Representations of *-Algebras, Locally Compact Groups, and Banach *-Algebraic Bundles

Volume 1
Basic Representation Theory of
Groups and Algebras

This is Volume 125 in
PURE AND APPLIED MATHEMATICS

Editors: Samuel Eilenberg and Hyman Bass

A complete list of titles in this series appears at the end of this volume.

Representations of *-Algebras, Locally Compact Groups, and Banach *-Algebraic Bundles

Volume 1
Basic Representation Theory of Groups and Algebras

J. M. G. Fell
Department of Mathematics
University of Pennsylvania
Philadelphia, Pennsylvania

R. S. Doran
Department of Mathematics
Texas Christian University
Fort Worth, Texas

ACADEMIC PRESS, INC.
Harcourt Brace Jovanovich, Publishers

Boston   San Diego   New York
Berkeley   London   Sydney
Tokyo   Toronto

ACADEMIC PRESS, INC.
1250 Sixth Avenue, San Diego, CA 92101

United Kingdom Edition published by
ACADEMIC PRESS, INC. (LONDON) LTD.
24-28 Oval Road, London NW1 7DX

Library of Congress Cataloging-in-Publication Data

Fell, J. M. G. (James Michael Gardner), Date
  Representations of *-algebras, locally compact
groups, and Banach *-algebraic bundles.

  (Pure and applied mathematics; 125-126)
  Includes bibliographies and indexes.
  Contents: v. 1. Basic representation theory of
groups and algebras — v, 2. Banach *-algebraic
bundles, induced representations, and the generalized
Mackey analysis.
  1. Representations of algebras. 2. Banach algebras.
3. Locally compact groups. 4. Fiber bundles
(Mathematics) I. Doran, Robert S., Date
II. Title. III. Series: Pure and applied mathematics
(Academic Press); 125-126.
QA3.P8 vol. 125-126 [QA326] 510 s [512'.55] 86-30222
ISBN 0-12-252721-6 (v. 1: alk. paper)
ISBN 0-12-252722-4 (v. 2: alk. paper)

88 89 90 91   9 8 7 6 5 4 3 2 1

Transferred to Digital Printing 2005

"All things come from thee, and of thy own have we given thee. For we are strangers before thee, and sojourners, as all our fathers were; our days on the earth are like a shadow, and there is no abiding. Thine, O Lord, is the greatness, and the power, and the glory, and the victory, and the majesty; for all that is in the heavens and in the earth is thine; thine is the kingdom, O Lord, and thou are exalted as head above all."

—I Chronicles 29:14, 15, 11
Revised Standard Version

Contents

Chapter III. Locally Compact Groups   163

Chapter IV. Algebraic Representation Theory   265

Topics in Volume 2:

Chapter VIII.   Banach *-Algebraic Bundles and Their Representations

Chapter IX.   Compact Groups

Chapter X.   Abelian Groups and Commutative Banach *-Algebraic
                Bundles

Chapter XI.   Induced Representations and the Imprimitivity Theorem

Chapter XII.   The Generalized Mackey Analysis

Preface

This two-volume work serves a double purpose, roughly (but not entirely) corresponding to the division into volumes. The first purpose (which occupies Volume 1 and part of Volume 2) is to provide a leisurely introduction to the basic functional analysis underlying the theory of infinite-dimensional unitary representations of locally compact groups and its generalization to representations of Banach algebras. Thus, the first volume covers generally well-known material on measure theory (Chapter II), topological groups and their group algebras (Chapter III), purely algebraic representation theory (Chapter IV), Banach algebras (Chapter V), C*-algebras and their *-representations (Chapter VI), and the topology of *-representations (Chapter VII). Chapters IX and X (in Volume 2), which deal mostly with classical material on compact groups and locally compact Abelian groups respectively, also serve primarily the first purpose.

The second purpose (which occupies Chapters VIII, XI, and XII of Volume 2) is to bring the reader to the frontiers of present-day knowledge in one limited but important area, namely, the Mackey normal subgroup analysis and its generalization to the context of Banach *-algebraic bundles. Here too our presentation is quite leisurely. Chapter VIII deals with Banach *-algebraic bundles and their representations. These objects can be thought of roughly as Banach *-algebras which are "over" a base group, and in which multiplication and involution are "covariant" with the operations of multiplication and inverse in the base group. The primary motivating example of a Banach *-algebraic bundle is in fact (very roughly speaking) just the group algebra of a locally compact group; the more general objects in this category can be regarded as "generalized group algebras." Chapters XI and XII are devoted to the "bundle generalization" (i.e., the generalization to Banach *-algebraic bundles) of the Mackey normal subgroup analysis. In its classical form developed by Frobenius for finite groups and by Mackey for separable locally compact groups, this is a powerful technique for classifying the unitary representations of a group H when we are given a closed normal subgroup N of H; under appropriate conditions it classifies the irreducible representations of H in terms of those of N and of subgroups of the quotient group $G = H/N$. More generally, suppose we are given a Banach *-algebra A, together with some suitable Banach *-algebraic bundle structure $\mathscr{B}$ for A over some locally compact group G. The "bundle generalization" of the normal subgroup analysis will then enable us to classify the *-representations of A in terms of those of subgroups of G and those of the so-called unit fiber of $\mathscr{B}$ (corresponding in the classical case to the normal subgroup N). The fundamental tools of the classical normal subgroup analysis are the construction of induced representations and the Imprimitivity Theorem. These will be developed in Chapter XI in the "bundle context" (in fact in an even more general, purely algebraic form). Finally, in Chapter XII we take up the Banach *-algebraic bundle version of the Mackey normal subgroup analysis itself and discuss numerous examples. (We remark that we use the phrase "generalized normal subgroup analysis" or "generalized Mackey analysis," even though in the generalized bundle version no normal subgroups of a group need appear!)

The choice of the material presented in Volume 1 is of course to a large extent governed by the needs of Volume 2. For example, little is said about von Neumann algebras and nothing about non-commutative direct integral decomposition theory, since it turns out (as mentioned later in this Preface) that these topics are hardly needed in the Loomis–Blattner–Glimm version of the normal subgroup analysis. On the other hand, in Chapter VII we go quite deeply into the topology of the space of all *-representations of a given *-algebra.

It is presumed that the reader has some prior knowledge of elementary group and ring theory, general topology, general linear topological spaces, and abstract measure theory (though a rapid introduction to the latter is given in Chapter II, from a slightly novel standpoint and much of it without proofs). With this equipment the reader should have no difficulty in reading the present work.

A word about the historical background and genesis of this work might be of interest here.

The story goes back to the summer of 1955, when the first-named author attended the summer course on group representations given at the University of Chicago by G. W. Mackey. The outcome of that course was a set of notes, prepared by him jointly with the late D. B. Lowdenslager, presenting the three stages of the program developed by Mackey in the course for the classification of separable group representations, namely: 1) Global multiplicity theory (the theory of von Neumann algebras); 2) direct integral decomposition theory; and 3) the normal subgroup analysis based on Mackey's version of the Imprimitivity Theorem. These notes made no pretense of giving complete proofs of the major results, because of the wealth of technical preliminaries (mostly measure-theoretic) which would have been necessary in order to do so. However, the first-named author had by this time become fascinated with the whole subject of symmetry and group representations and made up his mind to write an exhaustive account of the subject sooner or later. The present work is the outcome of that intention.

However, as even a nodding acquaintance with the earlier notes will show, the plan of those notes was quite different from the present work. Both aim at the normal subgroup analysis as their climax, but there the similarity ends. Two mathematical discoveries in the intervening years account for much of the difference of approach. First, a new approach to the theory of induced representations, the Imprimitivity Theorem and the normal subgroup analysis for groups was developed by L. H. Loomis and R. J. Blattner in the early 1960s, which dispenses entirely with Mackey's separability assumptions and substantially reduces the role of measure theory. A key factor in the Loomis–Blattner development was an observation by J. G. Glimm (in 1962) to the effect that just that portion of direct integral theory which is needed for the Mackey normal subgroup analysis can be developed without any separability assumptions whatever. In view of these discoveries we decided to orient the present work toward the Imprimitivity Theorem and the normal subgroup analysis in the non-separable context, using the results of Loomis, Blattner, and Glimm rather than Mackey's original highly measure-theoretic approach.

This is not to say that the Loomis–Blattner–Glimm approach supersedes that of Mackey. Indeed, Mackey's measure-theoretic proof of the Imprimitivity Theorem gives useful information in the separable context which does not emerge in the Loomis–Blattner proof. Also, his direct integral decomposition theory gives a beautiful insight into the structure of arbitrary unitary representations of separable Type I groups which is quite unavailable in the non-separable context that we have chosen to work with in Chapters XI and XII.

The second discovery that has molded the present work is the realization, hinted at earlier in this Preface, that the natural context for the normal subgroup analysis consists not merely in locally compact group extensions, but in the more general structures that we call saturated Banach *-algebraic bundles over locally compact groups. The first step in this direction (at least for infinite-dimensional representations) seems to have been taken by M. Takesaki, who in 1967 extended the normal subgroup analysis to crossed products, or semidirect products, of Banach *-algebras and groups. Not long after (in 1969) the first-named present author showed that the normal subgroup analysis is valid for the wider class of so-called homogeneous Banach *-algebraic bundles. (One should also mention the work of H. Leptin, who in 1972 demonstrated the same for his generalized $\mathscr{L}_1$ algebras, which are very closely related to homogeneous Banach *-algebraic bundles.) Finally, it became clear that the normal subgroup analysis actually requires only the property of *saturation* of the Banach *-algebraic bundle, which is considerably weaker than homogeneity. We therefore decided to present the whole development of Chapter XII (the culminating point of the present work) in the context of merely saturated Banach *-algebraic bundles.

In his Springer Lecture Notes Volume of 1977 (Fell [17]), the first-named author proved the Imprimitivity Theorem for saturated Banach *-algebraic bundles, but the further step of developing the generalized normal subgroup analysis in that context was not taken up there. *Chapter XII of the present work is therefore the first exposition of the Mackey normal subgroup analysis for saturated Banach *-algebraic bundles which has appeared in print.* (However, it must be pointed out that our methods are basically the same as those of Blattner in his 1965 paper on the normal subgroup analysis for the non-separable group case. No real advance in technique is needed in order to pass to the more general situation.)

Actually, the Imprimitivity Theorem for Banach *-algebraic bundles will be obtained in Chapter XI as (more or less) a corollary of a more general, purely algebraic Imprimitivity Theorem that makes no reference to groups at all. Not only does this generalized Imprimitivity Theorem have applications in areas outside of the theory of groups, but it suggests the hope of eventually "algebraizing" the entire Mackey normal subgroup analysis, i.e., of presenting it in the context of algebras, and banishing groups from their present crucial role as base spaces of the Banach *-algebraic bundles. Realization of this hope would be an important advance in the program of "algebraizing" all of the theory of group representations—a program begun in the earliest stages of the work of Gelfand and Naimark on Banach *-algebras, when they showed that the fact of existence of "enough" irreducible unitary representations of locally compact groups was only a special case of the corresponding fact for Banach *-algebras. However, the hope of completely "algebraizing" the normal subgroup analysis has not yet been fulfilled—a disappointment that is partly responsible for the delay in the appearance of the present work.

Although we have taken pains to present the whole theory of induced representations, the Imprimitivity Theorem, and the normal subgroup analysis in the generalized context of Banach *-algebraic bundles, there is no doubt that the "group case" (where we are interested in classifying the unitary representations of a group that is presented to us as a group extension) is the case of greatest interest. We have therefore taken every occasion to point out explicitly how our results specialize to the "group case." We hope thereby to minimize the annoyance felt by those (and we suspect they are quite a few!) whose aim is to learn from the work only so much of the subject as is relevant to groups.

Let us now say a few words about the exercises that have been placed at the ends of the various chapters. Some of these involve checking routine details of proofs or remarks that have been left to the reader. Others involve new results and actual extensions of the theory, while still others provide examples or counter-examples to various assertions in the text. The exercises range in difficulty from extremely easy to rather difficult. Some of them have been provided with hints for their solution.

Throughout both volumes of this work the reader will observe that certain propositions and results have been starred with an asterisk. Generally speaking, these propositions contain information and insights which may be of interest to the reader but which are not essential to the main flow of the material being discussed. Their proofs have therefore either been omitted entirely or else only sketched. As part of the exercises the reader has been asked to supply proofs (or to fill in details of proofs which have only been sketched) to many of these starred results.

An earlier draft of this work had been more or less completed by the first-named author in 1975 (the present second-named author was not involved in this effort except in a peripheral way with proof-reading and offering suggestions here and there). Because of various difficulties, the full manuscript was not published at that time (although in 1977 the material of Chapter XI and the first part of Chapter VIII was published as the Springer Lecture Notes Volume referred to above).

During the spring of 1981, the second-named author spent the semester as a Member of The Institute for Advanced Study. The close proximity of

Princeton to Philadelphia made it possible for the two of us to easily get together. It was decided at that time that the whole manuscript should be published, and, starting in 1982, the second-named author, assisted by the first-named, undertook a revision and updating of the entire manuscript. The present two-volume work is the result of these efforts.

This Preface would not be complete without a warm acknowledgement, on the part of the first-named author, of the debt which he owes to Professor G. W. Mackey for introducing him to the subject of group representations and guiding his subsequent study of it. Professor Mackey's dedication to and enthusiasm for the subject is contagious, and the first-named author's mathematical life was permanently shifted into new channels by his exposure to Mackey's 1955 summer course in Chicago and subsequent contacts with him over the years.

In closing, we want to thank the National Science Foundation for its financial support of the first-named author's sabbatical leave during 1970–71, when the original manuscript began to take shape. The first-named author also wishes to express his appreciation of the hospitality extended to him by the Natural Sciences Institute of The University of Iceland, Reykjavik, during his sabbatical leave there in the fall of 1985—a period spent largely in putting final touches to this manuscript. Special thanks are due to Professor Marc Rieffel for the many conversations with him in the early 1970s, which helped to give Chapter XI its present form. We also wish to thank Professor Robert C. Busby for carefully reading an early draft of Chapters II through VII and offering a number of useful suggestions. We also express our warm appreciation of the excellent work of Paula Rose in typing the entire first manuscript (during the early 1970s). Likewise our special thanks are due to Shirley Doran, who has carried through the endless revisions and additions the manuscript has undergone in the past several years, and to the Texas Christian University Department of Mathematics for making its office facilities and resources available to us during this period.

We want also to express our gratitude to our wives Daphne and Shirley (and our children also) for providing the encouragement and happy home environment that have added so much to the purely mathematical joys of seeing this long-delayed project to completion. But, most of all, we want to express, at least in some tiny measure, our gratitude to our Creator and God, who has given us the strength, the education, and the opportunity both to enjoy and to share with others the beauties of mathematical discovery.

J. M. G. Fell
R. S. Doran

Introduction to Volume 1 (Chapters I to VII)

1

This Introduction is designed to orient the reader toward the subject-matter not only of Volume 1 but of Chapters IX and X of Volume 2 as well. The reader who is new to the subject will find parts of it quite hard going at first. However, it is to be hoped that, in the process of assimilating the details of the subject, he will occasionally return to this Introduction and find it helpful in gaining a useful overview of the field.

The strongest motivation for the study of infinite-dimensional algebras and their representations comes from the representation theory of locally compact groups. For this reason, although the greater part of Volume 1 of this work deals with the representation theory of algebras, most of this Introduction will be spent in motivating the theory of group representations, from which many of the problems and achievements of the more general theory of representations of algebras took their origin.

The theory of group representations has its historical roots in two developments of nineteenth-century mathematics, both of the utmost importance in the history of mathematics and of mathematical physics as well.

The first of these is the theory of Fourier series, whose importance for applied mathematics stems from its application to problems of vibration and

1

heat conduction. Although at first sight this theory seems to have nothing to do with groups, in fact it fits beautifully into a group-theoretic framework. To see this, let us begin by recalling a fundamental fact about Fourier series: Let f be any function in the Hilbert space X of all complex-valued functions on $[-\pi, \pi[$ which are square-integrable with respect to Lebesgue measure; and for each integer n (positive, negative, or 0) put $g_n(t) = (2\pi)^{-1/2}e^{int}(-\pi \le t < \pi)$. Then

$$f = \sum_{n=-\infty}^{\infty} c_n g_n, \tag{1}$$

where

$$c_n = \int_{-\pi}^{\pi} f(t)\overline{g_n(t)}dt. \tag{2}$$

The equality in (1) means that the series on the right of (1) converges to f in the mean square sense, that is,

$$\lim_{p\to\infty} \int_{-\pi}^{\pi} |f(t) - \sum_{n=-p}^{p} c_n g_n(t)|^2 \, dt = 0.$$

Equations (1) and (2) are called the Fourier analysis of the function f. In terms of the Hilbert space X, they assert that the (orthonormal) collection of all the g_n is in fact a basis of X, and the process of Fourier analysis becomes, geometrically speaking, just the expansion of an arbitrary vector in X as a linear combination of the basis vectors g_n.

Why are we interested in the basis $\{g_n\}$ rather than in other possible bases of X? It is in the answer to this question that groups make their appearance. It will be convenient to think of $[-\pi, \pi[$ as identified by means of the bijection $t \mapsto u = e^{it}$ with the unit circle $\mathbb{E}$ consisting of all complex numbers u of absolute value 1. Then X becomes the Hilbert space of all complex functions on $\mathbb{E}$ which are square-summable with respect to the measure of length on $\mathbb{E}$, and g_n becomes the function $u \mapsto (2\pi)^{-1/2}u^n$ on $\mathbb{E}$. Now $\mathbb{E}$ is a compact Abelian group under the usual multiplication of complex numbers; it is called the circle group. Each element u of $\mathbb{E}$ gives rise, by rotation about the origin, to a unitary operator R_u on X:

$$(R_u f)(v) = f(u^{-1}v) \qquad\qquad (f \in X; v \in \mathbb{E}).$$

Once checks that R_1 is the identity operator and $R_u R_v = R_{uv}$ $(u, v \in \mathbb{E})$. It can be shown that $u \mapsto R_u$ is continuous in the strong operator topology. These three facts are summarized in the statement that $R: u \mapsto R_u$ is a unitary representation of $\mathbb{E}$; it is called the *regular* representation of $\mathbb{E}$. Now, considered as a vector in X, each g_n is an eigenvector for every operator R_u; indeed,

$$R_u g_n = u^{-n} g_n. \tag{3}$$

In fact it is easy to see that the g_n are (to within multiplication by a complex constant) the *only* vectors in X which are eigenvectors for every R_u. This fact, it turns out, is what gives to the basis $\{g_n\}$ its peculiar importance.

We have seen that X has an orthonormal basis consisting of vectors g_n each of which is an eigenvector for every R_u ($u \in \mathbb{E}$). This statement has a substantial generalization leading us straight into the theory of group representations. Suppose that G is any topological group (with unit e) and that Y is a Hilbert space. By a *unitary representation of G acting in Y* we mean a map T carrying each element x of G into a unitary operator T_x on Y and satisfying (i) T_e is the identity operator, (ii) $T_x T_y = T_{xy}$ ($x, y \in G$), and (iii) $x \mapsto T_x$ is continuous in the strong operator topology. The following result will emerge from IX.8.10: If G is any compact Abelian group and T is any unitary representation of G acting in a Hilbert space Y, then Y has an orthonormal basis consisting of vectors each of which is an eigenvector for every one of the operators T_x ($x \in G$).

We can express this generalization in a somewhat different manner. Let G be any topological group, and let us denote by $\Gamma(G)$ the set of all continuous group homomorphisms of G into the circle group $\mathbb{E}$. Given a unitary representation T of G, acting in a Hilbert space Y, let us associate to each γ in $\Gamma(G)$ the closed linear subspace Y_γ of Y consisting of those ξ in Y such that

$$T_x \xi = \gamma(x)\xi \qquad \text{for all } x \text{ in } G.$$

If β and γ are distinct elements of $\Gamma(G)$ then Y_β and Y_γ are orthogonal. (Indeed: If $\xi \in Y_\beta$ and $\eta \in Y_\gamma$, then $(\beta(x) - \gamma(x))(\xi, \eta) = (T_x \xi, \eta) - (\xi, T_x^{-1}\eta) = 0$ for all x in G. So, choosing x so that $\beta(x) \neq \gamma(x)$, we get $(\xi, \eta) = 0$.) Of course, Y_γ may be the zero space for some (or even all) of the γ in $\Gamma(G)$. We call Y_γ the γ-subspace of Y (with respect to T). Notice now that, if ξ is a non-zero eigenvector for all the T_x, the equation

$$T_x \xi = \gamma(x)\xi \qquad\qquad (x \in G)$$

determines $\gamma : x \mapsto \gamma(x)$ as an element of $\Gamma(G)$. Therefore the final assertion of the last paragraph can be restated as follows: If G happens to be compact and Abelian, and if Y is the space of a unitary representation T of G, then Y is the Hilbert direct sum of its orthogonal closed subspaces Y_γ ($\gamma \in \Gamma(G)$). Let P_γ be the operator of projection onto Y_γ. Then the P_γ are pairwise orthogonal, and if G is compact and Abelian $\sum_\gamma P_\gamma = \texttt{+}$ (identity operator). That is,

$$\xi = \sum_{\gamma \in \Gamma(G)} P_\gamma \xi \qquad \text{for all } \xi \text{ in } Y, \tag{4}$$

and

$$T_x = \sum_{\gamma \in \Gamma(G)} \gamma(x) P_\gamma \qquad \text{for all } x \text{ in } G \tag{5}$$

(the infinite sum on the right of (5) being taken in the weak topology).

Let us look for a moment at the example of Fourier series with which we began. Here G is the compact Abelian circle group $\mathbb{E}$, and $T = R$ (and $Y = X$). Now for each integer n (positive, negative, or 0) the function $\chi_n : u \mapsto u^n$ belongs to $\Gamma(\mathbb{E})$; in fact it can be shown that $\Gamma(\mathbb{E}) = \{\chi_n : n \text{ integral}\}$. By (3) g_{-n} (considered as a vector in X) belongs to the χ_n-subspace X_{χ_n}. In fact X_{χ_n} is simply the one-dimensional linear span of the vector g_{-n}. Thus the general equation (4) reduces in this case to the ordinary Fourier analysis (1). (It is important to realize that the functions g_n and χ_n, similar as they are, play quite different roles in the above discussion; the χ_n are elements of $\Gamma(\mathbb{E})$ while the g_n are vectors in X. Another special property of this example is that each χ_n-subspace is one-dimensional. This is far from true in general.)

In the light of this example we refer to equation (4) as the *generalized Fourier analysis* (or the *harmonic analysis*) *of ξ with respect to T*, and to equation (5) as the *harmonic analysis of T*.

2

What happens to (4) and (5) if we drop the assumption that G is compact and Abelian? In their present form, as we expect, they fail miserably. But this is not the end of the story. Much of modern harmonic analysis has consisted in finding sophisticated modifications of (4) and (5) that will hold for groups that need not be compact and Abelian but that are at least locally compact. (It seems that locally compact groups form the widest class of groups to which the processes of harmonic analysis can reasonably be expected to apply. Indeed, only on such groups do we have an invariant (Haar) measure; see the appendix of A. Weil [1]. Without an invariant measure most of harmonic analysis comes to a standstill.)

Our first step will be to indicate what modifications of the compact Abelian theory are necessary if G is Abelian and locally compact but not compact.

The most important non-compact Abelian group is the additive group $\mathbb{R}$ of the reals. Now, in formulae (1), (2) let us make the change of variable $t \mapsto \pi^{-1}at$

(where a is some large positive number), so that the interval $[-\pi, \pi]$ becomes the large interval $[-a, a]$. Fixing a continuous complex function f on $\mathbb{R}$, that vanishes outside some compact set, and passing (purely formally) to the limit $a \to \infty$, we find that (1) and (2) take the symmetrical form

$$f(t) = (2\pi)^{-1/2} \int_{-\infty}^{\infty} \hat{f}(s) e^{ist} \, ds \qquad (t \in \mathbb{R}), \qquad (6)$$

where

$$\hat{f}(s) = (2\pi)^{-1/2} \int_{-\infty}^{\infty} f(t) e^{-ist} \, dt \qquad (s \in \mathbb{R}), \qquad (7)$$

Formula (6) is the *Fourier integral formula* for f; it expresses f in terms of its *Fourier integral transform* $\hat{f}$ defined in (7). (The purely formal derivation of (6) and (7) sketched here was given by Fourier himself in his classic memoir [1] on heat conduction. The problem of characterizing those f for which (6) and (7) are valid will be taken up in greater generality in §X.3.)

Given f, let us apply (6) and (7) not to f itself but to the function h given by

$$h(t) = (2\pi)^{-1/2} \int_{-\infty}^{\infty} \overline{f(r)} f(t + r) dr.$$

One verifies from (7) that $\hat{h}(s) = |\hat{f}(s)|^2$. Hence, applying (6) with $t = 0$, we find that

$$\int_{-\infty}^{\infty} |f(t)|^2 \, dt = \int_{-\infty}^{\infty} |\hat{f}(s)|^2 \, ds. \qquad (8)$$

This is the famous Plancherel formula (first proved as a very special case of results in Plancherel [1]). Formula (8) suggests that the Fourier transform mapping $F: f \mapsto \hat{f}$ is an isometric linear operator on the Hilbert space X of all Lebesgue-square-summable complex functions on $\mathbb{R}$. It can be shown that $F: X \mapsto X$ is indeed a well defined bijection, and so is a unitary operator on X.

A very important step in the development of abstract harmonic analysis was the recognition by Weil [1] (see also Cartan and Godement [1]) that the theory of the Fourier integral sketched above is valid not merely for $\mathbb{R}$ but for any locally compact Abelian group. Let G be such a group; and notice that the $\Gamma(G)$ defined in §1 is also an Abelian group under pointwise multiplication. Equipped with the topology of uniform convergence on compact subsets of G, $\Gamma(G)$ turns out to be a locally compact Abelian group, the so-called *character group* of G. (Since G is Abelian, $\Gamma(G)$ can be shown to be large enough to reflect fully the properties of G. In fact, the Pontryagin Duality Theorem asserts that $\Gamma(\Gamma(G)) \cong G$; see X.3.11.) If λ and $\tilde{\lambda}$ are suitably normalized invariant measures on G and $\Gamma(G)$ respectively (generalizing Lebesgue measure on $\mathbb{R}$), one can show, in analogy with (6) and (7), that

$$f(x) = \int_{\Gamma(G)} \hat{f}(\gamma)\gamma(x)d\tilde{\lambda}\gamma \qquad (x \in G), \qquad (9)$$

where

$$\hat{f}(\gamma) = \int_G f(x)\overline{\gamma(x)}d\lambda x \qquad (\gamma \in \Gamma(G)), \qquad (10)$$

for "well-behaved" complex functions f on G. From (9) and (10) we again deduce a "Plancherel formula," namely

$$\int_G |f(x)|^2 \, d\lambda x = \int_{\Gamma(G)} |\hat{f}(\gamma)|^2 \, d\tilde{\lambda}\gamma, \qquad (11)$$

and the generalized Fourier integral transform (9), (10) again sets up a linear isometry $F: f \mapsto \hat{f}$ of the Hilbert space $\mathcal{L}_2(\lambda)$ onto the Hilbert space $\mathcal{L}_2(\tilde{\lambda})$. (The status of (6) and (7) as special cases of (9) and (10) becomes clear when we regard the variable s in (6) and (7) as indexing the functions $\gamma_s: t \mapsto e^{ist}$ which constitute $\Gamma(\mathbb{R})$. The map $s \mapsto \gamma_s$ is an isomorphism between $\mathbb{R}$ and $\Gamma(\mathbb{R})$, so that for $G = \mathbb{R}$ the spaces $\mathcal{L}_2(\lambda)$ and $\mathcal{L}_2(\tilde{\lambda})$ can be identified, and $F: f \mapsto \hat{f}$ can be regarded as a unitary operator on a single Hilbert space. For general locally compact Abelian groups G, however, G and $\Gamma(G)$ are not isomorphic.)

Again let G be an arbitrary locally compact Abelian group; and let $\mathcal{L}_2(\lambda)$ and $\mathcal{L}_2(\tilde{\lambda})$ be the Hilbert spaces introduced above. In analogy with §1 we denote by R the *regular* representation of G, that is, the unitary representation of G acting on $\mathcal{L}_2(\lambda)$ by translation:

$$(R_x f)(y) = f(x^{-1}y) \qquad (x, y \in G; f \in \mathcal{L}_2(\lambda)).$$

On transformation by the generalized Fourier integral transform map F, the representation R goes into a new unitary representation $R': x \mapsto FR_x F^{-1}$ of G acting on $\mathcal{L}_2(\tilde{\lambda})$; and one verifies from (9) and (10) that

$$(R'_x \hat{f})(\gamma) = \gamma(x)\hat{f}(\gamma). \qquad (12)$$

Thus R is unitarily equivalent (see the next paragraph) with the unitary representation R' of G whose operators R'_x act on $\mathcal{L}_2(\tilde{\lambda})$ by multiplying each function in $\mathcal{L}_2(\tilde{\lambda})$ by the function $\gamma \mapsto \gamma(x)$.

It turns out that this statement about R can be generalized to arbitrary unitary representations of G. To state this generalization in the conceptually simplest terms, we make two definitions, valid for arbitrary topological groups G. First, given two unitary representations T and T' of G (acting in

Hilbert spaces Y and Y' respectively), we say that T and T' are *unitarily equivalent* (in symbols $T \cong T'$) if there is a linear isometry L from Y onto Y' that carries T into T', that is, $T_x = L^{-1}T'_x L$ for all x in G. We call L a T, T' *unitary equivalence*. Secondly, suppose that μ is a measure on a measure space M, and let Z be the Hilbert space $\mathcal{L}_2(\mu)$ of all complex μ-square-integrable functions on M. Suppose further that with each point m of M is associated an element γ_m of $\Gamma(G)$. A unitary representation T of G that acts on Z according to the formula

$$(T_x f)(m) = \gamma_m(x)f(m) \qquad (f \in Z; x \in G; m \in M) \qquad (13)$$

is called a *multiplication representation* of G on M, μ. Now we saw in the preceding paragraph that, if G is locally compact and Abelian, the regular representation of G is unitarily equivalent to a multiplication representation of G (on $\Gamma(G), \tilde{\lambda}$). The generalization of this fact referred to above is the following: *Every* unitary representation T of the locally compact Abelian group G is unitarily equivalent to a multiplication representation T' of G on some (suitably chosen) measure space M, μ.

Notice what happens if the measure space M, μ is *discrete*. In that case $Z = \mathcal{L}_2(\mu)$ is the direct sum of the one-dimensional subspaces Z_m corresponding to the points m in M which have non-zero μ-measure; and formula (13) says that $T'_x(Z_m) \subset Z_m$ for all x in G and all m. Thus the space of T', and hence of T also, has an orthonormal basis consisting of eigenvectors of each of the representation operators. This, as we saw in §1, is exactly what happens if G is not only Abelian but compact. The extra latitude that comes of allowing M, μ to be non-discrete is just what is needed in order that the above generalization be true for arbitrary locally compact Abelian groups.

Just as in (5) we can rephrase this generalization in terms of projections. Let T now be itself a multiplication representation on the measure space M, μ, and let γ_m be as in (13). If W is any Borel subset of $\Gamma(G)$, let $P(W)$ be the projection on $Z = \mathcal{L}_2(\mu)$ consisting of multiplication by the characteristic function of $\{m \in M : \gamma_m \in W\}$. The mapping $P: W \mapsto P(W)$ is what we shall call in II.11.2 a projection-valued measure. In case G is compact, the topological space $\Gamma(G)$ turns out to be discrete, and for each γ in $\Gamma(G)$ the projection $P(\{\gamma\})$ coincides with the P_γ of §1. The projection-valued measure P now permits us to generalize summation to integration in (5); on doing this, we obtain the version of equation (5) which remains true for the arbitrary locally compact Abelian group G:

$$T_x = \int_{\Gamma(G)} \gamma(x)dP\gamma. \qquad (14)$$

The integral on the right of (14) is the spectral integral discussed in II.11.17.

Thus *every* unitary representation T of G, being unitarily equivalent to a multiplication representation, gives rise to a projection-valued measure P on the Borel σ-field of $\Gamma(G)$ such that (14) holds for all x in G, and P turns out to be unique. This is the Generalized Stone Theorem (see X.2.1.). It reduces the classification of unitary representations of G to the classification of projection-valued measures on $\Gamma(G)$. The P of (14) is called the *spectral measure* of T, and (14) itself is the *spectral decomposition*, or the *harmonic analysis*, of T.

The above introductory discussion has proceeded by way of multiplication representations because of the conceptual simplicity of the latter. Our approach to the Stone Theorem in the body of this work, however, is more direct, and makes no mention of multiplication representations.

Can (4), like (5), be generalized to the non-compact situation? Let T be a unitary representation (acting in a Hilbert space Y) of the locally compact Abelian group G, with spectral measure P. One might at first expect to generalize (4) by writing each vector ξ in Y as an integral over $\Gamma(G)$ of a function $\gamma \mapsto \xi_\gamma$ with values in Y—the value ξ_γ being somehow associated with the point γ in $\Gamma(G)$. In general, however, this is not possible. For one thing, if G is non-compact, $P(\{\gamma\})$ will in general be 0 for one-element sets $\{\gamma\}$. The Fourier integral formula (6) clearly illustrates the situation for the regular representation R of $\mathbb{R}$: According to (6) a vector f in the space X of R is the integral (over all real numbers s) of the functions $f_s: t \mapsto \hat{f}(s)e^{ist}$. Now f_s is naturally associated with the element $\gamma_s: t \mapsto e^{ist}$ of $\Gamma(\mathbb{R})$; in fact it is just a scalar multiple of γ_s. But, if $\hat{f}(s) \neq 0$, f_s is not square-integrable and so not in X. In general, for an arbitrary unitary representation T of G, we can regain (4) only if we are willing to introduce some extra structure permitting us to "complete" the space Y of T and so to capture certain "limit eigenvectors" associated with individual points of $\Gamma(G)$ but not lying in Y itself. Such extra structure is provided by the "rigged Hilbert spaces" of Gelfand and Vilenkin [1]. We shall not pursue this topic here, since for our purposes (5) is more important than (4).

As a matter of fact, the direct integral decomposition theory of representations (to be mentioned in §4) provides a "weak" version of (4) for all second-countable locally compact groups.

3

What happens to (4) and (5) when G is compact but not Abelian? In the non-compact Abelian case we have seen that the required modification of (5) consists in replacing summation by integration, while keeping the same base space $\Gamma(G)$. In the non-Abelian compact case, (4) and (5) fail chiefly because $\Gamma(G)$ is too small. (Indeed, there are many compact groups G for which $\Gamma(G)$ consists of just one element, namely the function identically 1 on G.) The required modification consists in replacing $\Gamma(G)$ by a certain bigger set $\hat{G}$ (while keeping the summation process as before).

Let us be precise. Given any topological group G, we say that a unitary representation T of G (acting on a Hilbert space X) is *irreducible* if there is no closed linear subspace Y of X, other than $\{0\}$ and X itself, which is *stable* under T (that is, $T_x(Y) \subset Y$ for all x in G). We define the *structure space* of G to be the family of all unitary equivalence classes of irreducible unitary representations of G, and denote it by $\hat{G}$. An element of $\Gamma(G)$ can be regarded as a unitary representation of G acting on the one-dimensional Hilbert space (and hence trivially irreducible), and two elements γ and γ' of $\Gamma(G)$, looked at in this light, are unitarily equivalent if and only if they are equal. Thus $\Gamma(G)$ becomes a subset of $\hat{G}$. If G is Abelian it turns out that every irreducible unitary representation of G is one-dimensional, so that $\hat{G} = \Gamma(G)$. But, in general, the elements of $\hat{G}$ act on multi-dimensional (in fact infinite-dimensional) spaces, and $\hat{G}$ is much bigger than $\Gamma(G)$.

Now let S be any unitary representation of G, acting on the Hilbert space X, and let τ be a class belonging to $\hat{G}$. If Y is a closed S-stable subspace of X, the unitary representation $x \mapsto S_x | Y$ ($x \in G$) is called the *subrepresentation of S acting on Y*, and is denoted by ${}^Y S$. If ${}^Y S$ is irreducible and of class τ, we shall say that Y is *irreducible* and *of class τ* (*with respect to S*). By the *τ-subspace of X* (*with respect to S*), denoted by X_τ, we shall mean the closed linear span of the set of all those subspaces Y of X which are irreducible and of class τ (with respect to S). If $\tau \in \Gamma(G)$, then X_τ is the same as the τ-subspace defined in §1. One can show (generalizing the corresponding observation in §1) that X_σ and X_τ are orthogonal whenever σ and τ are distinct elements of $\hat{G}$. If the sum of the X_τ (τ running over $\hat{G}$) is dense in X, that is, if X is the Hilbert direct sum of its subspaces X_τ, then S is called *discretely decomposable*. If S is discretely decomposable, X can be written as a Hilbert direct sum of stable *irreducible* closed subspaces Y_i:

$$X = \sum_{i \in I}^{\oplus} Y_i \tag{15}$$

(I being some index set). Usually the Y_i occurring in (15) are by no means unique. If G is Abelian, (15) is just the statement that X has an orthonormal basis consisting of eigenvectors of every S_x (see §1). If (15) holds, then for each fixed τ in $\hat{G}$ the τ-subspace X_τ is just the Hilbert direct sum of those Y_i which are of class τ. Furthermore, the cardinal number $m(\tau; S)$ of $\{i \in I : Y_i \text{ is of class}$

$\tau\}$ is independent of the particular mode of writing X in the form (15); this cardinal number is called the *multiplicity of τ in S*. If S is discretely decomposable, its unitary equivalence class is uniquely described as follows by the collection of all its multiplicities $m(\tau; S)$ $(\tau \in \hat{G})$:

$$S \cong \sum_{\tau \in \hat{G}}{}^{\oplus} m(\tau; S)\tau. \tag{16}$$

(Here $\sum^{\oplus}$ is the Hilbert direct sum of unitary representations; and $m(\tau; S)\tau$ means a direct sum of $m(\tau; S)$ copies of some irreducible unitary representation of G of class τ.)

The preceding two paragraphs were valid for any topological group G. Let us now suppose that G is a compact group. The fundamental result on the representations of compact groups is the Peter–Weyl Theorem (IX.2.11), from which the following two assertions (IX.8.10 and IX.8.6) follow as corollaries:

(A) Every unitary representation of G is discretely decomposable;
(B) Every irreducible unitary representation of G is finite-dimensional.

Let S be a unitary representation of G, acting on a Hilbert space X; and for each τ in $\hat{G}$ let P_τ be projection onto the τ-subspace X_τ of X. Thus the P_τ are pairwise orthogonal; and in virtue of (A) $\sum_\tau P_\tau = \mathbb{1}$, that is,

$$\xi = \sum_{\tau \in \hat{G}} P_\tau \xi \tag{17}$$

for every ξ in X. This equation of course generalizes (4). It is called the *harmonic analysis* of the vector ξ with respect to the unitary representation S of G; the term $P_\tau \xi$ of (17) is called the *τ-component* of ξ.

As regards (5), it has no complete generalization to the compact non-Abelian case; for the concrete operators S_x are not determined by the P_τ. The best we can do is to express the unitary equivalence class of S in terms of P by means of (16) together with the evident relation

$$m(\tau; S) = (\dim(\tau))^{-1}\dim(\text{range}(P_\tau)).$$

(Here $\dim(\tau)$ is the dimension of the space in which any representation of class τ acts. Notice from (B) that $\dim(\tau) < \infty$.)

Let us illustrate (17) with the *regular* representation R of the compact group G, that is, the unitary representation acting by left translation on the Hilbert space $X = \mathcal{L}_2(\lambda)$ (λ being normalized left-invariant measure on G):

$$(R_x f)(y) = f(x^{-1}y) \qquad (f \in X; x, y \in G).$$

It can be shown that every class τ in $\hat{G}$ occurs in R; in fact

$$m(\tau; R) = \dim(\tau) \qquad \text{for all } \tau \text{ in } \hat{G} \tag{18}$$

(see IX.9.1). If $f \in X$ and $\tau \in \hat{G}$, the τ-component f_τ of f (with respect to R) turns out to be given by the following formula (see IX.9.2):

$$f_\tau(x) = \dim(\tau)\mathrm{Tr}(T_f^\tau T_{x^{-1}}^\tau) \qquad (x \in G). \qquad (19)$$

(Here T^τ is any irreducible unitary representation of class τ; T_f^τ is the operator $\int_G f(x)T_x^\tau \, d\lambda x$; and Tr stands for trace. By assertion (B) the operator $T_f^\tau T_{x^{-1}}^\tau$ acts in a finite-dimensional space, and so has a trace.) The reader should convince himself that (19) as applied to the circle group $\mathbb{E}$ specializes to (1) and (2).

The argument that led from (6) and (7) to the Plancherel formula (8) also leads from (19) to the following Plancherel formula for complex functions on the compact group G:

$$\int_G |f(x)|^2 \, d\lambda x = \sum_{\tau \in \hat{G}} \dim(\tau)\mathrm{Tr}((T_f^\tau)^* T_f^\tau). \qquad (20)$$

The reader should compare this with (8).

Here is another specific illustration of harmonic analysis with respect to compact groups. Consider the compact group SO(3) of all proper rotations about the origin in three-dimensional space. It turns out (see IX.12.4) that $(SO(3))\hat{}$ contains just one element of each odd dimension, and no elements of even dimension. Let $\tau^{(n)}$ ($n = 0, 1, 2, \ldots$) be the unique element of $(SO(3))\hat{}$ whose dimension is $2n + 1$. Now SO(3) acts transitively by rotation on the surface M of the unit sphere about the origin; and this action gives rise to a unitary representation S of SO(3) on the Hilbert space X of all complex functions on M which are square-integrable with respect to the measure of area on M:

$$(S_u f)(q) = f(u^{-1}q) \qquad (f \in X; u \in SO(3); q \in M).$$

One can show $m(\tau^{(n)}; S) = 1$ for every $n = 0, 1, 2, \ldots$ (see IX.12.5), so that the $\tau^{(n)}$-subspace X_n of X is of dimension $2n + 1$. The functions belonging to X_n are called *spherical harmonics* on M of order n; and the harmonic analysis (with respect to S) of a function in X is called its *expansion in spherical harmonics*.

It was mentioned at the beginning of this Introduction that the theory of Fourier series grew out of practical problems of vibration and heat conduction. At our present juncture it is easy to indicate the relevance of group representations to the solution of such linear differential equations as arise in the theory of vibration and heat conduction. For this purpose we must suppose that the space M on which the vibration or heat conduction takes place is acted upon by a group G of symmetries. For simplicity let M be a compact manifold acted upon by a compact Lie group of transformations. Then there is a measure μ on M which is invariant under G; and the action of G on M defines a unitary representation S of G on $X = \mathscr{L}_2(\mu)$:

$$(S_x f)(m) = f(x^{-1}m) \qquad (f \in X; x \in G; m \in M).$$

Let X_τ ($\tau \in \hat{G}$) denote the τ-subspace of X with respect to S.

Now suppose we have a linear differential operator D on M with the following two properties: (i) D is invariant under G; (ii) D belongs to the algebra of operators generated by the "infinitesimal motions" of M under G. A typical initial value problem of the sort that arises in the theory of heat conduction is the following: Given a real-valued function F on M (belonging to X), find a real-valued function $f: \; < m, t > \; \mapsto f(m, t)$ $(m \in M; t \in \mathbb{R})$ which satisfies the differential equation

$$\frac{\partial f}{\partial t} = Df \qquad (D \text{ acting on the } m \text{ variable}) \qquad (21)$$

and the initial condition

$$f(m, 0) = F(m). \tag{22}$$

To solve this we first observe from condition (ii) on D that D leaves stable every closed linear subspace Y of X which is stable under S. If Y is irreducible under S, then in view of condition (i) on D it can be shown that D is a scalar operator on Y, that is, there is a real number r such that

$$Df = -rf \qquad\qquad (f \in Y)$$

(We insert a minus sign since in actual heat conduction problems $-r$ turns out to be negative.) Again invoking (ii) one can show that r depends only on the class in $\hat{G}$ to which the irreducible subspace Y belongs. Thus for each τ in $\hat{G}$ there is a real number r_τ such that

$$Dg = -r_\tau \rho \qquad \text{for all } g \text{ in } X_\tau. \tag{23}$$

Given any g in X_τ it follows from (23) that

$$\phi(m, t) = e^{-r_\tau t}g(m) \tag{24}$$

is a solution of (21). Now for each τ let F_τ denote the τ-component of the initial-value function F. Replacing g in (24) by F_τ and summing over τ the different functions ϕ so obtained, we find that

$$f(m, t) = \sum_{\tau \in \hat{G}} e^{-r_\tau t} F_\tau(m) \tag{25}$$

satisfies both (21) and (22), and so solves the given initial value problem.

Take the simple case that $G = M = \mathbb{E}$, $\mathbb{E}$ acting on itself by multiplication. If θ is the angular variable measuring position on $\mathbb{E}$, then $D = (\partial^2/\partial\theta^2)$ satisfies the above conditions (i) and (ii) on D. As we saw in §1, harmonic analysis with respect to $\mathbb{E}$ in this context is just expansion in Fourier series; and the solution (25) becomes in this case Fourier's classical solution of the problem of heat conduction on a circle.

Suppose next that M is the surface of a sphere in three-dimensional space and that G is the proper rotation group acting on M; and let D be the "angular component" of the Laplace operator $(\partial^2/\partial x^2) + (\partial^2/\partial y^2) + (\partial^2/\partial z^2)$. Then D satisfies the above conditions (i) and (ii). As we have seen, in this situation the F_τ occurring in (25) are the spherical harmonics. The solution (25) for this case is very useful in the discussion of boundary-value problems in which boundary conditions are specified on spheres.

4

The next natural question is: What remains of (4) and (5) if G is locally compact but neither compact nor Abelian? One might conjecture that the residue would be the intersection of what remains in the two cases already discussed—the non-compact Abelian and the non-Abelian compact cases; and this is (very roughly) correct. Recall that in both these cases, given any unitary representation S of G, we were able to define a projection-valued measure P on the Borel σ-field of $\hat{G}$ (which in the compact case consists of *all* subsets of $\hat{G}$), whose values were projections on the space of S. It turns out that this construction has a far-reaching generalization. Let G be any locally compact group. There is a natural topology (which we call the *regional topology*; see VII.1.3 and VIII.21.4) on $\hat{G}$ making the latter a locally compact space (though not in general Hausdorff). Given any unitary representation S of G, we can construct from S a canonical projection-valued measure P whose values are projections on the space of S and which is defined on the σ-field of Borel sets with respect to the regional topology of $\hat{G}$. This P is called the *spectral measure* of S (or the *Glimm projection-valued measure* of S, after its discoverer J. Glimm), and generalizes the P of (4) and (5) to the arbitrary locally compact situation. We shall construct it in §VII.8 (in the even more general context of C^*-algebras). Given a Borel subset W of $\hat{G}$, range $(P(W))$ can be regarded, roughly speaking, as the part of the space of S on which S acts like a "direct integral" of the irreducible representations whose classes lie in W.

Unfortunately, though the P of (4) and (5) generalizes naturally, neither of the equations (4) and (5) themselves can be generalized to the locally compact situation. This was to be expected, since (4) breaks down in the non-compact situation (see §2) and (5) in the non-Abelian situation (see §3). Some of the most interesting results of the modern theory of group representations have been efforts to extend (4) and (5) as far as possible to arbitrary locally compact groups. We will mention two of these efforts here, though their detailed development is beyond the scope of this work.

As regards (5), though no formula specifying the concrete operators S_x is available in the general case, one can at least hope to generalize the description (16) of the *equivalence class* of S. It turns out (see for example Mackey [7]) that, if G is second-countable and has a Type I representation theory (see VI.24.13), then every unitary representation S of G is unitarily equivalent to an essentially unique "direct integral"

$$S \cong \int_{\hat{G}}^{\oplus} m(\tau)\tau \, d\mu\tau \tag{26}$$

over the structure space $\hat{G}$. Relation (26) generalizes (16). The "direct integral" on the right of (26) is a generalization of the notion of a direct sum of representations. The values of the mapping $\tau \mapsto m(\tau)$ are cardinal numbers, $m(\tau)$ serving as the "generalized multiplicity" of τ in S; and μ is a numerical measure on $\hat{G}$ whose null sets coincide with the null sets of the spectral measure of S. The theory of relation (26) (direct integral decomposition theory) is an extremely important part of the modern theory of group representations. Unfortunately, it lies outside the scope of the present work.

Equation (4), as we have seen, loses its meaning even in the non-compact Abelian situation. In the important case of the regular representation of G, however, both (4) and the closely related Plancherel formula have natural generalizations provided G is either compact or Abelian (see (9), (11), (19), (20)), so it is natural to conjecture that they will have valid generalizations to the regular representation of an arbitrary locally compact group G. For simplicity we will confine our attention to the Plancherel formula. It turns out that whenever G is a unimodular locally compact group whose representation theory is of Type I, the following generalization of (11) and (20) holds:

$$\int_G |f(x)|^2 \, d\lambda x = \int_{\hat{G}} \mathrm{Tr}((T_f^\tau)^* T_f^\tau) d\mu\tau. \tag{27}$$

Here, as in (20), for each τ in $\hat{G}$ the symbol T^τ stands for any concrete irreducible unitary representation of G of class τ; and T_f^τ denotes the integrated operator $\int_G f(x)T_x^\tau \, d\lambda x$ (λ being left-invariant measure on G). The essential content of (27) is the existence of a unique measure μ on $\hat{G}$ such that (27) holds for all f in some dense subfamily of $\mathscr{L}_2(\lambda)$; this μ is called the

Plancherel measure on $\hat{G}$. (Notice that in the present context, unlike the compact and Abelian cases, the representation classes τ in $\hat{G}$ are in general infinite-dimensional, so that at first sight the traces $\text{Tr}((T_f^\tau)^* T_f^\tau)$ appear to have no meaning. However it turns out that, under the above hypotheses on G, there is a dense set of functions f in $\mathscr{L}_2(\lambda)$ such that $(T_f^\tau)^* T_f^\tau$ is of trace class, and hence has a trace, for μ-almost all τ.)

Comparing (27) with (11) and (20), we see that if G is Abelian the Plancherel measure is just the invariant measure on $\hat{G}$; while if G is compact the Plancherel measure assigns to each point τ of $\hat{G}$ the measure $\dim(\tau)$. A central problem of modern harmonic analysis is the determination of the Plancherel measure for various important classes of unimodular groups with Type I representation theory. (See G. Warner [1, 2] for an exposition of Harish-Chandra's work on the Plancherel measure for semi-simple Lie groups.)

To summarize these first four sections of our Introduction, we may say that the theory of group representations provides an important tool for the analysis of functions on a space which is acted upon by a group G of symmetries. Given a locally compact group G, in order actually to carry out this analysis for all spaces on which G might act, it is first necesssary to find out what the structure space of G looks like, that is, to classify the irreducible unitary representations of G up to unitary equivalence. Having done this, one would like to develop a technique for obtaining the spectral measure of an arbitrary (nonirreducible) unitary representation of G. Finally, the reader will recall that a certain "Type I" property was mentioned as being important both in direct integral theory and in the Plancherel formula for G. This deep and important property of the representation theory of G (which appears in Chapters VI, VII, and XII of the present work) finds its natural setting in the general theory of von Neumann algebras. It is a vital problem of harmonic analysis to determine which groups have the Type I property and which do not.

Now there is a very important technique by which these three questions—the classification of $\hat{G}$, the harmonic analysis of an arbitrary unitary representation of G, and the decision whether or not the Type I property holds for G—can be attacked all at once. It is the so-called Mackey normal subgroup analysis, which forms the main topic of the last part of this work. To be a litle more precise, suppose that we are given a closed normal subgroup N of G. What the Mackey normal subgroup analysis does (under appropriate general hypotheses) is to reduce the solution of these three problems for G to the solution of the corresponding problems for N and certain subgroups of G/N. Since N and G/N may well be substantially smaller and less complicated than G, the Mackey analysis is frequently very powerful and for many important groups answers the main questions concerning harmonic analysis with respect to G. (Of course, if G is simple in the sense of having no non-trivial closed normal subgroups, the Mackey analysis will be of no use.)

5

Up to now we have dealt only with *unitary* representations of groups, because it is for these that the deepest results have been obtained. Nevertheless, much more general categories of representations exist, for which one can pose (if not solve) problems similar to those already discussed in the unitary case. For future reference it will be useful to mention here a very general sense of the term "representation."

Let G be any group; and let X be a set carrying some mathematical structure (topological, algebraic, or of any other kind) which we denote by $\mathscr{S}$. Any group homomorphism of G into the group of all automorphisms of X, $\mathscr{S}$ can be called a *representation of G of the type of $\mathscr{S}$*. For example, if X is a linear space over a field F, a homomorphism of G into the group of all F-linear automorphisms of X is called an *F-linear representation* of G. If X is an associative algebra over a field F, a homomorphism of G into the group of all automorphisms of the algebra X is an *F-algebra representation* of G. If X is a topological space, a homomorphism of G into the group of all homeomorphisms of X onto itself is a *topological representation* of G. From this standpoint any left action of G on an abstract set X can be called a *set-representation* of G (X being considered merely as a set, without further structure). Similarly, if X is a locally convex linear topological space or a complex Hilbert space, we have the corresponding notions of a *locally convex representation* or a *unitary representation*. (We disregard here the continuity condition imposed on unitary representations in §1.)

Corresponding to each of these definitions we have a notion of equivalence: For example, two F-linear representations [F-algebra representations, set-representations] T and T' of the same group G, acting on X and X' respectively, are *F-linearly equivalent* [*F-algebra equivalent, set-theoretically equivalent*] if there is an F-linear [F-algebra, set-theoretic] isomorphism Φ of X onto X' satisfying $\Phi \circ T_x = T'_x \circ \Phi$ ($x \in G$). Such definitions can obviously be multiplied indefinitely.

For each of these definitions, we can pose the problem of classifying all the representations of G of the given category to within the appropriate equivalence relation. Notice that for set-representations this problem is trivial:

Every set-representation of G is set-theoretically equivalent to a set-theoretic union of transitive actions of G; and the latter are classified by the conjugacy classes of subgroups of G (see III.3.5). In general it is the structure $\mathscr{S}$ imposed on the set X which causes the difficulty in classifying the corresponding representations.

We observe parenthetically that, if G is compact, most of the theory of unitary representations of G (as sketched in §3) is valid in the more general category of locally convex representations (see Chapter IX).

The process of classifying representations within different categories offers a very general means of classifying functions on a G-space in the spirit of Fourier analysis. Suppose that T is an F-linear [F-algebra, set-theoretical] representation of G on an F-linear space [F-algebra, set] X; and let ξ be any element of X. Let Y be the subset of X *generated under T by* ξ; that is, Y is the smallest F-linear subspace [F-subalgebra, subset] of X which contains ξ and is stable under T. Then $^Y T : x \mapsto T_x | Y$ is an F-linear [F-algebra, set-theoretic] representation of G acting on Y; and the F-linear [F-algebraic, set-theoretic] equivalence class $(^Y T)^{\sim}$ of $^Y T$ is a property of ξ. We call $^Y T$ the *subrepresentation of T generated by* ξ. One can attempt to classify the elements ξ according to the properties of the class $(^Y T)^{\sim}$ so obtained. (Of course the same construction is possible for the category of representations determined by *any* kind of mathematical structure.)

Consider for example the analysis of the complex-valued functions on the group G itself. For simplicity we assume that G is finite and Albelian. Let X be the space of all complex functions on G; and let R be the regular representation of G on X:

$$(R_x f)(y) = f(x^{-1} y) \qquad\qquad (x, y \in G; f \in X).$$

One can now interpret R as a representation with respect to at least three different mathematical structures on X:

(A) Let us first think of X merely as a set, with no further structure. Given an element ξ of X, the subset Y of X generated by ξ is then simply the set of all translates of ξ; and the (transitive) action of G on Y (i.e., the set-representation generated by ξ) is determined to within set-theoretic equivalence by the stability subgroup $\{x \in G : R_x \xi = \xi\}$ of G, i.e., by the set of all *periods* of the function ξ.

(B) Now let us consider X as a complex associative algebra (under pointwise multiplication). Since each R_x is then an algebra automorphism of X, R becomes a complex algebra representation of G.

From the structure of subalgebras of X one can show that the subalgebra Y generated by ξ under R is just the set of all those functions in X whose group of periods contains the group of periods of ξ. From this it can be shown that the algebra equivalence class of the subrepresentation of R generated by ξ is determined (as in (A)) by the group $\{x \in G : R_x \xi = \xi\}$ of periods of ξ.

(C) Finally, let us regard X as a complex-linear space so that R becomes a complex-linear representation of G on X. The complex-linear equivalence class $({}^Y T)^{\sim}$ of the subrepresentation of R generated by ξ will of course determine the group $\{x \in G : R_x \xi = \xi\}$ of periods of ξ; but, in contrast to (A) and (B), the group of periods of ξ will no longer determine $({}^Y T)^{\sim}$. There will be many different functions in X, all having the same periods, but giving rise to quite different complex-linear equivalence classes $({}^Y T)^{\sim}$.

The reasoning in (B) and (C) will become clearer when the reader has mastered Chapter IV. But the conclusion is simple: The classification of complex functions ξ on a finite Abelian group by means of set-representation theory and algebra representation theory gives the same meager information about ξ—namely the group of its periods. But the classification by linear representation theory is much more informative; it distinguishes many different functions having the same period group. This gives us some inkling of the special importance of linear representations in analysis. It also illustrates a rather important fact of mathematical methodology—that the choice of structure to be imposed on a mathematical system in order that its study be fruitful is often a rather subtle matter. Here, in the analysis of functions on a group, we had a choice of at least three stuctures—sets, linear spaces, and algebras, in ascending order of complexity. The fruitful one for the analysis of functions turns out to be the middle one, in which neither too little nor too much structure is abstracted.

6

The second stream of mathematical development which contributed heavily to the theory of group representations was the circle of ideas centering around invariant theory, the Klein Erlanger program, and vector and tensor analysis. The reader will find a very stimulating historical account of these ideas in Hermann Weyl's book [2]. Included in these ideas is the mathematical background of Einstein's Theory of Relativity. In fact the theory of group representations is nothing but the precise mathematical formulation of the global "principle of relativity" in its most general sense. To substantiate this statement we shall conduct the reader on a brief semi-philosophical excursion into the ideas of "absolute" and "relative" as they occur in physics.

There are some physical phenomena of which no absolute, intrinsic description is possible, but only a description by "pointing." To describe such a phenomenon the observer must first *point* to himself and his own position, orientation, state of motion, etc.; only in relation to these will he be able to specify the given phenomenon in terms of numbers or other impersonal symbols. For example, the question: "What is the present velocity and direction of motion of the particle P?" is meaningless by itself (if we take for granted the classical ideas of the relativity of motion). It becomes meaningful only when we point to a particular observer, equipped with a particular coordinate system, and ask what is the velocity with respect to this coordinate system. Indeed, no observer can be intrinsically distinguished from any other observer on the basis of the absolute speed or direction of his motion.

Quite generally, then, by a *frame of reference*, or *coordinate system*, we shall mean a particular material system, attached to an observer, in terms of which we suppose that the observer can describe and measure exhaustively (in terms of numbers or other impersonal symbols) any object or event in the universe. The "principle of relativity" asserts, to begin with, that there will be many different frames of reference (attached to different observers) which are *indistinguishable* from each other by any absolute properties; that is, the laws of nature formulated from the standpoint of any two of them will be identical. Let us denote by $\mathscr{S}$ the set of all frames of reference which are thus intrinsically indistinguishable from some given one.

Now, although all elements of $\mathscr{S}$ are intrinsically indistinguishable from each other, it is not true that all *pairs* of elements of $\mathscr{S}$ are intrinsically indistinguishable from each other. Indeed, if $\tau \in \mathscr{S}$, an observer looking out on the universe from the standpoint of τ, and seeing another observer with a frame of reference σ in $\mathscr{S}$, will be able to give an impersonal description (verbal or numerical) of σ which will determine σ uniquely in terms of τ. (This follows from the assumption that each frame of reference τ is "complete," so that every feature of the universe can be described in terms of τ without further "pointing.") Thus, given two pairs $\langle \sigma, \tau \rangle$ and $\langle \sigma', \tau' \rangle$ of elements of $\mathscr{S}$, we can ask: "Does σ' have the same symbolic description from the standpoint of τ' that σ has from the standpoint of τ?" And the answer (since it involves only comparison of symbols) will be an absolute "yes" or "no".

Thus we envisage an *absolute* equivalence relation $\sim$ between pairs of elements of $\mathscr{S}$:

$$\langle \sigma, \tau \rangle \sim \langle \sigma', \tau' \rangle \text{ if and only if } \sigma' \text{ has}$$
$$\text{the same description in terms of } \tau' \qquad (28)$$
$$\text{that } \sigma \text{ has in terms of } \tau.$$

The general "principle of relativity" can be regarded as laying down the following rather plausible postulates for $\sim$ (valid for all $\rho, \sigma, \tau, \rho', \sigma', \tau'$ in $\mathscr{S}$):

(P1) $\langle \sigma, \sigma \rangle \sim \langle \tau, \tau \rangle$;

(P2) if $\langle \sigma, \tau \rangle \sim \langle \sigma', \tau' \rangle$, then $\langle \tau, \sigma \rangle \sim \langle \tau', \sigma' \rangle$;

(P3) if $\langle \rho, \sigma \rangle \sim \langle \rho', \sigma' \rangle$ and $\langle \sigma, \tau \rangle \sim \langle \sigma', \tau' \rangle$, then $\langle \rho, \tau \rangle \sim \langle \rho', \tau' \rangle$;

(P4) for any three elements σ, τ, τ' of $\mathscr{S}$, there is a unique σ' in $\mathscr{S}$ such that $\langle \sigma, \tau \rangle \sim \langle \sigma', \tau' \rangle$.

We shall now forget for the moment the physical interpretation (28) of $\sim$, and regard $\mathscr{S}$ as an abstract non-void set and $\sim$ as an abstract equivalence relation on $\mathscr{S} \times \mathscr{S}$ satisfying (P1)–(P4). These postulates imply that there is a natural group G acting transitively and freely on $\mathscr{S}$. Indeed, let us define G to be the family of all equivalence classes in $\mathscr{S} \times \mathscr{S}$ under $\sim$, and denote by $\langle \sigma, \tau \rangle^{\sim}$ the class to which $\langle \sigma, \tau \rangle$ belongs. Given elements g and h of G, choose first a pair $\langle \sigma, \tau \rangle$ belonging to h, and then a pair $\langle \rho, \sigma \rangle$ (with the second member σ equal to the first member of $\langle \sigma, \tau \rangle$) belonging to g (such exists by P4); then put $gh = \langle \rho, \tau \rangle^{\sim}$. It follows from (P3) that gh depends only on g and h. An easy verification based on (P1)–(P4) shows that G is a group under the product operation $\langle g, h \rangle \mapsto gh$, with unit $e = \langle \sigma, \sigma \rangle^{\sim}$ ($\sigma \in \mathscr{S}$) and inverse given by

$$(\langle \sigma, \tau \rangle^{\sim})^{-1} = \langle \tau, \sigma \rangle^{\sim}.$$

If $g \in G$ and $\tau \in \mathscr{S}$, let $g\tau$ denote the unique element σ of $\mathscr{S}$ such that $\langle \sigma, \tau \rangle \in g$. Then $\langle g, \tau \rangle \mapsto g\tau$ is a transitive free left action of G on $\mathscr{S}$.

Conversely, given a set $\mathscr{S}$ acted upon (to the left) transitively and freely by a group G, the statement

$$\langle \sigma, \tau \rangle \sim \langle \sigma', \tau' \rangle \Leftrightarrow \text{there is an element } g$$
$$\text{of } G \text{ such that } \sigma = g\tau \text{ and } \sigma' = g\tau'$$

defines an equivalence relation $\sim$ on $\mathscr{S} \times \mathscr{S}$ satisfying (P1)–(P4). Thus the system $\mathscr{S}, \sim$ satisfying (P1)–(P4) is essentially just a set acted upon transitively and freely by a group.

Returning now to the interpretation (28) of $\mathscr{S}$ and $\sim$, we shall refer to G as the *symmetry group* of the universe (or part of the universe) under discussion.

This symmetry group describes precisely the kind of relativity to which all descriptions and measurements in this universe must be subject. To see this, suppose we wish to measure some fixed object or quantity in the given universe. We choose a frame of reference σ in $\mathscr{S}$. With respect to this σ, the result of the measurement will be some member q of the set Q of all possible outcomes of this kind of measurement. The set Q is independent of σ; but the particular q in Q which represents the result of this individual measurement will vary with the frame of reference σ with respect to which the measurement was made; to indicate this dependence we write q_σ for q. (For instance, if we are measuring the velocity of a particular particle at a particular moment, Q will be $\mathbb{R}^3$; that is, the result of the measurement is an ordered triple of real numbers, the three components of the velocity vector. But the particular triple which we obtain will depend on the frame of reference which we are using.) We will suppose that the measurement is *complete* in the sense that, if σ and τ are any two elements of $\mathscr{S}$, q_τ determines q_σ. (For example, a measurement of merely the x-component of velocity of a particle with respect to the observer's frame of reference would not be "complete"; for a knowledge of the x-component of velocity with respect to τ would not determine the x-component of velocity with respect to a rotated coordinate system σ.) Since the measurement is "complete," for each pair σ, τ of elements of $\mathscr{S}$ there will be a well-defined permutation $T_{\sigma,\tau}$ of Q such that $T_{\sigma,\tau}(q_\tau) = q_\sigma$ for all measurements of this particular sort. That is, $T_{\sigma,\tau}$ carries the result of the measurement made with respect to τ into the result of the measurement of the same quantity made with respect to σ. Evidently

$$T_{\sigma,\sigma} = \text{identity}$$
$$T_{\rho,\sigma} \circ T_{\sigma,\tau} = T_{\rho,\tau}. \tag{29}$$

The intrinsic indistinguishability of the frames of reference requires that $T_{\sigma,\tau}$ depend only on $\langle \sigma, \tau \rangle^\sim$; that is,

$$T_{g\tau,\tau} \text{ depends only on } g, \text{ not on } \tau. \tag{30}$$

So, writing T_g instead of $T_{g\tau,\tau}$, we have by (29)

$$T_e = \text{identity}$$
$$T_g T_h = T_{gh} \qquad (g, h \in G). \tag{31}$$

That is, $\langle g, q \rangle \mapsto T_g q$ is a left action of G on Q; and T is a set-representation of G on Q (see §5).

In principle one can regard any left action of G on a well-defined set Q as arising in this way from the process of measuring some kind of physical quantity in a universe whose symmetry group is G.

In this context we can generalize the classical physical notions of vectors, tensors, and so forth, as *invariant* constituents of physical description. For example, when we measure the velocity of a particle in two coordinate systems σ and τ, obtaining the results $u = \langle u_1, u_2, u_3 \rangle$ and $v = \langle v_1, v_2, v_3 \rangle$ respectively, we think of u and v as being descriptions of the *same velocity vector* in the different coordinate systems. This suggests that, logically speaking, a vector ξ is nothing but an equivalence class of pairs $\langle u, \sigma \rangle$, where σ is a coordinate system in $\mathscr{S}$ and u is the triple of "components of ξ relative to σ." Quite generally, given a set-representation T of our symmetry group G, acting on a set Q, we define a *quantity of type T, Q* to be an equivalence class of the following equivalence relation $\approx$ on $Q \times \mathscr{S}$:

$$\langle q, \sigma \rangle \approx \langle q', \sigma' \rangle \text{ if and only if, for some } g$$
$$\text{in } G, \text{ we have } \sigma' = g\sigma \text{ and } q' = T_g q.$$

Let $\langle q, \sigma \rangle^0$ denote the $\approx$-class of $\langle q, \sigma \rangle$. By analogy with the special case of vectors, the equation $\langle q, \sigma \rangle^0 = \alpha$ is usually read as follows: *q is the value of the quantity α in the frame of reference σ.* Of course, once σ is fixed, the mapping

$$q \mapsto \langle q, \sigma \rangle^0 \tag{32}$$

is a one-to-one correspondence between Q and the family Q^0 of all quantities of type T, Q.

Up to now we have considered Q merely as a set. If Q has algebraic or topological or other mathematical structure, it will be important to know whether this structure has "physical significance." Suppose for definiteness that Q has the structure of a real linear space. For each fixed σ in $\mathscr{S}$ we can use the bijection (32) to transfer this linear space structure to Q^0. To say that this linear space structure of Q has physical significance is to say that the linear space structure on Q^0 so obtained is *independent of σ.* By the definition of $\approx$ this will be the case if and only if T_g is a real-linear automorphism of Q for every g in G, that is, T is a real-linear representation of G. In this case, as we have said, Q^0 has a well-defined real linear space structure. Two elements of Q^0 can be added (or multiplied by real numbers) by adding (or multiplying by real numbers) their values in any one coordinate system. Elements of Q^0 are referred to in this case as *real-linear quantities.*

In this context the problem, mentioned in §5, of classifying to within equivalence all possible real linear representations of G takes the following form: To classify all possible kinds of real-linear quantities which can occur in a universe whose symmetry group is G.

Of course the same comments apply to any other kind of mathematical structure on the set Q.

7

Let us give some examples of physically interesting quantities whose types are defined as in §6 by group representations. Most of the representations involved will be real-linear. Thus the corresponding physical quantities will be real-linear; they can be added and multiplied by real numbers independently of the coordinate system.

To begin with, we take our symmetry group G to be the group O(3) of all 3×3 real orthogonal matrices. We interpret this group by regarding our observers as looking out on three-dimensional Euclidean space from a fixed vantage-point θ, the origin, and with a fixed unit of length (both being the same for all observers). Each one chooses a coordinate system consisting of an ordered triple of mutually perpendicular unit vectors through θ. If $g \in O(3)$ and σ is one of these coordinate systems, then $g\sigma$ is the rotated coordinate system consisting of the ordered triple of vectors whose components relative to σ are $\langle g_{11}, g_{12}, g_{13} \rangle$, $\langle g_{21}, g_{22}, g_{23} \rangle$, and $\langle g_{31}, g_{32}, g_{33} \rangle$ respectively.

Notice that an element g of O(3) has determinant $\det(g) = \pm 1$. If $\det(g) = 1$, g preserves orientation and is called *proper*; if $\det(g) = -1$, g includes a reflection and is *improper*.

Here are some examples of types Q, T of real-linear quantities with respect to O(3).

Example 1. $Q = \mathbb{R}$; T_g is the identity map on $\mathbb{R}$ for all g. Quantities of this type T, Q are called *scalars*. They are just real numbers which are independent of the coordinate system in which they are measured.

Example 2. $Q = \mathbb{R}$; for $g \in O(3)$, T_g is multiplication by $\det(g)$. Quantities of this type are sometimes called *pseudo-scalars*. They are like scalars as long as we stick to proper rotations, but are changed in sign by a reflection. An example of a pseudo-scalar is the 3×3 determinant formed by the components of three vectors (see Example 3).

Example 3. $Q = \mathbb{R}^3$; $T_g(x_1, x_2, x_3) = \langle x'_1, x'_2, x'_3 \rangle$, where $x'_i = \sum_{j=1}^{3} g_{ij} x_j$ $(i = 1, 2, 3)$. Quantities of this type are called *vectors*. This definition reproduces the elementary notion of a vector as an entity having a direction and a magnitude. Physical examples of vectors include velocity, acceleration, force, and electric field strength.

Example 4. $Q = \mathbb{R}^3$; $T_g(x_1, x_2, x_3) = \langle x'_1, x'_2, x'_3 \rangle$,   where $x'_i = \det(g) \sum_{j=1}^{3} g_{ij} x_j$. Quantities of this type are *pseudo-vectors*. They bear the same relation to vectors as pseudo-scalars bear to scalars. If x and y are vectors, the familiar vector product $x \times y$ of elementary vector analysis is actually not a vector but a pseudo-vector.

Example 5. Let r be a positive integer, and V_r the real-linear space of all real functions on $\{1, 2, 3\} \times \{1, 2, 3\} \times \cdots \times \{1, 2, 3\}$ (r times). Let T be the real-linear representation of G on V_r defined by

$$(T_g \xi)(i_1, \ldots, i_r) = \sum_{j_1=1}^{3} \cdots \sum_{j_r=1}^{3} g_{i_1 j_1} \cdots g_{i_r j_r} \xi(j_1, \ldots, j_r). \tag{33}$$

The quantities of this type T, V_r are called *tensors of rank r*. Obviously a tensor of rank 1 is just a vector. A physical example of a tensor of rank 2 is the state of "stress" at a point within a solid medium.

If we were to multiply the right side of (33) by $\det(g)$, we would obtain the definition of a *pseudo-tensor of rank r*.

The physical applicability of these examples leads us, from a purely physical motivation, to wonder how many essentially different real-linear representations of O(3) actually exist. That is, what types of real-linear quantities are there with respect to this group? Interestingly enough, it turns out that the quantities which we have already constructed in a certain sense exhaust all the possibilities. All real-linear quantities with respect to O(3) are in a natural sense "contained in" those which we have already constructed.

Of course the rotation group O(3) is only a part of the total symmetry group G of the actual physical universe. The precise nature of G depends of course on what physical theories we adopt. Let us first take the standpoint of classical Newtonian mechanics. From this standpoint a frame of reference means a method of describing *events* (i.e., points in space-time) by ordered quadruples $\langle x_1, x_2, x_3, t \rangle$ of real numbers (x_1, x_2, x_3 being the coordinates of the *position* of the event and t its *time*). The symmetry group of Newtonian mechanics turns out to be the multiplicative group N of all 5×5 real matrices of the form

$$
g = \left(\begin{array}{ccc:cc}
 & & & -v_1 & -u_1 \\
 & \rho & & -v_2 & -u_2 \\
 & & & -v_3 & -u_3 \\
\hdashline
0 & 0 & 0 & 1 & -s \\
0 & 0 & 0 & 0 & 1
\end{array}\right), \tag{34}
$$

where $\rho \in O(3)$ and the v_i, u_i, and s are real numbers. The meaning of the matrix g of (34) is as follows: If one and the same event is described in the two frames of reference τ and $g\tau$ by the quadruples $\langle x_1, x_2, x_3, t \rangle$ and $\langle x'_1, x'_2, x'_3, t' \rangle$ respectively, then the two quadruples must be related by

$$
\begin{pmatrix} x'_1 \\ x'_2 \\ x'_3 \\ t' \\ 1 \end{pmatrix} = g \begin{pmatrix} x_1 \\ x_2 \\ x_3 \\ t \\ 1 \end{pmatrix} \qquad \text{(matrix multiplication).} \tag{35}
$$

To clarify the meaning of N let us mention some special kinds of elements of N. If the v_i, u_i, and s are all 0, then (35) reduces to

$$
x'_i = \sum_{j=1}^{3} \rho_{ij} x_j \qquad (i = 1, 2, 3),
$$

$$
t' = t,
$$

and this is just a *rotation of the space coordinates* (leaving the time coordinate fixed). If $\rho = 1$ (the unit matrix) and the v_i and s are all 0, (35) becomes

$$
x'_i = x_i - u_i \qquad (i = 1, 2, 3),
$$

$$
t' = t,
$$

which describes a *translation of the space coordinates* by the amount $\langle u_1, u_2, u_3 \rangle$. If $\rho = 1$ and the v_i and u_i are all 0, (35) takes the form

$$
x'_i = x_i \qquad (i = 1, 2, 3),
$$

$$
t' = t - s;
$$

and this is a *translation of the time coordinate* by the amount s. Finally, if $\rho = 1$ and the u_i and s are 0, (35) reduces to

$$x_i' = x_i - v_i t \qquad (i = 1, 2, 3),$$

$$t' = t;$$

and g then describes a *transition to a new frame of reference moving with uniform velocity* $v = \langle v_1, v_2, v_3 \rangle$ with respect to the old one. These four kinds of elements clearly generate all of N.

The fact that space-rotations belong to the fundamental symmetry group N simply means that the laws of nature do not distinguish one direction in space from another. The fact that space-translations and time-translations are in N says that the laws of nature are the same in all parts of the universe and at all times respectively. Lastly, the presence in N of transitions to moving frames of reference means that there is no such thing as absolute rest in our universe; the laws of nature give no indication of any absolute difference between rest and uniform motion.

Notice that the map

$$\pi : g \mapsto \rho \qquad (g \text{ being as in (34)})$$

is a homomorphism of N onto O(3), and so enables us to pass from quantities with respect to O(3) to quantities with respect to N. For example, by a *vector with respect to N* we mean a real-linear quantity of type $T \circ \pi$, $\mathbb{R}^3$, where T is the representation of O(3) given in the preceding Example 3. Scalars, pseudoscalars, pseudo-vectors and tensors with respect to N are defined similarly.

Example 6. The *velocity* of a particle, which is a vector when only space rotations are considered, is no longer a vector with respect to N. Indeed, if a particle P is observed to have a velocity $w = \langle w_1, w_2, w_3 \rangle$ with respect to some frame of reference σ, and if g is as in (34), then the velocity of P with respect to $g\sigma$ will be $\rho w - v$ (where $v = \langle v_1, v_2, v_3 \rangle$). The map $S_g : w \mapsto \rho w - v$ is an affine transformation of $\mathbb{R}^3$ but is not linear (unless $v = 0$). Thus, with respect to N, the velocity of a particle is what we should call (in the spirit of §6) a *real-affine* quantity (of type S, $\mathbb{R}^3$), but is not a real-linear quantity; in particular it is not a vector.

Acceleration and force, however, remain vectors when the symmetry group is extended to N.

***Example* 7.** Let P be a particle of mass m moving with velocity $w = \langle w_1, w_2, w_3 \rangle$ (with respect to some frame of reference). With respect to the rotation group O(3) its *momentum* $p = mw$ is a vector (with components $p_i = mw_i$), and its *kinetic energy* $E = \frac{1}{2}m(w_1^2 + w_2^2 + w_3^2)$ is a scalar. With respect to the larger symmetry group N, however, p is no longer real-linear (see Example 6), and E is no longer a scalar. However, we can build out of m, p, and E a five-component real-linear quantity with respect to N as follows: Let S be the five-dimensional real-linear representation of N (on $\mathbb{R}^5$) which sends the element g of (34) onto the 5×5 matrix

$$
S_g = \left(
\begin{array}{ccc|c}
1 & -v_1' \quad -v_2' \quad -v_3' & & \frac{1}{2}(v_1^2 + v_2^2 + v_3^2) \\
\hline
0 & & & -v_1 \\
0 & & \rho & -v_2 \\
0 & & & -v_3 \\
\hline
0 & 0 \quad 0 \quad 0 & & 1
\end{array}
\right),
$$

where $v_i' = \sum_{j=1}^{3} v_j \rho_{ji}$. One verifies that this is indeed a real-linear representation of N. If the momentum and energy of P with respect to some frame of reference τ are p and E respectively, and with respect to $g\tau$ are p' and E' respectively, one checks that

$$
\begin{pmatrix} E' \\ p_1' \\ p_2' \\ p_3' \\ m \end{pmatrix} = S_g \begin{pmatrix} E \\ p_1 \\ p_2 \\ p_3 \\ m \end{pmatrix} \quad \text{(matrix multiplication).} \tag{36}
$$

In view of this, the array

$$
\begin{pmatrix} E \\ p_1 \\ p_2 \\ p_3 \\ m \end{pmatrix} \tag{37}
$$

is a real-linear quantity (with respect to N) of type S, $\mathbb{R}^5$.

Example 8. Let W_g be the 4×4 matix obtained by omitting the first row and column of S_g of Example 7:

$$W_g = \begin{pmatrix} & & & \vdots & -v_1 \\ & \rho & & \vdots & -v_2 \\ & & & \vdots & -v_3 \\ ----&--&--&\vdots&---- \\ 0 & 0 & 0 & \vdots & 1 \end{pmatrix}.$$

Notice that $W: g \mapsto W_g$ is a real-linear representation of N on $\mathbb{R}^4$. Thus the four-component array

$$\begin{pmatrix} p_1 \\ p_2 \\ p_3 \\ m \end{pmatrix}$$

obtained by omitting the kinetic energy E from (37) is a real-linear quantity of type W, $\mathbb{R}^4$ with respect to N.

The last two examples are relevant to the question: What is the status of *energy* as an "absolute" entity in the external world? Let us digress briefly to discuss this, since it vividly illustrates the physical importance of group representations.

On the one hand we know from the principle of the conservation of energy that the latter is more than an artificial mathematical construct from the laws of Newtonian mechanics; it is a fundamental persistent entity in the world around us. One should therefore be able to make absolute statements describing the "amount of energy" present in a given isolated material system, which (like similar statements about "amount of mass") would be independent of the observer. On the other hand, the formula $\frac{1}{2}m(v_1^2 + v_2^2 + v_3^2)$ describing the kinetic energy of a moving particle P is clearly dependent on the observer; a change to a new frame of reference which is in motion with respect to the old one will change the observed kinetic energy of P. The solution of this paradox is that we must allow "energy" to be not a scalar but a quantity (in the strict sense of §6) of more general type. Now this quantity cannot be described in each frame of reference by the single number $\frac{1}{2}m(v_1^2 + v_2^2 + v_3^2)$; for this number does not determine either m or $v_1^2 + v_2^2 + v_3^2$ separately, and so cannot determine the value of the kinetic energy as observed from a new frame of reference moving with respect to the first one. (See the discussion of the *completeness* of a measurement, in §6.) There is another consideration: One must be able to *add* amounts of energy invariantly. The total energy of a system consisting of several independent subsystems should be obtained by an "absolute" process of addition of the separate energies of the sub-systems. These considerations suggest that "energy" ought to be a real-linear quantity, of dimension greater than 1, with respect to the fundamental symmetry group N. Example 7 provides a five-dimensional candidate (37) for the expression of energy as an "absolute" real-linear quantity. Here the classical kinetic energy E appears as only one component of the total quantity; the kinetic energy must be "completed" by the adjunction of four other components describing the momentum and mass.

The Newtonian group, as an expression of the symmetry of the physical world, depended on nineteenth-century pre-relativistic notions of space and time. With the advent of Einstein's special theory of relativity in 1905, it became clear that the Newtonian group N must be replaced by an entirely different group of symmetries, the so-called *Poincaré group P*. For the precise definition of P we refer the reader to XII.7.14. Though by no means isomorphic to N, P is obtained from N by a "small deformation" of the latter, so that in a small neighborhood of the unit element the two groups look very similar. This is why, so long as particles with only small velocity are considered, the behavior of N closely approximates the behavior of P.

We mention that, with respect to P, the proper "absolute" description of energy represents it as a four-component quantity (the so-called energy-momentum vector), in analogy with Example 8 rather than with Example 7.

8

The two great achievements of twentieth-century physics are the theory of relativity and the quantum theory. It is in the mathematics of quantum theory that unitary representations of groups, which were defined in §1 and which form the major motivation of this work, find their most important physical application. This application falls within the general framework sketched in §6; and in the paragraphs that follow we will try to suggest something of its special character and challenges.

Let X be a complex Hilbert space with inner product $(\ ,\ )$, and E the family of all unit vectors in X. Two vectors ξ, η in E will be called *equivalent* if $\eta = \lambda\xi$ for some complex number λ with $|\lambda| = 1$; and an equivalence class of vectors in E will be called a *ray of* X. Let S^0 be the set of all rays of X; and for each pair α, β of rays let $\rho^0(\alpha, \beta) = |(\xi, \eta)|$, where $\xi \in \alpha$, $\eta \in \beta$. We will call the pair S^0, ρ^0 the *ray space derived from* X. In general, by a *ray space* we mean a pair

S, ρ such that (i) S is a set, (ii) $\rho: S \times S \to \mathbb{R}$ (called the *ray inner product*), and (iii) there is a Hilbert space X such that S, ρ is isomorphic with the ray space S^0, ρ^0 derived from X (in the obvious sense that there is a bijection $\Phi: S \to S^0$ carrying ρ into ρ^0). If G is a group, a *ray-representation of G on a ray space S, ρ* is a set-representation W of G on S which preserves ρ (i.e., $\rho(W_g \alpha, W_g \beta) = \rho(\alpha, \beta)$ for $g \in G$ and $\alpha, \beta \in S$). The ray-equivalence of two ray-representations of G is defined in the obvious manner (cf. §5).

Now fix an isolated physical system M. By a *state* of M we mean, very roughly speaking, the outcome of a maximally complete and accurate observation of M. It is a fundamental postulate of (single-particle) quantum mechanics that the set S of all states of M forms a ray space under a certain physically meaningful ray inner product ρ. We call S, ρ the *state space* of M. Thus, if G is the symmetry group of the universe, we expect to find a ray-representation W of G acting on S, ρ and having the usual interpretation in terms of the physics of M, namely: If $g \in G$ and α is the state of M as observed from the frame of reference σ, then $W_g \alpha$ is the state of M as observed from the frame of reference $g\sigma$. This W will be called the ray-representation *characteristic* of M. Evidently the ray-equivalence class of W will be a property of the kind of physical system M with which we are dealing. Classifying ray-representations of G to within ray-equivalence should therefore be a significant step toward analyzing what kinds of physical systems can exist in a quantum-mechanical universe with symmetry group G.

Let S^0, ρ^0 be the ray space derived from the complex Hilbert space X. Any unitary representation T of the symmetry group G acting in X will in an obvious manner generate a ray-representation—which we call $T^{(r)}$—of G on S^0, ρ^0. In general it is *not true* that every ray-representation W of G on S^0, ρ^0 is of the form $T^{(r)}$ for some unitary representation T of G on X; and, even if $W = T^{(r)}$ for some T, it is *not true* in general that the ray-equivalence class of W determines the unitary equivalence class of T. In spite of this, there is a close connection (which we will not pursue further) between the problem of classifying the ray-equivalence classes of ray-representations of G and the problem of classifying the unitary equivalence classes of unitary representations of G. Indeed, for the purposes of the rest of this section the former problem will be regarded as replaced by the latter.

So let us assume that the state space S^0, ρ^0 of M is derived from the Hilbert space X, and that T is a unitary representation of the symmetry group G on X such that $T^{(r)}$ is the ray-representation characteristic of M. We remark that, if ξ and η are two unit vectors in X such that $T_g \xi = \eta$ for some g in G, then in a sense ξ and η (or rather the rays through them) describe the same "intrinsic state" of M; for the transition from one state to the other can be exactly duplicated by a change in the standpoint of the observer.

Now suppose that T is *irreducible*. What does this say about M? It would be tempting to conjecture that irreducibility of T implies that $T^{(r)}$ acts transitively on the state space S^0. This is not the case; but for purposes of rough interpretation let us suppose that it is "essentially" the case. Then by the remark in the preceding paragraph M has only one "intrinsic state." It can never undergo any intrinsic change. Any change which it *appears* to undergo (change in position, velocity, etc.) can be "cancelled out" by an appropriate change in the frame of reference of the observer. Such a material system is called an *elementary system* or an *elementary particle*. The word "elementary" reflects our preconception that, if a physical system undergoes an intrinsic change, it must be that the system is "composite," and that the change consists in some rearrangement of the "elementary parts."

The search for such elementary particles has been of absorbing interest throughout the whole history of natural philosophy, starting with the Vaiśeshika philosophers of India and the ancient Greek atomists. It is more vital now than ever because of the proliferation of the so-called elementary particles of modern high-energy physics. And it now appears, as we have seen, that *the search for elementary particles is closely connected with the search for irreducible unitary representations of appropriate symmetry groups*. In fact, one of the major starting-points of the modern theory of infinite-dimensional unitary group representations was an article by the physicist E. Wigner [1] in 1939, in which he investigated the irreducible representations of the Poincaré group with precisely the motivation sketched above.

In concluding this section we shall indicate one very significant development in recent physics. It might at first appear that only one group—namely the Poincaré group, the symmetry group of the theory of relativity—would be admissible as the group G underlying the preceding discussion. No such limitation is inherently called for, however, if we remember the abstract deductive nature of physical theory. Suppose for example that G is taken to be bigger than P, so that it contains the Poincaré group as a proper subgroup. If τ and σ are two frames of reference related by an element of P (i.e., $\sigma = g\tau$ for some g in P), we will say that τ and σ are *spatio-temporally related*. But, since G is now assumed strictly bigger than P, we must admit the possibility of two observers having frames of reference which are related by some element of $G \setminus P$, that is, which differ in some "internal," non-spatio temporal respect. If this seems hard to visualize, we must remember that the prime requisite for a physical concept is not that it can be visualized but that consequences agreeing with experiment can be deduced from it. And certainly some experimental consequences are suggested by this train of thought. Indeed, imagine for simplicity's sake that we are able to find an irreducible unitary representation W of the assumed "total" symmetry group G which, when restricted to the Poincaré subgroup P, splits into a direct sum of two irreducible representations U and V of P; and suppose further that U and V, considered as elementary particles with respect to the "space-time" group P, correspond to the observed properties of (say) protons and neutrons respectively. Then W, as an elementary particle with respect to G, contains within itself the possibility of both protons and neutrons. More precisely, protons and neutrons are just different states of the same "W-particle." As a concrete illustration of this approach, suppose that a proton collides with an atomic nucleus. It might emerge from the collision as a proton with altered velocity; or possibly the proton might disappear and a neutron might emerge from the collision. These two possibilities, at first sight qualitatively so different, become unified when described in terms of W-particles. In both cases a W-particle, initially in a proton state, collides with the nucleus and rebounds from it in an altered state—in the first case a proton state with altered velocity, in the second a neutron state.

This approach suggests a group-representational basis for classifying and unifying the multitudes of different so-called elementary particles observed in high-energy physics laboratories. It must be admitted, however, that the actual success achieved by this approach has so far been quite limited.

For a much more thorough discussion of the applications of group representations to physics we refer the reader to Mackey [21].

9

Up to now we have been motivating the study of representations of groups by showing its importance in two areas of mathematics—the harmonic analysis of functions and the analysis of "quantities" in a universe with given symmetry group. Our next step is to show that the theory of group representations itself is to a large extent only a special case of a more general theory, namely the theory of the structure and representations of algebras.

If A is a (complex associative) algebra, by a *representation* of A we mean a homomorphism T of A into the algebra of all linear operators on some complex-linear space $X(T)$ (called the *space of* T). (In the light of §5, "complex-linear representation" might be a less ambiguous phrase; but we prefer the shorter usage.) If in addition A has an involution operation * making it a *-algebra (see Chapter I), and if $X(T)$ is a Hilbert space and the T_a are bounded and satisfy $T_{a^*} = (T_a)^*$, then T is a *-representation* of A.

Now let G be a finite group. The complex-linear space $\mathscr{L}(G)$ of all complex functions on G becomes a *-algebra, called the *group *-algebra* of G, under the following operations of multiplication (here called *convolution*) and involution:

$$(fg)(x) = \sum_{y \in G} f(y)g(y^{-1}x),$$

$$f^*(x) = \overline{f(x^{-1})} \tag{38}$$

$(f, g \in \mathscr{L}(G); x \in G)$. It has been known since the work of Frobenius that the structure of $\mathscr{L}(G)$, as an abstract *-algebra, has the following simple description: $\mathscr{L}(G)$ is isomorphic with a direct sum of total matrix *-algebras

$$\mathscr{L}(G) \cong \sum_{i=1}^{r} {}^{\oplus} M_{n_i}. \tag{39}$$

Here M_{n_i} stands for the total matrix *-algebra of all $n_i \times n_i$ complex matrices (multiplication and involution being the ordinary multiplication and adjoint of matrices). Furthermore, the right side of (39) is very simply related to the representation theory of G. Indeed, every irreducible unitary representation S of G gives rise to an irreducible *-representation S' of $\mathscr{L}(G)$ by the simple formula

$$S'_f = \sum_{x \in G} f(x)S_x; \tag{40}$$

and every such S' comes from projecting $\mathscr{L}(G)$ onto one of the terms M_{n_i} in the direct sum on the right of (39). Thus the terms on the right of (39) are in natural one-to-one correspondence with the set $\hat{G}$ of all unitary equivalence classes of irreducible unitary representations of G—the element of $\hat{G}$ corresponding to M_{n_i} being of dimension n_i. The sequence $n_1, \ldots, n_r$ in (39) is thus just the sequence of the dimensions of the elements of $\hat{G}$.

From the algebraic standpoint it is desirable to characterize more generally those *-algebras which have the simple structure given by the right side of (39). Here is such a characterization: A finite-dimensional *-algebra A turns out to be *-isomorphic with a direct sum of total matrix *-algebras, that is,

$$A \cong \sum_{i=1}^{r} {}^{\oplus} M_{n_i} \qquad \text{for some } n_1, \ldots, n_r, \tag{41}$$

if and only if

$$a^*a = 0 \Rightarrow a = 0 \tag{42}$$

for all a in A. (This is proved in IV.7.4.) Such a *-algebra will be called a *Frobenius *-algebra*. It is easy to see directly that the group *-algebra of a finite group G has property (42); for, if $f \in \mathcal{L}(G)$, $(f^*f)(e) = \sum_{x \in G} |f(x)|^2$, so $f^*f = 0 \Rightarrow \sum_{x \in G} |f(x)|^2 = 0 \Rightarrow f = 0$. One should note however that not every Frobenius *-algebra A can be the group *-algebra of a finite group. Indeed, since every group has at least one one-dimensional unitary representation (namely the trivial one sending each group element into 1), the presence of the number 1 among the n_i of (41) is necessary for A to be a group *-algebra. Another necessary condition is that each n_i should divide the dimension $n_1^2 + n_2^2 + \cdots + n_r^2$ of A. As a matter of fact, we know of no simple condition on the sequence $n_1, n_2, \ldots, n_r$ which is both necessary and sufficient for A to be a group *-algebra.

One of the crucial motivations for Volume I of the present work is the desire to generalize the above observations to arbitrary locally compact groups. Can we find a class of (in general infinite-dimensional) *-algebras which will play the role of Frobenius *-algebras in relation to arbitrary locally compact groups? The *-algebras of this class should be defined by relatively simple postulates (generalizing (42)). They should have a relatively accessible structure theory intimately related to their *-representations and consisting (hopefully) in a generalized analogue of (41). Furthermore, from every locally compact group G it should be possible to construct canonically a "group *-algebra" belonging to this class, whose *-representation theory should be in one-to-one correspondence with the unitary representation theory of G (the correspondence being given by some generalization of (40)).

Such a class of *-algebras does exist; it is the class of so-called C^*-algebras. The theory of C^*-algebras and their *-representations will be the main topic of the present Volume. To place it in a clearer light, we shall in this Introduction approach the axiomatic definition of a C^*-algebra in three stages, taking up first Banach algebras, then Banach *-algebras, and finally C^*-algebras.

10

For the definition of a normed algebra and a Banach algebra we refer the reader to Chapter I.

A fundamental step in the derivation of the structure of Frobenius *-algebras is the fact that a finite-dimensional division algebra F over the complex field $\mathbb{C}$ is one-dimensional, that is, $F = \mathbb{C}$. This is no longer true if we omit the assumption of finite-dimensionality (as the field of complex rational functions shows). In 1938 and 1939, however, Gelfand [2] and Mazur [1] made the important observation that F is still one-dimensional if (without assuming F finite-dimensional) we suppose merely that it carries a norm making it a normed algebra. This has many consequences, among them the fact (see V.7.4) that any regular maximal ideal of a commutative Banach algebra A is the kernel of a unique complex homomorphism of A (that is, an algebra homomorphism of A into the algebra $\mathbb{C}$). Let A be a *commutative* Banach algebra, and $\check{A}$ the space of all complex homomorphisms of A. By the observation just made, $\check{A}$ can be identified with the space of all regular maximal ideals of A. Each element a of A gives rise to a complex function $\check{a}$: $\phi \mapsto \phi(a)$ on $\check{A}$, called the *Gelfand transform* of a. The topology on $\check{A}$ generated by the family of all the $\check{a}$ ($a \in A$) turns out to be locally compact and Hausdorff; and the functions $\check{a}$ vanish at infinity on $\check{A}$. Equipped with this topology $\check{A}$ is called the *Gelfand space* of A. If A is semisimple, that is, if the intersection of all its regular maximal ideals is $\{0\}$, then $a \mapsto \check{a}$ is one-to-one; in that case A has been presented as an algebra of continuous complex functions vanishing at infinity on its Gelfand space $\check{A}$. This is the well-known *Gelfand presentation* of A, discussed in §V.7. It should be compared with the fact (evident from (41)) that a commutative Frobenius *-algebra is simply the algebra of all complex functions on a finite set. The important point is that the latter statement has now been generalized to the infinite-dimensional normed context.

This commutative theory is very useful in studying even noncommutative Banach algebras, since it can be applied to the commutative subalgebras of A.

Our main interest lies in unitary representations of groups and their involutive generalizations; so it is appropriate to study Banach algebras which are also *-algebras. Adding the convenient assumption that $\|a^*\| = \|a\|$, we obtain the notion of a Banach *-algebra (see Chapter I). A very important technical fact about these is that a *-representation T of a Banach *-algebra A is automatically norm-continuous; in fact

$$\|T_a\| \le \|a\| \qquad\qquad (a \in A) \qquad (43)$$

(see VI.3.8). Another interesting result concerns the existence of irreducible *-representations of a Banach *-algebra A: If A is known to have enough *-representations to separate its points, then A has enough *irreducible* *-representations to separate its points (see VI.22.14).

Of vital importance to our subject is the following fact: Given any locally compact group G, there is a Banach *-algebra A, called the $\mathscr{L}_1$ *group algebra* of G, whose non-degenerate *-representations are in natural correspondence with the unitary representations of G. The construction of A depends essentially on the existence of a left-invariant (Haar) measure λ on G; A is in fact just the Banach *-algebra of all λ-summable complex functions on G, the operations of convolution and involution being defined by generalizations of (38) (see III.11.9). It is the possibility of constructing the $\mathscr{L}_1$ group algebra, more than any other fact, which accounts for the importance of the category of locally compact groups in functional analysis. (As a matter of fact, the correspondence between representations of A and of G will not be proved until §VIII.13 of Volume II, where it will be treated in the more general context of Banach *-algebraic bundles.)

Let G be any locally compact group, with left-invariant measure λ. It is easy to see that its $\mathscr{L}_1$ group algebra A has enough *-representations to distinguish its points; in fact the so-called *regular *-representation* of A, which acts by convolution on $\mathscr{L}_2(\lambda)$, is faithful. Hence, by the result mentioned above, A has enough *irreducible* *-representations to separate its points. In virtue of the natural correspondence between *-representations of A and unitary representations of G, it follows that *G has enough irreducible unitary representations to separate its points*. This important result, the existence of plenty of irreducible unitary representations of an arbitrary locally compact group, is an excellent illustration of the principle that theorems on group representations should find their natural context in the larger theory of *-representations of Banach *-algebras.

Thus the class of Banach *-algebras seems to satisfy some of the requirements laid down at the end of §9. In fact, however, it is still too wide a class. The norm of an arbitrary Banach *-algebra is too unrestricted. Even the $\mathscr{L}_1$ norm of the $\mathscr{L}_1$ group algebra of a locally compact group is not related naturally to its representation theory.

It turns out that the subclass that meets all our requirements is the class of C^*-algebras. A C^*-*algebra* is defined as a Banach *-algebra A satisfying the following analogue of (42):

$$\|a^*a\| = \|a\|^2 \qquad\qquad (a \in A). \qquad (44)$$

Notice that the Banach *-algebra $\mathcal{O}(X)$ of all bounded linear operators on a Hilbert space X satisfies (44) and so is a C^*-algebra. Furthermore, any Banach *-algebra A gives rise to a C^*-algebra as follows: In virtue of (43), for each a in A the supremum of the set of all the numbers $\|T_a\|$ (T running over all *-representations of A) is a finite number which we call $\|a\|_c$. The function $a \mapsto \|a\|_c$ is then a seminorm on A satisfying (44); and on factoring out the null space of $\| \ \|_c$ and completing with respect to $\| \ \|_c$ we obtain a

C^*-algebra, called the C^*-*completion* of A and denoted by A_c. By its very definition there is a natural correspondence between the *-representations of A and those of A_c. In particular, if A is the $\mathscr{L}_1$ group algebra of a locally compact group G, A_c is called the group C^*-*algebra* of G; and there is a natural correspondence between the non-degenerate *-representations of A_c and the unitary representations of G.

A norm-closed *-algebra of bounded linear operators on a Hilbert space X, being a *-subalgebra of $\mathcal{O}(X)$, is a C^*-algebra. It is a fundamental result due to Gelfand and Naimark [1] that, conversely, every C^*-algebra is isometrically *-isomorphic with some norm-closed *-subalgebra of $\mathcal{O}(X)$ for some Hilbert space X. (We shall prove this in VI.22.12.) This famous result, showing that the postulates of a C^*-algebra exactly characterize norm-closed *-subalgebras of $\mathcal{O}(X)$, is the primary motivation for the definition (44) of a C^*-algebra. Together with (43), it implies that for a C^*-algebra A we have $\|a\|_c = \|a\|$ $(a \in A)$; consequently the norm of a C^*-algebra is naturally related to its *-representations. It also implies that a C^*-algebra has enough irreducible *-representations to separate its points.

Does the structure theory of an arbitrary C^*-algebra in any sense generalize the right side of (41)? The answer is "yes", though the required generalization of (41) is somewhat sophisticated. To gain some insight into the question let us begin with *commutative* C^*-algebras. If M is a locally compact Hausdorff space, then the *-algebra $\mathscr{C}_0(M)$ of all continuous complex functions on M which vanish at infinity, equipped with pointwise multiplication and complex conjugation and the supremum norm, is a commutative C^*-algebra. Furthermore, every commutative C^*-algebra A is of this form. Indeed, let $\check{A}$ be the Gelfand space of the commutative Banach algebra underlying A. Because of the special properties of A, it turns out that the Gelfand presentation $a \mapsto \check{a}$ of A maps A isometrically and *-isomorphically onto $\mathscr{C}_0(\check{A})$ (see VI.4.3). This result completely solves the problem of the structure of commutative C^*-algebras, presenting them as a clear generalization of commutative Frobenius *-algebras (which are simply the *-algebras $\mathscr{C}_0(M)$ for finite sets M).

Now let us consider the question for an arbitrary C^*-algebra A. A comparison of (41) with the commutative situation described in the preceding paragraph suggests that we should try somehow to represent A as a *-algebra of continuous functions on some topological space M—the values of the functions lying in matrix *-algebras or some generalization thereof. The index set $\{1, \ldots, r\}$ on the right of (41) was in natural correspondence with the set of equivalence classes of irreducible *-representations of A. So, in the general case, let us denote by $\hat{A}$ the family of all unitary equivalence classes of irreducible *-representations of A; and let us try to carry through the above program taking M to be $\hat{A}$.

To discuss continuous functions on $\hat{A}$ we need a topology on $\hat{A}$. It turns out that the space $\mathscr{S}(A)$ of *all* unitary equivalence classes of *-representations of A admits several more or less natural topologies, one of which—the so-called *regional topology*—is discussed at length in Chapter VII. Noting that $\hat{A} \subset \mathscr{S}(A)$, let us equip $\hat{A}$ with the relativized regional topology; the resulting topological space $\hat{A}$ is called the *structure space* of A. If A is commutative, $\hat{A}$ coincides with the Gelfand space $\check{A}$ of A. If A is not commutative, $\hat{A}$ loses all of its separation properties; in general it is not even T_0, much less Hausdorff. However $\hat{A}$ does continue to be locally compact in the sense that every point of $\hat{A}$ has a basis of compact neighborhoods (see VII.6.9).

If A is the group C^*-algebra of a locally compact group G, then $\hat{A}$ can be identified with the space $\hat{G}$ of all unitary equivalence classes of irreducible unitary representations of G; and the regional topology of $\hat{A}$ becomes the regional topology of $\hat{G}$ mentioned in §4.

Now there is a straightforward way to regard A as a *-algebra of functions on $\hat{A}$. For each class τ in $\hat{A}$ let $\mathrm{Ker}(\tau)$ be the kernel of any *-representation T belonging to the class τ; and form the quotient C^*-algebra $A_\tau = A/\mathrm{Ker}(\tau)$. (The fact that $A/\mathrm{Ker}(\tau)$ is a C^*-algebra is proved in VI.8.7; it is of course *-isomorphic with the range of T.) An element a of A now gives rise to a mapping $\hat{a}: \tau \mapsto a + \mathrm{Ker}(\tau)$ associating to each class τ in $\hat{A}$ the element $a + \mathrm{Ker}(\tau)$ of A_τ. In the special case that A is commutative each τ is one-dimensional, so that $A_\tau \cong \mathbb{C}$; and $\hat{a}$ coincides with the Gelfand transform of a. Quite generally we refer to $\hat{a}$ as the *generalized Gelfand transform* of a. The image $\Phi(A)$ of A under the map $\Phi: a \mapsto \hat{a}$ is a *-algebra of functions on $\hat{A}$; the values of these functions at each point τ of $\hat{A}$ lie in a C^*-algebra A_τ which in general varies from point to point of $\hat{A}$. Because of the way in which it arose, each C^*-algebra A_τ is *primitive* in the sense that it has a faithful irreducible *-representation. If τ is of finite dimension n then $A_\tau \cong M_n$. Thus, if all the τ are of finite dimension, $\Phi(A)$ becomes a *-algebra of *matrix-valued* functions as in (41). The generalized Gelfand transform map $a \mapsto \hat{a}$ is always one-to-one, in fact an isometry with respect to the supremum norm of functions on $\hat{A}$.

Can we assert (as we could in the commutative case) that $\Phi(A)$ consists of *all* such functions which are "continuous" and vanish at infinity on $\hat{A}$? To the extent that this can be asserted, to that extent we have generalized the structure theory (41) of Frobenius *-algebras. It turns out that one can prove a rather profound theorem of Stone-Weierstrass type for C^*-algebras (see Glimm [2]), as a result of which a modified form of this assertion can be established. Without entering into details let us briefly indicate some of the features of the situation. In the first place it is not even clear what "continuity" of the functions $\hat{a}$ should mean, since their values at different points τ of their domain lie in quite unrelated algebras A_τ. A minimal requirement for continuity of the $\hat{a}$, it would seem, should be the continuity of the numerical functions $\tau \mapsto \|\hat{a}(\tau)\|_{A_\tau}$ on $\hat{A}$; but if $\hat{A}$ is not Hausdorff even this turns out to be false. Besides these negative comments, however, there are positive properties of $\Phi(A)$. First, it can be shown that the $\hat{a}$ vanish at infinity on $\hat{A}$ (see VII.6.7). Secondly, we have the following result of Dauns and Hofmann [1] (see VIII.1.23): If $a \in A$ and λ is any bounded continuous complex function on $\hat{A}$, then there is an element b of A such that $\hat{b}(\tau) = \lambda(\tau)\hat{a}(\tau)$ for all τ in $\hat{A}$. Thirdly, if $\hat{A}$ happens to be Hausdorff, then $\tau \mapsto \|\hat{a}(\tau)\|_{A_\tau}$ *is* continuous on $\hat{A}$ for each a in A. Assuming that $\hat{A}$ is Hausdorff and putting these three facts together, one can define a natural notion of continuity for functions on $\hat{A}$ with values in the A_τ (see II.13.16), and then show that $\Phi(A)$ must (as in the commutative case) consist of all those functions on $\hat{A}$ with values in the C^*-algebras A_τ which are continuous and vanish at infinity. If $\hat{A}$ is not Hausdorff, the negative features mentioned above indicate the need for rather drastic modification of $\hat{A}$ and $\Phi(A)$ before the latter can be described in terms of *continuous* functions with values in the A_τ. Such a modification is possible; it requires us to replace $\hat{A}$ by its so-called Hausdorff compactification $H(\hat{A})$, and to extend the generalized Gelfand transforms $\hat{a}$ to $H(\hat{A})$. In this work we will not enter upon the details of this aspect of the structure theory of C^*-algebras; the reader will find them in Fell [3].

To summarize, we may say that C^*-algebras form the correct infinite-dimensional generalization of the notion of Frobenius *-algebra. C^*-algebras bear the same relation to arbitrary locally compact groups that Frobenius *-algebras bear to finite groups.

11

In this work our interest in the topology of the structure space $\hat{A}$ of a C^*-algebra A comes primarily not from an interest in the structure of $\Phi(A)$, but from the application of $\hat{A}$ to the study of arbitrary *-representations of A, in particular to the construction of the so-called spectral measure of a *-representation.

Reverting for a moment to Frobenius *-algebras, let S be any non-degenerate *-representation (acting on X) of a Frobenius *-algebra A; and for each τ in $\hat{A}$ let P_τ be projection onto the τ-*subspace* of X—that is, the closed subspace of X spanned by the S-stable subspaces on which S acts equivalently to τ. The P_τ are mutually orthogonal and add up to the identity operator. For any subset W of $\hat{A}$, $P(W) = \sum_{\tau \in W} P_\tau$ is the projection onto that part of X on which S acts like a direct sum of elements of W.

Now suppose that A is an arbitrary C^*-algebra and that S is any non-degenerate $*$-representation of A. Even though S may not be a direct sum of irreducibles (indeed, S may have no irreducible subrepresentations at all), Glimm [5] showed how to define a canonical projection-valued measure P on the Borel σ-field of $\hat{A}$, whose values are projections on the space of S, and which exactly generalizes the projection-valued map $W \mapsto P(W)$ defined above for Frobenius $*$-algebras. This P is called the *spectral measure* of S. If W is a Borel subset of $\hat{A}$, range $(P(W))$ can be thought of roughly as that part of the space of S on which S behaves like a "direct integral" of elements of W. This spectral measure is developed in detail in §VII.9. If A is the group C^*-algebra of a locally compact group G, so that S corresponds to a unitary representation S' of G, the spectral measure of S is the same as the spectral measure of S' which we discussed in §4.

In conclusion it should be remarked (as in §4) that the spectral measure of a $*$-representation S is merely a fragment of the much stronger direct integral decomposition theory of S, which is not treated in this work. It is a fragment, however, which holds under much weaker hypotheses than those required for the full direct integral theory. In particular, unlike the theory of direct integrals, it requires no assumptions of second countability. Further, as Blattner [6] pointed out, the spectral measure is just that fragment of direct integral theory which is required for the Mackey normal subgroup analysis developed in Chapter XII. It is for these reasons that in this work we have treated the spectral measure in detail, while ignoring the more descriptive direct integral theory which holds in the second-countable situation.

I   Preliminaries

This introductory chapter is a list of well-known notations, definitions, and facts that are used without comment in the rest of this work.

1. Logical, Set, and Functional Notation

Square brackets are frequently used informally to denote alternatives. For example, we might assert that if S is a set of real numbers which is bounded above [below], then S has a least upper bound [greatest lower bound].

$P \Rightarrow Q$ is logical implication: If the statement P is true, then so is Q. $P \Leftrightarrow Q$ means that $P \Rightarrow Q$ and $Q \Rightarrow P$.

$\{a, b, c, \ldots\}$ is the set consisting of $a, b, c, \ldots$ $\{x: P(x)\}$ is the set of all x such that the statement $P(x)$ is true. The void set is $\emptyset$. Ordered pairs are denoted by $\langle \ , \ \rangle$; more generally, $\langle a_1, a_2, \ldots, a_n \rangle$ is an ordered n-tuple. Card (S) is the cardinality of the set S.

$x \in A [x \notin A]$ means that x is [is not] a member of A. $A \cup B$ and $A \cap B$ are the union and intersection respectively of the sets A and B. If $\mathscr{S}$ is a family of sets, then $\bigcup \mathscr{S}$, or $\bigcup_{A \in \mathscr{S}} A$, is the union $\{x: x \in A \text{ for some } A \text{ in } \mathscr{S}\}$; and $\bigcap \mathscr{S}$, or $\bigcap_{A \in \mathscr{S}} A$, is the intersection $\{x: x \in A \text{ for all } A \text{ in } \mathscr{S}\}$. $A \setminus B$ is the difference $\{x: x \in A, x \notin B\}$ of the sets A and B; and $A \ominus B$ is the symmetric difference

$(A \setminus B) \cup (B \setminus A)$. $A \subset B$ means that A is contained in B, that is, $x \in A \Rightarrow x \in B$.

If f is a function, domain(f) and range(f) are its domain and range respectively. The statement $f: A \to B$ (in words, f is *on* or *from A to* or *into B*) means that f is a function with domain A and with range contained in B. If range(f) $= B$, f is said to be *onto B* or is *surjective*. If f maps distinct elements in A to distinct elements in B, f is said to be *one-to-one* or *injective*. If f is both surjective and injective it is called *bijective*. The statement $f: x \mapsto y$ (the tail of the arrow having a vertical stroke) means that f is a function and that $f(x) = y$.

Ch_A is the characteristic function of the set A; that is, $\text{Ch}_A(x) = 1$ if $x \in A$ and $\text{Ch}_A(x) = 0$ if $x \notin A$. 1_A denotes the identity function on A; that is, $1_A(x) = x$ for all $x \in A$.

If f and g are two functions, their composition is denoted by $f \circ g$, or sometimes simply by fg. If A is a subset of the domain of f, then $f|A$ denotes the restriction of f to A.

Recall that the *Axiom of Choice* states that the Cartesian product of a nonempty family of nonempty sets is nonempty, and *Zorn's Lemma* states that if X is a nonempty partially ordered set such that every chain (totally ordered subset of X) in X has an upper bound in X, then X contains a maximal element.

2. Numerical Notation

$\mathbb{Z} = $ the ring of all integers (positive, negative, or 0).

$\mathbb{R} = $ the real field.

$\mathbb{R}_+ = \{t \in \mathbb{R} : t \geq 0\}$.

$\mathbb{C} = $ the complex field.

$\mathbb{E} = \{z \in \mathbb{C} : |z| = 1\}$.

We denote by $\mathbb{R}_{\text{ext}}$ the real field with the two points at infinity ∞ and $-\infty$ adjoined. Intervals in $\mathbb{R}_{\text{ext}}$ of the form $\{x : a \leq x \leq b\}$, $\{x : a < x < b\}$, $\{x : a \leq x < b\}$, and $\{x : a < x \leq b\}$ are written $[a, b]$, $]a, b[$, $[a, b[$, and $]a, b]$ respectively. $\mathbb{R}_{\text{ext}}$ is assumed to carry the topology which (i) reduces on $\mathbb{R}$ to the ordinary topology of the reals, and (ii) makes $\{]a, \infty] : a \in \mathbb{R}\}$ and $\{[-\infty, a[: a \in \mathbb{R}\}$ a base of neighborhoods of ∞ and $-\infty$ respectively.

Addition and multiplication are partially extended to $\mathbb{R}_{ext}$ in the usual way: $a + \infty = \infty$ if $a \neq -\infty$; $a + (-\infty) = -\infty$ if $a \neq \infty$; $a \cdot \infty = \infty$ if $a > 0$; $a \cdot \infty = -\infty$ if $a < 0$; $0 \cdot \infty = 0$.

If $a, b \in \mathbb{R}_{ext}$, $a \vee b$ and $a \wedge b$ are the maximum and minimum of a and b respectively. If $S \subset \mathbb{R}_{ext}$, $\sup(S)$ and $\inf(S)$ are the least upper bound and greatest lower bound of S respectively. (Notice that $\inf(\emptyset) = \infty$, $\sup(\emptyset) = -\infty$.) Similar definitions hold for $\mathbb{R}_{ext}$-valued functions: If f, g are functions on a set X with values in $\mathbb{R}_{ext}$, we put

$$(f \vee g)(x) = f(x) \vee g(x), \qquad (f \wedge g)(x) = f(x) \wedge g(x).$$

If $\mathscr{F}$ is a collection of functions on X to $\mathbb{R}_{ext}$, $\sup(\mathscr{F})$ and $\inf(\mathscr{F})$ are the functions on X given by

$$\sup(\mathscr{F})(x) = \sup\{f(x) : f \in \mathscr{F}\},$$

$$\inf(\mathscr{F})(x) = \inf\{f(x) : f \in \mathscr{F}\}$$

respectively.

We say that $f \leq g$ if $f(x) \leq g(x)$ for all x in the common domain of f and g.

If $t_1, t_2, \ldots$ is a sequence of numbers in $\mathbb{R}_{ext}$, we write as usual

$$\liminf_{n \to \infty} t_n = \sup_{p=1}^{\infty} \inf_{n=p}^{\infty} t_n$$

$$\limsup_{n \to \infty} t_n = \inf_{p=1}^{\infty} \sup_{n=p}^{\infty} t_n.$$

Similarly, if $f_1, f_2, \ldots$ is a sequence of $\mathbb{R}_{ext}$-valued functions on a set X, we put

$$\left(\liminf_{n \to \infty} f_n \right)(x) = \liminf_{n \to \infty} (f_n(x))$$

$$\left(\limsup_{n \to \infty} f_n \right)(x) = \limsup_{n \to \infty} (f_n(x))$$

for all x in X. Also

$$\left(\lim_{n \to \infty} f_n \right)(x) = \lim_{n \to \infty} f_n(x)$$

provided the right side exists for all x in X. If $f_1 \leq f_2 \leq f_3 \leq \ldots$ $[f_1 \geq f_2 \geq f_3 \geq \ldots]$, the sequence $\{f_n\}$ is called *monotone increasing* [*monotone decreasing*]; and, in that case, if $\lim_{n \to \infty} f_n = f$, we sometimes write $f_n \uparrow_n f$ or simply $f_n \uparrow f [f_n \downarrow_n f$ or simply $f_n \downarrow f]$.

The real and imaginary parts of a complex number z are denoted by $\operatorname{Re}(z)$ and $\operatorname{Im}(z)$. If $\operatorname{Re}(z)$ and $\operatorname{Im}(z)$ are both rational, z is called *complex rational*.

If n is a positive whole number and F is a field, $M(n, F)$ is the set of all $n \times n$ matrices with entries in F. If $A \in M(n, F)$, $\det(A)$ stands for the determinant of A and $\operatorname{Tr}(A)$ for its trace.

If $A \in M(n, \mathbb{C})$, A^* is the *adjoint matrix*: $(A^*)_{ij} = \overline{A_{ji}}$.

The symbol δ_{xy} is the *Kronecker delta*; it means 1 if $x = y$ and 0 if $x \neq y$. Sometimes the notation δ_x is used for the function $y \mapsto \delta_{xy}$.

3. Topology

For the definitions of standard topological terms the reader is referred to Kelley [1]. We shall mention here only a few points.

A *topology* $\mathcal{T}$ on a set S means the family of all subsets of S which are open with respect to this topology. Hence the statement that $\mathcal{T}_1$ *contains*, or *is bigger than*, $\mathcal{T}_2$ ($\mathcal{T}_1$ and $\mathcal{T}_2$ being topologies on S) means that every $\mathcal{T}_2$-open set is $\mathcal{T}_1$-open.

If A is a subset of a topological space, the symbol $\bar{A}$ means the closure of A. (Note however that, if z is a complex number, $\bar{z}$ means the complex conjugate of z.)

A topological space S is T_0 if the family of all open subsets of S distinguishes points of S, that is, if $\{s\}^- = \{t\}^- \Rightarrow s = t$ for all s, t in S. It is T_1 if all one-element subsets of S are closed.

A topological space S is (i) *separable* if S has a countable dense subset, (ii) *first-countable* if every point of S has a countable base of neighborhoods, and (iii) *second-countable* if there is a countable base for the open subsets of S. For metric spaces separability and second-countability are the same. Many writers use the terms "separable" and "second-countable" synonymously, but we prefer not to do this. (There are non-metric spaces that are compact, Hausdorff, separable, and first-countable, but not second-countable; see Bourbaki [13, pp. 229–230, Exercise 12].)

Let S be a Hausdorff topological space, and ∞ an object which is not a member of S; equip $S \cup \{\infty\}$ with the topology consisting of those subsets A of $S \cup \{\infty\}$ such that (i) $A \cap S$ is open in S, and (ii) if $\infty \in A$ then $S \setminus A$ is compact in S. With this topology $S \cup \{\infty\}$ is called the *one-point compactification* of S.

Throughout this work we use nets (see Kelley [1], Chapter 2) rather than filters in discussions of convergence. Convergence of nets is often indicated by plain arrows (a usage of arrows to be distinguished from functional notation, see §1). Thus, the statement that the net $\{x_i\}$ converges to x may be expresseed by $x_i \underset{i}{\to} x$ or simply by $x_i \to x$.

Let $S_1, S_2, \ldots, S_n, T$ be topological spaces. A function $f: S_1 \times S_2 \times \cdots \times S_n \to T$ is *separately continuous* if, for each $i = 1, \ldots, n$, the function $x_i \mapsto f(x_1, \ldots, x_{i-1}, x_i, x_{i+1}, \ldots, x_n)$ is continuous on S_i to T when the other variables $x_1, \ldots, x_{i-1}, x_{i+1}, \ldots, x_n$ are given arbitrary fixed values. The function $f: S_1 \times \cdots \times S_n \to T$ is *jointly continuous*, or simply *continuous*, if it is continuous with respect to the Cartesian product topology of $S_1 \times \cdots \times S_n$.

Let X, A be topological spaces. Then

$\mathscr{C}(X; A)$ is the space of all continuous functions $f: X \to A$;

$\mathscr{C}(X) = \mathscr{C}(X; \mathbb{C})$;

$\mathscr{C}_r(X) = \mathscr{C}(X; \mathbb{R})$;

$\mathscr{C}_+(X) = \mathscr{C}_r(X) \cap \{f: f(x) \geq 0 \text{ for all } x \text{ in } X\}$.

Assume in addition that X is locally compact and Hausdorff and that A is a linear topological space. Then

$\mathscr{C}_0(X; A) = \mathscr{C}(X; A) \cap \{f: f(x) \to 0 \text{ as } x \to \infty \text{ in } X\}$;

$\mathscr{C}_0(X) = \mathscr{C}_0(X; \mathbb{C})$;

$\mathscr{C}_{0r}(X) = \mathscr{C}_0(X) \cap \mathscr{C}_r(X)$;

$\mathscr{L}(X; A) = \mathscr{C}(X; A) \cap \{f: \text{there is a compact subset } D \text{ of } X \text{ such that } f(x) = 0 \text{ for all } x \in X \setminus D\}$;

$\mathscr{L}(X) = \mathscr{L}(X; \mathbb{C})$;

$\mathscr{L}_r(X) = \mathscr{L}(X) \cap \mathscr{C}_r(X)$;

$\mathscr{L}_+(X) = \mathscr{L}(X) \cap \mathscr{C}_+(X)$.

If X is compact, $\mathscr{C}(X; A) = \mathscr{C}_0(X; A) = \mathscr{L}(X; A)$.

If in addition A is a normed linear space, and if $f \in \mathscr{C}_0(X; A)$, then

$$\|f\|_\infty = \text{the } supremum \ norm \text{ of } f$$
$$= \sup\{\|f(x)\|: x \in X\}.$$

In particular, if $f \in \mathscr{C}_0(X)$,

$$\|f\|_\infty = \sup\{|f(x)|: x \in X\}.$$

4. Algebra

Our terminology as regards algebra is entirely conventional. We remind the reader of a few points.

Let H be a subgroup of a group G. By G/H we always mean the space of all *left* cosets xH $(x \in G)$ of G modulo H. A *transversal* for G/H is a subset W of G which intersects each left coset xH $(x \in G)$ in exactly one element.

If G_1 and G_2 are groups and $f: G_1 \to G_2$ is a group homomorphism, $\mathrm{Ker}(f)$ stands for the *kernel* of f: $\mathrm{Ker}(f) = \{x \in G_1 : f(x)$ is the unit element of $G_2\}$. If the groups are written additively (for example, if G_1 and G_2 are linear spaces), then of course $\mathrm{Ker}(f) = \{x \in G_1 : f(x)$ is the zero element of $G_2\}$.

A (possibly non-commutative) ring with unit in which every non-zero element has a (two-sided) inverse is a *division ring*. A commutative division ring is a *field*.

Let D be a division ring and X a D-linear space (D acting on X to the left). A *Hamel basis of X (over D)* is an indexed collection $\{x_i\}$ of elements of X such that each element y of X can be written in one and only one way in the form

$$y = \sum_i d_i x_i,$$

where the d_i are in D and $d_i = 0$ for all but finitely many i. (Thus a Hamel basis is simply a basis in the usual purely algebraic sense.) The *Hamel dimension of X (over D)* is the cardinal number of any Hamel basis over D.

Again let D be a division ring, and let X_1 and X_2 be two D-linear spaces. A map $f: X_1 \to X_2$ is *D-linear* if

$$f(dx) = df(x) \qquad\qquad (x \in X_1; d \in D).$$

The set of all D-linear maps $f: X_1 \to X_2$ is denoted by $\mathcal{O}'_D(X_1, X_2)$. If D is a field (i.e., commutative), $\mathcal{O}'_D(X_1, X_2)$ is itself a natural D-linear space. We abbreviate $\mathcal{O}'_D(X_1, X_1)$ to $\mathcal{O}'_D(X_1)$. An element of $\mathcal{O}'_D(X_1)$ is a *D-linear endomorphism* or *D-linear operator* on X_1.

Let X_1, X_2, X_3 be linear spaces over a field F. A map $\beta: X_1 \times X_2 \to X_3$ is *F-bilinear* if (i) $x_1 \mapsto \beta(x_1, x_2)$ is F-linear on X_1 to X_3 for each fixed x_2 in X_2, and (ii) $x_2 \mapsto \beta(x_1, x_2)$ is F-linear on X_2 to X_3 for each fixed x_1 in X_1. If $X_3 = F$, the F-bilinear map β is called an *F-bilinear form*.

An *algebra* over a field F is a linear space A together with a multiplication map $\langle a, b \rangle \mapsto ab$ of $A \times A$ into A which is (i) F-bilinear and (ii) associative (i.e., $(ab)c = a(bc)$ for all a, b, c in A). *We emphasize that algebras are always taken to be associative.*

Throughout this work, all linear spaces and algebras are over the complex number field $\mathbb{C}$ unless the contrary is explicitly stated.

Elements of an abstract linear space are often referred to as *vectors*.

Let X be a linear space. By a *subspace* of X we mean a linear subspace of X. If S is a subset of X, the *linear span of S in X* is the smallest linear subspace of X containing S, that is, the subspace of all finite linear combinations of elements of S. (The linear span of $\emptyset$ is thus $\{0\}$.)

A subset S of a real or complex linear space X is *convex* if $\lambda x + (1 - \lambda)y \in S$ whenever $x, y \in S$ and $\lambda \in \mathbb{R}$, $0 \leq \lambda \leq 1$. The *convex hull* of a subset S of X is the smallest convex subset of X containing S, that is, the family of all *convex combinations* $\lambda_1 x_1 + \lambda_2 x_2 + \cdots + \lambda_n x_n$ ($x_i \in S$; $0 \leq \lambda_i \in \mathbb{R}$; $\lambda_1 + \lambda_2 + \cdots + \lambda_n = 1$) of elements of S.

If $\mathscr{W}$ is a collection of subspaces of a linear space X, then $\sum \mathscr{W}$ denotes the linear span of $\mathscr{W}$.

Let X and Y be (complex) linear spaces. We abbreviate $\mathcal{O}'_{\mathbb{C}}(X, Y)$ and $\mathcal{O}'_{\mathbb{C}}(X)$ to $\mathcal{O}'(X, Y)$ and $\mathcal{O}'(X)$ respectively. Also we denote by $X^{\#}$ the linear space $\mathcal{O}'(X, \mathbb{C})$ of all complex-valued linear functionals on X.

A subset W of $X^{\#}$ is *total* if it separates the points of X, that is, if for any non-zero element x of X there is an element α of W such that $\alpha(x) \neq 0$.

An element of $\mathcal{O}'(X)$ of the form $x \mapsto \lambda x$, where $\lambda \in \mathbb{C}$, is called a *scalar operator* on X.

Let X be a (complex) linear space. We can form from X a new linear space $\bar{X}$ as follows: The underlying set and the addition operation of $\bar{X}$ are the same as for X; and the scalar multiplication in $\bar{X}$ is given by

$$(\lambda x)_{\text{in } \bar{X}} = (\bar{\lambda} x)_{\text{in } X} \qquad (x \in X; \lambda \in \mathbb{C}).$$

This $\bar{X}$ is called the linear space *complex-conjugate to X*.

Let X and Y be two (complex) linear spaces. A map $f: X \to Y$ is *conjugate-linear* if $f(x_1 + x_2) = f(x_1) + f(x_2)$ and $f(\lambda x) = \bar{\lambda} f(x)$ ($x, x_1, x_2 \in X; \lambda \in \mathbb{C}$), that is, if f is linear as a map of $\bar{X}$ into Y (or, equivalently, of X into $\bar{Y}$).

Let X, Y, and Z be three linear spaces. A map $\beta: X \times Y \to Z$ is *conjugate-bilinear* if it is bilinear as a map of $X \times \bar{Y}$ into Z, that is, if $\beta(x, y)$ is linear in x and conjugate-linear in y. If $Z = \mathbb{C}$, we then refer to β as a *conjugate-bilinear form* on $X \times Y$.

If $\beta: X \times X \to Z$ is conjugate-bilinear (X, Z being linear spaces), one verifies the following important identity:

$$4\beta(x, y) = \beta(x + y, x + y) - \beta(x - y, x - y) + i\beta(x + iy, x + iy)$$

$$- i\beta(x - iy, x - iy)$$

(for all x, y in X). This is called the *polarization identity*.

Let X be a linear space, and $\beta: X \times X \to \mathbb{C}$ a conjugate-bilinear form. We say that β is *positive* if $\beta(x, x) \geq 0$ for all x in X. In that case β is *self-adjoint*, that is,

$$\beta(y, x) = \overline{\beta(x, y)} \qquad\qquad (x, y \in X);$$

and we have the vitally important *Schwarz Inequality*:

$$|\beta(x, y)|^2 \leq \beta(x, x)\beta(y, y) \qquad\qquad (x, y \in X).$$

We come now to involutive algebras.

By an *involutive algebra*, or a **-algebra*, we mean an algebra A (over the complex field) which has a further unary operation $*: A \to A$ with the properties: (i) $*$ is conjugate-linear (on A to A); (ii) $(ab)^* = b^*a^*$ $(a, b \in A)$; (iii) $a^{**} = a$ $(a \in A)$. We refer to $*$ as the *operation of involution*, or the **-operation*, of A.

Let A be a **-algebra. A **-subalgebra* of A is a subalgebra B of A which is also closed under the *-operation. A left or right ideal I of A which is closed under the *-operation (and which is therefore a two-sided ideal of A) is called a **-ideal* of A.

Let A and B be two *-algebras. A map $f: A \to B$ is a **-homomorphism* if it is an algebra homomorphism and also $f(a^*) = (f(a))^*$ $(a \in A)$. (By a *homomorphism* of two *-algebras we mean a homomorphism of their underlying algebras, the *-operations being disregarded.) A one-to-one *-homomorphism is a **-isomorphism*. Sometimes a *-homomorphism which is one-to-one is called *faithful* (especially in the context of representation theory).

Let $\{X_i\}$ $(i \in I)$ be an indexed collection of linear spaces; and form the set-theoretic product $X = \prod_{i \in I} X_i$ consisting of all functions x on I such that $x_i \in X_i$ for each i in I. Clearly X is a linear space under the operations of pointwise addition and scalar multiplication; it is called the *direct product* of the X_i. Let Y denote the subset of X consisting of those x such that $x_i = 0$ for all but finitely many i. Thus Y is a linear subspace of X; it is called the *algebraic direct sum* of the X_i and is denoted by $\sum_{i \in I}^{\oplus} X_i$. If in addition each X_i is an algebra [*-algebra], then X is an algebra [*-algebra] under the pointwise operations of multiplication and involution, called the *direct product* of the algebras [*-algebras] X_i; and Y is a subalgebra [*-subalgebra] of X, called the *algebraic direct sum* of the algebras [*-algebras] X_i.

Let X and Y be two linear spaces. It is well known that there is an essentially unique linear space $X \otimes Y$, together with a bilinear map $\langle x, y \rangle \mapsto x \otimes y$ of $X \times Y$ into $X \otimes Y$, with the following property: For each linear space Z and each bilinear map $\beta: X \times Y \to Z$, there is a unique linear map $f: X \otimes Y \to Z$ such that

$$\beta(x, y) = f(x \otimes y) \qquad\qquad (x \in X; y \in Y).$$

This space $X \otimes Y$ is called the *(algebraic) tensor product* of X and Y.

Suppose that X and Y are not only linear spaces but algebras. Then the equation

$$(x_1 \otimes y_1)(x_2 \otimes y_2) = (x_1 x_2) \otimes (y_1 y_2)$$

$(x_i \in X; y_i \in Y)$ determines a multiplication on $X \otimes Y$ making the latter an algebra. With this multiplication, $X \otimes Y$ is the *(algebraic) algebra tensor product* of X and Y. If in addition X and Y are *-algebras, the equation

$$(x \otimes y)^* = x^* \otimes y^* \qquad\qquad (x \in X; y \in Y)$$

determines an involution operation on $X \otimes Y$ making the latter a *-algebra, which we call the *(algebraic)* *-algebra tensor product* of X and Y.

Let A be an algebra and X a linear space. A *left A-module structure* for X is a bilinear map $\langle a, x \rangle \mapsto ax$ of $A \times X$ into X satisfying the associative law $a(bx) = (ab)x$ $(a, b \in A; x \in X)$. Similarly, a *right A-module structure* for X is a bilinear map $\langle x, a \rangle \mapsto xa$ of $X \times A$ into X satisfying

$$(xa)b = x(ab) \qquad\qquad (a, b \in A; x \in X).$$

A linear space X equipped with a left [right] A-module structure is called a *left [right] A-module.*

If A is a *-algebra, by a *left [right] A-module* we mean a left [right] A_0-module, where A_0 is the algebra underlying A; in other words we disregard the involutive structure of A.

5. Seminorms, Normed Spaces, Normed Algebras

In this work we assume that the reader is familiar with basic facts about linear topological spaces, Banach spaces, Hilbert spaces, and operators on these. As sources for reference we may mention Kelley and Namioka [1], Halmos [2]. In this section we remind the reader of certain basic definitions and facts, in order to establish terminology. As was stated earlier, *all linear spaces (including Banach spaces, Banach algebras, and Hilbert spaces) are over the complex field.*

A *linear topological space* is a linear space X which is also a Hausdorff topological space, in which (i) the addition operation $\langle x, y \rangle \mapsto x + y$ is continuous (jointly in both variables) on $X \times X$ to X, and (ii) the scalar multiplication $\langle \lambda, x \rangle \mapsto \lambda x$ is (jointly) continuous on $\mathbb{C} \times X$ to X.

Let X be an abstract linear space. A function $N : X \to \mathbb{R}_+$ is called a *seminorm* on X if it satisfies the following two conditions: (i) $N(\lambda x) = |\lambda| N(x)$ $(x \in X; \lambda \in \mathbb{C})$; (ii) $N(x + y) \leq N(x) + N(y)$ (subadditivity). If in addition $N(x) = 0 \Rightarrow x = 0$, then N is called a *norm* on X.

Let $\mathcal{N}$ be a family of seminorms on a linear space X with the property that the only element x of X such that $N(x) = 0$ for all N in $\mathcal{N}$ is the zero element. Such an $\mathcal{N}$ is called a *defining family of seminorms on* X. (In particular, a single norm constitutes such a family.) Then X becomes a linear topological space when equipped with the topology $\mathcal{T}$ in which convergence of nets is defined by:

$$x_i \underset{i}{\to} x \Leftrightarrow N(x_i - x) \underset{i}{\to} 0 \qquad \text{for all } N \in \mathcal{N}.$$

The topology $\mathcal{T}$ is said to be *generated* by $\mathcal{N}$.

Let X be a linear topological space. We say that X is *locally convex* if its topology is generated by some defining family of seminorms on X. It can be shown that X is locally convex if and only if the zero element of X has a base of neighborhoods consisting of convex sets.

All the linear topological spaces encountered in this work will be locally convex. We abbreviate "locally convex linear topological space" to "*locally convex space*" or, more usually, to "LCS."

Two defining families of seminorms on X are *equivalent* if they generate the same topology on X. A defining family $\mathcal{N}$ of seminorms is *upward directed* if, given any two seminorms N_1 and N_2 in $\mathcal{N}$, there is a third seminorm N_3 in $\mathcal{N}$ such that $N_1 \leq N_3$ and $N_2 \leq N_3$. Every defining family $\mathcal{N}$ of seminorms is equivalent to an upward directed defining family (as we see by adjoining to $\mathcal{N}$ all finite sums of elements of $\mathcal{N}$).

Let X be an LCS, and $\{x_i\} (i \in I)$ a collection of vectors in X indexed by a set I. Let $\mathcal{F}$ be the directed set of all finite subsets of I (directed by the relation of inclusion); and for each F in $\mathcal{F}$ let us form the finite sum $s(F) = \sum_{i \in F} x_i$. If $y = \lim_{F \in \mathcal{F}} s(F)$ exists in X, we say that $\sum_{i \in I} x_i$ *converges unconditionally to* y (*in* X).

If S is a subset of an LCS X, by the *closed linear span* of S we mean the closure in X of the linear span of S; and by the *closed convex hull* of S we mean the closure of the convex hull of S.

Let X and Y be two LCS's. We denote by $\mathcal{O}(X, Y)$ the linear space of all continuous linear maps from X to Y, and abbreviate $\mathcal{O}(X, X)$ to $\mathcal{O}(X)$. Thus $\mathcal{O}(X)$ is an algebra with composition as multiplication.

The linear space $\mathcal{O}(X, \mathbb{C})$ is called the *adjoint space of* X, and is denoted by X^*; it consists of all continuous complex-valued linear functionals on X. Since X is assumed to be an LCS, the Hahn-Banach Theorem shows that X^* is total in $X^{\#}$, that is, it separates the points of X. Hence the collection of all seminorms on X of the form $x \mapsto |f(x)|$, where $f \in X^*$, is a defining family of seminorms; and the topology generated by this family is called the *weak topology* of X. The weak topology is of course contained in the original topology of X.

By the *strong topology* of $\mathcal{O}(X, Y)$ we mean the topology of pointwise convergence; that is, a net $\{a_i\}$ converges to a strongly in $\mathcal{O}(X, Y)$ if and only if, for each x in X, $a_i(x) \mapsto a(x)$ in the topology of Y.

The *weak topology* of $\mathcal{O}(X, Y)$ is obtained by replacing the original topology of Y in the preceding definition by the weak topology of Y. That is, a net $\{a_i\}$ converges to a weakly in $\mathcal{O}(X, Y)$ if and only if $f(a_i(x)) \mapsto f(a(x))$ for all x in X and f in Y^*.

Let X be an LCS. A net $\{x_i\}$ of elements of X is a *Cauchy net* if $\lim_{i, j} (x_i - x_j) = 0$. If every Cauchy net in X converges to some element of X, X is *complete*.

A *Fréchet space* is an LCS X which is (i) complete and (ii) first-countable. The condition of first-countability is equivalent to saying that the topology of X is generated by a countable defining family of seminorms. It is also equivalent to the assertion that the topology of X is obtained from some metric on X.

A linear space X equipped with a (single) norm is called a *normed linear space*. In the absence of any explicit mention to the contrary, the norm of X is denoted by $\| \ \|$. A vector x in X satisfying $\|x\| = 1$ is called a *unit vector*. The set $\{x \in X : \|x\| \leq 1\}$ is the *unit ball* of X. The topology generated by $\| \ \|$ is the *norm-topology* of X.

Let X and Y be two normed linear spaces. For each a in $\mathcal{O}(X, Y)$ the number $\|a\|_0 = \sup\{\|a(x)\| : x \in X, \|x\| \leq 1\}$ is finite; and the map $a \mapsto \|a\|_0$ is a norm on $\mathcal{O}(X, Y)$, called the *operator norm*. Thus we have three topologies for $\mathcal{O}(X, Y)$—the operator norm-topology, the strong topology, and the weak topology. Notice that the operator norm of $\mathcal{O}(X)$ satisfies the crucial *Banach inequality*:

$$\|ab\| \leq \|a\| \|b\| \qquad\qquad (a, b \in \mathcal{O}(X)).$$

An element a of $\mathcal{O}(X, Y)$ satisfying $\|a(x)\| = \|x\|$ for all x in X is an *isometry*.

A complete normed linear space is called a *Banach space*. Thus a Banach space is a special kind of Fréchet space.

Let X be a Banach space and $\{x_i\}$ ($i \in I$) a collection of elements of X indexed by a set I. If $\sum_{i \in I} \|x_i\| < \infty$, the completeness of X implies that $\sum_{i \in I} x_i$ converges unconditionally; in that case we say that $\sum_{i \in I} x_i$ *converges absolutely*. In finite-dimensional spaces, unconditional convergence and absolute convergence are the same; but in infinite-dimensional Banach spaces unconditional convergence does not imply absolute convergence.

Suppose that X is a normed linear space and Y is a Banach space. Then $\mathcal{O}(X, Y)$ with the operator norm is complete, hence a Banach space. In particular the adjoint space X^* of X is a Banach space under the operator norm.

Banach algebras and Banach *-algebras will be studied in detail from Chapter V onwards. Nevertheless we give their definition at this point, since they are referred to in Chapter III.

A *normed algebra* is a normed linear space A which is also an algebra, and in which the multiplication is connected with the norm by the inequality:

$$\|ab\| \le \|a\| \|b\| \qquad\qquad (a, b \in A).$$

A *Banach algebra* is a normed algebra which (considered as a normed linear space) is complete.

If X is any normed linear space, then $\mathcal{O}(X)$, equipped with the operator norm and the composition operation, is a normed algebra (in view of the Banach inequality). If X is a Banach space, $\mathcal{O}(X)$ is complete and hence a Banach algebra.

A *normed *-algebra* is a normed algebra A which is also a *-algebra under an involution * which satisfies

$$\|a^*\| = \|a\| \qquad\qquad (a \in A).$$

A *Banach *-algebra* is a normed *-algebra which (as a normed linear space) is complete.

Let $\{X_i\}_{i \in I}$ be a collection of normed linear spaces indexed by a set I. An element x of the direct product $\prod_{i \in I} X_i$ is said to *vanish at infinity* if for any $\varepsilon > 0$ there is a finite subset F of I such that $\|x_i\| < \varepsilon$ for all $i \in I \setminus F$. By the C_0 *direct sum* of the X_i—in symbols $\sum_{i \in I}^{\oplus 0} X_i$—we mean the normed linear space of all those elements of $\prod_{i \in I} X_i$ which vanish at infinity, equipped with the norm $\|x\|_\infty = \sup\{\|x_i\| : i \in I\}$. If each X_i is a Banach space, then so is their C_0 direct sum. If each X_i is a Banach algebra [Banach *-algebra] then so also is $\sum_{i \in I}^{\oplus 0} X_i$ under the pointwise operations of multiplication [and involution].

6. Hilbert Spaces

Let X be a (complex) linear space. A positive conjugate-bilinear form β on $X \times X$ is called an *inner product on X* if

$$\beta(x, x) = 0 \Rightarrow x = 0 \qquad\qquad (x \in X).$$

By a *pre-Hilbert space* we mean a linear space X equipped with an inner product. In the absence of any explicit mention to the contrary, the inner product on a pre-Hilbert space X is denoted by $(\ ,\ )_X$ or, more usually, simply by $(\ ,\ )$.

Let $X, (\ ,\ )$ be a pre-Hilbert space; and for each x in X define $\|x\|$ to be the non-negative square root of (x, x). Then $x \mapsto \|x\|$ is a norm on X. If X is complete with respect to this norm, $X, (\ ,\ )$ is a *Hilbert space*.

Any positive conjugate-bilinear form gives rise to a Hilbert space in a natural way. Let $\beta : X \times X \to \mathbb{C}$ be a positive conjugate-bilinear form, X being a linear space. Then the null space $N = \{x \in X : \beta(x, x) = 0\}$ of β is a linear subspace of X; and the quotient space $\tilde{X} = X/N$ inherits an inner product $\tilde{\beta}$:

$$\tilde{\beta}(x + N, y + N) = \beta(x, y) \qquad\qquad (x, y \in X).$$

Thus $\tilde{X}, \tilde{\beta}$ is a pre-Hilbert space. Let X_c be the completion of $\tilde{X}$ with respect to the norm $\tilde{x} \mapsto \tilde{\beta}(\tilde{x}, \tilde{x})^{1/2}$. Then $\tilde{\beta}$ can be extended to a (jointly continuous) inner product $(\ ,\ )$ on X_c. The Hilbert space $X_c, (\ ,\ )$ is said to be *generated* by the positive conjugate-bilinear form β. We sometimes refer to it loosely as the *completion* of X, β.

Let $X, (\ ,\ )$ be a Hilbert space. If $x, y \in X$, the notation $x \perp y$ means that x and y are orthogonal, that is, $(x, y) = 0$; and if $Y \subset X$ the symbol $Y^\perp$ stands for the orthogonal complement $\{x \in X : x \perp y \text{ for all } y \text{ in } Y\}$. If Y is a closed linear subspace of X we have $Y^{\perp\perp} = Y$.

For a Hilbert space X we have $X^* \cong \bar{X}$. More precisely, to every α in X^* there is a unique element y of X such that

$$\alpha(x) = (x, y) \qquad \text{for all } x \text{ in } X.$$

Thus the statement that $x_i \underset{i}{\to} x$ weakly in X means that $(x_i, y) \underset{i}{\to} (x, y)$ for all y in X. If X and Y are two Hilbert spaces and $\{a_i\}$ is a net of elements of $\mathcal{O}(X, Y)$, the assertion that $a_i \underset{i}{\to} a$ weakly in $\mathcal{O}(X, Y)$ means that

$$(a_i(x)), y) \underset{i}{\to} (a(x), y)$$

for all x in X and y in Y.

Let X and Y be two Hilbert spaces. Given any a in $\mathcal{O}(X, Y)$, there is a unique map a^* in $\mathcal{O}(Y, X)$ with the property that

$$(a(x), y)_Y = (x, a^*(y))_X$$

for all x in X and y in Y. This a^* is called the map *adjoint to a*. The correspondence $a \mapsto a^*$ is conjugate-linear and has the propertites:

$$(ba)^* = a^*b^*,$$

$$a^{**} = a,$$

$$\|a^*\| = \|a\|,$$

$$\|a^*a\| = \|a\|^2$$

$(a \in \mathcal{O}(X, Y); b \in \mathcal{O}(Y, Z)$, Z being a third Hilbert space; $\| \ \|$ is the operator norm).

Applying these properties to the case that $Y = X$, we conclude that $\mathcal{O}(X)$, equipped with the operations of composition and adjoint and the operator norm, is a Banach *-algebra. (In fact, since the identity $\|a^*a\| = \|a\|^2$ holds, $\mathcal{O}(X)$ is a C^*-algebra; see Chapter VI.)

We notice that the adjoint operation $a \mapsto a^*$ is continuous with respect to the norm-topology and the weak topology of $\mathcal{O}(X, Y)$, but not with respect to the strong topology. The operation of multiplication (composition) in $\mathcal{O}(X)$ is jointly continuous in the norm-topology and separately continuous in the strong and weak topologies; it is jointly continuous in the strong topology if we restrict our attention to the unit ball of $\mathcal{O}(X)$.

A subset A of $\mathcal{O}(X)$ is said to be *self-adjoint* if $a \in A \Rightarrow a^* \in A$. Suppose that A is a self-adjoint subset of $\mathcal{O}(X)$; and let us define $C = \{b \in \mathcal{O}(X) : ab = ba$ for all a in $A\}$. Then C is also self-adjoint; in fact, C is a *-subalgebra of $\mathcal{O}(X)$ and is closed in $\mathcal{O}(X)$ in the weak operator topology (hence also in the strong operator topology and the norm-topology). We refer to C as the *commuting algebra* of A.

It is useful to recall that elements of $\mathcal{O}(X, Y)$ are in correspondence with continuous conjugate-bilinear forms on $X \times Y$. Indeed, for every element a of $\mathcal{O}(X, Y)$ the equation

$$\beta(x, y) = (a(x), y)_Y$$

defines a conjugate-bilinear form β on $X \times Y$ which is continuous; that is, there is a constant k satisfying

$$|\beta(x, y)| \leq k\|x\|\|y\| \qquad (x \in X; y \in Y).$$

Conversely, every continuous conjugate-bilinear form β on $X \times Y$ arises in this manner from a unique element a of $\mathcal{O}(X, Y)$.

As before let X be a fixed Hilbert space. We shall have to deal with several important special classes of elements of $\mathcal{O}(X)$. An element a of $\mathcal{O}(X)$ is

(i)   *self-adjoint* or *Hermitian* if $a^* = a$,
(ii) *normal* if $a^*a = aa^*$,
(iii) *positive* if $a^* = a$ and $(a(x), x) \geq 0$ for all x in X,
(iv) *unitary* if $a^*a = aa^* = 1_X$,
(v)   a *projection* if $a^2 = a = a^*$.

One has the implications: projection $\Rightarrow$ positive $\Rightarrow$ self-adjoint $\Rightarrow$ normal.

There is a natural one-to-one correspondence $p \mapsto Y = \text{range}(p)$ between the set of all projections p in $\mathcal{O}(X)$ and the set of all closed linear subspaces Y of X. Given a closed linear subspace Y of X, the projection p whose range is Y is described by:

$$p(x) = \begin{cases} x & \text{if } x \in Y \\ 0 & \text{if } x \in Y^\perp. \end{cases}$$

Two projections p_1 and p_2 are *orthogonal* if $p_1 p_2 = 0$ (equivalently, if $p_2 p_1 = 0$; also equivalently, if $\text{range}(p_1) \perp \text{range}(p_2)$). If p_1 and p_2 are projections we say that $p_1 \leq p_2$ *if* $p_2 p_1 = p_1$ (equivalently, if $\text{range}(p_1) \subset \text{range}(p_2)$). If $\mathcal{P}$ is any family of projections on X, we denote by $r = \inf(\mathcal{P})$ and $q = \sup(\mathcal{P})$ the greatest lower bound of $\mathcal{P}$ and the least upper bound of $\mathcal{P}$ respectively with respect to the ordering $\leq$ just defined; that is, r is the projection onto $\bigcap \{\text{range}(p): p \in \mathcal{P}\}$, and q is the projection onto the closed linear span of $\bigcup \{\text{range}(p): p \in \mathcal{P}\}$. If the projections in $\mathcal{P}$ are pairwise orthogonal, we may write $q = \sum \mathcal{P}$, since in that case

$$q(x) = \sum_{p \in \mathcal{P}} p(x) \qquad \text{(unconditional convergence)}$$

for all x in X.

An operator u in $\mathcal{O}(X)$ is unitary if and only if $u^* = u^{-1}$; also it is unitary if and only if it is an isometry onto X.

Let $\{X_i\}(i \in I)$ be a collection of Hilbert spaces indexed by a set I. The algebraic direct sum $X^0 = \sum_{i \in I}^{\oplus} X_i$ becomes a pre-Hilbert space when equipped with the inner product $(\ ,\ )$ given by

$$(x, y) = \sum_{i \in I} (x_i, y_i)_{X_i} \qquad\qquad (x, y \in X^0).$$

(Notice that the sum here is finite by the definition of the algebraic direct sum.) The Hilbert space completion of $X^0, (\ ,\ )$ is called the *Hilbert direct sum* of the X_i. It is also denoted by $\sum_{i \in I}^{\oplus} X_i$; we distinguish it from the algebraic direct sum by qualifying it verbally as the Hilbert direct sum when this is needed to avoid ambiguity.

Suppose we are given a Hilbert space X and an indexed collection $\{X_i\}(i \in I)$ of pairwise orthogonal closed linear subspaces of X such that $\sum_{i \in I} X_i$ is dense in X. Then X is naturally isometrically isomorphic with the Hilbert direct sum $\sum_{i \in I}^{\oplus} X_i$; and we shall often speak of X as being the *Hilbert direct sum* of its subspaces X_i.

Let $\{X_i\}(i \in I)$ be an indexed collection of linear spaces, and $X^0 = \sum_{i \in I}^{\oplus} X_i$ their algebraic direct sum. For each i let a_i be a linear endomorphism of X_i. Then the equation

$$(a^0(x))_i = a_i(x_i) \qquad\qquad (x \in X^0; i \in I)$$

defines a linear endomorphism a^0 of X^0, which we refer to as the (*algebraic*) *direct sum* of the a_i: $a^0 = \sum_{i \in I}^{\oplus} a_i$. Next, suppose in addition that the X_i are Hilbert spaces, that the a_i are continuous, and that in fact there is a constant k such that $k \geq \|a_i\|$ (operator norm) for all i in I. Then the algebraic direct sum a^0 is continuous with respect to the inner product of X^0, and so extends to a bounded linear operator a on the Hilbert direct sum $X = \sum_{i \in I}^{\oplus} X_i$. This a is called the *Hilbert direct sum* of the a_i, and is also denoted by $\sum_{i \in I}^{\oplus} a_i$ (being distinguished, if necessary, from the algebraic direct sum by the word "Hilbert").

We come now to tensor products of Hilbert spaces.

Let X and Y be two linear spaces, and $Z = X \otimes Y$ their algebraic tensor product. Let α and β be conjugate-bilinear forms on $X \times X$ and $Y \times Y$ respectively. Then the equation

$$\gamma(x_1 \otimes y_1, x_2 \otimes y_2) = \alpha(x_1, x_2)\beta(y_1, y_2)$$

determines a conjugate-bilinear form γ on $Z \times Z$; γ is referred to as the *tensor product* $\alpha \otimes \beta$ of α and β. It can be shown that if α and β are positive then so is γ. Also, if α and β are inner products, then so is γ; that is, if X, α and Y, β are pre-Hilbert spaces, then Z, γ is a pre-Hilbert space. We call Z, γ, the *pre-Hilbert tensor product* of X, α and Y, β.

Now suppose that X and Y are Hilbert spaces; and form the pre-Hilbert tensor product Z^0 of X and Y (with inner product $(\, , \,)_Z = (\, , \,)_X \otimes (\, , \,)_Y$). The Hilbert space completion Z of Z^0 is called the *Hilbert tensor product* of X and Y, and is denoted by $X \otimes Y$ (together with the distinguishing phrase "Hilbert tensor product", if necessary).

Notice that tensor products (whether algebraic or Hilbert) are associative; that is,

$$(X \otimes Y) \otimes Z \cong X \otimes (Y \otimes Z).$$

In combination with the operation of forming direct sums they also satisfy the distributive law:

$$X \otimes \left(\sum_{i \in I}^{\oplus} Y_i \right) \cong \sum_{i \in I}^{\oplus} (X \otimes Y_i).$$

This holds if the tensor products and direct sums are all algebraic, and also if they are all Hilbert.

If a and b are linear endomorphisms of linear spaces X and Y respectively, the equation

$$c^0(x \otimes y) = a(x) \otimes b(y)$$

defines a linear endomorphism c^0 of $X \otimes Y$, called the (*algebraic*) *tensor product* of a and b and denoted by $a \otimes b$. Suppose now that X and Y are Hilbert spaces with Hilbert tensor product Z, and that $a \in \mathcal{O}(X)$ and $b \in \mathcal{O}(Y)$. Then the algebraic tensor product $c^0 = a \otimes b$ is continuous with respect to $(\, , \,)_Z$, and so can be extended uniquely to a continuous linear operator c on Z. This c is called the *Hilbert tensor product* of a and b, and as usual is also denoted by $a \otimes b$.

7. Exercises for Chapter I

1. Let X and Y be non-void sets and $f: X \to Y$ a function. Prove that the following are equivalent:
   (a) f is injective;
   (b) $f^{-1}(f(S)) = S$ for all $S \subset X$;
   (c) $f(S \cap T) = f(S) \cap f(T)$ for all $S, T \subset X$;

(d) $f(S) \cap f(T) = \emptyset$ for all $S, T \subset X$ with $S \cap T = \emptyset$;

(e) $f(X \setminus S) \subset Y \setminus f(S)$ for all $S \subset X$;

(f) $f(S \setminus T) = f(S) \setminus f(T)$ for all $S, T \subset X$.

2. Let X and Y be non-void sets and $f: X \to Y$ a function. Prove that the following are equivalent:

(a) f is surjective;

(b) $f(f^{-1}(E)) = E$ for all $E \subset Y$;

(c) $Y \setminus f(S) \subset f(X \setminus S)$ for all $S \subset X$.

3. Prove that if X is a non-void set and $f: X \to X$ is a function, then there exists a subset A in X such that $f|A$ is injective and $f(A) = f(X)$.

4. Prove that if $f: X \to Y$ is a surjective function from a non-void set X onto a non-void set Y, then there exists an injective function $g: Y \to X$ such that $f \circ g$ is the identity on Y.

5. Let $I = [0, 1]$ and suppose that $f, g: I \to I$ are continuous mappings. Set $A = \{x \in I : f(x) = g(x)\}$. Prove that:

(a) If f and g are surjective, then A is non-void.

(b) If $f(g(x)) = g(f(x))$ for all x in I, then A is non-void.

6. Show that every set can be given a compact Hausdorff topology.

7. Let X be a topological space. Construct a Hausdorff space $\tilde{X}$ and a continuous mapping $\phi: X \to \tilde{X}$ such that for every Hausdorff space Y and every continuous mapping $f: X \to Y$ there exists one and only one continuous mapping $\tilde{f}: \tilde{X} \to Y$ such that $\tilde{f} \circ \phi = f$.

8. Let S be an uncountable discrete topological space, and X its one point compactification. Prove that if f is a continuous complex-valued function on X, then $f(X)$ is countable.

9. Find a topological space X which is locally compact Hausdorff, and second countable, and an equivalence relation R on X such that the quotient space X/R is not locally compact.

10. Let X be a non-void compact Hausdorff space. Prove that if $f: X \to X$ is continuous, then there exists a non-void closed subset A of X such that $f(A) = A$.

11. Let X be a metric space, M a compact subset of X and $T: X \to X$ an isometry. Prove that if $T(M) \subset M$ or $M \subset T(M)$, then $T(M) = M$.

12. Let $X = [0, \infty[$. Define a linear operator T on $\mathscr{C}_r(X)$ by

$$(Tf)(x) = \begin{cases} \dfrac{1}{x} \displaystyle\int_0^x f(\xi)d\xi, & \text{if } 0 < x < \infty \\[2ex] f(0), & \text{if } x = 0. \end{cases}$$

Prove that, for all $x \in X$, $(T^n f)(x) \to f(0)$ as $n \to \infty$.

13. Let X be a compact Hausdorff space. Prove that the following are equivalent:

(a) X is metrizable;

(b) X is second countable;

(c) $\mathscr{C}(X)$ is separable in the uniform topology.

14. Let X be a metric space. Prove that if the space of bounded continuous complex valued functions on X is separable in the uniform topology, then X is compact.

15. If $f \in \mathscr{C}_r(\mathbb{R})$ can be uniformly approximated throughout $\mathbb{R}$ by polynomials, is f itself a polynomial? Supply a proof or give a counter example.

16. Give an example of a separable Hausdorff space X and a dense nonseparable subspace S of X.

17. Let f be a continuous function of the plane E into itself. Does there exist a closed proper non-void subset X of E such that $f(X) \subset X$?

18. Must the one point compactification of a locally compact metric space be metrizable?

19. Let X be a compact T_0 space and let A be the set of closed singletons in X. Show that every subset containing A is compact.

20. Prove that if a group G is the set theoretic union of a family of proper normal subgroups each two of which have only the identity in common, then G is Abelian.

21. Prove that if p is the smallest prime factor of the order of a finite group G, then any subgroup of index p is normal.

22. Let H be a proper subgroup of a group G and H' its complement. Show that $K = H'H'$ is a normal subgroup of G which equals either G or H.

23. Let $f: G \to G'$ be a homomorphism of groups G and G'. Let K be a normal subgroup of G' and $N = f^{-1}(K)$. If H is any normal subgroup of G, then there is a canonical isomorphism $G/HN \cong f(G)K/f(H)K$.

24. Let A be a ring. Suppose $a \in A$ is nilpotent and $b \in A$ is a right zero divisor (i.e., there exists nonzero d such that $bd = 0$). Prove that if a and b commute, then there exists a nonzero $c \in A$ such that $ac = bc = 0$.

25. Suppose that a ring R is the set-theoretic union of a finite number of fields having the same identity element. Prove that R must be a field.

26. To within isomorphism, determine the number of rings whose additive group is cyclic of order n.

27. Let A be an algebra with identity and let I, J be two distinct two-sided ideals of codimension 1. Show that:

(a) $IJ \cup JI$ spans $I \cap J$.

(b) Give an example to show that IJ need not span $I \cap J$.

28. Let A, B, and C be algebras. Let $\phi: A \to B$ and $\psi: A \to C$ be homomorphisms with ϕ surjective and such that $\mathrm{Ker}(\phi) \subset \mathrm{Ker}(\psi)$. Prove that there exists a unique homomorphism $f: B \to C$ such that $\psi = f \circ \phi$.

29. Let C be a closed convex subset of a normed linear space X such that $B_{1+\varepsilon} \subset B_1 + C$ for some $\varepsilon > 0$, where $B_r = \{x : \|x\| \le r\}$. Prove that C must have non-void interior.

30. Let X be a normed linear space such that $\mathcal{O}(X)$ is complete. Prove that X is complete.

31. Prove that if X is an infinite-dimensional normed linear space, then there exists a discontinuous linear mapping of X into X.

32. Prove that if X is a normed linear space of dimension greater than 1, then the algebra $\mathcal{O}(X)$ is noncommutative.

33. Give an example of a Hausdorff locally convex space which is separable and which contains a nonseparable linear subspace.

34. Let T be a linear (not necessarily continuous) mapping of a Hilbert space H to itself. Suppose that there exists a subset M of H such that $Tx \in M$ and $x - Tx \in M^{\perp}$, the orthogonal complement of M, for all $x \in H$. Prove that M is a closed linear subspace of H and that T is the (necessarily continuous) orthogonal projection of H onto M.

35. Show that if x and y are vectors in a (complex) Hilbert space and m, n are integers, $0 \leq m < 2n$, then

$$\sum_{k=1}^{2n} \| x + \lambda^k y \|^{2m} \leq n \binom{2m}{m} 2^{1-m} (\|x\|^2 + \|y\|^2)^m,$$

where $\lambda = \exp(i\pi/n)$. Show that equality holds if and only if $x = \mu y$, where μ is a complex number of unit modulus.

36. Let H be a Hilbert space and T an element of $\mathcal{O}(H)$ such that $T^2 = 0$. Prove that $T + T^*$ is invertible if and only if there exist operators U, V in $\mathcal{O}(H)$ such that $UT + TV = \mathbf{1}$, where $\mathbf{1}$ is the identity operator on H.

37. Let X be a normed linear space and M a closed linear subspace in X such that both M and the quotient space X/M are complete. Prove that X is complete.

38. Let $\{x_1, x_2, \ldots, x_n\}$ be a linearly independent set of vectors in a normed linear space X. Prove that there is a real number $c > 0$ such that for every choice of scalars $\alpha_1, \ldots, \alpha_n$ we have $\| \sum_{i=1}^{n} \alpha_i x_i \| \geq c(\sum_{i=1}^{n} |\alpha_i|)$.

39. Prove that every finite-dimensional linear subspace Y of a normed linear space X is complete.

40. Let X be a normed linear space. Prove that every finite-dimensional linear subspace Y of X is closed in X. Show by example that infinite-dimensional linear subspaces of X need not be closed.

41. (F. Riesz). Let X be a normed linear space. Suppose that Y and Z are linear subspaces of X and that Y is a closed proper subset of Z. Prove that for every real number $\alpha, 0 < \alpha < 1$, there is a vector z in Z such that $\|z\| = 1$ and $\|z - y\| \geq \alpha$ for all y in Y.

42. Prove that a normed linear space X is finite-dimensional if and only if the closed unit ball $B = \{x \in X : \|x\| \leq 1\}$ is compact in the norm topology.

43. Prove that if a normed linear space X is finite-dimensional, then every linear mapping $T : X \to X$ is continuous.

44. Let X be a normed linear space; and let $\phi : X \to \mathbb{C}$ be a linear functional on X which is *not* continuous. Prove that, for any non-void open subset U of X, we have $\phi(U) = \mathbb{C}$.

Hence, show that, if $\psi : X \to \mathbb{C}$ is a linear functional such that $\psi^{-1}(\{0\})$ is closed in X, then ψ is continuous.

45. Let X be a linear topological space; and B a convex subset of X with non-void interior. Prove that:

(a) $\bar{B}$ is convex, where the bar denotes closure in X.

(b) If $x \in \bar{B}$, $y \in \text{int}(B)$ (the interior of B) and $z = \alpha x + \beta y$, $\alpha > 0$, $\beta > 0$, $\alpha + \beta = 1$, then $z \in \text{int}(B)$.

(c) $\text{int}(B)$ is convex.

(d) $\bar{B} = [\text{int}(B)]^-$.

(e) $\text{int}(B) = \text{int}(\bar{B})$.

(f) The boundary of B is nowhere dense in X.

46. Let X be a linear topological space. Show that:

(a) The closed convex hull of a subset S of X is the intersection of all closed convex subsets of X containing S.

(b) The convex hull of an open subset of X is open.

(c) If $B_1, B_2, \ldots, B_n$ are compact convex subsets of X, then the convex hull of the union $B = \bigcup_{i=1}^n B_i$ is compact.

(d) If X is a normed linear space, then the diameter of the convex hull of a subset S of X is equal to the diameter of S.

47. Let X be an infinite-dimensional normed space; and let $C = \{x \in X : \|x\| = 1\}$. Show that 0 is in the closure of C relative to the weak topology of X.

Notes and Remarks

References for the preliminary material of this chapter are Berberian [2], Bourbaki [8, 9, 13, 14, 15], Conway [1], Dunford and Schwartz [1], R. E. Edwards [1], Halmos [2], Kelley [1], Kelley and Namioka [1], and Rudin [2]. Tensor products of Hilbert spaces are treated at length in Gaal [1], Kadison and Ringrose [1], and in Hewitt and Ross [2]. The general theory of normed and Banach algebras will be treated in Chapter V; basic references are Bonsall and Duncan [1], Naimark [8], and Rickart [2].

II Integration Theory and Banach Bundles

Like other writers on functional analysis we have felt it necessary to begin with an account of the theory of measure and integration. Indeed, if there is one property of locally compact groups more responsible than any other for the rich positive content of their representation theory, it is their possession of left-invariant and right-invariant (Haar) measures. The connection between the representation theory of a locally compact group and that of general Banach algebras proceeds directly from Lebesgue integration with respect to Haar measure on the group. The same is true for the more general objects to which much of this work is devoted, namely Banach *-algebraic bundles. For these objects, indeed, we need integration theory in an even more sophisticated form than for locally compact groups. We must study the integrability not merely of complex-valued functions but of "vector fields" whose values lie in Banach spaces varying from point to point of the domain. Since integration in this generality is certainly not part of the standard equipment even of abstract analysts, we are obliged to present it in some detail.

In the conflict between brevity and the desire to make this chapter as self-contained as possible, we have adopted a compromise. Definitions and theorems that relate to measure theory and integration of *complex-valued* functions are stated in full, but proofs of such theorems are often only sketched, or references to them from standard texts are given. This of course

makes the present chapter not very suitable for a first exposure to abstract
integration theory! Only in the sections on measurability and integration of
vector fields will the reader find complete proofs and a more leisurely
presentation.

The chapter is divided into sixteen sections. In §1 we define measure and
measurability of sets and complex functions. We have deviated from the usual
practice of considering measures as defined on σ-rings (which are closed
under countable unions). Instead, measures for us are defined on δ-rings,
which are closed merely under finite unions and countable intersections. This
permits us to dispense entirely with ∞ as a possible value of measures, and
(more importantly) to include from the beginning "unbounded complex
measures" within the framework of the general theory.

§2 is the standard theory of integration of complex-valued functions over a
general measure space. In §3 we begin vector-valued integration by constructing the "outsize $\mathscr{L}_p$-spaces" of vector fields, following the approach initiated
by Stone [3] and developed by Bourbaki [10]. These spaces are independent
of any measurability properties of the vector fields composing them. In §4 we
axiomatize the notion of a locally measurable vector field over a general
measure space, and combine this notion with §3 to obtain the appropriate
generalizations of the familiar $\mathscr{L}_p$-spaces of measurable complex functions. §5
is devoted to the vector-valued integral of a summable function whose values
lie in a single Banach space B, and §6 generalizes this to the case where B is an
arbitrary locally convex space. §7 centers around the classical version of the
Radon–Nikodym Theorem. §8 examines the special properties of regular
measures on locally compact spaces; these are in fact the only measures that
will appear in subsequent chapters of this work. In §9 we state the main
results on product measures and the Fubini Theorem for locally compact
spaces. §10 considers the transformations of measures that correspond to
mappings of sets.

One can of course develop a theory of integration with respect to
generalized measures taking values in arbitrary linear topological spaces.
One special instance of this is of great importance for representation theory,
namely, when the values of the measure are projections on a Hilbert
space. §11 presents the basic properties of these "projection-valued
measures"and of "spectral integrals" with respect to them. In §12 we apply
these spectral integrals to describe *-representations of commutative C^*-
algebras (to use the terminology of Chapter VI). This is the measure-theoretic
backbone of the Spectral Theorem for normal operators in Hilbert space (see
Chapter VI, §11).

The last four sections of the chapter are devoted to Banach bundles. §13 develops the basic properties of Banach bundles (including the Construction Theorem 13.16, from which almost all the specific Banach bundles occurring in this work will be derived). In §14 some important special properties of Banach bundles over locally compact spaces emerge. §15 applies the general notions of measurability and summability developed in §§3–5 to define and study the $\mathscr{L}_p$-spaces of cross-sections of Banach bundles over locally compact spaces. Finally, in §16 the Fubini Theorem is generalized to the context of Banach bundles over a product of two locally compact spaces.

1. δ-Rings, Measures, and Measurable Functions

δ-Rings and Measures

1.1. Take a fixed set X. A *ring of subsets of* X is a non-void family $\mathscr{S}$ of subsets of X such that $A \setminus B$ and $A \cup B$ are in $\mathscr{S}$ whenever A and B are in $\mathscr{S}$. This implies that $\emptyset \in \mathscr{S}$, and that $\mathscr{S}$ is also closed under finite intersections.

Consider a ring $\mathscr{S}$ of subsets of X. If $\bigcap \mathscr{W} \in \mathscr{S}$ whenever $\mathscr{W}$ is a non-void countable subfamily of $\mathscr{S}$, then $\mathscr{S}$ is a *δ-ring of subsets of* X. Equivalently, $\mathscr{S}$ is a δ-ring if and only if it is closed under "bounded" countable unions, that is, $\bigcup \mathscr{W} \in \mathscr{S}$ whenever $\mathscr{W}$ is a countable subfamily of $\mathscr{S}$ such that $\bigcup \mathscr{W} \subset A$ for some A in $\mathscr{S}$.

A ring $\mathscr{S}$ of subsets of X is a *field (of subsets of X)* if $X \in \mathscr{S}$. If it is both a field and a δ-ring, it is a *σ-field (of subsets of X)*; in that case it is closed under the taking of complements in X and of arbitrary countable unions and intersections.

The intersection of an arbitrary collection of δ-rings is again a δ-ring. Therefore, if $\mathscr{T}$ is any family of subsets of X, there is a unique smallest δ-ring $\mathscr{S}$ containing $\mathscr{T}$; $\mathscr{S}$ is called the δ-ring (of subsets of X) *generated* by $\mathscr{T}$. The same of course applies to rings, fields, and σ-fields.

1.2. Let $\mathscr{S}$ be a δ-ring of subsets of X, fixed for the rest of this section.

A *complex measure on* $\mathscr{S}$ is a complex-valued function μ on $\mathscr{S}$ which is countably additive, that is,

$$\mu\left(\bigcup_{n=1}^{\infty} A_n\right) = \sum_{n=1}^{\infty} \mu(A_n) \qquad \text{(absolute convergence)}$$

whenever $A_1, A_2, \dots$ is a sequence of pairwise disjoint sets in $\mathscr{S}$ such that $\bigcup_{n=1}^{\infty} A_n \in \mathscr{S}$. (In particular $\mu(\emptyset) = 0$). If the values of μ are all real and non-negative, μ is called simply a *measure on* $\mathscr{S}$.

If $\{A_n\}$ is a monotone decreasing sequence of sets in $\mathscr{S}$ and μ is a complex measure on $\mathscr{S}$, then $\mu(\bigcap_n A_n) = \lim_{n \to \infty} \mu(A_n)$. Similarly, if $\{A_n\}$ is monotone increasing and $\bigcup_n A_n \in \mathscr{S}$, then $\mu(\bigcup_n A_n) = \lim_{n \to \infty} \mu(A_n)$.

1.3. Proposition. *Take a complex measure μ on $\mathscr{S}$; and for each set A in $\mathscr{S}$, define $v(A)$ to be the supremum of $\{\sum_{r=1}^{n} |\mu(B_r)| : n$ is a positive integer, and the $B_1, \dots, B_n$ are pairwise disjoint sets in $\mathscr{S}$ such that $\bigcup_{r=1}^{n} B_r = A\}$. Then $v(A) < \infty$ for each A in $\mathscr{S}$, and in fact v is a measure on $\mathscr{S}$.*

The fact that $v(A) < \infty$ can be shown from the existence of the Hahn decomposition for real-valued complex measures (Halmos [1], Theorem 29A). The countable additivity of v then follows as an easy exercise (see Hewitt and Ross [1], Theorem 14.14).

The measure v is called the *total variation* of μ, and is denoted by $|\mu|$. It is the unique smallest measure on $\mathscr{S}$ which majorizes μ (that is, $|\mu(A)| \leq v(A)$ for every A in $\mathscr{S}$).

If μ is a measure, then of course $|\mu| = \mu$.

1.4. The set of all complex measures on the δ-ring $\mathscr{S}$ forms a linear space under the natural linear operations $(\mu + v)(A) = \mu(A) + v(A)$, $(c\mu)(A) = c\mu(A)$. Also, the measures on $\mathscr{S}$ are partially ordered in the obvious manner: $\mu \leq v \Leftrightarrow \mu(A) \leq v(A)$ for all A. We then have

$$|\mu + v| \leq |\mu| + |v| \tag{1}$$

$$|c\mu| = |c||\mu| \qquad (c \in \mathbb{C}). \tag{2}$$

1.5. A complex measure μ on $\mathscr{S}$ is *bounded* if $\sup\{|\mu|(A) : A \in \mathscr{S}\} < \infty$.

It follows from 1.4 that the set of all bounded complex measures on $\mathscr{S}$, which we will denote by $M_b(\mathscr{S})$, is a normed linear space under the linear operations of 1.4 and the *total variation norm*

$$\|\mu\| = \sup\{|\mu|(A) : A \in \mathscr{S}\}.$$

In fact $M_b(\mathscr{S})$ is complete, i.e., a Banach space. (The crux of the completeness proof lies in the fact that, if $\{\mu_n\}$ is a Cauchy sequence in $M_b(\mathscr{S})$, the numerical sequence $\{\mu_n(B)\}$ converges uniformly for all sets B contained in any given set A in $\mathscr{S}$.)

If $\mathscr{S}$ is a field, hence a σ-field of subsets of X, then every complex measure μ on $\mathscr{S}$ is of course bounded, and $\|\mu\| = |\mu|(X)$.

1.6. Take a complex measure μ on $\mathscr{S}$. A subset A of X is μ-*null* if there is a countable subfamily $\mathscr{W}$ of $\mathscr{S}$ such that $A \subset \bigcup \mathscr{W}$ and $|\mu|(B) = 0$ for every B in $\mathscr{W}$. Clearly subsets of μ-null sets and countable unions of μ-null sets are μ-null. If A is contained in some set belonging to $\mathscr{S}$, then A is μ-null if and only if there is a single set B in $\mathscr{S}$ for which $A \subset B$ and $|\mu|(B) = 0$.

A subset A of X is μ-*measurable* if $A \ominus B$ is μ-null for some B in $\mathscr{S}$. The family $\mathscr{M}$ of all μ-measurable sets is a δ-ring containing $\mathscr{S}$ and containing all μ-null sets; and μ can be extended in just one way to a complex measure μ' on $\mathscr{M}$ satisfying $\mu'(A) = 0$ for all μ-null sets A. We usually write simply $\mu(A)$ instead of $\mu'(A)$ for $A \in \mathscr{M}$.

A subset A of X is *locally μ-measurable* [*locally μ-null*] if $A \cap B$ is μ-measurable [μ-null] for every set B in $\mathscr{S}$. Thus X is always locally μ-measurable, and the family of all locally μ-measurable sets forms a σ-field of subsets of X containing $\mathscr{M}$. As with μ-null sets, subsets and countable unions of locally μ-null sets are again locally μ-null. Locally μ-null sets are locally μ-measurable.

A property of points of X which holds for all x outside some μ-null [locally μ-null] set is said to hold for μ-*almost all x* [*locally μ-almost all x*].

1.7. Let μ be a complex measure on $\mathscr{S}$. We have already defined boundedness of μ. Let us say that μ is σ-*bounded* if there exists a countable subfamily $\mathscr{W}$ of $\mathscr{S}$ such that $X \setminus \bigcup \mathscr{W}$ is locally μ-null.

We also define μ to be *parabounded* if there exists a (possibly uncountable) pairwise disjoint subfamily $\mathscr{W}$ of $\mathscr{S}$ such that (i) for each A in $\mathscr{S}$, $\{B \in \mathscr{W} : A \cap B \neq \emptyset\}$ is countable, and (ii) $X \setminus \bigcup \mathscr{W}$ is locally μ-null.

Notice that boundedness implies σ-boundedness. (Indeed, if μ is bounded we can take $\mathscr{W} = \{A_1, A_2, \ldots\}$, where $\{A_n\}$ is a sequence of sets in $\mathscr{S}$ for which $|\mu|(A_n) \to \sup\{|\mu|(A) : A \in \mathscr{S}\}$). Likewise σ-boundedness implies paraboundedness.

Notice also that if μ is parabounded, and if $\mathscr{W}$ is the disjoint family whose existence defines paraboundedness, a subset A of X is locally μ-measurable if and only if $A \cap B$ is μ-measurable for all B in $\mathscr{W}$.

Measurability of Complex Functions

1.8. Suppose that X and Y are two sets, and that $\mathscr{S}$ and $\mathscr{T}$ are σ-fields of subsets of X and Y respectively. A function $f : X \to Y$ is called an $\mathscr{S}, \mathscr{T}$ *Borel function* if $f^{-1}(A) \in \mathscr{S}$ whenever $A \in \mathscr{T}$. It is of course enough to require that $f^{-1}(A) \in \mathscr{S}$ for all A in some subfamily of $\mathscr{T}$ which generates $\mathscr{T}$.

Suppose that $Y = \mathbb{C}$ and that $\mathscr{T}$ is generated by the usual topology of $\mathbb{C}$. Then if f and g are $\mathscr{S}, \mathscr{T}$ Borel functions on X to $\mathbb{C}$, the functions $|f|$, $cf(c \in \mathbb{C})$, $f + g$, and fg are also $\mathscr{S}, \mathscr{T}$ Borel functions.

Next, assume that $Y = \mathbb{R}_{\text{ext}}$ and that $\mathscr{T}$ is generated by the usual topology of $\mathbb{R}_{\text{ext}}$. If $f, g, f_1, f_2, f_3, \ldots$ are any $\mathscr{S}, \mathscr{T}$ Borel functions on X to $\mathbb{R}_{\text{ext}}$, then the functions

$$f \wedge g,\ f \vee g,\ \inf_{n} f_n,\ \sup_{n} f_n,\ \liminf_{n \to \infty} f_n,\ \limsup_{n \to \infty} f_n$$

are all $\mathscr{S}, \mathscr{T}$ Borel.

These facts are proved in any standard text on Lebesgue integration.

1.9. Now let μ be a complex measure on a δ-ring $\mathscr{S}$ of subsets of a set X.

A function f on X to $\mathbb{C}$ (or to $\mathbb{R}_{\text{ext}}$) is said to be *locally μ-measurable* if it is $\mathscr{L}, \mathscr{T}$ Borel, where $\mathscr{L}$ is the σ-field of locally μ-measurable subsets of X, and $\mathscr{T}$ is the σ-field generated by the usual topology of $\mathbb{C}$ (or of $\mathbb{R}_{\text{ext}}$).

Thus the results of 1.8 apply to locally μ-measurable functions.

If two complex (or $\mathbb{R}_{\text{ext}}$-valued) functions on X coincide locally μ-almost everywhere, and one is locally μ-measurable, then so is the other.

If $X \in \mathscr{S}$, or if we are talking about the behavior of a function f only on a fixed set A in $\mathscr{S}$, we will omit the word "locally" and speak of f as μ-*measurable* instead of "locally μ-measurable."

1.10. Suppose f is a complex function on X such that, for each A in $\mathscr{S}$, there is a sequence $\{g_n\}$ of locally μ-measurable complex functions on X satisfying $\lim_{n \to \infty} g_n(x) = f(x)$ for μ-almost all x in A. Then evidently f is locally μ-measurable.

1.11. Egoroff's Theorem. *Let $\{f_n\}$ be a sequence of locally μ-measurable complex-valued functions on X, and f a complex-valued function on X such that $\lim_{n \to \infty} f_n(x) = f(x)$ for locally μ-almost all x. Then for any A in $\mathscr{S}$ and any positive number ε, there is a set B in $\mathscr{S}$ such that $B \subset A$, $|\mu|(A \setminus B) < \varepsilon$, and $f_n(x) \to f(x)$ uniformly for x in B.*

For the proof see Halmos [1], Theorem 21A.

Extension of Complex Measures to Larger δ-Rings

1.12. Let μ be a complex measure on $\mathscr{S}$. It may happen that μ has a natural extension to a δ-ring even larger than the $\mathscr{M}$ of 1.6.

Let us define $\mathscr{B}_\mu$ as the family of all those locally μ-measurable sets B such that

$$\sup\{|\mu|(A): A \in \mathscr{S}, A \subset B\} < \infty. \tag{3}$$

Evidently $\mathscr{B}_\mu$ is a δ-ring of subsets of X which contains all μ-measurable sets and all locally μ-null sets (and perhaps many more besides). Given a set B in $\mathscr{B}_\mu$, let us choose a sequence $\{A_n\}$ of pairwise disjoint subsets of B belonging to $\mathscr{S}$ such that $\sum_{n=1}^\infty |\mu|(A_n)$ equals the left side of (3), and define

$$\tilde{\mu}(B) = \sum_{n=1}^\infty \mu(A_n). \tag{4}$$

One verifies without difficulty that the right side of (4) is independent of the choice of $\{A_n\}$, and that the $\tilde{\mu}$ determined by (4) is a complex measure on $\mathscr{B}_\mu$ which reduces on $\mathscr{M}$ to the μ' of 1.6 (and of course assigns measure 0 to locally μ-null sets).

Definition. We shall call $\tilde{\mu}$ the *innate extension of* μ.

Evidently $\mathscr{B}_{|\mu|} = \mathscr{B}_\mu$, and

$$|\mu|^\sim = |\tilde{\mu}|. \tag{5}$$

If μ is bounded, then of course $\mathscr{B}_\mu$ consists of *all* locally μ-measurable sets, and $\|\mu\| = |\tilde{\mu}|(X)$.

2. Integration of Complex Functions

2.1. Throughout this section μ is a fixed (non-negative) measure on a δ-ring $\mathscr{S}$ of subsets of a set X.

In the linear space of all complex functions on X (with the pointwise linear operations) we form the linear span $\mathscr{F}$ of $\{\mathrm{Ch}_A: A \in \mathscr{S}\}$. Elements of $\mathscr{F}$ are called *simple* functions. We put $\mathscr{F}_+ = \{f \in \mathscr{F}: f(x) \geq 0 \text{ for all } x\}$. One verifies that there is a unique linear functional I on $\mathscr{F}$ satisfying $I(\mathrm{Ch}_A) = \mu(A)$ $(A \in \mathscr{S})$. If $f, g \in \mathscr{F}$ and $f \leq g$, then $I(f) \leq I(g)$; in particular $f \in \mathscr{F}_+ \Rightarrow I(f) \geq 0$.

Let $\mathscr{U}$ be the family of all locally μ-measurable functions $f: X \to [0, \infty]$ such that $\{x: f(x) \neq 0\}$ is contained in a countable union of sets in $\mathscr{S}$. A function $f: X \to [0, \infty]$ belongs to $\mathcal{U}$ if and only if there exists a monotone increasing sequence $\{g_n\}$ of functions in $\mathcal{F}_+$ such that

$$g_n(x) \uparrow f(x) \qquad \text{for } \mu\text{-almost all } x. \tag{1}$$

Proposition.   *Suppose f and f' are two functions in $\mathcal{U}$, and $\{g_n\}$ and $\{g'_n\}$ are any two monotone increasing sequences of functions in $\mathcal{F}_+$ such that $g_n(x) \uparrow f(x)$ and $g'_n(x) \uparrow f'(x)$ for μ-almost all x. Then, if $f \le f'$, we have $\lim_{n \to \infty} I(g_n) \le \lim_{n \to \infty} I(g'_n)$.*

Sketch of Proof.   For each fixed n, $(g_n \wedge g'_m)(x) \uparrow_m g_n(x)$ for μ-almost all x, whence by Egoroff's Theorem 1.11 $I(g_n \wedge g'_m) \uparrow_m I(g_n)$. It follows that $\lim_{m \to \infty} I(g'_m) \ge I(g_n)$ for every n, implying the desired conclusion. ∎

2.2.   Putting $f' = f$, we see from Proposition 2.1 that $\lim_{n \to \infty} I(g_n)$ depends only on f, being independent of the particular sequence $\{g_n\}$. We therefore extend I to $\mathcal{U}$ by defining

$$I(f) = \lim_{n \to \infty} I(g_n) \tag{2}$$

whenever $f \in \mathcal{U}$ and $\{g_n\}$ is a monotone increasing sequence of functions in $\mathcal{F}_+$ satisfying (1). Thus $I: \mathcal{U} \to [0, \infty]$, and we have

Proposition.   *If $f, g \in \mathcal{U}$ and $0 \le r \in \mathbb{R}$, then*

   (i) $I(f + g) = I(f) + I(g)$;
   (ii) $I(rf) = rI(f)$;
   (iii) $f \le g \Rightarrow I(f) \le I(g)$;
   (iv) $I(f) < \infty \Rightarrow f(x) < \infty$ *for μ-almost all x;*
   (v) $I(f) = 0 \Leftrightarrow f(x) = 0$ *for μ-almost all x;*
   (vi)   *if $\{f_n\}$ is a monotone increasing sequence of functions in $\mathcal{U}$, then $\lim_{n \to \infty} f_n \in \mathcal{U}$, and*

$$I(f_n) \uparrow I\left(\lim_{n \to \infty} f_n \right).$$

We leave the proof of these facts to the reader.

2.3.   Fatou's Lemma.   *For any sequence $\{f_n\}$ of functions in $\mathcal{U}$,*

$$I\left(\liminf_{n \to \infty} f_n \right) \le \liminf_{n \to \infty} I(f_n).$$

Proof.   Apply Proposition 2.2(vi) to the increasing sequence of functions $g_p = \inf\{f_n : n = p, p + 1, p + 2 \ldots\}$. ∎

2.4. Corollary. *Let* $f, f_1, f_2, \ldots$ *be functions in* $\mathcal{U}$ *such that* $f_n \leq f$ *for all* n *and* $\lim_{n \to \infty} f_n(x) = f(x)$ *for all* x. *Then*

$$I(f) = \lim_{n \to \infty} I(f_n).$$

Proof. $I(f) \leq \lim\inf_{n \to \infty} I(f_n)$ (by 2.3) $\leq \lim\sup_{n \to \infty} I(f_n) \leq I(f).$ ∎

2.5. Definition. Let $f: X \to \mathbb{C}$ be locally μ-measurable, and suppose that $\{x: f(x) \neq 0\}$ is contained in a countable union of sets in $\mathscr{S}$. Then $|f| \in \mathcal{U}$. We shall say that f is μ-*summable* if $I(|f|) < \infty$. This is equivalent to saying that f is μ-summable if and only if it is of the form

$$f = f_1 - f_2 + i(f_3 - f_4), \tag{3}$$

where, for each $i = 1, \ldots, 4$, $f_i \in \mathcal{U}$ (with $f_i(x) < \infty$ for all x) and $I(f_i) < \infty$.

We want to emphasize that, in our terminology, f can only be μ-summable if it vanishes outside a countable union of sets in the δ-ring $\mathscr{S}$ on which μ is defined.

If f is of the form (3), we define the *integral of* f *with respect to* μ as

$$\int f \, d\mu = I(f_1) - I(f_2) + i(I(f_3) - I(f_4)). \tag{4}$$

It must of course be verified that this definition does not depend on the particular choice of the f_i; we leave this to the reader. We often use the fuller notation $\int f(x) d\mu x$ for the integral $\int f \, d\mu$.

Clearly the space of all μ-summable complex functions on X is a linear space; and it is easy to see that $\int f \, d\mu$ is linear in f on this space.

Since $\int f \, d\mu$ coincides with $I(f)$ when $f \in \mathcal{U}$ and $I(f) < \infty$, we shall usually write $\int f \, d\mu$ instead of $I(f)$ *for all* f *in* $\mathcal{U}$.

By 2.2(v), altering a function f on a μ-null set does not alter its μ-summability, or the value of $\int f \, d\mu$ if the latter exists. Thus it is quite reasonable to speak of a function f as μ-summable, and of the number $\int f \, d\mu$, provided f coincides μ-almost everywhere with a μ-summable function (even if $f(x)$ is undefined or infinite on a μ-null set). For instance, in view of 2.2(iv), we shall regard f as μ-summable whenever $f \in \mathcal{U}$ and $\int f \, d\mu < \infty$.

2.6. Hölder's Inequality. *Let* p *and* q *be positive numbers with* $1/p + 1/q = 1$. *If* f *and* g *are in* $\mathcal{U}$,

$$\int fg \, d\mu \leq \left(\int f^p \, d\mu \right)^{1/p} \left(\int g^q \, d\mu \right)^{1/q}.$$

Imitate the proof in the classical context; see for example Hewitt and Ross [1], 12.4.

2.7. For real-valued functions we have:

Monotone Convergence Theorem. *Let* $\{f_n\}$ *be a sequence of μ-summable real-valued functions on X which is either monotone increasing or monotone decreasing; and put* $f(x) = \lim_{n\to\infty} f_n(x)$ $(x \in X)$. *Then the following two conditions are equivalent*:

   (I) f *is μ-summable*;
   (II) $\lim_{n\to\infty} \int f_n \, d\mu$ *is finite.*

If (I) *and* (II) *hold, then*

$$\int f \, d\mu = \lim_{n\to\infty} \int f_n \, d\mu.$$

Proof. Assume the $\{f_n\}$ are monotone increasing. Translating by the summable function $|f_1|$ we may also assume that $f_n \geq 0$. Then $f \in \mathcal{U}$, and by 2.2(vi)

$$\int f \, d\mu = \lim_{n\to\infty} \int f_n \, d\mu.$$

This implies both parts of the required conclusion.   ∎

2.8. Lebesgue Dominated-Convergence Theorem. *Let g be a non-negative-valued μ-summable function on X, and let $f, f_1, f_2, \ldots$ be locally μ-measurable complex-valued functions on X such that for μ-almost all x we have* (i) $|f_n(x)| \leq g(x)$ *for all n, and* (ii) $\lim_{n\to\infty} f_n(x) = f(x)$. *Then $f, f_1, f_2, \ldots$ are all μ-summable and*

$$\int f \, d\mu = \lim_{n\to\infty} \int f_n \, d\mu. \tag{5}$$

Proof. Splitting the f_n into their real and imaginary parts we may assume that f and the f_n are all real-valued. Altering the functions on a null set, we also suppose that $g(x) < \infty$ for all x, and that (i) and (ii) hold for all x.

The summability of f and f_n is evident from hypothesis (i). Since $f_n + g \geq 0$, Fatou's Lemma 2.3 implies

$$\int (f + g)d\mu = \int \lim_{n \to \infty} (f_n + g)d\mu \leq \liminf_{n \to \infty} \int (f_n + g)d\mu$$

$$= \liminf_{n \to \infty} \int f_n \, d\mu + \int g \, d\mu.$$

Cancelling $\int f \, d\mu$, we get

$$\int f \, d\mu \leq \liminf_{n \to \infty} \int f_n \, d\mu. \tag{6}$$

Replacing f_n and f by $-f_n$ and $-f$ in (6) and combining the resulting inequality with (6), we obtain (5). ∎

2.9. Proposition. *Suppose that $f: X \to \mathbb{C}$ is μ-summable. Then for each $\varepsilon > 0$ there exists a number $\delta > 0$ such that $\int (\mathrm{Ch}_A f)d\mu < \varepsilon$ whenever $A \in \mathcal{S}$ and $\mu(A) < \delta$.*

Proof. We may assume that $f \geq 0$, so that $f \in \mathcal{U}$ and $I(f) < \infty$. By 2.2(vi) there is a function $g \in \mathcal{F}_+$ such that $g \leq f$ and $I(f - g) < \frac{1}{2}\varepsilon$. Let $k = \sup\{g(x): x \in X\}$, $\delta = (2k)^{-1}\varepsilon$. Then, if $A \in \mathcal{S}$ and $\mu(A) < \delta$, $\int \mathrm{Ch}_A f \, d\mu = \int \mathrm{Ch}_A g \, d\mu + \int \mathrm{Ch}_A(f - g)d\mu \leq k\delta + I(f - g) < \varepsilon.$ ∎

Upper Integrals

2.10. Definition. Let $\mathcal{P}$ be the space of all functions on X to $[0, \infty]$ (whether locally μ-measurable or not). If $f \in \mathcal{P}$, we define $\bar{\int} f \, d\mu$ (or $\bar{\int} f(x)d\mu x$) to be the infimum of $\{\int g \, d\mu : g \in \mathcal{U}, g \geq f\}$. (Recall that the infimum of $\emptyset$ is ∞). This $\bar{\int} f \, d\mu$ is called the *upper integral* of f (*with respect to* μ). It lies in $[0, \infty]$ and coincides with $\int f \, d\mu$ if $f \in \mathcal{U}$; it is finite if and only if $f \leq g$ for some μ-summable function g. In particular, for $\bar{\int} f \, d\mu$ to be finite it is necessary that f vanish outside a countable union of sets in $\mathcal{S}$.

Suppose that $f \in \mathcal{P}$ and $\bar{\int} f \, d\mu < \infty$. Letting $\{g_n\}$ be a sequence of functions in $\mathcal{U}$ such that $\int g_n \, d\mu \downarrow \bar{\int} f \, d\mu$, and putting $g = \inf_{n=1}^{\infty} g_n$, we find that g is a μ-summable function in $\mathcal{U}$ such that $f \leq g$ and $\bar{\int} f \, d\mu = \int g \, d\mu$.

2.11.   Proposition.   *If* $f, g \in \mathscr{P}$ *and* $0 \leq r \in \mathbb{R}$, *then*

(i) $\bar{\int}(f + g)d\mu \leq \bar{\int} f \, d\mu + \bar{\int} g \, d\mu$;

(ii) $\bar{\int}(rf)d\mu = r \bar{\int} f \, d\mu$;

(iii) $f \leq g \Rightarrow \bar{\int} f \, d\mu \leq \bar{\int} g \, d\mu$;

(iv) $\bar{\int} f \, d\mu < \infty \Rightarrow f(x) < \infty$ *for* μ-*almost all* x;

(v) $\bar{\int} f \, d\mu = 0 \Leftrightarrow f(x) = 0$ *for* μ-*almost all* x;

(vi)   *if* $\{f_n\}$ *is a monotone increasing sequence of functions in* $\mathscr{P}$, *then* $\bar{\int} f_n \, d\mu \uparrow \bar{\int} (\lim_{n \to \infty} f_n) d\mu$.

The proof of these facts depends on 2.2 and is left to the reader. That of (v) and (vi) is especially facilitated by the last remark of 2.10.

The only formal difference between 2.11 and 2.2 lies in (i); the upper integral $\bar{\int} f \, d\mu$ is only subadditive in f, not additive.

3.   The Outsize $\mathscr{L}_p$ Spaces

3.1.   We shall now construct $\mathscr{L}_p$ spaces in a very general setting.

Let μ be a fixed measure on a δ-ring $\mathscr{S}$ of subsets of a set X. Assume that, attached to each x in X, we have a Banach space B_x. The linear operations and norm are denoted by the same symbols $+, \cdot, \| \ \|$ in all the spaces B_x (though the B_x for different x may be entirely unrelated to each other). The zero element of B_x is called 0_x. We refer to a function f on X such that $f(x) \in B_x$ for each x in X as a *vector field on* X.

3.2.   Fix a number p such that $1 \leq p < \infty$. By $\mathscr{L}_p(\mu; \{B_x\})$, or $\mathscr{L}_p$ for short, we shall mean the family of all those vector fields f on X such that

$$\bar{\int} \|f(x)\|^p \, d\mu x < \infty. \tag{1}$$

The number $(\bar{\int} \|f(x)\|^p \, d\mu x)^{1/p}$ is denoted by $\|f\|_p$.

Proposition. $\mathscr{L}_p$ *is a linear space; and* $\| \ \|_p$ *is a seminorm on* $\mathscr{L}_p$.

This is proved from Hölder's Inequality 2.6, just as in more classical contexts; see for example Hewitt and Ross [1], 12.6.

By 2.10 an element f of $\mathscr{L}_p$ must vanish outside some countable union C of sets in $\mathscr{S}$ (i.e., $f(x) = 0_x$ for $x \in X \setminus C$).

Strictly speaking, $\| \ \|_p$ is not a norm on $\mathscr{L}_p$, since by 2.11(v) $\|f\|_p = 0$ if and only if $f(x) = 0_x$ for μ-almost all x. However, if we factor $\mathscr{L}_p$ by the equivalence relation of "differing only on a μ-null set", the quotient space does become a normed linear space under $\| \ \|_p$. As usual, we shall not distinguish notationally between this quotient space and $\mathcal{L}_p$ itself, identifying two vector fields in $\mathcal{L}_p$ if they differ only on a μ-null set.

3.3. Actually $\mathcal{L}_p$ is complete, hence a Banach space. To see this, we begin with a lemma.

Lemma. *Let* $\{f_n\}$ *be a sequence of vector fields in* $\mathcal{L}_p$ *such that* $\sum_{n=1}^{\infty} \|f_n\|_p < \infty$. *Then:*

(A) *For* μ-almost all x, $\sum_{n=1}^{\infty} f_n(x)$ *converges absolutely in* B_x *to a vector* $g(x)$;
(B) *the resulting vector field* g *is in* $\mathcal{L}_p$;
(C) $\sum_{n=1}^{\infty} f_n = g$ *(absolute convergence in* $\mathcal{L}_p$).

Proof. Put $\phi_n(x) = \|f_n(x)\|$, $\phi = \sum_{n=1}^{\infty} \phi_n$. By 2.11(vi) and the subadditivity of $\| \ \|_p$,

$$\overline{\int} \phi^p \, d\mu = \lim_{m \to \infty} \overline{\int} \left(\sum_{n=1}^{m} \phi_n \right)^p d\mu$$

$$= \lim_{m \to \infty} \left\| \sum_{n=1}^{m} \phi_n \right\|_p^p \leq \lim_{m \to \infty} \left(\sum_{n=1}^{m} \|\phi_n\|_p \right)^p$$

$$= \left(\sum_{n=1}^{\infty} \|f_n\|_p \right)^p < \infty. \tag{2}$$

So by 2.11(v) $\sum_{n=1}^{\infty} \|f_n(x)\| = \phi(x) < \infty$ for μ-almost all x. This proves (A).

Put $g(x) = \sum_n f_n(x)$. (This defines $g(x)$ μ-almost everywhere. We can complete the definition to all of X in any way we want.) Since $\|g(x)\| \leq \phi(x)$ μ-almost everywhere, $\overline{\int} \|g(x)\|^p \, d\mu x < \infty$ by (2). So (B) holds.

Putting $g_m = \sum_{n=1}^{m} f_n$, and applying (2) to the series $\sum_{n=m+1}^{\infty} f_n = g - g_m$, we get

$$\|g - g_m\|_p^p \leq \left(\sum_{n=m+1}^{\infty} \|f_n\|_p \right)^p \to 0 \quad \text{as} \quad m \to \infty.$$

This establishes (C). ∎

3.4. Corollary. $\mathcal{L}_p$ *is a Banach space.*

Proof. If $\{f_n\}$ is a Cauchy sequence in $\mathscr{L}_p$, we can replace $\{f_n\}$ by a subsequence and assume that $\sum_{n=1}^{\infty} \| f_{n+1} - f_n \|_p < \infty$. But then $\{f_n\}$ converges in $\mathscr{L}_p$ by 3.3.   ■

Definition. $\mathscr{L}_p$ is called the *outsize $\mathscr{L}_p$-space of vector fields* on X.

The term "outsize" refers to the fact that in general $\mathscr{L}_p$ is far too big to be useful. Only after passing to a much smaller space of "measurable" vector fields will many of the familiar desirable properties of integration become available.

3.5.   The proof of 3.4, based on 3.3, also implies:

Corollary.   *If $\{f_n\}$ is any sequence converging to f in $\mathscr{L}_p$, there is a subsequence $\{f'_m\}$ of $\{f_n\}$ such that*

$$f'_m(x) \to f(x) \text{ in } B_x \text{ for } \mu\text{-almost all } x.$$

4.   Local Measurability Structures

4.1.   In this section we keep to the assumptions and notation of 2.1 and 3.1.

What should be meant by the term "measurability" as applied in 3.4 to a vector field? One important fragment of measurability is the requirement that $x \mapsto \| f(x) \|$ be locally μ-measurable in the sense of 1.9. But this requirement by itself is not enough; for the set of vector fields satisfying it is not in general closed under addition. (As an example illustrating this statement, let μ be Lebesgue measure on the unit interval $I = [0, 1]$, and take $B_x = \mathbb{C}$ for each x. Choose a non-Lebesgue-measurable subset A of I and let $f = \text{Ch}_A - \text{Ch}_{I \setminus A}$, $g = 1 = \text{Ch}_I$. Then $|f(x)| = |g(x)| = 1$ for all x; but $|f + g| = 2\text{Ch}_A$, which is not Lebesgue-measurable.) Since the spaces B_x are to begin with totally unrelated to each other, it seems clear that some more structure must be introduced before we shall be able to speak of passing "in a measurable manner" from one space B_x to another. This structure will consist in laying down what vector fields are to be measurable.

Definition. A local μ-measurability structure for the $\{B_x\}$ $(x \in X)$ is a non-void collection $\mathcal{M}$ of vector fields on X satisfying the following conditions:

(I)   If f and g are in $\mathcal{M}$ then $f + g \in \mathcal{M}$.

(II)   If $f \in \mathcal{M}$ and $\phi : X \to \mathbb{C}$ is locally μ-measurable (1.9), then $\phi f : x \mapsto \phi(x)f(x)$ is in $\mathcal{M}$.

(III) If $f \in \mathcal{M}$, the numerical function $x \mapsto \|f(x)\|$ is locally μ-measurable.

(IV)   Suppose f is a vector field such that, for each A in $\mathscr{S}$, there exists a sequence $\{g_n\}$ of elements of $\mathcal{M}$ (perhaps depending on A) such that $g_n(x) \to f(x)$ (in B_x) for μ-almost all x in A. Then $f \in \mathcal{M}$.

Condition (IV) is motivated by 1.10. Notice from (IV) that, if f is a vector field which coincides except on a locally μ-null set with a vector field g in $\mathcal{M}$, then $f \in \mathcal{M}$.

4.2.   Definition 4.1 is just a list of properties that we want for the set of locally μ-measurable vector fields. Our problem is to show that local μ-measurability structures exist. Evidently the intersection of any non-void family of local μ-measurability structures is again a local μ-measurability structure; so, if there exists one local μ-measurability structure containing a given family $\mathscr{Q}$ of vector fields, then there is a unique smallest one, which we call the local μ-measurability structure *generated* by $\mathscr{Q}$. The next result asserts that there are only two requirements that $\mathscr{Q}$ must satisfy in order to generate a local μ-measurability structure, namely linearity and 4.1(III). The importance of linearity was foreshadowed in a remark preceding Definition 4.1.

Proposition.   *Let $\mathscr{Q}$ be a linear space of vector fields on X such that $x \mapsto \|f(x)\|$ is locally μ-measurable on X for every f in $\mathscr{Q}$. Then $\mathscr{Q}$ generates a local μ-measurability structure $\mathcal{M}$ for the $\{B_x\}$.*

Proof.   If $\mathscr{T}$ is any family of vector fields, let us denote by $\mathscr{T}^{-}$ the family of all those vector fields f such that, for each A in $\mathscr{S}$, there is a sequence $\{g_n\}$ of elements of $\mathscr{T}$ such that $g_n(x) \to f(x)$ (in B_x) for μ-almost all x in A.

We claim that, if $\mathscr{T}$ is any linear space of vector fields such that $x \mapsto \|f(x)\|$ is locally μ-measurable for each f in $\mathscr{T}$, then $\mathscr{T}^{-}$ has the same properties; and in addition $(\mathscr{T}^{-})^{-} = \mathscr{T}^{-}$.

Indeed: Clearly $\mathscr{T}^{-}$ has the same properties. So we have only to take a vector field f in $(\mathscr{T}^{-})^{-}$ and show that $f \in \mathscr{T}^{-}$. Fix $A \in \mathscr{S}$; and choose a sequence $\{g_n\}$ of elements of $\mathscr{T}^{-}$ converging to f pointwise μ-almost everywhere in A. Since the numerical functions $x \mapsto \|f(x) - g_n(x)\|$ are locally μ-measurable, for each $m = 1, 2, \ldots$ Egoroff's Theorem (1.11) lets us choose a set D_m in $\mathscr{S}$, and one of the g_n—call it g^m— such that

$$\mu(D_m) < 2^{-m-1} \quad \text{and} \quad \|f(x) - g^m(x)\| < (2m)^{-1} \quad \text{for all } x \text{ in } A \setminus D_m. \quad (1)$$

Similarly, since $g^m \in \mathscr{T}^-$, for each m Egoroff's Theorem gives us a set E_m in $\mathscr{S}$ and a vector field h^m in $\mathscr{T}$ such that

$$\mu(E_m) < 2^{-m-1} \quad \text{and} \quad \|g^m(x) - h^m(x)\| < (2m)^{-1} \quad \text{for all } x \text{ in } A \setminus E_m. \quad (2)$$

Combining (1) and (2) we see that

$$\mu(D_m \cup E_m) < 2^{-m} \quad \text{and} \quad \|f(x) - h^m(x)\| < m^{-1} \quad \text{for } x \in A \setminus (D_m \cup E_m).$$

Hence, if $F_p = \bigcup_{m=p+1}^{\infty} (D_m \cup E_m)$, we have for each $p = 1, 2, \ldots$

$$\mu(F_p) < 2^{-p} \quad \text{and} \quad \|f(x) - h^m(x)\| < m^{-1}$$

$$\text{for all } m > p \quad \text{and} \quad x \in A \setminus F_p. \quad (3)$$

This implies that $\lim_{m \to \infty} h^m(x) = f(x)$ whenever $x \in \bigcup_p (A \setminus F_p) = A \setminus \bigcap_p F_p$. Since $\mu(\bigcap_p F_p) = 0$ (by (3)) and the h^m are in $\mathscr{T}$, this gives $f \in \mathscr{T}^-$, proving the claim.

Recall from 2.1 the definition of $\mathscr{F}$, and let $\mathscr{R}$ be the linear span of the collection of all products $\phi f : x \mapsto \phi(x) f(x)$, where $\phi \in \mathscr{F}$ and $f \in \mathscr{Q}$. We observe that, for any g in $\mathscr{R}$, $x \mapsto \|g(x)\|$ is locally μ-measurable. Indeed, if $g = \sum_{i=1}^{n} \phi_i f_i (\phi_i \in \mathscr{F}; f_i \in \mathscr{Q})$, we can find a finite pairwise disjoint sequence $A_1, \ldots, A_r$ of sets in $\mathscr{S}$ such that each ϕ_i is constant on each A_j, and the ϕ_i all vanish outside $\bigcup_j A_j$. It follows that, on each A_j, g coincides with a linear combination of the f_i with *constant* coefficients, that is, with an element of $\mathscr{Q}$ itself; and that g vanishes outside $\bigcup_j A_j$. This ensures the local μ-measurability of $x \mapsto \|g(x)\|$.

Thus we can apply the preceding claim with $\mathscr{T}$ replaced by $\mathscr{R}$. Define $\mathscr{M} = \mathscr{R}^-$. By the above claim $\mathscr{M}$ is linear, its elements have locally μ-measurable norm-functions, and $\mathscr{M}^- = \mathscr{M}$. So, to show that $\mathscr{M}$ is a local μ-measurability structure, we have only to prove 4.1(II). For this, we first observe that $\phi f \in \mathscr{M}$ whenever $\phi : X \to \mathbb{C}$ is locally μ-measurable and $f \in \mathscr{Q}$. Indeed, given any A in $\mathscr{S}$, we can find a sequence $\{\phi_n\}$ of elements of $\mathscr{F}$ with $\phi_n \to \phi$ pointwise μ-almost everywhere on A; since $\phi_n f \in \mathscr{R}$ and $\phi_n f \to \phi f$ μ-almost everywhere on A, it follows that $\phi f \in \mathscr{M}$. Now take any f in $\mathscr{M}$ and any locally μ-measurable $\phi : X \to \mathbb{C}$. Given $A \in \mathscr{S}$, there is a sequence $\{f_n\}$ of elements of $\mathscr{R}$ such that $f_n \to f$, hence also $\phi f_n \to \phi f$, pointwise μ-almost everywhere on A. But the preceding observation implies that $\phi f_n \in \mathscr{M}$. So $\phi f \in \mathscr{M}^- = \mathscr{M}$.

We have now shown that $\mathscr{M}$ is a local μ-measurability structure. It clearly contains $\mathscr{Q}$. So the proof of the Proposition is complete. ■

Remark.   Notice that the $\mathscr{M}$ of the above proof is actually the local μ-measurability structure generated by $\mathscr{Q}$.

4.3.   Here is an important technical fact relating to the proof of 4.2. We keep the notation of 4.2.

Lemma.   *Suppose that $f \in \mathscr{M}$ and that f vanishes (i.e., $f(x) = 0_x$) outside a countable union of sets in $\mathscr{S}$. Then there exists a sequence $\{g_n\}$ of vector fields in $\mathscr{R}$ such that*

(i) $g_n(x) \to f(x)$ in B_x for μ-almost all x, and
(ii) $\|g_n(x)\| \leq \|f(x)\|$ for all n and μ-almost all x.

Proof.   We shall first assume that $X \in \mathscr{S}$.
   By the proof of 4.2 there is a sequence $\{g_n'\}$ of elements of $\mathscr{R}$ satisfying (i). For each n consider the non-negative μ-measurable function

$$\rho_n : x \mapsto \min\{1, \|g_n'(x)\|^{-1} \|f(x)\|\}.$$

(Put $\rho_n(x) = 1$ if $g_n'(x) = 0_x$.) Thus $\rho_n(x) \to 1$ for μ-almost all x. Therefore by Egoroff's Theorem 1.11 there is a sequence $\{\phi_n\}$ of non-negative simple functions on X such that for μ-almost all x

$$\phi_n(x) \to 1 \quad \text{and} \quad \phi_n(x) \leq \rho_n(x) \qquad \text{for all } n. \tag{4}$$

Since the ϕ_n are simple, the vector fields $g_n = \phi_n g_n'$ belong to $\mathscr{R}$, and we see from (4) that properties (i) and (ii) hold for the g_n.
   Now drop the assumption that $X \in \mathscr{S}$, and choose a sequence $\{A_m\}$ of pairwise disjoint sets in $\mathscr{S}$ such that f vanishes outside $\bigcup_m A_m$. Applying the last paragraph inside each A_m, we obtain for each m a sequence $\{g_{mn}\}_n$ of vector fields in $\mathscr{R}$ satisfying (i) and (ii) inside A_m, such that g_{mn} vanishes outside A_m. If we set $g_n = \sum_{m=1}^{n} g_{mn}$, it is clear that the g_n belong to $\mathscr{R}$ and satisfy (i) and (ii) everywhere. ■

4.4.   We give here a useful criterion for local μ-measurability.

Proposition.   *Let $\mathscr{M}$ be a local μ-measurability structure for the $\{B_x\}$. Suppose that a vector field f satisfies the following two conditions: (I) $x \mapsto \|f(x) - g(x)\|$ is locally μ-measurable for every g in $\mathscr{M}$; (II) for each A in $\mathscr{S}$ there is a countable subfamily $\mathscr{N}$ of $\mathscr{M}$ such that, for μ-almost all x in A, $f(x)$ lies in the closed linear span (in B_x) of $\{g(x) : g \in \mathscr{N}\}$. Then $f \in \mathscr{M}$.*

Proof.   Fix A in $\mathcal{S}$, and let $\mathcal{N}$ be as in (II). Given $\varepsilon > 0$ it is enough by 4.1(IV) to find a vector field h in $\mathcal{M}$ such that

$$\|h(x) - f(x)\| < \varepsilon \quad \text{for } \mu\text{-almost all } x \text{ in } A. \tag{5}$$

Adjoining to $\mathcal{N}$ all linear combinations of elements of $\mathcal{N}$ with complex-rational coefficients, we may as well suppose that $f(x)$ lies in the closure of $\{g(x) : g \in \mathcal{N}\}$ for μ-almost all x in A. Choose a maximal disjoint family $\mathcal{T}$ of non-μ-null subsets of A in $\mathcal{S}$ such that for each B in $\mathcal{T}$ there is a vector field g_B in $\mathcal{N}$ for which $\|f(x) - g_B(x)\| < \varepsilon$ for all x in B. Since $\mathcal{T}$ is necessarily countable, $C = A \setminus \cup \mathcal{T}$ is in $\mathcal{S}$. Hypothesis (I) and the definition of $\mathcal{T}$ imply that $\|f(x) - g(x)\| \geq \varepsilon$ for all g in $\mathcal{N}$ and μ-almost all x in C; that is, for μ-almost all x in C, $f(x)$ is not in the closure of $\{g(x) : x \in \mathcal{N}\}$. Consequently C is μ-null. Hence, choosing the locally μ-measurable vector field h to coincide with g_B on each B in $\mathcal{T}$, we obtain (5).   ∎

4.5.   Fix a local μ-measurability structure $\mathcal{M}$ for the $\{B_x\}$.

Definition.   If $1 \leq p < \infty$, we shall denote by $\mathcal{L}_p(\mu; \{B_x\}; \mathcal{M})$, or $\mathcal{L}_p$ for short, the intersection $\bar{\mathcal{L}}_p(\mu; (B_x)) \cap \mathcal{M}$.

This $\mathcal{L}_p$ is a closed linear subspace of $\bar{\mathcal{L}}_p(\mu; \{B_x\})$, hence a Banach space in its own right. (The closedness of $\mathcal{L}_p$ follows immediately from 3.5 and 4.1(IV)). $\mathcal{L}_p$ is the useful "measurable" subspace of $\bar{\mathcal{L}}_p(\mu; \{B_x\})$ referred to in 3.4.

4.6. Proposition.   *Suppose $\mathcal{M}$ is the local μ-measurability structure generated (as in 4.2) by a linear space $\mathcal{D}$ of vector fields. Fix $1 \leq p < \infty$, and let $\mathcal{R}$ (as in the proof of 4.2) be the linear span of $\{\phi f : \phi \in \mathcal{F}, f \in \mathcal{D}\}$. Let us assume that for each f in $\mathcal{D}$, $\{\|f(x)\| : x \in X\}$ is bounded; then certainly $\mathcal{R} \subset \mathcal{L}_p$. In fact, $\mathcal{R}$ is dense in $\mathcal{L}_p$.*

Proof.   Take any vector field f in $\mathcal{L}_p$. Thus $f \in \mathcal{M}$ and f vanishes outside a countable union of sets in $\mathcal{S}$. Hence by 4.3 there is a sequence $\{g_n\}$ of vector fields in $\mathcal{R}$ such that 4.3(i), (ii) hold. Applying the Dominated-Convergence Theorem 2.8 to the sequence of numerical functions $x \mapsto \|f(x) - g_n(x)\|^p$, we conclude that $g_n \to f$ in $\mathcal{L}_p$.   ∎

4.7. Proposition.   *Assume that each B_x is a Hilbert space (with inner product
$(\ , \ )$). If $\mathcal{M}$ is a local μ-measurability structure for the $\{B_x\}$, then $\mathcal{L}_2 = \mathcal{L}_2(\mu; \{B_x\}; \mathcal{M})$ is a Hilbert space, whose inner product is given by:*

$$(f, g) = \int (f(x), g(x)) d\mu x \qquad (f, g \in \mathcal{L}_2). \qquad (6)$$

Proof.   Since the polarization identity (Chapter I) expresses $(f(x), g(x))$ as a
linear combination of four expressions of the form $\|h(x)\|^2$ (where $h \in \mathcal{L}_2$),
and since the functions $x \mapsto \|h(x)\|^2$ are μ-summable, it follows that $x \mapsto$
$(f(x), g(x))$ is μ-summable. Thus the right side of (6) defines an inner product
in $\mathcal{L}_2$ consistent with the norm of $\mathcal{L}_2$.   ∎

Remark.   This Proposition illustrates the importance of the "measurable"
subspaces $\mathcal{L}_p$ of $\bar{\mathcal{L}}_p$. Indeed, even if each B_x is a Hilbert space, the reader can
verify that the outsize space $\bar{\mathcal{L}}_2$ is not in general a Hilbert space.

5.   Integration of Functions with Values in a Banach Space

5.1.   An important special case arises when the B_x of 3.1 are all the same
Banach space B.

   In this section we assume as before that μ is a measure on the δ-ring $\mathcal{S}$ of
subsets of X; and we fix a Banach space B. By 4.2 the family $\mathcal{D}$ of all *constant*
functions on X to B generates a local μ-measurability structure $\mathcal{M}$ for
functions on X to B. Functions on X to B which belong to this $\mathcal{M}$ will be
called *locally μ-measurable*. Let us denote by $\mathscr{F}(B)$ the linear span of the
collection of all functions from X to B of the form $\text{Ch}_A \xi : x \mapsto \text{Ch}_A(x)\xi$, where
$A \in \mathcal{S}$ and $\xi \in B$. Elements of $\mathscr{F}(B)$ are called *simple (B-valued) functions*. In
our present context, $\mathscr{F}(B)$ is the $\mathcal{R}$ of the proof of 4.2. Hence by that proof we
have:

Proposition.   *A function $f : X \to B$ is locally μ-measurable if and only if, for
each A in $\mathcal{S}$, there is a sequence $\{g_n\}$ of simple (B-valued) functions such that
$g_n(x) \to f(x)$ in B for μ-almost all x in A.*

5.2.   Corollary.   *Let B' be another Banach space, and $\phi : B \to B'$ a continuous
map. If $f : X \to B$ is locally μ-measurable, then $\phi \circ f : X \to B'$ is locally μ-
measurable.*

5.3.   In the context of 5.1 we shall abreviate $\mathcal{L}_p(\mu; \{B_x\}; \mathcal{M})$ to $\mathcal{L}_p(\mu; B)$. By
4.6 one has:

Proposition.   *For every* $1 \leq p < \infty$, *$\mathscr{F}(B)$ is dense in $\mathscr{L}_p(\mu; B)$.*

5.4.   Functions f which belong to $\mathscr{L}_1(\mu; B)$ are called μ-*summable*. For these we shall define an integral $\int f \, d\mu$.

In fact, proceeding a little more generally, from here to the end of this section ν will denote a complex measure on $\mathscr{S}$ satisfying $|\nu| \leq \mu$. As usual, $\| \ \|_1$ is the norm in $\mathscr{L}_1(\mu; B)$. One verifies without difficulty that there is a (unique) linear map $J: \mathscr{F}(B) \to B$ such that

$$J(\mathrm{Ch}_A \xi) = \nu(A)\xi \qquad\qquad (A \in \mathscr{S}; \xi \in B)$$

(where $(\mathrm{Ch}_A \xi)(x) = \mathrm{Ch}_A(x)\xi$), and that J satisfies

$$\|J(f)\| \leq \|f\|_1 \qquad\qquad (f \in \mathscr{F}(B)). \qquad (1)$$

Therefore by 5.3 J extends in just one way to a linear map (also called J) of $\mathscr{L}_1(\mu; B)$ into B satisfying (1) for all f in $\mathscr{L}_1(\mu; B)$.

Definition.   If $f \in \mathscr{L}_1(\mu; B)$, we denote $J(f)$ by $\int f \, d\nu$, or (more fully) by $\int f(x) d\nu x$, and refer to it as the *B-valued vector integral of f with respect to ν*.

Evidently $\int f \, d\nu$ depends only on f and ν (being independent of the particular μ satisfying $|\nu| \leq \mu$), and exists if and only if $f \in \mathscr{L}_1(|\nu|; B)$.

We shall often write $\mathscr{L}_1(\nu; B)$ instead of $\mathscr{L}_1(|\nu|; B)$; and functions in $\mathscr{L}_1(\nu; B)$ will be called ν-*summable* instead of $|\nu|$-summable. Also, *local ν-measurability* will mean local $|\nu|$-measurability.

If we replace μ by $|\nu|$, (1) becomes the important inequality:

$$\left\| \int f \, d\nu \right\| \leq \int \|f(x)\| d|\nu|x. \qquad (2)$$

5.5.   If we take the B of this section to be $\mathbb{C}$, the definitions of local μ-measurability, μ-summability, and $\int f \, d\mu$ as given in this section clearly coincide with those of 1.9 and 2.5. *In this case we abbreviate $\mathscr{L}_p(\mu; \mathbb{C})$ to $\mathscr{L}_p(\mu)$.*

5.6. Lebesgue Dominated-Convergence Theorem.   *Let $1 \leq p < \infty$; let $g: X \to [0, \infty]$ be μ-summable; and let $f, f_1, f_2, \ldots$ be locally μ-measurable functions on X to B such that for μ-almost all x (i) $\|f_n(x)\|^p \leq g(x)$ for all n, and (ii) $f_n(x) \to f(x)$ in B. Then f and the f_n are all in $\mathscr{L}_p(\mu; B)$ and $f_n \to f$ in $\mathscr{L}_p(\mu; B)$. Further, if $p = 1$ and $|\nu| \leq \mu$,*

$$\int f_n \, d\nu \to \int f \, d\nu \text{ in } B. \qquad (3)$$

Proof. Applying 2.8 to the sequence of numerical functions $x \mapsto \|f(x) - f_n(x)\|^p$, we conclude that $f_n \to f$ in $\mathscr{L}_p(\mu; B)$. In case $p = 1$, the continuity (2) of $f \mapsto \int f \, dv$ gives (3). ∎

5.7. The following extremely useful result says that continuous linear maps can be "passed through an integral sign".

Proposition. *Let B' be another Banach space, and $\phi: B \to B'$ a continuous linear map. If $f \in \mathscr{L}_1(v; B)$, then $\phi \circ f \in \mathscr{L}_1(v; B')$, and*

$$\int (\phi \circ f) \, dv = \phi\left[\int f \, dv\right]. \tag{4}$$

Sketch of Proof. First note that, if $f \in \mathscr{F}(B)$, then $\phi \circ f \in \mathscr{F}(B')$, $\int \|\phi \circ f\| d|v| \leq \|\phi\| \int \|f\| d|v|$, and (4) holds. From this we pass by continuity to arbitrary elements of $\mathscr{L}_1(\mu; B)$. ∎

5.8. Corollary. *Let ξ be a vector in B and $f \in \mathscr{L}_1(v)$. Then $f\xi: x \mapsto f(x)\xi$ belongs to $\mathscr{L}_1(v; B)$; and*

$$\int f\xi \, dv = \left[\int f \, dv\right]\xi.$$

5.9. The proof of the following useful observation is left to the reader.

Proposition. *Let $\bar{B}$ be the Banach space complex-conjugate to B, and $\bar{v}: A \mapsto \overline{v(A)}$ the complex-conjugate of the complex measure v on $\mathscr{S}$. If $f \in \mathscr{L}_1(v; B)$, then $\int f \, d\bar{v}$, considered as a $\bar{B}$-valued integral, coincides with $\int f \, dv$ considered as a B-valued integral.*

In particular, if $f \in \mathscr{L}_1(v)$, then

$$\int \bar{f} \, d\bar{v} = \left(\int f \, dv\right)^{-}$$

($^{-}$ denoting complex conjugation).

5.10. Suppose that $\alpha: X \to Y$ is a bijection of X onto another set Y, and that $\mathscr{T}$ is the δ-ring of subsets of Y onto which α carries $\mathscr{S}$ (that is, $\alpha(A) \in \mathscr{T} \Leftrightarrow A \in \mathscr{S}$). Clearly α carries the complex measure v on $\mathscr{S}$ into a complex measure $\alpha \cdot v: A \mapsto v(\alpha^{-1}(A))$ on $\mathscr{T}$. Likewise, any function f on X to the Banach space B is transported by α into a function $\alpha \cdot f = f \circ \alpha^{-1}$ on Y to B. It is an important though rather evident fact that every property possessed by

f with respect to v is also possessed by $\alpha \cdot f$ with respect to $\alpha \cdot v$. For example, f is locally v-measurable [v-summable] if and only if $\alpha \cdot f$ is locally $(\alpha \cdot v)$-measurable [$(\alpha \cdot v)$-summable]; and if f is v-summable, then

$$\int f \, dv = \int (\alpha \cdot f) d(\alpha \cdot v)$$

$$= \int f(\alpha^{-1}(x)) d(\alpha \cdot v)x. \tag{5}$$

It is often useful to use the notation $dv(\alpha(x))$ for the transported measure $\alpha^{-1} \cdot v$. Then equation (5), applied to α^{-1}, becomes:

$$\int f \, dv = \int f(\alpha(x)) dv(\alpha(x)).$$

All this applies in particular, of course, if $Y = X$ and $\mathcal{T} = \mathcal{S}$, that is, if α is an $\mathcal{S}$-preserving permutation of X.

5.11.   Here is another useful and rather evident remark.

Suppose $\mathcal{T}$ is another δ-ring of subsets of X such that $\mathcal{T} \subset \mathcal{S}$, with the property that, if $A \in \mathcal{S}$ and $A \subset B \in \mathcal{T}$, then $A \in \mathcal{T}$. Put $\mu_0 = \mu|\mathcal{T}$, $v_0 = v|\mathcal{T}$ (where $|v| \leq \mu$). If $f: X \to B$ vanishes outside a countable union of sets in $\mathcal{T}$, we have:

(I) f is locally μ-measurable if and only if it is locally μ_0-measurable;

(II) f is μ-summable if and only if it is μ_0-summable, in which case

$$\int f \, dv_0 = \int f \, dv.$$

5.12.   Here is an interesting criterion for local μ-measurability in the context of this section.

Proposition*.   *A function $f: X \to B$ is locally μ-measurable if and only if it satisfies the following two conditions: (I) $\alpha \circ f$ is a locally μ-measurable (numerical) function for every α in B^*; (II) for each A in $\mathcal{S}$ there is a closed separable linear subspace D of B such that $f(x) \in D$ for μ-almost all x in A.*

To prove the 'if' part it is enough to show that (I) and (II) imply the local μ-measurability of $x \mapsto \|f(x)\|$ (for then by "translation" $x \mapsto \|f(x) - g(x)\|$ will be locally μ-measurable for every g in $\mathcal{M}$, implying by 4.4 that $f \in \mathcal{M}$). To show this, use the Hahn-Banach Theorem to obtain, for each separable subspace D of B, a countable subfamily W of B^* such that $\|\xi\| = \sup\{|\alpha(\xi)|:$ $\alpha \in W\}$ for every ξ in D.

6. Integration of Functions with Values in a Locally Convex Space

6.1. It is sometimes useful to be able to integrate functions whose values lie in an arbitrary locally convex space.

Let L be any fixed locally convex space, and P an upward directed defining family of seminorms for L (see Chapter I). If $p \in P$, let $N_p = \{\xi \in L : p(\xi) = 0\}$; form the normed linear space L/N_p, with norm

$$\|\xi + N_p\|_p = p(\xi);$$

and denote by $\tilde{L}_p$ the Banach space completion of L/N_p with respect to $\|\ \|_p$. Thus, if p and q are elements of P with $p \prec q$ (that is, for some constant k, $p(\xi) \leq kq(\xi)$ for all ξ in L), there are natural continuous linear maps $\gamma_p : L \to \tilde{L}_p$ and $\gamma_{pq} : \tilde{L}_q \to \tilde{L}_p$ given by:

$$\gamma_p(\xi) = \xi + N_p, \gamma_{pq}(\xi + N_q) = \xi + N_p \qquad (\xi \in L).$$

If $p \prec q \prec r$ $(p, q, r, \in P)$, we clearly have

$$\gamma_{pq} \circ \gamma_q = \gamma_p, \gamma_{pq} \circ \gamma_{qr} = \gamma_{pr}. \tag{1}$$

For each p in P suppose we are given a vector ξ_p in $\tilde{L}_p$. We shall say that the collection $\{\xi_p\}$ is *coherent* if $\gamma_{pq}(\xi_q) = \xi_p$ whenever $p, q \in P$ and $p \prec q$. Here is a useful fact:

Lemma. *If L is complete, and $\{\xi_p\}$ $(p \in P)$ is a coherent collection in the sense just defined, there exists a unique vector ξ in L such that*

$$\xi_p = \gamma_p(\xi) \text{ for every } p \text{ in } P.$$

Proof. Consider the directed set D of all $\langle p, \{\varepsilon_q\}_{q \in P} \rangle$, where $p \in P$ and $\varepsilon_q > 0$ for each q in P; the directing relation $\prec$ in D is given by: $\langle p, \{\varepsilon_q\} \rangle \prec \langle p', \{\varepsilon_q'\} \rangle$ if and only if $p \prec p'$ and $\varepsilon_q' \leq \varepsilon_q$ for each q. We define a net on D as follows: For each $\alpha = \langle p. \{\varepsilon_q\} \rangle$ in D let η_α be a vector in L such that

$$\|\gamma_p(\eta_\alpha) - \xi_p\|_p < \varepsilon_p. \tag{2}$$

We claim that $\{\eta_\alpha\}$ is a Cauchy net in L. To see this, we fix r in P and $\delta > 0$, and define $\delta_q = (3\|\gamma_{rq}\|)^{-1}\delta$ for $q \succ r$ (for $q \nsucc r$, δ_q can be any positive number we like). Then, if $\alpha = \langle p, \{\varepsilon_q\} \rangle \succ \langle r, \{\delta_q\} \rangle = \beta$ in D, we have $r(\eta_\alpha - \eta_\beta) = \|\gamma_{rp}\gamma_p(\eta_\alpha) - \gamma_r(\eta_\beta)\|_r \leq \|\gamma_{rp}(\gamma_p(\eta_\alpha) - \xi_p)\|_r + \|\gamma_r(\eta_\beta) - \xi_r\|_r +$

$\|\gamma_{rp}(\xi_p) - \xi_r\|_r < \|\gamma_{rp}\|\varepsilon_p + \delta_r$ (by (2) and the coherence condition) $\leq \delta/3 + \delta/3 < \delta$. By the arbitrariness of r and δ this implies that $\{\eta_\alpha\}$ is Cauchy, proving the claim. So, since L is complete, $\{\eta_\alpha\}$ converges to an element ξ of L; and it is easy to see that $\gamma_p(\xi) = \xi_p$ for every p in P. ∎

6.2.   As in §5, let v be a complex measure on a δ-ring $\mathscr{S}$ of subsets of X.

Definition.   A function $f: X \to L$ will be called v-*summable* if $\gamma_p \circ f: X \to \tilde{L}_p$ is v-summable in the sense of 5.4 for every p in P.

Assume that $f: X \to L$ is v-summable and put $\xi_p = \int (\gamma_p \circ f)dv$ $(p \in P)$. In view of 5.7 and (1), $\{\xi_p\}$ is a coherent collection. Hence, *if L is complete*, there exists by 6.1 an element ξ of L such that

$$\gamma_p(\xi) = \xi_p = \int (\gamma_p \circ f)dv \tag{3}$$

for all p in P. This ξ is called the *L-valued vector integral of f with respect to v*.

Even if L is not complete, there will be an element ξ in the completion of L satisfying (3). If it so happens that $\xi \in L$, ξ is still called the *L-valued vector integral of f with respect to v*, and is denoted by $\int f \, dv$.

Notice that the definition of v-summability and of the L-valued vector integral of f depends only on the topology of L, not on the particular defining family of seminorms. This follows from 5.7 and the fact that any upward directed defining family of seminorms is cofinal in the upward directed set of *all* continuous seminorms on L.

The definition does however depend strongly on the topology of L. For instance, a function $f: X \to L$ might well be v-summable with respect to the weak topology of L without being v-summable with respect to the original topology of L.

However, if L is a locally convex space under two different topologies T_1 and T_2, where T_2 is weaker than T_1, and if $\int f \, dv$ exists in L with respect to T_1, then it obviously exists with respect to T_2.

6.3.   Proposition 5.7 has an obvious generalization, whose proof we leave to the reader.

Proposition.   *Let L and L' be two complete locally convex spaces, and $\phi: L \to L'$ a continuous linear map. If $f: X \to L$ is v-summable, then $\phi \circ f: X \to L'$ is also v-summable; and*

$$\int (\phi \circ f)dv = \phi \left[\int f \, dv \right].$$

7. The Radon-Nikodym Theorem and Related Topics

Measures of the Form f dμ

7.1. In this section μ is a fixed complex measure on the δ-ring $\mathcal{S}$ of subsets of the set X.

If $f: X \to \mathbb{C}$, A is a locally μ-measurable set, and $\mathrm{Ch}_A f$ is μ-summable, one writes $\int_A f d\mu$ to mean $\int (\mathrm{Ch}_A f) d\mu$.

Suppose that $f: X \to \mathbb{C}$, and that $A_1, A_2, \ldots$ is a sequence of pairwise disjoint locally μ-measurable subsets of X with union A, such that $\mathrm{Ch}_A f$ is μ-summable. Then $\mathrm{Ch}_{A_n} f$ is μ-summable for each n, and by 5.6

$$\int_A f \, d\mu = \sum_{n=1}^{\infty} \int_{A_n} f \, d\mu. \tag{1}$$

7.2. Definition. A function $f: X \to \mathbb{C}$ is *locally μ-summable* if $\mathrm{Ch}_A f$ is μ-summable for every A in $\mathcal{S}$.

Thus local μ-summability implies local μ-measurability.

If $1 \le p < \infty$, any function in $\mathcal{L}_p(\mu)$ is locally μ-summable in virtue of 2.6. If $f: X \to \mathbb{C}$ is locally μ-summable, by 7.1 the equation

$$\rho(A) = \int_A f \, d\mu \qquad (A \in \mathcal{S})$$

defines a new complex measure ρ on $\mathcal{S}$. We denote this ρ briefly by the symbol $f \, d\mu$.

7.3. Proposition. *If $f: X \to \mathbb{C}$ is locally μ-summable, then $|f \, d\mu| = |f| \, d|\mu|$.*

Proof. By 5.4(2) $|(f \, d\mu)(A)| \le (|f| \, d|\mu|)(A)$ for $A \in \mathcal{S}$; hence by a remark in 1.3

$$|f \, d\mu| \le |f| \, d|\mu|. \tag{2}$$

Fix $A \in \mathcal{S}$. We wish to show that

$$|f \, d\mu|(A) = (|f| \, d|\mu|)(A). \tag{3}$$

The verification of (3) involves only subsets of A; so we may as well assume $A = X \in \mathcal{S}$. Then (3) is to be proved for all $f \in \mathcal{L}_1(\mu)$.

Both sides of (3), as functions of f, are seminorms on $\mathcal{L}_1(\mu)$ (see 1.4). By (2) these seminorms are both continuous. So it is enough to verify (3) for simple functions f (since these form a dense subset of $\mathcal{L}_1(\mu)$). But for simple functions (3) is evident. ∎

7.4.   Corollary.   *Given two locally μ-summable complex functions f and f' on X, we have $\int f\,d\mu = \int f'\,d\mu$ if and only if f and f' coincide locally μ-almost everywhere.*

7.5.   Theorem.   *Let $f: X \to \mathbb{C}$ be locally μ-summable, and put $v = f\,d\mu$. Let B be any Banach space, and $g: X \to B$ a function which vanishes outside a countable union of sets in $\mathscr{S}$. Then g belongs to $\mathscr{L}_1(v; B)$ if and only if fg belongs to $\mathscr{L}_1(\mu; B)$; and in that case*

$$\int g\,dv = \int fg\,d\mu. \tag{4}$$

Proof.   On the part of the space where f vanishes the result is trivially true. So we may assume that $f(x) \neq 0$ for all x. In that case both conditions whose equivalence is asserted imply that g is locally μ-measurable; so we shall assume this. Hence by 4.3 there is a sequence $\{g_n\}$ of simple B-valued functions on X such that, for μ-almost all x, $\|g_n(x)\| \leq \|g(x)\|$ for all n and $g_n(x) \to g(x)$.

Now it is clear from 7.3 that

$$\int \|g_n\| d|v| = \int |f| \|g_n\| d|\mu| \qquad \text{for each } n. \tag{5}$$

Since $\|g_n\| \leq \|g\|$ and $\|g_n\| \to \|g\|$ pointwise, it follows from (5) and 2.4 that

$$\int \|g\| d|v| = \int |f| \|g\| d|\mu|. \tag{6}$$

Thus the two sides of (6) are finite or infinite together, whence $g \in \mathscr{L}_1(v; B)$ if and only if $fg \in \mathscr{L}_1(\mu; B)$.

By 5.4(2) the left side of (4) is continuous in g on $\mathscr{L}_1(v; B)$. Likewise $\int h\,d\mu$ is continuous in h on $\mathscr{L}_1(\mu; B)$; so by (6) the right side of (4) is continuous in g on $\mathscr{L}_1(v; B)$. So it is enough to prove (4) for a dense set of functions g, for example for the simple functions (see 5.3)). But if g is simple, (4) holds in virtue of 5.8.   ∎

7.6.   Corollary.   *Let $f: X \to \mathbb{C}$ be locally μ-summable; put $v = f\,d\mu$; and suppose $g: X \to \mathbb{C}$ is locally v-summable. Then fg is locally μ-summable and*

$$g\,dv = (fg)d\mu.$$

If in addition $\{x: f(x) = 0\}$ is locally μ-null, then f^{-1} is locally v-summable, and

$$f^{-1}\,dv = \mu.$$

Absolute Continuity and the Radon-Nikodym Theorem

7.7.   Definition.   If v and μ are two complex measures on $\mathscr{S}$, we say that v is *absolutely continuous with respect to* μ, in symbols $v \ll \mu$, if every μ-null set is v-null. Equivalently, $v \ll \mu$ if and only if $|\mu|(A) = 0 \Rightarrow |v|(A) = 0$ for each A in $\mathscr{S}$. If both $v \ll \mu$ and $\mu \ll v$ (that is, if the μ-null sets and the v-null sets are the same), we say that μ and v are *(measure-theoretically) equivalent*, in symbols $v \sim \mu$.

Notice that $|\mu| \sim \mu$, and that $|v| \le |\mu| \Rightarrow v \ll \mu$.

If $v \sim \mu$, then evidently measurable sets and locally measurable sets and functions are the same with respect to both measures.

7.8.   If $f: X \to \mathbb{C}$ is locally μ-summable, then $f\, d\mu \ll \mu$. (Notice that $f\, d\mu \sim \mu$ if and only if $\{x : f(x) = 0\}$ is locally μ-null). The Radon-Nikodym Theorem states that, with some extra hypothesis on μ such as paraboundedness, the converse of this holds.

Radon-Nikodym Theorem.   *Let μ be a parabounded complex measure on the δ-ring $\mathscr{S}$; and let v be a complex measure on $\mathscr{S}$ such that $v \ll \mu$. Then there is a locally μ-summable function $f: X \to \mathbb{C}$ such that $v = f\, d\mu$.*

Proof.   Assume first that $X \in \mathscr{S}$. For that case, and for non-negative μ, we refer the reader to the proof in Halmos [1], Theorem 31B. Since $v \ll |\mu|$ and $\mu \ll |\mu|$, we thus obtain locally μ-summable complex functions g and h such that $v = g\, d|\mu|$ and $\mu = h\, d|\mu|$. Since $\mu \sim |\mu|$, $h(x) \ne 0$ for μ-almost all x (in fact, by 7.3, $|h(x)| \equiv 1$ μ-almost everywhere). Hence 7.6 gives $v = (h^{-1}g)d\mu$. This completes the proof in case $X \in \mathscr{S}$.

We now drop the assumption that $X \in \mathscr{S}$, and choose a subfamily $\mathscr{W}$ of $\mathscr{S}$ as in the definition of paraboundedness (1.7). By the last paragraph there is a function $f: X \to \mathbb{C}$ such that $v = f\, d\mu$ within each set A of $\mathscr{W}$. By the last remark of 1.7 f is locally μ-measurable. Let B be any set in $\mathscr{S}$. By 1.7(i), $\mathscr{W}_0 = \{A \in \mathscr{W} : B \cap A \ne \emptyset\}$ is countable. So (using 7.3) $|v|(B) = \sum_{A \in \mathscr{W}_0} |v|(B \cap A) = \sum_{A \in \mathscr{W}_0} \int_{B \cap A} |f|\, d|\mu|$. From this and 2.2(vi) we see that $\int_B |f|\, d|\mu| < \infty$. Hence f is locally μ-summable. Since v and $f\, d\mu$ coincide on subsets of each set in $\mathscr{W}$, they are equal.   ∎

Remark 1.   By 7.4 the f of the above theorem is unique up to a locally μ-null set. It is called the *Radon-Nikodym derivative of v with respect to μ.*

Remark 2.   The hypothesis of paraboundedness cannot simply be omitted. For an example illustrating this see Halmos [1], §31, Exercise 9.

7.9. Let μ be a measure on $\mathscr{S}$ (whether parabounded or not). If $f \in \mathscr{L}_1(\mu)$, then $f\,d\mu$ is bounded, and in fact by 7.3 the total variation norm of $f\,d\mu$ is equal to $\|f\|_1$. As a corollary of the Radon-Nikodym Theorem we have a converse of this:

Corollary. *Let v be any bounded complex measure on $\mathscr{S}$ such that $v \ll \mu$. Then there is a function f in $\mathscr{L}_1(\mu)$ such that $v = f\,d\mu$.*

Proof. By 1.7 there is a countable union E of sets in $\mathscr{S}$ such that $X \setminus E$ is locally v-null. We shall restrict our attention to subsets of E; that is, we assume $X = E$. Then μ becomes parabounded; and we can apply 7.8 to obtain a locally μ-summable function f such that $v = f\,d\mu$. Now 7.3 and 2.4 imply that $f \in \mathscr{L}_1(\mu)$. ∎

Remark. From this Corollary and the remark preceding it we see that the map $f \mapsto f\,d\mu$ is a linear isometry of $\mathscr{L}_1(\mu)$ into $M_b(\mathscr{S})$ (see 1.5) whose range is $\{v \in M_b(\mathscr{S}) : v \ll \mu\}$.

The Adjoint Space of $\mathscr{L}_1(\mu)$

7.10. Let μ be a measure on $\mathscr{S}$. A function $f : X \to \mathbb{C}$ is said to be *μ-essentially bounded* if there exists a number $k \geq 0$ such that $\{x \in X : |f(x)| > k\}$ is locally μ-null. In that case there is a *smallest* such k, which we denote by $\|f\|_\infty$. Let $\mathscr{L}_\infty(\mu)$ stand for the linear space of all μ-essentially bounded locally μ-measurable complex functions on X. Then $\|\ \|_\infty$ is a seminorm on $\mathscr{L}_\infty(\mu)$, whose null space consists of those functions which vanish except on a locally μ-null set. Let us identify functions in $\mathscr{L}_\infty(\mu)$ which differ only on a locally μ-null set. Then $\mathscr{L}_\infty(\mu)$ is a Banach space with $\|\ \|_\infty$ as norm.

7.11. Each f in $\mathscr{L}_\infty(\mu)$ gives rise to an element α_f of the adjoint space of $\mathscr{L}_1(\mu)$:

$$\alpha_f(g) = \int fg\,d\mu \qquad\qquad (g \in \mathscr{L}_1(\mu));$$

and it is easy to see that $\|\alpha_f\| = \|f\|_\infty$. Conversely, we have;

Theorem. *Assume that μ is a parabounded measure on $\mathscr{S}$. Then the correspondence $f \mapsto \alpha_f$ is a linear isometry of $\mathscr{L}_\infty(\mu)$ onto $(\mathscr{L}_1(\mu))^*$.*

Proof. We have only to fix an element α of $(\mathscr{L}_1(\mu))^*$ and show that $\alpha = \alpha_f$ for some f in $\mathscr{L}_\infty(\mu)$. For each A in $\mathscr{S}$ put $\nu(A) = \alpha(\mathrm{Ch}_A)$. By the continuity of α, ν is a complex measure on $\mathscr{S}$; and clearly $\nu \ll \mu$. So by 7.8 $\nu = \int f \, d\mu$ for some locally μ-summable complex function f. An easy estimate shows that f is μ-essentially bounded; in fact $\|f\|_\infty \leq \|\alpha\|$. By the definition of ν, α and α_f coincide on simple functions. Hence $\alpha = \alpha_f$. ∎

Remark. One can also prove—even without the hypothesis of para-boundedness—that if $1 < p < \infty$ and $p^{-1} + q^{-1} = 1$, then $(\mathscr{L}_p(\mu))^* \cong \mathscr{L}_q(\mu)$. The proof of this, like the proof given above, is based on the Radon-Nikodym Theorem, but requires more intricate norm-estimates. The proper techniques will be found, for example, in Hewitt and Ross [1], Theorem 12.18.

8. Measures on Locally Compact Hausdorff Spaces

8.1. The *Borel σ-field* $\mathscr{B}(X)$ of a topological space X means as usual the σ-field of subsets of X generated by the family of all open subsets of X. The elements of $\mathscr{B}(X)$ are called *Borel subsets* of X.

From here to the end of this section X is a fixed locally compact Hausdorff space.

Definition. We denote by $\mathscr{S}(X)$ the δ-ring of subsets of X which is generated by the family of all compact subsets of X; we shall refer to $\mathscr{S}(X)$ as the *compacted Borel δ-ring* of X.

Thus, $A \in \mathscr{S}(X)$ if and only if A is a Borel subset of X and the closure of A is compact.

8.2. A measure [complex measure] on $\mathscr{S}(X)$ will be called a *Borel measure* [*complex Borel measure*] *on* X. (Strictly speaking one ought to speak of "compacted" Borel measures; but we omit the "compacted" for brevity).

If μ is a complex Borel measure on X, any Borel subset of X is locally μ-measurable; hence any Borel (and so any continuous) function on X to $\mathbb{C}$ or $\mathbb{R}_\mathrm{ext}$ is locally μ-measurable.

Definition. A complex Borel measure μ on X is *regular* if one (hence both) of the following two equivalent conditions holds:

(I) For each A in $\mathscr{S}(X)$ and each $\varepsilon > 0$, there is an open set U in $\mathscr{S}(X)$ such that $A \subset U$ and $|\mu|(U \setminus A) < \varepsilon$.

(II) For each A in $\mathscr{S}(X)$ and each $\varepsilon > 0$, there is a compact subset C of X such that $C \subset A$ and $|\mu|(A \setminus C) < \varepsilon$.

The proof of the equivalence of (I) and (II) is the same as the proof of Halmos [1], Theorem 52E.

8.3. If μ is regular, then obviously so is $|\mu|$.

If v and μ are complex Borel measures on X, $v \ll \mu$, and μ is regular, then v is regular. This is clear when we express Condition (II) in the following form: For each A in $\mathscr{S}(X)$ there is a countable union D of compact sets such that $D \subset A$ and $|\mu|(A \setminus D) = 0$.

The family $M_{br}(\mathscr{S}(X))$ of all regular elements of $M_b(\mathscr{S}(X))$ (see 1.4) is a norm-closed linear subspace of $M_b(\mathscr{S}(X))$, and hence a Banach space in its own right.

Definition. We shall denote by $\mathscr{M}_r(X)$ the Banach space $M_{br}(\mathscr{S}(X))$ of all bounded regular complex Borel measures on X, with the total variation norm.

8.4. If μ is a regular complex Borel measure on X and B is any Borel subset of X, then $\mu_B : A \mapsto \mu(A \cap B)$ $(A \in \mathscr{S}(X))$ is also a regular complex Borel measure on X, called the *cut-down of μ to B*. Notice that

$$|\mu_B| = |\mu|_B.$$

8.5. If Y is a closed subspace of X and v is a regular complex Borel measure on Y (considered as a locally compact Hausdorff space in its own right), then $\mu : A \mapsto v(A \cap Y)$ is a regular complex Borel measure on X, called the *injection of v onto X*.

Notice that this remark could not be made if Y were for example open instead of closed; for in that case, if $A \in \mathscr{S}(X)$, $A \cap Y$ need not be in $\mathscr{S}(Y)$.

8.6. Proposition. *For a complex Borel measure μ on X the following three conditions are equivalent:*

(I)   *μ is regular.*

(II) *$|\mu|(U) = \sup\{|\mu|(C) : C \text{ compact}, C \subset U\}$ for every open set U in $\mathscr{S}(X)$.*

(III) *$\lim_\alpha |\mu|(U_\alpha) = |\mu|(U)$ whenever U is an open set in $\mathscr{S}(X)$ and $\{U_\alpha\}$ is an increasing net of open sets in $\mathscr{S}(X)$ such that $\bigcup_\alpha U_\alpha = U$.*

Proof. Assume (II); and let U, $\{U_\alpha\}$ be as in (III). Given $\varepsilon > 0$, there is by (II) a compact subset C of U such that $|\mu|(U \setminus C) < \varepsilon$. Since $\bigcup_\alpha U_\alpha \supset C$ and C is compact, there is an index α_0 such that $C \subset U_{\alpha_0}$. But then $\alpha \succ \alpha_0 \Rightarrow$ $|\mu|(U \setminus U_\alpha) \le |\mu|(U \setminus C) < \varepsilon$. So (III) holds.

Conversely, assume (III), and let U be an open set in $\mathscr{S}(X)$. The set $\mathscr{N}$ of all those open subsets of U whose (compact) closures are contained in U is upward directed, and $\bigcup \mathscr{N} = U$. So by (III)

$$|\mu|(U) = \sup\{|\mu|(V): V \in \mathscr{N}\} \le \sup\{|\mu|(\bar{V}): V \in \mathscr{N}\} \le |\mu|(U).$$

It follows that $\sup\{|\mu|(\bar{V}): V \in \mathscr{N}\} = |\mu|(U)$. Since the $\bar{V}$ ($V \in \mathscr{N}$) are compact, (II) must hold.

We have shown that (II) $\Leftrightarrow$ (III). For the proof that (I) $\Leftrightarrow$ (II) we refer the reader to the proof of Theorem 52F of Halmos [1]. ∎

8.7. Proposition. *Every regular complex Borel measure on X is parabounded (1.7).*

To prove this, we take for the $\mathscr{W}$ of 1.7 a maximal disjoint family of compact subsets B of X with the property that $|\mu|(U \cap B) > 0$ whenever U is open and $U \cap B \ne \emptyset$. For details see Hewitt and Ross [1], Theorem 11.39.

It follows that for a regular complex Borel measure μ on X the Radon-Nikodym Theorem 7.8 is valid, as well as the relation $(\mathscr{L}_1(\mu))^* \cong \mathscr{L}_\infty(\mu)$ (7.11).

8.8. Proposition. *If the locally compact Hausdorff space X satisfies the second axiom of countability, then every complex Borel measure on X is regular.*

Proof. This is proved by combining Halmos [1], Theorem 52G, with the observation that, by second countability, every compact set is a G_δ. ∎

8.9. Definition. Let μ be a complex Borel measure on X. By the *closed support of μ*—in symbols supp(μ)—we mean the set of all those ponts x of X such that $|\mu|(U) > 0$ for every open neighborhood U of x in $\mathscr{S}(X)$.

Evidently supp(μ) is a closed subset of X.

Now assume that μ is regular. Then supp(μ) can be described as the complement in X of the unique largest locally μ-null open subset of X. This equivalent description follows easily from Proposition 8.6(III).

Integration on Locally Compact Spaces

8.10. Recall that $\mathscr{L}(X)$ is the linear space of continuous complex functions with compact support on X; $\mathscr{L}_r(X)$ $[\mathscr{L}_+(X)]$ is the family of all real-valued [non-negative-real-valued] elements of $\mathscr{L}(X)$.

Let μ be a complex Borel measure on X. Each f in $\mathscr{L}(X)$ is locally μ-measurable (see 8.2) and bounded, and has compact support, and so is μ-summable. Thus μ gives rise to a linear functional I_μ on $\mathscr{L}(X)$:

$$I_\mu(f) = \int f \, d\mu \qquad\qquad (f \in \mathscr{L}(X)). \qquad (1)$$

8.11.   Notice that, *if μ is regular*, then it is uniquely determined by I_μ. To see this, fix a set A in $\mathscr{S}(X)$, and let D be the directed set of all pairs $\langle U, C \rangle$, where C is compact, U is an open set in $\mathscr{S}(X)$, and $C \subset A \subset U$; the directing relation $\prec$ is given by

$$\langle U, C \rangle \prec \langle U', C' \rangle \Leftrightarrow U' \subset U \quad \text{and} \quad C' \supset C.$$

For each pair $\langle U, C \rangle$ in D, we can choose an element $f_{U,C}$ of $\mathscr{L}(X)$ such that range$(f_{U,C}) \subset [0, 1]$, $f_{U,C}$ vanishes outside U, and $f_{U,C} \equiv 1$ on C. The regularity of μ now clearly implies that

$$\lim_{\langle U, C \rangle} \int f_{U,C} \, d\mu = \mu(A). \qquad (2)$$

Since D was defined independently of μ, (2) shows that $\mu(A)$ depends only on I_μ.

Remark.   Two distinct complex Borel measures may very well give the same I_μ if not both are regular. For an example see Halmos [1], p. 231, Example (10).

8.12.   For each compact subset K of X, let $\mathscr{L}_K(X)$ be the space of those f in $\mathscr{L}(X)$ which vanish outside K; and equip $\mathscr{L}_K(X)$ with the supremum norm: $\|f\|_\infty = \sup\{|f(x)| : x \in K\}$.

Generalized F. Riesz Representation Theorem.   *The map $\mu \mapsto I_\mu$ is a linear bijection from the space $\mathscr{M}$ of all regular complex Borel measures μ on X onto the space $\mathscr{I}$ of all those linear functionals I on $\mathscr{L}(X)$ which satisfy the following continuity condition: For each compact subset K of X, $I|\mathscr{L}_K(X)$ is continuous on $\mathscr{L}_K(X)$ with respect to $\| \ \|_\infty$.*

Proof.   By 8.11 $\mu \mapsto I_\mu$ is one-to-one. It is clear from 5.4(2) that I_μ must satisfy the above continuity condition. So it remains only to show that every I in $\mathscr{I}$ is of the form I_μ for some $\mu \in \mathscr{M}$.

Let $\mathscr{I}_r = \{I \in \mathscr{I} : I(f) \text{ is real for all } f \text{ in } \mathscr{L}_r(X)\}$; and let $\mathscr{I}_+ = \{I \in \mathscr{I} : I(f) \geq 0 \text{ whenever } f \in \mathscr{L}_+(X)\}$. Any I in $\mathscr{I}$ is of the form $I_1 + iI_2$, where $I_1, I_2 \in \mathcal{I}_r$. (Indeed, take $I_1(f) = \frac{1}{2}(I(f) + \overline{I(\bar{f})})$, $I_2(f) = (1/2i)(I(f) - \overline{I(\bar{f})})$, $^-$ denoting complex conjugation.) Again, take an element I of $\mathcal{I}_r$, and consider it as a real-linear function on $\mathcal{L}_r(X)$. By the continuity hypothesis I is relatively bounded in the sense of Hewitt and Ross [1], B.31 (p. 461). Hence by Theorem B.37 (p. 463) of that reference we have $I = I_+ - I_-$, where I_+ and I_- are in $\mathcal{I}_+$. Putting these facts together, we find that $\mathcal{I}$ is the linear span of $\mathcal{I}_+$. Hence it is sufficient to show that every I in $\mathcal{I}_+$ is of the form I_μ for some $\mu \in \mathcal{M}$. For this fact we refer the reader to 11.37 (p. 129) of Hewitt and Ross [1]. ∎

Remark 1. After Definition 14.3 we shall be able to rephrase the continuity condition in the above Theorem as follows: *I is continuous on $\mathcal{L}(X)$ in the inductive limit topology.*

Remark 2. We see from 8.11(2) that the set $\mathcal{I}_r$ [$\mathcal{I}_+$] corresponds under the bijection $\mu \mapsto I_\mu$ with the set of all μ in $\mathcal{M}$ whose values are all real [non-negative].

In this connection notice that any linear functional I on $\mathcal{L}(X)$ which satisfies $I(f) \geq 0$ for all f in $\mathcal{L}_+(X)$ automatically satisfies the continuity hypothesis of the above Theorem, and so belongs to $\mathcal{I}_+$.

Definition. We shall sometimes refer to elements of $\mathcal{I}$ as *complex integrals* on X, and to elements of $\mathcal{I}_+$ as *integrals* on X.

Thus, the Riesz Representation Theorem asserts a one-to-one correspondence between complex integrals on X and regular complex Borel measures on X, and also between integrals on X and regular Borel measures on X.

An element μ of $\mathcal{M}$ is bounded (i.e., belongs to $\mathcal{M}_r(X)$) if and only if the corresponding complex integral I is continuous with respect to the supremum norm of $\mathcal{L}(X)$ (and so can be extended to a continuous linear functional on $\mathscr{C}_0(X)$). (The verification of this is left as an exercise for the reader.) In that case, it follows from Hewitt and Ross [1], Theorems 14.6 and 14.4, that the $(\mathscr{C}_0(X))^*$-norm of I coincides with the total variation norm of μ. Thus *the correspondence $\mu \leftrightarrow I$ identifies $\mathcal{M}_r(X)$, as a Banach space, with $(\mathscr{C}_0(X))^*$.*

8.13. Lemma. *Let μ be a regular (non-negative) Borel measure on X, and f a lower semi-continuous function on X to $[0, \infty]$ which vanishes outside a countable union of compact sets. Then*

$$\int f \, d\mu = \sup\left\{\int g \, d\mu : g \in \mathcal{L}_+(X), g \leq f\right\}.$$

Proof. If f is the limit of a monotone increasing sequence of lower semi-continuous functions for which the Lemma holds, then it holds for f by 2.2(vi). Likewise, by 2.2(i), (ii), if the Lemma holds for $f_1, \ldots, f_n$, it holds for any non-negative linear combination of them.

Let f be as in the hypothesis. If n and r are positive integers, let $U_{n,r}$ be the open set $\{x \in X : f(x) > r2^{-n}\}$; and for each $n = 1, 2, \ldots$ put

$$g_n = \sum_{r=1}^{n2^n} 2^{-n} \mathrm{Ch}_{U_{n,r}}.$$

We check that $\{g_n\}$ is monotone increasing and $g_n \uparrow f$.

The last two paragraphs imply that it is sufficient to prove the Lemma for $f = \mathrm{Ch}_U$, U being open. But for such f the Lemma is an immediate consequence of the regularity of μ.   ∎

8.14. Proposition. *Let μ and f be as in Lemma 8.13; and suppose $\{g_\alpha\}$ is a monotone increasing net of functions in $\mathscr{L}_+(X)$ such that $\sup_\alpha g_\alpha = f$. Then $\int f \, d\mu = \sup_\alpha \int g_\alpha \, d\mu$.*

Proof. By 8.13 it is sufficient to show that $\sup_\alpha \int g_\alpha \, d\mu \geq \int g \, d\mu$ whenever $g \in \mathscr{L}_+(X)$ and $g \leq f$. Let $g \in \mathscr{L}_+(X)$, $g \leq f$, $\varepsilon > 0$. Since $\sup_\alpha g_\alpha \geq g$, a compactness argument shows that $g_\alpha \geq g - \varepsilon$ for some α; and this implies $\sup_\alpha \int g_\alpha \, d\mu \geq \int g \, d\mu$.   ∎

Remark. If $\{g_\alpha\}$ were a *sequence*, this Proposition would follow immediately from 2.2(vi). Its validity for *nets* is a special property of regular measures on locally compact Hausdorff spaces.

Extensions of Regular Borel Measures

8.15. In treating convolution of measures in Chapter III we shall find it necessary to extend a regular complex Borel measure on X to a larger δ-ring than $\mathscr{S}(X)$, using a modification of the innate extension mentioned in 1.12.

Let μ be a regular complex Borel measure on X, and $\tilde{\mu}$ its innate extension with domain $\mathscr{B}_\mu$, as in 1.12. We shall define a sub-δ-ring $\mathscr{E}_\mu$ of $\mathscr{B}_\mu$ as follows: $\mathscr{E}_\mu$ is to consist of those Borel subsets A of X such that $A \subset U$ for some *open* set belonging to $\mathscr{B}_\mu$. Evidently $\mathscr{E}_\mu$ contains $\mathscr{S}(X)$. The restriction of $\tilde{\mu}$ to $\mathscr{E}_\mu$ is thus an extension of μ; we denote it by μ_e. Evidently μ_e has regularity properties analogous to (I), (II) of 8.2, namely:

For each A in $\mathscr{E}_\mu$ and each $\varepsilon > 0$, there is an open set U in $\mathscr{E}_\mu$ and a compact set C such that $C \subset A \subset U$ and $|\mu|(U \setminus C) < \varepsilon$.

Definition.   The above μ_e is called the *maximal regular extension* of μ.
From 1.12(5) it follows immediately that $\mathscr{E}_{|\mu|} = \mathscr{E}_\mu$ and

$$|\mu|_e = |\mu_e|. \tag{3}$$

Remark 1.   By the regularity of μ, an open subset U of X belongs to $\mathscr{E}_\mu$ if and only if

$$\sup\{|\mu|(C): C \text{ compact} \subset U\} < \infty.$$

Furthermore, for an arbitrary set A belonging to $\mathscr{E}_\mu$ we have

$$\mu_e(A) = \lim_C \mu(C),$$

where $\{C\}$ runs over the set (directed by inclusion) of all compact subsets of A.

Remark 2.   The reason why we chose to define μ_e on $\mathscr{E}_\mu$, rather than on the (in general larger) δ-ring of *all* Borel sets in $\mathscr{B}_\mu$, will become apparent in the next section (see Remark 9.17).

Remark 3.   By 5.11, a function on X which is summable with respect to μ is also summable with respect to μ_e, and $\int f\, d\mu_e = \int f\, d\mu$. However, in general there will be many functions which are μ_e-summable but not μ-summable — for example the constant functions if μ is bounded and X is not σ-compact.

   In this connection, the reader will verify the following useful fact: If μ is a bounded regular complex Borel measure on X, A is a Banach space, and $f: X \to A$ is any bounded continuous function, then $f \in \mathscr{L}_1(\mu_e; A)$.

8.16.   Suppose the μ of 8.15 is bounded. Then clearly $\mathscr{E}_\mu = \mathscr{B}(X)$, the σ-field of all Borel subsets of X; and we have extended μ to the entire Borel σ-field of X.
   Assume that X is not compact; and let X_0 be any compactification of X; that is, X_0 is a compact Hausdorff space containing X as a dense open topological subspace. (For example, X_0 might be the one-point compactification; see Chapter I). Each Borel subset A of X_0 has a Borel intersection with X; so $\mu_0: A \mapsto \mu_e(A \cap X)\ (A \in \mathscr{B}(X_0))$ is a complex Borel measure on X_0. By the regularity property of μ_e, μ_0 is also regular. The correspondence $\mu \mapsto \mu_0$ is thus one-to-one from the family of all bounded regular complex Borel measures on X onto the family of all those regular complex Borel measures ν on X_0 which satisfy $\nu(X_0 \setminus X) = 0$.

8.17. *Definition.* We shall say that a complex measure v on the Borel σ-field $\mathscr{B}(X)$ is *regular* if it coincides with μ_e for some bounded regular complex Borel measure μ on X.

This definition, of course, adds nothing new unless X is non-compact.

Proposition. *For a complex measure v on $\mathscr{B}(X)$, the following three conditions are equivalent:*

(I)   *v is regular;*

(II) *For any Borel subset A of X and any $\varepsilon > 0$, there is a compact subset C of X such that $C \subset A$ and $|v|(A \setminus C) < \varepsilon$;*

(III) *$\lim_i |v|(U_i) = |v|(U)$ whenever $\{U_i\}$ is an increasing net of open subsets of X and $U = \bigcup_i U_i$.*

This follows easily from 8.16 and 8.6.

8.18. Applying 8.8 to the one-point compactification of X, and remembering 8.16, we obtain:

Proposition. *If X satisfies the second axiom of countability, every complex measure on $\mathscr{B}(X)$ is regular.*

9. Product Measures and Fubini's Theorem

9.1. The theory of product measures and Fubini's Theorem have two forms—one for δ-rings on general sets, and one for locally compact Hausdorff spaces. Since the latter is the only one needed in this work, and since it is not a special case of the usual version of the former, for brevity we confine ourselves to the theory on locally compact spaces.

9.2. Throughout this section X and Y are two fixed locally compact Hausdorff spaces.

A *Borel rectangle* [*compacted Borel rectangle*] in $X \times Y$ is a set of the form $A \times B$, where $A \in \mathscr{B}(X)$ and $B \in \mathscr{B}(Y)$ [$A \in \mathscr{S}(X)$ and $B \in \mathscr{S}(Y)$]. Denote by $\mathscr{T}$ the set of all compacted Borel rectangles in $X \times Y$. The δ-ring generated by $\mathscr{T}$ is a subfamily of $\mathscr{S}(X \times Y)$, but need not be equal to $\mathscr{S}(X \times Y)$ even if X and Y are compact.

Remark 1. As an example illustrating the last statement, let X be any set whose cardinality is greater than the power c of the continuum. Let $\mathscr{W}$ be the set of all "rectangles" $A \times B$, where A and B are any subsets of X; and define

$\mathscr{S}$ to be the σ-field of subsets of $X \times X$ generated by $\mathscr{W}$. We claim that the "diagonal" set $D = \{\langle x, x \rangle : x \in X\}$ does not belong to $\mathscr{S}$. To see this, set $E_x = \{y \in X : \langle x, y \rangle \in E\}$ $(E \subset X \times X; x \in X)$; and let $\mathscr{R} = \{E \subset X \times X : \mathrm{card}\{E_x : x \in X\} \leq c\}$. One checks that $\mathscr{R}$ is a σ-field of subsets of $X \times X$ which contains $\mathscr{W}$ but not D. This proves the claim.

Remark 2. Let X and Y satisfy the second axiom of countability; then so does $X \times Y$, and every compact subset of $X \times Y$ is a G_δ. It then follows by an argument similar to 9.3 that the δ-ring generated by $\mathscr{T}$ is equal to $\mathscr{S}(X \times Y)$.

9.3. Proposition. *Let ρ and σ be regular complex Borel measures on $X \times Y$ such that $\rho(D) = \sigma(D)$ for all D in $\mathscr{T}$. Then $\rho = \sigma$.*

Proof. Let $\mathscr{R}$ be the ring of subsets of $X \times Y$ generated by $\mathscr{T}$. Notice that every set in $\mathscr{R}$ is a finite *disjoint* union of sets in $\mathscr{T}$. Hence our hypothesis implies that ρ and σ coincide on $\mathscr{R}$.

Now fix a set A in $\mathscr{S}(X \times Y)$; and define the directed set D in terms of A as in 8.11. If $\langle U, C \rangle \in D$, an easy compactness argument enables us to pick an open set $E_{U,C}$ in $\mathscr{R}$ such that $C \subset E_{U,C} \subset U$. Then by regularity and the preceding paragraph

$$\rho(A) = \lim_{\langle U,C \rangle} \rho(E_{U,C}) = \lim_{\langle U,C \rangle} \sigma(E_U, C) = \sigma(A).$$

Hence by the arbitrariness of A, $\rho = \sigma$. ∎

9.4. The last Proposition tells us that a regular complex Borel measure on $X \times Y$ is determined by its values on compacted Borel rectangles. The argument of the proof tells us something more.

Proposition. *Let ρ be any regular Borel measure on $X \times Y$. For each $1 \leq p < \infty$, the linear span of $\{Ch_D : D \in \mathscr{T}\}$ is dense in $\mathscr{L}_p(\rho)$.*

Proof. Given $A \in \mathscr{S}(X \times Y)$ it is enough by 5.3 to approximate Ch_A in $\mathscr{L}_p(\rho)$ by functions Ch_E, where $E \in \mathscr{R}$. But this was done in the proof of 9.3. ∎

9.5. Proposition. *Let μ and v be (non-negative) regular Borel measures on X and Y respectively. Then there is a unique regular Borel measure ρ on $X \times Y$ satisfying*

$$\rho(A \times B) = \mu(A)v(B)$$

for all A in $\mathscr{S}(X)$ and B in $\mathscr{S}(Y)$.

Proof.   The uniqueness of ρ follows from 9.3. For the proof of its existence we refer the reader to Hewitt and Ross [1], Theorem 13.13.   ∎

Definition.   The ρ of the above Proposition is called the *regular product of* μ *and* v, and is denoted by $\mu \times v$.

Remark.   We shall see in 9.10 that this Proposition remains true when μ, v, ρ are allowed to be regular *complex* Borel measures.

9.6.   Proposition.   *If* μ *and* μ' *[*v *and* v'*] are regular Borel measures on* X *[*Y*] such that* $\mu \leq \mu'$ *and* $v \leq v'$, *then* $\mu \times v \leq \mu' \times v'$.

Proof.   This follows from the argument of 9.3.   ∎

9.7.   We now present the Fubini Theorem, in its various forms, for complex-valued functions. Generalizations to vector-valued functions will emerge in §16.

From here to 9.9 we fix regular Borel measures μ and v on X and Y respectively.

It is worth noting separately the following fundamental tool in the proof of the Fubini Theorem.

Proposition.   *If* D *is a* $(\mu \times v)$*-null subset of* $X \times Y$, *then* $\{y \in Y : \langle x, y \rangle \in D\}$ *is* v*-null for* μ*-almost all* x *in* X; *and similarly* $\{x \in X : \langle x, y \rangle \in D\}$ *is* μ*-null for* v*-almost all* y *in* Y.

For the proof, see for example Hewitt and Ross [1], Theorem 13.7.

Remark.   If the subset D of $X \times Y$ is merely locally $(\mu \times v)$-null, but is not contained in a countable union of compact sets, then it is false in general that $\{y \in Y : \langle x, y \rangle \in D\}$ is locally v-null for locally μ-almost all x. For example, let X be the unit interval $[0, 1]$ (with its usual topology) and μ Lebesgue measure on X; and let Y be the unit interval with the discrete topology, and v the "counting measure" on Y (assigning to each finite subset of Y its cardinality). Then the "diagonal" set $D = \{\langle x, x \rangle : x \in X\}$ is locally $(\mu \times v)$-null; but $\{y \in Y : \langle x, y \rangle \in D\}$ is a one-element set, and hence not locally v-null, for every x in X.

9.8.   Fubini's Theorem.   *Let* f *be a locally* $(\mu \times v)$*-measurable function on* $X \times Y$ *to* $\mathbb{C}$ *(or to* $\mathbb{R}_{ext}$*) which vanishes outside a union of countably many compact subsets of* $X \times Y$. *Then:*

(I) $y \mapsto f(x, y)$ is locally v-measurable on Y for μ-almost all x; and similarly $x \mapsto f(x, y)$ is locally μ-measurable on X for v-almost all y.

(II) Suppose that $\text{range}(f) \subset [0, \infty]$. Then: (i) by (I) $\int f(x, y)dvy$ exists (and belongs to $[0, \infty]$) for μ-almost all x; and $\int f(x, y)d\mu x$ exists for v-almost all y. (ii) $x \mapsto \int f(x, y)dvy$ and $y \mapsto \int f(x, y)d\mu x$ are locally μ-measurable and locally v-measurable $[0, \infty]$-valued functions respectively, and vanish outside a countable union of compact sets in X and Y respectively. (iii) We have the Fubini equality

$$\int_{X \times Y} f(x, y)d(\mu \times v)\langle x, y \rangle = \int_X \int_Y f(x, y)dvy \, d\mu x$$

$$= \int_Y \int_X f(x, y)d\mu x \, dvy \qquad (1)$$

(the three terms existing in view of (ii)).

(III) Suppose now that $f \in \mathscr{L}_1(\mu \times v)$. Then: (i) $y \mapsto f(x, y)$ is in $\mathscr{L}_1(v)$ for μ-almost all x, and $x \mapsto f(x, y)$ is in $\mathscr{L}_1(\mu)$ for v-almost all y. (ii) The functions $x \mapsto \int f(x, y)dvy$ and $y \mapsto \int f(x, y)d\mu x$ are μ-summable and v-summable respectively. (iii) The Fubini equality (1) holds.

Proof. As regards (I) and (II), for non-negative functions we refer the reader to Hewitt and Ross [1], Theorem 13.9. Part (I) for complex functions f is obtained by writing f as a linear combination of non-negative-valued functions and applying what we know for the latter. Part (III) is Theorem 13.8 of Hewitt and Ross [1]. (It can also be obtained directly from Part (II) by writing f as a linear combination of non-negative functions in $\mathscr{L}_1(\mu \times v)$.) ∎

Products of Complex Measures

9.9.   **Proposition.**   *Let $f : X \to \mathbb{C}$ be locally μ-summable (7.2), and let $g : Y \to \mathbb{C}$ be locally v-summable. Then $h : \langle x, y \rangle \mapsto f(x)g(y)$ is locally $(\mu \times v)$-summable. If we put $\rho = f \, d\mu$, $\sigma = g \, dv$, $\tau = h \, d(\mu \times v)$, then*

$$\tau(A \times B) = \rho(A)\sigma(B)$$

for all A in $\mathscr{S}(X)$ and B in $\mathscr{S}(Y)$.

Remark.   Notice from 8.3 that ρ, σ, and τ are regular.

Proof.   The local $(\mu \times v)$-measurability of h is an easy consequence of the definition of $\mu \times v$. The rest follows immediately from 9.8. ∎

9.10.   Corollary.   *Let μ' and v' be arbitrary regular complex Borel measures on X and Y respectively. Then there is a unique regular complex Borel measure, which we denote by $\mu' \times v'$, on $X \times Y$ satisfying:*

$$(\mu' \times v')(A \times B) = \mu'(A)v'(B)$$

for all A in $\mathscr{S}(X)$ and B in $\mathscr{S}(Y)$. Further

$$|\mu' \times v'| = |\mu'| \times |v'|. \tag{2}$$

Proof.   The uniqueness of $\mu' \times v'$ was proved in 9.3.

Put $\mu = |\mu'|, v = |v'|$. By 7.8 and 8.7 $\mu' = f\, d\mu$ and $v' = g\, dv$, where f and g are locally μ-summable and locally v-summable respectively. Putting $h(x, y) = f(x)g(y)$, we see from 9.9 that $h\, d(\mu \times v)$ has the property characterizing $\mu' \times v'$.

To prove the last statement, we use 7.3 to deduce that $\mu = |\mu'| = |f|\,d\mu$, whence $|f(x)| = 1$ for locally μ-almost all x. Similarly $|g(y)| = 1$ for locally v-almost all y. Hence $|h(x, y)| = 1$ for locally $(\mu \times v)$-almost all $\langle x, y\rangle$. It follows, again using 7.3, that

$$|\mu' \times v'| = |h\, d(\mu \times v)| = |h|\,d(\mu \times v) = \mu \times v. \qquad \blacksquare$$

Remark.   It follows easily from (2) that

$$\operatorname{supp}(\mu' \times v') = \operatorname{supp}(\mu') \times \operatorname{supp}(v').$$

9.11.   Proposition.   *If μ' and μ'' are regular complex Borel measures on X such that $\mu'' \ll \mu'$, and if v' is a regular complex Borel measure on Y, then $\mu'' \times v' \ll \mu' \times v'$. If $\mu'' \sim \mu'$, then $\mu'' \times v' \sim \mu' \times v'$.*

Proof.   Combine the Radon-Nikodym Theorem (see 8.7) with the proof of 9.10. $\blacksquare$

9.12.   Fubini Theorem for Complex Measures.   *Let μ', v', and $\mu' \times v'$ be as in 9.10. If $\phi \in \mathscr{L}_1(\mu' \times v')$, then (i) $y \mapsto \phi(x, y)$ is in $\mathscr{L}_1(v')$ for μ'-almost all x, and $x \mapsto \phi(x, y)$ is is $\mathscr{L}_1(\mu')$ for v'-almost all y. (ii) The functions $x \mapsto \int \phi(x, y)dv'y$ and $y \mapsto \int \phi(x, y)d\mu'x$ are μ'-summable and v'-summable respectively. (iii) We have:*

$$\int_{X \times Y} \phi(x, y)d(\mu' \times v')\langle x, y\rangle = \int_X \int_Y \phi(x, y)dv'y\, d\mu'x$$

$$= \int_Y \int_X \phi(x, y)d\mu'x\, dv'y.$$

Proof. Let $\mu = |\mu'|$, $v = |v'|$, $\mu' = f \, d\mu$, $v' = g \, dv$, and $\mu' \times v' = h(\mu \times v)$, where $h(x, y) = f(x)g(y)$, as in the proof of 9.10. In view of 7.5, the Theorem follows on applying 9.8(III) to the function $h\phi$. ∎

Regular Maximal Extensions and the Fubini Theorem

9.13. Let μ and v be regular complex Borel measures on X and Y respectively, with maximal regular extensions μ_e and v_e (see 8.15). Let us form the product regular complex Borel measure $\mu \times v$, and examine its maximal regular extension $(\mu \times v)_e$. As in 8.15 the domains of μ_e, v_e, and $(\mu \times v)_e$ are called $\mathscr{E}_\mu$, $\mathscr{E}_v$, and $\mathscr{E}_{\mu \times v}$ respectively.

Proposition. *If $A \in \mathscr{E}_\mu$ and $B \in \mathscr{E}_v$, then $A \times B \in \mathscr{E}_{\mu \times v}$ and*

$$(\mu \times v)_e(A \times B) = \mu_e(A)v_e(B). \tag{3}$$

Proof Assume first that A and B are open.

If C is a compact subset of $A \times B$, then the projections C_1 and C_2 of C on X and Y are compact subsets of A and B respectively, and $C \subset C_1 \times C_2$. So by (2)

$$\sup\{|\mu \times v|(C) : C \text{ compact} \subset A \times B\}$$

$$= \sup\{|\mu|(C_1)|v|(C_2) : C_1 \text{ compact} \subset A, C_2 \text{ compact} \subset B\}.$$

$$= \sup\{|\mu|(C_1) : C_1 \text{ compact} \subset A\}\sup\{|v|(C_2) : C \text{ compact} \subset B\}.$$

It follows (see Remark 1 of 8.15) that $A \times B \in \mathscr{E}_{\mu \times v}$.

Now drop the assumption that A and B are open. Then A and B are contained in open sets A' and B' belonging to $\mathscr{E}_\mu$ and $\mathscr{E}_v$ respectively; and so $A \times B$ is a Borel subset of the open set $A' \times B'$, which belongs to $\mathscr{E}_{\mu \times v}$ by the preceding paragraph. So $A \times B \in \mathscr{E}_{\mu \times v}$.

To prove (3), observe that the family of product compact sets $C_1 \times C_2$, where C_1 and C_2 are compact subsets of A and B respectively, is cofinal in the directed set of all compact subsets of $A \times B$. Therefore by Remark 1 of 8.15

$$(\mu \times v)_e(A \times B) = \lim(\mu \times v)(C_1 \times C_2)$$

$$= \lim \mu(C_1)v(C_2)$$

$$= \mu_e(A)v_e(B). \quad ∎$$

9.14. **Corollary.** *If the μ and v of 9.13 are bounded, then $\mu \times v$ is bounded, and*

$$\|\mu \times v\| = \|\mu\|\|v\|. \tag{4}$$

9.15. Keeping the notation of 9.13, suppose that the closed supports of μ and v (see 8.9) are all of X and Y respectively. Then we have the following weak converse of 9.13.

Proposition. *Let U be an open subset of $X \times Y$ belonging to $\mathcal{E}_{\mu \times v}$, and U_1 and U_2 the (open) projections of U on X and Y respectively. Then $U_1 [U_2]$ is a countable union of open sets belonging to $\mathcal{E}_\mu [\mathcal{E}_v]$.*

Proof. Consider the upward directed set F of all functions f in $\mathcal{L}_+(X \times Y)$ such that $f \leq \mathrm{Ch}_U$; and put $\phi_f(x) = \int f(x, y) d|v|y$ ($f \in F, x \in X$). An easy uniform continuity argument shows that ϕ_f is continuous, hence in $\mathcal{L}(X)$; and by 9.8 an 9.10(2)

$$\int \phi_f \, d|\mu| = \int f \, d|\mu \times v| \leq k < \infty \qquad (f \in F) \qquad (5)$$

where $k = |\mu \times v|_e(U)$.

Now for each positive integer n define

$$V_n = \{x \in X : \phi_f(x) > n^{-1} \text{ for some } f \text{ in } F\}.$$

Since the ϕ_f are continuous, V_n is an open subset of U_1; and since $\mathrm{supp}(v) = Y$, $\bigcup_n V_n = U_1$. We shall show that $V_n \in \mathcal{E}_\mu$. Indeed, for each f in F and each pair of positive integers, n, m, let $D^n_{f,m}$ be the compact set

$$\left\{ x \in X : \phi_f(x) \geq \frac{1}{n} + \frac{1}{m} \right\}.$$

Then $n^{-1}|\mu|(D^n_{f,m}) \leq \int \phi_f \, d|\mu|$; so by (5)

$$|\mu|(D^n_{f,m}) \leq nk. \qquad (6)$$

Now, for fixed n, the family of all $D^n_{f,m}$ ($f \in F; m = 1, 2, \ldots$) is cofinal in the upward directed set of all compact subsets of V_n. Hence $V_n \in \mathcal{E}_\mu$ by (6) and Remark 1, 8.15.

Since $\bigcup_n V_n = U_1$, we have shown that U_1 is a countable union of open sets in $\mathcal{E}_\mu$. Similarly, U_2 is a countable union of open sets in $\mathcal{E}_v$. ∎

9.16.   The Fubini Theorem for Maximal Regular Extensions.   *Let μ and v be regular complex Borel measures on X and Y respectively; and suppose that $f \in \mathcal{L}_1((\mu \times v)_e)$. Then: (i) $x \mapsto f(x, y)$ is μ_e-summable for v_e-almost all y, and*

$y \mapsto f(x, y)$ is v_e-*summable for* μ_e-*almost all* x. (ii) $y \mapsto \int f(x, y) d\mu_e x$ *is* v_e-*summable and* $x \mapsto \int f(x, y) dv_e y$ *is.* μ_e-*summable.* (iii) *We have*:

$$\int_{X \times Y} f \, d(\mu \times v)_e = \int_Y \int_X f(x, y) d\mu_e x \, dv_e y$$

$$= \int_X \int_Y f(x, y) dv_e y \, d\mu_e x. \tag{7}$$

Proof. We shall first prove this under the assumption that μ are v are bounded.

Let X_0 and Y_0 be compactifications of X and Y respectively. By 8.16 the equations $\mu_0(A) = \mu_e(A \cap X)$ and $v_0(B) = v_e(B \cap Y)$ $(A \in \mathscr{B}(X_0); B \in \mathscr{B}(Y_0))$ define regular complex Borel measures on X_0 and Y_0; and by 9.13 $\mu_0 \times v_0$ coincides with $(\mu \times v)_e$ on Borel subsets of $X \times Y$. Therefore the required conclusion is obtained by applying 9.12 to $\mu_0 \times v_0$.

We now drop the assumptions that μ and v are bounded.

Notice that it is sufficient to assume that $\mathrm{supp}(\mu) = X$ and $\mathrm{supp}(v) = Y$. Indeed, let $X' = \mathrm{supp}(\mu)$, $Y' = \mathrm{supp}(v)$. Then by 8.9, the open sets $X \setminus X'$, $Y \setminus Y'$, and $(X \times Y) \setminus (X' \times Y')$ are μ_e-null, v_e-null, and $(\mu \times v)_e$-null respectively; and we may throw away these sets, leaving the spaces X', Y', and $X' \times Y'$. So we shall henceforth assume that $\mathrm{supp}(\mu) = X$ and $\mathrm{supp}(v) = Y$.

By the definition of $\mathscr{E}_{\mu \times v}$, f must vanish outside a countable union U of open sets in $\mathscr{E}_{\mu \times v}$. By 9.15 and our assumption about the supports of μ and v, we have $U \subset V \times W$, where $V[W]$ is a countable union of open sets in $\mathscr{E}_\mu$ $[\mathscr{E}_v]$. Let us write $V = \bigcup_{n=1}^\infty V_n$, $W = \bigcup_{n=1}^\infty W_n$, where $\{V_n\}[\{W_n\}]$ is an increasing sequence of open sets in $\mathscr{E}_\mu[\mathscr{E}_v]$; and put $f_n = \mathrm{Ch}_{V_n \times W_n} f$ $(n = 1, 2, \ldots)$.

Now f_n vanishes except on $V_n \times W_n$, and μ and v are bounded on V_n and W_n respectively. So the first part of the proof implies that the conclusion of the Theorem holds for each f_n. To see that the Theorem holds for f, we can argue as follows: By a double application of 2.2(vi),

$$\iint |f_n(x, y)| d|\mu_e| x \, d|v_e| y \uparrow \iint |f(x, y)| d|\mu_e| x \, d|v_e| y. \tag{8}$$

Likewise

$$\int |f_n| d|(\mu \times v)_e| \uparrow \int |f| d|(\mu \times v)_e|. \tag{9}$$

By what we have already seen the left sides of (8) and (9) are equal for each n. So by (8) and (9)

$$\iint |f(x, y)| d|\mu_e|x \, d|v_e|y = \int |f| d|(\mu \times v)_e| < \infty. \tag{10}$$

In particular $\int |f(x, y)| d|\mu_e|x < \infty$ v_e-almost everywhere; so by 5.6

$$\int f_n(x, y) d\mu_e x \to \int f(x, y) d\mu_e x, \tag{11}$$

both expressions in (11) being dominated in absolute value by $\int |f(x, y)| d|\mu_e|x$. Combining this with (10) and 5.6, we deduce that

$$\iint f_n(x, y) d\mu_e x \, dv_e y \to \iint f(x, y) d\mu_e x \, dv_e y. \tag{12}$$

But the left side of (12) is $\int f_n \, d(\mu \times v)_e$, which, again by 5.6, approaches $\int f \, d(\mu \times v)_e$. So the latter integral coincides with the right side of (12), proving half of (7). The other half, of course, is obtained by reversing the roles of X and Y.   ∎

9.17.   Corollary.   *In the context of 9.16 suppose that A is a $(\mu \times v)_e$-null set. Then $\{x \in X : \langle x, y \rangle \in A\}$ is μ_e-null for v_e-almost all y. Similarly, $\{y \in Y : \langle x, y \rangle \in A\}$ is v_e-null for μ_e-almost all x.*

Remark.   Suppose that in 8.15 μ_e had been defined as the restriction of μ to the δ-ring of *all* Borel sets in $\mathscr{B}_\mu$. Then the μ_e-null Borel sets would have been exactly the locally μ-null Borel sets, and similarly for v_e and $(\mu \times v)_e$. The above Corollary would therefore have been false, since it would have contradicted the counter-example mentioned in Remark 9.7; and so 9.16 would have been false. This is why we defined μ_e as we did in 8.15.

Measures in Products of Several Spaces

9.18.   It is a very easy matter to generalize the theory of this section to products of an arbitrary finite number of locally compact Hausdorff spaces $X_1, X_2, \ldots, X_n$. The natural generalizations of the results of this section can be proved either by imitating the proofs given here for $n = 2$, or by induction in n.

Thus, suppose that for each $i = 1, \ldots, n$, μ_i is a regular complex Borel measure on X_i. Then, generalizing 9.10, we obtain a unique regular complex Borel measure μ on $X_1 \times \cdots \times X_n$, denoted by $\mu_1 \times \cdots \times \mu_n$, such that

$$\mu(A_1 \times \cdots \times A_n) = \prod_{i=1}^{n} \mu_i(A_i)$$

whenever $A_i \in \mathscr{S}(X_i)$ for each i. Furthermore

$$|\mu| = |\mu_1| \times \cdots \times |\mu_n|.$$

The product $\mu = \mu_1 \times \cdots \times \mu_n$ is associative; and it is easy to write down a Fubini Theorem for μ by building it up one step at a time and applying 9.12 inductively. The reader should verify the details.

10. Measure Transformations

10.1. Let $\mathscr{S}$ be a δ-ring of subsets of a set X, and $\mathscr{T}$ a δ-ring of subsets of a set Y; and let us fix a map $F: X \to Y$ which is $\mathscr{S}, \mathscr{T}$ *Borel* in the sense that $F^{-1}(B) \in \mathscr{S}$ whenever $B \in \mathscr{T}$.

If μ is a complex measure on $\mathscr{S}$, the equation

$$(F_*(\mu))(B) = \mu(F^{-1}(B)) \qquad\qquad (B \in \mathscr{T})$$

defines a complex measure $F_*(\mu)$ on $\mathscr{T}$, which we call the *F-transform of* μ. Of course $F_*(\mu)$ is a measure (i.e., non-negative) if μ is. Notice that $|F_*(\mu)(B)| \leq |\mu|(F^{-1}(B)) = F_*(|\mu|)(B)$, so that

$$|F_*(\mu)| \leq F_*(|\mu|) \tag{1}$$

by a remark in 1.3. In general equality does not hold in (1); for an example see Remark 10.7.

10.2. Proposition. *Let B be a Banach space, and μ a measure on $\mathscr{S}$; and suppose that $f: Y \to B$ is locally $F_*(\mu)$-measurable and vanishes outside a countable union of sets in $\mathscr{T}$. Then:*

(I) $f \in \mathscr{L}_1(F_*(\mu); B)$ *if and only if* $f \circ F \in \mathscr{L}_1(\mu; B)$.

(II)   *If the conditions of* (I) *hold, and if v is any complex measure on $\mathscr{S}$ with* $|v| \leq \mu$, *we have*

$$\int_Y f \, dF_*(v) = \int_X (f \circ F) dv. \tag{2}$$

Proof. By 4.3 there is a sequence $\{g_n\}$ of simple B-valued functions on Y such that, for $F_*(\mu)$-almost all y, $g_n(y) \to f(y)$ in B and $\|g_n(y)\| \leq \|f(y)\|$ for all n. It follows that $\{g_n \circ F\}$ is a sequence of simple B-valued functions on X such that, for μ-almost all x, $(g_n \circ F)(x) \to (f \circ F)(x)$ and $\|(g_n \circ F)(x)\| \leq \|(f \circ F)(x)\|$ for all n. Thus $f \circ F$ is locally μ-measurable and vanishes outside a countable union of sets in $\mathscr{S}$. Also it is clear from the simplicity of the g_n and $g_n \circ F$ that

$$\int_Y \|g_n\| \, dF_*(\mu) = \int_X \|g_n \circ F\| \, d\mu \qquad \text{for each } n. \tag{3}$$

By 2.4, (3) and the preceding statements imply that

$$\int_Y \|f\| \, dF_*(\mu) = \int \|f \circ F\| \, d\mu, \tag{4}$$

from which (I) follows.

To prove (II), notice from (4) and 5.4(2) that both sides of (2) are continuous (as functions of f) on $\mathscr{L}_1(F_*(\mu); B)$. Hence by 5.3 it is sufficient to prove (2) when f is a simple function. But for simple functions (2) is almost obvious. ■

Remark. A very special case of this was observed in 5.10.

Transformation of Regular Borel Measures

10.3. The rest of this section is motivated by the desire to apply 10.1, 10.2 when X and Y are locally compact Hausdorff spaces, $F: X \to Y$ is continuous, and μ is a regular complex Borel measure on X. This will be important when we come to study convolution of measures in Chapter III.

Let X and Y be fixed locally compact Hausdorff spaces, and $F: X \to Y$ a continuous map. Then, although F is $\mathscr{B}(X)$, $\mathscr{B}(Y)$ Borel, it need not be $\mathscr{S}(X)$, $\mathscr{S}(Y)$ Borel; that is, the inverses of compact sets under F need not be compact. Therefore, given an arbitrary regular complex Borel measure μ on X, we cannot form $F_*(\mu)$ directly.

We recall from 8.15 the definition of the maximal regular extension μ_e of μ, and of its domain $\mathscr{E}_\mu$.

Definition. Let μ be a regular complex Borel measure on X. The continuous map $F: X \to Y$ will be called μ-*proper* if F is $\mathscr{E}_\mu$, $\mathscr{S}(Y)$ Borel, that is, if $F^{-1}(B) \in \mathscr{E}_\mu$ whenever $B \in \mathscr{S}(Y)$.

Thus, F is μ-proper if and only if $\sup\{|\mu|(C) : C$ is a compact subset of $F^{-1}(D)\} < \infty$ for every compact subset D of Y.

If F is μ-proper, it is of course $|\mu|$-proper, and also ν-proper whenever ν is a regular complex Borel measure such that $|\nu| \leq |\mu|$. If F is μ-proper and also ν-proper, it is $(r\mu + s\nu)$-proper whenever $r, s \in \mathbb{C}$.

If μ is bounded, then F is μ-proper, for in that case $\mathscr{E}_\mu = \mathscr{B}(X)$.

10.4. Clearly a sufficient (though not a necessary) condition for F to be μ-proper is that $\mathrm{supp}(\mu) \cap F^{-1}(D)$ be compact for every compact subset D of Y. In particular, F is μ-proper if $F^{-1}(D)$ is compact for every compact subset D of Y.

10.5. Suppose in 10.3 that F is μ-proper. Then we can form the complex measure $F_*(\mu_e)$ on $\mathscr{S}(Y)$:

$$F_*(\mu_e)(B) = \mu_e(F^{-1}(B)) \qquad\qquad (B \in \mathscr{S}(Y)).$$

Notice by 10.1(1) and 8.15(5) that

$$|F_*(\mu_e)| \leq F_*(|\mu_e|) = F_*(|\mu|_e). \tag{5}$$

Now we claim that $F_*(\mu_e)$ is regular. Indeed: By (5) it is enough to prove this assuming that $\mu \geq 0$. Let $B \in \mathscr{S}(Y)$ and $\varepsilon > 0$. By the regularity property of μ_e (8.15) we can find a compact subset C of $F^{-1}(B)$ such that $\mu_e(F^{-1}(B) \setminus C) < \varepsilon$. Thus $F(C)$ is a compact subset of B and $F_*(\mu_e)(B \setminus F(C)) = \mu_e(F^{-1}(B \setminus F(C))) \leq \mu_e(F^{-1}(B) \setminus C) < \varepsilon$, proving the regularity of $F_*(\mu_e)$.

Definition. Assuming that F is μ-proper, we shall refer to $F_*(\mu_e)$ as the *regular F-transform of μ*, and denote it simply by $F_*(\mu)$.

The inequality (5) thus becomes, as in (1):

$$|F_*(\mu)| \leq F_*(|\mu|). \tag{6}$$

Remark. Sometimes we may say that "$F_*(\mu)$ exists" to mean that F is μ-proper.

10.6. Proposition. *If μ is bounded, then $F_*(\mu)$ (which exists by 10.3) is bounded, and*

$$\|F_*(\mu)\| \leq \|\mu\|,$$

equality holding if $\mu \geq 0$. (Here $\| \ \|$ is the total variation norm).

Proof. This follows easily from (6). ∎

10.7. Suppose that F is μ-proper as before. We can then form the maximal regular extension $F_*(\mu))_e$ of $F_*(\mu)$, with domain $\mathscr{E}_{F_*(\mu)}$. We claim that F is $\mathscr{E}_\mu$, $\mathscr{E}_{F_*(|\mu|)}$ Borel, and that

$$(F_*(\mu))_e(B) = \mu_e(F^{-1}(B)) \tag{7}$$

for all B in $\mathscr{E}_{F_*(|\mu|)}$.

Indeed: Let B be in $\mathscr{E}_{F_*(|\mu|)}$. If C is a compact subset of $F^{-1}(B)$, then

$$|\mu|(C) \le |\mu|_e(F^{-1}F(C)) = F_*(|\mu|)(F(C)) \le F_*(|\mu|)(B). \tag{8}$$

Therefore

$$\sup\{|\mu|(C): C \text{ compact} \subset F^{-1}(B)\} < \infty. \tag{9}$$

If B is open, then so is $F^{-1}(B)$; and by (9) and the definition of $\mathscr{E}_\mu$ we have $F^{-1}(B) \in \mathscr{E}_\mu$. Even if B is not open, it is contained in some open set B' belonging to $\mathscr{E}_{F_*(|\mu|)}$; and what we have proved for B' implies $F^{-1}(B) \in \mathscr{E}_\mu$.

Now $\mathscr{E}_{F_*(|\mu|)} \subset \mathscr{E}_{F_*(\mu)}$ by (6). Since $B \in \mathscr{E}_{F_*(|\mu|)}$, the regularity property of $(F_*(\mu))_e$ implies that

$$F_*(\mu)_e(B) = \lim_D F_*(\mu)(D), \tag{10}$$

where D runs over the upward directed set $\mathscr{D}$ of all compact subsets of B. Likewise, since $C \subset F^{-1}(F(C)) \subset F^{-1}(B)$ for every compact subset C of $F^{-1}(B)$, the regularity property of μ_e implies

$$\mu_e(F^{-1}(B)) = \lim_D \mu_e(F^{-1}(D)) \tag{11}$$

(D running over the same $\mathscr{D}$). Combining (10), (11), and the definition of $F_*(\mu)$, we obtain (7).

Remark. It is false in general that $|F_*(\mu)| = F_*(|\mu|)$ (see (1)); and we cannot assert in the above claim that F is $\mathscr{E}_\mu$, $\mathscr{E}_{F_*(\mu)}$ Borel. Here is an easy example: Suppose that $X = \{0, 1\} \times \mathbb{Z}$ and $Y = \mathbb{Z}$, and that $F: X \to Y$ is given by $F(r, n) = n$. For μ we take the complex Borel measure on X such that, for any n in $\mathbb{Z}$, $\mu(\{\langle 0, n \rangle\}) = 1$ and $\mu(\{\langle 1, n \rangle\}) = -1$. Then $F_*(\mu)$ exists and is the zero measure, so that $\mathscr{E}_{F_*(\mu)}$ consists of all subsets of Y. On the other hand $\mathscr{E}_\mu = \mathscr{S}(X)$ (the set of all finite subsets of X).

10.8. Proposition. *Let F be μ-proper. Then*

$$\operatorname{supp}(F_*(\mu)) \subset (F(\operatorname{supp}(\mu)))^- \tag{12}$$

($^-$ *being closure*). *If $\mu \ge 0$, equality holds in* (12).

Proof. Assume first that $\mu \geq 0$. If $y \in Y \setminus (F(\text{supp}(\mu)))^-$, we can choose an open neighborhood V of y with compact closure such that $V \cap (F(\text{supp}(\mu)))^- = \emptyset$. Then $F^{-1}(V) \cap \text{supp}(\mu) = \emptyset$, so that $F_*(\mu)(V) = 0$, whence $y \notin \text{supp}(F_*(\mu))$. Conversely, if $y \in (F(\text{supp}(\mu)))^-$, any open neighborhood V of y with compact closure will satisfy $F^{-1}(V) \cap \text{supp}(\mu) \neq \emptyset$, whence $F_*(\mu)(V) \neq 0$; so $y \in \text{supp}(F_*(\mu))$. Therefore equality holds in (12).

Now drop the assumption that $\mu \geq 0$. Applying the last paragraph to $|\mu|$, and using (6), we have

$$\text{supp}(F_*(\mu)) \subset \text{supp}(F_*(|\mu|)) = (F(\text{supp}(\mu)))^-. \qquad \blacksquare$$

10.9. The next two propositions show that the regular F-transform behaves reasonably under the operations of composition and direct product.

Proposition. *Suppose that X, Y, and Z are locally compact Hausdorff spaces, $F: X \to Y$ and $G: Y \to Z$ are continuous maps, and μ is a regular complex Borel measure on X. We shall assume that F is μ-proper and that G is $(F_*(|\mu|))$ —proper. Then $G \circ F$ is μ-proper, and*

$$(G \circ F)_*(\mu) = G_*(F_*(\mu)). \tag{13}$$

Proof. By 10.7 the hypotheses imply that F is $\mathscr{E}_\mu$, $\mathscr{E}_{F_*(|\mu|)}$ Borel and that G is $\mathscr{E}_{F_*(|\mu|)}$, $\mathscr{S}(Z)$ Borel. Therefore $G \circ F$ is $\mathscr{E}_\mu$, $\mathscr{S}(Z)$ Borel, whence $G \circ F$ is μ-proper. Furthermore, by (7), if $D \in \mathscr{S}(Z)$,

$$\mu_e[(G \circ F)^{-1}(D)] = \mu_e[F^{-1}(G^{-1}(D))]$$
$$= F_*(\mu)_e(G^{-1}(D))$$
$$= (G_*(F_*(\mu)))(D),$$

proving (13). $\blacksquare$

10.10. Proposition. *Let X_1, X_2, Y_1, Y_2 be locally compact Hausdorff spaces; let $F_i: X_i \to Y_i (i = 1, 2)$ be a continuous map; and let $\mu_i (i = 1, 2)$ be a regular complex Borel measure on X_i such that F_i is μ_i-proper. Then the product map $F: X_1 \times X_2 \to Y_1 \times Y_2$ given by $F(x_1, x_2) = \langle F_1(x_1), F_2(x_2) \rangle$ is $(\mu_1 \times \mu_2)$-proper, and*

$$F_*(\mu_1 \times \mu_2) = (F_1)_*(\mu_1) \times (F_2)_*(\mu_2).$$

Proof. If $D \in \mathscr{S}(Y_1)$ and $E \in \mathscr{S}(Y_2)$, then any compact subset of $F^{-1}(D \times E)$ $(= F_1^{-1}(D) \times F_2^{-1}(E))$ is contained in a product $A \times B$, where A and B are compact subsets of $F_1^{-1}(D)$ and $F_2^{-1}(E)$. It follows that

$$\sup\{(|\mu_1| \times |\mu_2|)(C): C \text{ compact} \subset F^{-1}(D \times E)\}$$

$$= \sup\{|\mu_1|(A): A \text{ compact} \subset F_1^{-1}(D)\}$$

$$\times \sup\{|\mu_2|(B): B \text{ compact} \subset F_2^{-1}(E)\}$$

$$= |\mu_1|_e(F_1^{-1}(D))|\mu_2|_e(F_2^{-1}(E))$$

$$= (F_1)_*(|\mu_1|)(D)(F_2)_*(|\mu_2|)(E) < \infty. \tag{14}$$

If in addition D and E are open, this implies that $F^{-1}(D \times E) \in \mathscr{E}_{|\mu_1| \times |\mu_2|}$. Since any element of $\mathscr{S}(Y_1 \times Y_2)$ is contained in some product $D \times E$, where D and E are open sets in $\mathscr{S}(Y_1)$ and $\mathscr{S}(Y_2)$ respectively, we deduce that F is $(|\mu_1| \times |\mu_2|)$-proper. Hence by 9.10(2) F is $(\mu_1 \times \mu_2)$-proper.

If $D \in \mathscr{S}(Y_1)$ and $\mathscr{E} \in \mathscr{S}(Y_2)$, the argument of (14), together with the regularity property (8.15) of $(\mu_1 \times \mu_2)_e$, shows that

$$F_*(\mu_1 \times \mu_2)(D \times E) = (F_1)_*(\mu_1)(D)(F_2)_*(\mu_2)(E)$$

$$= ((F_1)_*(\mu_1) \times (F_2)_*(\mu_2))(D \times E).$$

Thus the two regular complex Borel measures $F_*(\mu_1 \times \mu_2)$ and $(F_1)_*(\mu_1) \times (F_2)_*(\mu_2)$ coincide on all compacted Borel rectangles, and so by 9.3 are equal. ■

10.11.   Remark.   Let X be a locally compact Hausdorff space, Y a subspace of X which is locally compact in the relativized topology, $i: Y \to X$ the identity injection, and μ a regular complex Borel measure on Y. In general i need not be μ-proper. If however Y is closed in X, i is always μ-proper, and $i_*(\mu)$ is the injection of μ onto X mentioned in 8.5.

11.   Projection-Valued Measures and Spectral Integrals

11.1.   Up until now we have studied only complex-valued measures. But there is no reason why one should not try to integrate with respect to measures taking values in more general linear spaces. This section takes up the elementary theory of "measures" whose values are projections in a Hilbert space.

11.2.   The "measures" of this section will be defined on a σ-field (rather than a δ-ring). We fix a σ-field $\mathscr{S}$ of subsets of a set X, and a Hilbert space H.

Definition. An *H-projection-valued measure* on $\mathscr{S}$ is a function P on $\mathscr{S}$ assigning to each set A in $\mathscr{S}$ a projection $P(A)$ on H, and satisfying: (i) $P(X) = 1_H =$ identity operator on H; (ii) for any sequence $\{A_n\}$ of pairwise disjoint sets in $\mathscr{S}$, the $\{P(A_n)\}$ are pairwise orthogonal projections, and

$$P\left(\bigcup_{n=1}^{\infty} A_n\right) = \sum_{n=1}^{\infty} P(A_n), \tag{1}$$

(where the convergence on the right side of (1) is unconditional in the strong operator topology, and the sum is equal to the projection onto the closed linear span of $\{\text{range}(P(A_n)): n = 1, 2, 3, \ldots\}$).

11.3. For the rest of this section we fix an H-projection-valued measure P on $\mathscr{S}$.

From (ii) it follows in particular that $P(\emptyset) = 0$. So (1) holds for finite as well as infinite sequences.

If $A, B \in \mathscr{S}$ and $A \subset B$, then $P(A) \leq P(A) + P(B \setminus A) = P(B)$. Thus P is monotone increasing.

If $\{A_n\}$ is a monotone increasing [decreasing] sequence of sets in $\mathscr{S}$, then $P(\bigcup_n A_n) = \sup_n P(A_n) \, [P(\bigcap_n A_n) = \inf_n P(A_n)]$.

If $\xi, \eta \in H$, then $A \mapsto (P(A)\xi, \eta) \, (A \in \mathscr{S})$ is a complex measure on $\mathscr{S}$. If $\xi = \eta$ we obtain a (non-negative) measure $A \mapsto (P(A)\xi, \xi)$.

11.4. Proposition. *If* $A, B \in \mathscr{S}$, *then*

$$P(A \cup B) + P(A \cap B) = P(A) + P(B), \tag{2}$$

$$P(A \cap B) = P(A)P(B) = P(B)P(A). \tag{3}$$

Proof. By (1) $P(A \cup B) = P(A \cap B) + P(A \setminus B) + P(B \setminus A)$. Adding $P(A \cap B)$ to both sides and noticing that $P(A) = P(A \cap B) + P(A \setminus B)$, $P(B) = P(A \cap B) + P(B \setminus A)$, we get (2). Multiplying (2) by $P(A)$ on either side and recalling that $P(A \cap B) \leq P(A) \leq P(A \cup B)$, we get $P(A) + P(A \cap B) = P(A) + P(A)P(B) = P(A) + P(B)P(A)$, giving (3). ∎

11.5. (3) says that projection-valued measures, unlike ordinary complex measures, are not only additive but multiplicative. This has an important consequence.

Proposition. *Let* $\mathscr{T}$ *be a subfamily of* $\mathscr{S}$ *which generates* $\mathscr{S}$ *(i.e.,* $\mathscr{S}$ *is the smallest σ-field of subsets of X containing* $\mathscr{T}$). *If Q is an H-projection-valued measure on* $\mathscr{S}$ *such that* $Q(A) = P(A)$ *for all A in* $\mathscr{T}$, *then* $Q = P$.

Proof. It follows from (1) and (3) that $\{A \in \mathscr{S} : Q(A) = P(A)\}$ is a σ-field. Since it contains $\mathscr{T}$, it coincides with $\mathscr{S}$. ∎

11.6. *P-null sets, P-measurability,* etc. are defined just as for complex measures. Thus a subset A of X is *P-null* if $A \subset B$ for some set B in $\mathscr{S}$ for which $P(B) = 0$; it is *P-measurable* if $A \ominus B$ is P-null for some B in $\mathscr{S}$. A function $f : X \to \mathbb{C}$ is *P-measurable* if $f^{-1}(D)$ is P-measurable for every Borel subset D of $\mathbb{C}$; it is *P-essentially bounded* if there is a non-negative number k such that $\{x : |f(x)| > k\}$ is P-null; in this case the smallest such k is called $\|f\|_\infty$. The set of all P-measurable P-essentially bounded complex functions on X will be called $\mathscr{L}_\infty(P)$; two such functions are identified if they differ only on a P-null set. Thus $\mathscr{L}_\infty(P)$ with the norm $\| \ \|_\infty$ is a Banach space (in fact, a Banach algebra under pointwise multiplication).

Let $\mathscr{F}$ be the complex linear span of $\{\mathrm{Ch}_B : B \in \mathscr{S}\}$. Clearly $\mathscr{F}$ is dense in $\mathscr{L}_\infty(P)$. Elements of $\mathscr{F}$ are called *simple functions.*

11.7. Recall that $\mathcal{O}(H)$ is the linear space of all bounded linear operators on H. For each f in $\mathscr{L}_\infty(P)$ we are going to define a new kind of integral $\int f \, dP$ so that $f \to \int f \, dP$ will be a linear map from $\mathscr{L}_\infty(P)$ to $\mathcal{O}(H)$.

We begin by observing that there is a unique linear mapping $I : \mathscr{F} \to \mathcal{O}(H)$ defined by:

$$I\left(\sum_{i=1}^n \lambda_i \mathrm{Ch}_{B_i} \right) = \sum_{i=1}^n \lambda_i P(B_i) \tag{4}$$

$(n = 1, 2, \ldots; B_i \in \mathscr{S}; \lambda_i \in \mathbb{C})$. To verify that (4) is legitimate, consider a function $f = \sum_{i=1}^n \lambda_i \mathrm{Ch}_{B_i}$ in $\mathscr{F}$ (where $B_i \in \mathscr{S}$). It is easy to see that f can also be written in the form $\sum_{j=1}^m \lambda_j' \mathrm{Ch}_{B_j'}$, where the B_j' are pairwise disjoint sets in $\mathscr{S}$, and each B_j' is contained in some B_i; and one verifies that (i) $\sum_{i=1}^n \lambda_i P(B_i) = \sum_{j=1}^m \lambda_j' P(B_j')$, and (ii) $\sum_{j=1}^m \lambda_j' P(B_j')$ is unaltered on replacing the $\{B_j'\}$ by a "refinement." This shows that (4) is legitimate. In fact, replacing the $\{B_i\}$ by the disjoint $\{B_j'\}$, we also observe that I is an isometry:

$$\|I(f)\| = \|f\|_\infty \qquad (f \in \mathscr{F}) \tag{5}$$

($\|I(f)\|$ being the usual operator norm on $\mathcal{O}(H)$.)

We remarked in 11.6 that $\mathscr{F}$ is dense in $\mathscr{L}_\infty(P)$. So by (5) the above I extends to a linear isometry (also called I) of $\mathscr{L}_\infty(P)$ into $\mathcal{O}(H)$.

Definition. If $f \in \mathscr{L}_\infty(P)$, the operator $I(f)$ in $\mathcal{O}(H)$ is called the *spectral integral of f with respect to P,* and is denoted by $\int f \, dP$, or $\int_X f \, dP$, or $\int f(x) \, dPx$.

11.8. The following Proposition states some of the elementary properties of the spectral integral. The functions f, g in it are assumed to be in $\mathscr{L}_\infty(P)$.

Proposition.

(I) $\int f \, dP = \int g \, dP$ if and only if f and g differ only on a P-null set.

(II)   The spectral integral $\int f \, dP$ is linear in f.

(III) The spectral integral $\int f \, dP$ is multiplicative in f; that is, $\int (fg)dP = (\int f \, dP)(\int g \, dP)$.

(IV) $\int \bar{f} \, dP = (\int f \, dP)^*$ ($^-$ being complex conjugation).

(V) $\|\int f \, dP\| = \|f\|_\infty$.

(VI)   If $A \in \mathscr{S}$, $\int \mathrm{Ch}_A \, dP = P(A)$; in particular, $\int dP = P(X) = 1_H$.

(VII) If $\xi, \eta \in H$, let $\mu_{\xi,\eta}$ be the complex measure $A \mapsto (P(A)\xi, \eta) \, (A \in \mathscr{S})$. If $E = \int f \, dP$, we have

$$(E\xi, \eta) = \int f \, d\mu_{\xi,\eta}. \tag{6}$$

$$\|E\xi\|^2 = \int |f|^2 \, d\mu_{\xi,\xi}. \tag{7}$$

(VIII) $P(A)$ commutes with $\int f \, dP$ for all A in $\mathscr{S}$.

(IX) $\int f \, dP$ is a normal operator. It is Hermitian if and only if $f(x)$ is real for P-almost all x.

These properties are best proved by first checking them on $\mathscr{F}$ and then extending them by continuity to all of $\mathscr{L}_\infty(P)$. We omit the details.

Projection-Valued Measures on Locally Compact Hausdorff Spaces

11.9. We now suppose that X is a non-void locally compact Hausdorff space.

An H-projection-valued measure P on the Borel σ-field $\mathscr{B}(X)$ is called an H-projection-valued Borel measure on X. Take such a P. We call P regular if

$$P(A) = \sup\{P(C) : C \text{ is a compact subset of } A\} \tag{8}$$

for all Borel subsets A of X.

Applying (8) to $X \setminus A$, we see that (8) implies:

$$P(A) = \inf\{P(U) : U \text{ is open and } U \supset A\} \tag{9}$$

Let $\mu_{\xi,\xi}$ ($\xi \in H$) be the measure on $\mathscr{B}(X)$ defined in 11.8(VII). It is clear from (8) that P is regular if and only if $\mu_{\xi,\xi}$ is regular (in the sense of 8.17) for every ξ in H. Thus, by Proposition 8.17(III), we have:

Proposition. *P is regular if and only if*

$$P(\bigcup \mathscr{W}) = \sup\{P(U): U \in \mathscr{W}\} \tag{10}$$

for every family $\mathscr{W}$ of open subsets of X.

11.10. Proposition. *If X satisfies the second axiom of countability, then every H-projection-valued Borel measure on X is regular.*

Proof. Apply 8.18 to the measures $\mu_{\xi,\xi}$. ∎

11.11. By 11.5 an H-projection-valued Borel measure on X is determined by its values on *closed* sets.

11.12. By the *closed support* of an H-projection-valued Borel measure P on X we mean the (closed) set D of all those x in X such that $P(U) \neq 0$ for every open neighborhood U of x. We denote D by supp(P).

Evidently supp(P) is the closure in X of $\bigcup_{\xi \in H}$ supp($\mu_{\xi,\xi}$) (see 8.9).

11.13. The next result says that the closed support of a regular projection-valued measure whose range is $\{0, 1_H\}$ must be a single point.

Proposition. *Let P be a regular H-projection-valued Borel measure on X such that* range(P) = $\{0, 1_H\}$. *Then there exists a point x in X such that $P(\{x\}) = 1_H$.*

Proof. If $H = \{0\}$, the result is trivial. Assume that $H \neq \{0\}$, and that the proposition is false. By 11.9(9) every x in X has a P-null open neighborhood. Thus every compact set is P-null, whence by regularity $P(X) = 0$, a contradiction. ∎

The Spectral Integral of Unbounded Functions

11.14. We return to the case that X is any set and $\mathscr{S}$ is any σ-field of subsets of X.

In 11.7 we defined the spectral integral of bounded measurable complex functions f. The spectral theory of unbounded normal operators, as we shall present it in §VI.12, will require that we generalize the spectral integral to the case where f may be unbounded.

11.15. Fix an H-projection-valued measure P on $\mathscr{S}$; and let $f: X \to \mathbb{C}$ be P-measurable (but not necessarily P-essentially bounded). Then evidently we can find a sequence $\{A_n\}(n = 1, 2, \ldots)$ of pairwise disjoint sets in $\mathscr{S}$ such that

(i) $\bigcup_{n=1}^{\infty} A_n = X$ and (ii) f is P-essentially bounded on each A_n (i.e. $\mathrm{Ch}_{A_n} f$ is P-essentially bounded). Put $H_n = \mathrm{range}(P(A_n))$, $\mathscr{S}_n = \{A \in \mathscr{S} : A \subset A_n\}$ $(n = 1, 2, \ldots)$. Thus the restriction P_n of P to $\mathscr{S}_n$ is an H_n-projection-valued measure on $\mathscr{S}_n$. Since $f \,|\, A_n$ is P_n-essentially bounded, for each n we can define the spectral integral

$$T_n = \int_{A_n} (f \,|\, A_n) dP_n \qquad \left(\text{or } \int_{A_n} f \, dP \text{ for short} \right)$$

as a bounded normal operator on H_n, according to 11.7. Now the H_n are pairwise orthogonal and $H = \sum_n^{\oplus} H_n$. Therefore by Appendix B28 there is a unique (possibly unbounded) normal operator T in H—namely $T = \sum_n^{\oplus} T_n$—which coincides on H_n with T_n (for each n).

Definition. This T is defined as the *spectral integral* $\int f \, dP$ (or $\int f(x) \, dPx$) of f with respect to P.

In order for this definition to make sense, it must of course be shown that T is independent of the choice of the A_n. To see this, let $\{A_m'\}$ be another sequence with the same properties as the $\{A_n\}$; and assume first that $\{A_m'\}$ is a refinement of $\{A_n\}$ (that is, each A_m' is contained in some A_n). Let $T_n = \int_{A_n} f \, dP$, $T_m' = \int_{A_m'} f \, dP$. If $A_m' \subset A_n$, then obviously T_n and T_m' coincide on $\mathrm{range}(P(A_m'))$. Therefore, for each m, $\sum_n^{\oplus} T_n$ and $\sum_p^{\oplus} T_p'$ both coincide on range $(P(A_m'))$ with T_m'. It then follows from the uniqueness assertion in B28 that $\sum_n^{\oplus} T_n = \sum_p^{\oplus} T_p'$.

Now drop the assumption that $\{A_m'\}$ is a refinement of $\{A_n\}$. At any rate, there certainly exists a sequence $\{A_r''\}$, with the properties of $\{A_n\}$, which refines both $\{A_n\}$ and $\{A_m'\}$. Applying the preceding paragraph twice, we get

$$\sum_n^{\oplus} \int_{A_n} f \, dP = \sum_r^{\oplus} \int_{A_r''} f \, dP = \sum_m^{\oplus} \int_{A_m'} f \, dP,$$

showing that the definition of T is indeed independent of the particular choice of $\{A_n\}$.

Notice that $\int f \, dP$ is bounded if and only if f is P-essentially bounded. In that case the present definition of $\int f \, dP$ coincides of course with that of 11.7.

11.16. The routine proofs of the following several propositions are left to the reader. In all of them, P is an H-projection-valued measure on $\mathscr{S}$.

Proposition. *If $f: X \to \mathbb{C}$ and $g: X \to \mathbb{C}$ are P-measurable, then*

$$\int f \, dP = \int g \, dP$$

if and only if $f(x) = g(x)$ for P-almost all x.

11.17. Proposition. *If $f: X \to \mathbb{C}$ is P-measurable, and $X = \bigcup_{n=1}^{\infty} A_n$ (where the A_n are pairwise disjoint sets in $\mathscr{S}$), then*

$$\int_X f \, dP = \sum_n^{\oplus} \int_{A_n} f \, dP.$$

Here, of course, $\int_{A_n} f \, dP$ means $\int \mathrm{Ch}_{A_n} f \, dP$. The direct sum $\sum_n^{\oplus}$ is taken in the sense of Appendix B16. We do not of course assume that f is bounded on each A_n.

11.18. Proposition. *Let $f: X \to \mathbb{C}$ be P-measurable; and put $T = \int f \, dP$. A vector ξ in H belongs to* domain(T) *if and only if*

$$\int |f(x)|^2 \, d\|P\xi\|^2 x < \infty,$$

where $\|P\xi\|^2$ is the measure $\mu_{\xi, \xi}: A \mapsto \|P(A)\xi\|^2$ on $\mathscr{S}$; and in this case

$$\|T\xi\|^2 = \int |f(x)|^2 \, d\|P\xi\|^2 x. \tag{11}$$

If $\xi \in$ domain(T) and $\eta \in X$,

$$(T\xi, \eta) = \int f(x) d(P\xi, \eta)x, \tag{12}$$

where $(P\xi, \eta)$ is the measure $\mu_{\xi, \eta}: A \mapsto (P(A)\xi, \eta)$ on $\mathscr{S}$.

11.19. Proposition. *If $W \in \mathcal{O}(H)$, $WP(A) = P(A)W$ for all A in $\mathscr{S}$, and $f: X \to \mathbb{C}$ is P-measurable, then W comm $\int f \, dP$ (see B12).*

In particular, range$(P(A))$ *reduces $\int f \, dP$ (see B19) for each A in $\mathscr{S}$.*

11.20. For the next proposition let $\mathscr{S}_0$ be a δ-ring contained in $\mathscr{S}$, with the property that

$$A \in \mathscr{S}, A \subset B \in \mathscr{S}_0 \Rightarrow A \in \mathscr{S}_0. \tag{13}$$

Let μ be a measure on $\mathscr{S}_0$; form the Hilbert space $H = \mathscr{L}_2(\mu)$; and let P be the canonical H-projection-valued measure on $\mathscr{S}$:

$$P(A)\phi = \mathrm{Ch}_A \phi \qquad\qquad (A \in \mathscr{S}; \phi \in H).$$

(This makes sense by (13).)

Proposition. *Let* $f : X \to \mathbb{C}$ *be a P-measurable (hence locally μ-measurable) function. Then* $T = \int f \, dP$ *is just the operator of multiplication by f; that is,* domain$(T) = \{\phi \in H; \, f\phi \in H\}$, *and* $T(\phi) = f\phi \in$ domain(T).

11.21. Proposition. *Let* $f : X \to \mathbb{C}$ *and* $g : X \to \mathbb{C}$ *be P-measurable. Then:*

(I) $\quad \displaystyle\int f \, dP + \int g \, dP \subset \int (f + g) dP.$

(II) $\quad \displaystyle\lambda \int f \, dP = \int (\lambda f) dP \qquad\qquad\qquad\qquad (0 \ne \lambda \in \mathbb{C}).$

(III) $\displaystyle \left(\int f \, dP \right)\!\left(\int g \, dP \right) \subset \int fg \, dP.$

(IV) $\displaystyle \left(\int f \, dP \right)^{*} = \int \bar{f} \, dP.$

(The left sides of (I)–(IV) *are defined as in* B3 *and* B8.) *Equality holds in* (I) *provided at least one of f and g is P-essentially bounded. Equality holds in* (III) *provided that there exist positive numbers k_1, k_2 satisfying*

$$|g(x)| \ge k_1 \Rightarrow |f(x)| \ge k_2 \qquad\qquad (14)$$

for P-almost all x in X. In particular, equality holds in (III) *if g is P-essentially bounded.*

For proving (III), and the condition for equality in (III), one combines 11.18 with 7.5. The same combination yields the following useful fact:

(V) *If ξ belongs to the domains of both $\int g \, dP$ and $\int fg \, dP$, then $(\int g \, dP)(\xi)$ belongs to the domain of $\int f \, dP$, that is, ξ belongs to the domain of the left side of* (III).

It follows in particular that, if, $\{x \in X : f(x) = 0\}$ *is P-null, then* $\int f \, dP$ *is one-to-one and*

$$\left(\int f \, dP \right)^{-1} = \int (f(x))^{-1} \, dPx. \qquad\qquad (15)$$

11.22. Corollary. *Let* $f : X \to \mathbb{C}$ *be P-measurable. Then, for each positive integer n,*

$$\left(\int f \, dP \right)^{n} = \int (f(x))^{n} \, dPx.$$

Proof. By induction in n, using condition (14) for equality in 11.21 (III).   ∎

11.23.   Proposition. *Let Y be another set, and $\mathcal{T}$ a σ-field of subsets of Y. Let $\Phi: X \to Y$ be an $\mathcal{S}, \mathcal{T}$ Borel function, and Q the H-projection-valued measure on $\mathcal{T}$ given by:*

$$Q(B) = P(\Phi^{-1}(B)) \qquad\qquad (B \in \mathcal{T}). \qquad (16)$$

Then, if $f: Y \to \mathbb{C}$ is Q-measurable,

$$\int_Y f\, dQ = \int_X f(\Phi(x)) dPx.$$

In particular, if $Y = \mathbb{C}$ and $\mathcal{T}$ is the Borel σ-field of $\mathbb{C}$, then $\Phi: X \to \mathbb{C}$ is a complex-valued $\mathcal{S}$-Borel function, and

$$\int_X \Phi\, dP = \int_{\mathbb{C}} \lambda\, dQ\lambda.$$

12.   The Analogue of the Riesz Theorem for Projection-Valued Measures

12.1. The main theorem of this section (12.8) is the analogue, for projection-valued measures, of the Riesz Theorem 8.12 for complex measures on a locally compact Hausdorff space. Theorem 8.12 will play a critical role in the proof of 12.8.

12.2. From here to 12.7 X is a fixed *compact* Hausdorff space. Thus $\mathscr{C}(X)$ is a Banach *-algebra under pointwise multiplication, pointwise complex conjugation, and the supremum norm $\|f\|_\infty = \sup\{|f(x)|: x \in X\}$. Let $\mathbb{1}$ be the unit element of $\mathscr{C}(X)$, that is, the function with constant value 1.

12.3. Fix a Hilbert space H; and suppose we are given a linear map

$$T: \mathscr{C}(X) \to \mathcal{O}(H)$$

with the following properties: (i) $T_{fg} = T_f T_g$ $(f, g \in \mathscr{C}(X))$; (ii) $T_{\bar{f}} = (T_f)^*$ $(f \in \mathscr{C}(X)$; $\bar{}$ means complex conjugation); (iii) $T_{\mathbb{1}} = \mathbb{1}_H$. (Later in this work we shall refer to such a T as a *non-degenerate *-representation* of $\mathscr{C}(X)$).

Notice that, if $f \in \mathscr{C}_+(X)$, then T_f is a positive operator. Indeed: Such an f is of the form $\bar{g}g$, where $g \in \mathscr{C}(X)$; so by (i),(ii), $(T_f \xi, \xi) = (T_{gg} \xi, \xi) = (T_g \xi, T_g \xi) \geq 0$ for every ξ in H.

12.4. We have just seen that if $\xi \in H$, the linear functional $f \mapsto (T_f \xi, \xi)$ is non-negative on $\mathscr{C}_+(X)$; hence by Remark 2 of 8.12 it is continuous on $\mathscr{C}(X)$. So, by the polarization identity, $f \mapsto (T_f \xi, \eta)$ must be continuous on $\mathscr{C}(X)$ for every ξ, η in H. It follows by 8.12 that each pair of vectors ξ, η in H gives rise to a unique regular complex Borel measure $\mu_{\xi,\eta}$ on X satisfying

$$(T_f \xi, \eta) = \int_X f \, d\mu_{\xi,\eta} \tag{1}$$

for all f in $\mathscr{C}(X)$. Clearly (1) implies that $\mu_{\xi,\eta}$ is linear in ξ and conjugate-linear in η.

By 12.3

$$\mu_{\xi,\xi} \geq 0 \qquad\qquad (\xi \in H). \tag{2}$$

This implies of course that $\mu_{\xi,\xi}$ is real-valued and hence, by the polarization identity, that

$$\mu_{\eta,\xi} = \overline{\mu_{\xi,\eta}}. \tag{3}$$

Substituting $f = 1$ in (1), we have by 12.3(iii)

$$\mu_{\xi,\eta}(X) = (\xi, \eta) \qquad\qquad (\xi, \eta \in H). \tag{4}$$

In view of (2) we can apply Schwarz's Inequality to the function $\langle \xi, \eta \rangle \mapsto \mu_{\xi,\eta}(A)$, getting by (4):

$$|\mu_{\xi,\eta}(A)|^2 \leq \mu_{\xi,\xi}(A)\mu_{\eta,\eta}(A)$$
$$\leq \mu_{\xi,\xi}(X)\mu_{\eta,\eta}(x)$$
$$\leq \|\xi\|^2 \|\eta\|^2 \tag{5}$$

for each Borel subset A of X and all ξ, η in H.

It follows from (5) and the conjugate-bilinearity of $\mu_{\xi,\eta}$ in ξ and η that for each Borel subset A of X there is a unique operator $P(A)$ in $\mathscr{O}(H)$ satisfying $\|P(A)\| \leq 1$ and

$$(P(A)\xi, \eta) = \mu_{\xi,\eta}(A) \qquad\qquad (\xi, \eta \in H). \tag{6}$$

By (3) $P(A)$ is Hermitian.

Our purpose now is to show that $P: A \mapsto P(A)$ is a regular projection-valued measure, and that, for each f in $\mathscr{C}(X)$, $T_f = \int_X f \, dP$ (the spectral integral).

12.5. Lemma. *For any Borel subsets A and B of X,*

$$P(A \cap B) = P(A)P(B) = P(B)P(A).$$

Proof.   Fix vectors ξ, η in H. For each g in $\mathscr{C}(X)$ let $v_g = g \, d\mu_{\xi,\eta}$ (see 7.2). We claim that

$$v_g = \mu_{T_g\xi,\eta}. \tag{7}$$

Indeed: For all f, g in $\mathscr{C}(X)$,

$$\int f \, d\mu_{T_g\xi,\eta} = (T_f T_g \xi, \eta) \qquad \text{(by (1))}$$

$$= (T_{fg}\xi, \eta) \qquad \text{(by 12.3(i))}$$

$$= \int fg \, d\mu_{\xi,\eta} \qquad \text{(by (1))}$$

$$= \int f \, dv_g \qquad \text{(by 7.6)} \tag{8}$$

By 8.1B (8) implies (7).

Now fix a Borel set $A \subset X$, and define

$$\rho(B) = \mu_{\xi,\eta}(A \cap B)$$

(for Borel subsets B of X). Thus ρ is a regular complex Borel measure (see 8.4); and for any g in $\mathscr{C}(X)$

$$\int g \, d\rho = \int_A g \, d\mu_{\xi,\eta} = v_g(A)$$

$$= (P(A)T_g\xi, \eta) \qquad \text{(by (6) and (7))}$$

$$= (T_g\xi, P(A)\eta) \qquad \text{(since $(P(A)$ is Hermitian)}$$

$$= \int g \, d\mu_{\xi,P(A)\eta} \qquad \text{(by (1))}.$$

By the arbitrariness of g and 8.11 this implies

$$\rho = \mu_{\xi,P(A)\eta}. \tag{9}$$

Now, for any Borel set B,

$$(P(A \cap B)\xi, \eta) = \mu_{\xi,\eta}(A \cap B) \qquad \text{(by (6))}$$

$$= \rho(B) = \mu_{\xi,P(A)\eta}(B) \qquad \text{(by (9))}$$

$$= (P(B)\xi, P(A)\eta) \qquad \text{(by (6))}$$

$$= (P(A)P(B)\xi, \eta) \qquad \text{(since $P(A)$ is Hermitian)}.$$

Since the vectors ξ, η were arbitrary, this gives

$$P(A \cap B) = P(A)P(B). \tag{10}$$

Interchanging A and B in (10), we obtain the conclusion of the Lemma   ∎

12.6. Corollary. *For each Borel subset A of X, $P(A)$ is a projection operator. If A, B are disjoint Borel sets, $P(A)P(B) = 0$. Also $P(X) = 1_H$.*

Proof. The first two statements follow from 12.5 and the Hermitian property of $P(A)$. The last is a consequence of (4).   ∎

12.7. Proposition. *P is a regular H-projection-valued Borel measure on X; and*

$$T_f = \int f \, dP \qquad (spectral\ integral) \tag{11}$$

for every f in $\mathscr{C}(X)$.

Proof. Let $\{A_n\}$ be a sequence of pairwise disjoint Borel subsets of X. By 12.6 the $P(A_n)$ are pairwise orthogonal projections. Putting $Q = \sum_n P(A_n)$, we have for each ξ in H

$$(Q\xi, \xi) = \sum_n (P(A_n)\xi, \xi)$$

$$= \sum_n \mu_{\xi, \xi}(A_n) \qquad \text{(by (6))}$$

$$= \mu_{\xi, \xi}\left(\bigcup_n A_n\right)$$

$$= \left(P\left(\bigcup_n A_n\right)\xi, \xi\right). \qquad \text{(by (6))}$$

It follows that $\sum_n P(A_n) = P(\bigcup_n A_n)$; and P is a projection-valued measure. Its regularity follows from that of the $\mu_{\xi, \xi}$ (see 11.9).

To prove (11), take $f \in \mathscr{C}(X)$, and form the spectral integral $E = \int f \, dP$. For each ξ, η in H we have by 11.8(VII) and (1)

$$(E\xi, \eta) = \int f \, d\mu_{\xi, \eta} = (T_f \xi, \eta).$$

So $E = T_f$; and the proof is complete.   ∎

12.8. We now drop the hypothesis that X is compact.

Theorem. *Let X be a locally compact Hausdorff space, and H a Hilbert space. Suppose that $T: \mathscr{C}_0(X) \to \mathcal{O}(H)$ is a linear map with the following properties:* (i) $T_{fg} = T_f T_g$ $(f, g \in \mathscr{C}_0(X))$; (ii) $T_{\bar{f}} = (T_f)^*$ $(f \in \mathscr{C}_0(X))$; (iii) $\{T_f \xi : f \in \mathscr{C}_0(X);$ $\xi \in H\}$ *spans a dense subspace of H. Then there exists a unique regular H-projection-valued Borel measure P on X such that*

$$T_f = \int_X f \, dP \qquad\qquad \text{(spectral integral)}$$

for all f in $\mathscr{C}_0(X)$.

Proof. We first prove the existence of P.

If X is compact, hypothesis (iii) clearly coincides with 12.3(iii); so 12.7 asserts the existence of P.

Assume that X is not compact, and let $X_\infty = X \cup \{\infty\}$ be its one-point compactification. Since every f in $\mathscr{C}(X_\infty)$ can be written in just one way in the form $\lambda \cdot \mathbf{1} + g$ $(\lambda \in \mathbb{C}; g \in \mathscr{C}_0(X); \mathbf{1} = \text{unit function on } X_\infty)$ we can extend T to a linear map $T': \mathscr{C}(X_\infty) \to \mathcal{O}(H)$ by setting

$$T'_{\lambda \cdot \mathbf{1} + g} = \lambda \cdot \mathbf{1}_H + T_g$$

$(\lambda \in \mathbb{C}; g \in \mathscr{C}_0(X))$. Conditions (i) and (ii) of the Theorem imply the same for T'; and obviously $T'_{\mathbf{1}} = \mathbf{1}_H$. So by the compact case there is a regular H-projection-valued Borel measure P' on X_∞ such that

$$T'_f = \int_{X_\infty} f \, dP' \qquad \text{for all } f \text{ in } \mathscr{C}(X_\infty). \qquad (12)$$

We claim that

$$P'(\{\infty\}) = 0. \qquad (13)$$

Indeed: Using 11.8(III),(VI), we have by (12) for all g in $\mathscr{C}_0(X)$

$$P'(\{\infty\})T_g = \left(\int \mathrm{Ch}_{\{\infty\}} dP' \right)\left(\int g \, dP' \right)$$

$$= \int \left(\mathrm{Ch}_{\{\infty\}} g \right) dP'$$

$$= 0 \qquad\qquad (\text{since } g(\infty) = 0).$$

By hypothesis (iii) of the Theorem this implies (13).

In view of (13) $P'(X) = P'(X_\infty \setminus \{\infty\}) = \mathbf{1}_H$. Therefore the restriction P of P' to the Borel σ-field of X is a regular H-projection-valued Borel measure on X. By (12) $\int_X f \, dP = T_f$ for all f in $\mathscr{C}_0(X)$.

Thus we have proved the existence of the required P.

To prove the uniqueness of P, suppose that Q is another regular H-projection-valued Borel measure on X with the same property. Putting $\mu_{\xi,\eta}(A) = (P(A)\xi, \eta), \nu_{\xi,\eta}(A) = (Q(A)\xi, \eta)$ $(\xi, \eta \in H; A$ a Borel subset of $X)$, we obtain from 11.8(VII)

$$\int f \, d\mu_{\xi,\eta} = (T_f \xi, \eta) = \int f \, d\nu_{\xi,\eta} \tag{14}$$

for all f in $\mathscr{C}_0(X)$. By 8.11, (14) implies that $\mu_{\xi,\eta} = \nu_{\xi,\eta}$ for all ξ, η in H. From this we deduce that $P(A) = Q(A)$ for all Borel sets A, or $P = Q$. ∎

12.9. Corollary. *Let X be a locally compact Hausdorff space and H a Hilbert space. The relation*

$$T_f = \int f \, dP \qquad (f \in \mathscr{C}_0(X)) \tag{15}$$

sets up a one-to-one correspondence between the set of all regular H-projection-valued Borel measures P on X and the set of all linear maps $T: \mathscr{C}_0(X) \to \mathcal{O}(H)$ which satisfy 12.8(i), (ii), (iii).

Proof. In view of 12.8 it remains only to show that for each such P the relation (15) defines a map T satisfying 12.8(i), (ii), (iii). But 12.8(i), (ii) follow from 11.8(III),(IV). To prove 12.8(iii), observe that if C is a compact subset of X, $f \in \mathscr{C}_0(X)$, and $f \equiv 1$ on C, then by 11.8(III), (VI) range$(P(C)) \subset$ range(T_f). Also, by the regularity of P, $\bigcup \{\text{range}(P(C)): C \text{ compact}\}$ is dense in H. The last two statements imply that $\bigcup \{\text{range } (T_f): f \in \mathscr{C}_0(X)\}$ is dense in H, proving 12.8(iii). ∎

13. Banach Bundles

13.1. Definitions. Let X be a fixed Hausdorff topological space. A *bundle over X* is a pair $\langle B, \pi \rangle$, where B is a Hausdorff topological space and $\pi: B \to X$ is a continuous open surjection.

Let $\mathscr{B} = \langle B, \pi \rangle$ be a bundle over X. We call X the *base space*, B the *bundle space*, and π the *bundle projection* of $\mathscr{B}$. For each x in X, $\pi^{-1}(x)$ is the *fiber over x*; we shall usually denote it by B_x (omitting explicit reference to π). Sometimes it is natural to refer to the bundle $\mathscr{B}$ itself as $\langle B, \{B_x\}_{x \in X} \rangle$. *We should like to emphasize that all bundles in this work are over Hausdorff base spaces.*

A *cross-section of $\mathcal{B}$* is a function $f: X \to B$ such that $\pi \circ f = 1_X$, that is, $f(x) \in B_x$ for each x. If $s \in \text{range}(f)$, we say that f *passes through s*. Continuous cross-sections will play a very important role. If for every s in B there exists a continuous cross-section of $\mathcal{B}$ passing through s, we say that $\mathcal{B}$ *has enough continuous cross-sections*.

13.2. Because of the importance of bundles, and hence of open maps, throughout this work, we remind the reader of the following useful criterion for openness of a map.

Proposition. *Let Y and Z be topological spaces and $f: Y \to Z$ a surjection. A necessary and sufficient condition for f to be open is that, whenever $y \in Y$ and $\{z_i\}$ is a net of elements of Z converging to $f(y)$, there should exist a subnet $\{z'_j\}$ of $\{z_i\}$, based on some directed set J, and a net $\{y_j\}$ (based on the same J) of elements of Y, such that (i) $f(y_j) = z'_j$ for all j in J, and (ii) $y_j \to y$ in Y.*

Proof. The proof of sufficiency is very easy and is left to the reader. As regards the necessity, let I be the directed set domain of $\{z_i\}$. Form the new directed set J consisting of all pairs $\langle i, U \rangle$, where $i \in I$ and U is a Y-neighborhood of y, ordered as follows: $\langle i, U \rangle \succ \langle i', U' \rangle \Leftrightarrow i \succ i'$ and $U \subset U'$. If $j = \langle i, U \rangle \in J$, let i_j be an element of I such that $i_j \succ i$ and $z_{i_j} \in f(U)$; such an i_j exists since f is assumed open. Thus for each $j = \langle i, U \rangle$ in J we can choose an element y_j in U for which $f(y_j) = z_{i_j}$. The y_j and $z'_j = z_{i_j}$ then satisfy properties (i) and (ii), and verify the necessity of the condition. ∎

13.3. Let $\mathcal{B} = \langle B, \pi \rangle$ be a bundle over the Hausdorff space X, Y another Hausdorff space, and $\phi: Y \to X$ a continuous map. Let C be the topological subspace $\{\langle y, s \rangle : y \in Y, s \in B, \phi(y) = \pi(s)\}$ of the Cartesian product $Y \times B$; and define $\rho(y, s) = y$ for $\langle y, s \rangle \in C$. Evidently $\rho: C \to Y$ is a continuous surjection. Also ρ is open. To see this, take a point $\langle y, s \rangle$ in C and a net $\{y_i\}$ converging to y in Y. Thus $\phi(y_i) \to \phi(y) = \pi(s)$; so by 13.2 and the openness of π we can replace $\{y_i\}$ by a subnet (without change of notation) and find a net $\{s_i\}$ in B such that $\pi(s_i) = \phi(y_i)$ for all i and $s_i \to s$. This says that $\langle y_i, s_i \rangle \in C$ and $\langle y_i, s_i \rangle \to \langle y, s \rangle$. Since $\rho(y_i, s_i) = y_i$, another application of 13.2 shows that ρ is open. So $\langle C, \rho \rangle$ is a bundle over Y.

Definition. The bundle $\mathcal{C} = \langle C, \rho \rangle$ just constructed is called the *retraction of $\mathcal{B}$ by ϕ*.

Notice that, for each y in Y, the fiber C_y is homeomorphic with the fiber $B_{\phi(y)}$ under the bijection $s \mapsto \langle y, s \rangle$ $(s \in B_{\phi(y)})$.

In the special case that Y is a topological subspace of X and $\phi: Y \to X$ is the identity injection, C can be identified with $\pi^{-1}(Y)$ and ρ with $\pi|\pi^{-1}(Y)$. In this case we call $\mathscr{C}$ the *reduction of $\mathscr{B}$ to Y*, and denote it by $\mathscr{B}_Y$.

13.4. Definition. A *Banach bundle* $\mathscr{B}$ over X is a bundle $\langle B, \pi \rangle$ over X, together with operations and norms making each fiber B_x ($x \in X$) into a Banach space, and satisfying the following conditions:

(i) $s \mapsto \|s\|$ is continuous on B to $\mathbb{R}$.
(ii) The operation $+$ is continuous as a function on $\{\langle s, t \rangle \in B \times B:$
      $\pi(s) = \pi(t)\}$ to B.
(iii) For each complex λ, the map $s \mapsto \lambda \cdot s$ is continuous on B to B.
(iv) If $x \in X$ and $\{s_i\}$ is any net of elements of B such that $\|s_i\| \to 0$ and
      $\pi(s_i) \to x$ in X, then $s_i \to 0_x$ in B.

Remark. Here and in the sequel we write $+, \cdot,$ and $\| \ \|$ for the operations of addition, scalar multiplication, and norm in each fiber B_x. In connection with (ii) we emphasize that $s + t$ is defined only if s and t belong to the same fiber, i.e., $\pi(s) = \pi(t)$. The symbol 0_x in (iv) stands for the zero element of the Banach space B_x.

To avoid multiplicity of symbols we shall often throughout this work deliberately confuse a bundle $\mathscr{B}$ with its bundle space B. For example, by a function with domain $\mathscr{B}$ we will mean a function defined on the bundle space B.

Remark. Notice that condition (iv) is equivalent to:

(iv') If $x \in X$, the collection of all subsets of B of the form $\{b \in B: \pi(b) \in U,$
      $\|b\| < \varepsilon\}$, where U is a neighborhood of x in X and $\varepsilon > 0$, is a basis of
      neighborhoods of 0_x in B.

It is clear from (iv) that the zero cross-section $x \mapsto 0_x$ is continuous.

13.5. Definition. A Banach bundle $\mathscr{B}$ each of whose fibers is a Hilbert space is called a *Hilbert bundle*.

If $(\ , \ )$ denotes the inner product on each fiber of the Hilbert bundle $\mathscr{B}$, then $\langle s, t \rangle \mapsto (s, t)$ is continuous on $\{\langle s, t \rangle \in B \times B: \pi(s) = \pi(t)\}$. This follows from the polarization identity together with the continuity of addition, scalar multiplication, and the norm.

13.6. As a first example, let A be any Banach space; put $B = X \times A$ and $\pi(x, \xi) = x$ ($\langle x, \xi \rangle \in B$); and equip each fiber $B_x = \pi^{-1}(x)$ with the Banach space structure derived from A via the bijection $\xi \mapsto \langle x, \xi \rangle$. Then $\mathscr{B} = \langle B, \pi \rangle$

is obviously a Banach bundle over X. It is called the *trivial Banach bundle with constant fiber A*. Cross-sections of this $\mathscr{B}$ will often be deliberately confused with the corresponding functions $f: X \to A$. Obviously a trivial Banach bundle has enough continuous cross-sections.

13.7.   Let $\mathscr{B}$ be a Banach bundle over the Hausdorff space X, Y another Hausdorff space, and $\phi: Y \to X$ a continuous map. Let us form the retraction $\mathscr{C}$ of the bundle $\mathscr{B}$ by ϕ as in 13.3; and for each y in Y let us equip C_y with the Banach space structure under which the bijection $s \mapsto \langle y, s \rangle$ of $B_{\phi(y)}$ onto C_y becomes a linear isometry. One then verifies immediately that $\mathscr{C}$ satisfies 13.4(i)–(iv), and so is a Banach bundle.

If $\mathscr{B}$ has enough continuous cross-sections, so does $\mathscr{C}$.

Definition.   This $\mathscr{C}$ is called the *Banach bundle retraction of $\mathscr{B}$ by ϕ*.

If Y is a topological subspace of X and ϕ is the identity injection, $\mathscr{C}$ can be identified as in 13.3 with the subspace $\pi^{-1}(Y)$ of B. Then $\mathscr{C}$ is called the *reduction of $\mathscr{B}$ to Y*, and is denoted by $\mathscr{B}_Y$.

13.8.   Definition.   Let $\mathscr{B} = \langle B, \pi \rangle$ and $\mathscr{B}' = \langle B', \pi' \rangle$ be Banach bundles over base spaces X and X'; and let $F: X \to X'$ be a surjective homeomorphism. By *an isomorphism of $\mathscr{B}$ and $\mathscr{B}'$ covariant with F* we mean a surjective homeomorphism $\Phi: B \to B'$ such that (i) $\pi' \circ \Phi = F \circ \pi$ (i.e., $\Phi(B_x) = B'_{F(x)}$ for each x in X), and (ii) $\Phi | B_x$ is linear (onto $B'_{F(x)}$) for each x in X. This implies (see 13.11) that $\Phi | B_x$ is a topological isomorphism of the Banach spaces B_x and $B'_{F(x)}$. If in addition Φ preserves norm ($\|\Phi(b)\| = \|b\|$ for $b \in B$), Φ is an *isometric isomorphism*.

Suppose $X = X'$. Then $\mathscr{B}$ and $\mathscr{B}'$ will be said to be *isometrically isomorphic as Banach bundles* if there is an isometric isomorphism of $\mathscr{B}$ and $\mathscr{B}'$ covariant with the identity map on X.

13.9.   Remark.   Let us say that a Banach bundle $\mathscr{B}$ over X is *locally trivial* if for every x in X there is a neighborhood U of x such that the reduced Banach bundle $\mathscr{B}_U$ over U is isomorphic with a trivial Banach bundle.

Evidently the crucial postulate 13.4(iv) can be looked upon as a "fragment" of the property of local triviality.

Of course not all Banach bundles are locally trivial. For example, let $\mathscr{B} = \langle B, \pi \rangle$ be a trivial Banach bundle over X whose constant fiber A is finite-dimensional. Take a point u of X and a proper linear subspace D of A; and set $B' = B \setminus (\{u\} \times (A \setminus D))$, $\pi' = \pi | B'$. Then $\mathscr{B}' = \langle B', \pi' \rangle$ is a Banach bundle (coinciding with $\mathscr{B}$ except that its fiber over u is D instead of A), but it is not locally trivial around u.

It can be shown that a Banach bundle $\mathscr{B}$ over a locally compact Hausdorff space, whose fibers are all of the same *finite* dimension, is necessarily locally trivial. If however the fibers are infinite-dimensional, $\mathscr{B}$ can fail to be locally trivial even when all the fibers are isometrically isomorphic to each other; see the example cited in 13.22.

Elementary Properties of Banach Bundles

13.10. For the rest of this section $\mathscr{B} = \langle B, \pi \rangle$ is a fixed Banach bundle over X. Let us refer to postulates 13.4(i)–(iv) simply as (i)–(iv).

Proposition. *The scalar multiplication map $\langle \lambda, s \rangle \mapsto \lambda \cdot s$ (of $\mathbb{C} \times B$ into B) is continuous.*

Proof. Let $\lambda_i \to \lambda$, $s_i \to s$. Then $\pi(\lambda_i s_i - \lambda s_i) = \pi(s_i) \to \pi(s)$ and $\|\lambda_i s_i - \lambda s_i\| = |\lambda_i - \lambda| \|s_i\| \to 0$ by (i). So $(\lambda_i - \lambda)s_i \to 0_{\pi(s)}$ by (iv). By (iii) $\lambda s_i \to \lambda s$. Combining the last two statements with (ii) we get $\lambda_i s_i \to \lambda s$. ∎

13.11. Proposition. *The topology of B relativized to a fiber B_x is the norm-topology of B_x.*

Proof. Let $s_i \to s$ in B, where the s_i and s are all in B_x. Then by (ii) and (i) $\|s_i - s\| \to 0$. The converse follows from (iv) and (ii). ∎

13.12. Proposition. *Let $\{s_i\}(i \in I)$ be a net of elements of B, and s an element of B such that $\pi(s_i) \to \pi(s)$. Suppose further that for each $\varepsilon > 0$ we can find a net $\{u_i\}$ of elements of B (indexed by the same I) and an element u of B such that (a) $u_i \to u$ in B, (b) $\pi(u_i) = \pi(s_i)$ for each i (and hence $\pi(u) = \pi(s)$), (c) $\|s - u\| < \varepsilon$ and (d) $\|s_i - u_i\| < \varepsilon$ for all large enough i. Then $s_i \to s$ in B.*

Proof. Since we could replace $\{s_i\}$ by any subnet of it, it suffices to show that some subnet of $\{s_i\}$ converges to s. Since π is open, by 13.2 we can pass to a subnet (without change of notation) and find for each i an element t_i of B such that $\pi(t_i) = \pi(s_i)$ and $t_i \to s$. Now, given $\varepsilon > 0$, choose $\{u_i\}$ and u as in the hypothesis. Since $t_i \to s$ and $u_i \to u$, postulates (ii), (i), and (c) above give $\|t_i - u_i\| \to \|s - u\| < \varepsilon$; so $\|t_i - u_i\| < \varepsilon$ for all large i. Combining this with hypothesis (d) we have $\|t_i - s_i\| < 2\varepsilon$ for large enough i. But ε was arbitrary; so $\|t_i - s_i\| \to 0$, whence by (iv) $t_i - s_i \to 0_{\pi(s)}$. Since $t_i \to s$, this and (ii) imply that $s_i \to s$. ∎

13.13.   Corollary.   *Let* $\{f_i\}$ *be a net of continuous cross-sections of* $\mathcal{B}$, *and* f *a cross-section of* $\mathcal{B}$ *such that* $f_i \to f$ *uniformly on* X *(i.e.,* $\| f_i(x) - f(x) \| \to 0$ *uniformly for* x *in* X). *Then* f *is continuous.*

13.14.   For the moment let Δ be the family of all cross-sections of $\mathcal{B}$. Evidently Δ is a complex linear space when addition and scalar multiplication are carried out pointwise on X. The zero element of Δ is the *zero cross-section* $x \mapsto 0_x$.

Now define $\mathscr{C}(\mathcal{B})$ to be the set of all *continuous* cross-sections of $\mathcal{B}$.

Proposition. $\mathscr{C}(\mathcal{B})$ *is a linear subspace of* Δ. *Furthermore, if* $f \in \mathscr{C}(\mathcal{B})$ *and* $\lambda : X \to \mathbb{C}$ *is continuous, then* $\lambda \cdot f : x \mapsto \lambda(x) f(x)$ *is in* $\mathscr{C}(\mathcal{B})$.

Proof.   That $\mathscr{C}(\mathcal{B})$ is closed under addition and under multiplication by continuous complex functions follows from (ii) and 13.10. That the zero cross-section is in $\mathscr{C}(\mathcal{B})$ follows from (iv).   ∎

13.15.   Proposition.   *Suppose that for each* s *in* B *and each* $\varepsilon > 0$ *there exists an element* f *of* $\mathscr{C}(\mathcal{B})$ *such that* $\| f(\pi(s)) - s \| < \varepsilon$. *Then* $\mathcal{B}$ *has enough continuous cross-sections.*

Proof.   Fix $s \in B$; and set $x = \pi(s)$. We must find a continuous cross-section passing through s.

First note that, for any f in $\mathscr{C}(\mathcal{B})$, we can find another f' in $\mathscr{C}(\mathcal{B})$ such that $f'(x) = f(x)$ and $\| f'(y) \| \leq \| f(x) \|$ for all y in X. Indeed, we have only to set $f'(y) = \min(1, \| f(x) \| \| f(y) \|^{-1}) f(y)$; this f' is continuous by 13.14.

Now by the hypothesis we can find a series $\sum_{n=1}^{\infty} t_n$ of elements of B_x converging absolutely (i.e., $\sum \| t_n \| < \infty$) to s, and elements $\{f_n\}$ of $\mathscr{C}(\mathcal{B})$ such that $f_n(x) = t_n$ for all n. By the preceding paragraph we can assume that $\| f_n(y) \| \leq \| t_n \|$ for all n and y. Then the series $\sum_{n=1}^{\infty} f_n$ converges uniformly on X to a cross-section f satisfying $f(x) = s$; and by 13.13 and 13.14 f is continuous.   ∎

13.16.   Next follow two propositions which often help in proving continuity of maps from one Banach bundle to another.

Proposition.   *Let* $\mathcal{B} = \langle B, \pi \rangle$ *and* $\mathcal{B}' = \langle B', \pi' \rangle$ *be two Banach bundles over the same Hausdorff base space* X. *Let* $\Phi : B \to B'$ *be a map such that*: (i) *for each* $x \in X$, $\Phi(B_x) \subset B'_x$ *and* $\Phi | B_x$ *is linear, and* (ii) *there is a constant* k *such that* $\| \Phi(b) \| \leq k \| b \|$ *for all* $b \in B$. *Suppose further that there is a family* Γ *of* *continuous cross-sections of B such that:* (iii) $\{\gamma(x): \gamma \in \Gamma\}$ *has dense linear span in* B_x *for each* $x \in X$, *and* (iv) $\Phi \circ \gamma$ *is a continuous cross-section of* B' *for all* γ *in* Γ. *Then* $\Phi: B \to B'$ *is continuous.*

Proof. Postulates 13.4(ii), (iii) permit us to replace Γ by its linear span and so to assume that: (iii') $\{\gamma(x): \gamma \in \Gamma\}$ is dense in B_x for all $x \in X$.

Suppose now that $b_i \to b$ in B. We must show that $\Phi(b_i) \to \Phi(b)$ in B'. Let $x = \pi(b)$, $x_i = \pi(b_i)$. Given $\varepsilon > 0$, we use (iii') to choose $\gamma \in \Gamma$ so that $\|\gamma(x) - b\| < \varepsilon k^{-1}$. Since $b_i \to b$ and $\gamma(x_i) \to \gamma(x)$, the continuity of the norm and of subtraction shows that $\|\gamma(x_i) - b_i\| < \varepsilon k^{-1}$ for all large enough i. By (ii) these facts imply that $\|(\Phi \circ \gamma)(x) - \Phi(b)\| < \varepsilon$ and $\|(\Phi \circ \gamma)(x_i) - \Phi(b_i)\| < \varepsilon$ for all large enough i. But by (iv) $(\Phi \circ \gamma)(x_i) \to (\Phi \circ \gamma)(x)$ in B'. Hence by Proposition 13.12 $\Phi(b_i) \to \Phi(b)$. ∎

Remark. At first glance one might wonder, by analogy with the Banach space situation, whether (i) and (ii) would imply continuity of Φ, without the assumption of a family Γ satisfying (iii) and (iv). But this is not so. One can in general clearly choose Φ so that (i) and (ii) hold, while the $\Phi|B_x$ vary with x in such a discontinuous manner that $\Phi(b)$ changes abruptly as b passes continuously from one fiber to neighboring ones.

13.17. The following proposition is useful in establishing isomorphisms of Banach bundles.

Proposition. *Let* $\mathscr{B} = \langle B, \pi \rangle$ *and* $\mathscr{B}' = \langle B', \pi' \rangle$ *be two Banach bundles over the same (Hausdorff) base space* X. *Let* $\Phi: B \to B'$ *be a map such that:* (i) *for each* $x \in X$, $\Phi(B_x) \subset B'_x$, *and* $\Phi|B_x$ *is linear,* (ii) $\Phi: B \to B'$ *is continuous, and* (iii) *there is a positive constant* k *such that* $\|\Phi(b)\| \geq k\|b\|$ *for all* $b \in B$. *Then* $\Phi^{-1}: \Phi(B) \to B$ *is continuous.*

Proof. Notice that Φ is injective by (iii), so that Φ^{-1} exists.

Let $\{b_i\}$ be a net in B and b an element of B such that $\Phi(b_i) \to \Phi(b)$ in B'. Thus in particular $\pi(b_i) \to \pi(b)$ in X. We must show that $b_i \to b$, that is, that every subnet of $\{b_i\}$ has a subnet converging to b. Let us replace $\{b_i\}$ by an arbitrary subnet of itself (without change of notation). By 13.2, passing again to a subnet of $\{b_i\}$ (without change of notation), we can find a net $\{c_i\}$ of elements of B (based on the same directed set $\{i\}$) such that $c_i \to b$ and $\pi(c_i) = \pi(b_i)$ for all i. Since Φ is continuous, this implies that $\Phi(c_i) \to \Phi(b)$. At the same time $\Phi(b_i) \to \Phi(b)$, and $\pi'(\Phi(b_i)) = \pi(b_i) = \pi(c_i) = \pi'(\Phi(c_i))$. The last two sentences imply, by the continuity of subtraction and of the norm, that

$$\|\Phi(b_i) - \Phi(c_i)\| \to 0.$$

By hypothesis (iii) this gives

$$\|b_i - c_i\| \to 0.$$

whence (since $\pi(b_i - c_i) \to \pi(b)$)

$$b_i - c_i \to 0_{\pi(b)}.$$

This, together with the continuity of addition and the fact that $c_i \to b$, shows that $b_i \to b$.

Thus any subnet of the original $\{b_i\}$ has a subnet converging to b. Hence the original $\{b_i\}$ converges to b; so Φ^{-1} is continuous. ∎

Construction of Banach Bundles

13.18. The construction of specific Banach bundles usually takes place in the following sort of situation. Let X be a Hausdorff space as before.

Suppose we are given an untopologized set A and a surjection $\rho: A \to X$; and suppose that for each x in X the set $A_x = \rho^{-1}(x)$ is a complex Banach space (with operations $+$, $\cdot$, and norm $\| \ \|$). Further assume that Γ is a complex linear space of cross-sections of $\langle A, \rho \rangle$ such that (a) for each f in Γ the numerical function $x \mapsto \|f(x)\|$ is continuous on X, and (b) for each x in X the set $\{f(x): f \in \Gamma\}$ is dense in A_x.

Theorem. *Under these hypotheses, there is a unique topology for A making $\langle A, \rho \rangle$ a Banach bundle over X such that all the elements of Γ are continuous cross-sections of $\langle A, \rho \rangle$.*

Proof. If such a topology $\mathscr{T}$ for A exists, then 13.12 tells us that a net $\{s_i\}$ of elements of A converges to s (relative to $\mathscr{T}$) if and only if (a) $\rho(s_i) \to \rho(s)$, and (b) for each f in Γ, $\|s_i - f(\rho(s_i))\| \to \|s - f(\rho(s))\|$. Therefore $\mathscr{T}$, if it exists, is unique.

To see that such a topology exists, let $\mathscr{W}$ be the family of all subsets of A of the form

$$W(f, U, \varepsilon) = \{s \in A: \rho(s) \in U, \|s - f(\rho(s))\| < \varepsilon\},$$

where f runs over Γ, U over all open subsets of X, and ε over all positive numbers. (This definition is motivated by the easy observation that, if the required topology for A exists, $\mathscr{W}$ must be a basis of open sets in that topology). Let $\mathscr{T}$ be the family of all unions of elements of $\mathscr{W}$.

We claim that $\mathscr{T}$ is the family of all open sets for some topology of A. To see this it is enough to show that the intersection of two elements of $\mathscr{W}$ is in

$\mathcal{T}$. So, given sets $W(f, U, \varepsilon)$ and $W(g, V, \delta)$ in $\mathcal{W}$ and a point s in their intersection, we must find a set $W(h, Z, \sigma)$ in $\mathcal{W}$ satisfying

$$s \in W(h, Z, \sigma) \subset W(f, U, \varepsilon) \cap W(g, V, \delta). \tag{1}$$

Put $x = \rho(s)$; choose ε' and δ' so that $\|s - f(x)\| < \varepsilon' < \varepsilon$ and $\|s - g(x)\| < \delta' < \delta$; and put $\sigma = \frac{1}{2}\min\{\varepsilon - \varepsilon', \delta - \delta'\}$. If we choose $h \in \Gamma$ so that $\|s - h(x)\| < \sigma$, we have

$$\left.\begin{array}{l} \|h(x) - f(x)\| \le \|h(x) - s\| + \|s - f(x)\| < \sigma + \varepsilon', \\ \|h(x) - g(x)\| \le \|h(x) - s\| + \|s - g(x)\| < \sigma + \delta'. \end{array}\right\} \tag{2}$$

Now $h - f \in \Gamma$; so $y \mapsto \|h(y) - f(y)\|$ is continuous; and similarly for $h - g$. So by (2) we can take an open neighborhood Z of x with $Z \subset U \cap V$ such that

$$\|h(y) - f(y)\| < \sigma + \varepsilon' \quad \text{and} \quad \|h(y) - g(y)\| < \sigma + \delta' \tag{3}$$

for all y in Z. Note that $s \in W(h, Z, \sigma)$ (since $\|s - h(x)\| < \sigma$). Further, if $t \in W(h, Z, \sigma)$ and $y = \pi(t)$, then $y \in Z \subset U$, and by (3) $\|t - f(y)\| \le \|t - h(y)\| + \|h(y) - f(y)\| < \sigma + \sigma + \varepsilon' \le \varepsilon$; so $t \in W(f, U, \varepsilon)$; and similarly $t \in W(g, V, \delta)$. Thus (1) has been proved for these h, Z, σ; and the claim is established.

We have now defined a topology $\mathcal{T}$ for A. It is very easy to see that ρ is continuous and open with respect to $\mathcal{T}$, and that 13.4(i), (iii) hold.

Let us prove 13.4(ii). Let s, t be elements of A with $x = \rho(s) = \rho(t)$; and suppose $s + t \in W(f, U, \varepsilon)$. Choose ε' so that

$$\|s + t - f(x)\| < \varepsilon' < \varepsilon; \tag{4}$$

put $\delta = \frac{1}{4}(\varepsilon - \varepsilon')$; and take g, h in Γ to satisfy

$$\|s - g(x)\| < \delta, \; \|t - h(x)\| < \delta. \tag{5}$$

Then by (4) and (5)

$$\|g(x) + h(x) - f(x)\| < 2\delta + \varepsilon'.$$

Since $g + h - f \in \Gamma$, we can therefore find an open neighborhood V of x with $V \subset U$ such that

$$\|g(y) + h(y) - f(y)\| < 2\delta + \varepsilon' \quad \text{for all } y \text{ in } V. \tag{6}$$

By (5) $W(g, V, \delta)$ and $W(h, V, \delta)$ are $\mathcal{T}$-neighborhoods of s and t respectively. Now assume that $p \in W(g, V, \delta)$, $q \in W(h, V, \delta)$, and $\rho(p) = \rho(q) = y \in V$. Then $\|p - g(y)\| < \delta$ and $\|q - h(y)\| < \delta$; and so by (6)

$$\|p + q - f(y)\| \leq \|p - g(y)\| + \|q - h(y)\| + \|g(y) + h(y) - f(y)\|$$
$$< \delta + \delta + 2\delta + \varepsilon' = \varepsilon,$$

whence $p + q \in W(f, U, \varepsilon)$. Since $W(f, U, \varepsilon)$ was an arbitrary element of $\mathcal{W}$ containing $s + t$, 13.4(ii) holds.

It remains only to prove 13.4(iv). Assume that $\{s_i\}$ is a net of elements of A such that $\|s_i\| \to 0$ and $\pi(s_i) \to x \in X$. We must show that $s_i \to 0_x$ relative to $\mathcal{T}$. Let $0_x \in W(f, U, \varepsilon)$. So $x \in U$; and we can choose ε' so that $\|f(x)\| < \varepsilon' < \varepsilon$. By the hypotheses on $\{s_i\}$, we have $\pi(s_i) \in U$, $\|f(\pi(s_i))\| < \varepsilon'$ and $\|s_i\| < \varepsilon - \varepsilon'$, whence $s_i \in W(f, U, \varepsilon)$, for all large enough i. Since $W(f, U, \varepsilon)$ was any element of $\mathcal{W}$ containing 0_x, we have shown that $s_i \to 0_x$ in $\mathcal{T}$.

Thus A, with the topology $\mathcal{T}$ and the bundle projection ρ, satisfies all the postulates for a Banach bundle. This completes the proof of the Proposition. ∎

13.19.   Remark. By 13.15, the construction of 13.18 always leads to Banach bundles having enough continuous cross-sections. As a matter of fact, all the specific Banach bundles encountered in this work will be over locally compact base spaces; and for these we have the beautiful result of A. Douady and L. Dal Soglio-Hérault proved in Appendix C:

Every Banach bundle over a base space which is either locally compact or paracompact has enough continuous cross-sections.

The reader will be assumed henceforth to recall, without explicit mention, that all Banach bundles over locally compact Hausdorff base spaces have enough continuous cross-sections.

13.20.   As a first example of the use of 13.18 we construct the C_0 direct sum of Banach bundles.

For each i in an index set I let $\mathcal{B}^i = \langle B^i, \pi^i \rangle$ be a Banach bundle over a Hausdorff base space X (the same X for all i). We shall assume that all the $\mathcal{B}^i$ have enough continuous cross-sections; by 13.19 this is automatically the case if X is locally compact.

For each $x \in X$ let $B_x = \sum_{i \in I}^{\oplus 0} B^i_x$ be the C_0 direct sum Banach space (Chap. I, §5); and let B be the disjoint union $\bigcup_{x \in G} B_x$, and $\pi : B \to X$ the surjection given by $\pi^{-1}(x) = B_x$. If $i \in I$ and $\phi \in \mathscr{C}(\mathcal{B}^i)$, then (since $B^i_x \subset B_x$) ϕ is a cross-section of $\langle B, \pi \rangle$. Let Γ be the linear span of the set of all such cross-sections (both i and ϕ varying). One easily verifies that $x \mapsto \|\psi(x)\|_B$ is continuous on X for each $\psi \in \Gamma$ and that $\{\psi(x): \psi \in \Gamma\}$ is dense in B_x for each $x \in G$. So by 13.18 there is a unique topology for B making $\langle B, \pi \rangle$ a Banach bundle over X and such that the cross-sections in Γ are continuous.

Definition. The $\mathscr{B} = \langle B, \pi \rangle$ just constructed is called the C_0 *direct sum* of the $\mathscr{B}^i$, in symbols,

$$\mathscr{B} = \sum_{i \in I}^{\oplus 0} \mathscr{B}^i.$$

13.21. For arguments involving separability, the following Proposition is important.

Proposition. *Assume that $\mathscr{B}$ has enough continuous cross-sections. Then, for the bundle space B to be second-countable, it is necessary and sufficient that the following two conditions hold: (a) X is second-countable; (b) there is a countable subfamily Δ of $\mathscr{C}(\mathscr{B})$ such that $\{f(x): f \in \Delta\}$ is dense in B_x for every x in X.*

Proof. *Sufficiency.* Assume (a) and (b). Let Δ be as in (b); and let C be a countable base of open subsets of X. Using the notation of the proof of 13.18, it is enough to show that the countable set of all $W(f, U, n^{-1})$, where $f \in \Delta$, $U \in C$, and $n = 1, 2, \ldots$, is a basis of open subsets of B.

To do this, let $s \in W(g, V, \varepsilon)$, where $g \in \mathscr{C}(\mathscr{B})$, V is an open subset of X, and $\varepsilon > 0$. By the proof of 13.18 these sets form a basis of open sets in B. So it is enough to find f in Δ, U in C, and $n = 1, 2, \ldots$ satisfying

$$s \in W(f, U, n^{-1}) \subset W(g, V, \varepsilon). \tag{7}$$

Put $x = \pi(s)$, so that $x \in V$ and $\|s - g(x)\| < \varepsilon$; let $\delta = \varepsilon - \|s - g(x)\|$. Choose $f \in \Delta$ and a positive integer n so that

$$\|s - f(x)\| < n^{-1} < \tfrac{1}{2}\delta. \tag{8}$$

It follows that $\|f(x) - g(x)\| \le \|f(x) - s\| + \|s - g(x)\| < \varepsilon - \delta/2$. Thus we can pick $U \in C$ so that $x \in U \subset V$ and

$$\|f(y) - g(y)\| < \varepsilon - \tfrac{1}{2}\delta \qquad \text{for all } y \text{ in } U. \tag{9}$$

Now from (8) we deduce that $s \in W(f, U, n^{-1})$. If $t \in W(f, U, n^{-1})$, then $y = \pi(t) \in U \subset V$, and by (8) and (9)

$$\|t - g(y)\| \le \|t - f(y)\| + \|f(y) - g(y)\|$$

$$< n^{-1} + \varepsilon - \tfrac{1}{2}\delta < \varepsilon;$$

so $t \in W(g, V, \varepsilon)$. Thus (7) has been proved.

Necessity. We assume that B is second-countable. This and the continuity and openness of π imply that X is second-countable. Furthermore, any basis of open sets in B must have a countable subfamily which is also a basis. Therefore, using the notation and results of the proof of 13.18, we can choose a basis for B consisting of a sequence of sets of the form $W(f_n, U_n, \varepsilon_n)$ ($f_n \in \mathscr{C}(\mathscr{B})$; U_n open in X; $\varepsilon_n > 0$). It is then easy to see that $\Delta = \{f_n\}$ satisfies (b). ∎

13.22. *Remark.* If the bundle space B is to be second-countable, each fiber B_x must of course be separable. One might conjecture that the B of 13.21 will be second-countable provided X is second-countable and each fiber B_x is separable. This is true if $\mathscr{B}$ is locally trivial (13.9); but in general it is false. An instructive counter-example is given in Maréchal [1], p. 140.

14. Banach Bundles over Locally Compact Base Spaces

14.1. The Banach bundles which we shall come across in this work will almost always have locally compact base spaces.

Throughout this section $\mathscr{B} = \langle B, \pi \rangle$ is a Banach bundle over a *locally compact Hausdorff* base space X, having therefore (by 13.19) enough continuous cross-sections.

By the *topology of uniform convergence on compact sets* for $\mathscr{C}(\mathscr{B})$ we mean of course that topology in which $f_i \to f$ if and only if $\| f_i(x) - f(x) \| \to 0$ uniformly in x on compact subsets of X.

Proposition. *Let Γ be a linear subspace of $\mathscr{C}(\mathscr{B})$ with the following two properties: (a) If $f \in \Gamma$ and $\lambda: X \to \mathbb{C}$ is continuous, then $\lambda f: x \mapsto \lambda(x) f(x)$ is in Γ; and (b) for each x in X, $\{ f(x): f \in \Gamma \}$ is dense in B_x. Then Γ is dense in $\mathscr{C}(\mathscr{B})$ in the topology of uniform convergence on compact sets.*

Proof. Let K be any compact subset of X, g any element in $\mathscr{C}(\mathscr{B})$, and $\varepsilon > 0$. Hypothesis (b), together with the theory of partitions of unity (Kelley [1], p. 171), gives us (i) a finite open covering $U_1, \dots, U_n$ of K, (ii) for each $i = 1, \dots, n$ an element f_i of Γ such that $\| f_i(x) - g(x) \| < \varepsilon$ for all x in U_i, and (iii) for each i a continuous non-negative function λ_i on X vanishing outside U_i such that $\sum_{i=1}^{n} \lambda_i(x) = 1$ for all x in K.

We now form the cross-section $f = \sum_{i=1}^{n} \lambda_i f_i$ which by hypothesis (a) belongs to Γ. Fixing $x \in K$, we have

$$\| f(x) - g(x) \| = \left\| \sum_{i=1}^{n} \lambda_i(x)(f_i(x) - g(x)) \right\|$$

$$\leq \sum_{i=1}^{n} \lambda_i(x) \| f_i(x) - g(x) \|. \tag{1}$$

Now, for each i either $x \in U_i$, in which case $\| f_i(x) - g(x) \| < \varepsilon$, or $x \notin U_i$, in which case $\lambda_i(x) = 0$. So in all cases $\lambda_i(x) \| f_i(x) - g(x) \| \leq \varepsilon \lambda_i(x)$, and (1) gives

$$\| f(x) - g(x) \| \leq \varepsilon \sum_{i=1}^{n} \lambda_i(x) = \varepsilon$$

for $x \in K$. By the arbitrariness of g, K, and ε, this completes the proof. ∎

Remark. Let Δ be any subfamily of $\mathscr{C}(X)$ which is dense in $\mathscr{C}(X)$ in the topology of uniform convergence on compact sets. The above Proposition remains true if hypothesis (a) is weakened to (a'): If $f \in \Gamma$ and $\lambda \in \Delta$, then $\lambda f \in \Gamma$.

To see this, we form the closure $\bar{\Gamma}$ of Γ (with respect to the topology of uniform convergence on compact sets). By the denseness of Δ and the evident continuity of $\langle \lambda, f \rangle \mapsto \lambda f$, hypothesis (a') implies that $\bar{\Gamma}$ satisfies the original hypothesis (a). So the Proposition asserts that $\bar{\Gamma}$ is dense in, hence equal to, $\mathscr{C}(\mathscr{B})$; and this implies that Γ is dense in $\mathscr{C}(\mathscr{B})$.

The Inductive Limit Topology of $\mathscr{L}(\mathscr{B})$

14.2. We denote by $\mathscr{L}(\mathscr{B})$ the subspace of $\mathscr{C}(\mathscr{B})$ consisting of those f such that the closure in X of $\{x \in X : f(x) \neq 0_x\}$ is compact. This closure is called the *compact support* of f, and is denoted by $\text{supp}(f)$.

Notice that there are enough elements of $\mathscr{L}(\mathscr{B})$ to pass through every point of B. Indeed, if $s \in B$, we can choose a continuous cross-section g passing through s and a continuous complex function λ on X with compact support such that $\lambda(\pi(s)) = 1$. Then by 13.14 λg belongs to $\mathscr{L}(\mathscr{B})$ and passes through s.

If $\mathscr{B}$ is the trivial Banach bundle whose constant fiber is $\mathbb{C}$, $\mathscr{L}(\mathscr{B})$ coincides with $\mathscr{L}(X)$.

14.3. It is often useful to equip $\mathscr{L}(\mathscr{B})$ with the so-called inductive limit topology. To define this, consider a compact set $K \subset X$, and denote by $\mathscr{L}^K$ (or $\mathscr{L}^K(\mathscr{B})$) the subspace of $\mathscr{L}(\mathscr{B})$ consisting of those f such that $\text{supp}(f) \subset$

K. Thus $\mathscr{L}^K$ can be identified with a linear subspace of $\mathscr{C}(\mathscr{B}_K)$ (the space of all continuous cross-sections of the reduced bundle $\mathscr{B}_K$); and $\mathscr{C}(\mathscr{B}_K)$ is a normed linear space (in fact a Banach space, by 13.13) under the supremum norm $\|f\|_\infty = \sup\{\|f(x)\| : x \in K\}$. We give to $\mathscr{L}^K$ the relativized topology of $\mathscr{C}(\mathscr{B}_K)$. Also, let $i_K : \mathscr{L}^K \to \mathscr{L}(\mathscr{B})$ be the identity injection; and let $p_K : \mathscr{L}(\mathscr{B}) \to \mathscr{C}(\mathscr{B}_K)$ be the restriction map $f \mapsto f \,|\, K$.

Proposition.   *There is a unique largest topology $\mathscr{I}$ on $\mathscr{L}(\mathscr{B})$ with respect to which* (i) *$\mathscr{L}(\mathscr{B})$ is a locally convex space, and* (ii) *the map i_K is continuous for every compact subset K of X.*

  This $\mathscr{I}$ has the following universal Property (P): *For any locally convex space M and any linear map $F : \mathscr{L}(\mathscr{B}) \to M$, F is continuous (with respect to $\mathscr{I}$) if and only if $F \circ i_K : \mathscr{L}^K \to M$ is continuous for every compact subset K of X.*

Proof.   Let us denote by $\mathscr{U}$ the family of all convex subsets U of $\mathscr{L}(\mathscr{B})$ with the following property: For every compact subset K of X, $U \cap \mathscr{L}^K$ is a neighborhood of 0 in the normed linear space $\mathscr{L}^K$. We leave it to the reader to verify that $\mathscr{U}$ satisfies the postulates for a "local basis" as given in Theorem 5.1 of Kelley and Namioka [1]. Furthermore, $\mathscr{U}$ separates 0 from all non-zero points of $\mathscr{L}(\mathscr{B})$; indeed, $\{f \in \mathscr{L}(\mathscr{B}) : \|f\|_\infty < \varepsilon\} \in \mathscr{U}$ for all $\varepsilon > 0$. Therefore, since the elements of $\mathscr{U}$ are convex, Theorem 5.1 of Kelley and Namioka [1] tells us that there is a unique topology $\mathscr{I}$ making $\mathscr{L}(\mathscr{B})$ a locally convex space and having $\mathscr{U}$ as a basis of neighborhoods of 0.

  Evidently $i_K : \mathscr{L}^K \to \mathscr{L}(\mathscr{B})$ is continuous with respect to $\mathscr{I}$ for every compact set K.

  Suppose that M is any other locally convex space and $F : \mathscr{L}(\mathscr{B}) \to M$ is a linear map. If $F \circ i_K$ is continuous for all compact sets K, then $F^{-1}(V)$ is in $\mathscr{U}$ for every convex M-neighborhood V of 0, and so F is continuous with respect to $\mathscr{I}$. Conversely, if F is continuous with respect to $\mathscr{I}$, the continuity of the i_K shows that the $F \circ i_K$ are continuous. Thus $\mathscr{I}$ has the universal Property (P).

  Suppose that $\mathscr{J}$ is another topology making $\mathscr{L}(\mathscr{B})$ a locally convex space and making all the i_K continuous. Property (P) applied to the identity map $i : \mathscr{L}(\mathscr{B}) \to \mathscr{L}(\mathscr{B})$ (taking $\mathscr{I}$ and $\mathscr{J}$ for the topologies of the domain and range respectively) shows that i is continuous, hence that $\mathscr{J}$ is weaker than $\mathscr{I}$. So $\mathscr{I}$ is the unique largest such topology, and the proof is complete.   ∎

Definition.   The topology $\mathscr{I}$ of the above theorem is called the *inductive limit topology of $\mathscr{L}(\mathscr{B})$*.

  If $\mathscr{B}$ is the trivial Banach bundle with constant fiber $\mathbb{C}$, then $\mathscr{L}(\mathscr{B})$ becomes $\mathscr{L}(X)$, and we have defined the *inductive limit topology of $\mathscr{L}(X)$*.

Remark. The injections $i_K: \mathscr{L}^K \to \mathscr{L}(\mathscr{B})$ (K being compact) are actually homeomorphisms with respect to $\mathscr{I}$. Indeed, we have already seen that they are continuous. Obviously $p_K \circ i_L: \mathscr{L}^L \to \mathscr{L}^K$ is continuous for every pair of compact sets K, L; so by Property (P) the p_K are continuous. Since $p_K \circ i_K$ is the identity of $\mathscr{L}^K$, i_K^{-1} is continuous.

14.4. If $\lambda \in \mathscr{C}(X)$ and $f \in \mathscr{L}(\mathscr{B})$, the pointwise product $\lambda f: x \mapsto \lambda(x)f(x)$ belongs to $\mathscr{L}(\mathscr{B})$. The map $\langle \lambda, f \rangle \mapsto \lambda f$ of $\mathscr{C}(X) \times \mathscr{L}(\mathscr{B})$ into $\mathscr{L}(\mathscr{B})$ is bilinear; and one deduces immediately from the universal Property (P) that it is *separately* continuous with respect to the topology of uniform convergence on compact sets for $\mathscr{C}(X)$, and the inductive limit topology for $\mathscr{L}(\mathscr{B})$. In particular, $\langle \lambda, f \rangle \mapsto \lambda f$ is separately continuous on $\mathscr{L}(X) \times \mathscr{L}(\mathscr{B})$ to $\mathscr{L}(\mathscr{B})$ with respect to the inductive limit topologies.

As regards the *joint* continuity of $\langle \lambda, f \rangle \mapsto \lambda f$, see 14.15.

14.5. Proposition. *Suppose S is a subset of $\mathscr{L}(\mathscr{B})$ with the property that for each f in $\mathscr{L}(\mathscr{B})$ there is a compact subset K of X such that f lies in the $\mathscr{L}^K$-closure of $S \cap \mathscr{L}^K$. Then S is dense in $\mathscr{L}(\mathscr{B})$ in the inductive limit topology.*

This is clear.

14.6. In analogy with 14.1 we have:

Proposition. *Let Γ be a linear subspace of $\mathscr{L}(\mathscr{B})$ with the following two properties: (I) If λ is a continuous complex function on X and $f \in \Gamma$, then $\lambda f \in \Gamma$; (II) for each x in X, $\{f(x): f \in \Gamma\}$ is dense in B_x. Then Γ is dense in $\mathscr{L}(\mathscr{B})$ in the inductive limit topology.*

Proof. By 14.1 Γ is dense in $\mathscr{L}(\mathscr{B})$ in the topology of uniform convergence on compact sets. Let $g \in \mathscr{L}(\mathscr{B})$, and let λ be an element of $\mathscr{L}(X)$ which takes the value 1 on supp(g). We can then choose a net $\{f_i\}$ of elements of Γ converging to g uniformly on $K = \text{supp}(\lambda)$. Then $\lambda f_i \to \lambda g = g$ uniformly on all of X; and all the λf_i vanish outside K. So $\lambda f_i \to g$ in the inductive limit topology, and $\lambda f_i \in \Gamma$. ∎

Remark. Let Δ be any subfamily of $\mathscr{C}(X)$ which is dense in $\mathscr{C}(X)$ in the topology of uniform convergence on compact sets. Analogously with Remark 14.1, we observe that the above Proposition continues to hold if hypothesis (I) is weakened to (I'): If $\lambda \in \Delta$ and $f \in \Gamma$, then $\lambda f \in \Gamma$.

Indeed, let $\bar{\Gamma}$ be the inductive limit closure of Γ. Hypothesis (I'), together with the denseness of Δ and the separate continuity of $\langle \lambda, f \rangle \mapsto \lambda f$ (see 14.4), imply that $\bar{\Gamma}$ satisfies hypothesis (I) of the Proposition. Therefore the Proposition asserts that $\bar{\Gamma}$ is dense in, hence equal to, $\mathscr{L}(\mathscr{B})$; hence Γ is inductively dense in $\mathscr{L}(\mathscr{B})$.

14.7.   Definition.   We shall denote by $\mathscr{C}_0(\mathscr{B})$ the linear subspace of $\mathscr{C}(\mathscr{B})$ consisting of those f which "vanish at ∞" (that is, for each $\varepsilon > 0$ there is a compact subset E of X such that $\| f(x) \| < \varepsilon$ for all $x \in X \setminus E$). It is a familiar fact that $\mathscr{C}_0(\mathscr{B})$ is a Banach space under the supremum norm $\| f \|_\infty = \sup_{x \in X} \| f(x) \|$. We call it the C_0 *cross-sectional Banach space* of $\mathscr{B}$.

Obviously $\mathscr{L}(\mathscr{B}) \subset \mathscr{C}_0(\mathscr{B}) \subset \mathscr{C}(\mathscr{B})$; and $\mathscr{L}(\mathscr{B})$ is dense in $\mathscr{C}_0(\mathscr{B})$ with respect to the supremum norm.

If X is compact, then of course $\mathscr{L}(\mathscr{B}) = \mathscr{C}_0(\mathscr{B}) = \mathscr{C}(\mathscr{B})$.

If X is discrete, then $\mathscr{C}_0(\mathscr{B})$ coincides with the $\sum_{x \in X}^{\oplus 0} B_x$ defined in Chapter I, §5.

If $\mathscr{B}$ is the trivial Banach bundle with constant fiber A, $\mathscr{C}_0(\mathscr{B})$ is just the $\mathscr{C}_0(X; A)$ of Chapter I, §3. In particular, if $A = \mathbb{C}$, then $\mathscr{C}_0(\mathscr{B}) = \mathscr{C}_0(X)$.

Here is a useful corollary of 14.1:

Corollary.   *Let Γ be a linear subspace of $\mathscr{L}(\mathscr{B})$ with the properties* (I) *and* (II) *of Proposition* 14.6. *Then Γ is dense in $\mathscr{C}_0(\mathscr{B})$* (*In the supremum norm of the latter*).

Proof.   Evidently the supremum norm topology relativized to $\mathscr{L}(\mathscr{B})$ is weaker than (i.e. contained in) the inductive limit topology of $\mathscr{L}(\mathscr{B})$. So by 14.1 Γ is supremum-norm dense in $\mathscr{L}(\mathscr{B})$. Since the latter is dense in $\mathscr{C}_0(\mathscr{B})$, Γ must be dense in $\mathscr{C}_0(\mathscr{B})$. ∎

14.8.   As another application of 14.1 we obtain a bundle version of the Tietze Extension Theorem.

Theorem.   *Let Y be any closed subset of X. Then every element g of $\mathscr{L}(\mathscr{B}_Y)$ is the restriction $f | Y$ to Y of some f in $\mathscr{L}(\mathscr{B})$.*

Proof.   Let $g \in \mathscr{L}(\mathscr{B}_Y)$; and choose an open subset U of X with compact closure, such that U contains the compact support C of g. By the Tietze Extension Theorem for numerical functions, the family $\mathscr{S} = \{ f | Y : f \in \mathscr{L}(\mathscr{B}) \}$ satisfies the hypotheses of 14.1 relative to $\mathscr{C}(\mathscr{B}_Y)$. Hence by 14.1 we can find a sequence $\{ f_n \}$ $(n = 1, 2, \ldots)$ of elements of $\mathscr{L}(\mathscr{B})$ such that

$$f_n | Y \to g \text{ uniformly on } \bar{U} \cap Y. \tag{2}$$

Multiplying the f_n by a fixed element of $\mathscr{L}(X)$ which is 1 on C and 0 outside U, we may assume that

$$\text{all the } f_n \text{ vanish outside } U. \qquad (3)$$

Next, passing to a subsequence of $\{f_n\}$, we may by (2) assume without loss of generality that, for each $n \geq 2$,

$$\sup\{\|f_n(y) - f_{n-1}(y)\| : y \in \bar{U} \cap Y\} < 2^{-n}. \qquad (4)$$

Now put $h'_n = f_n - f_{n-1}$ $(n \geq 2)$, and define

$$h_n(x) = \begin{cases} h'_n(x) & \text{if } \|h'_n(x)\| \leq 2^{-n}, \\ \|h'_n(x)\|^{-1} 2^{-n} h'_n(x) & \text{if } \|h'_n(x)\| \geq 2^{-n}. \end{cases}$$

Evidently $h_n \in \mathscr{L}(\mathscr{B})$, and $\|h_n(x)\| \leq 2^{-n}$ for all x. Thus

$$f(x) = f_1(x) + \sum_{n=2}^{\infty} h_n(x) \qquad (x \in X)$$

converges absolutely uniformly in x; and f is continuous by 13.13. By (3) f vanishes outside U, and so is in $\mathscr{L}(\mathscr{B})$. By (4), $h_n(y) = f_n(y) - f_{n-1}(y)$ for $y \in \bar{U} \cap Y$, whence $f(y) = f_1(y) + \sum_{n=2}^{\infty}(f_n(y) - f_{n-1}(y)) = g(y)$ for $y \in \bar{U} \cap Y$. Since both f and g vanish on $Y - \bar{U}$, this implies $f | Y = g$, completing the proof. ∎

14.9. Given a continuous map $\phi: X \to Y$ of locally compact base spaces, the retraction process (13.7) carries us from a Banach bundle over Y to one over X. Conversely, if ϕ is open, the formation of C_0 partial cross-sectional bundles, which we present next, carries us in the opposite direction—from a Banach bundle $\mathscr{B}$ over X to one over Y. The Generalized Tietze Extension Theorem is an essential tool here.

Proposition. *Let Y be another locally compact Hausdorff space, and $\phi: X \to Y$ a continuous open surjection. For each $y \in Y$ let $\mathscr{B}^y$ be the reduction of $\mathscr{B}$ to the closed subset $\phi^{-1}(y)$ of X; and let D_y be the C_0 cross-sectional Banach space of $\mathscr{B}^y$ (see 14.7). There is a unique Banach bundle $\mathscr{D}$ over Y such that: (i) For each $y \in Y$, D_y is the fiber of $\mathscr{D}$ over y; and (ii) for each $f \in \mathscr{L}(\mathscr{B})$, the cross-section $y: \mapsto f | \phi^{-1}(y)$ of $\mathscr{D}$ is continuous.*

Proof. Let Γ be the family of all cross-sections of $\mathscr{D}$ of the form $g: y \mapsto f | \phi^{-1}(y)$ $(f \in \mathscr{L}(\mathscr{B}))$. (We notice, since $\phi^{-1}(y)$ is closed, that $g(y) \in \mathscr{L}(\mathscr{B}^y) \subset D_y$). Thus Γ is linear, and a routine proof based on the openness of ϕ shows that $y \mapsto \|g(y)\|_\infty$ is continuous on Y for each $g \in \Gamma$. Furthermore, for each fixed $y \in Y$, the Generalized Tietze Theorem (14.8) shows that $\{g(y): g \in \Gamma\}$ is equal to $\mathscr{L}(\mathscr{B}^y)$ and hence dense in D_y. So the Proposition follows from 13.18.   ∎

Definition.   We refer to the $\mathscr{D}$ constructed in this Proposition as the C_0 *partial cross-sectional bundle over Y derived from $\mathscr{B}$ and ϕ.*

14.10.   Proposition.   *Suppose that the bundle space B of $\mathscr{B}$ is second-countable. Then $\mathscr{L}(\mathscr{B})$ is separable (that is, there is a countable subset of $\mathscr{L}(\mathscr{B})$ which is dense in $\mathscr{L}(\mathscr{B})$ in the inductive limit topology).*

Proof.   We shall first prove that $\mathscr{L}(X)$ is separable.

By 13.21 X is second-countable. It follows easily that we can find a countable family $\mathscr{F}$ of real-valued functions in $\mathscr{L}(X)$ such that (i) $\mathscr{F}$ separates points of X (that is, if $x, y \in X$ and $x \neq y$, then $\phi(x) \neq \phi(y)$ for some ϕ in $\mathscr{F}$), and (ii) for each x in X, $\phi(x) \neq 0$ for some ϕ in $\mathscr{F}$. Replacing $\mathscr{F}$ by $\{r\phi: \phi \in \mathscr{F}, r \in \mathbb{C}_{\text{rat}}\}$ (where $\mathbb{C}_{\text{rat}}$ is the set of all complex-rational numbers) we may assume that $\mathscr{F}$ is countable and closed under multiplication by complex-rationals. Again, replacing $\mathscr{F}$ by the set of all finite sums of finite products of functions in $\mathscr{F}$, we may assume that $\mathscr{F}$ is countable and closed under pointwise addition and multiplication, as well as under pointwise complex conjugation and scalar multiplication by complex-rationals. Now let $\mathscr{G}$ be the inductive limit closure of $\mathscr{F}$. From the continuity properties of addition, multiplication (see 14.4) and pointwise conjugation with respect to the inductive limit topology, we conclude that $\mathscr{G}$ is linear and is closed under pointwise multiplication and complex conjugation. Also $\mathscr{G}$ shares properties (i) and (ii) of $\mathscr{F}$. So by the Stone–Weierstrass Theorem (Appendix A) $\mathscr{G}$ is dense in $\mathscr{C}(X)$ in the topology of uniform convergence on compact sets. Combining this with the fact that $\mathscr{G}$ is closed under pointwise multiplication and closed in the inductive limit topology, we conclude that $\lambda\phi \in \mathscr{G}$ whenever $\lambda \in \mathscr{C}(X)$ and $\phi \in \mathscr{G}$. From this and 14.6 it follows that $\mathscr{G}$ is inductively dense in $\mathscr{L}(X)$ and hence equal to $\mathscr{L}(X)$. Since $\mathscr{G}$ is the closure of a countable set, $\mathscr{L}(X)$ is separable.

By 13.21 there is a countable subset Δ of $\mathscr{C}(\mathscr{B})$ such that $\{f(x): f \in \Delta\}$ is dense in B_x for each x in X. Now form the family $\mathscr{H}$ of all finite sums of cross-sections of the form λf where $f \in \Delta$ and $\lambda \in \mathscr{F}$ ($\mathscr{F}$ being the above countable dense subset of $\mathscr{L}(X)$). Thus $\mathscr{H}$ is a countable subset of $\mathscr{L}(\mathscr{B})$. Let $\mathscr{K}$ be the inductive limit closure of $\mathscr{H}$. From the denseness of $\mathscr{F}$ and the continuity of $\lambda \mapsto \lambda f$ for each fixed f in $\mathscr{C}(\mathscr{B})$ (see 14.4), we conclude that $\mathscr{K}$ contains the set Γ of all finite sums of cross-sections of the form λf, where

$f \in \Delta$ and λ is any element of $\mathscr{L}(X)$. By the denseness property of Δ, this Γ satisfies the hypotheses of 14.6, and so is inductively dense in $\mathscr{L}(\mathscr{B})$. Thus $\mathscr{K}$ is inductively dense in $\mathscr{L}(\mathscr{B})$, and so equal to $\mathscr{L}(\mathscr{B})$. Therefore $\mathscr{L}(\mathscr{B})$ has a countable dense subset $\mathscr{H}$, completing the proof. ∎

Further Properties of the Inductive Limit Topology

14.11. The inductive limit topology, though convenient for the elegant formulation of results, will not be of primary importance in this work. We could well manage without it. However, some of its properties are quite interesting; and we shall mention a few of them, without proofs. The reader may treat them as exercises. A more detailed discussion (for the case of $\mathscr{L}(X)$) will be found in R. E. Edwards [1].

14.12. Proposition*. *Suppose that X is discrete (so that each element of $\mathscr{L}(\mathscr{B})$ vanishes except at finitely many points of X). Let P be the set of all functions on X to $\mathbb{R}_+$; and for each ϕ in P let $p_\phi(f) = \sum_{x \in X} \phi(x) \| f(x) \|$ ($f \in \mathscr{L}(\mathscr{B})$). Then $\{p_\phi : \phi \in P\}$ is a defining family of seminorms for the inductive limit topology of $\mathscr{L}(\mathscr{B})$.*

14.13. Proposition*. *Suppose that X is second-countable. Let Q be the family of all those functions on X to $\mathbb{R}_+$ which are bounded on each compact subset of X; and for each ϕ in Q put $q_\phi(f) = \sup\{\phi(x) \| f(x) \| : x \in X\}$ ($f \in \mathscr{L}(\mathscr{B})$). Then $\{q_\phi : \phi \in Q\}$ is a defining family of seminorms for the inductive limit topology of $\mathscr{L}(\mathscr{B})$.*

Remark. One cannot remove the hypothesis of second countability in the above Proposition. For example, if X is discrete and uncountable, the collection of all the above q_ϕ defines a strictly weaker topology than the inductive limit topology defined by the p_ϕ of 14.12.

14.14. Remark*. Suppose that X is either second-countable or discrete, and that $\{f_n\}$ is a *sequence* of elements of $\mathscr{L}(\mathscr{B})$ converging to g in the inductive limit topology of $\mathscr{L}(\mathscr{B})$. Then it is not hard to show that there is a compact subset K of X such that $\mathrm{supp}(f_n) \subset K$ for all n.

However, if $\{f_n\}$ is merely a net, not a sequence, then, even if it converges in the inductive limit topology, the $\mathrm{supp}(f_n)$ need not all be contained in a single compact set. Indeed, any inductive limit neighborhood of 0 contains elements of $\mathscr{L}(\mathscr{B})$ having arbitrarily large compact support; and from this, if X is not compact, it is easy to construct a net $\{f_i\}$ converging to 0 in the inductive limit topology, such that the $\mathrm{supp}(f_i)$ expand indefinitely.

14.15.   We shall now briefly examine the (joint) continuity of certain *bilinear* maps with respect to the inductive limit topology.

We have noted in 14.4 that the pointwise multiplication map $\langle \lambda, f \rangle \mapsto \lambda f$ ($\lambda \in \mathscr{C}(X)$; $f \in \mathscr{L}(\mathscr{B})$) is separately continuous with respect to the topology of uniform convergence on compact sets for $\mathscr{C}(X)$ and the inductive limit topology for $\mathscr{L}(\mathscr{B})$. In general however it is not jointly continuous with respect to these topologies. For example, let $X = \mathbb{Z}$. As in 14.12 let P be the directed set of all functions on $\mathbb{Z}$ to $\mathbb{R}_+$, ordered by the usual pointwise ordering. If $n = 1, 2, \ldots$ and $p \in P$, let $f_{n,p} = (p(n))^{-1}\delta_n$, $\lambda_{n,p} = p(n)\delta_n$ (δ_n being the element of $\mathscr{L}(\mathbb{Z})$ which has the value 1 at n and 0 elsewhere). Then, as $n \to \infty$ and p increases, $f_{n,p} \to 0$ in the inductive limit topology (see 14.12) and $\lambda_{n,p} \to 0$ in the topology of uniform convergence on compact sets, but $f_{n,p}\lambda_{n,p}(= \delta_n)$ does not approach 0 in the inductive limit topology.

However, if the λ are restricted to lie below a suitable fixed function on X, then the continuity discussed in the previous paragraph does hold. Indeed, we have:

Proposition*.   *Let* $p\colon X \to \mathbb{R}_+$ *be a fixed function, bounded on each compact subset of* X. *Let* $\mathscr{M} = \{\lambda \in \mathscr{C}(X) : |\lambda(x)| \le p(x) \text{ for all } x \text{ in } X\}$. *Then the pointwise multiplication map* $\langle \lambda, f \rangle \mapsto \lambda f$ *is continuous on* $\mathscr{M} \times \mathscr{L}(\mathscr{B})$ *with respect to the topology of uniform convergence on compact sets for* $\mathscr{M}$, *and the inductive limit topology for* $\mathscr{L}(\mathscr{B})$. *In particular,* $\langle \lambda, f \rangle \mapsto \lambda f$ *is continuous on* $\mathscr{L}(X) \times \mathscr{L}(\mathscr{B})$ *with respect to the inductive limit topologies of* $\mathscr{L}(X)$ *and* $\mathscr{L}(\mathscr{B})$.

In proving this, the crucial observation is the following: If $U \in \mathscr{U}$ (see the proof of 14.3), then $V = \{f \in \mathscr{L}(\mathscr{B}) : \lambda f \in U \text{ whenever } \lambda \in \mathscr{C}(X) \text{ and } |\lambda| \le p\}$ is also in $\mathscr{U}$.

14.16.   Consider two locally compact Hausdorff spaces X and Y. If $f \in \mathscr{L}(X)$ and $g \in \mathscr{L}(Y)$ let $f \times g$ be the element of $\mathscr{L}(X \times Y)$ given by $(f \times g)(x, y) = f(x)g(y)$. The bilinear map $\langle f, g \rangle \mapsto f \times g$ is certainly separately continuous with respect to the inductive limit topologies of $\mathscr{L}(X)$, $\mathscr{L}(Y)$, $\mathscr{L}(X \times Y)$. We ask whether it is jointly continuous.

Proposition*.   *If* X *and* Y *are second-countable, the map* $\langle f, g \rangle \mapsto f \times g$ *is jointly continuous with respect to the inductive limit topologies of* $\mathscr{L}(X)$, $\mathscr{L}(Y)$, *and* $\mathscr{L}(X \times Y)$.

To prove this, the essential step is to observe that, if $r\colon X \times Y \to \mathbb{R}_+$ is bounded on compact sets, then there exist functions $p\colon X \to \mathbb{R}_+$ and $q\colon Y \to$

$\mathbb{R}_+$, both bounded on compact sets, such that $r(x, y) \le p(x)q(y)$ for all x, y. One then completes the proof by invoking 14.13.

Remark 1. If X and Y are not both second-countable, the conclusion of this Proposition need not hold, even if X and Y are discrete. Assume that X and Y are discrete. By 14.12 and the argument of the last paragraph, a counter-example will be obtained if we exhibit a function $r: X \times Y \to \mathbb{R}_+$ such that there do not exist functions $p: X \to \mathbb{R}_+$ and $q: Y \to \mathbb{R}_+$ satisfying $r(x, y) \le p(x)q(y)$ for all x, y. To exhibit such a situation, let $X = \mathbb{Z}$; let Y be the family of all functions $\phi: \mathbb{Z} \to \mathbb{R}_+$; and define $r: X \times Y \to \mathbb{R}_+$ by: $r(n, \phi) = \phi(n)$ $(n \in \mathbb{Z}; \phi \in Y)$.

Remark 2. The above proposition can be generalized to the context of Banach bundles, provided we know how to form the "tensor product" (over $X \times Y$) of two bundles over X and Y.

15. Integration in Banach Bundles Over Locally Compact Spaces

15.1. In this section, as in the last, $\mathscr{B} = \langle B, \pi \rangle$ is a Banach bundle over the locally compact Hausdorff base space X (and hence having enough continuous cross-sections). Let μ be a fixed regular Borel measure on X (8.2).

Locally Measurable Cross-Sections

15.2. Having associated a Banach space B_x with each point x of X, we are in a context where the theory of local measurability structures (§4) can be applied. If $f \in \mathscr{C}(\mathscr{B})$ then of course the numerical function $x \to \| f(x) \|$ is continuous, hence (8.2) locally μ-measurable. So by 4.2 $\mathscr{C}(\mathscr{B})$ generates a local μ-measurability structure $\mathscr{M}$ for the cross-sections of $\mathscr{B}$.

Definition. A cross-section of $\mathscr{B}$ is said to be *locally μ-measurable* if it belongs to the above $\mathscr{M}$.

Notice that the local μ-measurability structure $\mathscr{M}'$ generated by $\mathscr{L}(\mathscr{B})$ is the same as the $\mathscr{M}$ generated by $\mathscr{C}(\mathscr{B})$. To see this we have only to observe from 4.1(IV) that $\mathscr{C}(\mathscr{B}) \subset \mathscr{M}'$.

15.3. Suppose that $\mathscr{B}$ is the trivial bundle with constant fiber A (see 13.6). Then the definition of local μ-measurability just given coincides with that of 5.1. This follows from the fact that, if K is a compact subset of X and $f: X \to A$

is continuous, then f can be approximated pointwise (in fact uniformly) on K by simple functions (in the sense of 5.1).

15.4.   Proposition.   *A cross-section f of $\mathscr{B}$ is locally μ-measurable if and only if it satisfies the following condition* (C): *For each compact subset K of X there exists a sequence $\{g_n\}$ of continuous cross-sections of $\mathscr{B}$ such that $g_n(x) \to f(x)$ (in B_x) for μ-almost all x in K.*

Proof.   Let $\mathscr{R}$ stand for the linear span of $\{\mathrm{Ch}_A f : A \in \mathscr{S}(X), f \in \mathscr{C}(\mathscr{B})\}$; and define $^-$ as in the proof of Proposition 4.2. We claim that

$$\mathscr{R} \subset \mathscr{C}(\mathscr{B})^-. \tag{1}$$

Indeed: Since $\mathscr{C}(\mathscr{B})^-$ is linear, we need only take a set A in $\mathscr{S}(X)$ and a continuous cross-section f, and show that $\mathrm{Ch}_A f \in \mathscr{C}(\mathscr{B})^-$. Now by the regularity of μ we can find a decreasing sequence $\{U_n\}$ of open sets in $\mathscr{S}(X)$, and an increasing sequence $\{C_n\}$ of compact sets, such that

$$C_n \subset A \subset U_n, \lim_n \mu(U_n \setminus C_n) = 0. \tag{2}$$

For each n choose a continuous function $\lambda_n : X \to [0, 1]$ such that $\lambda_n \equiv 1$ on C_n and λ_n vanishes outside U_n. It follows from (2) that $\lambda_n(x) \to \mathrm{Ch}_A(x)$, and hence that $(\lambda_n f)(x) \to (\mathrm{Ch}_A f)(x)$, for μ-almost all x. Since $\lambda_n f \in \mathscr{C}(\mathscr{B})$, we have $\mathrm{Ch}_A f \in \mathscr{C}(\mathscr{B})^-$, proving (1).

Now by the proof of Proposition 4.2 $\mathscr{M} = \mathscr{R}^-$. From the first part of the same proof we get $\mathscr{C}(\mathscr{B})^{--} = \mathscr{C}(\mathscr{B})^-$. Combining these facts with (1) we see that

$$\mathscr{M} = \mathscr{R}^- \subset \mathscr{C}(\mathscr{B})^{--} = \mathscr{C}(\mathscr{B})^- \subset \mathscr{M}.$$

Therefore $\mathscr{M} = \mathscr{C}(\mathscr{B})^-$. Since $\mathscr{C}(\mathscr{B})^-$ is just the set of cross-sections satisfying Condition (C), the proof is finished.   ■

15.5.   Generalized Lusin's Theorem.   *Let f be any locally μ-measurable cross-section of $\mathscr{B}$. For each compact subset K of X and each $\varepsilon > 0$, there is a Borel subset D of K with $\mu(K \setminus D) < \varepsilon$ such that $f|D$ is continuous on D.*

Proof.   By 15.4 there is a sequence $\{g_n\}$ of continuous cross-sections of $\mathscr{B}$ such that $g_n(x) \to f(x)$ (in B_x) for μ-almost all x in K. By Egoroff's Theorem 1.11 (applied to the numerical functions $x \mapsto \|g_n(x) - f(x)\|$) we can find a Borel subset D of K with $\mu(K \setminus D) < \varepsilon$ such that $g_n(x) \to f(x)$ uniformly for x in D. Applying 13.13 on the reduction of $\mathscr{B}$ to D we conclude that $f|D$ is continuous on D.   ■

15.6. Here is another useful condition for local measurability of a cross-section.

Proposition. *For a cross-section f of $\mathcal{B}$ to be locally μ-measurable, it is necessary and sufficient that the following two conditions hold:* (I) $x \mapsto \|f(x) - g(x)\|$ *is locally μ-measurable for every continuous cross-section g;* (II) *for each compact subset K of X, there is a sequence $\{H_n\}$ of compact subsets of B such that, for μ-almost all x in K, $f(x)$ belongs to the closed linear span (in B_x) of $\bigcup_{n=1}^{\infty} (H_n \cap B_x)$.*

Proof. *Necessity.* (I) is obviously necessary. If f is locally μ-measurable and K is a compact subset of X, by 15.4 we can choose continuous cross-sections $\{g_n\}$ such that $g_n(x) \to f(x)$ for μ-almost all x in K. The compact sets $H_n = g_n(K)$ then satisfy condition (II), showing that (II) is necessary.

Sufficiency. We need the following assertion:

For every compact subset H of B, we can find a countable subset Γ of $\mathcal{C}(\mathcal{B})$ such that every s in H belongs to the closure of $\{g(\pi(s)): g \in \Gamma\}$.

To prove this, fix $\varepsilon > 0$ and $x \in X$. Since $H \cap B_x$ is compact and metric, we can choose a finite subfamily Γ_x^ε of $\mathcal{C}(\mathcal{B})$ such that for each s in $H \cap B_x$ we have $\|g(x) - s\| < \varepsilon$ for some g in Γ_x^ε. Now we claim that there exists an open X-neighborhood U_x of x such that if $t \in H$ and $\pi(t) \in U_x$ then $\|g(\pi(t)) - t\| < \varepsilon$ for some g in Γ_x^ε. Indeed: If this were false we could find a net $\{t_i\}$ of elements of H with $\pi(t_i) \to x$ such that

$$\|g(\pi(t_i)) - t_i\| \geq \varepsilon \tag{3}$$

for all i and all g in Γ_x^ε. By the compactness of H we can pass to a subnet and suppose that $t_i \to t$ in H. Passing to the limit in (3) we obtain $\|g(x) - t\| \geq \varepsilon$ for all g in Γ_x^ε, contradicting the definition of Γ_x^ε. So the claim is proved. Now cover the compact set $K = \pi(H)$ with finitely many of the U_x:

$$K \subset U_{x_1} \cup \cdots \cup U_{x_n}.$$

Then $\Gamma^\varepsilon = \Gamma_{x_1}^\varepsilon \cup \cdots \cup \Gamma_{x_n}^\varepsilon$ is a finite subfamily of $\mathcal{C}(\mathcal{B})$ such that every s in H satisfies $\|g(\pi(s)) - s\| < \varepsilon$ for some g in Γ^ε. Consequently the countable set $\Gamma = \bigcup_{m=1}^{\infty} \Gamma^{m^{-1}}$ satisfies the requirement of the assertion.

We now return to the proof of sufficiency. Condition (II) of the present Proposition together with the above assertion imply that hypothesis (II) of Proposition 4.4 holds for f. Hypothesis (I) of 4.4 follows from Condition (I) and 15.4. So by 4.4 Conditions (I) and (II) imply that f is locally μ-measurable. ∎

The $\mathscr{L}_p$ Spaces

15.7. Definition. Let $1 \leq p < \infty$. We define $\mathscr{L}_p(\mu; \mathscr{B})$ to be the Banach space $\mathscr{L}_p(\mu; \{B_x\}; \mathscr{M})$ of 4.5 in our present context. That is, $\mathscr{L}_p(\mu; \mathscr{B})$ consists of all those locally μ-measurable cross-sections f of $\mathscr{B}$ which vanish outside a countable union of compact sets and for which

$$\|f\|_p^p = \int \|f(x)\|^p \, d\mu x < \infty. \tag{4}$$

It carries the norm $\| \ \|_p$ of (4); and two elements of $\mathscr{L}_p(\mu; \mathscr{B})$ are identified if they differ only on a μ-null set.

We call $\mathscr{L}_p(\mu; \mathscr{B})$ the $\mathscr{L}_p$ *cross-sectional space of $\mathscr{B}$ (with respect to μ).*

15.8. Evidently $\mathscr{L}(\mathscr{B}) \subset \mathscr{L}_p(\mu; \mathscr{B})$. It follows from the definition of the inductive limit topology (14.3) that the identity map $\mathscr{L}(\mathscr{B}) \rightarrow \mathscr{L}_p(\mu; \mathscr{B})$ is continuous with respect to the inductive limit topology of $\mathscr{L}(\mathscr{B})$ and the norm-topology of $\mathscr{L}_p(\mu; \mathscr{B})$.

15.9. Proposition. *If $1 \leq p < \infty$, $\mathscr{L}(\mathscr{B})$ is dense in $\mathscr{L}_p(\mu; \mathscr{B})$.*

Proof. If A, f, and λ_n are as in the proof of 15.4, then evidently $\lambda_n f \rightarrow \mathrm{Ch}_A f$ in $\mathscr{L}_p(\mu; \mathscr{B})$. Therefore the $\mathscr{L}_p$-closure of $\mathscr{L}(\mathscr{B})$ contains the linear span of $\{\mathrm{Ch}_A f : A \in \mathscr{S}(X), \ f \in \mathscr{C}(\mathscr{B})\}$. From this and 4.6 the Proposition follows. ∎

15.10. Combining 14.6, 15.8, and 15.9, we get:

Proposition. *Let Γ be a linear subspace of $\mathscr{L}(\mathscr{B})$ such that (I) if $\lambda \in \mathscr{C}(X)$ and $f \in \Gamma$ then $\lambda f \in \Gamma$, and (II) for each x in X, $\{f(x) : f \in \Gamma\}$ is dense in B_x. Then Γ is dense in $\mathscr{L}_p(\mu; \mathscr{B})$ for each $1 \leq p < \infty$.*

15.11 Similarly, combining 14.10, 15.8, and 15.9, we obtain:

Proposition. *If the bundle space B is second-countable, the Banach spaces $\mathscr{L}_p(\mu; \mathscr{B})$ $(1 \leq p < \infty)$ are all separable.*

Hilbert Bundles

15.12. Suppose that $\mathscr{B}$ is a *Hilbert* bundle. By 4.7 $\mathscr{L}_2(\mu; \mathscr{B})$ is a Hilbert space, with inner product

$$(f, g) = \int \, (f(x), g(x))_{B_x} \, d\mu x.$$

It is called the *cross-sectional Hilbert space of $\mathscr{B}$ with respect to μ.*

15.13. Let $\mathscr{B}$ be a Hilbert bundle; and put $H = \mathscr{L}_2(\mu; \mathscr{B})$. For each Borel subset A of X let $P(A)$ be the projection on H given by

$$(P(A)f)(x) = \mathrm{Ch}_A(x)f(x) \qquad (f \in H; x \in X).$$

It is easily verified that P is a regular H-projection-valued Borel measure on X (11.9); its regularity follows from that of μ.

Proposition. *Suppose $\phi \in \mathscr{L}_\infty(P)$ (11.6); and form the spectral integral $E = \int \phi \, dP$ as in 11.7. Then, if $f \in H$,*

$$(Ef)(x) = \phi(x)f(x) \qquad \text{for μ-almost all x.} \tag{5}$$

Proof. This is true by 11.8(VI) if ϕ is a linear combination of characteristic functions of Borel sets; and such ϕ are dense in $\mathscr{L}_\infty(P)$. Also (5) remains true under passage to limits in $\mathscr{L}_\infty(P)$. So it holds for all ϕ in $\mathscr{L}_\infty(P)$. Details are left to the reader. ∎

Direct Sums of Hilbert Bundles

15.14. For each i in an index set I let $\mathscr{B}^i = \langle B^i, \{B^i_x\}_{x \in X} \rangle$ be a Hilbert bundle over the locally compact Hausdorff base space X. Thus for each x in X we can form the Hilbert space direct sum $B_x = \sum^{\oplus}_{i \in I} B^i_x$. Let B be the disjoint union of the B_x ($x \in X$). For each i in I let $\gamma^i \colon B^i \to B$ be the injection which coincides on each B^i_x with the natural injection of B^i_x into B_x; and let $p^i \colon B \to B^i$ be the surjection which coincides on each B_x with the projection of B_x onto B^i_x.

Proposition. *There is a unique topology $\mathscr{T}$ for B making $\mathscr{B} = \langle B, \{B_x\} \rangle$ a Hilbert bundle over X, and such that for each i the map $\gamma^i \colon B^i \to B$ is continuous. As a matter of fact, with this topology the maps γ^i are all homeomorphisms. Also, the maps $p^i \colon B \to B^i$ are continuous.*

Proof. Let Γ be the linear span of the set of all cross-sections of $\langle B, \{B_x\} \rangle$ of the form $\gamma^i \circ f$, where i runs over I and $f \in \mathscr{C}(\mathscr{B}^i)$. It is easy to verify that Γ satisfies the hypotheses of 13.18, and so determines a topology $\mathscr{T}$ for B making $\langle B, \{B_x\} \rangle$ a Hilbert bundle $\mathscr{B}$ such that $\Gamma \subset \mathscr{C}(\mathscr{B})$.

   We claim that γ^i is a homeomorphism with respect to $\mathscr{T}$. Indeed, let $\{t_j\}$ be a net of elements of B^i and t an element of B^i such that $\pi(t_j) \to \pi(t)$ (π denoting all bundle projections); and choose f in $\mathscr{C}(\mathscr{B}^i)$ so that $f(\pi(t)) = t$. By 13.12

$t_j \to t$ if and only if $\lim_j \|t_j - f(\pi(t_j))\| = 0$. Also, since $\gamma^i \circ f$ is $\mathcal{T}$-continuous, 13.12 implies that $\gamma^i(t_j) \underset{j}{\to} \gamma^i(t)$ (in $\mathcal{T}$) if and only if $\lim_j \|\gamma^i(t_j) - \gamma^i(f(\pi(t_j)))\| \to 0$. But γ^i is an isometry; so the two norm-limit statements are equivalent. Thus $t_j \to t$ if and only if $\gamma^i(t_j) \underset{j}{\to} \gamma^i(t)$, proving the claim.

Next we shall show that each p^i is $\mathcal{T}$-continuous. Let $s_j \to s$ in B. Given $\varepsilon > 0$, choose f in Γ so that $\|s - f(\pi(s))\| < \varepsilon$, whence $\|s_j - f(\pi(s_j))\| < \varepsilon$ for all large enough j. Since p^i is norm-decreasing, this gives

$$\|p^i(s) - (p^i \circ f)(\pi(s))\| < \varepsilon,$$
$$\|p^i(s_j) - (p^i \circ f)(\pi(s_j))\| < \varepsilon \tag{6}$$

for large j. Now it follows from the definition of Γ that $p^i \circ f \in \mathcal{C}(\mathcal{B}^i)$. This, (6), and 13.12 imply that $p^i(s_j) \to p^i(s)$, showing that p^i is continuous.

We have shown that $\mathcal{T}$ has all the properties mentioned in the Proposition. It remains only to see that any other topology $\mathcal{T}'$ making $\mathcal{B}$ a Hilbert bundle and making the γ^i continuous must coincide with $\mathcal{T}$. But the continuity of the γ^i implies that the elements of Γ are all $\mathcal{T}'$-continuous. Hence $\mathcal{T}' = \mathcal{T}$ by the uniqueness assertion in 13.18. ∎

Definition. The Hilbert bundle $\mathcal{B}$ constructed above is called the *Hilbert bundle direct sum of the* $\mathcal{B}^i$, and is denoted by $\sum_{i \in I}^{\oplus} \mathcal{B}^i$.

Remark 1. Local compactness was of course used in this Proposition only to ensure the existence of enough continuous cross-sections.

Remark 2. The above topology $\mathcal{T}$ can also be characterized as the smallest topology for B under which the norm-function $s \mapsto \|s\|$ and all the projections p^i are continuous. The reader can verify this for himself.

15.15. Keep the notations of 15.14; and equip B with the above topology $\mathcal{T}$ so that $\mathcal{B} = \langle B, \{B_x\} \rangle$ is a Hilbert bundle.

Suppose now that we have a regular Borel measure μ on X. The continuity assertions of 15.14 together with 15.4 imply the next two statements:

A cross-section f of $\mathcal{B}^i$ is locally μ-measurable if and only if $\gamma^i \circ f$ is locally μ-measurable as a cross-section of $\mathcal{B}$.

A cross-section f of $\mathcal{B}$ is locally μ-measurable if and only if for each i in I, $p^i \circ f$ is a locally μ-measurable cross-section of $\mathcal{B}^i$.

From these statements it is a short step to the following:

Proposition. *For each i in I, the map $\tilde{\gamma}^i: f \mapsto \gamma^i \circ f (f \in \mathcal{L}_2(\mu; \mathcal{B}^i))$ is a linear isometry of $\mathcal{L}_2(\mu; \mathcal{B}^i)$ onto the closed subspace M^i of $\mathcal{L}_2(\mu; \mathcal{B})$ consisting of those g whose range is contained in $\gamma^i(B^i)$. The M^i are pairwise orthogonal and* *span a dense subspace of $\mathcal{L}_2(\mu; \mathcal{B})$. The projection of $\mathcal{L}_2(\mu; \mathcal{B})$ onto M^i is $\tilde{\gamma}^i \circ \tilde{p}^i$ (where $\tilde{p}^i(f) = p^i \circ f$).*

Regarding $\tilde{\gamma}^i$ as an identification, we can summarize this Proposition by the equivalence:

$$\mathcal{L}_2(\mu; \mathcal{B}) = \sum_{i \in I}^{\oplus} \mathcal{L}_2(\mu; \mathcal{B}^i) \qquad \text{(Hilbert direct sum)} \qquad (7)$$

Tensor Products of Hilbert Bundles

15.16. A program similar to 15.14 and 15.15 can be carried out for tensor products.

Let X and Y be two locally compact Hausdorff spaces, and let $\mathcal{B} = \langle B, \{B_x\}_{x \in X} \rangle$ and $\mathcal{D} = \langle D, \{D_y\}_{y \in Y} \rangle$ be Hilbert bundles over X and Y respectively. (Hence $\mathcal{B}$ and $\mathcal{D}$ have enough continuous cross-sections by Appendix C). For each x in X and y in Y form the Hilbert tensor product $E_{x,y} = B_x \otimes D_y$.

Proposition. *There is a unique Hilbert bundle $\mathcal{E} = \langle E, \{E_{x,y}\} \rangle$ over $X \times Y$ whose fiber over $\langle x, y \rangle$ is $E_{x,y}$, such that $\langle x, y \rangle \mapsto f(x) \otimes g(y)$ belongs to $\mathcal{C}(\mathcal{E})$ whenever $f \in \mathcal{C}(\mathcal{B})$ and $g \in \mathcal{C}(\mathcal{D})$.*

Proof. Form the linear span Γ of the set of cross-sections of the form $\langle x, y \rangle \mapsto f(x) \otimes g(y)$ $(f \in \mathcal{C}(\mathcal{B}); g \in \mathcal{C}(\mathcal{D}))$; and invoke 13.18.   ∎

Definition. $\mathcal{E}$ is called the *(outer) tensor product* of $\mathcal{B}$ and $\mathcal{D}$.

15.17. Now let μ and ν be regular Borel measures on X and Y respectively; and form the product regular Borel measure $\mu \times \nu$ on $X \times Y$ (9.5). From the definition of $\mathcal{E}$ along with 15.4 we conclude that, if f and g are locally measurable cross-sections of $\mathcal{B}$ and $\mathcal{D}$ respectively (with respect to μ and ν), then $\langle x, y \rangle \mapsto f(x) \otimes g(y)$ is a locally $(\mu \times \nu)$-measurable cross-section of $\mathcal{E}$. We shall call this cross-section $f \times g$. If in addition $f \in \mathcal{L}_2(\mu; \mathcal{B})$ and $g \in \mathcal{L}_2(\nu; \mathcal{D})$, then by Fubini's Theorem 9.8(II)

$$\int \|(f \times g)(x, y)\|^2 \, d(\mu \times \nu)\langle x, y \rangle = \iint \|f(x)\|^2 \|g(y)\|^2 \, d\mu x \, d\nu y \qquad (8)$$

$$= \|f\|^2_{\mathcal{L}_2(\mu; \mathcal{B})} \|g\|^2_{\mathcal{L}_2(\nu; \mathcal{D})} < \infty,$$

so that $f \times g \in \mathcal{L}_2(\mu \times \nu; \mathcal{E})$. If $f, f' \in \mathcal{L}_2(\mu; \mathcal{B})$ and $g, g' \in \mathcal{L}_2(\nu; \mathcal{D})$, the same calculation (8) (based this time on Fubini's Theorem in the form 9.8(III)) gives:

$$(f \times g, f' \times g')_{\mathcal{L}_2(\mu \times \nu; \mathcal{E})} = (f, f')_{\mathcal{L}_2(\mu; \mathcal{B})}(g, g')_{\mathcal{L}_2(\nu; \mathcal{D})}. \tag{9}$$

From (9) we conclude that the equation

$$\Phi(f \otimes g) = f \times g \qquad (f \in \mathcal{L}_2(\mu; \mathcal{B}); g \in \mathcal{L}_2(\nu; \mathcal{D}))$$

defines a linear isometry Φ of the Hilbert tensor product $\mathcal{L}_2(\mu; \mathcal{B}) \otimes \mathcal{L}_2(\nu; \mathcal{D})$ into $\mathcal{L}_2(\mu \times \nu; \mathcal{E})$.

Proposition.   *The range of Φ is all of $\mathcal{L}_2(\mu \times \nu; \mathcal{E})$.*

Proof.   Let $\mathcal{W}_0$ be the linear span of $\{f \times g : f \in \mathcal{L}(\mathcal{B}), g \in \mathcal{L}(\mathcal{D})\}$; and let $\mathcal{W}$ be the closure of $\mathcal{W}_0$ in $\mathcal{L}(\mathcal{E})$ with respect to the inductive limit topology. Let $\mathcal{T}$ be the linear span in $\mathcal{L}(X \times Y)$ of $\{\phi \times \psi : \phi \in \mathcal{L}(X), \psi \in \mathcal{L}(Y)\}$. Then $\mathcal{T}$ is closed under pointwise multiplication and complex conjugation and separates points of $X \times Y$, and so by the Stone-Weierstrass Theorem (Appendix A) is dense in $\mathcal{C}(X \times Y)$ in the topology of uniform convergence on compact sets. Notice also that $\mathcal{W}_0$ is closed under multiplication by functions in $\mathcal{T}$. These two facts, together with the separate continuity of χh in χ and h $(\chi \in \mathcal{C}(X \times Y); h \in \mathcal{L}(\mathcal{E}))$, imply that $\mathcal{W}$ is closed under multiplication by arbitrary continuous complex functions on $X \times Y$. Evidently $\{h(x, y) : h \in \mathcal{W}_0\}$ is dense in $E_{x,y}$ for each x, y. Thus we can apply 15.10 to conclude that $\mathcal{W}$ is dense in $\mathcal{L}_2(\mu \times \nu; \mathcal{E})$. By 15.8 $\mathcal{W}_0$ is $\mathcal{L}_2(\mu \times \nu; \mathcal{E})$-dense in $\mathcal{W}$. So $\mathcal{W}_0$ is dense in $\mathcal{L}_2(\mu \times \nu; \mathcal{E})$. On the other hand $\mathcal{W}_0 \subset \text{range}(\Phi)$. Therefore range($\Phi$) is dense in, and hence equal to, $\mathcal{L}_2(\mu \times \nu; \mathcal{E})$.   ∎

Remark.   In view of this result we can regard $\mathcal{L}_2(\mu \times \nu; \mathcal{E})$ as identified with the Hilbert tensor product $\mathcal{L}_2(\mu; \mathcal{B}) \otimes \mathcal{L}_2(\nu; \mathcal{D})$ via Φ.

In particular, if $\mathcal{B}$, $\mathcal{D}$, and hence $\mathcal{E}$ are the trivial Banach bundles with constant fiber $\mathbb{C}$, we obtain the well-known fact that

$$\mathcal{L}_2(\mu \times \nu) \cong \mathcal{L}_2(\mu) \otimes \mathcal{L}_2(\nu).$$

Vector Integrals and a Continuity Lemma

15.18.   Suppose now that $\mathcal{B}$ is a trivial Banach bundle over X with constant fiber A (13.6). Let μ be a regular complex Borel measure on X. Since the definitions 5.1 and 15.2 of local $|\mu|$-measurability coincide (see 15.3), we are at liberty to take the A-valued vector integral $\int f \, d\mu$ (see 5.4) of any function in

$\mathcal{L}_1(\mu; A)$. In particular $\int f \, d\mu$ exists (in A) whenever $f \in \mathcal{L}(\mathcal{B})$, i.e., whenever $f: X \to A$ is continuous with compact support.

15.19. One obtains an interesting result by applying 15.18 to the following particular context. As in 15.1 let $\mathcal{B} = \langle B, \pi \rangle$ be an arbitrary Banach bundle over the locally compact Hausdorff space X. Take another locally compact Hausdorff space Y; let $p: X \times Y \to X$ be the surjection $\langle x, y \rangle \mapsto x$; and form the Banach bundle retraction $\mathcal{D} = \langle D, \rho \rangle$ of $\mathcal{B}$ by p as in 13.7. Thus $\mathcal{D}$ is a Banach bundle over $X \times Y$ whose bundle space D can be identified with $B \times Y$, the bundle projection being $\rho: \langle s, y \rangle \mapsto \langle \pi(s), y \rangle$ $(s \in B; y \in Y)$.

Let v be a regular complex Borel measure on Y. Take an element f of $\mathcal{L}(\mathcal{D})$; we can regard f as a continuous function with compact support on $X \times Y$ to B, with $f(x, y) \in B_x$ for each $\langle x, y \rangle$. Thus, choosing a compact subset K of X such that $K \times Y$ contains the compact support of f, for each y in Y we obtain an element $F_y: x \mapsto f(x, y)$ of the space $\mathcal{L}^K(\mathcal{B})$ defined in 14.3.

We claim that the map $F: y \mapsto F_y$ of Y into $\mathcal{L}^K(\mathcal{B})$ is continuous. Indeed: Take $y_0 \in Y$ and $\varepsilon > 0$. Since the numerical function $\langle x, y \rangle \mapsto \| f(x, y) - f(x, y_0) \|$ is continuous on $X \times Y$ and vanishes for $y = y_0$, there must exist a neighborhood U of y_0 such that $\| f(x, y) - f(x, y_0) \| < \varepsilon$ whenever $y \in U$ and x is in the compact set K. But this says that $\| F_y - F_{y_0} \|_\infty \leq \varepsilon$ for $y \in U$, proving the claim.

Now notice that $\mathcal{L}^K(\mathcal{B})$ is a closed subspace of $\mathcal{C}(\mathcal{B}_K)$, hence a Banach space in its own right under the supremum norm $\| \; \|_\infty$ (see 14.3). Since F evidently has compact support, the preceding paragraph, together with the last statement of 15.18, permits us to form the $\mathcal{L}^K(\mathcal{B})$-valued vector integral

$$g = \int F_y \, dvy \in \mathcal{L}^K(\mathcal{B}). \tag{10}$$

Now notice that for any x in X the "evaluation map" $h \mapsto h(x)$ on $\mathcal{L}^K(\mathcal{B})$ to B_x is continuous. On applying this map to (10), we obtain by 5.7

$$g(x) = \int f(x, y) dvy \qquad (B_x\text{-valued integral}). \tag{11}$$

Equation (11) shows what we certainly would suspect—that the g of (10) is independent of the particular compact set K. Thus, recalling the definition 6.2 of vector integrals with respect to arbitrary locally convex spaces, we can write (10) in a manner independent of K:

$$g = \int F_y \, dvy, \tag{12}$$

the right side being a vector integral with respect to $\mathscr{L}(\mathscr{B})$ equipped with the inductive limit topology.

Equations (11) and (12) imply the following continuity assertion:

Proposition. *As above, let f be any element of $\mathscr{L}(\mathscr{D})$. For each x in X we can form the B_x-valued integral $\int f(x, y)dvy$. The cross-section*

$$x \longmapsto \int f(x, y)dvy$$

of $\mathscr{B}$ is then continuous.

This proposition will be vital at several points in this work (beginning with the generalized Fubini Theorems which follow).

16. Fubini Theorems for Banach Bundles

16.1.   The context of Banach bundles provides rather elegant generalizations of the Fubini Theorem.

Let X and Y be two locally compact Hausdorff spaces, and $\mathscr{B} = \langle B, \pi \rangle$ a Banach bundle over $X \times Y$. Each point y of Y gives rise naturally to a Banach bundle $\mathscr{B}^y = \langle B^y, \pi^y \rangle$ over X, namely the Banach bundle retraction of $\mathscr{B}$ by the injection $x \mapsto \langle x, y \rangle$ of X into $X \times Y$ (see 13.7). We can identify B^y with the subspace $\bigcup_x B_{x,y}$ of B, the fiber $(B^y)_x$ of $\mathscr{B}^y$ being the same as the fiber $B_{x,y}$ of $\mathscr{B}$. If f is a cross-section of $\mathscr{B}$ and $y \in Y$, f^y will denote the cross-section $x \mapsto f(x, y)$ of $\mathscr{B}^y$.

Let μ be a regular Borel measure on X; and fix a number $1 \le p < \infty$. There is a natural way to build a Banach bundle $\mathscr{L}^p$ over Y whose fiber at y is $\mathscr{L}_p(\mu; \mathscr{B}^y)$. Indeed: Let $\| \ \|_p$ be the norm of $\mathscr{L}_p(\mu; \mathscr{B}^y)$. If $f \in \mathscr{L}(\mathscr{B})$, notice that $y \mapsto \|f^y\|_p$ is continuous on Y (by 15.19 applied to numerical functions). Also, by 15.9 and 14.8, $\{f^y : f \in \mathscr{L}(\mathscr{B})\}$ is dense in $\mathscr{L}_p(\mu; \mathscr{B}^y)$ for each y. Therefore by 13.18 there is a unique Banach bundle $\mathscr{L}^p$ over Y whose fiber $(\mathscr{L}^p)_y$ at y is $\mathscr{L}_p(\mu; \mathscr{B}^y)$ and relative to which the cross-section $y \mapsto f^y$ is continuous for every f in $\mathscr{L}(\mathscr{B})$.

Now, in addition, suppose that v is a regular Borel measure on Y, and form the regular product $\mu \times v$ as in 9.5. Our first variant of Fubini's Theorem will assert that $\mathscr{L}_p(\mu \times v; \mathscr{B})$ is essentially the same Banach space as $\mathscr{L}_p(v; \mathscr{L}^p)$.

Theorem.

(I)   Let $f \in \mathscr{L}_p(\mu \times v; \mathscr{B})$. Then $f^y \in \mathscr{L}_p(\mu; \mathscr{B}^y)$ for v-almost all y. Let $\tilde{f}$ be the cross-section $y \mapsto f^y$ of $\mathscr{L}^p$ so obtained (defined except on a v-null set).

(II) $\tilde{f} \in \mathscr{L}_p(v; \mathscr{L}^p)$ for each f in $\mathscr{L}_p(\mu \times v; \mathscr{B})$.

(III) The map $f \mapsto \tilde{f}$ is a linear isometry of $\mathscr{L}_p(\mu \times v; \mathscr{B})$ onto $\mathscr{L}_p(v; \mathscr{L}^p)$.

Proof.   (I) By 15.9 there is a sequence $\{f_n\}$ of elements of $\mathscr{L}(\mathscr{B})$ converging to f in $\mathscr{L}_p(\mu \times v; \mathscr{B})$. Passing to a subsequence, we can assume by 3.5 that $f_n(x, y) \to f(x, y)$ (in $B_{x, y}$) for $(\mu \times v)$-almost all $\langle x, y \rangle$. By 9.7 there is a μ-null subset N of Y such that, for all y in $Y \setminus N$, $f_n(x, y) \to f(x, y)$ for μ-almost all x. By the continuity of the f_n, this implies by 15.4 that f^y is locally μ-measurable for all $y \in Y \setminus N$. Also, by the ordinary Fubini Theorem 9.8(III) applied to the numerical function $\langle x, y \rangle \mapsto \|f(x, y)\|^p$, we deduce that $x \mapsto \|f^y(x)\|^p$ is μ-summable for v-almost all y. From these facts and Definition 4.5 (I) follows.

   (II) Let f and the $\{f_n\}$ be as in the proof of (I); and put $\phi_n(x, y) = \|f_n(x, y) - f(x, y)\|^p$, $\psi_n(y) = \int \phi_n(x, y) d\mu x$. By the definition of $\{f_n\}$ we have $\phi_n \to 0$ in $\mathscr{L}_1(\mu \times v)$; and so $\psi_n \to 0$ in $\mathscr{L}_1(v)$ by 9.8(III). Thus by 3.5 we can pass to a subsequence and assume that $\psi_n(y) \to 0$ for v-almost all y. But this says that $(f_n)^y \to f^y$ in $\mathscr{L}_p(\mu; \mathscr{B}^y)$ for v-almost all y. Now $y \mapsto (f_n)^y$ is a continuous cross-section of $\mathscr{L}^p$ (by the definition of the latter). Therefore by 15.4 $\tilde{f}: y \mapsto f^y$ is locally v-measurable. Further, Fubini's Theorem 9.7(II) gives

$$\int_Y \|f^y\|_p^p \, dvy = \int \|f(x, y)\|^p \, d(\mu \times v) \langle x, y \rangle < \infty. \tag{1}$$

By Definition 4.4 these facts imply (II).

   (III) By (II) and (1) the map $f \mapsto \tilde{f}$ is a linear isometry. It remains only to show that it is surjective. Now we have already observed (while constructing $\mathscr{L}^p$) that $\{f^y : f \in \mathscr{L}(\mathscr{B})\}$ is dense in $(\mathscr{L}^p)_y = \mathscr{L}_p(\mu; \mathscr{B}^y)$ for each y. Obviously $\{\tilde{f} : f \in \mathscr{L}(\mathscr{B})\}$ is closed under multiplication by continuous complex functions on Y. Therefore by 15.10 the range of $f \mapsto \tilde{f}$ is dense in $\mathscr{L}_p(v; \mathscr{L}^p)$. Since $f \mapsto \tilde{f}$ is an isometry and its domain is complete, its range is in fact all of $\mathscr{L}_p(v; \mathscr{L}^p)$.   ∎

16.2.   We now specialize the context of 16.1. Let X, Y, μ, v be as in 16.1; and suppose that the $\mathscr{B}$ of 16.1 is the Banach bundle retraction of $\mathscr{D}$ by the surjection $\langle x, y \rangle \mapsto y$ ($\langle x, y \rangle \in X \times Y$), where $\mathscr{D} = \langle D, \rho \rangle$ is some Banach bundle over Y. In that case the $\mathscr{B}^y$ of 16.1 is just the trivial Banach bundle over X with constant fiber D_y.

Let μ' be any regular complex Borel measure on X with $|\mu'| \le \mu$.

Proposition. *Suppose that $f \in \mathcal{L}_1(\mu \times v; \mathcal{B})$. By Theorem 16.1 there is a v-null subset N of Y such that $f^y \in \mathcal{L}_1(\mu; D_y)$ for all $y \in Y \setminus N$. If we put*

$$g(y) = \int_X f^y \, d\mu' \in D_y \qquad \text{for } y \in Y \setminus N,$$

then $g \in \mathcal{L}_1(v; \mathcal{D})$.

Proof. We shall first show that g is locally v-measurable. As in the proof of Theorem 16.1(II) we can find a sequence $\{f_n\}$ of elements of $\mathcal{L}(\mathcal{B})$ such that $(f_n)^y \to f^y$ in $\mathcal{L}_1(\mu; D_y)$ for v-almost all y. Since the integration process $h \mapsto \int h \, d\mu'$ is continuous on $\mathcal{L}_1(\mu; D_y)$ to D_y, this implies that $\int (f_n)^y \, d\mu' \to \int f^y \, d\mu' = g(y)$ in D_y for v-almost all y. On the other hand, by 15.19 $y \mapsto \int (f_n)^y \, d\mu'$ is a continuous cross-section of $\mathcal{D}$ for each n. Hence g is locally v-measurable by 15.4.

Since $h \mapsto \int h \, d\mu'$ is norm-decreasing by 5.4(2), we have

$$\|g(y)\| \le \|f^y\|_{\mathcal{L}_1(\mu; D_y)} \qquad \text{for } v\text{-almost all } y. \tag{2}$$

Combining (2) with the local v-measurability of g and the fact that $y \mapsto f^y$ is in $\mathcal{L}_1(v; \mathcal{L}^1)$ (16.1), we conclude that $g \in \mathcal{L}_1(v; \mathcal{D})$. ∎

16.3. The next Theorem is the classical Fubini Theorem for Banach bundles. Here we assume that X, Y, μ, and v are as in 16.1, and that $\mathcal{B}$ is the trivial Banach bundle over $X \times Y$ with constant fiber A. Let μ' and v' be regular complex Borel measures on X and Y respectively with $|\mu'| \le \mu$, $|v'| \le v$. By 9.10 and 9.6

$$|\mu' \times v'| \le \mu \times v. \tag{3}$$

Fubini Theorem. *Suppose that $f \in \mathcal{L}_1(\mu \times v; A)$. Then:*

(I) *For v-almost all y in Y the function $x \mapsto f(x, y)$ belongs to $\mathcal{L}_1(\mu; A)$.*
(II) *The function $y \mapsto \int_X f(x, y) d\mu' x$ belongs to $\mathcal{L}_1(v; A)$.*
(III) $\int_Y \int_X f(x, y) d\mu' x \, dv' y = \int_{X \times Y} f(x, y) d(\mu' \times v')\langle x, y \rangle.$

Similar statements hold with X and Y interchanged, ending with the equality

$$\int_X \int_Y f(x, y) dv' y \, d\mu' x = \int_{X \times Y} f(x, y) d(\mu' \times v')\langle x, y \rangle.$$

Proof. (I) and (II) follow from 16.1 and 16.2. Hence by (3) both sides of (III) exist; and we have only to prove their equality. Since A^* separates points in A, it is enough to take an element α of A^* and show that application of α to both sides of (III) gives the same result. By 5.7 this amounts to showing that

$$\int_Y \int_X \alpha(f(x, y))d\mu'x\, dv'y = \int_{X \times Y} \alpha(f(x, y))d(\mu' \times v')\langle x, y\rangle. \qquad (4)$$

But this is the numerical Fubini Theorem 9.12. ■

16.4. *Remark.* In applications of Fubini's Theorem it is sometimes a nontrivial matter to decide whether or not a given function $f: X \times Y \to A$ is locally $(\mu \times v)$-measurable. As a first step in this direction, notice that, if $\phi: X \to A$ is locally μ-measurable, then $f: \langle x, y\rangle \mapsto \phi(x)$ is locally $(\mu \times v)$-measurable. This follows easily from 15.4.

16.5. Theorem 16.3 remains true when μ', v', and $\mu' \times v'$ are replaced by their regular maximal extensions (8.15). This is proved in the same way as the corresponding fact for numerical functions (Theorem 9.16). Keeping the notation of 16.3, we thus obtain:

Fubini Theorem for Maximal Regular Extensions.

Suppose that $f \in \mathscr{L}_1((\mu \times v)_e; A)$. Then:

(I)   *For v_e-almost all y in Y the function $x \mapsto f(x, y)$ belongs to $\mathscr{L}_1(\mu_e; A)$.*
(II) *The function $y \mapsto \int f(x, y)d(\mu')_e x$ belongs to $\mathscr{L}_1(v_e; A)$.*
(III) $\int_Y \int_X f(x, y)d(\mu')_e x\, d(v')_e y = \int_{X \times Y} f(x, y)d(\mu' \times v')_e \langle x, y\rangle.$

The same holds, of course, when X and Y are interchanged.

17.   Exercises for Chapter II

1. Give a detailed proof of Proposition 1.3. Show that the total variation $|\mu|$ is the unique smallest measure on the δ-ring $\mathscr{S}$ which majorizes μ.

2. Prove that the normed linear space $\mathscr{M}_b(\mathscr{S})$ of bounded complex measures on a δ-ring $\mathscr{S}$ under the total variation norm is complete; i.e., $\mathscr{M}_b(\mathscr{S})$ is a Banach space (see 1.5).

3. Prove that if μ is a σ-bounded complex measure, then μ is parabounded (see 1.7).

4. Let μ be a complex measure on a δ-ring $\mathscr{S}$ of subsets of a set X. Verify that if two complex valued functions on X coincide locally μ-almost everywhere, and one is locally μ-measurable, then so is the other (see 1.9).

5. Show that $\tilde{\mu}$, as defined in 1.12 by equation (4), is a complex measure which is independent of the sequence $\{A_n\}$ used to define it.

6. Show that if X is the space of bounded Lebesgue measurable real functions on the unit interval $[0, 1]$, then convergence almost everywhere is not equivalent to convergence in any topology on X.

7. Give detailed proofs of Proposition 2.1 and of Proposition 2.2.

8. Show that the integral $\int f\, d\mu$, as defined by (4), in 2.5 does not depend on the particular choice of the functions f_i.

9. Prove Hölder's inequality 2.6.

10. Prove Proposition 2.11.

11. Fill in the details of the proof of Proposition 5.7.

12. Prove Proposition 5.9.

13. Give a complete proof of Proposition 5.12.

14. Verify that Proposition 6.3 is true.

15. Let $f: X \to \mathbb{C}$ be locally μ-summable. Show that $f\, d\mu \ll \mu$ (see 7.8).

16. Show that the hypothesis of paraboundedness cannot be omitted from the statement of the Radon-Nikodym Theorem (write out the details of the example referred to in Remark 2 of 7.8.)

17. Show that $\mathscr{L}_\infty(\mu)$ is a Banach space with respect to the essential supremum norm $\| \; \|_\infty$ (see 7.10).

18. Show that if $\mu(X) < \infty$ and $f \in \mathscr{L}_\infty(\mu)$ (see 7.10), then $f \in \mathscr{L}_p(\mu)$ for every $p > 0$ and $\|f\|_\infty = \lim_{p \to \infty}(\int_x |f|^p\, d\mu)^{1/p}$.

19. Prove that the Banach space $\mathscr{L}_\infty(\mu)$ (7.10) is not separable unless it is finite-dimensional.

20. Verify in the proof of 7.11 that $\|f\|_\infty \leq \|\alpha\|$.

21. Prove that if $1 < p < \infty$ and $p^{-1} + q^{-1} = 1$, then $(\mathscr{L}_p(\mu))^* \cong \mathscr{L}_q(\mu)$ (see Remark 7.11).

22. Let X be a locally compact Hausdorff space and $\mathscr{S}(X)$ the compacted Borel δ-ring of X (see 8.2). Show that if μ is a complex measure on $\mathscr{S}(X)$, then the following are equivalent:

(a) For each $A \in \mathscr{S}(X)$ and each $\varepsilon > 0$, there is a compact subset C of X such that $C \subset A$ and $|\mu|(A \setminus C) < \varepsilon$.

(b) For each $A \in \mathscr{S}(X)$ there is a countable union D of compact sets such that $D \subset A$ and $|\mu|(A \setminus D) = 0$.

23. Give a detailed proof of Proposition 8.7.

24. Show that the closed support of a regular complex Borel measure μ on a locally compact Hausdorff space X (see 8.9) is the unique largest locally μ-null open subset of X.

25. Verify that a measure μ belongs to $\mathscr{M}_r(X)$ if and only if the corresponding complex integral I is continuous with respect to the supremum norm of $\mathscr{L}(X)$ (see the paragraph preceding Lemma 8.13).

26. Prove that if μ is a bounded regular complex Borel measure on X, A is a Banach space, and $f: X \to A$ is any bounded continuous function, then $f \in \mathscr{L}_1(\mu_e; A)$ (see 8.15, Remark 3).

27. Prove Proposition 8.17.

28. Verify that R, as defined in Remark 1 of 9.2, is a σ-field of subsets of $X \times X$ which contains W but not the diagonal D.

29. Show that if X and Y are locally compact Hausdorff spaces which satisfy the second axiom of countability, then $X \times Y$ also has these properties. Further, show that every compact subset of $X \times Y$ is a G_δ-set.

30. Show in the proof of Proposition 9.15 that the function ϕ_f defined by $\phi_f(x) = \int f(x, y)d|v|y$ is continuous.

31. Extend the results of §9 to products of an arbitrary finite number of locally compact Hausdorff spaces as described in 9.18.

32. Verify the statements made in Remark 10.11.

33. Show that the linear mapping I defined by equation (4) in 11.7 is legitimate by verifying the details of the argument sketched there.

34. Give a complete proof of all parts of Proposition 11.8.

35. Prove Propositions 11.16 through 11.23.

36. Prove the sufficiency in Proposition 13.2.

37. Let $\mathscr{B}$ be a Banach bundle over a locally compact Hausdorff space X, whose fibers are all of the same *finite* dimension.

 (i) Prove that $\mathscr{B}$ is locally trivial.
 (ii) Is B necessarily *isometrically locally trivial* (in the sense that each x in X has a neighborhood U such that the reduction of B to U is isometrically isomorphic with a trivial bundle over U)?

38. Show, in the proof of 13.18, that the mapping ρ is continuous and open with respect to the topology $\mathscr{T}$ for A, and that 13.4(i), (iii) hold.

39. Verify in 13.20 that the mapping $x \mapsto \|\psi(x)\|_B$ is continuous on X for each $\psi \in \Gamma$ and that $\{\psi(x): \psi \in \Gamma\}$ is dense in B_x for each $x \in G$.

40. Let X and S be two topological spaces, and $\pi: S \to X$ a continuous surjection. Show that π is open if and only if $\pi^{-1}(\bar{Y}) = \pi^{-1}(Y)^-$ for every subset Y of X.

41. Let $\mathscr{B} = \langle B, \pi \rangle$ be a Banach bundle over X having enough continuous cross-sections. For each $x \in X$ let D_x be a closed linear subspace of the fiber B_x over x; and let us denote by Γ the set of all those continuous cross-sections f of $\mathscr{B}$ such that $f(x) \in D_x$ for all $x \in X$. We shall say that $\{D_x\}$ is a *lower semi-continuous choice of subspaces* if $\{f(x): f \in \Gamma\}$ is dense in D_x for each $x \in X$.

Assume that $\{D_x\}$ is a lower semi-continuous choice of subspaces, and let $D = \{b \in B: b \in D_{\pi(b)}\}$ carry the relativized topology of B. Show that $\langle D, \pi | D \rangle$ is a Banach bundle over X. It is called the *Banach subbundle of $\mathscr{B}$ with fibers* $\{D_x\}$.

42. Show that the family U of convex subsets in $\mathscr{L}(\mathscr{B})$ described in the proof of Proposition 14.3 is a local basis.

43. Prove that the mapping $y \mapsto \|g(y)\|_\infty$ in the proof of Proposition 14.9 is continuous on Y for each $g \in \Gamma$.

44. Show the existence of the countable family F of real-valued functions in $\mathscr{L}(X)$ described in the proof of 14.10.

45. Prove Propositions 14.12 through 14.16 concerning the inductive limit topology.

46. Complete the proof of Proposition 15.13.

47. Verify Remark 2 of 15.14.

48. Prove, in 15.15, the two statements concerning local μ-measurability of cross-sections, and also prove the proposition stated there.

49. Let $f \in \mathscr{L}(\mathscr{D})$, and, for each x in X, form the B_x-valued integral $\int f(x, y)dvy$. Prove that the cross-section

$$x \mapsto \int f(x, y)dvy$$

of $\mathscr{B}$ is continuous (see Proposition 15.19).

50. Prove the assertion made in Remark 16.4.

51. Let $\mathscr{B} = \langle B, \{B_x\}_{x \in X}\rangle$ be a Banach bundle over the locally compact Hausdorff space X; and let μ be a regular Borel measure on X. Let us define L to be the linear space of all those locally μ-measurable cross-sections f of $\mathscr{B}$ which are μ-essentially bounded, i.e., such that there is a non-negative number k such that $\{x \in X : \|f(x)\| > k\}$ is locally μ-null. For each $f \in L$ the smallest k will be denoted by $\|f\|_\infty$.

Prove that L, $\| \ \|_\infty$ is a Banach space (provided we identify two cross-sections which differ only on a locally μ-null set).

This L, $\| \ \|_\infty$ is denoted by $\mathscr{L}_\infty(\mu; \mathscr{B})$; it is called the $\mathscr{L}_\infty$ cross-sectional space of $\mathscr{B}$. If $\mathscr{B}$ is the trivial Banach bundle with fiber $\mathbb{C}$, then of course $\mathscr{L}_\infty(\mu; \mathscr{B})$ coincides with the $\mathscr{L}_\infty(\mu)$ of 7.10.

52. Let $\mathscr{B} = \langle B, \pi\rangle$ and $\mathscr{E} = \langle E, \rho\rangle$ be two Banach bundles over the same locally compact Hausdorff space X (having fibers B_x and E_x respectively), such that:

(i) For each $x \in X$, E_x is the adjoint space $(B_x)^*$ of B_x;

(ii) The map $\langle \alpha, b\rangle \mapsto \alpha(b)$, defined on $D = \{\langle \alpha, b\rangle \in E \times B : \rho(\alpha) = \pi(b)\}$, is continuous on D in the relativized topology of $E \times B$.

Let μ be a regular Borel measure on X. Prove:

(A) If $f \in \mathscr{L}_\infty(\mu; \mathscr{E})$ and $\phi \in \mathscr{L}_1(\mu; \mathscr{B})$, then the function $x \mapsto f(x)[\phi(x)](x \in X)$ is μ-summable, and in fact

$$\left| \int_X f(x)[\phi(x)]d\mu x \right| \le \|f\|_\infty \|\phi\|_1.$$

(B) In view of (A) the equation

$$F_f(\phi) = \int_X f(x)[\phi(x)]d\mu x \qquad (f \in \mathscr{L}_\infty(\mu; \mathscr{E}), \phi \in \mathscr{L}_1(\mu; \mathscr{B}))$$

defines a norm-decreasing linear map $F : \mathscr{L}_\infty(\mu; \mathscr{E}) \to (\mathscr{L}_1(\mu; \mathscr{B}))^*$. Prove that F is actually an isometry. [Hint: Use 15.5.]

(C) Show that (unlike the "classical case" 8.7) the F of (B) is *not* in general *onto* $(\mathscr{L}_1(\mu; \mathscr{B}))^*$. [Hint: Let $X = \mathbb{R}$, let $\mathscr{B}$ be the trivial bundle over $\mathbb{R}$ with constant fiber $\mathscr{L}_1(\lambda)$, where λ is Lebesgue measure on $\mathbb{R}$, and let $\mathscr{E}$ be the trivial bundle over $\mathbb{R}$ with constant fiber $\mathscr{L}_\infty(\lambda)$. Then conditions (i) and (ii) above are satisfied. Use Fubini's Theorem to identify $\mathscr{L}_1(\mu; \mathscr{B})$ with $\mathscr{L}_1(\lambda \times \lambda)$, where

$\lambda \times \lambda$ is Lebesgue measure on the plane. Let g be the "diagonal" function on the plane given by

$$g(s, t) = \begin{cases} 1 & \text{if } |s - t| \leq 1 \\ 0 & \text{if } |s - t| > 1, \end{cases}$$

so that $g \in \mathscr{L}_\infty(\lambda \times \lambda)$, and let γ be the corresponding element of $(\mathscr{L}_1(\lambda \times \lambda))^* \cong (\mathscr{L}_1(\lambda; \mathscr{B}))^*$. Show that γ cannot be of the form $\gamma = F_f$ for any $f \in \mathscr{L}_\infty(\lambda; \mathscr{E})$.]

53. One might wonder whether Condition (II) could be omitted in Proposition 15.6, i.e., whether Condition (I) by itself is sufficient for local μ-measurability of f. (A) Give an example showing that in general (II) cannot be omitted. (B) Prove that, if the bundle space B is second-countable, then (II) is automatically satisfied by all cross-sections f, and so can be omitted.

Notes and Remarks

The purpose of this chapter has been to develop the general measure-theoretic tools which are necessary for the remainder of this work. There are many expositions of measure and integration in the literature. General references are Bourbaki [10, 11], Dinculeanu [1, 2], Dunford and Schwartz [1, 2], R. E. Edwards [1], Halmos [1], Hewitt and Ross [1], Hille and Phillips [1], Loomis [2], Nachbin [1], Naimark [8], Stone [3], and Williamson [2]. Excellent historical notes and further references on integration theory are contained in Bourbaki [10], Dunford and Schwartz [1], Hawkins [1], and Hewitt and Ross [1].

The terminology "δ-ring" introduced in §1 was taken from a forthcoming book on measure theory by Kelley and Srinavasan [1] where the same approach is adopted. δ-rings, under the name "semi-tribe" were utilized extensively in 1967 by Dinculeanu [1] in his treatment of vector measures. The basic techniques and methods of sections 1 through 11 are essentially a synthesis of known techniques which are developed in one form or another in the above references.

Projection-valued measures and spectral integrals have been treated extensively in the literature; see, for example, Dunford and Schwartz [2], Halmos [2], and Rudin [2]. Dunford and Schwartz [2] give historical notes and further references.

Among the first authors to consider the idea of Banach spaces which vary continuously over a topological space were Godement [6, 7], and Kaplansky [6]. Godement introduced what he called "continuous fields of Banach spaces" in order to study group representations. In essence he axiomatized the notion of a sub-direct continuous representation of Banach spaces. In 1961 Fell [3] studied so-called "algebras of operator fields" and used them to determine the structure of the group C^*-algebra of $SL(2, \mathbb{C})$. This paper essentially contained the definition of Banach bundle as given in §13 although the term "Banach bundle" was not explicitly used. The papers of Tomiyama and Takesaki [1], Tomiyama [1], and Dixmier and Douady [1] also contain significant bundle-theoretic results.

In 1968-1969 Dauns and Hofmann [1, 2] made important advances both in the theory of Banach bundles themselves as well as in the representation of various topological-algebraic structures in the space of cross-sections of these bundles. Hofmann [2, 3, 4] continued and extended this work in several ways. The main difference between the bundles considered by Hofmann and those defined in §13 is that he merely assumes that the function $s \mapsto \|s\|$ in (i) of 13.4 is upper-semi-continuous rather than continuous (the important result of A. Douady and L. Dal Soglio-Hérault discussed in Appendix C remains valid for such bundles; see C18).

Tensor products of Hilbert bundles were treated in 15.16; tensor products of general Banach bundles have been studied by Kitchen and Robbins [1]. Gierz [1] has given a comprehensive introduction to bundles of topological linear spaces and their duality. A closed graph theorem for Banach bundles is proved by Baker [1]. Further results dealing with various aspects of bundle theory and Banach bundles are given in Dupré [1, 2, 3, 4], Dupré and Gillette [1], Evans [1], Fell [13, 14, 15, 17], Fourman, Mulvey, and Scott [1], Gelbaum [1, 2, 3], Gelbaum and Kyriazis [1], Hofmann and Liukkonen [1], Schochetman [9, 10], Seda [4, 5, 6], and Varela [1, 2].

III Locally Compact Groups

The first four sections of this chapter are devoted to the basic properties and constructions of topological groups.

§5 deals mostly with central group extensions, whose importance to the theory of group representations will emerge in the last chapter of the work.

Some very interesting topological groups come from topological fields, whose elementary properties are given in §6. At the end of §6 we sketch the construction of the most important disconnected locally compact fields, namely, the p-adic fields.

The importance of the class of locally compact groups stems from the fact that on locally compact groups there exists a unique left-invariant (Haar) measure. In §7 we give the classical proof of this fact; and in §8 we relate left-invariant and right-invariant measures by means of the modular function. §9 studies the behavior of Haar measure in semidirect products and in topological fields.

As we pointed out in the Introduction, the most fruitful structures on which to do harmonic analysis and group representation theory are linear spaces. Thus the first step in the application of harmonic analysis to any group or group action is usually to "linearize" the situation. With this in mind, in §10 we study the space $\mathscr{M}(G)$ of all regular complex Borel measures on a locally compact group G. This space can be regarded as a (rather large) "linearization" of G, since it is itself a linear space and contains the group elements (if we identify each x in G with the unit mass at x). We introduce into $\mathcal{M}(G)$ the operations of convolution and involution; these are the natural extensions to $\mathcal{M}(G)$ of the product and inverse in G. The subspace $\mathcal{M}_r(G)$ of *bounded* complex measures becomes a Banach *-algebra under these operations and the total variation norm; it is the so-called measure algebra of G.

For many purposes the measure algebra is too big. It does not adequately reflect the topology of G; for one thing, the operation of translating a measure is not in general continuous (see 11.11). For harmonic analysis a much more useful algebra is the ideal of $\mathcal{M}_r(G)$ consisting of those μ which are absolutely continuous with respect to Haar measure λ. This can also be described in terms of $\mathcal{L}_1(\lambda)$. It is called the $\mathcal{L}_1$ group algebra, and is studied in §11. In Chapter VIII we shall show that the unitary representation theories of G and of its $\mathcal{L}_1$ group algebra exactly coincide—a fact which is far from being true for the measure algebra.

If H is a closed subgroup of the locally compact group G, great interest attaches to the invariance properties of measures on the quotient space G/H. In §13 we shall see how to construct certain measures on G/H from so-called rho-functions on G; and we shall find a necessary and sufficient condition for the existence of a measure on G/H which is invariant under the action of G. In §14 we shall prove a result which is fundamental to the theory of induced representations: Whether or not there exists an invariant measure on G/H, there always exists a unique invariant equivalence class of measures on G/H (two measures being called equivalent if they have the same null sets).

1.   Topological Groups and Subgroups

1.1.   Definition.   A *topological group* is a triple $\langle G, \cdot, \mathcal{T} \rangle$, where (a) $\langle G, \cdot \rangle$ is a group, (b) $\langle G, \mathcal{T} \rangle$ is a Hausdorff topological space, and (c) the map $\langle x, y \rangle \mapsto x \cdot y^{-1}$ of $G \times G$ into G is continuous with respect to $\mathcal{T}$.

We shall normally refer to a topological group by a single letter, say G, without explicitly naming either the group operation or the topology. It is then understood that the group product of x and y is to be denoted by xy, and the inverse of x by x^{-1}. The unit element is usually called e. However, when speaking of an *additive* group, we call the group operation $+$, the inverse of x is $-x$, and the unit element is 0.

Condition (c) of the above definition can be replaced by the following two conditions: (c') the product map $\langle x, y \rangle \mapsto xy$ is continuous on $G \times G$ to G; (c'') the inverse map $x \mapsto x^{-1}$ is continuous on G to G.

1.2. Fix a topological group G, with unit e.

The inverse map $x \mapsto x^{-1}$, being continuous and of order 2, is a homeomorphism of G onto itself. Likewise, for each fixed y in G, the left translation $x \mapsto yx$ and the right translation $x \mapsto xy$ are continuous and have continuous inverses (namely $x \mapsto y^{-1}x$ and $x \mapsto xy^{-1}$), and so are homeomorphisms of G onto itself.

If S and T are subsets of G, we write S^{-1} for $\{x^{-1} : x \in S\}$ and ST for $\{xy : x \in S, y \in T\}$. If n is a positive integer, $S^n = SS \dots S$ (n times).

A subset S of G is *symmetric* if $S^{-1} = S$.

Since U^{-1} is a neighborhood of e whenever U is, and since $U \cap U^{-1}$ is symmetric, it follows that e has a basis of symmetric open neighborhoods.

If U is any neighborhood of e, we can find a neighborhood V of e such that $V^2 \subset U$. This follows from the continuity of the group product and the fact that $e^2 = e$. Likewise we can choose a neighborhood W of e to satisfy either (or both) of the inclusions $W^{-1}W \subset U$, $WW^{-1} \subset U$.

Notice that for any subset A of G we have $\bar{A} = \bigcap_V (VA)$, the intersection running over all neighborhoods V of e. Indeed: $x \in \bar{A} \Leftrightarrow A \cap V^{-1}x \neq \phi$ for all neighborhoods V of $e \Leftrightarrow x \in VA$ for all neighborhoods V of e. Similarly $\bar{A} = \bigcap_V (AV)$ (V being as before).

In particular, if V is a neighborhood of e, then $\bar{V} \subset V^2$. It follows that to every neighborhood U of e there is a neighborhood V of e such that $\bar{V} \subset U$.

1.3. If $\langle G, \cdot, \mathcal{T} \rangle$ is a topological group and $:$ is the reversed product (given by $x : y = y \cdot x$), then $\langle G, : \mathcal{T} \rangle$ is also a topological group. It is called the *reverse topological group* of $\langle G, \cdot, \mathcal{T} \rangle$.

This observation is often useful in proving statements about multiplication in both orders; see for example the proofs of 1.6, 1.8, 7.17.

1.4. Remark. Definition 1.1 is unaltered if condition (b) is replaced by the apparently weaker assertion that the topological space $\langle G, \mathcal{T} \rangle$ is T_0 (i.e., the open sets separate points); see for example Hewitt and Ross [1], Theorem 4.8, p. 19. In the other direction, it can be shown that a topological group is necessarily completely regular; see Hewitt and Ross [1], Theorem 8.4, p. 70.

1.5. When $\langle G, \cdot, \mathcal{T} \rangle$ is asserted to have some topological property, this means of course that the underlying topological space $\langle G, \mathcal{T} \rangle$ has this property.

A topological group which is locally compact [compact, discrete] will be called simply a *locally compact [compact, discrete] group.*

We recall that a topological space is *second-countable* if it has a countable base of open sets. Many writers use the term "separable" synonymously with "second-countable"; but we prefer to distinguish them, reserving "separable" for spaces having a countable dense subset. (For metric spaces separability and second-countability are the same; but a non-metric space may be separable without being second-countable; see Kelley [1], p. 49).

A topological space is *σ-compact* if it is the union of countably many compact subsets. Notice that a second-countable locally compact space is σ-compact.

If a topological group is locally compact, σ-compact, and first-countable, then it is second-countable. The reader can verify this as an exercise.

1.6. *From here to the end of this section G is a fixed topological group, with unit e.*

Proposition. *Let A and B be subsets of G, A being closed and B compact. Then AB and BA are closed. If A and B are both compact, then AB and BA are compact.*

Proof. Suppose that $x_i y_i \to z$ in G, where $\{x_i\}$ and $\{y_i\}$ are nets of elements of A and B respectively. Since B is compact, we can pass to a subnet and assume that $y_i \to y$ in B. But then $x_i = (x_i y_i) y_i^{-1} \to z y^{-1}$ in G. Since A is closed this implies that $z y^{-1} \in A$, whence $z = (z y^{-1})y \in AB$. So AB is closed. Applied to the reverse group (1.3), this shows that BA is closed.

If A and B are both compact, AB is the image of the compact space $A \times B$ under the product map, and so is compact. Similarly for BA.   ■

Remark. Simple examples show that, if A and B are merely closed subsets of G, then AB need not be closed.

1.7. Proposition. *If A and B are connected subsets of G, then A^{-1} and AB are connected.*

Proof. Since the inverse map is a homeomorphism, the connectedness of A^{-1} is obvious. To prove the connectedness of AB, we may assume without cost that $e \in B$; (otherwise replace B by Bu^{-1}, where $u \in B$). Then AB is the union of the connected sets xB, where $x \in A$; and each such set xB intersects the connected set A. Therefore AB is connected.   ■

1.8. Proposition. *Let C be a compact subset of G and U an open subset of G with $C \subset U$. Then there is a neighborhood V of e such that $VC \subset U$ and $CV \subset U$.*

Proof. Assume that there is no neighborhood V of e satisfying $VC \subset U$. Then we can choose nets $\{x_i\}$ and $\{y_i\}$ of elements of G such that $x_i \to e$, $y_i \in C$, and $x_i y_i \notin U$. Since C is compact we pass to a subnet and assume $y_i \to y$ in C. But then $x_i y_i \to ey = y$; and since the $x_i y_i$ belong to the closed set $G \setminus U$, this implies $y \in G \setminus U$, a contradiction. Therefore $VC \subset U$ for some neighborhood V of e. Applying this conclusion to the reverse group (1.3), we see that $CV \subset U$ for some neighborhood V of e. ■

1.9. Proposition. *Let $f \in \mathscr{C}(G)$ and $\varepsilon > 0$; and let C be a compact subset of G. There is a neighborhood V of e such that $|f(x) - f(y)| < \varepsilon$ whenever x, $y \in C$ and either $x^{-1}y \in V$ or $yx^{-1} \in V$.*

Proof. If this is false there are two nets $\{x_i\}$ and $\{y_i\}$ of elements of C such that

$$|f(x_i) - f(y_i)| \geq \varepsilon \qquad \text{for each } i, \tag{1}$$

and

$$\text{either } x_i^{-1}y_i \to e \quad \text{or} \quad y_i x_i^{-1} \to e. \tag{2}$$

By the compactness of C we can pass to a subnet and assume that $x_i \to x$ and $y_i \to y$ in C. Passing to the limit in i, (1) then implies that $|f(x) - f(y)| \geq \varepsilon$, and (2) implies that $x = y$. But this is a contradiction. ■

1.10. Corollary. *Let G be locally compact. If $f \in \mathscr{L}(G)$ and $\varepsilon > 0$, there is a neighborhood U of e such that $|f(x) - f(y)| < \varepsilon$ whenever x, $y \in G$ and either $x^{-1}y \in U$ or $yx^{-1} \in U$.*

Subgroups

1.11. Definition. If H is a subgroup of the underlying group of G, then, equipped with the relativized topology of G, H is a topological group in its own right. As such, it is called a *topological subgroup* of G. If in addition H is closed as a subset of the topological space G, we call it a *closed subgroup* of G (meaning thereby a closed topological subgroup).

1.12. Proposition. *If H is a subgroup of G, so is its closure $\bar{H}$ in G. If H is Abelian, so is $\bar{H}$. If H is normal, so is $\bar{H}$.*

We leave the proof to the reader.

1.13. Associated with the topological group G are certain important closed normal subgroups. For example:

The *center* $Z = \{x \in G : xy = yx \text{ for all } y \text{ in } G\}$ is easily seen to be a closed normal subgroup of G.

Let C be the closure of the subgroup of G generated by $\{xyx^{-1}y^{-1} : x, y \in G\}$. Thus by 1.12 C is a closed normal subgroup of G called the *commutator subgroup* of G.

Let N be the (unique) largest connected subset of the topological space G containing e. By 1.7 together with the properties of connected sets, N is a closed normal subgroup of G. It is called the *connected component of the unit* in G.

1.14. The next few numbers describe special conditions under which closed subgroups arise.

Proposition. *A subgroup H of G which is open in G is also closed in G.*

Proof. Since H is open each coset xH $(x \in G)$ is open. So $G \setminus H = \bigcup \{xH : x \in G \setminus H\}$ is open, and H is closed. ∎

1.15. If V is a non-void open subset of G, any subgroup of G containing V must be open (since it is a union of sets of the form xV), and hence closed by 1.14. In particular the subgroup generated by V is both open and closed. Thus we have:

Corollary. *If G is connected and V is a non-void open subset of G, the subgroup of G generated by V is G itself.*

1.16. Proposition. *A locally compact topological subgroup H of G is closed in G.*

Proof. We may assume without cost that H is dense in G; otherwise we replace G by $\bar{H}$.

We claim that H is open in G. Indeed: Let $x \in H$. Since H is locally compact, there is an open H-neighborhood V of x whose H-closure $\bar{V}$ is compact. Thus $\bar{V}$ is also closed in G, and so coincides with the G-closure of V. Now V, being open in H, is of the form $U \cap H$, where U is open in G. Since H is dense in G, V must be dense in U, and so its G-closure $\bar{V}$ contains U. But $\bar{V} \subset H$. Therefore $U \subset H$; and the claim is proved.

Thus, by 1.14, H is closed in G, and the proof is complete. ∎

Remark. It follows in particular from this Proposition that a discrete topological subgroup of G is closed.

Suppose that G itself is locally compact. Then a closed subgroup of G is certainly locally compact. Thus an arbitrary topological subgroup of G is locally compact if and only if it is closed.

1.17. It is well known that a non-void subset of a finite group which is closed under multiplication is also closed under the inverse operation and is therefore a subgroup. For compact groups an analogous statement is true.

Proposition. *Let G be a compact group. A non-void closed subset H of G which is closed under multiplication is also closed under the taking of inverses, and so is a closed subgroup of G.*

Proof. Let $x \in H$. Since G is compact, the sequence $x, x^2, x^3, \ldots$, has a limit point y in G. Now, given a neighborhood U of e, choose a neighborhood V of e such that $V^{-1}V \subset U$. Since y is a limit point of $\{x^n\}$, there are two positive powers of x, say x^p and x^q, such that $p > q + 1$ and x^p and x^q are both in yV. Setting $n(U) = p - q$, we have $x^{n(U)} \in (yV)^{-1}yV = V^{-1}y^{-1}yV = V^{-1}V \subset U$. So $x^{n(U)} \to e$ as U shrinks to e. Consequently

$$x^{n(U)-1} \to x^{-1} \tag{3}$$

as U shrinks to e. Since $p > q + 1$, $n(U) - 1 > 0$; and (3) says that x^{-1} belongs to the closure of H, hence to H itself. Thus H is closed under the inverse operation. ∎

Examples

1.18. Here are a few simple examples of topological groups.

Any (untopologized) group becomes a topological group when equipped with the discrete topology. Perhaps the most important infinite discrete group is the additive group $\mathbb{Z}$ of the integers (the infinite cyclic group).

The real number field $\mathbb{R}$, with its usual topology and the operation of addition, forms a second-countable connected locally compact group called the *additive group of the reals.* The set of all non-zero real numbers, with the relativized topology of $\mathbb{R}$ and the operation of multiplication, forms a second-countable locally compact group $\mathbb{R}_*$ called the *multiplicative group of non-zero reals.* This group is not connected; its connected component of the unit is the multiplicative subgroup $\mathbb{R}_{++}$ of all *positive* real numbers.

The same of course can be said of the complex field $\mathbb{C}$, except that the multiplicative group $\mathbb{C}_* = \mathbb{C} \setminus \{0\}$ is connected.

Let us denote $\{z \in \mathbb{C} : |z| = 1\}$ by $\mathbb{E}$. With the operation of multiplication, and with the relativized topology of the complex plane, $\mathbb{E}$ becomes a second-countable connected compact group; it is called the *circle group*. It plays an enormous role in the theory of unitary group representations.

1.19. All the above groups are of course Abelian. An extremely important non-Abelian group is the $n \times n$ *general linear group* $GL(n, \mathbb{C})$ $(n = 2, 3, \ldots)$. This is defined as the group, under multiplication, of all non-singular $n \times n$ complex matrices. We give to $GL(n, \mathbb{C})$ the topology of entry-wise convergence; that is, a net $\{a^r\}$ converges to a in $GL(n, \mathbb{C})$ if and only if $a^r_{ij} \to a_{ij}$ in $\mathbb{C}$ for all $i, j = 1, \ldots, n$. Since the non-singularity of a means the non-vanishing of $\det(a)$, $GL(n, \mathbb{C})$ is an open subset of $\mathbb{C}^{n^2}$, hence locally compact. The formulae for the product and inverse of matrices show that these operations are continuous. Hence $GL(n, \mathbb{C})$ is a second-countable locally compact group.

1.20. Many of the closed subgroups of $GL(n, \mathbb{C})$ are important. Here are a few of them:

(1) $SL(n, \mathbb{C}) = \{a \in GL(n, \mathbb{C}) : \det(a) = 1\}$, the $n \times n$ *special linear group.*
(2) $GL(n, \mathbb{R}) = \{a \in GL(n, \mathbb{C}) : a_{ij}$ is real for all $i, j\}$, the $n \times n$ *general real linear group.*
(3) $SL(n, \mathbb{R}) = GL(n, \mathbb{R}) \cap SL(n, \mathbb{C})$, the $n \times n$ *special real linear group.*
(4) $U(n) = \{a \in GL(n, \mathbb{C}) : a^* = a^{-1}\}$, where $(a^*)_{ij} = \overline{a_{ji}}$. This is the $n \times n$ *unitary group.*
(5) $SU(n) = U(n) \cap SL(n, \mathbb{C})$, the $n \times n$ *special unitary group.*
(6) $O(n) = U(n) \cap GL(n, \mathbb{R})$, the $n \times n$ *orthogonal group.*
(7) $SO(n) = O(n) \cap SL(n, \mathbb{R})$, the $n \times n$ *special orthogonal group.*

Of these, the subgroups (4), (5), (6), and (7) are compact. Indeed, let us show that $U(n)$ is compact. If $a \in U(n)$, we have $1 = (a^*a)_{ii} = \sum_j |a_{ji}|^2$ for each i, showing that $|a_{ij}| \le 1$ for all i, j. It follows that $U(n)$ is contained in the compact subset $D = \{a \in \mathbb{C}^{n^2} : |a_{ij}| \le 1$ for all $i, j\}$ of $\mathbb{C}^{n^2}$. Thus $U(n)$, being the intersection of D with the closed subset $\{a : a^*a = aa^* = 1\}$ of $\mathbb{C}^{n^2}$, is compact. Consequently $SU(n)$, $O(n)$, and $SO(n)$, being closed subgroups of $U(n)$, are also compact.

2. Quotient Spaces and Homomorphisms

2.1. From here to 2.6 G is a topological group with unit e, and H is a fixed closed subgroup of G. Let $\pi: G \to G/H$ be the natural surjection $x \mapsto xH$. We equip G/H with the quotient topology deduced from π; that is, a subset A of G/H is open if and only if $\pi^{-1}(A)$ is open in G.

Proposition. *G/H is a Hausdorff space. The map π is continuous and open. Hence, G/H is discrete if and only if H is open in G.*

Proof. π is continuous by definition. If U is an open subset of G, then $\pi^{-1}(\pi(U)) = UH = \bigcup \{Ux : x \in H\}$, which is open. So $\pi(U)$ is open in G/H, and π is an open map.

To prove that G/H is Hausdorff, let x and y be elements of G with $\pi(x) \neq \pi(y)$. Then $G \setminus yH$ is an open neighborhood of x; so by 1.2 we can find a neighborhood U of e such that

$$U^{-1}Ux \subset G \setminus yH. \tag{1}$$

This implies that $UxH \cap UyH = \emptyset$; for, if $uxh = vyk$ ($u, v \in U$; $h, k \in H$), we would have $y = v^{-1}uxhk^{-1} \in U^{-1}UxH$, contradicting (1). So $\pi(Ux)$ and $\pi(Uy)$ are disjoint neighborhoods of $\pi(x)$ and $\pi(y)$. $\blacksquare$

2.2. Corollary. *If G is locally compact [compact], then G/H is locally compact [compact].*

Proof. If G is locally compact and $x \in G$, there is a compact neighborhood U of x. Then by the continuity and openness of π (2.1) $\pi(U)$ is a compact neighborhood of $\pi(x)$. So G/H is locally compact. $\blacksquare$

2.3. Lemma. *Assume that* (i) *U is a closed symmetric neighborhood of e such that $[(U^3) \cap H]^-$ is compact,* (ii) *C is a compact subset of G/H such that $C \subset \pi(U)$. Then $U \cap \pi^{-1}(C)$ is compact in G.*

Proof. We begin with a general observation: Suppose B is any compact subset of G, and $\{x_i\}$ is any net of elements of G such that no subnet of $\{x_i\}$ converges to any point of B. Then there exists a neighborhood V of e such that

$$x_i \notin VB \qquad i\text{-eventually}. \tag{2}$$

Indeed: By the hypotheses, there is a finite covering

$$B \subset W_1 \cup W_2 \cup \cdots \cup W_n$$

of B by open sets W_k such that for each $k = 1, \ldots, n$, the net $\{x_i\}$ is i-eventually outside W_k. It follows that

$$\{x_i\} \quad \text{is eventually outside} \quad W = \bigcup_{k=1}^{n} W_k.$$

By 1.8 we can find an open neighborhood V of e such that $VB \subset W$; and this gives (2).

Returning to the proof of the lemma, we note that $U \cap \pi^{-1}(C)$ is closed. Assume it is not compact. Then there is a net $\{x_i\}$ of elements of $U \cap \pi^{-1}(C)$, no subnet of which converges to any element of G. Since C is compact, we may pass to a subnet (without change of notation) and assume that $\pi(x_i) \to \gamma \in C$; and so (since $C \subset \pi(U)$)

$$\pi(x_i) \to \pi(u_0), \qquad \text{where } u_0 \in U.$$

Now $u_0([(U^3) \cap H]^-)$ is compact by hypothesis. So by the general observation with which we began this proof, there is a symmetric neighborhood V of e such that $V \subset U$ and

$$x_i \notin Vu_0([(U^3) \cap H]^-) \qquad i - \text{eventually.} \tag{3}$$

On the other hand, since $\pi(x_i) \to \pi(u_0)$, *the openness of π* permits us by II.13.2 to pass to a subnet (again without change of notation) and suppose

$$x_i \in y_i H, \qquad \text{where} \quad y_i \to u_0 \text{ in } G.$$

Thus $y_i \in Vu_0$ i-eventually, or

$$x_i \in Vu_0 H \qquad i\text{-eventually.} \tag{4}$$

Now if i is any index for which $x_i \in Vu_0 H$, we have

$$x_i = vu_0 h \qquad\qquad (v \in V, h \in H), \tag{5}$$

whence

$$h = u_0^{-1} v^{-1} x_i \in U^3 \tag{6}$$

(since $u_0 \in U = U^{-1}$, $v \in V = V^{-1} \subset U$ and $x_i \in U$). So by (4), (5), and (6),

$$x_i \in Vu_0((U^3) \cap H) \qquad i\text{-eventually,}$$

contradicting (3).   ■

2.4. Proposition. *If H and G/H are compact [locally compact], then G is compact [locally compact].*

Proof. Assume that H and G/H are compact. Applying 2.3 with $U = G$, $C = G/H$, we conclude that $G = U \cap \pi^{-1}(C)$ is compact.

Now assume that H and G/H are locally compact. Since H is locally compact, we can choose a neighborhood Z of e in G such that $Z \cap H$ has compact closure. Then by 1.2 we can choose a closed symmetric neighborhood U of e in G so that $U^3 \subset Z$, and hence $U^3 \cap H$ has compact closure. Now, since π is open, $\pi(U)$ is a neighborhood of eH in G/H. So, since the latter is locally compact, we can find a compact neighborhood C of eH in G/H such that $C \subset \pi(U)$. Lemma 2.3 now states that $U \cap \pi^{-1}(C)$ is compact in G. But $U \cap \pi^{-1}(C)$ is a neighborhood of e. ■

2.5. The following fact is sometimes useful.

Proposition. *Let G be locally compact. If K is a compact subset of G/H, there exists a compact subset D of G such that $\pi(D) = K$.*

Proof. By the openness of π, the compactness of K, and the local compactness of G, there are finitely many open subsets $U_1, \ldots, U_n$ of G, each having compact closure, such that $\bigcup_{i=1}^{n} \pi(U_i) \supset K$. Thus $E = \bigcup_{i=1}^{n} \bar{U}_i$ is compact and $\pi(E) \supset K$. Hence $D = E \cap \pi^{-1}(K)$ is compact and $\pi(D) = K$. ■

2.6. Proposition. *Assume that the closed subgroup H of G is normal. Then G/H, with the quotient group structure and the quotient topology of 2.1, is a topological group.*

Proof. G/H is Hausdorff by 2.1. So we have only to show that $\langle r, s \rangle \mapsto rs^{-1}$ is continuous ($r, s \in G/H$).

Let $r, s \in G/H$; and let W be a neighborhood of rs^{-1}. Choose x, y in G such that $\pi(x) = r$, $\pi(y) = s$; then $\pi^{-1}(W)$ is a neighborhood of xy^{-1}. By the continuity of the operations in G we can therefore find open neighborhoods U and V of x and y respectively such that $UV^{-1} \subset \pi^{-1}(W)$. This implies by the openness of π that $\pi(U)$ and $\pi(V)$ are neighborhoods of r and s satisfying $\pi(U)(\pi(V))^{-1} \subset W$. So $\langle r, s \rangle \mapsto rs^{-1}$ is continuous. ■

Definition. The topological group G/H of the above Proposition is called the *quotient group of G modulo H*.

Homomorphisms and Isomorphisms

2.7. Let G and K be two topological groups, and $f\colon G \to K$ a homomorphism (in the purely algebraic sense that $f(xy) = f(x)f(y)$ for $x,\ y \in G$). We shall be very interested in homomorphisms which are continuous. Observe that f is continuous if and only if f is continuous at the unit element e of G. Indeed: If f is continuous at e and $x_i \to x$ in G, then $x^{-1}x_i \to e$, whence $f(x^{-1}x_i) \to f(e)$; so $f(x_i) = f(x)f(x^{-1}x_i) \to f(x)f(e) = f(x)$.

2.8. *Definition.* Let G and K be two topological groups. A group isomorphism f of G onto K which is also a homeomorphism is called an *isomorphism of topological groups*. If such an isomorphism f exists, we say that G and K are *isomorphic (as topological groups)*—in symbols $G \cong K$.

An isomorphism of the topological group G with itself is called an *automorphism*.

If G is a normal topological subgroup of another topological group L, then every element x of L gives rise to the automorphism $y \mapsto xyx^{-1}$ ($y \in G$) of G. If x itself is in G, the automorphism $y \mapsto xyx^{-1}$ of G is called *inner*.

2.9. *Examples.* The additive group $\mathbb{R}$ of the reals is isomorphic (as a topological group) with the multiplicative group $\mathbb{R}_{++}$ of the positive reals, under the isomorphism $f\colon t \mapsto e^t$.

The circle group $\mathbb{E}$ is isomorphic with $SO(2)$ (see 1.20) under the isomorphism

$$e^{it} \mapsto \begin{pmatrix} \cos t & -\sin t \\ \sin t & \cos t \end{pmatrix} \qquad (t \in \mathbb{R}).$$

2.10. Let G and K be topological groups, and $f\colon G \to K$ a continuous homomorphism. Then the kernel $H = \mathrm{Ker}(f)$ of f is a closed normal subgroup of G; and by 2.6 we can form the quotient topological group $\tilde{G} = G/H$. The equation

$$\tilde{f}(xH) = f(x) \qquad (x \in G)$$

then defines a one-to-one homomorphism $\tilde{f}\colon \tilde{G} \to K$. One sees immediately that $\tilde{f}$ is continuous. Furthermore:

Proposition. *Let f be onto K. Then $\tilde{f}$ is an isomorphism of the topological groups $\tilde{G}$ and K if and only if $f\colon G \to K$ is open.*

2.11.   *Examples.*   The map $t \mapsto e^{2\pi i t} (t \in \mathbb{R})$ is a continuous open homomorphism of $\mathbb{R}$ onto $\mathbb{E}$ whose kernel is $\mathbb{Z}$. It therefore follows from 2.10 that

$$\mathbb{E} \cong \mathbb{R}/\mathbb{Z}.$$

Similarly, the determinent map $a \mapsto \det(a)$ is a continuous open homomorphism of $GL(n, \mathbb{C})$ onto $\mathbb{C}_*$, of $GL(n, \mathbb{R})$ onto $\mathbb{R}_*$, of $U(n)$ onto $\mathbb{E}$, and of $O(n)$ onto $\{1, -1\}$ (see 1.18–1.20). So by 2.10, if $n = 2, 3, \ldots$,

$$GL(n, \mathbb{C})/SL(n, \mathbb{C}) \cong \mathbb{C}_*,$$

$$GL(n, \mathbb{R})/SL(n, \mathbb{R}) \cong \mathbb{R}_*,$$

$$U(n)/SU(n) \cong \mathbb{E},$$

$$O(n)/SO(n) \cong \{1, -1\} \qquad \text{(the two-element group)}.$$

In 3.11 we shall find general conditions under which a continuous surjective homomorphism is automatically open.

3.   Topological Transformation Spaces

3.1.   *Definition.*   Let G be a group (without topology) with unit e. A *left G-transformation space* (or simply a *left G-space*) is a pair $\langle M, \alpha \rangle$, where M is a set and $\alpha: G \times M \to M$ is a function satisfying:

(i) $\alpha(e, m) = m$,

(ii) $\alpha(y, \alpha(x, m)) = \alpha(yx, m)$

for all x, y in G and m in M. We call $\alpha(x, m)$ the *(left) action* of x on m.

A *right G-transformation space* is defined in the same way, except that (ii) is replaced by

(ii′) $\alpha(y, \alpha(x, m)) = \alpha(xy, m)$ $(x, y \in G; m \in M)$.

If in addition G is a topological group, M is a topological space, and $\alpha: G \times M \to M$ is continuous, then $\langle M, \alpha \rangle$ is a *left [right] topological G-transformation space* (or simply a *left [right] topological G-space*). We may also express this fact by saying that G *acts as a left [right] topological transformation group on M*.

A *G-transformation space*, without mention of "left" or "right", will always mean a *left* G-transformation space.

In referring to a left or right G-space $\langle M, \alpha \rangle$, we almost always omit explicit mention of α, writing simply xm $[mx]$ instead of $\alpha(x, m)$ for a left [right] G-space. Conditions (i) and (ii) for a left G-space then become $em = m$ and $y(xm) = (yx)m$; while for a right G-space conditions (i) and (ii') become $me = m$ and $(mx)y = m(xy)$.

In the topological context, notice that $m \mapsto xm[m \mapsto mx]$ is a homeomorphism of M onto itself for each x in G. Indeed, in the left case, the continuity of α implies that both $m \mapsto xm$ and its inverse $m \mapsto x^{-1}m$ are continuous.

A right G-space $\langle M, \alpha \rangle$ is automatically a left $\tilde{G}$-space, where $\tilde{G}$ is the reverse group of G (see 1.3), and conversely. So any statement or definition made for left G-spaces has an automatic analogue for right G-spaces.

3.2. Fix a group G and a (left) G-space M.

A *G-subspace* of M is a subset N of M which is closed under the action of G, that is, $x \in G$, $m \in N \Rightarrow xm \in N$. Each m in M belongs to a unique smallest non-void G-subspace, namely the *orbit* $Gm = \{xm : x \in G\}$ of m. Notice that two orbits are either equal or disjoint; in fact the orbits in M are just the equivalence classes of the equivalence relation $\sim$ in M given by

$$m_1 \sim m_2 \Leftrightarrow m_2 = xm_1 \qquad \text{for some } x \text{ in } G.$$

Definition. M is *transitive* if it is non-void and contains only one orbit, that is, for every pair m_1, m_2 of elements of M there is an element x of G such that $xm_1 = m_2$.

3.3. Here is a vitally important example. Let K be any subgroup of a group G, and let M be the left coset space G/K. If $m \in G/K$ and $x \in G$, then xm is also a left K coset and hence in G/K. Evidently $em = m$ and $x(ym) = (xy)m$; and every coset is of the form xm_0, m_0 being the coset eH. So the action $\langle x, m \rangle \mapsto xm$ makes M into a transitive G-space.

Now let G be a topological group and K a closed subgroup of G; and topologize M as in 2.9. Then:

Proposition. *M is a transitive topological G-space.*

Proof. The continuity of the action $\langle x, m \rangle \mapsto xm$ is proved by almost the same argument as was Proposition 2.6. ∎

3.4. Let M and M' be two G-spaces. A function $\phi: M \to M'$ is said to *intertwine* M and M', or to be *G-equivariant*, if

$$x(\phi(m)) = \phi(xm)$$

for all m in M and x in G. If ϕ is a bijection (onto M'), we call it an *isomorphism of the G-spaces* M and M', and say that M and M' are *isomorphic as G-spaces* (in symbols $M \cong M'$).

If in addition M and M' are topological G-spaces and ϕ is a homeomorphism of M onto M', then ϕ is an *isomorphism of topological G-spaces*, and M and M' are *isomorphic as topological G-spaces* (in symbols $M \cong M'$).

3.5. Let M be a G-space, and m an element of M. The set G_m of those elements x of G for which $xm = m$ is a subgroup of G. (Indeed: If $xm = m$ and $ym = m$, then $(yx)m = m$ by 3.1(ii). If $xm = m$, applying x^{-1} to both sides and using 3.1(i), (ii), we get $m = x^{-1}m$). We call G_m the *stabilizer of m in G*. Now form the left coset space G/G_m, and let $\pi: G \to G/G_m$ be the natural quotient map. If $x, y \in G$, then $ym = xm$ if and only if $x^{-1}y \in G_m$, that is, $\pi(y) = \pi(x)$. Hence the map $x \mapsto xm$ is a composition $\phi_m \circ \pi$, where $\phi_m: G/G_m \to M$ is the injection given by:

$$\phi_m(\pi(x)) = xm \qquad\qquad (x \in G).$$

The range of ϕ_m is the orbit of m. Notice that ϕ_m intertwines the two G-spaces G/G_m and M; indeed, if $x, y \in G$,

$$\phi_m(y\pi(x)) = \phi_m(\pi(yx)) = (yx)m = y\phi_m(\pi(x)).$$

If M itself is transitive, then range $(\phi_m) = M$; and we obtain:

Proposition. *A transitive G-space M is isomorphic (as a G-space) with G/G_m, where m is any element of M.*

3.6. Notice that, if two elements m and p of the G-space M belong to the same orbit under G, then their stabilizers differ only by an inner automorphism. In fact

$$G_{xm} = xG_m x^{-1} \qquad\qquad (x \in G, m \in M).$$

Indeed: $y \in G_{xm} \Leftrightarrow yxm = xm \Leftrightarrow x^{-1}yxm = m \Leftrightarrow x^{-1}yx \in G_m \Leftrightarrow y \in xG_m x^{-1}$.

3.7. Proposition. *Let G be a topological group, and K and L two closed subgroups of G. Then G/K and G/L are isomorphic as topological G-spaces if and only if $L = xKx^{-1}$ for some x in G.*

Proof. (A) Assume that $L = xKx^{-1}$. Then, if $m = yK \in G/K$, we have $mx^{-1} = (yx^{-1})xKx^{-1} = yx^{-1}L \in G/L$; so that $m \mapsto mx^{-1}$ ($m \in G/K$) is a mapping of G/K into G/L. It is easy to see that this mapping is a homeomorphism onto G/L and is G-equivariant. So G/K and G/L are isomorphic as topological G-spaces.

(B) Assume that G/K and G/L are isomorphic as G-spaces under an isomorphism f; and let $f(K) = xL \in G/L$. (Here K is considered as a coset belonging to G/K). Since K and xL correspond under f, they have the same stabilizers. But the stabilizer of K in G/K is K; and the stabilizer of xL in G/L is xLx^{-1} (see 3.6). So $K = xLx^{-1}$. ■

3.8. Our next step is to try to carry through 3.5 in a topological context.

Let G be a topological group and M a topological G-space. The first question is: When are the stabilizers G_m closed in G?

If M is a T_1 space (i.e., one-element sets are closed) it is almost obvious that the stabilizers G_m are closed. A stronger result holds:

Proposition. *If M is a T_0 space (i.e., the open sets separate points of M), the stabilizers G_m are closed.*

Proof. Fix $m \in M$; put $H = G_m$; and denote closure by bars. Since $H\{m\} = \{m\}$, the continuity of the action gives

$$\bar{H}\{m\}^- \subset \{m\}^-. \qquad\qquad (1)$$

Now let $x \in \bar{H}$. Since $p \mapsto xp$ is a homeomorphism of M, we have by (1)

$$\{xm\}^- = x\{m\}^- \subset \{m\}^-. \qquad\qquad (2)$$

By 1.12 $\bar{H}$ is a subgroup; so $x^{-1} \in \bar{H}$ and by (1) $x^{-1}\{m\}^- \subset \{m\}^-$, whence $\{m\}^- \subset x\{m\}^-$. Combining this with (2) we get

$$\{xm\}^- = \{m\}^-. \qquad\qquad (3)$$

Since M is a T_0 space, (3) implies that $xm = m$, or $x \in H$. But x was any element of $\bar{H}$; so H is closed.   ∎

Transitive Transformation Spaces

3.9. Let G, M be as in 3.8; and suppose that M is a T_0-space, so that by 3.8 the stabilizers G_m are closed. Fix an element m of M; and construct $\phi_m : G/G_m \to M$ as in 3.5. By 2.1, G/G_m is a Hausdorff space, and the natural surjection $\pi : G \to G/G_m$ is open. Since $x \mapsto xm$ is continuous, it follows from the openness of π that ϕ_m is also continuous.

If ϕ_m were necessarily a homeomorphism, one could immediately translate Proposition 3.5 into a topological context, and assert that every transitive T_0 topological G-space is isomorphic (as a topological G-space) to G/H for some closed subgroup H. However this is false. For example let G be the additive group of rational numbers (with the discrete topology), and let M be the same set of rational numbers, but carrying the non-discrete topology relativized from $\mathbb{R}$. The action of G on M by addition is topological and transitive, but M is not homeomorphic with any G/H.

3.10. There are certain rather general conditions, however, under which the ϕ_m of 3.9 has to be a homeomorphism if M is transitive.

Theorem. *Let G be a σ-compact locally compact group, and M a transitive locally compact Hausdorff topological G-space. Then, for each m in M, the map $\phi_m : G/G_m \to M$ of 3.9 is bicontinuous, so that M is isomorphic as a topological G-space with G/G_m.*

Proof. Fix $m \in M$. To prove that ϕ_m is open, hence bicontinuous, it is enough to show that the map $x \mapsto xm\ (x \in G)$ is open. To do this we shall first show that Vm is a neighborhood of m for each neighborhood V of the unit element e.

Let V be a neighborhood of e, and choose a compact neighborhood W of e such that $W^{-1}W \subset V$. Since G is σ-compact, G can be covered by countably many left translates $\{x_n W\}$ of $W(n = 1, 2, \ldots)$. Since W is compact, the $\{x_n Wm\}$ form a sequence of compact, hence closed, subsets of M; and, since M is transitive, $\bigcup_n x_n Wm = M$. So by the local compactness of M and the Baire Category Theorem, we can choose an n such that $x_n Wm$ has non-void interior. Pick an element y of W such that $x_n ym$ is an interior point of $x_n Wm$. Then $m = y^{-1}x_n^{-1}(x_n ym)$ is an interior point of $y^{-1}x_n^{-1}(x_n Wm) = y^{-1}Wm \subset W^{-1}Wm \subset Vm$. So we have shown that m is an interior point of Vm.

Now, if U is a neighborhood of the point x in G, $x^{-1}U$ is a neighborhood of e. So by the last paragraph m is an interior point of $x^{-1}Um$, whence xm is an interior point of Um. Thus $x \mapsto xm$ is open. ∎

3.11. Corollary. *Let G and H be locally compact groups, G being σ-compact; and let f be a continuous homomorphism of G onto H. Then f is open; and, if K is the kernel of f, H and G/K are isomorphic (as topological groups) under the isomorphism $\tilde{f}$ induced by f:*

$$\tilde{f}(xK) = f(x) \qquad\qquad (x \in G).$$

Proof.   Apply 3.10 to the transitive action $\langle x, h \rangle \mapsto f(x)h$ of G on H. ∎

3.12. Corollary (Closed Graph Theorem). *Let G and H be two σ-compact locally compact groups, and $f: G \to H$ a group homomorphism whose graph is a closed subset of $G \times H$. Then f is continuous.*

Proof.   The graph F of f, being closed, is itself a σ-compact locally compact subgroup of $G \times H$. So we may apply 3.11 to the projection $p_1: F \to G$, and conclude that $p_1^{-1}: G \to F$ is continuous. ∎

3.13. Corollary. *Let G be a locally compact group; and let H, K be two closed subgroups of G, such that K is σ-compact and $KH = G$. Then the natural bijection $\alpha: k(H \cap K) \mapsto kH$ ($k \in K$) of $K/(H \cap K)$ onto G/H is a homeomorphism.*

Proof. G/H is a locally compact Hausdorff K-space (on restricting the action of G to K). Since $KH = G$, G/H is transitive under K; and the stabilizer for eH is $H \cap K$. So α is a homeomorphism by 3.10. ∎

3.14. Corollary. *Let G be a locally compact group, and H, K two closed subgroups of G such that (i) one of H and K is σ-compact, (ii) $KH = G$, and (iii) $K \cap H = \{e\}$. Then the bijection $\langle k, h \rangle \mapsto kh$ of $K \times H$ onto G is a homeomorphism.*

Examples

3.15.   Here are some examples of quotient spaces and quotient groups, and of the usefulness of Theorem 3.10 and its Corollary 3.11.

(A) Let S_{n-1} be the unit sphere $\{t \in \mathbb{R}^n : t_1^2 + t_2^2 + \cdots + t_n^2 = 1\}$ in $\mathbb{R}^n$ ($n = 2, 3, \ldots$); and consider S_{n-1} as a topological $SO(n)$-space (with the ordinary action of matrices on $\mathbb{R}^n$). Since any unit vector in $\mathbb{R}^n$ can be transformed into any other unit vector by a matrix in $SO(n)$, S_{n-1} is transitive. Let $s = (0, 0, \ldots, 0, 1) \in S_{n-1}$. Clearly the stabilizer of s in $SO(n)$ consists of all

$$a = \begin{pmatrix} a' & | & 0 \\ \hline 0 & | & 1 \end{pmatrix} \qquad (a' \in SO(n-1)),$$

and so may be identified with $SO(n-1)$. Thus by Theorem 3.10 S_{n-1}, as a topological $SO(n)$-space, is isomorphic with the quotient space $SO(n)/SO(n-1)$.

(B) Let n and r be positive integers with $r < n$; and let $P_{n,r}$ be the family of all r-dimensional linear subspaces of $\mathbb{C}^n$. Each a in $GL(n, \mathbb{C})$, considered as acting on $\mathbb{C}^n$, carries elements of $P_{n,r}$ onto elements of $P_{n,r}$, and so permutes $P_{n,r}$. Thus $P_{n,r}$ is a $GL(n, \mathbb{C})$-space—in fact, as we easily see, a transitive one. Let p be the particular element $\{\langle t_1, \ldots, t_r, 0, \ldots, 0 \rangle : t_1, \ldots, t_r \in \mathbb{C}\}$ of $P_{n,r}$. The stabilizer of p in $GL(n, \mathbb{C})$ is the closed subgroup G_p consisting of all a such that $a_{ij} = 0$ whenever $i > r$ and $j \leq r$. Thus, since $P_{n,r}$ is transitive, the map $\phi_p : aG_p \mapsto ap$ is an isomorphism of the (untopologized) G-spaces $GL(n, \mathbb{C})/G_p$ and $P_{n,r}$.

Now, since G_p is closed, $GL(n, \mathbb{C})/G_p$ has the natural locally compact Hausdorff quotient topology. Let us give to $P_{n,r}$ the topology making ϕ_p a homeomorphism. Thus $P_{n,r}$ becomes a transitive locally compact Hausdorff $GL(n, \mathbb{C})$-space isomorphic with $GL(n, \mathbb{C})/G_p$.

Notice that by 3.13 we would have obtained the same topology for $P_{n,r}$ had we started, not with $GL(n, \mathbb{C})$, but with any closed subgroup of $GL(n, \mathbb{C})$ whose action on $P_{n,r}$ is transitive. For example, it is easy to see that $U(n)$ acts transitively on $P_{n,r}$; and the stabilizer of p in $U(n)$ is the subgroup $U(r) \times U(n-r)$ consisting of all $\begin{pmatrix} u & 0 \\ 0 & v \end{pmatrix}$, where $u \in U(r)$ and $v \in U(n-r)$. So $P_{n,r}$ is homeomorphic with $U(n)/(U(r) \times U(n-r))$. In particular, since $U(n)$ is compact, $P_{n,r}$ is compact.

4. Direct and Semidirect Products

Direct Products

4.1. For each i in an index set I let G_i be a topological group. We form the Cartesian product topological space $G = \prod_{i \in I} G_i$ (consisting of all functions x on I such that $x_i \in G_i$ for each i, with the topology of pointwise convergence), and equip G with the pointwise product operation: $(xy)_i = x_i y_i$ (the right side being the product in G_i). One verifies immediately that G is then a topological group. It is called the *Cartesian* or *direct product* of the G_i, and is denoted by $\prod_{i \in I} G_i$.

If I is finite—say $I = \{1, \ldots, n\}$—we may write $G_1 \times \cdots \times G_n$ instead of $\prod_{i=1}^{n} G_i$.

If $I = \emptyset$, $\prod_{i \in I} G_i$ will mean the one-element group.

If $\mathscr{I}$ is a family of pairwise disjoint subsets of I with $\cup \mathscr{I} = I$, the following associative law is easily verified:

$$\prod_{i \in I} G_i \cong \prod_{J \in \mathscr{I}} \prod_{i \in J} G_i. \tag{1}$$

4.2. We recall some elementary topological facts about the direct product. If all the G_i are compact, then $\prod_{i \in I} G_i$ is compact (Tihonov's Theorem).

If all the G_i are locally compact, $\prod_{i \in I} G_i$ will not in general be locally compact. But, if all the G_i are locally compact and all but finitely many are compact, then $\prod_{i \in I} G_i$ is locally compact. Indeed: One first verifies this when I is finite, and then extends it by (1) and Tihonov's Theorem to the case that I is infinite.

$\prod_{i \in I} G_i$ is connected if and only if all the G_i are connected.

4.3. For example, the additive group of $\mathbb{R}^n$ (real n-space) is just $\mathbb{R} \times \mathbb{R} \times \cdots \times \mathbb{R}$ (n times), where $\mathbb{R}$ is the additive group of the reals. Likewise $\mathbb{C}^n = \mathbb{C} \times \mathbb{C} \times \cdots \times \mathbb{C}$ (n times).

The direct product $\mathbb{E}^n = \mathbb{E} \times \mathbb{E} \times \cdots \times \mathbb{E}$ (n times) is called the *n-dimensional torus*.

Note that $\mathbb{C}_* \cong \mathbb{R}_{++} \times \mathbb{E}$ under the isomorphism $z \mapsto \langle |z|, |z|^{-1}z \rangle$ ($z \in \mathbb{C}_*$).

Semidirect Products

4.4. The semidirect product is an extremely useful generalization of the direct product of two groups.

Let G_1 and G_2 be two topological groups, and suppose we are given a homomorphism τ of G_2 into the group of all group automorphisms of G_1 such that the map $\langle x, y \rangle \mapsto \tau_y(x)$ is continuous on $G_1 \times G_2$ to G_1. (In particular each τ_y is bicontinuous). We shall construct a new topological group G as follows: The topological space underlying G is the Cartesian product $G_1 \times G_2$ of the topological spaces underlying G_1 and G_2; and the product in G is given by:

$$\langle x, y \rangle \cdot \langle x', y' \rangle = \langle x \cdot \tau_y(x'), yy' \rangle$$

$(x, x' \in G_1; y, y' \in G_2)$. One easily verifies that this operation makes G into a group. The unit element is $e = \langle e_1, e_2 \rangle$ (e_i being the unit of G_i); and inverses in G are given by

$$\langle x, y \rangle^{-1} = \langle \tau_{y^{-1}}(x^{-1}), y^{-1} \rangle.$$

Product and inverse are clearly continuous in the topology of G. So G is a topological group.

Definition. G is called the τ-*semidirect product* of G_1 and G_2, and is denoted by $G_1 \underset{\tau}{\times} G_2$.

If G_1 and G_2 are locally compact, then of course $G_1 \underset{\tau}{\times} G_2$ is locally compact.

If τ is trivial, that is, $\tau_y(x) = x$ for all x, y, then of course $G_1 \underset{\tau}{\times} G_2$ is just the direct product $G_1 \times G_2$.

4.5. We notice in 4.4 that $x \mapsto \langle x, e_2 \rangle (x \in G_1)$ is an isomorphism of G_1 onto the closed subgroup $G_1' = \{\langle x, e_2 \rangle : x \in G_1\}$ of G; while $y \mapsto \langle e_1, y \rangle (y \in G_2)$ is an isomorphism of G_2 onto the closed subgroup $G_2' = \{\langle e_1, y \rangle : y \in G_2\}$ of G. The subgroup G_1' is normal in G; indeed

$$\langle e_1, y \rangle \langle x, e_2 \rangle \langle e_1, y \rangle^{-1} = \langle \tau_y(x), e_2 \rangle \tag{2}$$

$(x \in G_1; y \in G_2)$. Also

$$G_1'G_2' = G, \qquad G_1' \cap G_2' = \{e\}. \tag{3}$$

It is sometimes convenient to identify x with $\langle x, e_1 \rangle$ $(x \in G_1)$ and y with $\langle e_1, y \rangle$ $(y \in G_2)$, hence G_1 with G_1' and G_2 with G_2'. When we do this, the pair $\langle x, y \rangle = \langle x, e_2 \rangle \langle e_1, y \rangle$ becomes just xy; and (2) becomes

$$yxy^{-1} = \tau_y(x) \qquad\qquad (x \in G_1; y \in G_2),$$

showing that the original τ_y has become the action of y on G_1 by inner automorphism in G.

The map $\langle x, y \rangle \mapsto y$ $(\langle x, y \rangle \in G)$ is an open continuous homomorphism of G onto G_2 with kernel G_1'. Hence by 2.10

$$G/G_1' \cong G_2. \tag{4}$$

4.6. Relations (3) suggest a converse problem. Given a topological group G with closed subgroups G_1 and G_2, how do we recognize whether $G \cong G_1 \underset{\tau}{\times} G_2$ (for suitable τ)?

Proposition.   *Let G be a topological group, G_1 a closed normal subgroup of G, and G_2 a closed subgroup of G such that* (i) $G_1 G_2 = G$, (ii) $G_1 \cap G_2 = \{e\}$, *and* (iii) *the bijection*

$$\langle x, y \rangle \mapsto xy \qquad\qquad (x \in G_1, y \in G_2) \qquad (5)$$

of $G_1 \times G_2$ onto G is a homeomorphism (equivalently, that the bijection $y \mapsto yG_1$ of G_2 onto G/G_1 is open). If we define $\tau_y(x) = yxy^{-1}(x \in G_1; y \in G_2)$, then the topological groups $G_1 \underset{\tau}{\times} G_2$ and G are isomorphic under the isomorphism (5).

If G is locally compact and one of G_1 and G_2 is σ-compact, hypothesis (iii) *is automatically satisfied and may be omitted.*

Proof.   The proof of all but the last statement is quite routine and is left to the reader. The last statement is just 3.14.   ∎

Definition.   Under the conditions of this Proposition we often say that *G is the semidirect product of its normal subgroup G_1 and the complementary subgroup G_2.*

Remark.   The hypothesis (iii) is not in general superfluous. For example, let $\mathbb{R}$ be the additive group of the reals with the ordinary topology, and $\mathbb{R}_d$ the same group with the discrete topology; and put $G = \mathbb{R} \times \mathbb{R}_d$, $G_1 = \{\langle 0, t \rangle : t \in \mathbb{R}_d\}$, $G_2 = \{\langle t, t \rangle : t \in \mathbb{R}\}$. With these ingredients, the hypotheses (i) and (ii) hold; but hypothesis (iii) and the conclusion of the Proposition fail.

4.7.   Here is an important general example of a semidirect product.

Let H be any topological subgroup of $GL(n, \mathbb{R})$ $(n = 1, 2, \ldots)$; and define $\tau_a(t) = at$ for $a \in H$, $t \in \mathbb{R}^n$ (at being the ordinary action of an $n \times n$ matrix on $\mathbb{R}^n$). Thus τ is a homomorphism of H into the automorphism group of the additive group of $\mathbb{R}^n$, and is continuous in both variables. So we may form the semidirect product $G = \mathbb{R}^n \underset{\tau}{\times} H$, consisting of all pairs $\langle t, a \rangle (t \in \mathbb{R}^n, a \in H)$, with multiplication

$$\langle t, a \rangle \langle t', a' \rangle = \langle t + at', aa' \rangle.$$

This has $\mathbb{R}^n$ as a closed Abelian normal subgroup, and by (4) $G/\mathbb{R}^n \cong H$.

Notice that G also can be represented as a matrix group. Indeed, the map

$$\langle t, a \rangle \mapsto \begin{pmatrix} & & \vdots & t_1 \\ & a & \vdots & t_2 \\ & & \vdots & \vdots \\ & & \vdots & t_n \\ \hline 0 & \dots & 0 & \vdots & 1 \end{pmatrix} \qquad (6)$$

$(t \in \mathbb{R}^n; a \in H)$ is a (homeomorphic) isomorphism of G onto a topological subgroup of $GL(n + 1, \mathbb{R})$.

If we momentarily agree to identify the vector x of $\mathbb{R}^n$ with the vector $\langle x_1, \dots, x_n, 1 \rangle$ of $\mathbb{R}^{n+1}$, (6) shows us that G acts as a topological transformation group on $\mathbb{R}^n$, the action being:

$$\langle t, a \rangle x = ax + t \qquad\qquad (\langle t, a \rangle \in G; x \in \mathbb{R}^n).$$

This action is faithful; that is, if $\langle t, a \rangle x = x$ for all x in $\mathbb{R}^n$, then $\langle t, a \rangle$ is the unit of G. The range of this action is the subgroup of the group of all permutations of $\mathbb{R}^n$ which is generated by H together with the set of all translations $x \mapsto x + t$ (t being an element of $\mathbb{R}^n$).

4.8. Here are two special cases of 4.7.

(A) The "$ax + b$" group. We take $n = 1$ and $H = \mathbb{R}_{++} = $ the multiplicative group of positive reals. Thus G consists of all pairs $\langle b, a \rangle$ where a, $b \in \mathbb{R}$ and $a > 0$, multiplication being given by:

$$\langle b, a \rangle \langle b', a' \rangle = \langle b + ab', aa' \rangle.$$

$\mathbb{R}$ is a G-transformation space under the action

$$\langle b, a \rangle \cdot x = ax + b.$$

The latter formula of course accounts for the name '"$ax + b$" group.'

(B) The Euclidean groups. Let n be any positive integer, and take $H = O(n)$. Then the G of 4.7 is the group of transformations of $\mathbb{R}^n$ generated by all translations, rotations and reflections; that is, G is the group of all those transformations of $\mathbb{R}^n$ which leave invariant the Euclidean distance between points. This G is called the *Euclidean group of n-space*.

5. Group Extensions

5.1. Fix two topological groups N and G.

Definition. An *extension of N by G* is a triple $\langle H, i, j \rangle$, where: (i) H is a topological group, (ii) i is a (homeomorphic) isomorphism of N onto a closed subgroup N' of H, and (iii) j is an open continuous homomorphism of H onto G whose kernel coincides with N'.

The extension is usually denoted schematically by

$$N \underset{i}{\to} H \underset{j}{\to} G.$$

5.2. This definition implies that N' is normal in H and (by 2.10) that $G \cong H/N'$. Conversely, let H be any topological group and N a closed normal subgroup of H; then

$$N \underset{i}{\to} H \underset{\pi}{\to} H/N,$$

where i is the identity injection and π is the continuous open surjection of 2.1, is an extension of N by H/N.

5.3 *Definition.* Two extensions

$$N \underset{i}{\to} H \underset{j}{\to} G$$

and

$$N \underset{i'}{\to} H' \underset{j'}{\to} G$$

of N by G are *isomorphic* if there exists an isomorphism $f : H \to H'$ of the topological groups H and H' such that

$$i' = f \circ i, \qquad j = j' \circ f; \tag{1}$$

that is, such that the diagram

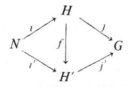

commutes.

This relation of isomorphism is clearly an equivalence relation.

Remark. It is useful to observe that the preceding definition is unaltered if we assume merely that there is a continuous homomorphism $f: H \to H'$ satisfying (1); for this implies that f must be a surjective homeomorphism.

Indeed: Suppose $f(h) = e'$ (unit of H'). Then by (1) $j(h) = (j' \circ f)(h) =$ unit of G; so $h = i(n)$ $(n \in N)$, and by (1) $f(h) = i'(n)$. Since i' is one-to-one and $i'(n) = f(h) = e'$, n must be the unit of N; so $h = e$ (unit of H). Consequently f is one-to-one. Now let $h' \in H'$. Since $j = j' \circ f$ there are an element h of H and an element n of N such that $h' = f(h)i'(n)$. Since $i' = f \circ i$, we have $h' = f(h)f(i(n)) = f(hi(n))$; so f is onto H'.

It remains only to show that f^{-1} is continuous. Assume that $f(h_\alpha) \to f(h)$ in H', where $\{h_\alpha\}$ is a net of elements of H and $h \in H$. We must show that $h_\alpha \to h$. Since we can replace $\{h_\alpha\}$ by any subnet of it, it is enough to show that some subnet of $\{h_\alpha\}$ converges to h. Now, since $f(h_\alpha) \to f(h)$, (1) implies that $j(h_\alpha) \to j(h)$. By the openness of j and II.13.2 we can pass to a subnet and choose elements k_α of H and n_α of N such that $h_\alpha = k_\alpha i(n_\alpha)$ and $k_\alpha \to h$. Applying f and using (1), we find that $f(k_\alpha) \to f(h)$ and $f(k_\alpha)i'(n_\alpha) = f(k_\alpha i(n_\alpha)) = f(h_\alpha) \to f(h)$. From these two facts we deduce that $i'(n_\alpha) \to$ unit of H', whence $n_\alpha \to$ unit of N, so that $h_\alpha = k_\alpha i(n_\alpha) \to h$. This completes the proof that f^{-1} is continuous.

5.4. The most obvious extension of N by G is the *direct product extension*

$$N \underset{i}{\to} N \times G \underset{j}{\to} G,$$

where $i(n) = \langle n, e \rangle$ $(n \in N; e =$ unit of G), and $j(n, x) = x$ $(\langle n, x \rangle \in N \times G)$.

5.5. More generally, let H_τ be the τ-semidirect product $N \underset{\tau}{\times} G$ defined as in 4.4. Then

$$N \underset{i}{\to} H_\tau \underset{j}{\to} G, \tag{2}$$

where $i(n) = \langle n, e \rangle$ $(n \in N; e =$ unit of G) and $j(n, x) = x$ $(\langle n, x \rangle \in H_\tau)$, is an extension of N by G. We call it the *τ-semidirect product extension*.

An arbitrary extension $N \underset{i}{\to} H \underset{j}{\to} G$ of N by G is called a *semidirect product extension* if it is isomorphic to (2) for some τ. By 4.6, this is the case if and only if there is a closed subgroup G' of H which is "complementary" to $N' = i(N)$; that is, $N'G' = H$, $N' \cap G' = \{e\}$, and the bijection $\langle n', x' \rangle \mapsto n'x'$ of $N' \times G'$ onto H is a homeomorphism.

Central Extensions

5.6. Quite different from the semidirect product extensions are the so-called central extensions, with which we will be much concerned in the last chapter of this book.

Definition. An extension

$$N \underset{i}{\to} H \underset{j}{\to} G$$

of N by G is called *central* if $i(N)$ is contained in the center of H. This of course implies that N is Abelian.

An extension which is isomorphic to a central extension is itself central. If N is Abelian, the direct product extension of N by G is central.

Definition. If N and G are topological groups, N being Abelian, let us denote by $\Gamma(N, G)$ the family of all isomorphism classes of central extensions of N by G.

5.7. Remark. Let N and G be topological groups, N being Abelian. It is very interesting to observe (though we shall not need the fact in this work) that $\Gamma(N, G)$ has a natural Abelian group structure. Indeed, let

$$N \underset{i_1}{\to} H_1 \underset{j_1}{\to} G \tag{3}$$

and

$$N \underset{i_2}{\to} H_2 \underset{j_2}{\to} G \tag{4}$$

be two central extensions. Let H_0 be the closed subgroup $\{\langle s, t\rangle \in H_1 \times H_2 : j_1(s) = j_2(t)\}$ of $H_1 \times H_2$, and let M be the closed central subgroup $\{\langle i_1(n), i_2(n^{-1})\rangle : n \in N\}$ of H_0. Denote the quotient topological group H_0/M by H, and define $i: N \to H$ and $j: H \to G$ as follows:

$$i(n) = \langle i_1(n), e_2\rangle M \qquad (n \in N; e_2 = \text{unit of } H_2),$$

$$j(\langle s, t\rangle M) = j_1(s) = j_2(t) \qquad\qquad (\langle s, t\rangle \in H_0).$$

Then

$$N \underset{i}{\to} H \underset{j}{\to} G \tag{5}$$

is a central extension. If ξ_1 and ξ_2 stand for the isomorphism classes of the extensions (3) and (4) respectively, then the isomorphism class ξ of (5) depends only on ξ_1 and ξ_2; we denote ξ by $\xi_1\xi_2$. Under this product

$\langle \xi_1, \xi_2 \rangle \mapsto \xi_1 \xi_2$, $\Gamma(N, G)$ turns out to be an Abelian group. The inverse operation in $\Gamma(N, G)$ is that which sends the isomorphism class of

$$N \xrightarrow[i]{} H \xrightarrow[j]{} G$$

into the isomorphism class of

$$N \xrightarrow[i^{-1}]{} H \xrightarrow[j]{} G,$$

where $i^{-1}(n) = i(n^{-1})$ $(n \in N)$. The unit element of $\Gamma(N, G)$ is the isomorphism class of the direct product extension.

The proof of these facts is quite routine, and we leave it to the reader.

Remark. The inverse of the isomorphism class of $N \xrightarrow[i]{} H \xrightarrow[j]{} G$ can also be described as the isomorphism class of

$$N \xrightarrow[i]{} \tilde{H} \xrightarrow[\tilde{j}]{} G.$$

Here $\tilde{H}$ is the reverse group of H (see 1.3) and $\tilde{j}(h) = (j(h))^{-1}$ $(h \in \tilde{H})$. (The isomorphism of these two versions of the inverse is proved by noting the isomorphism $h \mapsto h^{-1}$ of H and $\tilde{H}$.)

5.8. Proposition. *Let* $\gamma : N \xrightarrow[i]{} H \xrightarrow[j]{} G$ *and* $\gamma' : N \xrightarrow[i']{} H' \xrightarrow[j']{} G$ *be two central extensions; and form the closed subgroup*

$$H_0 = \{ \langle h, h' \rangle \in H \times H' : j'(h') = j(h) \}$$

of $H \times H'$. *A necessary and sufficient condition for* γ *and* γ' *to be isomorphic is that there should exist a continuous homomorphism* $\phi : H_0 \to N$ *satisfying*

$$\phi(i(n)h, h') = n^{-1}\phi(h, h'),$$
$$\phi(h, i'(n)h') = n\phi(h, h') \tag{6}$$

for all $\langle h, h' \rangle$ *in* H_0 *and* n *in* N.

Proof. *Necessity.* Suppose that γ and γ' are isomorphic under $F : H \to H'$. Then $F(h)^{-1}h' \in i'(N)$ for all $\langle h, h' \rangle \in H_0$, and we can define $\phi : H_0 \to N$ by means of the equation:

$$\phi(h, h') = i'^{-1}[F(h)^{-1}h'] \qquad (\langle h, h' \rangle \in H_0).$$

One verifies without difficulty that ϕ is a continuous homomorphism satisfying (6).

Sufficiency. Let $\phi: H_0 \to N$ be a continuous homomorphism satisfying (6). We define $F: H \to H'$ as follows: Given $h \in H$, we can find an element h' of H' such that $j(h) = j'(h')$. Now set

$$F(h) = (i'(\phi(h, h')))^{-1} h' \in H'. \tag{7}$$

One verifies from (6) that the right side of (7) is unaltered on replacing h' by $i'(m)h'$ ($m \in N$); hence it depends only on h, and (7) is a legitimate definition. Since ϕ and i' are homomorphisms and $i'(N)$ is central, $F: H \to H'$ is a homomorphism. It also follows easily from (6) that

$$F \circ i = i', \qquad j' \circ F = j. \tag{8}$$

We claim that F is continuous. Indeed, let $h_\alpha \to h$ in H. Then $j(h_\alpha) \to j(h)$, and by the openness of j' and II.13.2 we can pass to a subnet and choose elements $\{h'_\alpha\}$ and h' of H' such that $\langle h_\alpha, h'_\alpha \rangle$ and $\langle h, h' \rangle$ are in H_0 and $\langle h_\alpha, h'_\alpha \rangle \to \langle h, h' \rangle$ in H_0. Since ϕ is continuous, it then follows from (7) that $F(h_\alpha) \to F(h)$, proving the claim.

We have shown that $F: H \to H'$ is a continuous homomorphism satisfying (8). So by Remark 5.3 γ and γ' are isomorphic. ∎

5.9. Corollary. *A central extension $\gamma: N \xrightarrow{i} H \xrightarrow{j} G$ is isomorphic to the direct product extension if and only if there exists a continuous homomorphism $f: H \to N$ such that $f(i(n)) = n$ for all n in N.*

Proof. Assume that γ is equal to the direct product extension. Then for f we can take the projection homomorphism $\langle n, x \rangle \mapsto n$ ($\langle n, x \rangle \in N \times G$). Thus the "only if" part holds.

Assume that such an f exists; and let $\gamma': N \xrightarrow{i'} H' = N \times G \xrightarrow{j'} G$ be the direct product extension. Then, if H_0 is defined as in 5.8, the equation

$$\phi(h, \langle n, x \rangle) = nf(h)^{-1} \qquad (n \in N; h \in H; x = j(h))$$

defines a continuous homomorphism $\phi: H_0 \to N$ satisfying (6). Therefore by 5.8 γ and γ' are isomorphic. ∎

5.10. Remark. 5.8 is an easy consequence of 5.7 and 5.9. Indeed, if ξ and ξ' are the isomorphism classes of the γ and γ' of 5.8, the condition in 5.8 asserts by 5.9 that $\xi\xi'^{-1}$ (constructed as in 5.7) is the unit class of $\Gamma(N, G)$. We have preferred, however, not to assume 5.7.

Cases where $\Gamma(N, G)$ *is Trivial*

5.11. It is an important fact that in certain cases $\Gamma(N, G)$ is *trivial*, that is, contains only one element (the class of the direct product extension).

Proposition. $\Gamma(N, G)$ *is trivial in the following cases:*

(I) $G = \mathbb{Z}$ *and* N *is any Abelian topological group.*

(II) G *is the finite cyclic group* $\mathbb{Z}_p$ *of order* p, *and* N *is any Abelian topological group such that for any* n *in* N *there exists an element* m *of* N *satisfying* $m^p = n$. *(For example,* N *could be* $\mathbb{R}^q$ *or* $\mathbb{E}^q$.)

(III) $G = \mathbb{R}$ *and* N *is any locally compact Abelian group.*

(IV) $G = \mathbb{E}$ *and* $N = \mathbb{E}^p$ *for some positive integer* p.

Proof. (I) Let $\gamma: N \underset{i}{\to} H \underset{j}{\to} \mathbb{Z}$ be a central extension. If we choose any element h of $j^{-1}(1)$, then $\langle n, p \rangle \mapsto i(n)h^p$ is an isomorphism of γ with the direct product extension.

(II) Let $\gamma: N \underset{i}{\to} H \underset{j}{\to} \mathbb{Z}_p$ be a central extension, a a generator of $\mathbb{Z}_p$, and h any element of $j^{-1}(a)$. Then $h^p \in i(N)$; so that by the hypothesis of (II) there is an element n of N such that $(hi(n))^p = e$ (unit of H). Putting $k = hi(n)$, we see that $\langle m, a^q \rangle \mapsto i(m)k^q$ is an isomorphism of γ with the direct product extension.

(III) Let $\gamma: N \underset{i}{\to} H \underset{j}{\to} \mathbb{R}$ be a central extension. We claim that H is Abelian. Indeed: Let h be an element of H and L its (closed) centralizer $\{h' \in H: hh' = h'h\}$. We shall show that $L = H$. This is obvious if $h \in i(N)$ since $i(N)$ is central. So we assume that $h \notin i(N)$, that is, $t = j(h) \neq 0$. If p is a positive integer and k is an element of H such that $j(k) = p^{-1}t$, then $j(k^p) = j(h)$, whence $h = k^p i(n)$ for some n in N, implying (since $i(N)$ is central) that $hk = kh$, or $k \in L$. Since L is a subgroup and $i(N) \subset L$, it follows that L contains $j^{-1}(p^{-1}t)$ for every $p = 1, 2, \ldots,$ and hence contains $j^{-1}(rt)$ for every rational r. But $\{rt: r \text{ rational}\}$ is dense in $\mathbb{R}$ (recall that $t \neq 0$); and so (since j is open) $\bigcup\{j^{-1}(rt): r \text{ rational}\}$ is dense in H. Thus L is dense in H, and so equal to H, proving the claim.

Now by 2.4 H is locally compact. We shall first make the assumption that H is compactly generated (i.e., there exists a compact subset C of H such that the smallest subgroup of H containing C is H itself). The structure theory of locally compact compactly generated Abelian groups (see for example Hewitt and Ross [1], Theorem 9.8) tells us that

$$H = E \times \mathbb{R}^n \times \mathbb{Z}^m,$$

where E is a compact Abelian group and n and m are non-negative integers. Since $\mathbb{R}$ has no non-trivial compact subgroups, we have $j(E) = \{0\}$, that is, $E \subset i(N)$. It follows that

$$i(N) = E \times L,$$

where L is a closed subgroup of $\mathbb{R}^n \times \mathbb{Z}^m$. Now it is easy to see that any continuous group-homomorphism $\phi \colon \mathbb{R}^n \to \mathbb{R}$ is actually linear, so that there is a closed subgroup (in fact a linear subspace) of $\mathbb{R}^n$ complementary (see 4.6) to $\mathrm{Ker}(\phi)$; and clearly the same is true for any continuous surjective group-homomorphism $\mathbb{R}^n \times \mathbb{Z}^m \to \mathbb{R}$. Taking the direct product with E, we see that there is a closed subgroup of H complementary to $\mathrm{Ker}(j) = i(N)$. So γ is the direct product extension.

Now drop the assumption that H is compactly generated, and let H_0 be the open subgroup of H generated by a compact neighborhood of e in H. If $N_0 = i^{-1}(H_0)$, $i_0 = i|N_0$ and $j_0 = j|H_0$, then

$$\gamma_0 \colon N_0 \underset{i_0}{\to} H_0 \underset{j_0}{\to} \mathbb{R}$$

is a central extension; and H_0 is compactly generated. So by the preceding paragraph γ_0 is the direct product extension, that is, there is a closed subgroup K complementary to $i_0(N_0)$ in H_0. It is now easy to see that K is complementary to $i(N)$ in H. So γ is the direct product extension.

(IV) Let $\gamma \colon \mathbb{E}^p \underset{i}{\to} H \underset{j}{\to} \mathbb{E}$ be a central extension. Form the closed subgroup $H_0 = \{\langle t, x \rangle \in \mathbb{R} \times H \colon e^{it} = j(x)\}$ of $\mathbb{R} \times H$; and put $i_0(u) = \langle 0, i(u) \rangle (u \in \mathbb{E}^p)$, $j_0(t, x) = t \, (\langle t, x \rangle \in H_0)$. We verify that $\gamma_0 \colon \mathbb{E}^p \underset{i_0}{\to} H_0 \underset{j_0}{\to} \mathbb{R}$ is a central extension of $\mathbb{E}^p$ by $\mathbb{R}$. By (III) γ_0 is isomorphic to the direct product extension, and so by 5.9 there is a continuous homomorphism $\phi_0 \colon H_0 \to \mathbb{E}^p$ satisfying

$$\phi_0(i_0(u)) = u \qquad\qquad (u \in \mathbb{E}^p). \qquad (9)$$

Since $\langle 2\pi n, e \rangle \in H_0$ for all integers n (e being the unit of H), $n \mapsto \phi_0(2\pi n, e)$ is a homomorphism of $\mathbb{Z}$ into $\mathbb{E}^p$; and there exists a continuous homomorphism $\chi \colon \mathbb{R} \to \mathbb{E}^p$ satisfying

$$\chi(2\pi n) = \phi_0(2\pi n, e) \qquad\qquad (n \in \mathbb{Z}). \qquad (10)$$

We now put $\phi_0'(t, x) = (\chi(t))^{-1}\phi_0(t, x) \, (\langle t, x \rangle \in H_0)$, so that ϕ_0' is again a continuous homomorphism of H_0 into $\mathbb{E}^p$. By (10) $\langle 2\pi n, e \rangle \in \mathrm{Ker}(\phi_0')$. Since $\{\langle 2\pi n, e \rangle \colon n \in \mathbb{Z}\}$ is the kernel of the open continuous homomorphism $\langle t, x \rangle \mapsto x$ of H_0 onto H, ϕ_0' gives rise to the continuous homomorphism $\phi \colon H \to \mathbb{E}^p$:

$$\phi(x) = \phi_0'(t, x) \qquad\qquad (\langle t, x \rangle \in H_0);$$

and, if $u \in E^p$,

$$\phi(i(u)) = \phi'_0(0, i(u)) = \phi_0(0, i(u)) = u$$

by (9). So by 5.9 γ is isomorphic to the direct product extension. ∎

Remark. The most important assertion in the above Proposition, from the point of view of group representations, is (III). It is unfortunate that for its proof we had to call upon Theorem 9.8 of Hewitt and Ross [1]. However, this Proposition will be used merely for illustrative purposes, not in the main stream of the exposition in this work.

The Description of Central Extensions by Cocycles

5.12. In the past literature of the theory of group representations, central extensions have usually been discussed in terms of cocycles. For topological reasons it is difficult to apply the cocycle description if the groups are not second-countable, and in this work we shall manage to avoid it entirely in the formal development. However, to clarify the connection between the work of other authors and our development, we shall present the cocycle description in a context where topological difficulties are avoided—namely, when G is discrete.

Let G be a fixed discrete group with unit e, and N an Abelian (multiplicative) topological group with unit $\mathbb{1}$. By a *cocycle* (*for N, G*) we shall mean a function $\sigma : G \times G \to N$ such that:

$$\sigma(e, x) = \sigma(x, e) = \mathbb{1}, \tag{11}$$

$$\sigma(xy, z)\sigma(x, y) = \sigma(x, yz)\sigma(y, z) \tag{12}$$

(for all $x, y, z \in G$). Condition (12) is called the *cocycle identity*. If we take $y = x^{-1}$, $z = x$, conditions (11) and (12) combine to give

$$\sigma(x, x^{-1}) = \sigma(x^{-1}, x). \tag{13}$$

The set $C = C(N, G)$ of all cocycles is clearly an Abelian group under pointwise multiplication.

If $\rho : G \to N$ is a function satisfying $\rho(e) = \mathbb{1}$, the equation

$$\sigma_\rho(x, y) = \rho(xy)(\rho(x)\rho(y))^{-1} \tag{14}$$

defines a cocycle σ_ρ. The set of all cocycles of the form σ_ρ for some such ρ is denoted by B or $B(N, G)$; and its elements are called *coboundaries*. Clearly B is a subgroup of C. The quotient group $Z(N, G) = Z = C/B$ is the *cocycle class group* (or the *second cohomology group of G with values in N*); and its elements are *cocycle classes*. Two cocycles belonging to the same cocycle class are *cohomologous*.

5.13.   Let σ be a fixed cocycle. There is a canonical way to construct from σ a central extension

$$\gamma_\sigma : N \underset{i_\sigma}{\to} H_\sigma \underset{j}{\to} G \tag{15}$$

of N by G.

Indeed: Let H_σ, as a topological space, be $N \times G$; and define multiplication in H_σ as follows:

$$\langle n, x \rangle \langle m, y \rangle = \langle nm(\sigma(x, y))^{-1}, xy \rangle \tag{16}$$

$(n, m \in N; x, y \in G)$. One verifies (using (13)) that H_σ is a topological group under (16), the unit element being $\langle 1, e \rangle$ and $\langle n, x \rangle^{-1}$ being $\langle \sigma(x, x^{-1})n^{-1}, x^{-1} \rangle$. If we now set $i_\sigma(n) = \langle n, e \rangle$ and $j_\sigma(n, x) = x$, it is clear that (15) is a central extension.

It turns out that all central extensions of N by G are essentially obtained in this way from cocycles. In fact we have:

Proposition.   *Every central extension of N by G is isomorphic with γ_σ for some σ in C. if σ, τ are two cocycles, then γ_σ and γ_τ are isomorphic if and only if σ and τ are cohomologous.*

Proof.   Let $\gamma : N \underset{i}{\to} H \underset{j}{\to} G$ be a central extension. For each x in G pick out an element τ_x of the coset $j^{-1}(x)$, making sure that τ_e is the unit of H. Thus for every pair x, y of elements of G, there is a unique element $\sigma(x, y)$ of N such that

$$\tau_{xy} = \tau_x \tau_y i(\sigma(x, y)). \tag{17}$$

Now σ is a cocycle. Indeed, (11) is obvious since τ_e is the unit. By (17) and the centrality of γ,

$$(\tau_x \tau_y)\tau_z i(\sigma(xy, z)\sigma(x, y)) = \tau_{xy}\tau_z i(\sigma(xy, z))$$

$$= \tau_{xyz} = \tau_x \tau_{yz} i(\sigma(x, yz))$$

$$= \tau_x(\tau_y \tau_z)i(\sigma(x, yz)\sigma(y, z)), \tag{18}$$

whence (12) follows. If we now construct the central extension $\gamma_\sigma : N \to H_\sigma \to G$ corresponding to this σ, it is easy to verify that $\phi : \langle n, x \rangle \mapsto \tau_x i(n)$ $(\langle n, x \rangle \in H_\sigma)$ sets up an isomorphism between γ_σ and γ. So the first statement of the Proposition is proved.

Now let σ and α be two cocycles.

If σ and α are cohomologous, there is a function $\rho: G \to N$ with $\rho(e) = 1$ such that $\alpha(x, y) = \sigma(x, y)\rho(x)\rho(y)(\rho(xy))^{-1}(x, y \in G)$. The map $\phi: \langle n, x \rangle \mapsto \langle n\rho(x), x \rangle$ is then easily seen to be an isomorphism of γ_σ and γ_α.

Conversely, suppose γ_σ and γ_α are isomorphic under an isomorphism $\phi: H_\sigma \to H_\alpha$. For each x in G ϕ carries $\langle N, x \rangle$ onto $\langle N, x \rangle$; so there is an element $\rho(x)$ of N such that $\phi(1, x) = \langle \rho(x), x \rangle$. Since $\phi \circ i_\sigma = i_\alpha$, we have $\phi(n, e) = (n, e)$. In particular $\rho(e) = 1$. Also $\phi(n, x) = \phi(\langle n, e \rangle \langle 1, x \rangle) = \langle n, e \rangle \langle \rho(x), x \rangle = \langle n\rho(x), x \rangle$. Consequently, for any x, y in G,

$$\langle \sigma(x, y)^{-1}\rho(xy), xy \rangle = \phi(\sigma(x, y)^{-1}, xy)$$

$$= \phi[\langle 1, x \rangle \langle 1, y \rangle] \qquad \text{(multiplication in } H_\sigma)$$

$$= \phi(1, x)\phi(1, y) \qquad \text{(multiplication in } H_\alpha)$$

$$= \langle \rho(x)\rho(y)(\alpha(x, y))^{-1}, xy \rangle,$$

whence $\sigma(x, y)^{-1}\rho(xy) = \rho(x)\rho(y)(\alpha(x, y))^{-1}$. So σ and α are cohomologous, and the Proposition is proved. ∎

Remark. This Proposition implies of course a one-to-one correspondence between the cocycle classes and the isomorphism classes in $\Gamma(N, G)$ (see 5.6). One verifies that under this correspondence the Abelian group structure of $\Gamma(N, G)$ mentioned in 5.7 goes into the quotient group structure of $Z(N, G)$. The unit element of $Z(N, G)$ corresponds to the direct product extension.

6. Topological Fields

6.1. Definition. A *topological ring* is a ring A with a Hausdorff topology such that the ring operations $x \mapsto -x$, $\langle x, y \rangle \mapsto x + y$, and $\langle x, y \rangle \mapsto xy$ are continuous with respect to this topology.

In particular, the additive group of a topological ring is an Abelian topological group.

6.2. Definition. A *topological division ring* is a division ring F with a topology such that F is a topological ring and, in addition, the multiplicative inverse map $x \mapsto x^{-1}$ is continuous on $F \setminus \{0\}$.

Thus, if F is a topological division ring, $F_* = F \setminus \{0\}$ is a topological group under multiplication. In this case each element x of F_* gives rise to a (homeomorphic) automorphism $y \mapsto xy$ ($y \in F$) of the additive topological group of F.

A commutative topological division ring is called a *topological field*.

Remark. The zero and unit elements of any topological division ring will always be called 0 and 1.

6.3. The real field $\mathbb{R}$ and the complex field $\mathbb{C}$ are of course connected locally compact topological fields. It can be shown (see for example Chapter IV, Pontryagin [6]) that $\mathbb{R}$, $\mathbb{C}$, and the quaternion division ring are the only connected locally compact topological division rings.

6.4. From here to 6.10 F is a fixed locally compact topological division ring whose topology is *not discrete*. We shall obtain a few simple general facts about F which are basic to its detailed structure theory (as given for example in Chapter IV, Pontryagin [6]), and which will be useful in Chapter X.

6.5. Proposition. *If $\{x_i\}$ is a net of non-zero elements of F converging to 0, then $x_i^{-1} \to \infty$ in F. In particular F cannot be compact.*

Proof. Suppose the first statement is false. Then there is a subnet of $\{x_i^{-1}\}$ which stays inside some compact set, and hence a subnet which is convergent. So, replacing $\{x_i\}$ by a subnet, we may assume that $x_i^{-1} \to y$ in F. Thus $1 = x_i x_i^{-1} \to 0y = 0$, a contradiction. This proves the first statement.

Since F is not discrete, there does exist a net of non-zero elements converging to 0. Therefore the second statement follows from the first. ∎

6.6. Proposition. *If K is a compact subset of F and U is a neighborhood of 0, there exists a neighborhood W of 0 such that $KW \subset U$.*

Proof. If not, there would be a net $\{x_i\}$ approaching 0 in F and a net $\{y_i\}$ of elements of K such that $y_i x_i \notin U$ for all i. Passing to a subnet we can assume that $y_i \to y$ in K. But then $y_i x_i \to y0 = 0$, a contradiction. ∎

6.7. Corollary. *F satisfies the first axiom of countability.*

Proof. Since F is non-discrete it has an infinite compact set, hence a countably infinite subset C with a limit point u not in C. Translating additively we may assume that $u = 0$. Thus if U is any open neighborhood of 0 with compact closure, it follows from 6.6 that $\{Ux : x \in C\}$ is a basis of neighborhoods of 0. ∎

6.8. Lemma. *There exists a non-zero element t of F such that $t^n \to 0$ as $n \to \infty$ (n being a positive integer).*

Proof. Let U be an open neighborhood of 0 with compact closure such that $1 \notin \bar{U}$; and put $K = \bar{U} \cup \{1\}$. By 6.6 we have $Kt \subset U$ for some $t \neq 0$. By induction this implies $Kt^n \subset U$ for all $n = 1, 2, \ldots$. We claim $t^n \to 0$ as $n \to \infty$. Indeed: Since $t^n \in \bar{U}$ for all $n = 1, 2, \ldots$, it is enough to derive a contradiction from the assumption that some subnet $\{t^{n_i}\}$ of $\{t^n\}$ converges to $c \neq 0$ (and hence $t^{-n_i} \to c^{-1}$). But this implies that $1 = cc^{-1}$ is a limit point of $\{t^m : m = 1, 2, \ldots\}$, contradicting the fact that $t^m \in \bar{U}$, $1 \notin \bar{U}$. ∎

6.9. Proposition. *If $\{x_i\}$ is a net of elements of F such that $x_i \to \infty$, then $x_i^{-1} \to 0$.*

Proof. Choose t as in 6.8, and let U be an open neighborhood of 0 with compact closure. By 6.5 and 6.8 we have:

$$t^n \to 0 \quad \text{as} \quad n \to \infty,$$
$$t^n \to \infty \quad \text{as} \quad n \to -\infty. \tag{1}$$

Consequently, for each $0 \neq x \in F$, there are unique integers $r(x)$ and $s(x)$ such that

$$xt^n \in \bar{U} \quad \text{for all } n \geq r(x),$$
$$t^n x \in \bar{U} \quad \text{for all } n \geq s(x), \tag{2}$$
$$xt^{r(x)-1} \notin \bar{U}, \qquad t^{s(x)-1}x \notin \bar{U}.$$

Multiplying the last equations by t, we find

$$xt^{r(x)} \in \bar{U} \setminus \bar{U}t, \qquad t^{s(x)}x \in \bar{U} \setminus t\bar{U}. \tag{3}$$

Now suppose that $x_i \to \infty$ in F. We claim that $r(x_i) \to \infty$ and $s(x_i) \to \infty$. Indeed: if $r(x_i) \leq k < \infty$ for arbitrarily large i, then by (2) $x_i t^k \in \bar{U}$ or $x_i \in \bar{U}t^{-k}$ for arbitrarily large i, contradicting $x_i \to \infty$. So $r(x_i) \to \infty$. Similarly $s(x_i) \to \infty$.

We shall now show that $x_i^{-1} \to 0$. Since $\{x_i^{-1}\}$ cannot approach a non-zero limit in F, it is enough to assume $x_i^{-1} \to \infty$ and obtain a contradiction. Define $b_i = t^{s(x_i^{-1})}x_i^{-1}$, $c_i = x_i t^{r(x_i)}$. By the above claim and (1), $b_i c_i = t^{s(x_i^{-1})+r(x_i)} \to 0$. On the other hand, by (3), we can pass to subnets and assume $b_i \to b$ in the compact set $\bar{U} \setminus tU$, and $c_i \to c$ in the compact set $\bar{U} \setminus Ut$. Thus $b \neq 0$, $c \neq 0$, and $b_i c_i \to bc \neq 0$, a contradiction. ∎

6.10. Corollary. *F satisfies the second axiom of countability.*

Proof. Since F is first-countable we have only to show (see 1.5) that F is σ-compact. Take a decreasing basis $\{U_n\}(n = 1, 2, \ldots)$ of open neighborhoods of 0; and let $V_n = (U_n \setminus \{0\})^{-1}$, $C_n = F \setminus V_n$. Thus C_n is closed for each n. If it were not compact, we could find a net $\{x_i\}$ converging to ∞ in C_n; and this would contradict 6.9. Clearly $\bigcup_n C_n = F$. ∎

The p-adic Fields

6.11. There are important classes of non-discrete locally compact topological fields F other than $\mathbb{R}$ and $\mathbb{C}$. Notice that such an F must be totally disconnected. (Indeed, by 6.3 it cannot be connected. So there is a non-trivial open-closed subset A of F; and $\{yA + x : x, y \in F, y \neq 0\}$ is a family of open-closed subsets of F which separates points.) The most important such topological fields are the p-adic fields, whose construction we now sketch.

6.12. Let Q be the rational number field, and fix a prime number $p = 2, 3, 5, \ldots$

Every non-zero number r in Q can be written in the form

$$r = m^{-1}np^t, \tag{4}$$

where n, m are non-zero integers not divisible by p, and t is an integer. Clearly t depends only on r (not on the particular choice of n and m); so we are free to define

$$w(r) = p^{-t}.$$

Making the supplementary definition $w(0) = 0$, we obtain a function $w : Q \to [0, \infty)$ with the following easily verifiable properties:

$$w(r) = 0 \Leftrightarrow r = 0; \tag{5}$$

$$w(-r) = w(r); \tag{6}$$

$$w(r + s) \leq \max(w(r), w(s)); \tag{7}$$

$$w(rs) = w(r)w(s) \tag{8}$$

$(r, s \in Q)$. In particular, the equation

$$d(r, s) = w(r - s) \qquad (r, s \in Q)$$

defines a metric d on Q. The completion of Q under the metric d will be denoted by F_p.

If $\{r_n\}$ and $\{s_n\}$ are Cauchy sequences of elements of Q with respect to d, the properties (6), (7), (8) imply that $\{-r_n\}$, $\{r_n + s_n\}$, and $\{r_n s_n\}$ are also Cauchy sequences. Thus if r and s are the limits of $\{r_n\}$ and $\{s_n\}$ in F_p respectively, we are at liberty to define $-r, r + s$, and rs as the limits in F_p of $\{-r_n\}, \{r_n + s_n\}$, and $\{r_n s_n\}$ respectively. If in addition $r \neq 0$, we have by (7), (8),

$$w(r_n^{-1} - r_m^{-1}) = (w(r_n)w(r_m))^{-1}w(r_m - r_n); \tag{9}$$

and since $w(r_n) = d(r_n, 0) \to d(r, 0) \neq 0$, the right side of (9) approaches 0 as $n, m \to \infty$; therefore $\{r_n^{-1}\}$ is a Cauchy sequence, whose limit in F_p we define to be r^{-1}. Under these extended operations F_p becomes a field; the algebraic properties required of them are easily verified from the corresponding properties in Q by passing to the limit.

The function w extends uniquely to a continuous function, also called w, on F_p to $[0, \infty)$; indeed, we have only to define

$$w(r) = d(r, 0) \qquad\qquad (r \in F_p).$$

Properties (5)–(8) continue to hold for all r, s in F_p; and in addition

$$d(r, s) = w(r - s) \qquad\qquad (r, s \in F_p). \tag{10}$$

From (6), (7), (8), (10) we deduce that the field operations in F_p are continuous with respect to d, so that F_p (equipped with the topology of the metric d) becomes a topological field.

Definition.   The topological field F_p is called the *p-adic field*; and its elements are called *p-adic numbers*.

6.13. Set $I_p = \{r \in F_p : w(r) \leq 1\}$. By (6), (7), (8), I_p is closed under addition, negation, and multiplication; it is therefore a subring of F_p. Its elements are called *p-adic integers*.

We shall show that I_p is compact (in the topology of d). For this we need a Lemma:

Lemma.   *For each r in I_p and each positive integer n, there is a unique integer $s = 0, 1, 2, \ldots, p^n - 1$ such that*

$$w(r - s) \leq p^{-n}. \tag{11}$$

Proof.   To prove the existence of s, we shall first assume that $r \in Q$, so that $r = uv^{-1}$, where u and v are integers and p does not divide v. Elementary number theory then assures us that there are integers s and k satisfying $u = sv + kp^n$; translating s by a suitable multiple of p^n, we may suppose that $0 \leq s < p^n$. This implies that $r - s = uv^{-1} - s = kv^{-1}p^n$, whence (11) holds. Having now proved the existence of s for $r \in Q \cap I_p$ (which is dense in I_p), we obtain its existence for arbitrary r in I_p by an easy passage to the limit.

To see the uniqueness of s, suppose that s' is another integer such that $0 \leq s' < p^n$, $w(r - s') \leq p^{-n}$. Then by (7) $w(s' - s) \leq \max(w(r - s), w(r - s')) \leq p^{-n}$, so that p^n divides $s' - s$, an impossibility unless $s' = s$. ∎

6.14. Corollary. I_p is compact. Thus F_p is a non-discrete totally disconnected locally compact topological field.

Proof. By Lemma 6.13 I_p is totally bounded with respect to the metric d. Being a closed subset of a complete metric space, it is therefore compact. This proves the first statement. Since I_p is a neighborhood of 0 in F_p and compact, its translate $x + I_p$ is a compact neighborhood of the arbitrary point x of F_p. So F_p is locally compact. Since I_p is infinite and compact its topology is certainly not discrete. Finally, from the fact that w is continuous on F_p and its range is $\{p^t : t \in \mathbb{Z}\} \cup \{0\}$, we see that I_p is open-closed in F_p; so by the argument of 6.11 F_p is totally disconnected. ∎

7. Haar Measure

7.1. Throughout this section we fix a locally compact group G, with unit e. As usual $\mathscr{L}(G)$ is the space of continuous complex functions on G with compact support; and $\mathscr{L}_+(G)$ is the set of all non-negative real-valued functions in $\mathscr{L}(G)$. If $f \in \mathscr{L}(G)$, put $\|f\|_\infty = \sup_{x \in G} |f(x)|$; and, for $x \in G$, let f_x and f^x be the left and right translates given by $f_x(y) = f(xy)$, $f^x(y) = f(yx)$.

Our goal is to construct an integral I on G (see II.8.12) which will be *left-invariant* (or *right-invariant*), that is, which will satisfy $I(f_x) = I(f)$ (or $I(f^x) = I(f)$) for all f in $\mathscr{L}(G)$ and x in G.

7.2. Lemma. *Assume that $f, g \in \mathscr{L}_+(G)$ and $g \neq 0$. Then*

$$f \leq \sum_{i=1}^{r} c_i g_{u_i} \tag{1}$$

for some positive integer r, non-negative numbers $c_1, \ldots, c_r$, and elements $u_1, \ldots, u_r$ of G.

Proof. Since $g \neq 0$, to each x in G there is a number $c \geq 0$ and an element u of G such that $f \leq cg_u$ everywhere on some neighborhood of x. Covering the compact support of f with finitely many such neighborhoods, and adding the corresponding functions cg_u, we obtain (1). ∎

7.3. Definition   If $f, g \in \mathcal{L}_+(G)$ and $g \neq 0$, set

$$(f; g) = \inf\left\{ \sum_{i=1}^{r} c_i \right\},$$

where $\langle c_1, \ldots, c_r \rangle$ runs over all finite sequences of non-negative numbers for which group elements $u_1, \ldots, u_r$ exist satisfying (1).

By Lemma 7.2, $0 \leq (f; g) < \infty$.

Remark. $(f; g)$ might be called the "approximate integral" of f relative to g. It is analogous to approximating the volume of a region V in space by finding the smallest number $(V; U)$ of translates of a given "test region" U which are required to cover V.

7.4. Lemma.   *Assume that $f, f_1, f_2, g, h \in \mathcal{L}_+(G)$ and $g \neq 0, h \neq 0$. Then*:

(i) $(f_x; g) = (f; g)$ for all x in G;
(ii) $(f_1 + f_2; g) \leq (f_1; g) + (f_2; g)$;
(iii) $(cf; g) = c(f; g)(0 \leq c \in \mathbb{R})$;
(iv)   if $f_1 \leq f_2$, then $(f_1; g) \leq (f_2; g)$;
(v) $(f; h) \leq (f; g)(g; h)$;
(vi) $(f; g) \geq \|g\|_\infty^{-1} \|f\|_\infty$; in particular $(f; g) > 0$ if $f \neq 0$.

Proof.   The proof of (i)–(iv) is quite routine and is left to the reader. In the context of (1) we have $\|f\|_\infty \leq \sum_{i=1}^{r} c_i \|g\|_\infty$. So (vi) holds. To prove (v), let

$$f \leq \sum_{i=1}^{r} c_i g_{u_i}, \qquad g \leq \sum_{j=1}^{s} d_j h_{v_j},$$

where $c_i, d_j \geq 0$ and $u_i, v_j \in G$. Then $g_{u_i} \leq \sum_j d_j (h_{v_j})_{u_i}$; so

$$f \leq \sum_{i=1}^{r} \sum_{j=1}^{s} c_i d_j (h_{v_j})_{u_i}.$$

It follows that

$$(f; h) \leq \sum_{i=1}^{r} \sum_{j=1}^{s} c_i d_j = \left(\sum_{i=1}^{r} c_i \right)\left(\sum_{j=1}^{s} d_j \right).$$

Letting $\sum_i c_i$ and $\sum_j d_j$ sink toward their respective infima, we obtain (v).   ∎

7.5.   *Definition.* Let us fix a non-zero element g of $\mathscr{L}_+(G)$. If f, $\phi \in \mathscr{L}_+(G)$ and $\phi \neq 0$, define

$$I_\phi(f) = (g;\phi)^{-1}(f;\phi).$$

Remark. In the analogy of volume measurement given in Remark 7.3, the ratio $(W;U)^{-1}(V;U)$ is the approximate ratio of the volumes of V and W, when the latter are estimated by means of the "test region" U. One expects that as the test region U becomes smaller and smaller (hence more "discriminating"), the ratio $(W;U)^{-1}(V;U)$ will approach the true ratio of the volumes of V and W. Similarly we shall find here that as ϕ becomes smaller and smaller $I_\phi(f)$ will approach the ratio of the Haar integrals of f and g.

7.6.   Lemma.   *Let* $f, f_1, f_2 \in \mathscr{L}_+(G)$. *Then:*

(i)   *If* $f \neq 0$, $(g;f)^{-1} \leq I_\phi(f) \leq (f;g)$;
(ii) $I_\phi(f_x) = I_\phi(f)\ (x \in G)$;
(iii) $I_\phi(f_1 + f_2) \leq I_\phi(f_1) + I_\phi(f_2)$;
(iv) $I_\phi(cf) = cI_\phi(f)\ (c \in \mathbb{R}_+)$.

Proof.   By (v), (i), (ii), (iii) of Lemma 7.4.   ∎

7.7. We next show that, if ϕ has small compact support, I_ϕ is "nearly additive."

Lemma.   *Given* $f_1, f_2 \in \mathscr{L}_+(G)$ *and* $\varepsilon > 0$, *we can find a neighborhood* V *of* e *such that, if* $0 \neq \phi \in \mathscr{L}_+(G)$ *and* $\mathrm{supp}(\phi) \subset V$, *then*

$$|I_\phi(f_1) + I_\phi(f_2) - I_\phi(f_1 + f_2)| \leq \varepsilon. \tag{2}$$

Proof.   Fix a non-zero function f' in $\mathscr{L}_+(G)$ which is strictly positive everywhere on the compact support of $f_1 + f_2$; and let δ be such a positive number that

$$(f';g)\delta(1 + 2\delta) + 2\delta(f_1 + f_2; g) < \varepsilon. \tag{3}$$

Now put $f = f_1 + f_2 + \delta f'$, and

$$h_i(x) = \begin{cases} \dfrac{f_i(x)}{f(x)} & \text{if } f(x) \neq 0, \\[2mm] 0 & \text{if } f(x) = 0 \end{cases}$$

$(i = 1, 2; x \in G)$. One verifies easily that $h_i \in \mathscr{L}_+(G)$.

By the uniform continuity of h_i (see 1.10) we can choose a neighborhood V of e so that

$$|h_i(x) - h_i(y)| < \delta \qquad \text{for } i = 1, 2 \text{ and } x^{-1}y \in V. \tag{4}$$

Now let ϕ be any non-zero function in $\mathscr{L}_+(G)$ with compact support contained in V. We claim that (2) holds.

Let $f \le \sum_{j=1}^r c_j \phi_{u_j}$ $(0 \le c_j \in \mathbb{R}; u_j \in G)$. If $\phi(u_j x) \ne 0$, then $u_j x \in V$ so that by (4)

$$|h_i(u_j^{-1}) - h_i(x)| < \delta.$$

Hence for $i = 1, 2$, and for *every* x in G, we have

$$f_i(x) = h_i(x)f(x) \le \sum_{j=1}^r c_j \phi(u_j x)h_i(x)$$

$$\le \sum_{j=1}^r c_j \phi(u_j x)(h_i(u_j^{-1}) + \delta);$$

whence

$$(f_i; \phi) \le \sum_{j=1}^r c_j(h_i(u_j^{-1}) + \delta). \tag{5}$$

But $h_1 + h_2 \le 1$; so by (5)

$$(f_1; \phi) + (f_2; \phi) \le \sum_{j=1}^r c_j(1 + 2\delta).$$

Using the arbitrariness in the c_j, we thus obtain:

$$(f_1; \phi) + (f_2; \phi) \le (1 + 2\delta)(f; \phi),$$

or

$$I_\phi(f_1) + I_\phi(f_2) \le (1 + 2\delta)I_\phi(f). \tag{6}$$

By Lemma 7.6 and the definition of f,

$$I_\phi(f) \le I_\phi(f_1 + f_2) + \delta I_\phi(f').$$

Combining this with (6), (3), and 7.6(i), we find:

$$I_\phi(f_1) + I_\phi(f_2) \le I_\phi(f_1 + f_2) + 2\delta I_\phi(f_1 + f_2) + \delta(1 + 2\delta)I_\phi(f')$$
$$< I_\phi(f_1 + f_2) + \varepsilon.$$

This together with 7.6(iii) completes the proof of (2) and hence of the Lemma. ∎

7.8. We are now in a position to prove the existence of a left-invariant integral.

Proposition. *There exists a non-zero left-invariant integral on G.*

Proof. For each non-zero f in $\mathscr{L}_+(G)$, let $S_f = [(g; f)^{-1}, (f; g)]$; and put

$$S = \prod_{0 \neq f \in \mathscr{L}_+(G)} S_f$$

(with the Cartesian product topology). By Tihonov's Theorem S is compact.

Let $\{\phi_i\}$ be a net of non-zero elements of $\mathscr{L}_+(G)$ such that, for each neighborhood V of e, $\mathrm{supp}(\phi_i) \subset V$ for all sufficiently large i. By Lemma 7.6(i) $I_{\phi_i} \in S$ for each i. Since S is compact, we can replace $\{\phi_i\}$ by a subnet, and assume that $I_{\phi_i} \to I$ in S. Putting $I(0) = 0$, and passing to the i-limit in Lemmas 7.6 and 7.7, we then get:

$$(g; f)^{-1} \leq I(f) \leq (f; g) \qquad \text{if } 0 \neq f \in \mathscr{L}_+(G), \tag{7}$$

$$I(f_x) = I(f) \, (f \in \mathscr{L}_+(G); x \in G), \tag{8}$$

$$I(cf) = cI(f) \, (c \in \mathbb{R}_+; f \in \mathscr{L}_+(G), \tag{9}$$

$$I(f_1 + f_2) = I(f_1) + I(f_2) \, (f_1, f_2 \in \mathscr{L}_+(G)). \tag{10}$$

Now any f in $\mathscr{L}(G)$ is of the form

$$f = f_1 - f_2 + i(f_3 - f_4), \tag{11}$$

where the f_i are in $\mathscr{L}_+(G)$. If (11) holds, set $I(f) = I(f_1) - I(f_2) + i(I(f_3) - I(f_4))$. The legitimacy of this extension of I is assured by (10). It is linear on $\mathscr{L}(G)$ by (9) and (10), left-invariant by (8), and non-zero by (7). ∎

7.9. Notice that the integral I of 7.8 has the property that $I(f) > 0$ whenever $0 \neq f \in \mathscr{L}_+(G)$.

7.10. Let μ be a regular complex Borel measure on G and I_μ the corresponding complex integral on G (see II.8.10). If $x \in G$ let μ_x and μ^x be the translated regular complex Borel measures:

$$\mu_x(W) = \mu(xW), \qquad \mu^x(W) = \mu(Wx)$$

($W \in \mathscr{S}(G)$). We say that μ is *left-invariant* [*right-invariant*] if $\mu_x = \mu \, [\mu^x = \mu]$ for every x in G.

Since μ is regular, it will be left-invariant provided $\mu(xW) = \mu(W)$ for all x in G and all compact sets W (or all open sets W in $\mathscr{S}(G)$). The same holds for right invariance.

Notice that I_μ is left-invariant if and only if μ is. Indeed, we verify that

$$I_{\mu_x}(f) = I_\mu(f_{x^{-1}}) \qquad\qquad (f \in \mathscr{L}(G)).$$

Therefore, since I_ν determines ν (II.8.11), I_μ is left-invariant $\Leftrightarrow I_\mu = I_{\mu_x}$ for all $x \Leftrightarrow \mu = \mu_x$ for all $x \Leftrightarrow \mu$ is left-invariant.

Similarly, of course, I_μ is right-invariant if and only if μ is.

7.11. From now on, let μ be the unique regular (non-negative) Borel measure on G such that I_μ coincides with the I of 7.8; such a μ exists in virtue of II.8.12 (see especially Remark 2). By 7.8 and 7.10 μ is *left-invariant*.

7.12. Our next aim is to prove that up to a multiplicative constant, μ is the *only* left-invariant regular complex Borel measure on G.

Let ν be any left-invariant regular complex Borel measure in G; and set $J = I_\nu$.

We shall anticipate §11 by defining convolution with respect to μ. If $f, g \in \mathscr{L}(G)$, set

$$(f * g)(x) = \int_G f(xy)g(y^{-1})d\mu y$$

$$= \int_G f(y)g(y^{-1}x)d\mu y \qquad\qquad (x \in C).$$

7.13. Lemma. *If $f, g \in \mathscr{L}(G)$, then $f * g \in \mathscr{L}(G)$.*

Proof. Given $\varepsilon > 0$, choose a neighborhood U of e so that $xz^{-1} \in U \Rightarrow |f(x) - f(z)| < \varepsilon$ $(x, z \in G)$ (see Corollary 1.10). Then, if $x, z \in G$ and $xz^{-1} \in U$, we have

$$|(f * g)(x) - (f * g)(z)| \le \int_G |f(xy) - f(zy)||g(y^{-1})|d\mu y \le \varepsilon \int_G |g(y^{-1})|d\mu y$$

(since $(xy)(zy)^{-1} \in U$ for all y). So $f * g$ is continuous.

Let C and D be the compact supports of f and g. In order to have $(f * g)(x) \ne 0$, there must exist at least one y in G such that $xy \in C$ and $y^{-1} \in D$; and that implies $x \in CD$, which is compact by 1.6. So $f * g$ has compact support. ∎

7.14. Let $\{\phi_i\}$ be a net of non-zero elements of $\mathscr{L}_+(G)$ with supports shrinking down to e; that is, for each neighborhood V of e, $\mathrm{supp}(\phi_i) \subset V$ for all large enough i. By 7.9 we may (and shall) normalize the ϕ_i so that $\int \phi_i(y^{-1})d\mu y = 1$ for each i. Then

Lemma. *For each f in $\mathscr{L}(G)$, $J(f * \phi_i) \to J(f)$.*

Proof. We have

$$\left| \int ((f * \phi_i)(x) - f(x)) dvx \right| = \left| \iint (f(xy) - f(x)) \phi_i(y^{-1}) d\mu y \, dvx \right|. \quad (12)$$

Fix a compact neighborhood V of e. If C is the compact support of f, and $\phi_i \equiv 0$ outside V, the x-integral in (12) need be taken only over CV^{-1}. Given $\varepsilon > 0$, we can find (by 1.10) a neighborhood U of e such that $|f(xy) - f(x)| < \varepsilon$ whenever $y \in U$. Thus, if i is so large that $\phi_i \equiv 0$ outside $U \cap V$, the right side of (12) is majorized by

$$\int_{CV^{-1}} \int \varepsilon \phi_i(y^{-1}) d\mu y \, d|v|x = \varepsilon |v|(CV^{-1})$$

(see II.5.4(2)). So the Lemma is proved. ∎

Notice that the Lemma remains true if J is replaced by I.

7.15. Proposition. *There is a complex constant k such that $J = kI$, hence $v = k\mu$.*

Proof. Let f_1 and f_2 be any two non-zero elements of $\mathscr{L}_+(G)$; and let $\{\phi_i\}$ be as in 7.14. By Fubini's Theorem (II.9.12) and the left-invariance of v,

$$J(f_j * \phi_i) = \iint f_j(y) \phi_i(y^{-1}x) d\mu y \, dvx$$

$$= \iint f_j(y) \phi_i(x) dvx \, d\mu y$$

$$= I(f_j) J(\phi_i). \quad (13)$$

Similarly

$$I(f_j * \phi_i) = I(f_j) I(\phi_i). \quad (14)$$

Combining (13) and (14) (and remembering 7.9), we get

$$\frac{J(f_1 * \phi_i)}{I(f_1 * \phi_i)} = \frac{J(\phi_i)}{I(\phi_i)} = \frac{J(f_2 * \phi_i)}{I(f_2 * \phi_i)}. \quad (15)$$

In (15) let $i \to \infty$. By Lemma 7.14 we get

$$\frac{J(f_1)}{I(f_1)} = \frac{J(f_2)}{I(f_2)}. \quad (16)$$

By the arbitrariness of f_1 and f_2, (16) asserts that there is a complex constant k such that

$$\frac{J(f)}{I(f)} = k \qquad \text{whenever } 0 \neq f \in \mathcal{L}_+(G) \tag{17}$$

Clearly (17) implies that $J = kI$. From this it follows by II.8.11 that $v = k\mu$. ∎

7.16. Propositions 7.8 and 7.15 combine to give the fundamental result:

Theorem. *Let G be any locally compact group. There exists a non-zero left-invariant regular Borel measure μ on G. Any other left-invariant regular complex Borel measure v on G is of the form $k\mu$ for some complex constant k.*

Remark. If the above v is non-negative and non-zero, then of course the k will be positive.

7.17. Whatever we have proved about left-invariance applies to right-invariance also. Indeed, let $\tilde{G}$ be the reverse group of G (see 1.3). Then a left-invariant regular Borel measure for $\tilde{G}$ is right-invariant for G, and conversely. So by 7.16 there is a non-zero right-invariant regular Borel measure μ on G, and any other right-invariant regular complex Borel measure is a constant multiple of μ.

7.18. *Definition*. A non-zero left-invariant [right-invariant] regular Borel measure on G is called a *left [right] Haar measure on G.*

Remark. In general a left Haar measure will not be right-invariant, nor a right Haar measure left-invariant. See §8.

7.19. Proposition. *If μ is a left (or right) Haar measure on the locally compact group G, and U is a non-void open set in $\mathcal{S}(G)$, then $\mu(U) > 0$. (That is, the closed support of μ is all of G.)*

Proof. This follows from 7.9. It can also be proved directly. For, if $\mu(U) = 0$, then any compact set, being covered by finitely many left (or right) translates of U, must have μ measure 0; and so $\mu = 0$. ∎

7.20. Proposition. *The locally compact group G is compact if and only if its left (or right) Haar measure is bounded.*

Proof.   Any Borel measure on a compact space is bounded by definition. So we have only to prove the "if" direction.

Assume that G is not compact; and let V be a compact neighborhood of e. No finite set of left translates of V covers G. Hence there is a sequence $\{x_n\}$ of points of G such that for each n

$$x_n \notin \bigcup_{m=1}^{n-1} x_m V.$$

Choose an open neighborhood U of e such that $UU^{-1} \subset V$. Since $x_n \notin x_m UU^{-1}$ whenever $m < n$, it follows that the $\{x_n U\}$ are pairwise disjoint. So, if μ is a left Haar measure on G, we have

$$\mu\left(\bigcup_{m=1}^{n} x_m U\right) = \sum_{m=1}^{n} \mu(x_m U) = n\mu(U) \to \infty$$

as $n \to \infty$ (by 7.19). Thus μ is not bounded. An exactly similar argument holds if μ is right-invariant.   ∎

7.21.   Proposition.   *A necessary and sufficient condition for G to be discrete is that the left (or right) Haar measure of some (hence every) one-element subset of G be greater than 0.*

Proof.   If G is discrete, the "counting measure" on $\mathscr{S}(G)$ (assigning to each set its cardinality) is both left and right Haar measure. This proves the necessity. On the other hand, if G is not discrete, it has an infinite compact subset C; and the Haar measure of C could not be finite if all its one-element subsets had the same positive Haar measure. This establishes the sufficiency.   ∎

7.22.   If G_0 is an open (and hence also closed) subgroup of a locally compact group G, notice that the restriction to $\mathscr{S}(G_0)$ of a left Haar measure μ on G is a left Haar measure μ_0 on G_0. Further, if we know μ_0 we know μ; for the image of μ_0 under the translation $y \mapsto xy \, (y \in G_0)$ is just the restriction of μ to the (open) coset xG_0.

8.   The Modular Function

8.1.   In this section we study the relationship between left and right Haar measure.

Fix a locally compact group G with unit e and left Haar measure μ. By $\text{Aut}(G)$ we mean the group of all automorphisms of the topological group G.

8.2. The following definition and lemma are stated in a generality which will be useful not only in this section but in later work.

Each α in $\text{Aut}(G)$ carries $\mathscr{S}(G)$ onto $\mathscr{S}(G)$, and so carries a regular complex Borel measure v on G into another regular complex Borel measure αv given by

$$(\alpha v)(A) = v(\alpha^{-1}(A)) \qquad\qquad (A \in \mathscr{S}(G)).$$

Observe that $\alpha\mu$ is again left-invariant. Indeed, $(\alpha\mu)(xA) = \mu(\alpha^{-1}(xA)) = \mu(\alpha^{-1}(x)\alpha^{-1}(A)) = \mu(\alpha^{-1}(A)) = (\alpha\mu)(A)$ $(x \in G; A \in \mathscr{S}(G))$. So by the uniqueness of Haar measure (7.16), there is a positive number $\Gamma(\alpha)$ satisfying

$$\alpha\mu = (\Gamma(\alpha))^{-1}\mu; \qquad\qquad (1)$$

that is, $\mu(\alpha(A)) = \Gamma(\alpha)\mu(A)$ for all A in $\mathscr{S}(G)$.

Definition. We call $\Gamma(\alpha)$ the *expansion factor* of α.

Evidently $\Gamma(\alpha_1\alpha_2) = \Gamma(\alpha_1)\Gamma(\alpha_2)$ $(\alpha_1, \alpha_2 \in \text{Aut}(G))$; that is, Γ is a homomorphism of $\text{Aut}(G)$ into the multiplicative group of positive reals.

8.3. Lemma. *Let $s \mapsto \alpha_s$ be a homomorphism of a topological group S into $\text{Aut}(G)$ which is continuous in the sense that $\langle s, x\rangle \mapsto \alpha_s(x)$ is continuous on $S \times G$ to G. Then $s \mapsto \Gamma(\alpha_s)$ $(s \in S)$ is a continuous homomorphism of S into the positive reals.*

Proof. By 8.2 $s \mapsto \Gamma(\alpha_s)$ is a homomorphism. By 2.7 we need only show that it is continuous at the unit e of S.

Fix a non-zero element f of $\mathscr{L}_+(G)$, so that $\int f \, d\mu > 0$ (by 7.9). For $s \in S$ we have by II.5.10 and (1)

$$\int f \, d\mu = \int (\alpha_s^{-1} f) d(\alpha_s^{-1}\mu)$$

$$= \Gamma(\alpha_s) \int f(\alpha_s(x)) d\mu x. \qquad\qquad (2)$$

Now let U be a compact set in G whose interior contains the compact support C of f. By the joint continuity of α there is a neighborhood W of e in S such that $\alpha_s^{-1}(C) \subset U$ whenever $s \in W$. It follows that

$$f(\alpha_s(x)) = 0 \qquad \text{for } s \in W \text{ unless } x \in U. \qquad\qquad (3)$$

Again, by the joint continuity of α, if $\varepsilon > 0$ there is a neighborhood V of e in S such that

$$|f(\alpha_s(x)) - f(x)| < \varepsilon \qquad \text{for } x \in U, s \in V. \qquad\qquad (4)$$

From (3) and (4) we have for $s \in V \cap W$:

$$\left| \int_G f(\alpha_s(x)) d\mu x - \int_G f(x) d\mu x \right| = \left| \int_U (f(\alpha_s(x)) - f(x)) d\mu x \right| \leq \varepsilon \mu(U).$$

It follows that the integral in (2) is continuous in s at e. Hence by (2) $\Gamma(\alpha_s)$ is continuous in s at e.   ∎

8.4. For each y in G let $\alpha_y(x) = yxy^{-1} (x \in G)$; and put $\Delta(y) = (\Gamma(\alpha_y))^{-1}$, the reciprocal of the expansion factor of the inner automorphism α_y.

By Lemma 8.3 Δ is a continuous homomorphism of G into the multiplicative group of positive reals.

Definition. Δ is called the *modular function of G*.

8.5. Proposition. $\int_G f(x) d\mu x = \Delta(y) \int_G f(xy) d\mu x$ *for* $f \in \mathscr{L}_1(\mu)$, $y \in G$.

Proof.   By II.5.10 and the left-invariance of μ,

$$\int f(xy) d\mu x = \int f(\alpha_{y^{-1}}(x)) d\mu x = \int f(x) d(\alpha_{y^{-1}}\mu)x$$

$$= \Delta(y^{-1}) \int f(x) d\mu x.   ∎$$

8.6.   Proposition 8.5 can be expressed in the form

$$d\mu(xy) = \Delta(y) d\mu x$$

or

$$\mu^y = \Delta(y)\mu \qquad\qquad (y \in G). \qquad (5)$$

(see II.5.10 and 7.10). It gives the effect of right translation on the left Haar measure μ. In particular we have:

Corollary.   *For a left Haar measure to be also a right Haar measure it is necessary and sufficient that* $\Delta(y) = 1$ *for all y in G.*

8.7. Definition. G is *unimodular* if $\Delta(y) = 1$ for all y in G.

Thus the locally compact group G is unimodular if and only if there is a non-zero regular Borel measure on G which is both left- and right-invariant.

Remark.   Perhaps the simplest example of a non-unimodular group is the "$ax + b$" group; see Corollary 9.8.

8.8.   Proposition.   *A locally compact group G which is either Abelian, compact, or discrete is unimodular.*

Proof.   The Abelian case is obvious. If G is compact, then the range of Δ is a compact subgroup of the positive reals, hence equal to $\{1\}$; so G is unimodular. If G is discrete, we saw in the proof of 7.21 that the "counting measure" on $\mathscr{S}(G)$ is both left- and right-invariant; hence G is unimodular. ■

8.9.   The proof of the compactness part of Proposition 8.8 shows the more general fact that, if K is any compact subgroup of the arbitrary locally compact group G, then $\Delta(x) = 1$ for all x in K.

8.10.   If G is unimodular, we shall refer to a left (hence also right) Haar measure on G simply as a *Haar measure* on G.

If G is compact, *normalized Haar measure* on G is that Haar measure μ for which $\mu(G) = 1$.

In general, different (left) Haar measures on G, differing by multiplicative constants, are often referred to as different *normalizations* of (left) Haar measure on G.

8.11.   The next Proposition will tell us how left Haar measure behaves under the inverse map.

Proposition. $d\mu(x^{-1}) = \Delta(x^{-1})d\mu x$. *Thus, if $f \in \mathscr{L}_1(\mu)$,*

$$\int f(x)d\mu x = \int f(x^{-1})\Delta(x^{-1})d\mu x. \tag{6}$$

Proof.   Putting $J(f) = \int f(x^{-1})\Delta(x^{-1})d\mu x$ for $f \in \mathscr{L}(G)$, we have if $y \in G$:

$$J(f_y) = \int f(yx^{-1})\Delta(x^{-1})d\mu x$$

$$= \Delta(y)\int f(x^{-1})\Delta(y^{-1}x^{-1})d\mu x \qquad \text{(by 8.5)}$$

$$= \int f(x^{-1})\Delta(x^{-1})d\mu x = J(f).$$

So J is left-invariant on $\mathscr{L}(G)$. By the uniqueness of Haar measure there is a positive constant k such that

$$\int f(x^{-1})\Delta(x^{-1})d\mu x = k \int f(x)d\mu x \qquad (7)$$

for all f in $\mathscr{L}(G)$. Applying (7) to the function $g(x) = f(x^{-1})\Delta(x^{-1})$, we get

$$\int f(x)d\mu x = \int g(x^{-1})\Delta(x^{-1})d\mu x = k \int g(x)d\mu x = k^2 \int f(x)d\mu x$$

for all f in $\mathscr{L}(G)$, whence $k^2 = 1$, or $k = 1$. We have now shown that (6) holds for all f in $\mathscr{L}(G)$.

This, II.5.10, II.7.5, and II.8.11 imply that $d\mu(x^{-1}) = \Delta(x^{-1})d\mu x$, and hence that (6) holds for all f in $\mathscr{L}_1(\mu)$. ∎

8.12. Since the inverse map $x \mapsto x^{-1}$ clearly carries a left-invariant into a right-invariant measure, 8.11 tells us that $d\mu_r x = \Delta(x^{-1})d\mu x$ is a right Haar measure.

In analogy with (5) we have

$$d\mu_r(yx) = \Delta(y^{-1})d\mu_r x.$$

8.13. The following proposition is left as an easy exercise to the reader.

Proposition. *Let $\phi: G_1 \to G_2$ be an isomorphism of the two locally compact groups G_1 and G_2. Then $\Delta_1(x) = \Delta_2(\phi(x))\ (x \in G)$, where Δ_1 and Δ_2 are the modular functions of G_1 and G_2 respectively.*

8.14. Applying 8.13 to an automorphism of G onto itself, we obtain:

Corollary. *Any automorphism α of the locally compact group G leaves Δ invariant; that is, $\Delta \circ \alpha = \Delta$.*

8.15. Again, applying 8.13 to the isomorphism $x \mapsto x^{-1}$ of G with its reverse group $\tilde{G}$ (see 1.3), we find:

Corollary. *The modular function $\tilde{\Delta}$ of $\tilde{G}$ is given by*

$$\tilde{\Delta}(x) = \Delta(x^{-1}) \qquad\qquad (x \in G).$$

8.16. Here is an interesting description of Haar measure on a compact group.

Assume that G is a compact group. For each f in $\mathscr{L}(G)$, let D_f be the $\| \ \|_\infty$-closure (in $\mathscr{L}(G)$) of the set of all convex linear combinations g of left translates of f (that is, $g = \sum_{i=1}^r c_i f_{u_i}$, where $u_1, \ldots, u_r \in G$, $0 \le c_i \in \mathbb{R}$, and $\sum_{i=1}^r c_i = 1$).

Proposition.* *For each f in $\mathscr{L}(G)$, D_f contains exactly one constant function h on G; and the constant value of h is $\int f \, d\mu$ (μ being normalized Haar measure on G).*

The main step in proving this is to show that D_f does contain the function h with constant value $\int f \, d\mu$. To show this, notice that $h(x) = \int f(yx) d\mu y$, and approximate $\int f(yx) d\mu y$ by finite sums uniformly in x.

9. Examples of Haar Measure and the Modular Function

9.1. We have seen in Proposition 8.8 (and its proof) that a discrete group G is unimodular, and that its Haar measure is the "counting measure" μ on G, assigning to each finite set A the number of elements of A.

9.2. Lebesgue measure λ is of course Haar measure on the additive group $\mathbb{R}$ of the reals. More generally, n-dimensional Lebesgue measure λ^n is Haar measure for the additive group $\mathbb{R}^n$. This is a special case of the next Proposition.

9.3. Proposition. *If $\mu_1, \ldots, \mu_n$ are left [right] Haar measures on locally compact groups $G_1, \ldots, G_n$ respectively, and if $\Delta_1, \ldots, \Delta_n$ are their respective modular functions, then: (a) The product measure $\mu = \mu_1 \times \cdots \times \mu_n$ (see II.9.18) is a left [right] Haar measure on $G = G_1 \times \cdots \times G_n$, and (b) the modular function Δ of G is given by:*

$$\Delta(x_1, \ldots, x_n) = \Delta_1(x_1)\Delta_2(x_2)\ldots\Delta_n(x_n).$$

Proof. Suppose the μ_i are left-invariant; and let $x = \langle x_1, \ldots, x_n \rangle \in G$. The image $x\mu$ of μ under left translation by x is again a regular Borel measure on G; and if $A_i \in \mathscr{S}(G_i)$ for each i,

$$(x\mu)(A_1 \times \cdots \times A_n) = \mu(x_1^{-1}A_1 \times \cdots \times x_n^{-1}A_n) = \prod_{i=1}^{n} \mu_i(x_i^{-1}A_i)$$

$$= \prod_{i=1}^{n} \mu_i(A_i).$$

But by II.9.18 the product μ is that unique regular Borel measure satisfying $\mu(A_1 \times \cdots \times A_n) = \prod_{i=1}^{n} \mu_i(A_i)$ for all such A_i. Thus $x\mu = \mu$; and μ is left-invariant.

Similarly, if the μ_i are right-invariant μ is right-invariant.

Let the μ_i be left-invariant; and for each i let A_i be a set in $\mathscr{S}(G_i)$ with $0 < \mu_i(A_i)$. Then by 8.6, if $y = \langle y_1, \ldots, y_n \rangle \in G$ and $A = A_1 \times \cdots \times A_n$,

$$\Delta(y)\mu(A) = \mu(Ay) = \mu(A_1 y_1 \times \cdots \times A_n y_n)$$

$$= \mu_1(A_1 y_1) \ldots \mu_n(A_n y_n)$$

$$= \Delta_1(y_1) \ldots \Delta_n(y_n)\mu_1(A_1) \ldots \mu_n(A_n)$$

$$= \Delta_1(y_1) \ldots \Delta_n(y_n)\mu(A).$$

So $\Delta(y) = \Delta_1(y_1) \ldots \Delta_n(y_n)$.   ∎

Haar Measure on Semidirect Products

9.4. What are the Haar measures and modular functions of the semidirect products of §4?

Let G_1, G_2, τ be as in 4.4, and let G be the semidirect product $G_1 \underset{\tau}{\times} G_2$. We shall assume now that G_1 and G_2 (and hence G) are locally compact. Let μ_1 and μ_2 be left Haar measures for G_1 and G_2; and let Δ_1, Δ_2, and Δ be the modular functions of G_1, G_2, and G respectively. Further, for each y in G_2 let $\Gamma(y)$ be the expansion factor (see 8.2) for the automorphism τ_y of G_1; that is,

$$d\mu_1(\tau_y(x)) = \Gamma(y)d\mu_1 x. \tag{1}$$

By Lemma 8.3, Γ is a continuous homomorphism of G_2 into the multiplicative group of positive reals.

9.5. Proposition. *The regular product $\mu = \mu_1 \times (\Gamma(y))^{-1} \, d\mu_2 y$ is a left Haar measure on G. Further*

$$\Delta(\langle x, y \rangle) = \Delta_1(x)\Delta_2(y)(\Gamma(y))^{-1}. \tag{2}$$

Proof. As in 4.5, let us embed G_1 and G_2 in G, writing xy for $\langle x, y \rangle$. If $g_0 = x_0 y_0 \in G$ and $f \in \mathscr{L}(G)$, we have (using II.7.5, Fubini's Theorem, and the left-invariance of μ_1 and μ_2):

$$\int f(g_0 g)d\mu g = \iint f(x_0 y_0 xy)(\Gamma(y))^{-1} \, d\mu_1 x \, d\mu_2 y$$

$$= \iint f(x_0(y_0 xy_0^{-1})y)\Gamma(y^{-1}y_0)d\mu_2 y \, d\mu_1 x$$

$$= \iint f(x_0 xy)\Gamma(y^{-1}y_0)d\mu_2 y \, d\mu_1(y_0^{-1}xy_0) \qquad \text{(see II.5.10)}$$

$$= \iint f(x_0 xy)(\Gamma(y))^{-1} \, d\mu_1 x \, d\mu_2 y \qquad \text{(by (1))}$$

$$= \iint f(xy)(\Gamma(y))^{-1} \, d\mu_1 x \, d\mu_2 y = \int f(g)d\mu g.$$

So μ is a left Haar measure on G.

To calculate Δ we shall use Proposition 8.5. Let f be an element of $\mathscr{L}(G)$ with $\int f \, d\mu \neq 0$. Then

$$\int f(gx_0^{-1})d\mu g = \iint f(xyx_0^{-1})(\Gamma(y))^{-1} \, d\mu_1 x \, d\mu_2 y$$

$$= \iint f(x(yx_0^{-1}y^{-1})y)(\Gamma(y))^{-1} \, d\mu_1 x \, d\mu_2 y$$

$$= \iint f(xy)\Delta_1(yx_0 y^{-1})(\Gamma(y))^{-1} \, d\mu_1 x \, d\mu_2 y \qquad \text{(by 8.5)}$$

$$= \Delta_1(x_0) \int f(g)d\mu g$$

(since $\Delta_1(yx_0 y^{-1}) = \Delta_1(x_0)$ by 8.14). So by 8.5

$$\Delta(x_0) = \Delta_1(x_0). \tag{3}$$

Also, if $y_0 \in G_2$,

$$\int f(gy_0^{-1})d\mu g = \iint f(xyy_0^{-1})(\Gamma(y))^{-1}\,d\mu_2 y\,d\mu_1 x$$

$$= \iint f(xy)(\Gamma(yy_0))^{-1}\Delta_2(y_0)d\mu_2 y\,d\mu_1 x \qquad \text{(by 8.5)}$$

$$= (\Gamma(y_0))^{-1}\Delta_2(y_0)\int f d\mu.$$

So by 8.5

$$\Delta(y_0) = (\Gamma(y_0))^{-1}\Delta_2(y_0). \tag{4}$$

Putting (3) and (4) together, we obtain (2).   ∎

9.6. Applying 8.12 in the context of 9.5, we see that the right Haar measure μ^r on G is given by

$$d\mu^r(xy) = \Delta(xy)^{-1}(\Gamma(y))^{-1}\,d\mu_1 x\,d\mu_2 y$$

$$= \Delta_1(x)^{-1}\Delta_2(y)^{-1}\,d\mu_1 x\,d\mu_2 y$$

$$= d\mu_1^r x\,d\mu_2^r y,$$

or

$$\mu^r = \mu_1^r \times \mu_2^r,$$

where μ_1^r and μ_2^r are right Haar measures on G_1 and G_2 respectively.

9.7. In the general example 4.7 of a semidirect product, notice that the expansion factor $\Gamma(a)$ of τ_a is $|\det(a)|$. So we have:

Corollary. *In* 4.7 *let* H *be a closed unimodular subgroup of* $GL(n, \mathbb{R})$ *such that* $|\det(a)| = 1$ *for all* a *in* H. *Then* $\mathbb{R}^n \underset{\tau}{\times} H$ *is unimodular. In particular, the Euclidean groups (see* 4.8(B)) *are unimodular.*

9.8 Corollary. *The "$ax + b$" group of* 4.8(A) *is not unimodular. Its modular function is given by*:

$$\Delta(b, a) = a^{-1}.$$

Modules of locally compact fields

9.9. As another illustration of 8.2 and 8.3, let us consider a locally compact topological field F (see §6). This gives rise to two locally compact Abelian groups, the additive group F_+ of F, and the multiplicative group $F_* = F \setminus \{0\}$.

Let μ be a Haar measure on F_+. Each element y of F_* gives rise to an automorphism $x \mapsto yx$ of F_+. Let us denote by $\mathrm{mod}_F(y)$ the expansion factor of this automorphism (see 8.2). By 8.3 mod_F is a continuous homomorphism of F_* into the multiplicative group of the positive reals. We now complete the definition of mod_F on all of F by setting $\mathrm{mod}_F(0) = 0$. Evidently the multiplicative property

$$\mathrm{mod}_F(yz) = \mathrm{mod}_F(y)\mathrm{mod}_F(z)$$

now holds for all y, z in F.

Proposition. mod_F *is continuous on all of* F.

Proof. We have only to show that mod_F is continuous at 0. For this we may as well assume that F is not discrete. Choose a compact subset C of F with $k = \mu(C) > 0$; and let $\varepsilon > 0$. By 7.21 and the regularity of μ, there is an open F-neighborhood U of 0 such that $\mu(U) < \varepsilon k$. By 6.6 we can find a neighborhood W of 0 such that $WC \subset U$. But then, for any point x of W,

$$\mathrm{mod}_F(x)\mu(C) = \mu(xC) \leq \mu(U) < \varepsilon k = \varepsilon\mu(C),$$

whence $\mathrm{mod}_F(x) < \varepsilon$. This proves the continuity of mod_F at 0.   ∎

Definition. The function mod_F is called the *module of* F.

9.10. Proposition. *In the context of* 9.9, *the Haar measure of* F_* *is* $(\mathrm{mod}_F(x))^{-1} d\mu x$.

Proof. If $a \in F_*$, we have $d\mu(ax) = \mathrm{mod}_F(a)d\mu x$ by the definition of mod_F. So $(\mathrm{mod}_F(ax))^{-1} d\mu(ax) = (\mathrm{mod}_F(x))^{-1} d\mu x$, proving the invariance of $(\mathrm{mod}_F(x))^{-1} d\mu x$ under multiplication by non-zero elements of F.   ∎

9.11. It is evident that

$$\mathrm{mod}_{\mathbb{R}}(x) = |x| \qquad\qquad (x \in \mathbb{R});$$

$$\mathrm{mod}_{\mathbb{C}}(z) = |z|^2 \qquad\qquad (z \in \mathbb{C}).$$

9.12. Let us find mod_{F_p} for the p-adic fields F_p introduced in 6.12. We adopt the notation of 6.12.

Proposition. $\operatorname{mod}_{F_p}(r) = w(r)$ *for all r in* F_p.

Proof. For each non-negative integer n let $U_n = \{r \in F_p : w(r) \le p^{-n}\}$. By Lemma 6.13 then p^n additive translates $s + U_n$, where s runs over $\{0, 1, 2, \ldots, p^n - 1\}$, are disjoint and have $U_0 = I_p$ for their union.

Now we saw in 6.14 that I_p is a compact neighborhood of 0. Let μ be the Haar measure of the additive group of F_p normalized so that $\mu(I_p) = 1$. Then by the preceding paragraph

$$\mu(U_n) = p^{-n} \qquad\qquad (n = 0, 1, 2, \ldots). \qquad (5)$$

Now suppose $r \in F_p$ and $w(r) = p^{-n}$ $(n = 0, 1, 2, \ldots)$. By the multiplicative property of w,

$$rI_p = U_n. \qquad (6)$$

Combining (5) and (6) with the definition of mod_{F_p}, we obtain $\operatorname{mod}_{F_p}(r) = p^{-n} = w(r)$. Thus $\operatorname{mod}_{F_p}(r) = w(r)$ whenever $w(r) \le 1$. If $w(r) > 1$, then $w(r^{-1}) = (w(r))^{-1} < 1$; so

$$(\operatorname{mod}_{F_p}(r))^{-1} = \operatorname{mod}_{F_p}(r^{-1}) = w(r^{-1}) = (w(r))^{-1},$$

whence $\operatorname{mod}_{F_p}(r) = w(r)$. Therefore mod_{F_p} and w coincide everywhere. ∎

10. Convolution and Involution of Measures on G

10.1. Let G be a locally compact group. We are going to extend the product operation on G to a product operation (called convolution) defined for certain pairs of measures on G.

We shall denote by $\mathcal{M}(G)$ the linear space of all regular complex Borel measures on G, and by $\mathcal{M}_+(G)$ the set of (non-negative) measures in $\mathcal{M}(G)$. Let $p: G \times G \to G$ stand for the product operation $\langle x, y \rangle \mapsto xy$.

10.2. *Definition.* Given two elements μ, ν of $\mathcal{M}(G)$, we shall say that μ and ν (in that order) are *convolvable*, or that $\mu * \nu$ *exists*, if p is $(\mu \times \nu)$-proper in the sense of II.10.3. If this is the case, the regular p-transform $p_*(\mu \times \nu)$ of $\mu \times \nu$ (see II.10.5) is called the *convolution* of μ and ν and is denoted by $\mu * \nu$. Thus, if μ and ν are convolvable, $\mu * \nu$ is again an element of $\mathcal{M}(G)$.

10.3. Since p is $(\mu \times \nu)$-proper if and only if it is $|\mu \times \nu|$-proper, it follows from II.9.10(2) that $\mu * \nu$ exists if and only if $|\mu| * |\nu|$ exists; and in this case we see from II.10.5(6):

$$|\mu * \nu| \le |\mu| * |\nu|. \qquad (1)$$

10.4. If μ and v are convolvable and non-negative, then $\mu * v$ is non-negative.

If $\mu \in \mathcal{M}(G)$, the set of those v for which $\mu * v$ $[v * \mu]$ exists is a linear subspace of $\mathcal{M}(G)$; and the map $v \mapsto \mu * v$ $[v * \mu]$ is linear on that space.

Further, if μ, μ', v, v' are in $\mathcal{M}(G)$ and $|\mu'| \leq |\mu|$ and $|v'| \leq |v|$, then the existence of $\mu * v$ implies that of $\mu' * v'$. This follows from II.9.6 and II.10.3.

Remark. The order of convolution is of course essential. It is even possible for $\mu * v$ to exist while $v * \mu$ fails to exist; see Bourbaki [11], Chapter 8, §3, Exercise 12.

10.5. Proposition. *If $\mu, v \in \mathcal{M}(G)$ and $\mu * v$ exists, then*

$$\mathrm{supp}(\mu * v) \subset (\mathrm{supp}(\mu)\mathrm{supp}(v))^- \tag{2}$$

($^-$ denoting closure). If $\mu \geq 0$ and $v \geq 0$, equality holds in (2).

Proof. This follows from Remark II.9.10 and II.10.8. ■

10.6. Proposition. *If μ, v are in $\mathcal{M}(G)$ and one of $\mathrm{supp}(\mu)$ and $\mathrm{supp}(v)$ is compact, then $\mu * v$ exists. If $\mathrm{supp}(\mu)$ and $\mathrm{supp}(v)$ are both compact, then $\mathrm{supp}(\mu * v)$ is compact.*

Proof. If $K = \mathrm{supp}(\mu)$ is compact, then, since $\mathrm{supp}(\mu \times v) \subset K \times G$ by Remark II.9.10, it follows that $p^{-1}(D) \cap \mathrm{supp}(\mu \times v)$ is compact for every compact subset D of G. Hence $\mu * v$ exists by II.10.4. Similarly, $\mu * v$ exists if $\mathrm{supp}(v)$ is compact.

The last statement now follows from 10.5. ■

10.7. Proposition. *If μ and v are bounded elements of $\mathcal{M}(G)$, then $\mu * v$ exists and is bounded; and*

$$\|\mu * v\| \leq \|\mu\| \|v\| \tag{3}$$

($\| \ \|$ being the total variation norm). Equality holds in (3) if μ and v are non-negative.

Proof. By II.9.14 $\mu \times v$ is bounded and $\|\mu \times v\| = \|\mu\| \|v\|$. Hence the conclusion follows from II.10.6. ■

10.8. The next proposition tells us how to integrate with respect to $\mu * v$. Suppose that μ, v are two elements of $\mathcal{M}(G)$ such that $\mu * v$ exists; and let A be a Banach space. We recall from II.8.15 the definition of the maximal regular extension μ_e of μ.

Proposition. *Suppose that $f \in \mathcal{L}_1((|\mu| * |v|)_e; A)$. Then the map $\langle x, y \rangle \mapsto$ $f(xy)$ belongs to $\mathcal{L}_1((|\mu| \times |v|)_e; A)$; and*

$$\int_G f \, d(\mu * v)_e = \int_G \int_G f(xy) d\mu_e x \, dv_e y = \int_G \int_G f(xy) dv_e y \, d\mu_e x. \qquad (4)$$

Proof. By II.10.2 combined with II.10.7, the map $\langle x, y \rangle \mapsto f(xy)$ belongs to $\mathcal{L}_1((|\mu| \times |v|)_e; A)$, and

$$\int_G f \, d(\mu * v)_e = \int_{G \times G} f(xy) d(\mu \times v)_e \langle x, y \rangle.$$

Now (4) follows from this and Fubini's Theorem in the form II.16.5. ∎

Remark. Suppose it happens that μ and v are σ-bounded (see II.1.7), and that the f of the above Proposition vanishes outside a countable union of compact sets. Thus there is a countable union E of compact subsets of G such that $G \setminus E$ is locally μ-null and locally v-null, and f vanishes outside E. By II.5.11 the subscript e can be removed from the integral $\int f \, d(\mu * v)_e$ and the inner integrals $\int f(xy) d\mu_e x$ and $\int f(xy) dv_e y$ occurring in (4). Furthermore, $\int f(xy) d\mu x$ clearly vanishes unless $y \in E^{-1}E$; and $\int f(xy) dvy$ vanishes unless $x \in EE^{-1}$. Since $E^{-1}E$ and EE^{-1} are also countable unions of compact sets, it follows, again by II.5.11, that we can also remove the subscripts e from the outer integrals in (4). So, with the above hypotheses, (4) becomes

$$\int_G f \, d(\mu * v) = \int_G \int_G f(xy) d\mu x \, dvy = \int_G \int_G f(xy) dvy \, d\mu x. \qquad (5)$$

10.9. Corollary *Let μ and v be elements of $\mathcal{M}(G)$ whose convolution $\mu * v$ exists. Then for each D in $\mathcal{S}(G)$:*

(I) *The function $x \mapsto v(x^{-1}D)$ is μ_e-summable, and*

$$(\mu * v)(D) = \int_G v(x^{-1}D) d\mu_e x. \qquad (6)$$

(II) *The function $y \mapsto \mu(Dy^{-1})$ is v_e-summable, and*

$$(\mu * v)(D) = \int_G \mu(Dy^{-1}) dv_e y. \qquad (7)$$

As in Remark 10.8, if μ and v are σ-bounded we can remove the subscripts e from the integrals in (6) and (7).

To illustrate this Corollary, let μ be any element of $\mathcal{M}(G)$, and let δ_x be the unit mass at the element x of G. Thus $\delta_x * \mu$ and $\mu * \delta_x$ exist by 10.6, and by the above Corollary

$$(\delta_x * \mu)(D) = \mu(x^{-1}D), \qquad (\mu * \delta_x)(D) = \mu(Dx^{-1}) \tag{8}$$

for all D in $\mathcal{S}(G)$. That is, $\delta_x * \mu$ and $\mu * \delta_x$ are just the results of translating μ by x on the left and right respectively.

In particular, (8) implies that

$$\delta_x * \delta_y = \delta_{xy} \qquad\qquad (x, y \in G),$$

justifying the statement that convolution is the extension to measures of the multiplication on the group.

10.10. We shall next prove the associativity of convolution.

Proposition. *Let μ, ν, ρ be three elements of $\mathcal{M}(G)$ such that $\mu * \nu$, $\nu * \rho$, $(|\mu| * |\nu|) * |\rho|$, and $|\mu| * (|\nu| * |\rho|)$ all exist. Then*

$$\mu * (\nu * \rho) = (\mu * \nu) * \rho$$

(and both sides exist).

Proof. Recall that $p: G \times G \to G$ is the group product; and let $i: G \to G$ be the identity map. Thus $(p \times i)(x, y, z) = \langle xy, z \rangle$ and $(i \times p)(x, y, z) = \langle x, yz \rangle$; and by the associativity of multiplication in G,

$$p \circ (p \times i) = p \circ (i \times p). \tag{9}$$

Now by hypothesis p is $(|\nu| \times |\rho|)$-proper; and i is obviously μ-proper. So by II.10.10 $i \times p$ is $(|\mu| \times |\nu| \times |\rho|)$-proper, and

$$(i \times p)_*(|\mu| \times |\nu| \times |\rho|) = |\mu| \times (|\nu| * |\rho|). \tag{10}$$

Again, p is $(|\mu| \times (|\nu| * |\rho|))$-proper by hypothesis; so by (10) and II.10.9, $p \circ (i \times p)$ is $(|\mu| \times |\nu| \times |\rho|)$-proper. Combining these facts with another application of II.10.9, II.10.10, we get

$$(p \circ (i \times p))_*(\mu \times \nu \times \rho) = \mu * (\nu * \rho); \tag{11}$$

and similarly

$$(p \circ (p \times i))_*(\mu \times \nu \times \rho) = (\mu * \nu) * \rho. \tag{12}$$

From (9), (11), and (12), we obtain the required result. ∎

Involution of Measures

10.11. There is another important unary operation on $\mathcal{M}(G)$.

Definition. If $\mu \in \mathcal{M}(G)$, let μ^* denote the element of $\mathcal{M}(G)$ given by

$$\mu^*(A) = \overline{\mu(A^{-1})} \qquad\qquad (A \in \mathcal{S}(G))$$

($^{-}$ denoting complex conjugation). The operation $\mu \mapsto \mu^*$ is called *involution of measures*.

If $x \in G$, evidently $(\delta_x)^* = \delta_{x^{-1}}$; so the involution operation is an extension to measures of the inverse operation on the group.

10.12. Here are some elementary properties of involution.

Proposition. *Let* $\mu, v \in \mathcal{M}(G)$. *Then:*

(i) $(\mu + v)^* = \mu^* + v^*$;

(ii) $(c\mu)^* = \bar{c}\mu^* \ (c \in \mathbb{C})$;

(iii) $\mu^{**} = \mu$;

(iv) *if* $\mu * v$ *exists, then* $v^* * \mu^*$ *exists and equals* $(\mu * v)^*$;

(v) $|\mu^*| = |\mu|^*$;

(vi) $\operatorname{supp}(\mu^*) = (\operatorname{supp}(\mu))^{-1}$.

Proof. We shall content ourselves with proving (iv).

Let $w: G \times G \to G \times G$ be the homeomorphism given by $w(x, y) = \langle y^{-1}, x^{-1} \rangle$; and let $D \in \mathcal{S}(G)$. Then the images of $p^{-1}(D)$, $v^* \times \mu^*$, and $(v^* \times \mu^*)_e$ under w are clearly $p^{-1}(D^{-1})$, $\bar{\mu} \times \bar{v}$, and $(\bar{\mu} \times \bar{v})_e$ respectively ($^{-}$ denoting complex conjugation). From this it follows that $v^* * \mu^*$ exists, and that

$$(v^* * \mu^*)(D) = (v^* \times \mu^*)_e(p^{-1}(D))$$

$$= (\bar{\mu} \times \bar{v})_e(p^{-1}(D^{-1}))$$

$$= ((\mu \times v)_e(p^{-1}(D^{-1})))^{-}$$

$$= ((\mu * v)(D^{-1}))^{-} = (\mu * v)^*(D).$$

So (iv) is proved. ∎

10.13. From 10.12(v) we obtain:

Corollary. *If* μ *is a bounded element of* $\mathcal{M}(G)$, *then* μ^* *is bounded, and*

$$\|\mu^*\| = \|\mu\| \qquad (\text{total variation norm}).$$

10.14.   *Remark.* Notice that μ^* is the complex conjugate of the image of μ under the inverse map. Therefore a Borel set A belongs to domain $((\mu^*)_e)$ if and only if $A^{-1} \in$ domain (μ_e), in which case $(\mu^*)_e(A) = \overline{\mu_e(A^{-1})}$.

The Measure Algebra

10.15. In II.8.3 we defined $\mathscr{M}_r(G)$ as the Banach space of all bounded elements of $\mathscr{M}(G)$. In view of 10.7, 10.10, 10.12, and 10.13, $\mathscr{M}_r(G)$ is a Banach *-algebra under the operations of convolution and involution.

Definition. The Banach *-algebra $\mathscr{M}_r(G)$ (with the operations of convolution and involution, and with the total variation norm) is called the *measure algebra of G*.

10.16. Let $\mathscr{M}_{cr}(G) = \{\mu \in \mathscr{M}_r(G): \operatorname{supp}(\mu)$ is compact$\}$.

Proposition. $\mathscr{M}_{cr}(G)$ *is a dense *-subalgebra of* $\mathscr{M}_r(G)$.

Proof. $\mathscr{M}_{cr}(G)$ is a *-subalgebra by 10.6 and 10.12(vi). Its denseness is easily verified. ∎

We shall refer to $\mathscr{M}_{cr}(G)$ as the *compacted measure algebra of G*.

Convolution as a Vector Integral

10.17. By the *weak topology of* $\mathscr{M}(G)$ we mean the topology of pointwise convergence on $\mathscr{L}(G)$; that is, given a net $\{\mu_i\}$ of elements sof $\mathscr{M}(G)$, we have $\mu_i \to \mu$ weakly if and only if $I_{\mu_i}(f) \underset{i}{\to} I_\mu(f)$ for every f in $\mathscr{L}(G)$. (Here I_μ is the complex integral corresponding to μ as in II.8.10.)

Evidently $\mathscr{M}(G)$ is a locally convex space under the weak topology; and the functions $\mu \mapsto |\int f \, d\mu|$, as f runs over $\mathscr{L}(G)$, provide a defining family of seminorms for the weak topology.

10.18. If $v \in \mathscr{M}(G)$ and $x \in G$, let xv stand for the left translate of v by x:

$$(xv)(A) = v(x^{-1}A) \qquad\qquad (A \in \mathscr{S}(G)).$$

Proposition. *Let* μ, v *be two elements of* $\mathscr{M}(G)$ *such that* $\mu * v$ *exists. Then*

$$\mu * v = \int_G (xv) d\mu_e x, \qquad\qquad (13)$$

the right side being an $\mathscr{M}(G)$*-valued vector integral (defined as in Chapter II, §6) with respect to the weak topology of* $\mathscr{M}(G)$.

Proof. In view of the last remark of 10.17, we have only to verify that

$$I_{\mu \ast v}(f) = \int_G I_{xv}(f)d\mu_e x$$

$$= \int_G \int_G f(xy)dvy \, d\mu_e x$$

for every f in $\mathcal{L}(G)$. But this follows from 10.8 (when we recall from II.5.11 that the subscript e can be dropped from $\int f \, d(\mu \ast v)_e$ and $\int f(xy)dv_e y$).   ∎

10.19. Assuming σ-boundedness of the measures, we shall obtain a converse of 10.18.

Proposition. *Let μ and v be σ-bounded elements of $\mathcal{M}(G)$. If $\int_G (x|v|)d|\mu|_e x$ exists as an $\mathcal{M}(G)$-valued vector integral with respect to the weak topology of $\mathcal{M}(G)$, then $\mu \ast v$ exists.*

Proof. Let D be any compact subset of G. Let E be a countable union of compact subsets of G such that $G \setminus E$ is locally μ-null and locally v-null. Let f be an element of $\mathcal{L}(G)$ such that $f \geq \text{Ch}_D$. Then $\langle x, y \rangle \mapsto f(xy)\text{Ch}_E(x)\text{Ch}_E(y)$ is a non-negative Borel function vanishing outside a countable union of compact sets in $G \times G$. So by Fubini's Theorem in the form II.9.8(II)

$$\int_{G \times G} f(xy)\text{Ch}_E(x)\text{Ch}_E(y)d(|\mu| \times |v|)\langle x, y \rangle$$

$$= \int_G \int_G f(xy)\text{Ch}_E(x)\text{Ch}_E(y)d|v|y \, d|\mu|x \qquad (14)$$

(where the value of each side might a priori be ∞). Now by the hypothesis of the Proposition, the right side of (14) has a finite value which is clearly independent of E; call it k. Let K be any compact subset of $p^{-1}(D)$. If we choose E big enough, Ch_K will be majorized by $\langle x, y \rangle \mapsto f(xy)\text{Ch}_E(x)\text{Ch}_E(y)$; and so by (14)

$$(|\mu| \times |v|)(K) \leq k. \qquad (15)$$

Since (15) holds for every compact subset K of $p^{-1}(D)$, $p^{-1}(D)$ must belong to the domain of $(|\mu| \times |\nu|)_e$. By the arbitrariness of D this says that p is $(|\mu| \times |\nu|)$-proper, hence that $\mu * \nu$ exists.   ∎

11. Convolution of Functions, the $\mathscr{L}_1$ Group Algebra

11.1.   Again G will be a locally compact group, with unit e; let λ be a fixed left Haar measure on G, and Δ the modular function of G.

11.2.   We recall from II.7.2 and II.7.8 that every locally λ-summable function $f: G \to \mathbb{C}$ gives rise to a complex Borel measure $f\, d\lambda$ on G satisfying $f\, d\lambda \ll \lambda$ and therefore regular (II.8.3.). Conversely, by the Radon–Nikodym Theorem II.7.8 (see II.8.7), every regular complex Borel measure ν on G satisfying $\nu \ll \lambda$ is of the form $f\, d\lambda$, where $f: G \to \mathbb{C}$ is a locally λ-summable function uniquely determined up to a locally λ-null set.

Notation.   We shall write ρ_f instead of $f\, d\lambda$ for the rest of this section.

11.3.   We saw in II.7.2 that if $1 \le p < \infty$ and $f \in \mathscr{L}_p(\lambda)$, then f is locally λ-summable. If $f \in \mathscr{L}_1(\lambda)$ then ρ_f is bounded, and by II.7.9 $\rho | \mathscr{L}_1(\lambda))$ is a linear isometry of $\mathscr{L}_1(\lambda)$ onto the closed subspace $\{\mu \in \mathscr{M}_r(G): \mu \ll \lambda\}$ of the measure algebra $\mathscr{M}_r(G)$.

11.4.   Suppose μ is a regular complex Borel measure on G, $f: G \to \mathbb{C}$ is locally λ-summable, and $\mu * \rho_f$ exists. Then it follows from 10.9(6) and the left-invariance of λ that $\mu * \rho_f \ll \lambda$, and hence that

$$\mu * \rho_f = \rho_g \tag{1}$$

for some locally λ-summable function $g: G \to \mathbb{C}$. One of our principal goals in this section is to express g in terms of μ and f. To do this we shall need the following useful lemma:

Lemma.   *Let H be a closed subgroup of G and σ any regular complex Borel measure on H. Let B be a Banach space, and $f: G \to B$ a locally λ-measurable function (see II.5.1). Then the three functions $\langle x, h \rangle \mapsto f(x)$, $\langle x, h \rangle \mapsto f(xh)$, and $\langle x, h \rangle \mapsto f(hx)$ (on $G \times H$ to B) are all locally $(\lambda \times \sigma)$-measurable.*

Proof.   The local $(\lambda \times \sigma)$-measurability of the function $\phi\colon \langle x, h \rangle \mapsto f(x)$ was observed in II.16.4.

Consider the "shear transformations" U and V given by $U(x, h) = \langle h^{-1}x, h \rangle$ and $V(x, h) = \langle xh^{-1}, h \rangle$. These are homeomorphisms of $G \times H$ onto itself. So by II.5.10 the local $(\lambda \times \sigma)$-measurability of ϕ implies the local $(U(\lambda \times \sigma))$-measurability of $U\phi$ and the local $(V(\lambda \times \sigma))$-measurability of $V\phi$. Now one checks that $(U\phi)(x, h) = f(hx)$ and $(V\phi)(x, h) = f(xh)$. Next we claim that $U(\lambda \times \sigma) = \lambda \times \sigma$. Indeed: If $C \in \mathscr{S}(G \times H)$, $(U(\lambda \times \sigma))(C) = (\lambda \times \sigma)(U^{-1}(C)) = \iint \mathrm{Ch}_{U^{-1}(C)}(x, h)d\lambda x\, d\sigma h$ (by II.9.12) $= \iint \mathrm{Ch}_C(h^{-1}x, h)d\lambda x\, d\sigma h = \iint \mathrm{Ch}_C(x, h)d\lambda x\, d\sigma h$ (by the left-invariance of λ) $= (\lambda \times \sigma)(C)$ (by II.9.12). So $U(\lambda \times \sigma)$ and $\lambda \times \sigma$ coincide, proving the claim. Similarly one proves (recalling II.7.5) that $V(\lambda \times \sigma) = \Delta'\, d(\lambda \times \sigma)$, where $\Delta'(x, h) = \Delta(h)$.

The preceding paragraph shows that $\langle x, h \rangle \mapsto f(hx)$ is locally $(\lambda \times \sigma)$-measurable, and $\langle x, h \rangle \mapsto f(xh)$ is locally $(\Delta'\, d(\lambda \times \sigma))$-measurable. But, since $\lambda \times \sigma$ and $\Delta'\, d(\lambda \times \sigma)$ have the same null sets (see II.7.8), local measurability with respect to $\lambda \times \sigma$ and $\Delta'\, d(\lambda \times \sigma)$ is the same. This completes the proof. ∎

11.5. Theorem.   *Let μ be a regular complex Borel measure on G, and $f\colon G \to \mathbb{C}$ a locally λ-summable function vanishing outside a countable union of compact sets.*

(I)   *Assume that $\mu * \rho_f$ exists. Then $\int_G |f(y^{-1}x)|d|\mu|y < \infty$ for locally λ-almost all x, and the equation*

$$g(x) = \int_G f(y^{-1}x)d\mu y \tag{2}$$

(for locally λ-almost all x) defines a locally λ-summable function $g\colon G \to \mathbb{C}$ satisfying

$$\mu * \rho_f = \rho_g. \tag{3}$$

(II)   *Assume that $\rho_f * \mu$ exists. Then $\int_G |f(xy^{-1})|\Delta(y^{-1})d|\mu|y < \infty$ for locally λ-almost all x, and the equation*

$$h(x) = \int_G f(xy^{-1})\Delta(y^{-1})d\mu y \tag{4}$$

(*for locally λ-almost all x*) *defines a locally λ-summable function* $h: G \to \mathbb{C}$ *satisfying*

$$\rho_f * \mu = \rho_h. \tag{5}$$

Proof. (I) Let $D \in \mathscr{S}(G)$. By 10.8(5) (noting that the second half of 10.8(5) did not require the σ-boundedness of μ), we have

$$(\mu * \rho_f)(D) = \iint \mathrm{Ch}_D(xy)d\rho_f y\, d\mu x$$

$$= \iint \mathrm{Ch}_D(xy)f(y)d\lambda y\, d\mu x$$

$$= \iint \mathrm{Ch}_D(y)f(x^{-1}y)d\lambda y\, d\mu x. \tag{6}$$

It follows from (6) that if $\lambda(D) = 0$, then $(\mu * \rho_f)(D) = 0$. So $\mu * \rho_f \ll \lambda$, as we asserted at the beginning of 11.4; and there exists a locally λ-summable function g satisfying (3).

To prove (2) we must apply Fubini's Theorem to the right side of (6). By Lemma 11.4 $\langle x, y \rangle \mapsto f(x^{-1}y)$ is locally $(\mu \times \lambda)$-measurable. So $\langle x, y \rangle \mapsto \mathrm{Ch}_D(y)f(x^{-1}y)$ is locally $(\mu \times \lambda)$-measurable and also vanishes outside a countable union of compact subsets of $G \times G$. The same of course is true when f and μ are replaced by $|f|$ and $|\mu|$; and with this replacement (6) implies that

$$\iint \mathrm{Ch}_D(y)|f(x^{-1}y)|d\lambda y\, d|\mu|x < \infty.$$

Hence from Fubini's Theorem in the form II.9.8(II) we conclude that $\langle x, y \rangle \mapsto \mathrm{Ch}_D(y)f(x^{-1}y)$ belongs to $\mathscr{L}_1(\mu \times \lambda)$. Thus by (6) and Fubini's Theorem II.9.8(III) we get

$$(\mu * \rho_f)(D) = \int_D d\lambda y \int f(x^{-1}y)d\mu x. \tag{7}$$

The derivation of (7) assures us that $\int f(x^{-1}y)d\mu x$ exists for λ-almost all y in D, and that $y \mapsto \int f(x^{-1}y)d\mu x$ is λ-summable over D. From the arbitrariness of D we thus conclude that $\int f(x^{-1}y)d\mu x$ exists for locally λ-almost all y, and that $y \mapsto \int f(x^{-1}y)d\mu x$ is locally λ-summable. Comparing (7) with the definition of g, we obtain (2).

(II) The proof of (II) proceeds quite similarly. ∎

11.6.   Corollary.   *Suppose f and g are locally λ-summable functions on G to* ℂ *such that at least one of f and g vanishes outside a countable union of compact sets; and assume that $\rho_f * \rho_g$ exists. Then $\rho_f * \rho_g = \rho_h$, where h: G → ℂ is the locally λ-summable function given by*

$$h(x) = \int f(y)g(y^{-1}x)d\lambda y. \tag{8}$$

(The integrand $y \mapsto f(y)g(y^{-1}x)$ on the right of (8) is λ-summable for locally λ-almost all x.)

Proof.   This follows easily from Parts (I) and (II) of 11.5, when we remember II.7.5.   ∎

11.7.   The next proposition bears the same relation to involution as 11.6 does to convolution.

Proposition.   *Let f: G → ℂ be locally λ-summable. Then the equation*

$$g(x) = \overline{f(x^{-1})}\Delta(x^{-1}) \qquad\qquad (x \in G) \tag{9}$$

defines a locally λ-summable function g: G → ℂ; and we have

$$(\rho_f)^* = \rho_g. \tag{10}$$

Proof.   By 8.11 g is locally λ-summable; and

$$\rho_g(A) = \int_A \overline{f(x^{-1})}\Delta(x^{-1})d\lambda x = \int_{A^{-1}} \overline{f(x)}d\lambda x$$
$$= \overline{\rho_f(A^{-1})} \qquad\qquad (A \in \mathscr{S}(G)).$$

11.8.   *Notation.*   It is usually convenient to identify two locally λ-summable functions f if they differ only on a locally λ-null set, and then to identify f with the corresponding complex measure ρ_f. Thus, if $\mu \in \mathscr{M}(G)$ and $f: G \to ℂ$ is a locally λ-summable function such that $\mu * \rho_f$ exists, we will usually write $\mu * f$

instead of $\mu * \rho_f$; and similarly for $f * \mu$. With this convention, equations (2)–(5), (8), and (10) become:

$$(\mu * f)(x) = \int_G f(y^{-1}x)d\mu y, \tag{11}$$

$$(f * \mu)(x) = \int_G f(xy^{-1})\Delta(y^{-1})d\mu y, \tag{12}$$

$$(f * g)(x) = \int f(y)g(y^{-1}x)d\lambda y, \tag{13}$$

$$f^*(x) = \overline{f(x^{-1})}\Delta(x^{-1}). \tag{14}$$

The $\mathscr{L}_1$ Group Algebra

11.9. Using the convention 11.8, we have the following important result:

Theorem. *$\mathscr{L}_1(\lambda)$ is a closed (two-sided)*-ideal of the measure algebra $\mathscr{M}_r(G)$. Thus $\mathscr{L}_1(\lambda)$ is a Banach *-algebra in its own right under the operations of convolution and involution given by (13) and (14).*

Proof. We have already seen in II.7.9 that the norms of $\mathscr{L}_1(\lambda)$ and $\mathscr{M}_r(G)$ coincide on $\mathscr{L}_1(\lambda)$. So $\mathscr{L}_1(\lambda)$ is a closed linear subspace of $\mathscr{M}_r(G)$. If $\mu \in \mathscr{M}_r(G)$ and $f \in \mathscr{L}_1(\lambda)$, then $\mu * f$ and $f * \mu$ are in $\mathscr{M}_r(G)$ by 10.7 and are absolutely continuous with respect to λ by 11.5, and so are in $\mathscr{L}_1(\lambda)$ by II.7.9. Thus $\mathscr{L}_1(\lambda)$ is a two-sided ideal of $\mathscr{M}_r(G)$. It is a *-ideal by (14).

If $f, g \in \mathscr{L}_1(\lambda)$, it follows from 11.6 that the element $f * g$ of $\mathscr{L}_1(\lambda)$ is given locally λ-almost everywhere by (13). But since f and g vanish outside a countable union of compact sets, the same is true of the right side of (13). Hence $(f * g)(x)$ coincides λ-almost everywhere with the right side of (13). ■

Definition. The Banach *-algebra $\mathscr{L}_1(\lambda)$, with the usual $\mathscr{L}_1$ norm and the operations of convolution (13) and involution (14), is called the $\mathscr{L}_1$ *group algebra of* G.

11.10. From 7.13 and the denseness of $\mathscr{L}(G)$ we conclude that $\mathscr{L}(G)$ is a dense *-subalgebra of $\mathscr{L}_1(\lambda)$.

The Continuity of Translation

11.11. If $\mu \in \mathcal{M}_r(G)$, the translation operation $x \mapsto x\mu$ (defined as in 10.18) need not be continuous on G with respect to the norm of $\mathcal{M}_r(G)$. For example, if $\mu = \delta_e$, then $x\mu = \delta_x$ (the unit mass at x); and $\|\delta_x - \delta_y\| = 2$ whenever $x \neq y$. In view of this observation, we ask which are the μ in $\mathcal{M}_r(G)$ for which $x \mapsto x\mu$ is norm-continuous on G. It will turn out (11.20) that these are exactly the elements of $\mathcal{L}_1(\lambda)$.

11.12. Proposition. *If $f \in \mathcal{L}(G)$, the translation maps $x \mapsto f_x$ and $x \mapsto f^x$ (defined as in 7.1) are continuous on G to $\mathcal{L}(G)$ with respect to the inductive limit topology of the latter.*

Proof.   Fix a point x in G. Let $K = \mathrm{supp}(f_x)$, and let L be a compact subset of G whose interior contains K. For any $\varepsilon > 0$, 1.10 permits us to find a neighborhood U of x such that $|f_u(y) - f_x(y)| < \varepsilon$ for all u in U and all y. Narrowing U if necessary, we may also suppose that $\mathrm{supp}(f_u) \subset L$ for all u in U (by 1.8). Thus $f_u \to f_x$ in $\mathcal{L}^L$ as $u \to x$ (see II.14.3), and so $x \mapsto f_x$ is continuous in the inductive limit topology. The same holds for the right translation map $x \mapsto f^x$. ∎

11.13. Corollary. *Let $1 \leq p < \infty$. If $f \in \mathcal{L}_p(\lambda)$, the translation maps $x \mapsto f_x$ and $x \mapsto f^x$ are continuous on G to $\mathcal{L}_p(\lambda)$ with respect to the norm-topology of the latter.*

Proof. Since the $\mathcal{L}_p$ topology relativized to $\mathcal{L}(G)$ is weaker than the inductive limit topology, the required conclusion holds if $f \in \mathcal{L}(G)$ by 11.12. Now for any f in $\mathcal{L}_p(\lambda)$ and x in G observe that

$$\|f_x\|_p = \|f\|_p, \qquad \|f^x\|_p = (\Delta(x))^{-(1/p)}\|f\|_p. \tag{15}$$

Since $\mathcal{L}(G)$ is dense in $\mathcal{L}_p(\lambda)$ (II.15.9), we can choose a sequence $\{f_n\}$ in $\mathcal{L}(G)$ converging in $\mathcal{L}_p(\lambda)$ to f. By the first statement of this proof the functions $x \mapsto (f_n)_x$ and $x \mapsto (f_n)^x$ are continuous for each n; and by (15) $(f_n)_x \to f_x$ and $(f_n)^x \to f^x$ in $\mathcal{L}_p(\lambda)$ uniformly on compact sets. So $x \mapsto f_x$ and $x \mapsto f^x$ are continuous. ∎

11.14. As a first application of 11.13 we will prove the existence of convolution in a new context.

One should bear in mind that, since λ is left-invariant, the left translate $f_{x^{-1}}$ of a function f corresponds, under the convention 11.8, with the left-translated measure $xf \cong x\rho_f$. (In a non-unimodular group, this will not be true for right translation.)

Proposition. *Let $1 \leq p < \infty$. If $\mu \in \mathcal{M}_r(G)$ and $f \in \mathcal{L}_p(\lambda)$, then $\mu * f$ exists and belongs to $\mathcal{L}_p(\lambda)$.*

Proof. By 11.13 (and (15)), the map $x \mapsto xf$ is continuous and bounded on G to $\mathcal{L}_p(\lambda)$. So $\int (xf)d\mu_e x$ exists as an $\mathcal{L}_p(\lambda)$-valued integral in view of Remark 3 of II.8.15. Now the weak topology of $\mathcal{M}(G)$, then relativized to $\mathcal{L}_p(\lambda)$, is weaker than the $\mathcal{L}_p$-topology (in view of Hölder's Inequality II.2.6); and so $\int (xf)d\mu_e x$ exists as an $\mathcal{M}(G)$-valued integral in the weak topology (see II.6.2). Since μ and f are both σ-bounded, it follows from 10.19 that $\mu * f$ exists. By 10.18, therefore,

$$\mu * f = \int (xf)d\mu_e x \in \mathcal{L}_p(\lambda). \tag{16}$$

11.15. Here is another application of 11.12:

Proposition. *If $\mu \in \mathcal{M}(G)$ and $f \in \mathcal{L}(G)$, then $\mu * f$ and $f * \mu$ exist and belong to $\mathcal{C}(G)$. If $\mu \in \mathcal{M}(G)$ and $\mu * f = 0$ for all f in $\mathcal{L}(G)$ (or $f * \mu = 0$ for all f in $\mathcal{L}(G)$), then $\mu = 0$.*

Proof. $\mu * f$ and $f * \mu$ exist by 10.6.
  By (11) and (12),

$$(\mu * f)(x) = I_\mu(\phi_x), \qquad (f * \mu)(x) = I_\mu(\psi_x), \tag{17}$$

where $\phi_x(y) = f(y^{-1}x)$, $\psi_x(y) = f(xy^{-1})\Delta(y^{-1})$, and I_μ is the complex integral corresponding to μ. It is an easy consequence of 11.12 that $x \mapsto \phi_x$ and $x \mapsto \psi_x$ are continuous with respect to the inductive limit topology. Since I_μ is continuous with respect to the inductive limit topology, it follows from (17) that $\mu * f$ and $f * \mu$ are continuous, hence in $\mathcal{C}(G)$.
  Suppose that $\mu * f = 0$ for all f in $\mathcal{L}(G)$. Since $\mu * f$ is continuous and by (17) $(\mu * f)(e) = I_\mu(\tilde{f})$ (where $\tilde{f}(y) = f(y^{-1})$), we see that $I_\mu(\tilde{f}) = 0$ for all f in $\mathcal{L}(G)$. So $I_\mu = 0$, whence $\mu = 0$ (by II.8.11). Similarly, if $f * \mu = 0$ for all f in $\mathcal{L}(G)$, then $\mu = 0$. ∎

11.16. Corollary. *$\mathcal{L}(G)$ is a (two-sided) *-ideal of the compacted measure algebra $\mathcal{M}_{cr}(G)$.*

Approximate Units

11.17. The measure algebra of g has a unit element, namely the unit mass δ_e at e. On the other hand $\mathcal{L}_1(\lambda)$ in general has no unit. As we shall see, however, it has elements which are "close" to being units.

Definition. By an *approximate unit on G* we mean a net $\{\psi_i\}$ of elements of $\mathscr{L}_+(G)$ such that (i) $\int \psi_i \, d\lambda = 1$ for each i, and (ii) for each neighborhood U of e we have supp$(\psi_i) \subset U$ for all large enough i.

Approximate units obviously exist in abundance. In fact, replacing ψ_i by $\frac{1}{2}(\psi_i + \psi_i^*)$, we can obtain approximate units $\{\psi_i\}$ satisfying $\psi_i^* = \psi_i$ for all i.

11.18.   Lemma.   *Let $\{\psi_i\}$ be an approximate unit on G. If A is a Banach space and $f: G \to A$ is a continuous map, then*

$$\int_G \psi_i(x) f(x) d\lambda x \xrightarrow{i} f(e) \quad \text{in} \quad A \tag{18}$$

(the left side of (18) being an A-valued integral).

Proof.   Given $\varepsilon > 0$, we choose a neighborhood U of e such that $\|f(x) - f(e)\| < \varepsilon$ for $x \in U$. Then, if i is large enough so that supp$(\psi_i) \subset U$,

$$\left\| \int \psi_i(x) f(x) d\lambda x - f(e) \right\| = \left\| \int \psi_i(x)(f(x) - f(e)) d\lambda x \right\| \quad \text{(since } \int \psi_i \, d\lambda = 1\text{)}$$

$$\leq \int_U \psi_i(x) \|f(x) - f(e)\| d\lambda x \qquad \text{(by II.5.4(2))}$$

$$\leq \varepsilon \int \psi_i \, d\lambda = \varepsilon. \quad \blacksquare$$

11.19.   Proposition.   *If $\{\psi_i\}$ is an approximate unit on G, then:*

(I)   *For each f in $\mathscr{L}(G)$, $\psi_i * f \to f$ and $f * \psi_i \to f$ in the inductive limit topology.*

(II)   *For each f in $\mathscr{L}_1(\lambda)$, $\psi_i * f \to f$ and $f * \psi_i \to f$ in the $\mathscr{L}_1$ norm.*

Proof.   Let f be in $\mathscr{L}_1(\lambda)$. Then by 11.13 $x \mapsto xf$ is continuous; and

$$\psi_i * f = \int \psi_i(x)(xf) d\lambda x \tag{19}$$

($\mathscr{L}_1(\lambda)$-valued integral) by (16) and II.7.5. Hence Lemma 11.18 implies that $\psi_i * f \to f$ in $\mathscr{L}_1(\lambda)$.

If f is not only in $\mathscr{L}_1(\lambda)$ but in $\mathscr{L}(G)$, then $x \mapsto xf$ is continuous in the inductive limit topology by 11.12. So, applying 11.18 to the Banach space $\mathscr{L}^K$ (with suitably large compact K; see II.14.3), we see from (19) that $\psi_i * f \to f$ in the inductive limit topology.

Notice that $\{\psi_i^*\}$ is also an approximate unit. Hence, since the * operation preserves both the $\mathscr{L}_1$ and the inductive limit topologies, we have

$$f * \psi_i = (\psi_i^* * f^*)^* \to f^{**} = f$$

in the contexts of both (I) and (II). This completes the proof. ∎

11.20. Here is an interesting characterization, independent of Haar measure, of the subspace $\mathscr{L}_1(\lambda)$ of $\mathscr{M}_r(G)$. We recall the convention 11.8.

Corollary. *An element μ of $\mathscr{M}_r(G)$ belongs to $\mathscr{L}_1(\lambda)$ if and only if $x \mapsto x\mu$ is continuous on G with respect to the norm topology of $\mathscr{M}_r(G)$.*

Proof. We proved the "only if" part in 11.13.

Suppose that μ is an element of $\mathscr{M}_r(G)$ for which $x \mapsto x\mu$ is continuous; and let $\{\psi_i\}$ be an approximate unit on G. By 11.18

$$\int \psi_i(x)(x\mu)d\lambda x \to \mu \quad \text{in} \quad \mathscr{M}_r(G). \tag{20}$$

But by 10.18 the left side of (20) is $\psi_i * \mu$, which belongs to $\mathscr{L}_1(\lambda)$ since by 11.9 the latter is an ideal of $\mathscr{M}_r(G)$. So (20) says that μ lies in the closure of $\mathscr{L}_1(\lambda)$, hence in $\mathscr{L}_1(\lambda)$ itself. This proves the "if" part. ∎

11.21. The reader should now have little difficulty in verifying the following equivalence.

Proposition*. *The following three conditions are equivalent:*

(I) *G is discrete.*
(II) *$\mathscr{L}_1(\lambda) = \mathscr{M}_r(G)$.*
(III) *$\mathscr{L}_1(\lambda)$ has a unit element.*

11.22. Here is another characterization, independent of Haar measure, of $\mathscr{L}_1(\lambda)$ as a subspace of $\mathscr{M}_r(G)$.

Proposition*. *$\mathscr{L}_1(\lambda)$ is the unique smallest non-zero norm-closed left (or right) ideal I of $\mathscr{M}_r(G)$ having the property: If $\mu \in I$, $v \in \mathscr{M}_r(G)$, and $v \ll \mu$, then $v \in I$.*

12. Relations between Measure and Topology on G

12.1. Let G be a locally compact group, with unit e and left Haar measure λ; and let $\tilde{\lambda}$ be the "reverse" of λ, that is, $\tilde{\lambda}(A) = \lambda(A^{-1})$ $(A \in \mathscr{S}(G))$.

12.2. The following lemma is given in somewhat greater generality than we need. The reader should recall the convention of 11.8.

Lemma. *Suppose that $1 < p < \infty$ and $p^{-1} + q^{-1} = 1$. Let $f \in \mathscr{L}_p(\lambda)$ and $g \in \mathscr{L}_q(\tilde{\lambda})$. Then $f * g$ exists and coincides λ-almost everywhere with a continuous function vanishing at ∞ on G.*

Proof. Let $\tilde{g}(x) = g(x^{-1})$, so that $\tilde{g} \in \mathscr{L}_q(\lambda)$; and let $x\phi$ be the left x-translate of the function ϕ on G: $(x\phi)(y) = \phi(x^{-1}y)$.

By Hölder's Inequality II.2.6, the equation

$$\alpha_f(\phi) = \int f(y)\phi(y)d\lambda y \qquad\qquad (\phi \in \mathscr{L}_q(\lambda))$$

defines a continuous linear functional α_f on $\mathscr{L}_q(\lambda)$. Thus by 11.13 the function

$$h: x \mapsto \alpha_f(x\tilde{g}) = \int f(y)g(y^{-1}x)d\lambda y \qquad\qquad (1)$$

is continuous on G (and of course the integral in (1) is finite). By Fubini's Theorem II.9.8(II), (III) (together with 11.4), we have

$$\int h\psi \, d\lambda = \iint f(y)(yg)(x)\psi(x)d\lambda x \, d\lambda y$$

$$= \int f(y)\left[\int (yg)\psi \, d\lambda \right]d\lambda y$$

for all ψ in $\mathscr{L}(G)$, whence by 10.19 and 10.18 $f * g$ exists and equals h.

To show that h vanishes at ∞, let $\varepsilon > 0$, and choose functions f' and g', with compact supports C and D respectively, such that

$$\|f' - f\|_p < \varepsilon, \quad \|\tilde{g}' - \tilde{g}\|_q < \varepsilon.$$

It follows that $\|x\tilde{g}' - x\tilde{g}\|_q < \varepsilon$ for all x; hence by (1) and 2.6, putting $h' = f' * g'$, we have

$$|h'(x) - h(x)| = |\alpha_{f'}(x\tilde{g}') - \alpha_f(x\tilde{g})|$$

$$\leq |\alpha_{f'}(x\tilde{g}' - x\tilde{g})| + |\alpha_{f'-f}(x\tilde{g})|$$

$$\leq (\|f\|_p + \varepsilon)\varepsilon + \varepsilon\|\tilde{g}\|_q \qquad\qquad (2)$$

for all x in G. On the other hand, by (1), h' vanishes outside the compact set CD. Hence, by (2) and the arbitrariness of ε, h vanishes at ∞. ∎

12.3. Proposition. *Let A and B be sets in $\mathscr{S}(G)$ with $\lambda(A) > 0$, $\lambda(B) > 0$. Then the function $x \mapsto \lambda(A \cap xB)$ is continuous and not identically 0 on G.*

Proof. If $f = \mathrm{Ch}_A$ and $g = \mathrm{Ch}_{B^{-1}}$, the h of (1) becomes the function $x \mapsto \lambda(A \cap xB)$. So the continuity of this function follows from 12.2; and we have by Fubini's Theorem applied to (1)

$$\int \lambda(A \cap xB) d\lambda x = \iint \mathrm{Ch}_A(y) \, \mathrm{Ch}_{B^{-1}}(y^{-1}x) d\lambda y \, d\lambda x$$

$$= \iint \mathrm{Ch}_A(y) \, \mathrm{Ch}_{B^{-1}}(x) d\lambda x \, d\lambda y$$

$$= \lambda(A)\lambda(B^{-1}). \tag{3}$$

Since $\lambda(B) > 0$, $\lambda(B^{-1}) > 0$ by 8.11. Hence the right side of (3) is non-zero, showing that $\lambda(A \cap xB)$ cannot vanish for all x. ∎

12.4. Corollary. *Let A and B be locally λ-measurable subsets of G neither of which is locally λ-null. Then:*

(i) *AB has an interior point;*
(ii) *e is an interior point of AA^{-1}.*

Proof. Reducing A and B, we may as well assume that both are in $\mathscr{S}(G)$. Now

$$AB \supset \{x \in G : \lambda(A \cap xB^{-1}) > 0\}, \tag{4}$$

and by 12.3 the right side of (4) is a non-void open set. This proves (i).

Now $x \mapsto \lambda(A \cap xA)$ is continuous by 12.3, and is obviously positive for $x = e$. So, replacing B by A^{-1} in (4), we conclude that e is an interior point of AA^{-1}. ∎

Remark. By this Corollary, the topology of a locally compact group G is determined once we know the locally Haar-measurable sets and the locally Haar-null sets of G; indeed, (ii) of the Corollary implies that a basis of neighborhoods of e consists of all sets of the form AA^{-1}, where A is a locally Haar measurable set which is not locally Haar null.

12.5. Corollary. *Let H be a locally λ-measurable subgroup of G. Then either H is locally λ-null or H is open in G.*

Proof. If H is not locally λ-null, then by 12.4(ii) e is an interior point of $HH^{-1} = H$. ∎

12.6. Corollary. *Let G_1 and G_2 be locally compact groups, G_2 being second-countable. Let λ be left Haar measure on G_1. We suppose that $f : G_1 \to G_2$ is a group homomorphism which is locally λ-measurable (that is, $f^{-1}(U)$ is locally λ-measurable for every Borel subset U of G_2). Then f is continuous.*

Proof. Let V be any open neighborhood of the unit e_2 of G_2. We claim that $f^{-1}(V)$ cannot be locally λ-null. Indeed, suppose it were. Then, since $f^{-1}(f(x)V) = xf^{-1}(V)$ and left translation preserves local λ-nullness,

$$f^{-1}(f(x)V) \qquad \text{is locally } \lambda\text{-null for all } x \text{ in } G_1. \tag{5}$$

Now the open subset $f(G_1)V$ of G_2, like G_2 itself, is a second-countable locally compact Hausdorff space, and therefore has the Lindelöf property (that every open covering of it contains a countable open covering). Hence there are countably many elements $\{x_n\}$ ($n = 1, 2, \ldots$) of G_1 such that $\bigcup_{n=1}^{\infty} f(x_n)V = f(G_1)V$. Therefore $G_1 = f^{-1}(f(G_1)V) = \bigcup_{n=1}^{\infty} f^{-1}(f(x_n)V)$. and so by (5) G_1 is a countable union of locally λ-null sets, hence itself locally λ-null. But this is impossible; and the claim is proved.

Now take an arbitrary neighborhood U of e_2; and let V be an open neighborhood of e_2 such that $VV^{-1} \subset U$. By the above claim, the locally λ-measurable set $f^{-1}(V)$ is not locally λ-null; and so by 12.4 e_1 (the unit of G_1) is an interior point of $f^{-1}(V)f^{-1}(V)^{-1}$. But the latter set is a subset of $f^{-1}(VV^{-1})$, which is contained in $f^{-1}(U)$. So e_1 is an interior point of $f^{-1}(U)$ for every neighborhood U of e_2; and f is continuous. ∎

13. Invariant Measures on Coset Spaces

Rho-Functions and Measures on Coset Spaces

13.1 In this and the next sections G is a fixed locally compact group with unit e; and K is a fixed closed subgroup of G. Let M be the left coset space G/K (which by 2.1 and 2.3 is locally compact and Hausdorff), and $\pi : G \to M$ the quotient map. Let λ and ν be fixed left Haar measures, and Δ and δ the modular functions, of G and K respectively.

13.2.   Definition.   A *K-rho-function on G* is a locally λ-summable complex-valued function ρ on G satisfying

$$\rho(xk) = \frac{\delta(k)}{\Delta(k)}\,\rho(x) \qquad (1)$$

for all x in G and k in K.

Thus ρ is determined on each coset xK by its value at one point of that coset.

Since K has been fixed, we shall generally omit mention of K, and speak simply of a "rho-function."

A rho-function is said to have *compact support in M* if there is a compact subset D of M such that ρ vanishes outside $\pi^{-1}(D)$.

13.3.   The next Proposition will show that continuous rho-functions exist in abundance.

Proposition.   *If $f \in \mathscr{L}(G)$, the function ρ defined by*

$$\rho(x) = \int_K \Delta(k)(\delta(k))^{-1} f(xk)dvk \qquad (x \in G)$$

is a continuous rho-function with compact support in M.

Proof.   This   is   easily   verified.   Notice   that $\rho(x) = 0$ when $x \notin \pi^{-1}(\pi(\mathrm{supp}(f)))$.   ∎

13.4.   Definition.   If $f \in \mathscr{L}(G)$, we define f^0 to be the complex function on G given by

$$f^0(x) = \int_K f(xk)dvk \qquad (x \in G).$$

It is an easy consequence of 1.10 that f^0 is continuous on G. By the left invariance of v, f^0 is constant on each coset xK, and so gives rise to a continuous function f^{00} on $M: f^{00}(xK) = f^0(x)$ $(x \in G)$. Evidently f^{00} vanishes outside $\pi(\mathrm{supp}(f))$, and so belongs to $\mathscr{L}(M)$.

13.5.   Lemma.   *Every g in $\mathscr{L}(M)$ is of the form f^{00} for some f in $\mathscr{L}(G)$. If $g \geq 0$, then f can be taken to be non-negative.*

Proof. Let $g \in \mathcal{L}(M)$, $D = \text{supp}(g)$. By 2.5 we can find a compact subset C of G such that $\pi(C) = D$. Choose a non-negative element h of $\mathcal{L}(G)$ such that $h(x) > 0$ for all x in C. Then evidently $h^{00}(m) > 0$ for all m in D. Let r be the continuous function defined on M by:

$$r(m) = \begin{cases} (h^{00}(m))^{-1} g(m) & \text{if } h^{00}(m) \neq 0, \\ 0 & \text{if } h^{00}(m) = 0. \end{cases}$$

If we set

$$f(x) = r(\pi(x))h(x) \qquad\qquad (x \in G),$$

then $f \in \mathcal{L}(G)$, and we have

$$f^{00}(m) = r(m)h^{00}(m) = g(m)$$

for all m in M. So $g = f^{00}$.

If $g \geq 0$, the above construction yields a non-negative f. ∎

Thus $f \mapsto f^{00}$ is a linear map of $\mathcal{L}(G)$ onto $\mathcal{L}(M)$.

13.6. Now follows the most important fact about rho-functions.

Proposition. *Let ρ be a rho-function on G, and f an element of $\mathcal{L}(G)$ such that $f^0 = 0$. Then*

$$\int_G \rho(x)f(x)d\lambda x = 0. \qquad\qquad (2)$$

Proof. By 8.11 and the hypothesis, $\int_K f(xk^{-1})\delta(k^{-1})dvk = \int_K f(xk)dvk = 0$ for every x in G. So, for any g in $\mathcal{L}(G)$, we have by Fubini's Theorem and 11.4

$$0 = \int_G \int_K \rho(x)g(x)f(xk^{-1})\delta(k^{-1})dvk\, d\lambda x$$

$$= \int_K \int_G \rho(x)g(x)f(xk^{-1})\delta(k^{-1})d\lambda x\, dvk$$

$$= \int_K \int_G \rho(xk)g(xk)f(x)\Delta(k)\delta(k^{-1})d\lambda x\, dvk \qquad \text{(using 8.5)}$$

$$= \int_G \int_K \rho(x)f(x)g(xk)dvk\, d\lambda x \qquad\qquad \text{(by (1))}$$

$$= \int_G \rho(x)f(x)g^0(x)d\lambda x.$$

Now by 13.5 we can choose g so that $g^0 \equiv 1$ on supp(f). Then the preceding equation becomes (2).   ∎

13.7.   There is a useful near-converse of 13.6:

Lemma.   *Let ρ be a locally λ-summable complex function on G such that*

$$\int_G \rho(x)f(x)d\lambda x = 0 \qquad \text{whenever } f \in \mathcal{L}(G) \text{ and } f^0 = 0.$$

Then, for each k in K,

$$\rho(xk) = \frac{\delta(k)}{\Delta(k)}\rho(x) \qquad \text{for locally } \lambda\text{-almost all } x.$$

Proof.   Let $k \in K$, $f \in \mathcal{L}(G)$, and $f^k(x) = f(xk)\delta(k)$ $(x \in G)$. Then by 8.5 $(f^k)^0 = f^0$; and so by 8.5 and the hypothesis

$$0 = \int_G \rho(x)(f(xk)\delta(k) - f(x))d\lambda x$$

$$= \int_G f(x)(\rho(xk^{-1})\delta(k)(\Delta(k))^{-1} - \rho(x))d\lambda x.$$

Since f is an arbitrary element of $\mathcal{L}(G)$, this implies that

$$\rho(xk^{-1})\delta(k)(\Delta(k))^{-1} - \rho(x) = 0$$

locally λ-almost everywhere.   ∎

13.8.   Let ρ be a rho-function on G. In view of Proposition 13.6, if $f \in \mathcal{L}(G)$ the number $\int \rho(x)f(x)d\lambda x$ depends only on f^{00}. Hence (and in view of 13.5) the following definition is legitimate:

Definition.   We denote by $\rho^{\#}$ the linear functional on $\mathcal{L}(M)$ given by:

$$\rho^{\#}(f^{00}) = \int_G \rho(x)f(x)d\lambda x \qquad (f \in \mathcal{L}(G)). \qquad (3)$$

13.9.   Proposition. $\rho^{\#}$ *is a complex integral on M.*

Proof.   Suppose that $\{g_i\}$ is a net of elements of $\mathcal{L}(M)$, vanishing outside a common compact set and converging uniformly to 0. To show that $\rho^{\#}$ is continuous in the inductive limit topology, it must be shown that $\rho^{\#}(g_i) \to 0$.

But the construction in the proof of Lemma 13.5 shows that we can write $g_i = f_i^{00}$, where the f_i all vanish outside the same compact subset of G and converge to 0 uniformly on G. Hence

$$\rho^\#(g_i) = \int \rho(x) f_i(x) d\lambda x \to 0. \quad \blacksquare$$

13.10. In the context of 13.8 and 13.9 we shall also use the symbol $\rho^\#$ to denote the regular complex Borel measure on M associated (according to II.8.12) with the complex integral $\rho^\#$. Thus we have the important double integration formula:

$$\int_G \rho(x) f(x) d\lambda x = \int_M d\rho^\#(xK) \int_K f(xk) d\nu k, \qquad (4)$$

valid for all f in $\mathscr{L}(G)$. (Later on, in 14.17, we will show that (4) holds for a wider class of functions f on G.)

If $\rho \geq 0$ and $g \in \mathscr{L}_+(M)$, then $\rho^\#(g) \geq 0$ by (3) and the last statement of Lemma 13.5. So, if $\rho \geq 0$, $\rho^\#$ is a (non-negative) regular Borel measure on M.

Remark 1. Clearly $\rho^\#$ depends on the normalization of the Haar measures λ and ν. If λ and ν are multiplied by positive constants a and b respectively, $\rho^\#$ becomes multiplied by ab^{-1}.

Remark 2. By (3) $\rho^\# = 0$ if and only if $\rho(x) = 0$ for locally λ-almost all x. So the linear mapping $\rho \mapsto \rho^\#$ is one-to-one if we identify rho-functions which coincide locally λ-almost everywhere. The range of this mapping will be characterized in 14.19.

Remark 3. If D is a closed subset of M and ρ vanishes outside $\pi^{-1}(D)$, evidently $\text{supp}(\rho^\#) \subset D$. Thus, if D is compact, $\rho^\#$ has compact support.

13.11. Proposition. *Let f be in $\mathscr{L}(G)$; and construct from it a rho-function $\rho = \rho_f$ as in 13.3. Then*

$$\int_M g \, d(\rho_f)^\# = \int_G f(x) g(\pi(x)) d\lambda x$$

for each g in $\mathscr{L}(M)$. Thus in the terminology of II.10.5, $(\rho_f)^\# = \pi_*(f \, d\lambda)$.

The proof is left to the reader.

13.12. The following covariance property is easily verified and very useful.

Proposition. *Let ρ be a rho-function and $x \in G$. Then the left translate $x\rho$ of ρ (defined by $(x\rho)(y) = \rho(x^{-1}y)$) is a rho-function; and $(x\rho)^{\#}$ is the translate $x(\rho^{\#})$ of $\rho^{\#}$ (given by $(x(\rho^{\#}))(A) = \rho^{\#}(x^{-1}A)$ for $A \in \mathcal{S}(M)$).*

13.13. Corollary. *Suppose that ρ is a rho-function, and S is some subgroup of G such that $\rho(sx) = \rho(x)$ for all s in S and x in G. Then $\rho^{\#}$ is invariant under the action of S on M.*

Invariant Measures on Coset Spaces

13.14. Corollary 13.13 leads naturally into the question of the existence of measures on M invariant under the action of G. We take up this question a little more generally.

If μ is a regular Borel measure on M and ϕ is a continuous homomorphism of G into the multiplicative group of positive numbers, we shall say that μ is ϕ-*variant* if $\mu \neq 0$ and $x\mu = \phi(x^{-1})\mu$ for all x in G (the x-translate $x\mu$ of μ being defined as in 13.12). In particular, if $\phi \equiv 1$, μ is *invariant* (under the action of G).

13.15. Theorem. *Let ϕ be a continuous homomorphism of G into the multiplicative group of positive numbers. A necessary and sufficient condition for the existence of a ϕ-variant regular Borel measure on M is that*

$$\phi(k) = \frac{\delta(k)}{\Delta(k)} \qquad \text{for all } k \text{ in } K. \tag{5}$$

If (5) holds, the ϕ-variant regular Borel measure is unique to within a positive multiplicative constant.

Proof. Assume that (5) holds. Then ϕ is a rho-function on G; and by Proposition 13.12, $x(\phi^{\#}) = (x\phi)^{\#} = \phi(x^{-1})\phi^{\#}$. So $\phi^{\#}$ is ϕ-variant; and the sufficiency is proved.

Now assume that μ is a ϕ-variant regular Borel measure on M. Let σ be the regular Borel measure on G associated with the integral

$$f \mapsto \int_{M} f^{00}(m)d\mu m$$

$(f \in \mathscr{L}(G))$. For each x in G and f in $\mathscr{L}(G)$,

$$(x\sigma)(f) = \sigma(x^{-1}f) = \int_M (x^{-1}f)^{00}(m)d\mu m$$

$$= \int_M (x^{-1}(f^{00}))(m)d\mu m = (x\mu)(f^{00})$$

$$= \phi(x^{-1})\mu(f^{00}) = \phi(x^{-1})\sigma(f).$$

Thus $x\sigma = \phi(x^{-1})\sigma$ for all x in G; in other words,

$$d\sigma(xy) = \phi(x)d\sigma y. \tag{6}$$

If we put $d\tau y = \phi(y^{-1})d\sigma y$, (6) says that τ is left-invariant. Hence, by the uniqueness of left Haar measure, $\tau = c\lambda$ for some positive constant c; that is, $d\sigma y = c\phi(y)d\lambda y$. Substituting this in the definition of σ, we get

$$c\int_G \phi(y)f(y)d\lambda y = \int_M f^{00}(m)d\mu m. \tag{7}$$

for $f \in \mathscr{L}(G)$. It follows that ϕ satisfies the hypothesis on ρ in Lemma 13.7; so, by that Lemma and the fact that ϕ is a continuous homomorphism, we have (5). This proves the necessity of (5).

The argument just given for the necessity of (5), in particular equation (7), shows that any ϕ-variant measure must equal $c\phi^{\#}$ for some positive constant c. So the last statement of the Theorem holds.   ∎

13.16. Corollary.   *A necessary and sufficient condition for the existence of a non-zero G-invariant regular Borel measure μ on M is that $\delta(k) = \Delta(k)$ for all k in K. If this holds, μ is unique to within a positive multiplicative constant.*

Proof.   Take $\phi \equiv 1$ in 13.15.   ∎

13.17.   By the proof of 13.15, the G-invariant regular Borel measure μ on M, if it exists, is equal to $1^{\#}$ (1 being the function identically 1). Hence in that case 13.10(4) becomes the double integral formula:

$$\int_G f\, d\lambda = \int_M d\mu(xK)\int_K f(xk)d\nu k \quad (f \in \mathscr{L}(G)). \tag{8}$$

13.18. Corollary.   *If the subgroup K is compact, there exists a G-invariant regular Borel measure on M.*

Proof.   By 13.16 and 8.9.   ∎

13.19.   Corollary. *If K is normal in G, then* $\delta(k) = \Delta(k)$ *for all k in K.*

Proof.   In that case M is a locally compact group (2.3, 2.6), and so has a left Haar measure μ which is clearly G-invariant. Now apply 13.16.   ∎

13.20.   The formula (8) permits us to generalize formula 9.5(2) to arbitrary locally compact group extensions.

Proposition.   *Assume that K is normal in G. Let* Δ' *be the modular function of the locally compact quotient group* $M = G/K$; *and let* Γ *be the continuous homomorphism of G into the positive reals such that* $d\nu(x^{-1}kx) = \Gamma(x)d\nu k$ *for* $x \in G$ *(see 8.2, 8.3). Then*

$$\Delta(x) = \Gamma(x)\Delta'(\pi(x)) \qquad \text{for } x \in G. \tag{9}$$

Proof.   Notice from 8.6 that

$$\Gamma(k) = \delta(k) \qquad\qquad (k \in K). \tag{10}$$

Let μ be the left Haar measure of M normalized so that (8) holds. Applying (8) with f replaced by $x \mapsto f(x^{-1})$, we have

$$\int f(x)\Delta(x^{-1})d\lambda x = \int f(x^{-1})d\lambda x \qquad\qquad \text{(by 8.11)}$$

$$= \int d\mu(xK) \int f(k^{-1}x^{-1})d\nu k$$

$$= \int d\mu(xK) \int f(kx^{-1})\delta(k^{-1})d\nu k \qquad\qquad \text{(by 8.11)}$$

$$= \int d\mu(xK) \int f(x^{-1}k)\delta(k^{-1})\Gamma(x)d\nu k \qquad\qquad \text{(recall 8.14)}$$

$$= \int d\mu(xK) \int f(xk)(\Delta'(\pi(x))\Gamma(xk))^{-1} \, d\nu k$$

$$\text{(by 8.11 and (10)).} \tag{11}$$

If we now apply (8) again, with f replaced by $f \cdot (\Gamma \cdot (\Delta' \circ \pi))^{-1}$, we obtain from (11)

$$\int f(x)(\Delta(x^{-1}) - \Gamma(x^{-1})(\Delta' \circ \pi)(x^{-1}))d\lambda x = 0;$$

and this holds for all f in $\mathscr{L}(G)$. It follows that (9) holds (since both sides of (9) are continuous in x).   ∎

Examples

13.21.   As an application of 13.20 we observe:

Proposition.   *The groups* GL(n, $\mathbb{C}$), SL(n, $\mathbb{C}$), GL(n, $\mathbb{R}$), SL(n, $\mathbb{R}$) *(see* 1.20) *are unimodular for all* $n = 2, 3, \ldots$

Proof.   Let us first show that SL(n, $\mathbb{R}$) is unimodular.

Let C be the commutator subgroup of SL(n, $\mathbb{R}$) (see 1.13). We claim that $C = $ SL(n, $\mathbb{R}$). We shall merely sketch the proof of this fact. One first recalls that every matrix in SL(n, $\mathbb{R}$) can be converted into the unit matrix by a finite succession of "elementary operations" which do not lead outside of SL(n, $\mathbb{R}$). These "elementary operations" are of three kinds: (I) multiplication of the i-th row (column) by λ and the j-th row (column) by λ^{-1} ($\lambda \in \mathbb{R}$); (II) interchanging the i-th and j-th rows (columns) and multiplying one of them by -1; (III) adding a constant multiple of the i-th row (column) to the j-th row (column). (In all of these, $i \neq j$.) Now each such elementary operation consists in multiplying the matrix, either to the left or to the right, by a matrix U in SL(n, $\mathbb{R}$) which, as a linear transformation on $\mathbb{R}^n$, differs from the identity in only two of the dimensions of $\mathbb{R}^n$. Thus every matrix in SL(n, $\mathbb{R}$) is a product of matrices of the above form U; and so, to prove the claim, it is enough to show that these U are in C. Since each of these U acts essentially only on two dimensions, it will therefore be enough if we establish the claim for $n = 2$. But for $n = 2$ the claim can be verified by direct calculation.

Now the modular function Δ of any locally compact group G must be identically 1 on the commutator subgroup C of G (since $\Delta(xyx^{-1}y^{-1}) = \Delta(x)\Delta(y)\Delta(x)^{-1}\Delta(y)^{-1} = 1$). Hence the above claim implies the unimodularity of SL(n, $\mathbb{R}$).

Let Γ be defined as in 13.20 when we take GL(n, $\mathbb{R}$) for G and SL(n, $\mathbb{R}$) for K; and let S be the "scalar" subgroup of GL(n, $\mathbb{R}$) consisting of all the matrices

$$\begin{pmatrix} \lambda & 0 & & 0 \\ 0 & \lambda & & 0 \\ & & \ddots & \\ 0 & 0 & & \lambda \end{pmatrix} \qquad (0 \neq \lambda \in R).$$

Since S is central in GL(n, $\mathbb{R}$), it is clear that $\Gamma(s) = 1$ for $s \in S$. Further, the unimodularity of SL(n, $\mathbb{R}$) implies that Γ is identically 1 on SL(n, $\mathbb{R}$) (see (10)). So Γ is identically 1 on the product P of S and SL(n, $\mathbb{R}$). Now $P = $ GL (n, $\mathbb{R}$) if n is odd; while if n is even P consists of all x in GL(n, $\mathbb{R}$) for which det(x) > 0, so that $GL(n, \mathbb{R})/P$ is of order 2. In either case, the fact that Γ is positive-valued and identically 1 on P implies that $\Gamma \equiv 1$ on $GL(n, \mathbb{R})$.

Now by 2.11 $GL(n, \mathbb{R})/SL(n, \mathbb{R})$ is Abelian, hence unimodular. So, in view of the constancy of Γ, the unimodularity of $GL(n, \mathbb{R})$ follows from 13.20.

Exactly the same argument shows that $SL(n, \mathbb{C})$ and $GL(n, \mathbb{C})$ are unimodular. ∎

13.22.   Remark.   We have seen in 4.7 that the "$ax + b$" group of 4.8, which is non-unimodular by 9.8, can be represented as a (closed) subgroup of $GL(2, \mathbb{R})$, which is unimodular by 13.21. Thus a closed subgroup of a unimodular group need not be unimodular. In particular, by 13.16, G-invariant measures on coset spaces G/H will not in general exist.

13.23.   It is instructive to apply 13.16 and 13.21 to the Example 3.15(B). In that Example we saw that the space $P_{n,r}$ $(r < n)$ of all r-dimensional linear subspaces of $\mathbb{C}^n$ is acted upon transitively by $GL(n, \mathbb{C})$; in fact it can be identified, as a topological $GL(n, \mathbb{C})$-space, with the space $GL(n, \mathbb{C})/G_p$, where $G_p = \{a \in GL(n, \mathbb{C}): a_{ij} = 0 \text{ whenever } i > r \text{ and } j \leq r\}$.

Now G_p is *not* unimodular. To see this, one represents G_p as a semidirect product of the normal Abelian subgroup $N = \{a \in G_p: a_{ii} = 1 \text{ for all } i \text{ and } a_{ij} = 0 \ (i \neq j) \text{ unless } i \leq r \text{ and } j > r\}$, and the closed subgroup $S(\cong GL(r, \mathbb{C}) \times GL(n - r, \mathbb{C}))$ of all matrices a such that $a_{ij} = 0$ whenever either $i > r$ and $j < r$ or $i < r$ and $j > r$. By 13.21 and 9.3 S is unimodular. Likewise N, being Abelian, is unimodular. But it is easy to check that the action of S on N by inner automorphisms does not preserve the Haar measure (Lebesgue measure) on N. So by 9.5 G_p is not unimodular.

Since $GL(n, \mathbb{C})$ is unimodular (13.21) while G_p is not, 13.16 says that there is no $GL(n, \mathbb{C})$-invariant regular Borel measure on $GL(n, \mathbb{C})/G_p$, and hence none on $P_{n,r}$.

We mentioned in 3.15(B) that $P_{n,r}$ is also transitive under the action of $U(n)$; and that $P_{n,r}$ coincides as a $U(n)$-space with $U(n)/(U(r) \times U(n - r))$. Since $U(n)$ is compact (1.20), and compact groups are all unimodular (8.8), it follows from 13.16 that there is a unique regular Borel measure on $P_{n,r}$ which is invariant under the action of the unitary group $U(n)$.

13.24.   Remark.   In analogy with 7.20 one might conjecture that if there exists a (non-zero) *bounded* G-invariant regular Borel measure on G/K, then G/K has to be compact. But this is false. For example, if $n \geq 2$ and $SL(n, \mathbb{Z}) = \{a \in SL(n, \mathbb{R}): a_{ij} \in \mathbb{Z} \text{ for all } i, j\}$, it follows from 13.16, 13.21 and the discreteness of $SL(n, \mathbb{Z})$ that there is a (non-zero) $SL(n, \mathbb{R})$-invariant regular Borel measure μ on $M = \mathrm{SL}(n, \mathbb{R})/\mathrm{SL}(n, \mathbb{Z})$. It can be shown that μ is bounded, but that M is not compact (see N. Bourbaki [11], Chap. 7, §3, Exer. 7).

14. Quasi-Invariant Measures on Coset Spaces

Existence and Uniqueness of the Quasi-Invariant Measure Class

14.1. We keep the notation of 13.1.

A regular Borel measure μ on M will be called *G-quasi-invariant* if $\mu \neq 0$ and $x\mu \sim \mu$ for all x in G. (Here as usual $x\mu$ is the x-translate $A \mapsto \mu(x^{-1}A)$ of μ; and $\sim$ is the equivalence relation of II.7.7.) If μ is G-quasi-invariant and $\mu \sim \mu'$, then clearly μ' is also G-quasi-invariant.

For brevity let us say simply "quasi-invariant" without explicit mention of G.

Notice that a quasi-invariant regular Borel measure μ must have all of M for its closed support. Indeed, if $\mu(U) = 0$ for some non-void open subset U of M, then $\mu(xU) = 0$ for every x in G, whence $\mu(D) = 0$ for every compact set $D \subset M$; thus $\mu = 0$.

14.2. We have seen in §13 that a G-invariant measure on M is essentially unique if it exists, but that in general such a measure will not exist. In this section we are going to show that there always exists a quasi-invariant regular Borel measure on M; and that any two quasi-invariant regular Borel measures are equivalent. We shall show, in other words, that there is a unique G-invariant *equivalence class* of non-zero regular Borel measures on M.

14.3. The problem of constructing a quasi-invariant Borel measure on M is essentially that of finding a rho-function which is nowhere 0. We shall in fact find *continuous* rho-functions which are nowhere 0.

14.4. Lemma. *Let U be an open symmetric neighborhood of e with compact closure. Then there exists a subset A of G with the following properties:* (i) *Every coset xK intersects Uy for some y in A;* (ii) *If C is any compact subset of G, then CK intersects Uy for only finitely many y in A.*

Proof. Using Zorn's Lemma, define A to be a maximal subset of G such that:

$$\text{If } y, z \in A \text{ and } y \neq z, \quad \text{then } z \notin UyK. \tag{1}$$

If some coset xK is disjoint from Uy for all y in A, we can adjoin x to A, contradicting the maximality of the latter. So (i) holds. To verify (ii), let C be a compact subset of G and put $A' = \{y \in A : CK \text{ intersects } Uy\}$. Then $\bar{U}C \cap yK \neq \emptyset$ for all y in A'. Suppose A' is infinite; and for each y in A' let $u_y \in \bar{U}C \cap yK$. Since $\bar{U}C$ is compact and A' is infinite, the u_y cluster at some point x of $\bar{U}C$; that is, if V is a neighborhood of x so small that $VV^{-1} \subset U$, there are distinct elements y and z of A' such that $u_y \in V, u_z \in V$. But then we have $u_y u_z^{-1} \in VV^{-1} \subset U$, whence $u_y \in Uu_z \subset UzK$; so $y \in UzK$, contradicting (1). Thus A' is finite, and so (ii) holds. ∎

14.5. Proposition. *There exists an everywhere positive continuous rho-function on G.*

Proof. Choose a non-negative function f in $\mathscr{L}(G)$ with $f(e) > 0$ and $f(x^{-1}) = f(x)$ for all x; put $U = \{x : f(x) > 0\}$, so that U is open, symmetric, and has compact closure; and let A have properties (i) and (ii) of Lemma 14.4. For each y in A let $f^y(x) = f(xy^{-1})$, and put

$$\rho^y(x) = \int_K \Delta(k)(\delta(k))^{-1} f^y(xk) d\nu k \qquad (x \in G).$$

By Proposition 13.3 ρ^y is a continuous rho-function; it clearly vanishes outside UyK. Now define $\rho(x) = \sum_{y \in A} \rho^y(x)$. By property (ii) of 14.4, for each compact subset C of G only finitely many terms contribute to this sum for $x \in C$. Hence ρ is a continuous rho-function. By property (i) of 14.4 $\rho(x) > 0$ for all x. So the Proposition is proved. ∎

14.6. The next Proposition is proved, for the present, only for continuous rho-functions.

Proposition. *Let ρ_1 and ρ_2 be continuous rho-functions on G such that $\rho_2(x) \neq 0$ for all x. Thus the equation $\sigma(xK) = \rho_1(x)(\rho_2(x))^{-1}$ $(x \in G)$ defines a continuous complex function σ on M. We have:*

$$d\rho_1^{\#} m = \sigma(m) d\rho_2^{\#} m.$$

Proof. Let $f \in \mathscr{L}(G)$. We have by definition:

$$\int f^{00}(m)d\rho_1^{\#}m = \int \rho_1(x)f(x)d\lambda x$$

$$= \int \rho_2(x)\sigma(xK)f(x)d\lambda x$$

$$= \int \sigma(m)f^{00}(m)d\rho_2^{\#}m. \tag{2}$$

Now f^{00} is the generic element of $\mathscr{L}(M)$ (by 13.5). Therefore (2) asserts that $d\rho_1^{\#}m = \sigma(m)d\rho_2^{\#}m$. ∎

14.7. Corollary. *If ρ is an everywhere positive continuous rho-function on G, then $\rho^{\#}$ is quasi-invariant.*

Proof. Let $x \in G$. By 13.12 $x(\rho^{\#}) = (x\rho)^{\#}$; so by 14.6

$$d(x(\rho^{\#}))m = \sigma(m; x)d\rho^{\#}m, \tag{3}$$

where

$$\sigma(yK; x) = \rho(x^{-1}y)(\rho(y))^{-1} \qquad (x, y \in G). \tag{4}$$

Since $\sigma(yK; x) > 0$, it follows from (3) that $x(\rho^{\#}) \sim \rho^{\#}$. So $\rho^{\#}$ is quasi-invariant. ∎

Remark. We notice that the function σ of (4) is continuous on $M \times G$. Thus the $\rho^{\#}$ of this Corollary is not only quasi-invariant, but has the further property that its G-translates have *continuous* Radon–Nikodym derivatives with respect to $\rho^{\#}$. (See 14.10.)

14.8. The next Proposition shows what the null sets for a quasi-invariant measure must be, and so proves that any two quasi-invariant measures are equivalent.

Proposition. *Let μ be a quasi-invariant regular Borel measure on M. A set D in $\mathscr{S}(M)$ is μ-null if and only if $\pi^{-1}(D)$ is locally λ-null.*

Proof. Let $D \in \mathscr{S}(M)$, and let $f \in \mathscr{L}_{+}(G)$. Thus $\langle x, m \rangle \mapsto f(x)\mathrm{Ch}_D(xm)$ is a bounded Borel function with compact support on $G \times M$; so we can apply Fubini's Theorem to it, getting:

$$\int_M \int_G f(x)\mathrm{Ch}_D(xm)d\lambda x\, d\mu m = \int_G \int_M f(x)\mathrm{Ch}_D(xm)d\mu m\, d\lambda x. \tag{5}$$

Now assume that $\pi^{-1}(D)$ is locally λ-null. Then for all y in G we have (putting $\pi(y) = m$):

$$0 = \int f(xy^{-1})\Delta(y^{-1})\text{Ch}_{\pi^{-1}(D)}(x)d\lambda x$$

$$= \int f(x)\text{Ch}_{\pi^{-1}(D)}(xy)d\lambda x$$

$$= \int f(x)\text{Ch}_D(xm)d\lambda x.$$

So the left side of (5) is 0. Therefore so is the right side; that is, for λ-almost all x,

$$0 = \int_M f(x)\text{Ch}_D(xm)d\mu m = f(x)\mu(x^{-1}D). \tag{6}$$

Suppose $f \neq 0$. It then follows from (6) that $\mu(x^{-1}D) = 0$ for some x. Since μ is quasi-invariant, this implies that $\mu(D) = 0$.

Conversely, assume that $\mu(D) = 0$. Then by quasi-invariance $\mu(x^{-1}D) = 0$ for all x, so the right side of (5) is 0. Therefore the left side is 0, giving:

$$\int f(x)\text{Ch}_D(xm)d\lambda x = 0 \qquad \text{for } \mu\text{-almost all } m. \tag{7}$$

Now let C be a compact subset of G and U a neighborhood of e with compact closure; and suppose $f \in \mathcal{L}_+(G)$ is chosen so that $f \geq 1$ on CU^{-1}. Since π is open and the closed support of μ is M (see 14.1), we can choose y in U so that (7) holds for $m = \pi(y)$. Thus:

$$0 = \Delta(y)\int f(x)\text{Ch}_D(\pi(xy))d\lambda x$$

$$= \int f(xy^{-1})\text{Ch}_{\pi^{-1}(D)}(x)d\lambda x. \tag{8}$$

Now if $x \in C$, then $xy^{-1} \in CU^{-1}$. So $f(xy^{-1})\text{Ch}_{\pi^{-1}(D)}(x) \geq \text{Ch}_{\pi^{-1}(D) \cap C}(x)$ for all x. Hence by (8) $\lambda(\pi^{-1}(D) \cap C) = 0$. Since C was an arbitrary compact set, this says that $\pi^{-1}(D)$ is locally λ-null.

We have now shown that $\pi^{-1}(D)$ is locally λ-null if and only if $\mu(D) = 0$, and the proof is complete. ∎

14.9. Putting together 14.5, 14.7, and 14.8, we obtain:

Theorem.   *There exists a quasi-invariant regular Borel measure μ on M; and any two such are equivalent. A set D in $\mathscr{S}(M)$ is μ-null if and only if $\pi^{-1}(D)$ is locally λ-null.*

Translation-Continuous Quasi-Invariant Measures

14.10.   In 14.7 we pointed out that there exist quasi-invariant measures which are translation-continuous in the following sense:

Definition.   A quasi-invariant regular Borel measure μ on M is *translation-continuous* if there exists a positive-real-valued continuous function σ on $M \times G$ such that

$$d(x\mu)m = \sigma(m, x)d\mu m \tag{9}$$

for all $x \in G$.

In particular, if μ is translation-continuous, the Radon–Nikodym derivatives of the translates of μ with respect to μ are continuous.

Notice that (9) implies the following identities for σ:

$$\sigma(m, e) = 1, \tag{10}$$

$$\sigma(m, xy) = \sigma(x^{-1}m, y)\sigma(m, x) \tag{11}$$

$(m \in M; x, y \in G)$. Indeed: (10) is evident. By (9) (see II.5.10, II.7.6) we have:

$$\sigma(x^{-1}m, y)\sigma(m, x)d\mu m = \sigma(x^{-1}m, y)d\mu(x^{-1}m)$$

$$= d(y\mu)(x^{-1}m)$$

$$= d(xy\mu)m = \sigma(m, xy)d\mu m.$$

Since σ is continuous and $\mathrm{supp}(\mu) = M$ (by 14.1), this implies (11).

14.11.   To prove our main result on translation-continuity, we begin with the case that $K = \{e\}$, $M = G$.

Lemma.   *Every translation-continuous quasi-invariant regular Borel measure μ on G is of the form $\gamma(x)d\lambda x$, where γ is a continuous function on G to the positive reals.*

Proof.   Let σ be the continuous function (defined now of course on $G \times G$) satisfying (9). From the identity (11) we deduce that $v = (\sigma(e; x^{-1}))^{-1}\,d\mu x$ is left-invariant. Hence by the uniqueness of Haar measure $v = c\lambda$ for some positive constant c. Therefore μ coincides with $c\sigma(e; x^{-1})d\lambda x$, proving the Lemma.   ∎

Remark. Even without translation-continuity, the Radon–Nikodym Theorem together with 14.9 shows that $\mu = \gamma(x)d\lambda x$ for some locally λ-summable function γ. The continuity of γ is the essential content of the above Lemma.

14.12. Returning to an arbitrary closed subgroup K, we have:

Proposition. *Let μ be a regular Borel measure on M. Then μ is quasi-invariant and translation-continuous if and only if there is an everywhere positive continuous rho-function ρ on G such that $\mu = \rho^{\#}$.*

Proof. Assume that μ is quasi-invariant and translation-continuous; and let σ be as in (9). One verifies (via II.8.12) that the equation

$$\int_G f \, d\tau = \int_M f^{00} \, d\mu \qquad (f \in \mathcal{L}(G)) \qquad (12)$$

defines a regular Borel measure τ on G. Since $(xf)^{00} = x(f^{00})$, (12) and (9) imply that for any x in G and f in $\mathcal{L}(G)$,

$$\int_G f \, d(x\tau) = \int_G (x^{-1}f)d\tau = \int_M f^{00} \, d(x\mu)$$

$$= \int_M f^{00}(m)\sigma(m, x)d\mu m$$

$$= \int_G f(y)\sigma(\pi(y), x)d\tau y.$$

By the arbitrariness of f, this gives

$$d(x\tau)y = \sigma(\pi(y), x)d\tau y. \qquad (13)$$

Now (13) implies by 14.11 that $d\tau y = \rho(y)d\lambda y$ for some positive-valued continuous function ρ on G. By 13.7 and the definition (12) of τ, ρ is a rho-function; and from (12) we see that $\mu = \rho^{\#}$. Thus we have proved the "only if" part of the Proposition.

The "if" part was established in the proof of 14.7. ■

Remark. By the above Proposition together with 14.6, any two quasi-invariant translation-continuous regular Borel measures μ and μ' on M are related by the equation $d\mu'm = g(m)d\mu m$ for some positive-valued everywhere continuous function g on M.

14.13. Proposition 14.12 leads to an illuminating interpretation of the rather mysterious ratio $\delta(k)(\Delta(k))^{-1}$ occurring crucially in such places as 13.2(1), 13.15(5). Let μ be any quasi-invariant translation-continuous regular Borel measure on M. By 14.12 $\mu = \rho^{\#}$ for some continuous positive rho-function ρ, and so by 14.7(3), (4) and 13.2(1) we have for $k \in K$

$$d\mu(km) = d(k^{-1}\mu)m = \sigma(m; k^{-1})d\mu m, \tag{14}$$

where

$$\sigma(eK; k^{-1}) = \rho(k)(\rho(e))^{-1}$$

$$= \delta(k)(\Delta(k))^{-1}. \tag{15}$$

Thus translation by an element k of K leaves the point eK of M fixed and *multiplies the μ-measures of small regions around eK by a factor very close to* $\delta(k)\Delta(k))^{-1}$.

From this geometric interpretation of $\delta(k)(\Delta(k))^{-1}$ it is easy to see why the ratio $\delta(k)(\Delta(k))^{-1}$ must equal 1 if there is to exist a G-invariant measure on M.

14.14. Corollary. *Let S be another closed subgroup of G such that $SK = G$, and such that the canonical bijection $S/(S \cap K) \to M$ is a homeomorphism. (The latter condition will follow from the fact that $SK = G$ if G is σ-compact; see 3.13). Let δ_S and Δ_S be the modular functions of $S \cap K$ and S respectively. Then*

$$\delta_S(k)(\Delta_S(k))^{-1} = \delta(k)(\Delta(k))^{-1} \tag{16}$$

for all k in $S \cap K$. In particular, if ρ is a continuous K-rho-function on G, $\rho|S$ is an $(S \cap K)$-rho-function on S.

Proof. Since G/K and $S/(S \cap K)$ essentially coincide as locally compact S-spaces, the S-quasi-invariant translation-continuous regular Borel measures on both are essentially the same. Hence (16) follows from (14) and (15). The last statement follows from (16) and the characteristic identity 13.2(1) for rho-functions. ∎

Double Integration of Summable Vector-Valued Functions

14.15. Fix an everywhere positive continuous rho-function ρ on G. Replacing f by $\rho^{-1}f$ in 13.10(4), we have

$$\int_G f d\lambda = \int_M d\rho^{\#}(xK) \int_K \rho(xk)^{-1}f(xk)dvk \tag{17}$$

for all f in $\mathcal{L}(G)$. We propose to generalize (17) to the case where f is any λ-summable function on G to a Banach space.

14.16. First we prove the corresponding fact about null sets.

Proposition. *Let N be a λ-null subset of G. Then for ρ^*-almost all cosets m in M, the set $\{k \in K : xk \in N\}$ (x being any fixed element of m) is ν-null.*

Proof. Let U be any open subset of G with compact closure. We claim that (17) holds for $f = \mathrm{Ch}_U$.

Indeed: Let $\{f_i\}$ be a monotone increasing net of elements of $\mathcal{L}_+(G)$ such that $f_i \uparrow \mathrm{Ch}_U$ pointwise. Then by II.8.14

$$\int f_i \, d\lambda \uparrow \lambda(U). \tag{18}$$

Further, for each x in G, applying II.8.14 to the $f_i|(xK)$, we have

$$\int \rho(xk)^{-1} f_i(xk) dvk \uparrow \int \rho(xk)^{-1} \mathrm{Ch}_U(xk) dvk. \tag{19}$$

Now for each i the left side of (19) is continuous in xK (see 13.4). Hence the right side of (19) is lower-semicontinuous in xK; and II.8.14 applied to (19) gives

$$\int_M d\rho^*(xK) \int_K \rho(xk)^{-1} f_i(xk) dvk \uparrow \int_M d\rho^*(xK) \int_K \rho(xk)^{-1} \mathrm{Ch}_U(xk) dvk. \tag{20}$$

By (17) the left sides of (18) and (20) are equal for each i. Hence the right sides of (18) and (20) are equal, proving the claim.

To prove the Proposition it is enough to assume that the λ-null set N is contained in a compact set. Then there is a decreasing sequence $\{U_n\}$ of open subsets of G with compact closure such that $N \subset U_n$ for each n and $\lambda(U_n) \downarrow 0$. Hence by the above claim

$$\int d\rho^*(xK) \int \rho(xk)^{-1} \mathrm{Ch}_N(xk) dvk \le \int d\rho^*(xK) \int \rho(xk)^{-1} \mathrm{Ch}_{U_n}(xk) dvk \downarrow_n 0.$$

So

$$\int d\rho^*(xK) \int \rho(xk)^{-1} \mathrm{Ch}_N(xk) dvk = 0,$$

implying that $\int \rho(xk)^{-1} \mathrm{Ch}_N(xk) dvk = 0$ for ρ^*-almost all xK. Since $\rho > 0$ everywhere, this says that for ρ^*-almost all xK, $\mathrm{Ch}_N(xk) = 0$ for ν-almost all k; and the Proposition is proved. ∎

14.17. We now have the following Fubini-like result.

Theorem. *Let A be any fixed Banach space; and let $f: G \to A$ be a λ-summable function (see II.5.4). Then:*

(I)   *For $\rho^{\#}$-almost all m in M, the function $k \mapsto \rho(xk)^{-1}f(xk)$ (where $x \in m$) is v-summable.*

(II) *The function $xK \mapsto \int_K \rho(xk)^{-1}f(xk)dvk$ is $\rho^{\#}$-summable.*

(III) $\int_M d\rho^{\#}(xK) \int_K \rho(xk)^{-1}f(xk)dvk = \int f(x)d\lambda x.$

Remark. In connection with (I) and (II), notice that the K-left-invariance of v implies that, for a fixed coset m, the existence and value of $\int \rho(xk)^{-1}f(xk)dvk$ is independent of which x in m we take.

Proof. For each m in M, the formula

$$v_m(W) = v(x^{-1}W) \qquad\qquad (W \in \mathscr{S}(m)),$$

where x is a fixed element of m, defines a regular Borel measure v_m on m which is independent of the particular x. Put $B_m = \mathscr{L}_1(v_m; A)$. We shall define a Banach bundle $\mathscr{B}$ over M whose fiber at m is B_m.

Given $\phi \in \mathscr{L}(G; A)$, let $\tilde{\phi}$ be the function on M whose value at m is the element $\tilde{\phi}(m)$ of $\mathscr{L}(m; A)$ given by

$$\tilde{\phi}(m)(y) = \rho(y)^{-1}\phi(y) \qquad\qquad (y \in m).$$

Notice from II.14.8 that every element of $\mathscr{L}(m; A)$ is of the form $\tilde{\phi}(m)$ for some ϕ in $\mathscr{L}(G; A)$; and that by II.15.9 $\mathscr{L}(m; A)$ is dense in B_m. Hence, putting $\Phi = \{\tilde{\phi} : \phi \in \mathscr{L}(G; A)\}$, we see that $\{\tilde{\phi}(m) : \tilde{\phi} \in \Phi\}$ is dense in B_m for every m. Furthermore, as was mentioned in 13.4, $m \mapsto \|\tilde{\phi}(m)\|_{B_m} = \int \rho(xk)^{-1}\|\phi(xk)\|dvk$ (where $x \in m$) is continuous in m for each $\tilde{\phi}$ in Φ. Thus we can apply the Construction Theorem II.13.18 to Φ, obtaining a Banach bundle $\mathscr{B}$ over M whose fibers are the B_m ($m \in M$), and having the $\tilde{\phi}$ ($\phi \in \mathscr{L}(G; A)$) as continuous cross-sections.

By (17) the map $I : \phi \mapsto \tilde{\phi}$ extends to a linear isometry of $\mathscr{L}_1(\lambda; A)$ into $\mathscr{L}_1(\rho^{\#}; \mathscr{B})$. Furthermore, by II.15.10 Φ is dense in $\mathscr{L}_1(\rho^{\#}; \mathscr{B})$. Therefore I is onto $\mathscr{L}_1(\rho^{\#}; \mathscr{B})$. We claim that, for each f in $\mathscr{L}_1(\lambda; A)$, $I(f)(m)$ coincides v_m-almost everywhere with $(\rho^{-1}f)|m$ for $\rho^{\#}$-almost all m. Indeed: Let

$f \in \mathscr{L}_1(\lambda; A)$; and choose a sequence $\{\phi_n\}$ of elements of $\mathscr{L}(G; A)$ approaching f in $\mathscr{L}_1(\lambda; A)$. Then we also have $I(\phi_n) \to I(f)$ in $\mathscr{L}_1(\rho^\#; \mathscr{B})$. By II.3.5 we can pass to a subsequence and assume that:

$$\phi_n(x) \to f(x) \qquad \text{in } A \text{ for } \lambda\text{-almost all } x, \tag{21}$$

$$I(\phi_n)(m) \to I(f)(m) \qquad \text{in } B_m \text{ for } \rho^\#\text{-almost all } m. \tag{22}$$

Now by the definition of I the claim holds for $f \in \mathscr{L}(G; A)$. Hence, applying 14.16 to the exceptional null set of (21), we conclude that, for $\rho^\#$-almost all m, $I(\phi_n)(m) \to (\rho^{-1}f)|m$ pointwise ν_m-almost everywhere. From this and (22) the claim follows; and with it follows (I) of the Theorem.

In view of this claim, for each f in $\mathscr{L}_1(\lambda; A)$ the function

$$f^* : m \mapsto \int_m I(f)(m)d\nu_m \in A$$

is defined $\rho^\#$-almost everywhere on M. By the argument leading to (22) we can find a sequence $\{\phi_n\}$ of elements of $\mathscr{L}(G; A)$ such that $(\phi_n)^*(m) \to f^*(m)$ for $\rho^\#$-almost all m. By an easy generalization of the argument of 13.4, $(\phi_n)^*$ is continuous on M to A. So f^* is locally $\rho^\#$-measurable. In fact, since

$$\|f^*(m)\| \le \|I(f)(m)\| \qquad \text{(by II.5.4(2))} \tag{23}$$

and $I(f) \in \mathscr{L}_1(\rho^\#; \mathscr{B})$, f^* is $\rho^\#$-summable. This (together with the above claim) proves (II) of the Theorem.

By (23) the linear map

$$J_1 : f \mapsto \int_M f^* d\rho^\# \qquad\qquad (f \in \mathscr{L}_1(\lambda; A))$$

is continuous on $\mathscr{L}_1(\lambda; A)$ to A. To prove (III), and thereby to complete the proof of the Theorem, it suffices to show that J_1 coincides with $J_2 : f \mapsto \int_G f d\lambda$. Since both J_1 and J_2 are continuous, it is enough to take an arbitrary α in A^* and show that $\alpha \circ J_1$ coincides with $\alpha \circ J_2$ on the dense subset $\mathscr{L}(G; A)$ of $\mathscr{L}_1(\lambda; A)$. But by II.5.7,

$$\alpha(J_1(\phi)) = \int_M d\rho^\#(xK) \int_K \rho(xk)^{-1}(\alpha \circ \phi)(xk)d\nu k,$$

$$\alpha(J_2(\phi)) = \int_G (\alpha \circ \phi)(x)d\lambda x;$$

and, if $\phi \in \mathscr{L}(G; A)$ (so that $\alpha \circ \phi \in \mathscr{L}(G)$), the right sides of these two equalities coincide by (17). ∎

14.18. Corollary. *Let ρ be a continuous everywhere positive rho-function on G. A function $f: M \to \mathbb{C}$ is locally $\rho^{\#}$-measurable [locally $\rho^{\#}$-summable] if and only if the function $\phi: G \to \mathbb{C}$ given by $\phi(x) = f(xK)$ is locally λ-measurable [locally λ-summable].*

Proof. First we prove the statement about local measurability. For this it is enough to take a subset D of M with compact closure, and show that D is $\rho^{\#}$-measurable if and only if $\pi^{-1}(D)$ is locally λ-measurable.

Assume that D is $\rho^{\#}$-measurable. Then there is a Borel subset E of M such that $D \ominus E$ is $\rho^{\#}$-null. Since $\pi^{-1}(D \ominus E) = \pi^{-1}(D) \ominus \pi^{-1}(E)$ is locally λ-null by 14.8, and since $\pi^{-1}(E)$ is a Borel subset of G, it follows that $\pi^{-1}(D)$ is locally λ-measurable.

Conversely, assume that $\pi^{-1}(D)$ is locally λ-measurable. By 13.5 there is an element g of $\mathscr{L}(G)$ such that $g^{00} \equiv 1$ on D. Then $h = \rho g \, \mathrm{Ch}_{\pi^{-1}(D)}$ is λ-summable, and by 14.17(II) the function

$$xK \mapsto \int_K \rho(xk)^{-1} h(xk) dvk = \mathrm{Ch}_{\pi^{-1}(D)}(x) = \mathrm{Ch}_D(xK)$$

is $\rho^{\#}$-summable. This implies that D is $\rho^{\#}$-measurable.

We have now proved the statement about local measurability. It remains only to take a locally $\rho^{\#}$-measurable function $f: M \to \mathbb{C}$, and show that f is locally $\rho^{\#}$-summable if and only if $\phi: x \mapsto f(xK)$ (which is locally λ-measurable by the preceding part of the proof) is locally λ-summable.

Assume that ϕ is locally λ-summable. Given $g \in \mathscr{L}(M)$, use 13.5 to choose an h in $\mathscr{L}(G)$ satisfying $h^{00} = g$. By hypothesis $\rho h \phi$ is λ-summable. So 14.17(II) implies that gf is $\rho^{\#}$-summable. Since g is arbitrary in $\mathscr{L}(M)$, this implies that f is locally $\rho^{\#}$-summable.

Now assume that f is locally $\rho^{\#}$-summable; and suppose without loss of generality that $f \geq 0$. For each $n = 1, 2, \ldots$ let $f_n = \min(f, n)$, $\phi_n = \min(\phi, n)$; and take an arbitrary function h in $\mathscr{L}_+(G)$. Applying 14.17(III) to the λ-summable function $\rho h \phi_n$, we get

$$\int \rho h \phi_n \, d\lambda = \int_M h^{00}(m) f_n(m) d\rho^{\#} m. \tag{24}$$

Since f is locally $\rho^{\#}$-summable, as $n \to \infty$ the right side of (24) approaches the finite number $\int h^{00} f \, d\rho^{\#}$ (see II.2.7). Hence, applying II.2.7 to the left side of

(24), we conclude that $\int \rho h \phi_n \, d\lambda \uparrow \int \rho h \phi \, d\lambda < \infty$. By the arbitrariness of h this implies that ϕ is locally λ-summable. ∎

14.19. Corollary*. *For a regular complex Borel measure μ on M, the following two conditions are equivalent:*

(I) $\mu = \sigma^\#$ *for some rho-function σ on G;*

(II) μ *is absolutely continuous with respect to any given quasi-invariant regular Borel measure on M.*

15. Exercises for Chapter III

1. Let G be a topological group, where it is not assumed that the topology on G is Hausdorff. Prove that the following are equivalent:

(a) G is a T_0-space;

(b) G is a T_1-space;

(c) G is a Hausdorff space;

(d) $\{e\}$ is a closed subset of G;

(e) $\bigcap\{U : U \in \mathcal{N}\} = \{e\}$, where $\mathcal{N}$ is the family of neighborhoods of e in G.

2. Let G be a topological group and $\mathcal{N}$ the family of neighborhoods of e. Prove that for any subset B of G, $\bar{B} = \bigcap\{UBU : U \in \mathcal{N}\}$.

3. Give an example of two closed subsets A and B of a topological group G such that AB is not closed.

4. Let A and B be subsets of a topological group G. Prove that:

(a) $\bar{A}\bar{B} \subset (AB)^-$;

(b) $(\bar{A})^{-1} = (A^{-1})^-$;

(c) $x\bar{A}y = (xAy)^-$ for all $x, y \in G$.

(d) Give an example to show that the inclusion in part (a) can be proper.

5. Prove that a topological group G with more than one element has the discrete topology if and only if there exists a compact open subset V of G such that Vx is not contained in V for all x in G with $x \neq e$.

6. Show that a countable locally compact group G must have the discrete topology. [Hint: Baire Category.]

7. Show that the commutator subgroup C (i.e., the subgroup generated by all elements of the form $x^{-1}y^{-1}xy$) of a connected topological group G is connected.

8. Give an example of a group G and a compact Hausdorff topology on G such that G with this topology is not a topological group.

9. Give an example to show that the two axioms (c') and (c'') in the definition of a topological group (see Definition (1.1)) are independent (that is, neither can be proved from the other).

10. Show that a subgroup H of a topological group G is open if and only if the interior of H is non-void.

11. Let G be a topological group and H a subgroup of G such that $\bar{U} \cap H$ is closed in G for some neighborhood U of e in G. Prove that H is closed.

12. Prove that a subgroup H of a topological group G is closed if and only if there exists an open set U in G such that $U \cap H \neq \emptyset$ and $U \cap H = U \cap \bar{H}$.

13. Let H be a dense subgroup of a topological group G, and let K be a normal subgroup of H. Prove that the closure $\bar{K}$ of K in G is a normal subgroup of G.

14. Prove that if H is a nonclosed subgroup of a topological group G, then $\bar{H} \cap H'$ is dense in $\bar{H}$, where H' denotes the complement of H in G.

15. Prove that the intersection of all open subgroups and the intersection of all closed subgroups of a topological group G are closed normal subgroups of G.

16. Prove Proposition 1.12.

17. Let H be a subgroup of a topological group G. In 1.14 it was shown that if H is open in G, then it is also closed. Prove that every closed subgroup H of finite index of G is open.

18. Show that if a subgroup H of a topological group G has an isolated point, then H is closed. Does the result remain true if the topology on G is non-Hausdorff?

19. Let x be an element in a compact group G. Prove that if $A = \{x^n : n = 0, 1, 2, \ldots\}$, then $\bar{A}$ is a subgroup of G.

20. Prove that if G is a compact group and H is a closed subgroup of G, then $xHx^{-1} = H$ if and only if $xHx^{-1} \subset H$.

21. Prove that if G is a compact group, then every neighborhood U of e in G contains a neighborhood V of e which is invariant under conjugation.

22. Let G be a separable topological group and H a subgroup of G. Prove that the quotient space G/H is separable.

23. Let G be a topological group and H a subgroup of G. If H and G/H are separable, is G separable?

24. Let G be a topological group. Prove that if H is a closed normal subgroup of G and A is a closed subgroup of G such that $H \subset A$, then the image $\pi(A)$ of A under the natural mapping is a closed subgroup of G/H.

25. Let G and H be topological groups and f a homomorphism of G onto H. Prove that if $f(U)$ has nonempty interior for all U in an open basis at e, then f is an open mapping.

26. Prove that if U is a symmetric neighbourhood of the identity e in a topological group G, then $H = \bigcup_{i=1}^{n} U^n$ is an open and closed subgroup of G.

27. Prove that every connected locally compact group is Lindelöf (i.e., every open covering of X admits a countable subcovering).

28. Let G be a locally compact group which is σ-compact (i.e., G is a countable union of compact subsets of G). Show that there exists a compact normal subgroup H of G such that G/H satisfies the second axiom of countability. (See Kakutani and Kodaira [1].)

29. Show that the topological space underlying a locally compact group is normal.

30. Let G be a topological group and A a connected subset of G which contains e. Prove that the smallest subgroup of G containing A is also connected.

31. Let G be a topological group, and H a closed subgroup of G. Prove that:

(a) If G is connected, then G/H is connected.

(b) If H and G/H are connected, then G is connected.

32. Let G be a locally compact group; and let K be a closed normal subgroup of G such that K and G/K are σ-compact. Prove that G is σ-compact.

33. Let G be a locally compact group; and let N denote the connected component of e. Prove that the following are equivalent:

(a) G is σ-compact;

(b) G is a Lindelöf space;

(c) If V is a neighborhood of e, then G can be covered by countably many left [right] translates of V;

(d) For any closed subgroup H of G, H and G/H are σ-compact;

(e) There is a closed subgroup K of G such that K and G/K are σ-compact;

(f) G/N is σ-compact.

34. Let G be a topological group, C a compact subset of G, and U an open subset of G containing C. Prove that there exists a neighborhood V of e such that $(CV) \cup (VC) \subset U$. Show that if G is locally compact, then V can be chosen so that $((CV) \cup (VC))^-$ is compact.

35. A topological group G is said to be *compactly generated* if it contains a compact subset C for which the subgroup generated by C is G. Prove that assertions (a), (b), and (c) are equivalent:

(a) G is compactly generated;

(b) There is an open subset U of G such that $\bar{U}$ is compact and U generates G;

(c) There is a neighborhood U of e in G such that $\bar{U}$ is compact and U generates G.

36. Let G be a locally compact group and H a closed normal subgroup of G. Prove that if both H and G/H are compactly generated, then G is compactly generated.

37. Let $\langle G, \cdot, d \rangle$ be a triple such that $\langle G, \cdot \rangle$ is a group, $\langle G, d \rangle$ is a metric space and the function $\langle x, y \rangle \mapsto xy$ is continuous from $G \times G$ to G. Prove that if d is left-invariant (i.e., $d(ax, ay) = d(x, y)$ for all $a, x, y \in G$), then $\langle G, \cdot, d \rangle$ is a topological group under the metric topology of d. Is this true without the assumption that d is left-invariant?

38. Let G be a topological group such that $\bigcup_{n=1}^{\infty} V^n = G$ for every neighborhood V of e. Prove that if H is a discrete normal subgroup of G, then H is contained in the center of G.

39. Give an example to show that the image of a closed subgroup under an open continuous homomorphism need not be closed.

40. Let G, H, and K be topological groups and let $f: G \to H$, $g: H \to K$ be continuous surjective homomorphisms. Prove that if $g \circ f: G \to K$ is a topological isomorphism, then f and g are topological isomorphisms.

41. Let G, H, and K be topological groups. Let $f: G \to H$ be a continuous open surjective homomorphism, and $g: H \to K$ a surjective homomorphism. Prove that if $g \circ f: G \to K$ is a topological isomorphism, then f and g are topological isomorphisms.

42. Verify the details left to the reader in the proof of part (III) of Proposition 5.11.

43. Let G be a locally compact totally disconnected topological group. Prove that if H is a closed subgroup of G, then the space G/H is totally disconnected.

44. Let G and H be topological groups and $f: G \to H$ a continuous open surjective homomorphism. Prove that if G is locally compact and N is the component of e in G, then $\overline{f(N)}$ is the component of the identity in H.

45. Let $\{G_i\}_{i \in I}$ be a family of topological groups and consider the direct product topological group $G = \prod_{i \in I} G_i$ (see 4.1). Let H be the set of all $x = (x_i)_{i \in I}$ in G such that x_i is the unit element of G_i for all but finitely many i. Prove that H is a normal subgroup of G which is dense in G.

46. Let $\{G_i\}_{i \in I}$ be a family of topological groups and, for each i, let H_i be a closed normal subgroup of G_i. Prove that $\prod_{i \in I} H_i$ is a closed normal subgroup in $\prod_{i \in I} G_i$ and that the topological groups $\prod_{i \in I}(G_i/H_i)$ and $(\prod_{i \in I} G_i)/(\prod_{i \in I} H_i)$ are topologically isomorphic.

47. Suppose that, for each i in an index set I, G_i is a locally compact group having a compact open subgroup H_i. Let G be the (untopologized) group of all those x in $\prod_{i \in I} G_i$ such that $x_i \in H_i$ for all but finitely many i in I.

Show that there is one and only one topology on G making G a topological group such that: (i) $H = \prod_{i \in I} H_i$ is an open subgroup of G; (ii) the topology of G relativized to H is the Cartesian product topology of H (and hence compact).

(Note: As an example of this construction, let I be the set of positive prime numbers; and for each prime p take G_p and H_p to be the additive group of p-adic numbers and p-adic integers respectively. The resulting G is important in algebraic number theory.)

48. Let N and G be topological groups, with N Abelian. Show that the family $\Gamma(N, G)$ of isomorphism classes of central extensions of N by G is an Abelian group (see 5.6). Verify the assertions in 5.7.

49. Show that the mapping $\phi: H_0 \to N$ defined by $\phi(h, h') = i'^{-1}[F(h)^{-1}h']$ in the proof of 5.8 is a continuous homomorphism which satisfies (6). Furthermore, show that the relations in (8) hold in the proof of sufficiency.

50. Prove parts (i)–(iv) of Lemma 7.4.

51. Prove Proposition 8.13.

52. Give a complete proof of Proposition 8.16.

53. Let $G = GL(n, \mathbb{R})$. Show that the Haar measure λ of G is given by the following differential formula:

$$d\lambda x = \frac{\prod_{i,j=1}^n dx_{ij}}{|\det(x)|^n},$$

where of course

$$x = \begin{pmatrix} x_{11} & x_{12} & \cdots & x_{1n} \\ x_{21} & x_{22} & \cdots & x_{2n} \\ \vdots & \vdots & & \vdots \\ x_{n1} & x_{n2} & \cdots & x_{nn} \end{pmatrix}.$$

54. Let $G = GL(n, \mathbb{C})$. Show that the Haar measure λ of G is given by:

$$d\lambda x = \frac{\prod_{i,j=1}^{n} dx_{ij}' \, dx_{ij}''}{|\det(x)|^{2n}},$$

where

$$x = \begin{pmatrix} x_{11} & x_{12} & \cdots & x_{1n} \\ x_{21} & x_{22} & \cdots & x_{2n} \\ \vdots & \vdots & & \vdots \\ x_{n1} & x_{n2} & \cdots & x_{nn} \end{pmatrix}, \qquad x_{ij}' = \mathrm{Re}(x_{ij}), \ x_{ij}'' = \mathrm{Im}(x_{ij}).$$

55. Establish properties (i)–(iii) and (v), (vi) of Proposition 10.12.

56. Show that the compacted measure algebra $\mathcal{M}_{cr}(G)$ is dense in the measure algebra $\mathcal{M}_r(G)$ (see 10.16).

57. Prove part (II) of Theorem 11.5.

58. Prove Propositions 11.21 and 11.22.

59. Let K, M be as in 13.1; let ρ be a K-rho-function on G, and let W be a Borel subset of M. Define $\sigma: G \to \mathbb{C}$ by:

$$\sigma(x) = \begin{cases} \rho(x), & \text{if } xK \in W \\ 0, & \text{if } xK \notin W. \end{cases}$$

Show that σ is a K-rho-function and that $\sigma^*(A) = \rho^*(A \cap W)$ for $A \in \mathcal{S}(M)$.

60. Let G be a locally compact group; and K a closed subgroup. Show that if $f \in \mathcal{L}(G)$, then the function ρ defined on G by

$$\rho(x) = \int_K \Delta(k)(\delta(k))^{-1} f(xk) dv$$

is a continuous K-rho-function with compact support in G/K (see 13.3).

61. Show that if $f \in \mathcal{L}(G)$, then the function f defined in 13.4 is continuous on the locally compact group G.

62. Prove Proposition 13.11.

63. Prove Proposition 13.12.

64. Prove Corollary 14.19.

65. Let G be a locally compact group with left Haar measure λ. By a *norm-function* on G we shall mean a function $\sigma: G \to \mathbb{R}$ such that $\sigma(x) > 0$ for all $x \in G$ and the following two conditions hold:

(i) $\sigma(xy) \leq \sigma(x)\sigma(y)$ $(x, y \in G)$;

(ii) σ is lower semi-continuous on G (i.e., $\lim \inf_{x \to u} \sigma(x) \geq \sigma(u)$ for all $u \in G$).

(A) Show that a norm-function σ on G is bounded on each compact subset of G. [Hint: Apply the Baire Category Theorem to the countable covering of G by the closed sets $\{x: \sigma(x) \leq n\}$ for $n = 1, 2, \ldots$]

(B) Let σ be a norm-function on G. Show that $\mathscr{L}(G)$ is a normed algebra under convolution and the norm $\| \ \|_\sigma$ given by

$$\|f\|_\sigma = \int_G |f(x)| \sigma(x) d\lambda x.$$

The completion of $\mathscr{L}(G)$ with respect to $\| \ \|_\sigma$ is called the σ-*group algebra* of G.

(C) Let T be a Banach representation of G (see VIII.8.2, 8.5). Show that $x \mapsto \|T_x\|$ $(x \in G)$ is then a norm-function in the above sense.

(D) Give an example of a Banach representation T of $\mathbb{R}$ such that $t \mapsto \|T_t\|$ is *not* continuous on $\mathbb{R}$. [Hint for part (D): Let X be the two-dimensional Hilbert space $\mathbb{C}^2$ with inner product $(\xi, \eta) = \xi_1 \bar{\eta}_1 + 2\xi_2 \bar{\eta}_2$; and for each integer n let $S^{(n)}$ be the repesentation of $\mathbb{R}$ on X given by

$$S_t^{(n)} = \begin{pmatrix} \cos nt & \sin nt \\ -\sin nt & \cos nt \end{pmatrix}.$$

Let S be the direct sum of the $S^{(n)}$ $(n = 1, 2, \ldots)$, acting on the Hilbert space direct sum $\sum_{n=1}^{\oplus \infty} X$. Then $\limsup_{t \to 0} \|S_t\| = \sqrt{2}$.]

Notes and Remarks

Basic references for topological groups are Bourbaki [13], Hewitt and Ross [1], Higgins [1], Husain [1], Pontryagin [1], Montgomery and Zippin [1], and Roelcke and Dieroff [1]. The axiomatic definition of a topological group as given in 1.1 was introduced by the Polish mathematician F. Leya [1] in 1927. O. Schreier [1] also gave a set of axioms for groups that were Fréchet L-spaces in which the group operations were continuous. The roots of topological groups can be traced back to the early 1870's with the work of Lie and Klein on continuous transformation groups. A nice account of this is given in Bourbaki [16].

In the 1930's the theory of topological groups emerged quickly as a subject in its own right. The monographs of Pontryagin [1] and Weil [1] summarized what was known up to 1938. Modern treatments of the subject are given in Bourbaki [13], Hewitt and Ross [1], Husain [1], and Roelcke and Dieroff [1]. For information on topological transformation groups we refer the reader to Bourbaki [13], Bredon [1], and Montgomery and Zippin [1]. Group extensions are discussed, for example, in Bourbaki [15], Curtis and Reiner [1, 2], and Kirillov [6].

Invariant integration on continuous groups was known as early as 1897 (see Hurwitz [1]). Schur and Frobenius, during the period between 1900 and 1920, frequently made use of averages over finite groups; see the notes in

Weyl [2]. During 1925–1926 Weyl [1] computed the invariant integrals for the groups $O(n)$ and $U(n)$. For *Lie* groups the existence of an invariant integral is very easily proved; one simply translates the infinitesimal volume element at the unit e to all points of the group. In 1927 Peter and Weyl [1] made use of this invariant integral to prove the well-known Peter-Weyl Theorem (see X.2) for compact *Lie* groups.

The crucial and decisive step was taken by A. Haar [1] in 1933 who constructed (directly!) a left-invariant integral on a second countable locally compact group. His construction was extended, independently, to arbitrary locally compact groups by Weil and Kakutani (see Hewitt and Ross [1]). The uniqueness of Haar measure was proved by von Neumann [5, 6], first for compact groups, and then for arbitrary second countable locally compact groups. In 1940 H. Cartan [1] gave a proof which simultaneously demonstrated the existence and uniqueness of Haar measure.

With the establishment of invariant measure on arbitrary locally compact groups, it followed automatically that (for example) the Peter-Weyl Theorem was valid for arbitrary compact groups.

Further information on Haar measure is given in Bourbaki [11], Dinculeanu [2], Gaal [1], Halmos [1], Hewitt and Ross [1], Loomis [2], Nachbin [1], Naimark [8], and Weil [1].

In 1952 Mackey [5] established the existence of quasi-invariant measures for second countable locally compact groups. The fact that a continuous rho-function exists in general was first proved in 1956 by Bruhat [1, 5]. In 1960 Loomis [3] gave an independent proof of the existence of a quasi-invariant measure with continuous rho-function. Additional results on quasi-invariant measures can be found, for example, in Bourbaki [11], Katznelson and Weiss [1], Krieger [1], and G. Warner [1].

IV   Algebraic Representation Theory

From a purely logical point of view, much of the present, purely algebraic, chapter is unnecessary for the rest of the book. We have seen fit to include it mostly because of its value in motivating the development of infinite-dimensional topological representation theory. Much of the modern topological representation theory of locally compact groups and Banach algebras grew out of the purely algebraic work of Frobenius, Wedderburn, and others on finite groups and finite-dimensional algebras. Often the main goals (if not the detailed features) of the topological theory emerge quite clearly and simply in the purely algebraic theory, uncluttered by topological and measure-theoretic technicalities. To give one example, the guiding purpose of all the processes known as "harmonic analysis" appears quite clearly and forcibly in the simple algebraic context of §2 (see 2.13 and 2.23). Another example (to be studied in Chapter XI) is the theory of induced representations, which even for finite groups is quite non-trivial, and whose generalization from the finite to the locally compact situation has been one of the most fruitful achievements of modern harmonic analysis.

In this chapter, except for the last section, linear spaces and algebras are over an arbitrary ground field. But the reader will lose little if he takes the ground field to be always either the reals or complexes.

265

§1 and §3 contain some of the basic definitions from algebraic repesenta-tion theory. §2 develops the theory of multiplicity for completely reducible representations; here the word "representation" is understood in the (superfi-cially) very general sense of an "operator set" (see 1.2).

In §4 we present the Extended Jacobson Density Theorem on irreducible representations and several of its corollaries (including the classical Burnside Theorem). In §5 we apply §4 to obtain the structure of finite-dimensional semisimple algebras (Wedderburn's Theorem); and from this in §6 we derive a few of the fundamental results on the representations of finite groups. Finally, in §7 the ground field is assumed to be the complex field. We obtain the structure of a large class of finite-dimensional *-algebras, which includes the group *-algebras of finite groups; and we show that for these the *-representations and the algebraic representations are essentially the same.

1. Fundamental Definitions

1.1. Let F be a fixed (commutative) field. Throughout this chapter it is understood that linearity of spaces and maps means linearity with respect to F, unless something is said to the contrary.

1.2. *Definition.* Let X be a linear space. By an *operator set on X (indexed by* a set E) we shall mean simply a map T on E to the family of all linear operators on X.

Usually X is called the *space of T*, and is denoted by $X(T)$.

One often speaks of the *dimension* $\dim(T)$ *of T*, meaning thereby the dimension of $X(T)$. Similarly T is called *finite-dimensional* [*infinite-dimensional*] if $X(T)$ is finite-dimensional [infinite-dimensional].

Remark. Let $M(n, F)$ be the space of all $n \times n$ matrices with entries in F. A matrix a in $M(n, F)$ can be identified with a linear operator on the n-dimensional linear space F^n in the usual way:

$$a\langle \xi_1, \ldots, \xi_n \rangle = \langle \eta_1, \ldots, \eta_n \rangle, \qquad \text{where } \eta_i = \sum_{j=1}^{n} a_{ij} \xi_j.$$

Thus a map $T: E \to M(n, F)$ can be considered as an operator set on F^n. Such a map is called an *operator set in matrix form*.

In this and the next section all operator sets are indexed by the fixed set E.

1.3. Definition. Let T be an operator set. A linear subspace Y of $X(T)$ is T-stable if $T_u(Y) \subset Y$ for all u in E.

The intersection of any family of T-stable linear subspaces is T-stable. Likewise the algebraic sum of any family of T-stable subspaces is T-stable. Obviously $\{0\}$ and $X(T)$ are T-stable; T-stable subspaces other than these two are called *non-trivial*.

If Z is any subset of $X(T)$, the smallest T-stable subspace of $X(T)$ containing Z (i.e., the intersection of all the T-stable subspaces containing Z) is said to be *generated by Z*.

1.4. Definition. Let T be an operator set, and Y a T-stable subspace of $X(T)$. Then

$$T': u \mapsto T_u | Y \qquad\qquad (u \in E)$$

is an operator set whose space is Y; it is called the *restriction* of T to Y, and is denoted by $^Y T$. Let $\tilde{X}$ be the quotient linear space $X(T)/Y$. Each T_u, since it leaves Y stable, defines a quotient linear operator $\tilde{T}_u$ on $\tilde{X}$:

$$\tilde{T}_u(\xi + Y) = (T_u \xi) + Y \qquad\qquad (\xi \in X(T)).$$

The operator set $\tilde{T}: u \mapsto \tilde{T}_u$ is called the *quotient of T on $\tilde{X}$*.

If Y and Z are two T-stable subspaces of $X(T)$ with $Y \subset Z$, the quotient operator set derived from T whose space is Z/Y is called a *subquotient* of T.

1.5. Definition. An operator set T is *irreducible* if $X(T) \neq \{0\}$ and there are no T-stable subspaces of $X(T)$ other than $\{0\}$ and $X(T)$.

A one-dimensional operator set is obviously irreducible.

A T-stable linear subspace Y of $X(T)$ is said to be *irreducible under T* (or *T-irreducible*) if $^Y T$ is irreducible.

1.6. Definition. Let T and T' be two operator sets (both of course indexed by E). We denote by $\mathrm{Hom}(T, T')$ the linear space of all those linear maps $f: X(T) \to X(T')$ which satisfy

$$f \circ T_u = T'_u \circ f \qquad \text{for all } u \text{ in } E.$$

Elements of $\mathrm{Hom}(T, T')$ are called *T, T' intertwining operators*, and are said to *intertwine T and T'*.

If there exists a bijection $f: X(T) \to X(T')$ belonging to $\mathrm{Hom}(T, T')$, we say that T and T' are *equivalent*—in symbols $T \cong T'$; the equivalence is said to be *implemented by f*. This relation of equivalence is clearly an equivalence relation in the usual sense.

Remark. Every finite-dimensional operator set T is equivalent to one in matrix form (see Remark 1.2). To see this, we have only to introduce a basis of $X(T)$ and consider the matrix of T_u with respect to this basis.

1.7. In the context of 1.4, the identity injection $i: Y \to X(T)$ intertwines $^Y T$ and T; and the quotient map $\pi: X(T) \to \tilde{X}$ intertwines T and $\tilde{T}$.

1.8. Schur's Lemma. *Let T and T' be operator sets; and suppose that $f \in \mathrm{Hom}(T, T')$, $Y = \mathrm{Ker}(f)$, $Y' = \mathrm{range}(f)$. Then: (i) Y is T-stable, (ii) Y' is T'-stable, and (iii) the quotient $\tilde{T}$ of T on $\tilde{X} = X(T)/Y$ is equivalent to $^{Y'}(T')$.*

Proof. If $u \in E$ and $\xi \in Y$, $f(T_u \xi) = T'_u(f(\xi)) = 0$; so $T_u \xi \in Y$, and (i) is proved. If $u \in E$ and $\eta = f(\xi) \in Y'$, then $T'_u \eta = f(T_u \xi) \in Y'$; so (ii) is proved. The bijection $\tilde{f}: \tilde{X} \to Y'$ defined by f clearly implements the equivalence of $\tilde{T}$ and $^{Y'}(T')$. ∎

1.9. Corollary. *If T and T' are irreducible operator sets, then either $\mathrm{Hom}(T, T') = \{0\}$ or $T \cong T'$.*

1.10. Definition. For each α in some index set A let T^α be an operator set on $X_\alpha = X(T^\alpha)$. Let us form the (algebraic) direct sum $X = \sum_{\alpha \in A}^{\oplus} X_\alpha$ (consisting of all functions ξ on A such that $\xi_\alpha \in X_\alpha$ for each α and $\xi_\alpha = 0$ for all but finitely many α, the linear operations being pointwise on A); and for each u in E let T_u be the linear operator on X given by $(T_u \xi)_\alpha = T_u^\alpha \xi_\alpha$ ($\xi \in X; \alpha \in A$). Then $T: u \mapsto T_u$ is called the *(algebraic) direct sum of the T^α*, and is denoted by $\sum_{\alpha \in A}^{\oplus} T^\alpha$.

1.11. If T is an operator set, and $\{Y_\alpha\}$ ($\alpha \in A$) is a linearly independent indexed family of T-stable subspaces of $X(T)$ such that $\sum_{\alpha \in A} Y_\alpha = X(T)$, then evidently

$$T \cong \sum_{\alpha \in A}^{\oplus (Y_\alpha)} T. \qquad (1)$$

We say that (1) is a *direct sum decomposition* of T.

Suppose now that $B \subset A$; and let Y be the T-stable subspace $\sum_{\alpha \in B} Y_\alpha$, and $\tilde{T}$ the quotient of T on $X(T)/Y$. Then clearly

$$Y_T \cong \sum_{\alpha \in B}^{\oplus (Y_\alpha)} T \qquad (2)$$

and

$$\tilde{T} \cong \sum_{\alpha \in A \setminus B}^{\oplus (Y_\alpha)} T. \qquad (3)$$

2. Complete Reducibility and Multiplicity for Operator Sets

2.1. *Definition*. An operator set T is *completely reducible* if it is equivalent to an algebraic direct sum of irreducible operator sets.

This amounts to saying that we can find a linearly independent indexed collection $\{Y_\alpha\}(\alpha \in A)$ of linear subspaces of $X(T)$ such that $X(T) = \sum_{\alpha \in A}^{\oplus} Y_\alpha$ and such that each Y_α is stable and irreducible under T.

The extent to which these subspaces Y_α are unique (if they exist) will be discussed later in this section (in 2.12 and 2.20).

2.2. *Remark*. Complete reducibility is definitely the exception rather than the rule for arbitrary operator sets. For example, the single operator with matrix $\begin{pmatrix} 0 & 1 \\ 0 & 0 \end{pmatrix}$, acting on the two-dimensional space F^2, is not completely reducible.

2.3. Suppose that T is an operator set, and that there exists a (Hamel) basis $\{\xi_\alpha\}$ of $X(T)$ and numbers $\lambda_{u,\alpha}$ in F such that $T_u \xi_\alpha = \lambda_{u,\alpha} \xi_\alpha$ for each u in E and each α. Then T is completely reducible; for $X(T) = \sum_{\alpha \in A}^{\oplus} F\xi_\alpha$, and the $F\xi_\alpha$, being T-stable and one-dimensional, are irreducible under T.

2.4. *Lemma*. *Let T be an operator set, Z a T-stable linear subspace of $X(T)$, and*

$$X(T) = \sum_{\alpha \in A}^{\oplus} Y_\alpha \tag{1}$$

a direct sum decomposition of $X(T)$ into T-stable T-irreducible subspaces Y_α. Then there is a subset B of A such that

$$X(T) = Z \oplus \sum_{\alpha \in B}^{\oplus} Y_\alpha \quad \text{(direct sum)}. \tag{2}$$

Proof. By Zorn's Lemma there is a maximal subset B of A such that $Z, \{Y_\alpha\}_{\alpha \in B}$ are linearly independent. Thus the sum $X' = Z + \sum_{\alpha \in B} Y_\alpha$ is direct. We have only to show that $X' = X(T)$. Assume that $X' \neq X(T)$. By (1) we have $Y_\beta \not\subset X'$ for some $\beta \in A \setminus B$. Now $X' \cap Y_\beta$ is T-stable. Since Y_β is T-irreducible, it follows that $X' \cap Y_\beta$ is either Y_β or $\{0\}$. If the former holds we have $Y_\beta \subset X'$, which is false. If the latter holds, Y_β is linearly independent of $Z, \{Y_\alpha\}_{\alpha \in B}$, contradicting the maximality of B. So $X' = X(T)$. ∎

2.5. *Corollary*. *If T is a completely reducible operator set, then every T-stable subspace of $X(T)$ has a complementary T-stable subspace in $X(T)$.*

Remark. Actually the condition of this Corollary is not only necessary but sufficient for complete reducibility. See 3.12.

2.6. Lemma. *In the context of Lemma 2.4, let $\tilde{T}$ be the quotient of T on $X(T)/Z$. Then $^{Z}T \cong \sum^{\oplus}_{\alpha \in A \setminus B} {}^{Y_{\alpha}}T$ and $\tilde{T} \cong \sum^{\oplus}_{\alpha \in B} {}^{Y_{\alpha}}T$.*

Proof. The second of these statements follows on combining 1.11(3) and 2.4(2). The first follows on applying 1.11(3) to 2.4(2) and also to the equation $\sum^{\oplus}_{\alpha \in B} Y_{\alpha} \oplus \sum^{\oplus}_{\alpha \in A \setminus B} Y_{\alpha} = X(T)$. ∎

2.7. Corollary. *If T is a completely reducible operator set and Y is a T-stable subspace of $X(T)$, then ^{Y}T and the quotient of T on $X(T)/Y$ are both completely reducible.*

2.8. It follows from 2.7 that a completely reducible operator set T has the following Property (P): Every non-zero T-stable linear subspace of $X(T)$ contains a T-irreducible T-stable subspace.

Property (P) certainly holds for all finite-dimensional operator sets. Not all of the latter, however, are completely reducible (see 2.2); so Property (P), while necessary, is not sufficient for complete reducibility.

Remark. As an example of the failure of Property (P), take F to be $\mathbb{C}$, let $X(T)$ be the space of all complex polynomials in one variable, and let range(T) consist of the single operator τ sending p into $t \mapsto tp(t)$. For any T-stable subspace Y, and any non-zero element p of Y, the linear span Z of $\{\tau^{n}(p): n = 1, 2, \ldots\}$ is T-stable, non-zero, and strictly less than Y (since $p \notin Z$). Thus there are no T-irreducible subspaces.

Remark. In this purely algebraic context, given an arbitrary operator set T, it is easier to find irreducible *quotients* of T than to find irreducible *subspaces* (see the argument of 3.8).

2.9. Proposition. *Let T be an operator set, and $\mathscr{I}$ the family of all T-stable T-irreducible subspaces of $X(T)$. Then T is completely reducible if and only if $\sum \mathscr{I} = X(T)$.*

Proof. The "only if" part is obvious. To prove the "if" part, let $\mathscr{I}'$ be a maximal linearly independent subfamily of $\mathscr{I}$, and argue as in the proof of 2.4. ∎

2.10. In general we refer to the $\sum \mathscr{I}$ of Proposition 2.9 as the *completely reducible part* of $X(T)$. By 2.9 the restriction of T to this subspace is completely reducible.

Harmonic Analysis

2.11. Suppose T is a completely reducible operator set, with

$$X(T) = \sum_{\alpha \in A}^{\oplus} Y_\alpha, \tag{3}$$

each Y_α being T-stable and T-irreducible. In what sense, if any, is the decomposition (3) *unique*?

To begin with, let us define $\mathscr{C}$ to be the family of all equivalence classes of irreducible operator sets (indexed by E). If T is any operator set and $\sigma \in \mathscr{C}$, we shall define $X_\sigma(T)$ to be the sum of all those T-stable T-irreducible subspaces Y of $X(T)$ such that ${}^Y T$ is of class σ. $X_\sigma(T)$ is called the *σ-subspace* of $X(T)$. Obviously $\sum_{\sigma \in \mathscr{C}} X_\sigma(T)$ is the completely reducible part of $X(T)$.

2.12. If T is an operator set, $\sigma \in \mathscr{C}$, and Y is a T-stable subspace, we write $Y \cong \sigma$ to mean that ${}^Y T$ is irreducible and of class σ.

Proposition. *Let T be an operator set. Then the σ-subspaces $X_\sigma(T)$ are linearly independent as σ runs over $\mathscr{C}$. (Some of course may be $\{0\}$). Thus, if T is completely reducible, $X(T) = \sum_{\sigma \in \mathscr{C}}^{\oplus} X_\sigma(T)$.*

Assume that T is completely reducible, and that we are given a direct sum decomposition (3) of $X(T)$ into irreducible subspaces. Then, for each σ in $\mathscr{C}$,

$$X_\sigma(T) = \sum \{Y_\alpha : Y_\alpha \cong \sigma\}. \tag{4}$$

Proof. Replacing T by ${}^Z T$, where Z is the completely reducible part of $X(T)$ (see 2.10), we may as well assume that T is completely reducible.

Take the decomposition (3), and let X'_σ be the right side of (4). Clearly the X'_σ ($\sigma \in \mathscr{C}$) are linearly independent and $\sum_\sigma X'_\sigma = X(T)$. Thus we have only to show that $X'_\sigma = X_\sigma(T)$ for each σ. Let $\sigma \in \mathscr{C}$. Since obviously $X'_\sigma \subset X_\sigma(T)$, it suffices to show that $X_\sigma(T) \subset X'_\sigma$, that is, that $Y \subset X'_\sigma$ for any T-stable subspace Y for which $Y \cong \sigma$. Take such a Y.

For each index α, let p_α be the idempotent operator on $X(T)$ with range Y_α which annihilates Y_β for all $\beta \neq \alpha$. Clearly p_α intertwines T and ${}^{Y_\alpha} T$; so $p_\alpha | Y$ intertwines ${}^Y T$ and ${}^{Y_\alpha} T$. Consequently, by 1.9, for each α either $p_\alpha(Y) = \{0\}$ or ${}^Y T \cong {}^{Y_\alpha} T$. Since $Y \cong \sigma$, the latter alternative means that $Y_\alpha \cong \sigma$. But $\xi = \sum_\alpha p_\alpha \xi$ for every ξ; consequently, if $\xi \in Y$, $\xi = \sum \{p_\alpha \xi : Y_\alpha \cong \sigma\} \in X'_\sigma$; that is, $Y \subset X'_\sigma$. ∎

2.13. *Remark.* Proposition 2.12 gives a partial answer to the question raised in 2.11. Let T be a completely reducible operator set. If $\sum_\alpha^\oplus Y_\alpha$ and $\sum_\beta^\oplus Y'_\beta$ are two direct sum decompositions of $X(T)$ into irreducible parts, it may happen that no Y_α coincides with any Y'_β; nevertheless, for each σ in $\mathscr{C}$ we shall have 2.12 $\sum\{Y_\alpha : Y_\alpha \cong \sigma\} = \sum\{Y'_\beta : Y'_\beta \cong \sigma\} = X_\sigma(T)$.

Let us say that a vector ξ in $X(T)$ is *of class* σ (*with respect to* T) if $\xi \in X_\sigma(T)$. By 2.12 every ξ in $X(T)$ has a unique decomposition

$$\xi = \sum_{\sigma \in \mathscr{C}} \xi_\sigma, \tag{5}$$

where ξ_σ is of class σ (and only finitely many of the ξ_σ are non-zero). We call ξ_σ the *σ-component of* ξ (*relative to* T). Equation (5) can be regarded as the *harmonic analysis of ξ with respect to the operator set T.*

Consider for example the situation of 2.3. Each function $\tau : E \to F$ can be identified with the element of $\mathscr{C}$ containing the one-dimensional irreducible operator set which sends u into multiplication by τ_u. For each $\tau : E \to F$, the τ-subspace $X_\tau(T)$ is by 2.12(4) the linear span of $\{\xi_\alpha : \lambda_{u,\alpha} = \tau_u$ for all u in $E\}$ (that is, the "eigenspace" of T for the set of "simultaneous eigenvalues" $\{\tau_u\}$).

Notice that, within each "eigenspace" $X_\tau(T)$, any Hamel basis of $X_\tau(T)$ will serve to decompose $X_\tau(T)$ into (one-dimensional) T-irreducible subspaces. This illustrates the fact that, if T is an arbitrary completely reducible operator set, there are in general many ways to decompose each $X_\sigma(T)$ into T-irreducible subspaces.

2.14. **Corollary.** *If T is an operator set, and $\mathscr{F}$ is a family of T-irreducible T-stable subspaces of $X(T)$ such that the ${}^Y T$ ($Y \in \mathscr{F}$) are pairwise inequivalent, then $\mathscr{F}$ is linearly independent.*

2.15. **Corollary.** *Let T be an operator set; let $\xi \in X(T)$; and let Y be the T-stable subspace generated by $\{\xi\}$ (see 1.3). Then ξ is of class σ if and only if ${}^Y T$ is the direct sum of irreducible operator sets all of class σ.*

Remark. If the condition of the Corollary holds, ${}^Y T$ will clearly be the direct sum of *finitely* many irreducible operator sets of class σ.

2.16. **Proposition.** *Let T and T' be two operator sets, and $f : X(T) \to X(T')$ a T, T' intertwining map. Then $f(X_\sigma(T)) \subset X_\sigma(T')$ for every σ in $\mathscr{C}$.*

Proof. It is sufficient to show that $f(Y) \subset X_\sigma(T')$ whenever Y is T-stable and $Y \cong \sigma$. Now $f | Y$ intertwines ${}^Y T$ and T'. Since ${}^Y T$ is irreducible, Schur's Lemma (1.9) says that either $f(Y) = \{0\}$ or ${}^Y T \cong {}^{Y'}(T')$, where $Y' = f(Y)$. Since $Y \cong \sigma$, both these alternatives assert that $f(Y) \subset X_\sigma(T')$. ∎

2.17. **Corollary.** *Let T be a completely reducible operator set, and Y a T-stable subspace of $X(T)$. For $\sigma \in \mathscr{C}$, let Y_σ be the σ-subspace for $^Y T$. Then:* (i) $Y_\sigma = Y \cap X_\sigma(T)$ *for each σ in $\mathscr{C}$; and* (ii) $Y = \sum_{\sigma \in \mathscr{C}}^{\oplus} Y_\sigma$.

Proof. Proposition 2.16 applied to the identity injection $Y \to X(T)$ gives $Y_\sigma \subset X_\sigma(T)$ for each σ. On the other hand, 2.7 and 2.12 imply that $Y = \sum_\sigma^{\oplus} Y_\sigma$. The conclusion now follows immediately. ∎

Multiplicity

2.18. We have seen that the decomposition of a completely reducible operator set T into its σ-subspaces ($\sigma \in \mathscr{C}$) is intrinsic to the structure of T. However, the further direct sum decomposition of each σ-subspace into irreducible subspaces is in general by no means unique (see 2.13). In spite of this, we shall now show that at least the *cardinal number* of irreducible subspaces appearing in the direct sum decomposition of each $X_\sigma(T)$ is independent of the particular decomposition.

2.19. **Lemma.** *Let T be an operator set. Let $\mathscr{F}$ be a family of T-stable T-irreducible subspaces of $X(T)$ whose sum (not necessarily direct) is $X(T)$; and let $\mathscr{G}$ be any linearly independent family of T-irreducible T-stable subspaces of $X(T)$. Then the cardinal number q of $\mathscr{G}$ is no greater than the cardinal number p of $\mathscr{F}$.*

Proof. First consider the case that p is finite. Assume that the conclusion is false, and take a sequence $Z_1, \ldots, Z_{p+1}$ of distinct elements of $\mathscr{G}$. We shall show by induction that for each $r = 0, 1, \ldots, p$ there is a set $\mathscr{F}_r$ consisting of $p - r$ subspaces belonging to $\mathscr{F}$ such that $\{Z_1, \ldots, Z_r\} \cup \mathscr{F}_r$ spans $X(T)$. For $r = 0$ this is obvious (take $\mathscr{F}_0 = \mathscr{F}$). Assume it for $r - 1$ ($r = 1, \ldots, p$). Then the family of subspaces $\{Z_1, \ldots, Z_r\} \cup \mathscr{F}_{r-1}$ cannot be linearly independent. Now write down the elements of $\mathscr{F}_{r-1}$ in some order: $\mathscr{F}_{r-1} = \{Y_1, Y_2, \ldots, Y_{p-r+1}\}$; and take the first element of the sequence

$$Z_1, Z_2, \ldots, Z_r, Y_1, Y_2, \ldots, Y_{p-r+1} \tag{6}$$

which is *not* linearly independent of the sum of its predecessors. This must be one of the Y_i, say Y_t. Thus

$$Y_t \cap (Z_1 + \cdots + Z_r + Y_1 + \cdots + Y_{t-1}) \tag{7}$$

is a non-zero T-stable subspace of Y_t. Since Y_t is T-irreducible, (7) equals Y_t. Therefore Y_t is contained in the sum of the subspaces remaining in (6) after Y_t

is removed. Thus, if we put $\mathscr{F}_r = \{Y_1, \ldots, Y_{t-1}, Y_{t+1}, \ldots, Y_{p-r+1}\}$, the family $\{Z_1, \ldots, Z_r\} \cup \mathscr{F}_r$ will span $X(T)$. Since $\mathscr{F}_r$ has $p - r$ elements, the inductive step is complete.

Applying the result of the preceding paragraph with $r = p$ (so that $\mathscr{F}_p$ is void), we find that p of the elements of $\mathscr{G}$ span $X(T)$. Since $\mathscr{G}$ is linearly independent, this leaves no room for Z_{p+1} to exist; and we have a contradiction. Thus $q \leq p$ if p is finite.

Next we consider the case that p is infinite. For each finite subset $\mathscr{E}$ of $\mathscr{F}$, define $\mathscr{G}(\mathscr{E})$ to be the family of those Z in $\mathscr{G}$ which are contained in $\sum \mathscr{E}$. Since $\sum \mathscr{F} = X(T)$, for each Z in $\mathscr{G}$ we can find a finite subset $\mathscr{E}$ of $\mathscr{F}$ such that Z and $\sum \mathscr{E}$ have non-zero intersection; since Z is irreducible and $\sum \mathscr{E}$ is T-stable, this implies that $Z \subset \sum \mathscr{E}$, or $Z \in \mathscr{G}(\mathscr{E})$. Further, applying the preceding part of the proof of the restriction of T to $\sum \mathscr{E}$, we conclude that the family $\mathscr{G}(\mathscr{E})$ is finite for each finite subset $\mathscr{E}$ of $\mathscr{F}$. It follows that $\mathscr{G} = \bigcup_{\mathscr{E}} \mathscr{G}(\mathscr{E})$, where each $\mathscr{G}(\mathscr{E})$ is finite and $\mathscr{E}$ runs over a set of cardinality p (namely, the family of all finite subsets of $\mathscr{F}$). So the cardinality q of $\mathscr{G}$ is no greater than p. ∎

2.20. Theorem. *Let T be a completely reducible operator set; and let*

$$X(T) = \sum_{\alpha \in A}{}^{\oplus} Y_\alpha = \sum_{\beta \in B}{}^{\oplus} Z_\beta$$

be two direct sum decompositions of $X(T)$ into T-stable T-irreducible sub-spaces $\{Y_\alpha\}$ and $\{Z_\beta\}$. Then, for each σ in $\mathscr{C}$, the sets $\{\alpha \in A : Y_\alpha \cong \sigma\}$ and $\{\beta \in B : Z_\beta \cong \sigma\}$ have the same cardinal number.

Note. For the meaning of "$Y_\alpha \cong \sigma$" see 2.12.

Proof. By 2.12, $X_\sigma(T) = \sum \{Y_\alpha : Y_\alpha \cong \sigma\} = \sum \{Z_\beta : Z_\beta \cong \sigma\}$. Thus, the Theorem is obtained on applying 2.19 twice to the $\{Y_\alpha\}$ and $\{Z_\beta\}$ considered as irreducible subspaces of $X_\sigma(T)$. ∎

Remark. The proof of this Theorem is a close imitation of the proof of invariance of the dimension of a linear space (which of course is a special case of it).

2.21. Definition. The cardinal number of $\{\alpha \in A : Y_\alpha \cong \sigma\}$ in 2.20 is called the *(algebraic) multiplicity of σ in T*.

In the general case, when T is not completely reducible, the *(algebraic) multiplicity of σ in T* is defined to mean the multiplicity of σ in the completely reducible part of T (see 2.10).

We say that *σ occurs in T* if the multiplicity of σ in T is not 0, that is, if $X_\sigma(T) \neq \{0\}$.

Remark. If σ is one-dimensional, then of course the multiplicity of σ in T is the (Hamel) dimension of $X_\sigma(T)$.

2.22. Theorem 2.20 gives us a complete classification of the equivalence classes of completely reducible operator sets in terms of $\mathscr{C}$ and cardinal numbers. Indeed, if p is any cardinal number and $\sigma \in \mathscr{C}$, let $p \cdot \sigma$ be the equivalence class of $\sum_{\alpha \in A}^{\oplus} S^\alpha$, where A is an index set of cardinality p and S^α is of class σ for each α. Then it is clear from Theorem 2.20 that *the map*

$$\pi \mapsto \sum_{\sigma \in \mathscr{C}}^{\oplus} \pi(\sigma) \cdot \sigma$$

is a one-to-one correspondence between the family of all cardinal-valued functions π on $\mathscr{C}$ and the family of all equivalence classes of completely reducible operator sets.

Among all operator sets the completely reducible ones are of course rather exceptional. Interestingly enough, however, we shall see in §5 and §6 that there are rather important classes of operator sets which are always completely reducible.

The Harmonic Analysis of Functions on a Domain With Maps

2.23. *Remark.* Suppose that S is any set, and that E is any collection of maps of S into itself. The results of this section suggest an interesting program for the study and classification of functions on S relative to E. This program is a relatively simple version of generalized harmonic analysis.

Let X stand for the linear space of all functions on S to the ground field F. Each u in E gives rise to a linear operator T_u on X:

$$T_u(f)(s) = f(u(s)) \qquad\qquad (f \in X; s \in S).$$

Thus $T : u \to T_u$ is an operator set on X indexed by E; and, given any f in X, we can ask how it behaves with respect to T. For example, does f belong to the completely reducible part of T? If it does, what is its harmonic analysis with respect to T, in the sense of 2.13?

As a specific illustration, suppose that $F = \mathbb{C}$, $S = \mathbb{R}$, and $E = \{u_r : r \in \mathbb{R}\}$, where u_r is the translation $x \mapsto x + r$ ($x \in \mathbb{R}$). Define T as above; take a continuous function $f : R \to \mathbb{C}$; and let Y be the T-stable subspace generated by $\{f\}$ (i.e., the linear span of the set of all translates of f). One can show:

Proposition*.

(I) Y *is finite-dimensional if and only if f is a linear combination of functions of the form $g(x) = e^{kx}x^n$ ($k \in \mathbb{C}$; $n = 0, 1, 2, \ldots$).*

(II) *Assume that Y is finite-dimensional. Then (A) Y is T-irreducible if and only if f is an exponential function $x \mapsto ce^{kx}$ (where $c, k \in \mathbb{C}$ and $c \neq 0$). This implies of course that Y is one-dimensional. (B) $^Y T$ is completely reducible if and only if f is a finite linear combination of exponentials:*

$$f(x) = c_1 e^{k_1 x} + \cdots + c_n e^{k_n x} \qquad (8)$$

($c_1, \ldots, c_n, k_1, \ldots, k_n \in \mathbb{C}$, the $k_1, \ldots, k_n$ being pairwise distinct). If this is the case, (8) is the harmonic analysis of f. (C) f is a polynomial function if and only if every irreducible subquotient of $^Y T$ is trivial.

Note. By a *trivial operator set* we mean a one-dimensional operator set sending every u in E into the identity operator.

To prove the "only if" part of (I), we show that, if Y is finite-dimensional, f must satisfy a linear differential equation with constant coefficients, and then use the theory of such equations.

Unfortunately we do not know whether or not the T-irreducibility of Y implies that Y is finite-dimensional, hence one-dimensional. Thus the above Proposition falls short of describing the completely reducible part even of the subspace of continuous functions in X.

3. Representations of Groups and Algebras

3.1. An operator set, as defined in 1.2, is too general an object for most purposes. Usually the indexing set E has some structure of its own; and the operator set T, as a map of E into operators, is required to preserve this structure in some sense. The most common special situations are the following:

3.2. Definition. Let G be a group with unit e. A *representation of G* is an operator set T indexed by G such that: (i) T_e is the identity operator on $X(T)$, (ii) $T_{xy} = T_x T_y$ for all elements x, y of G.

Thus, a representation T of G is a homomorphism of G into the group (under composition) of all linear bijections $X(T) \to X(T)$. We have in particular $T_{(x^{-1})} = (T_x)^{-1}$ for $x \in G$.

3.3. By "algebra" we always mean in this chapter an associative algebra over F.

Definition. Let A be an algebra. A *representation of A* is an operator set T indexed by A such that: (i) $T_{\lambda a} = \lambda T_a$, (ii) $T_{a+b} = T_a + T_b$, and (iii) $T_{ab} = T_a T_b$ for all a, b in A and λ in F.

Thus, a representation of A is a homomorphism T of A into the algebra of all linear endomorphisms of $X(T)$.

If T is one-to-one, i.e., $\mathrm{Ker}(T) = \{0\}$, T is called *faithful*.

3.4. A linear operator on a one-dimensional linear space is of course just scalar multiplication by an element of F. Thus an equivalence class of one-dimensional operator sets (indexed by E) can be identified with a map of E into F.

In particular, an equivalence class of one-dimensional representations of a group G can be identified with a group homomorphism of G into the multiplicative group $F \setminus \{0\}$. An equivalence class of one-dimensional representations of an algebra A can be identified with an algebra homomorphism of A into F.

3.5. If T is a representation of a group G [algebra A] and Y is a T-stable linear subspace of $X(T)$, notice that ${}^{Y}T$ and the quotient of T on $X(T)/Y$ (defined as operator sets by 1.4) are also representations of G [A]. We call ${}^{Y}T$ the *subrepresentation of T acting on Y*.

Notice also that a direct sum of representations of G [A] (in the sense of 1.10) is also a representation.

3.6. The concepts of intertwining map, equivalence, and complete reducibility as applied to representations will have the same meanings as defined in §1 and §2. However, the term "irreducible" as applied to a representation of an algebra A will have a slightly stronger sense.

Definition. A representation T of the algebra A is a *zero representation* if $T_a = 0$ for all a in A.

Definition. A representation T of A is *irreducible* if it is irreducible as an operator set (see 1.5) and also is not a zero representation.

Thus the only representation of A which is irreducible as an operator set but not as a representation is the one-dimensional zero representation.

With this definition of irreducibility, a representation T of the algebra A will be completely reducible (in the former sense of 2.1) if and only if it is a direct sum of representations each of which is either irreducible or a zero representation.

3.7. Proposition. *Let T be a representation of an algebra A, with $X(T) \neq \{0\}$. A necessary and sufficient condition for T to be irreducible is that, if ξ, η are any two vectors in $X(T)$ with $\xi \neq 0$, there should exist an element a of A satisfying $T_a \xi = \eta$.*

Proof. The sufficiency is almost evident, and is left to the reader.

Assume that T is irreducible. First notice that $N = \{\xi \in X(T): T_a\xi = 0$ for all a in $A\}$ is T-stable; indeed, if $\xi \in N$ and $b \in A$, then $T_a(T_b\xi) = T_{ab}\xi = 0$ for all a, so $T_b\xi \in N$. Thus N is either $\{0\}$ or $X(T)$. If $N = X(T)$ then T is a zero representation, which we ruled out in 3.6. So $N = \{0\}$. Now let ξ be a fixed non-zero vector in $X(T)$; and put $Y = \{T_a\xi : a \in A\}$. Since $\xi \notin N$, $Y \neq \{0\}$. Also Y is T-stable (by the same argument used above for N). So $Y = X(T)$, showing that every vector η in $X(T)$ is of the form $T_a\xi$ for some a in A. ∎

3.8. Definition. Given an algebra A [group G], we shall denote by $\overline{A}\,[\overline{G}]$ the family of all equivalence classes of irreducible representations of $A\,[G]$.

Remark 1. Sometimes, if no clarity is lost, we will deliberately confuse a representation with the equivalence class to which it belongs, and speak, for example, of a particular irreducible representation of A as if it were an element of $\overline{A}$.

Remark 2. Given any group G, there is always a one-dimensional element of $\overline{G}$, namely the *trivial representation*, sending every element of G into the number 1 (see 3.4).

3.9. One of the most important representations of an algebra A is its *regular representation* R, defined as the action of A on itself by left multiplication. More precisely: $X(R) = A$, $R_a(b) = ab$ $(a, b \in A)$.

A linear subspace of A is of course R-stable if and only if it is a left ideal of A. Given a left ideal J of A, the quotient $\tilde{R}$ of R on A/J is referred to as the *natural representation of A on A/J*. Clearly $\tilde{R}$ is irreducible as an operator set if and only if J is a maximal proper left ideal of A.

These "natural" representations are not too far from being exhaustive. Indeed, let T be any representation of A such that there is a vector ξ in $X(T)$ satisfying $X(T) = \{T_a\xi : a \in A\}$. If J is the left ideal $\{a \in A: T_a\xi = 0\}$, the map $a \mapsto T_a\xi$ lifts to a bijection $A/J \to X(T)$ which intertwines $\tilde{R}$ (the natural representation of A on A/J) and T. So $\tilde{R} \cong T$.

3.10. Actually, operator sets are not really more general than representations of algebras.

Given any set E, let $A(E)$ be the free associative algebra with unit generated by E. Thus $A(E)$ is an algebra with unit 1, containing E as a subset, such that every map f of E into an algebra B with unit $1'$ can be uniquely extended to an algebra homomorphism $\phi: A(E) \to B$ satisfying $\phi(1) = 1'$. Then in particular every operator set T indexed by E can be uniquely extended to a representation T' of $A(E)$ on the same space, carrying $\mathbb{1}$ into the identity operator. The T-stable and T'-stable subspaces are the same; and the map $T \mapsto T'$ preserves direct sums, irreducibility, complete reducibility, and intertwining operators.

3.11. The observation 3.10 is sometimes quite useful. Thus it enables us to prove:

Lemma. *Let T by any operator set (indexed by a set E). If $X(T) \neq \{0\}$, T has an irreducible subquotient.*

Proof. By 3.10 it is enough to prove this for the corresponding representation T' of $A(E)$. Choose a non-zero vector ξ in $X(T)$, and let J be the proper left ideal $\{a \in A(E): T'_a \xi = 0\}$ of $A(E)$. By Zorn's Lemma there is a *maximal* proper left ideal K of $A(E)$ containing J. As we remarked in 3.9, the natural representation Q of $A(E)$ on $A(E)/K$ is irreducible. Evidently Q is a quotient of the natural representation $\tilde{R}$ of $A(E)$ on $A(E)/J$; and by 3.9 $\tilde{R}$ is equivalent to the subrepresentation of T' acting on $T'_{A(E)}\xi$. Thus Q is equivalent to some subquotient S of T'. Since Q is irreducible, so is S. ∎

3.12. Corollary. *Let T be any operator set. A necessary and sufficient condition for T to be completely reducible is that, for every T-stable linear subspace Y of $X(T)$, there should exist a T-stable linear subspace Z complementary to Y in $X(T)$ (i.e., $Y + Z = X(T)$, $Y \cap Z = \{0\}$).*

Proof. The necessity was observed in 2.5.

To prove the sufficiency, we assume the condition, and denote by Y the completely reducible part of $X(T)$ (2.10). We must show that $Y = X(T)$. Suppose the contrary. By hypothesis there is a (non-zero) T-stable subspace Z complementary to Y. Now by 3.11 ^{Z}T has an irreducible subquotient Q. But it is easy to see that the property of having complementary stable subspaces is inherited by ^{Z}T from T, and that therefore Q is equivalent to some *subrepresentation* of ^{Z}T, acting (say) on W. Since ^{W}T is irreducible, it follows that $W \subset Y$, contradicting the fact that $\{0\} \neq W \subset Z$. So $Y = X(T)$. ∎

Remark. If T is finite-dimensional, the proof of 3.11 and therefore of the above Corollary becomes much simpler. We leave the details to the reader.

Remark. This Corollary will serve later as a vivid illustration of the difference between the purely algebraic representation theory of this chapter and the topological representation theory with which most of this work is concerned. See VI.10.13.

3.13. There is also a close relation between representations of groups and algebras.

Let G be a group with unit e. Let $B(G)$ be the linear space of all functions $f: G \to F$ such that $f(x) = 0$ for all but finitely many x in G; and introduce into $B(G)$ the associative multiplication given by

$$(fg)(x) = \sum_{y \in G} f(y)g(y^{-1}x) \tag{1}$$

($f, g \in B(G)$; $x \in G$; notice that the sum on the right is finite). We observe that (1) is the unique (bilinear) multiplication satisfying $\delta_x \delta_y = \delta_{xy}$ for all $x, y \in G$; here δ_x is the element $y \mapsto \delta_{xy}$ of $B(G)$. As an algebra with this multiplication, $B(G)$ is called the *discrete group algebra of G*. It has a unit element $\mathbb{1} = \delta_e$.

If S is a representation of G, then

$$T: f \mapsto \sum_{x \in G} f(x)S_x \tag{2}$$

is a representation of $B(G)$ on $X(S)$ sending $\mathbb{1}$ into the identity operator. Conversely, if T is a representation of $B(G)$ sending $\mathbb{1}$ into the identity operator, then

$$S: x \mapsto T_{\delta_x} \qquad\qquad (x \in G) \tag{3}$$

is a representation of G. Evidently the two constructions (2) and (3) are inverse to each other, and thus set up a one-to-one correspondence between representations of G and representations of $B(G)$ sending $\mathbb{1}$ into the identity operator. This correspondence preserves stable subspaces, intertwining operators, irreducibility, and complete reducibility.

Notice that if T is the regular representation of $B(G)$ (see 3.9), the representation S of G corresponding to T via (3) is given by:

$$X(S) = B(G), (S_x f)(y) = f(x^{-1}y)$$

($x, y \in G$; $f \in B(G)$). This S is called the *(algebraic) regular representation* of G.

3.14. The correspondences mentioned in 3.10 and 3.13 emphasize the special importance of representations of algebras. We end this section with a few elementary facts about these.

3.15. Proposition.   *An irreducible representation of a finite-dimensional algebra or of a finite group is finite-dimensional.*

Proof.   By 3.7 and 3.13.   ■

3.16. Proposition.   *Let A be an algebra and J a two-sided ideal of A. If T is an irreducible representation of A, then either $J \subset \mathrm{Ker}(T)$ or $T|J$ is irreducible.*

Proof. Assume that $J \not\subset \mathrm{Ker}(T)$. Then there are vectors ξ_0, η_0 in $X(T)$ and an element b of J such that $\eta_0 = T_b \xi_0 \neq 0$. Let ξ, η be arbitrary vectors in $X(T)$ with $\xi \neq 0$. By 3.7 there are elements a, c of A such that $T_a \eta_0 = \eta$, $T_c \xi = \xi_0$. Thus $T_{abc} \xi = T_a(T_b(T_c \xi)) = \eta$. Since $abc \in J$, the arbitrariness of ξ and η implies by 3.7 that $T|J$ is irreducible. ∎

3.17. *Definition.* A representation T of an algebra A is *non-degenerate* if the following two conditions hold: (I) The linear span of $\{T_a \xi : a \in A,$ $\xi \in X(T)\}$ is $X(T)$. (II) If $\xi \in X(T)$ and $T_a \xi = 0$ for all a in A, then $\xi = 0$. If T is not non-degenerate, it is *degenerate*.

A direct sum of non-degenerate representations is evidently non-degenerate. An irreducible representation is non-degenerate (by 3.7 and its proof); hence so is any direct sum of irreducible representations. However, a subrepresentation or a quotient representation of a non-degenerate representation may be degenerate. The reader should be able to construct examples illustrating this without too much difficulty.

If A has a unit element $\mathbf{1}$, a representation T of A is non-degenerate if and only if $T_{\mathbf{1}}$ is the identity operator.

3.18. Proposition. *Suppose that J is a two-sided ideal of the algebra A, and that S is a non-degenerate representation of J. Then S can be extended in one and only one way to a representation T of A on $X(S)$.*

Proof. Let $a \in A$. We claim that if $\xi_1, \ldots, \xi_r \in X(S)$, $b_1, \ldots, b_r \in J$, and $\sum_{i=1}^r S_{b_i} \xi_i = 0$, then $\sum_{i=1}^r S_{ab_i} \xi_i = 0$. Indeed: For all c in J we then have $S_c(\sum_{i=1}^r S_{ab_i} \xi_i) = \sum_{i=1}^r S_{c(ab_i)} \xi_i = S_{ca}(\sum_{i=1}^r S_{b_i} \xi_i) = 0$. By Condition (II) of Definition 3.17 this implies the claim.

The above claim, together with Condition (I) of 3.17, permits us to define a unique linear operator T_a on $X(S)$ satisfying

$$T_a\left(\sum_{i=1}^r S_{b_i} \xi_i \right) = \sum_{i=1}^r S_{ab_i} \xi_i \tag{4}$$

$(b_1, \ldots, b_r \in J; \xi_1, \ldots, \xi_r \in X(S))$. One verifies that $T : a \mapsto T_a$ is a representation of A on $X(S)$. If $a \in J$, then of course (4) holds on replacing T_a by S_a; so T is an extension of S. Further, any representation of A which extends S must satisfy (4); and this shows that the extension T is unique. ∎

3.19. Suppose that J is a two-sided ideal of the algebra A; and set $(\overline{A})_J = \{T \in \overline{A} : J \not\subset \mathrm{Ker}(T)\}$.

Corollary. *The map $T \mapsto T|J$ is a bijection of $(\overline{A})_J$ onto $\overline{J}$.*

Proof. If $T \in (\overline{A})_J$, then $T|J \in \overline{J}$ by 3.16. That $T \mapsto T|J$ maps $(\overline{A})_J$ biuniquely onto J follows immediately from 3.18. ∎

Remark. Thus $\overline{J}$ can be identified with the subset $(\overline{A})_J$ of $\overline{A}$.

Let $\pi: A \to A/J$ be the quotient homomorphism. If $R \in (A/J)^{-}$, then $T = R \circ \pi \in \overline{A}$; and it is easy to see that the map $R \mapsto R \circ \pi$ is a bijection of $(A/J)^{-}$ onto $\{T \in \overline{A} : J \subset \mathrm{Ker}(T)\}$, that is, onto $\overline{A} \setminus (\overline{A})_J$. Thus the two complementary subsets $(\overline{A})_J$ and $\overline{A} \setminus (\overline{A})_J$ are naturally identified with $\overline{J}$ and $(A/J)^{-}$ respectively.

3.20. Proposition. *Let $A_1, \ldots, A_n$ be algebras; and form their algebra direct sum $A = \sum_i^{\oplus} A_i$. (Thus $A = \sum_i^{\oplus} A_i$ as a linear space; the multiplication in A is given by $\langle a_1, \ldots, a_n \rangle \langle b_1, \ldots, b_n \rangle = \langle a_1 b_1, \ldots, a_n b_n \rangle$). If $T \in \overline{A}$, there is a (unique) $i = 1, \ldots, n$, and a (unique) element S of $(A_i)^{-}$ such that*

$$T_a = S_{a_i} \qquad \text{for all } a \text{ in } A.$$

Proof. We regard each A_i as a two-sided ideal of A. There must be an $i = 1, \ldots, n$ such that $A_i \not\subset \mathrm{Ker}(T)$. By 3.16 $S = T|A_i$ is irreducible. Suppose $j \neq i$. Given $\xi \in X(T)$ and $a \in A_j$, we can choose (by 3.7) an element b of A_i such that $T_b \xi = \xi$; but then $T_a \xi = T_{ab} \xi = 0$ since $ab = 0$. This shows that $A_j \subset \mathrm{Ker}(T)$ for every $j \neq i$. This essentially completes the proof. ∎

Remark. This Proposition enables us to identify $\overline{A}$ with the disjoint union of the $(A_i)^{-}$ $(i = 1, \ldots n)$.

3.21. Let $A = \sum_{i=1}^{n \oplus} A_i$ as in 3.20; and consider each A_i as a two-sided ideal of A.

Proposition. *A representation T of A is completely reducible if and only if $T|A_i$ is completely reducible for each i.*

Proof. To prove this by induction in n it is enough to prove it assuming $n = 2$.

The "only if" part follows immediately from 3.16.

For the "if" part, assume that $T|A_1$ and $T|A_2$ are completely reducible. We can then write $X(T) = \sum_{\alpha}^{\oplus} Y_{\alpha} \oplus Z$, where each Y_{α} is stable and irreducible under $T|A_1$, and $Z = \{\xi \in X(T): T_a \xi = 0 \text{ for all } a \text{ in } A_1\}$. If $\xi \in Z$ and $b \in A_2$, then $T_a(T_b \xi) = T_{ab} \xi = T_0 \xi = 0$ for all a in A_1, so $T_b \xi \in Z$. Thus Z is stable under $T|A_2$. By 2.7 $^Z(T|A_2)$ is completely reducible. So $Z = \sum_{\beta}^{\oplus} Z_{\beta}$, where each Z_{β} is stable and either irreducible or zero under $T|A_2$. Now the argument of the proof of 3.20 shows that $T|A_2$ is zero on each Y_α. Hence in the direct sum decomposition $X(T) = \sum_\alpha^\oplus Y_\alpha \oplus \sum_\beta^\oplus Z_\beta$, every Y_α and Z_β is T-stable and is either irreducible or zero under T. So T is completely reducible.   ∎

4. The Extended Jacobson Density Theorem

4.1. In this section we shall make a deeper study of irreducible representations of algebras. The fundamental fact about their structure is the Extended Jacobson Density Theorem.

4.2. Suppose for the moment that T is any irreducible operator set, indexed by a set E. Thus $\mathrm{Hom}(T, T)$ (defined in 1.6) is closed under operator composition, and so is an algebra (over F). In fact $\mathrm{Hom}(T, T)$ is a division algebra over F. To see this, we have only to show that each non-zero element α of $\mathrm{Hom}(T, T)$ has an inverse in $\mathrm{Hom}(T, T)$. Now by 1.8 $\mathrm{Ker}(\alpha)$ and $\mathrm{range}(\alpha)$ are T-stable. Since $\alpha \neq 0$ and T is irreducible, this says that $\mathrm{Ker}(\alpha) = \{0\}$, $\mathrm{range}(\alpha) = X(T)$. So α is a bijection. Since $\alpha^{-1} \circ T_u = \alpha^{-1} \circ (T_u \circ \alpha) \circ \alpha^{-1} = \alpha^{-1} \circ (\alpha \circ T_u) \circ \alpha^{-1} = T_u \circ \alpha^{-1}$ for each u in E, α^{-1} belongs to $\mathrm{Hom}(T, T)$.

Definition. $\mathrm{Hom}(T, T)$ is called the *commuting division algebra* of the irreducible operator set T.

Notice that $X(T)$ can be regarded as a linear space over the division algebra $\mathrm{Hom}(T, T)$, and that the operators T_u are then $\mathrm{Hom}(T, T)$-linear, that is, $T_u \circ \alpha = \alpha \circ T_u$ for all α in $\mathrm{Hom}(T, T)$.

4.3. *For the rest of this section we fix an algebra A.*

Extended Jacobson Density Theorem. *Let r be a positive integer, and let $T^1, \ldots, T^r$ be r pairwise inequivalent irreducible representations of A. Put $D^i = \mathrm{Hom}(T^i, T^i)$. For each $i = 1, \ldots, r$ let W_i be a D^i-linear subspace of $X(T^i)$ of finite D^i-dimension, and let g^i be any D^i-linear operator on $X(T^i)$. Then there exists an element a of A such that*

$$T_a^i(\xi) = g^i(\xi) \tag{1}$$

for all $i = 1, \ldots, r$ and all ξ in W_i.

Proof.   Let X_i stand for $X(T^i)$. We first make the following claim:

Suppose that, for each i, U_i is a D^i-linear subspace of X_i of finite D^i-dimension s_i. For some j let $\xi \in X_j \setminus U_j$. Then there exists an a in A such that

$$T_a^i \equiv 0 \text{ on } U_i \text{ for each } i, \text{ and } T_a^j \xi \neq 0. \qquad (2)$$

This claim is proved by induction in $s = \sum_{i=1}^r s_i$. For $s = 0$ it is obvious. Let $s > 0$; and assume that it is true for $s - 1$, but that there are subspaces U_i as above, and a vector $\xi \in X_j \setminus U_j$, such that no a satisfying (2) exists. Choose any integer k for which $U_k \neq \{0\}$, and write

$$U_k = V + D^k \eta, \qquad (3)$$

where V is a D^k-subspace of U_k of D^k-dimension $s_k - 1$, and $\eta \in U_k \setminus V$. Let J be the left ideal of A consisting of all those a such that $T_a^k(V) = \{0\}$ and $T_a^i(U_i) = \{0\}$ for all $i \neq k$. By the inductive hypothesis on $s - 1$, $T_J^k \eta \neq \{0\}$; hence by 3.7 and the irreducibility of T^k, $T_J^k \eta = X_k$. Furthermore, if $a \in J$ and $T_a^k \eta = 0$, by (3) a annihilates all the U_i and so $T_a^j \xi = 0$ (by the assumed failure of (2)). It follows that there is a unique F-linear map $g : X_k \to X_j$ satisfying

$$g(T_a^k \eta) = T_a^j \xi \qquad \text{for all } a \text{ in } J. \qquad (4)$$

This g intertwines T^k and T^j. Indeed let $a \in A$; then for $b \in J$ we have $ab \in J$, and so by (4) $g T_a^k (T_b^k \eta) = g(T_{ab}^k \eta) = T_{ab}^j \xi = T_a^j g(T_b^k \eta)$. Since $T_J^k \eta = X_k$, this implies $g \circ T_a^k = T_a^j \circ g$. So $g \in \text{Hom}(T^k, T^j)$.

Now we shall consider separately the two possibilities, $k = j$ and $k \neq j$. Assume first that $k = j$. Since $g \in \text{Hom}(T^j, T^j) = D^j$, we have by (4)

$$T_a^j(\xi - g\eta) = 0 \qquad \text{for all } a \text{ in } J.$$

Combining this with the inductive hypothesis on $s - 1$, we conclude that $\xi - g\eta \in V$, whence $\xi \in V + D^j \eta = U_j$ (by (3)), contradicting the definition of ξ.

Now assume that $k \neq j$. If $g = 0$, then by (4) $T_a^j \xi = 0$ for all a in J, contradicting the inductive hypothesis for the spaces $U_1, \ldots, V, \ldots, U_r$ (of total dimension $s - 1$). Thus $0 \neq g \in \text{Hom}(T^k, T^j)$, whence $T^k \cong T^j$ by 1.9, contradicting the hypothesis that the T^i are pairwise inequivalent.

Thus in either case we have reached a contradiction. So the original claim is proved.

To prove the Theorem, we choose for each i a D^i-basis $\xi_1^i, \ldots, \xi_{m_i}^i$ of W_i. For each of these vectors ξ_j^i the above claim shows that there is an element a_{ij} of A such that $T_{a_{ij}}^i \xi_j^i \neq 0$ while $T_{a_{ij}}^p \xi_q^p = 0$ for all pairs $\langle p, q \rangle$ other than $\langle i, j \rangle$.

Then we use 3.7 to obtain an element b_{ij} of A such that $T^i_{b_{ij}}(T^i_{a_{ij}}\xi^i_j) = g^i(\xi^i_j)$. The element $a = \sum_{i=1}^r \sum_{j=1}^{m_i} b_{ij}a_{ij}$ then satisfies (1).   ∎

4.4. Corollary. *Let T be an irreducible representation of A, with commuting division algebra D. If g is any D-linear operator on $X(T)$, and if U is any D-linear subspace of $X(T)$ of finite D-dimension, there is an element a of A such that $T_a(\xi) = g(\xi)$ for all ξ in U.*

Proof.   This is just the case $r = 1$ of 4.3.   ∎

4.5. Definition.   A representation T of A is *totally irreducible* if (i) $X(T) \neq \{0\}$, and (ii) for every linear operator g on $X(T)$ and every linear subspace U of $X(T)$ of finite (F-)dimension, there is an element a of A such that $T_a(\xi) = g(\xi)$ for all ξ in U.

A totally irreducible representation T is obviously irreducible. Furthermore, its commuting division algebra is F; that is, $\text{Hom}(T, T)$ consists of the "scalar" operators $\xi \mapsto c\xi$ $(c \in F)$ only. Indeed: Suppose $g \in \text{Hom}(T, T)$, $\xi \in X(T)$, and ξ and $g\xi$ are linearly independent. Since T is totally irreducible, there is an a in A such that $T_a\xi = \xi$ and $T_a g\xi = 0$. But then $0 = T_a g\xi = gT_a\xi = g\xi$, a contradiction. Therefore for each ξ in $X(T)$ we have $g\xi = c_\xi \xi$ for some $c_\xi \in F$. If $\xi \neq 0$, $\eta \neq 0$, and $c_\xi \neq c_\eta$, then $\xi + \eta$ and $c_\xi \xi + c_\eta \eta = g(\xi + \eta)$ are linearly independent, contradicting what was proved above. So there is a single number c in F satisfying $g(\xi) = c\xi$ for all ξ.

4.6.   In view of 4.4, the properties mentioned in 4.5 are sufficient for total irreducibility.

Corollary.   *A representation T of A is totally irreducible if and only if* (i) T *is irreducible, and* (ii) $\text{Hom}(T, T)$ *consists of the scalar operators only.*

Remark.   In the argument of 4.5 showing that total irreducibility implies $\text{Hom}(T, T) \cong F$, we actually only needed the special case of the definition of total irreducibility in which $\dim(U) = 2$. It follows that an irreducible representation T of A is totally irreducible if and only if, given any four vectors ξ, ξ', η, η' in $X(T)$ such that ξ and ξ' are linearly independent, there exists an element a of A such that $T_a\xi = \eta$, $T_a\xi' = \eta'$.

4.7. Proposition.   *If A is Abelian, any totally irreducible representation T of A is one-dimensional.*

Proof. Suppose $X(T)$ contains two linearly independent vectors ξ and η. By total irreducibility there are elements a, b of A such that $T_a\xi = \xi$, $T_a\eta = 0$, $T_b\xi = \eta$. But then $T_{ba}\xi = \eta \neq 0 = T_{ab}\xi$, which is impossible since $ab = ba$. ∎

Remark. If A is Abelian, a representation T of A which is merely irreducible need not be one-dimensional. As an example, let D be any commutative division algebra over F which is of dimension greater than 1, and let T be the regular representation of D.

4.8. Of particular interest is the case when each of the T^i in 4.3 is finite-dimensional. In that case we can take $W_i = X(T^i)$ in 4.3, getting:

Corollary. *Let $T^1, \ldots, T^r$ be a finite sequence of pairwise inequivalent finite-dimensional irreducible representations of A. Let D^i be the commuting division algebra of T^i; and for each $i = 1, \ldots, r$ let g_i be a D^i-linear operator on $X(T^i)$. Then there is an element a of A satisfying*

$$T_a^i = g_i$$

for all $i = 1, \ldots, r$.

4.9. Corollary 4.8 says in particular that, if T is a finite-dimensional irreducible representation of A, with commuting division algebra D, then the range of T consists of all D-linear operators.

Remark. If T is irreducible but not finite-dimensional, range(T) need no longer consist of *all* D-linear operators. For example, let X be an infinite-dimensional linear space, and A the algebra of all linear operators on X *of finite rank* (i.e., whose range is finite-dimensional). Then the identity representation T of A (sending each a in A into itself) is evidently totally irreducible. By 4.5, Hom(T, T) consists of the scalar operators only; so *all* linear operators on X are Hom(T, T)-linear.

4.10. Suppose for the moment that F is algebraically closed—for example the complex field. Then it is well known that the only finite-dimensional division algebra D over F is F itself. (To see this, take any element u of D, and let $\lambda_u(v) = uv$ ($v \in D$). Since F is algebraically closed, the equation $\det(\lambda_u - c \cdot 1) = 0$ has a solution c in F. This implies that the operation of left multiplication in D by the element $u - c$ is singular, and hence that $u = c$, since D is a division algebra.)

Now let T be a finite-dimensional irreducible representation of A. Clearly the commuting division algebra D of T is finite-dimensional over F; and so, if F is algebraically closed, it follows from the preceding paragraph that $D = F$. Putting this information into 4.8 we get:

Corollary. *Assume that F is algebraically closed. Let $T^1, \ldots, T^r$ be finitely many pairwise inequivalent irreducible finite-dimensional representations of A. For each $i = 1, \ldots, r$ let g_i be a linear operator on $X(T^i)$. Then there is an element a of A such that*

$$T_a^i = g_i \quad \textit{for each } i = 1, \ldots, r.$$

In particular, every finite-dimensional irreducible representation T of A is totally irreducible, and range(T) *consists of all linear operators on $X(T)$.*

The last statement of the Corollary is usually known as Burnside's Theorem.

4.11. Corollary. *If F is algebraically closed and A is Abelian, a finite-dimensional irreducible representation of A is one-dimensional.*

Proof. By 4.10 and 4.7.   ∎

4.12. Remark. A classical theorem of Frobenius states that there are, to within isomorphism, only three finite-dimensional division algebras over $\mathbb{R}$, namely $\mathbb{R}$, $\mathbb{C}$, and the quaternion division algebra Q. Thus, if $F = \mathbb{R}$, the irreducible finite-dimensional representations T of A split into three classes, namely those whose commuting division algebras are $\mathbb{R}$, $\mathbb{C}$, or Q respectively. By 4.6, the first of these classes is just the class of totally irreducible finite-dimensional representations of A.

4.13. Here is an interesting generalization of Burnside's Theorem:

Proposition*. *Suppose that F is algebraically closed, and that T is a (perhaps infinite-dimensional) irreducible representation of A such that T_a is of finite rank for every a in A. Then T is totally irreducible.*

To prove this we observe that the hypothesis of finite rank compels Hom(T, T) to be finite-dimensional over F.

Characters of Finite-Dimensional Representations

4.14. **Definition.** If T is a finite-dimensional representation of A, the linear functional $a \mapsto \mathrm{Tr}(T_a)$ is called the *character of T*, and will be denoted by χ_T.

By the properties of traces of operators,

$$\chi_T(ab) = \chi_T(ba) \qquad\qquad (a, b \in A).$$

If g and h are linear operators on $X(T)$, g being non-singular, then $\mathrm{Tr}(g^{-1}hg) = \mathrm{Tr}(h)$. It follows that $\chi_{T'} = \chi_T$ whenever T' and T are equivalent representations. Thus we can write χ_τ instead of χ_T, where τ is the equivalence class to which T belongs.

4.15. **Proposition.** *Assume that F has characteristic 0. If $T^1, \ldots, T^r$ are pairwise inequivalent finite-dimensional irreducible representations of A, the functionals $\chi_{T^1}, \ldots, \chi_{T^r}$ are linearly independent.*

Proof. Suppose that $\sum_{i=1}^r c_i \chi_{T^i} = 0 \ (c_i \in F)$; and fix $j = 1, \ldots, r$. By 4.8 there is an element a of A such that T_a^j is the identity operator on $X(T^j)$ and $T_a^i = 0$ for $i \neq j$. Thus $0 = \sum_{i=1}^r c_i \chi_{T^i}(a) = c_j \dim(T^j)$; and this implies $c_j = 0$ since F is of characteristic 0. ∎

4.16. This has an important corollary:

Corollary. *Assume that F has characteristic 0; and let T be a completely reducible non-degenerate finite-dimensional representation of A. Then T is determined to within equivalence by χ_T.*

Proof. The hypotheses imply that

$$T \cong \sum_{S \in \hat{A}}^{\oplus} n_S S, \qquad\qquad (5)$$

where the n_S are non-negative integers, and all but finitely many are 0. (For the notation see 2.22 and 3.8.) Thus

$$\chi_T = \sum_{S \in \hat{A}} n_S \chi_S. \qquad\qquad (6)$$

Since the χ_S are linearly independent by 4.15, there can only be one collection of integers n_S satisfying (6). So (5) implies the Corollary. ∎

Jordan–Hölder Series

4.17. Consider a finite-dimensional operator set T (with $X(T) \neq \{0\}$), indexed by a set E.

Definition. A *Jordan-Hölder series for* T is a finite strictly increasing sequence

$$\{0\} = X_0 \subset X_1 \subset X_2 \subset \cdots \subset X_m = X(T) \tag{7}$$

of T-stable subspaces of $X(T)$ such that, for each $i = 1, \ldots, m$, the subquotient $\tilde{T}^i$ of T acting on X_i/X_{i-1} is irreducible. The $\tilde{T}^i$ are called the *quotients* of the Jordan–Hölder series (7).

A Jordan–Hölder series for T clearly exists. Indeed, let X_1 be a T-stable subspace of minimal non-zero dimension; let X_2 be a T-stable subspace properly containing X_1 and of minimal dimension consistent with this condition; let X_3 be a T-stable subspace properly containing X_2 and of minimal dimension; and so forth. Then (7) will be a Jordan–Hölder series.

4.18. In general there will be many different Jordan-Hölder series for the same T. So it is somewhat remarkable that all of them have essentially the same quotients.

Theorem. *Assume that F has characteristic 0; and let T be a finite-dimensional operator set. Let (7) and*

$$\{0\} = X_0' \subset X_1' \subset \cdots \subset X_p' = X(T) \tag{8}$$

be two Jordan–Hölder series for T. Then $p = m$, and there is a permutation π of $\{1, \ldots, m\}$ such that $\tilde{T}^{\pi(j)} \cong \tilde{T}'^j$ for all $j = 1, \ldots, m$. (Here the $\tilde{T}^i$ and $\tilde{T}'^j$ are the quotients of (7) and (8) respectively.)

Proof. The observation 3.10 enables us to assume without loss of generality that T is a representation of the algebara A with unit $\mathbf{1}$, sending $\mathbf{1}$ into the identity operator.

Take a basis of $X(T)$ consisting of first a basis of X_1, then a basis of X_2 modulo X_1, then a basis of X_3 modulo X_2, and so forth. For any a in A the matrix of T_a with respect to this basis will be of upper triangular block form, the diagonal "blocks" being the matrices of the quotients $\tilde{T}^i$. It follows that

$$\chi_T(a) = \sum_{i=1}^{m} \chi_{\tilde{T}^i}(a) \qquad (a \in A). \tag{9}$$

Similarly

$$\chi_T(a) = \sum_{j=1}^{p} \chi_{\tilde{T}^{\prime j}}(a) \qquad\qquad (a \in A). \qquad (10)$$

Applying 4.15 to the two expressions (9) and (10) for χ_T, we conclude that each S in $\bar{A}$ must occur as many times among the $\tilde{T}^{\prime j}$ as it does among the $\tilde{T}^i$. From this the required conclusion follows immediately. ∎

Remark. This Theorem is actually true without the hypothesis that F is of characteristic 0. See for example Curtis and Reiner [1], Theorem 13.7.

Notice the similarity between the proofs of this Theorem and of 4.16.

4.19. Given a finite-dimensional operator set T (indexed by E) and an irreducible operator set S, Theorem 4.18 enables us to define the *multiplicity of S in T* as the number of distinct i such that $S \cong \tilde{T}^i$, where $\tilde{T}^1, \ldots, \tilde{T}^m$ are the quotients of some Jordan-Hölder series for T.

If T is completely reducible, this definition of multiplicity clearly coincides with that of 2.21.

5. Finite-dimensional Semisimple Algebras

The Radical of an Algebra

5.1. Let A be a fixed algebra.

The extent to which the irreducible representations of A are adequate to describe A is measured by the so-called radical of A.

Definition. The *radical* R of A is the intersection of the kernels of all the irreducible representations of A; that is, $R = \cap\{\mathrm{Ker}(T): T \in \bar{A}\}$.

Thus the radical of A is a two-sided ideal of A.

If $R = \{0\}$, we say that A is *semisimple*; if $R = A$, A is a *radical algebra*.

A radical algebra is thus one which has no irreducible representations at all. A semisimple algebra A is one which has enough irreducible representations to separate points of A.

5.2. Proposition. *If R is the radical of A, then A/R is semisimple, and R is a radical algebra. In fact, R is the largest radical two-sided ideal of A.*

Proof. The semisimplicity of A/R is obvious.

Any irreducible representation S of R would extend by 3.18 to an irreducible representation T of A, and we would have $R \not\subset \text{Ker}(T)$, contradicting the definition of R. So R is a radical algebra.

Let J be a two-sided ideal of A and a radical algebra. If T is an irreducible representation of A and $J \not\subset \text{Ker}(T)$, then by 3.16 $T|J$ is irreducible, contradicting the radical property of J; so $J \subset \text{Ker}(T)$. Since this holds for all T in $\bar{A}$, we have $J \subset R$. ■

5.3. We shall say that an algebra A is *nilpotent* if there is a positive integer n such that $A^n = \{0\}$ (i.e., any product of n elements of A is 0).

Proposition. *Suppose A is finite-dimensional. Then A is a radical algebra if and only if A is nilpotent.*

Proof. Suppose that $A^n = \{0\}$ ($n = 1, 2, \ldots$); and let T be an irreducible representation of A. Let $0 \neq \xi \in X(T)$. By 3.7 there is an element a of A such that $T_a \xi = \xi$. But this implies that $T_{(a^n)} \xi = (T_a)^n \xi = \xi \neq 0$, contradicting the fact that $a^n = 0$. So there is no such T; and A is a radical algebra.

Conversely, assume that A is a radical algebra. By the Remark following this proof, A has a finite-dimensional faithful representation T. Let

$$\{0\} = X_0 \subset X_1 \subset \cdots \subset X_m = X(T)$$

be a Jordan-Hölder series for T (4.17). Since A is a radical algebra, the quotients of this series are all zero representations; and this says that

$$T_a(X_i) \subset X_{i-1} \qquad (a \in A; i = 1, \ldots, m) \qquad (1)$$

If $a_1, \ldots, a_m$ are any elements of A, an m-fold application of (1) gives $T_{a_1 \ldots a_m}(X_m) \subset T_{a_1 \ldots a_{m-1}}(X_{m-1}) \subset \cdots \subset X_0 = \{0\}$, or $T_{a_1 \ldots a_m} = 0$; so, since T is faithful, $a_1 \ldots a_m = 0$. Thus A is nilpotent. ■

Remark. In the above proof we used the fact that every finite-dimensional algebra A has a faithful finite-dimensional representation T. To see this, let A_1 be A if A has a unit element; otherwise let A_1 be the result of adjoining a unit to A (see V.4.10). Then the restriction to A of the regular representation of A_1 has the properties required of T.

5.4. Corollary. *If* A *is finite-dimensional, its radical is the largest nilpotent two-sided ideal of* A.

Proof. By 5.2 and 5.3.   ∎

Remark. If A is finite-dimensional and Abelian, the radical of A is just the set J of all elements a of A which are nilpotent (i.e., $a^n = 0$ for some $n = 1, 2, \ldots$). Indeed: Clearly J is an ideal in A, and the radical R is contained in J by 5.4. On the other hand, the first part of the proof of 5.3 suffices to show that J can have no irreducible representations; and so by 5.2 $J \subset R$.

Representations of Total Matrix Algebras

5.5. It will now be necessary to study the representations of those algebras which, by 4.9, can arise as the range of an irreducible finite-dimensional representation. These are the total matrix algebras.

5.6. Fix a division algebra D (not necessarily finite-dimensional) over F; and let X be a D-linear space of finite dimension n over D. When X is regarded as an F-linear space (by restricting the field of scalars to F), we call it X_F. Let M now stand for the algebra (over F) of all D-linear operators on X. The identity operator $\mathbf{1}$ is thus the unit element of M.

Fix a D-basis $e^1, e^2, \ldots, e^n$ of X. If $\xi \in X$ let $\xi_1, \ldots, \xi_n$ be the elements of D giving $\xi = \sum_{i=1}^{n} \xi_i e^i$. Then every a in M gives rise to an $n \times n$ matrix $\{\alpha_{ij}\}$ with entries in D, as follows:

$$\alpha_{ij} = (ae^j)_i. \tag{2}$$

Conversely, every $n \times n$ matrix $\{\alpha_{ij}\}$ with entries in D arises via (2) from the element a of M defined as follows:

$$(a\xi)_i = \sum_{j=1}^{n} \xi_j \alpha_{ij} \qquad (\xi \in X). \tag{3}$$

So M can be identified with the family $M(n, D)$ of all $n \times n$ matrices with entries in D. When we make this identification, addition and scalar multiplication (by elements of F) in M go over into the corresponding obvious operations in $M(n, D)$, and multiplication becomes:

$$(\alpha\beta)_{ij} = \sum_{k=1}^{n} \beta_{kj} \alpha_{ik}. \tag{4}$$

Definition. With these operations $M(n, D)$ is called the $n \times n$ *total matrix algebra* (*over* F) *with entries in* D.

Let e^{rs} and m^u ($r, s = 1, \ldots, n$; $u \in D$) be the matrices in $M(n, D)$ given by: $(e^{rs})_{ij} = \delta_{ir}\delta_{js}$; $(m^u)_{ij} = \delta_{ij}u$. We have:

$$e^{rs}e^{pq} = \delta_{sp}e^{rq}, \tag{5}$$

$$m^u + m^v = m^{u+v}, m^u m^v = m^{vu}, \tag{6}$$

$$e^{rs}m^u = m^u e^{rs}, \tag{7}$$

$$\alpha = \sum_{r,s=1}^{n} m^{\alpha_{rs}} e^{rs} \qquad (\alpha \in M(n, D)). \tag{8}$$

5.7. Proposition. *There is, to within equivalence, only one irreducible representation of the algebra M, namely the identity representation W of M on X_F. More generally, every non-degenerate representation of M is a direct sum of copies of W.*

Proof. Since any non-zero vector ξ in X can be made the first element e^1 of a D-basis of X, the arbitrariness in the first column of a matrix in $M(n, D)$ shows (by (3)) that $M\xi = X$. Thus W is irreducible.

Now let T be any representation of M with $T_{\mathbb{1}}$ equal to the identity. We shall identify M with $M(n, D)$ as in 5.6; and shall abbreviate $T_a\xi$ to $a\xi$ ($a \in M$; $\xi \in X(T)$). By (6) $X(T)$ becomes a right linear space over D as follows:

$$\xi u = m^u \xi \qquad (\xi \in X(T); u \in D).$$

If $u \in F$, then of course ξu and $u\xi$ are the same.

Since $e^{11}e^{11} = e^{11}$, $T_{e^{11}}$ is idempotent; let V be its range. By (7) V is a D-subspace of $X(T)$. Let B be a D-basis of V; and for each ξ in B and $i = 1, \ldots, n$ set

$$\xi^i = e^{i1}\xi.$$

We claim that $\{\xi^i : \xi \in B; i = 1, \ldots, n\}$ is a D-basis of $X(T)$.

Indeed: Suppose $\sum_{\xi \in B} \sum_{i=1}^{n} \xi^i u_{\xi, i} = 0$ ($u_{\xi, i} \in D$). Then by (5) and (7), for each $k = 1, \ldots, n$

$$0 = e^{1k} \sum_{\xi \in B} \sum_{i=1}^{n} (\xi^i u_{\xi, i}) = \sum_{\xi \in B} \xi u_{\xi, k}.$$

Since B is D-independent, it follows that all the $u_{\xi, k}$ are 0. So the ξ^i form a D-independent set.

Let η be any vector in $X(T)$. For each k, $e^{1k}\eta \in V$; so

$$e^{1k}\eta = \sum_{\xi \in B} \xi u_{\xi,k} \qquad\qquad (u_{\xi,k} \in D).$$

It follows that

$$\eta = \sum_{k=1}^{n} e^{kk}\eta = \sum_{k=1}^{n} e^{k1}e^{1k}\eta$$

$$= \sum_{k=1}^{n} \sum_{\xi \in B} e^{k1}(\xi u_{\xi,k}) = \sum_{k=1}^{n} \sum_{\xi \in B} \xi^{k}u_{\xi,k}$$

(by (7)). So the D-linear span of the ξ^i is $X(T)$; and the claim is proved.

Now for each ξ in B let X_ξ be the D-subspace of $X(T)$ spanned by $\xi^1, \ldots, \xi^n$. In view of the above claim, the Proposition will be proved if we show that each X_ξ is T-stable, and that the subrepresentation of T on each X_ξ is equivalent to W. Fix $\xi \in B$.

Notice that

$$e^{ij}\xi^k = \delta_{jk}\xi^i \qquad (i,j,k = 1, \ldots, n). \qquad (9)$$

Let G be the F-linear bijection $X \to X_\xi$ given by

$$G\left(\sum_{i=1}^{n} u_i e^i\right) = \sum_{i=1}^{n} \xi^i u_i \qquad\qquad (u_i \in D).$$

If $a \in M$ and $u_1, \ldots, u_n \in D$ we have by (8), (7), (9), (3),

$$a\left(G\left(\sum_i u_i e^i\right)\right) = \sum_i a(\xi^i u_i) = \sum_{i,r,s} e^{rs}m^{ars}(\xi^i u_i)$$

$$= \sum_{i,r,s} (e^{rs}\xi^i)u_i a_{rs}$$

$$= \sum_{i,r} \xi^r u_i a_{ri}$$

$$= G\left(\sum_{i,r} u_i a_{ri} e^r\right)$$

$$= G\left(a\left(\sum_r u_r e^r\right)\right).$$

So G intertwines W and T, and the Proposition is proved.   ∎

5.8.   Corollary.   *Every representation T of the algebra M of 5.6 is completely reducible.*

Proof. M has a unit element $\mathbf{1}$, and we can write $X(T) = X_1 \oplus X_2$, where $X_1 = \text{range}(T_1)$ and $X_2 = \{\xi: T_\mathbf{1}\xi = 0\}$. Evidently X_1 and X_2 are T-stable; $X_2 T$ is a zero representation; and by 5.7 $X_1 T$ is a direct sum of copies of the irreducible representation W. ∎

5.9. Corollary. *The algebra M of 5.6 is simple (i.e., has no two-sided ideals except $\{0\}$ and M).*

Proof. Let I be a proper two-sided ideal of M. By Remark 5.3 M/I has a non-zero representation, giving rise to a non-zero representation of M whose kernel J contains I. But 5.7 implies that every non-zero representation of M is faithful. So $J = \{0\}$, whence $I = \{0\}$. ∎

5.10. Corollary. *If α is an automorphism of the algebra $M(n, F)$ ($n = 2, 3, \ldots$), there exists a non-singular matrix g in $M(n, F)$ such that*

$$\alpha(a) = g^{-1}ag \qquad\qquad (a \in M(n, F)).$$

Proof. If W is the unique irreducible representation of $M(n, F)$ (see 5.7), then $W \circ \alpha$ is also an irreducible representation of $M(n, F)$ and so is equivalent to W. The operator which implements this equivalence is the required g. ∎

The Structure of Finite-dimensional Semisimple Algebras

5.11. Theorem. *Let A be a finite-dimensional algebra. Then the following four conditions are equivalent:*

(I)   *A is semisimple.*
(II) *A has a faithful completely reducible representation.*
(III) *Every representation of A is completely reducible.*
(IV) *A is isomorphic to an algebra direct sum $\sum_{i=1}^{r \oplus} M(n_i, D^i)$, where r is a non-negative integer, the n_i are positive integers, the D^i are finite-dimensional division algebras over F, and $M(n_i, D^i)$ is the total matrix algebra of 5.6.*

Proof. Each total matrix algebra $M(n, D)$ has property (III) by 5.8. By 3.21 a finite direct sum of algebras with property (III) also has property (III). This shows that (IV) $\Rightarrow$ (III).

We have seen in Remark 5.3 that A has a faithful representation T. If (III) holds, T is completely reducible. So (III) $\Rightarrow$ (II).

If (II) holds, the irreducible representations occurring in a faithful completely reducible representation of A must separate points of A. So (II) $\Rightarrow$ (I).

It remains only to show that (I) $\Rightarrow$ (IV). Suppose that $T^1, \ldots, T^r$ is a finite sequence of pairwise inequivalent irreducible representations of A (necessarily finite-dimensional by 3.15). Let $D^i = \mathrm{Hom}(T^i, T^i)$, and let n_i be the D^i-dimension of $X(T^i)$. Identifying $M(n_i, D^i)$ with the algebra of all D^i-linear operators on $X(T^i)$ as in 5.6, we see from 4.8 that

$$\Phi: a \mapsto T_a^1 \oplus \cdots \oplus T_a^r \tag{10}$$

is a homomorphism of A *onto* the algebra direct sum $B = \sum_{i=1}^{r \oplus} M(n_i, D^i)$.

Now the dimension of B is at least r. Since Φ is onto, $\dim(B) \leq \dim(A)$. It follows that $r \leq \dim(A)$; and we have shown that *there can be no more that* $\dim(A)$ *inequivalent irreducible representations of* A.

In view of this, we may assume that the $T^1, \ldots, T^r$ are *all* the irreducible representations of A. Now suppose that A is semisimple. Then the $T^1, \ldots, T^r$ separate points of A, and so the map Φ of (10) is one-to-one, hence an isomorphism of A and B. Thus (I) $\Rightarrow$ (IV). ∎

5.12. Corollary. *Every finite-dimensional semisimple algebra has a unit element.*

Proof. Use (IV) of 5.11. ∎

5.13. The ideal structure of a semisimple finite-dimensional algebra is quite transparent in view of 5.11.

Proposition. *Let A be the algebra direct sum $\sum_{i=1}^{r \oplus} A_i$, where A_i is the total matrix algebra $M(n_i, D^i)$. Then every two-sided ideal of A is of the form $\sum_{i \in N} A_i$, where $N \subset \{1, \ldots, r\}$.*

Proof. Given a two-sided ideal J of A, put $N = \{i: A_i \subset J\}$. Then $\sum_{i \in N} A_i \subset J$.

Let e_i be the unit element of A_i, and suppose that $\sum_{i=1}^{r} a_i = a \in J$ $(a_i \in A_i)$. For any $i = 1, \ldots, r$, we have $a_i = e_i a \in J \cap A_i$. So, since A_i is simple (5.9), $a_i \neq 0 \Rightarrow J \cap A_i = A_i \Rightarrow i \in N$. Thus $a \in \sum_{i \in N} A_i$. Since a was an arbitrary element of J, this gives $J \subset \sum_{i \in N} A_i$. Therefore $J = \sum_{i \in N} A_i$. ∎

5.14. It follows from 5.13 that the A_i of 5.13 are just the minimal non-zero two-sided ideals of A. So 5.11 asserts that a finite-dimensional semisimple algebra is the direct sum of its minimal two-sided ideals, and that each of the latter is isomorphic to a total matrix algebra.

5.15. Combining the proof of 5.11 with 5.14, we see that card($\overline{A}$) (the number of inequivalent irreducible representations of A) is equal to card($\mathscr{J}$), where $\mathscr{J}$ is the set of minimal two-sided ideals of A. There is of course a natural correspondence between $\mathscr{J}$ and $\overline{A}$: Given $J \in \mathscr{J}$, the corresponding element T^J of $\overline{A}$ is the one which annihilates every element of $\mathscr{J}$ other than J, and coincides on J with the unique irreducible representation of J (see 5.7).

Harmonic Analysis of Representations of A

5.16. Let A be a semisimple finite-dimensional algebra. For each σ in $\overline{A}$ let J_σ be the minimal two-sided ideal of A corresponding to σ as in 5.15, and denote by e_σ the unit element of J_σ. If S is a representation of class σ, we observe that $S_{e_\tau} = 0$ for $\tau \in \overline{A}$, $\tau \neq \sigma$, while S_{e_σ} is the identity operator on $X(S)$. From this we conclude (recalling from 2.11 the definition of the σ-subspace):

Proposition. *Let T be any representation of the above algebra A. For each σ in $\overline{A}$, the range of the idempotent operator T_{e_σ} is equal to the σ-subspace of $X(T)$; and T_{e_σ} annihilates the τ-subspace for all $\tau \neq \sigma$. Thus, if $\xi \in X(T)$, $T_{e_\sigma}\xi$ is the σ-component of ξ (see 2.13); and (if T is non-degenerate) the equation*

$$\xi = \sum_{\sigma \in \overline{A}} T_{e_\sigma}\xi$$

is the harmonic analysis of ξ.

5.17. Some of the consequences of 5.11 take on a particularly simple form when F is algebraically closed, so that the D^i of 5.11(IV) are equal to F (see 4.10). *From here to the end of this section we take F to be algebraically closed, and A to be a finite-dimensional algebra over F.* If $\sigma \in \overline{A}$, put $d_\sigma = \dim(\sigma)$. Results 5.18 to 5.21 are easy consequences of 5.11, and are left to the reader.

5.18. Proposition. *If A is semisimple, then*

$$A \cong \sum_{\sigma \in \overline{A}}^{\oplus} M(d_\sigma, F). \tag{11}$$

In particular,

$$\dim(A) = \sum_{\sigma \in \overline{A}} d_\sigma^2. \tag{12}$$

The Abelian Semisimple Case

5.19. Proposition. *If A is semisimple, a necessary and sufficient condition for it to be Abelian is that all its irreducible representations be one-dimensional. In this case* card$(\overline{A})$ = dim(A); *and A is isomorphic to the algebra, under pointwise multiplication, of all* dim(A)-*termed sequences of elements of F.*

The Structure of the Regular Representation

5.20. In 3.9 we defined the regular representation of A. Since $\dim(M(n, F)) = n^2$ and the unique irreducible representation W of $M(n, F)$ is of dimension n, the multiplicity of W in the regular representation of $M(n, F)$ is n. From this and 5.11 (or 5.18) we easily obtain:

Proposition. *Suppose that A is semisimple, and that J_σ is the minimal two-sided ideal of A associated (as in 5.14, 5.15) with the element σ of A. If R is the regular representation of A, then:* (I) J_σ *is the σ-subspace of A with respect to R:* (II) *the multiplicity of σ in R is* dim(σ).

The Center of a Semisimple Algebra

5.21. Proposition. *Let A be semisimple; and let e_J be the unit element of the minimal two-sided ideal J of A. The center C of A is semisimple and of dimension equal to* card$(\overline{A})$. *The e_J (J running over all minimal two-sided ideals of A) form a basis of C.*

Traces on Semisimple Algebras

5.22. Sometimes a finite-dimensional semisimple algebra A is presented to us in a form which makes its decomposition into minimal ideals far from transparent. This is the case with the group algebras of finite groups, to be studied in the next section. In such a case it is useful to have an explicit relation between the matrices of an irreducible representation of A and the elements of the corresponding minimal two-sided ideal of A.

We continue to assume that F is algebraically closed. In addition, *from here to the end of the section, F is supposed to have characteristic 0.* Let A be a fixed finite-dimensional semisimple algebra. If $\sigma \in \overline{A}$, let $d_\sigma = \dim(\sigma)$.

5.23. *Definition.* A *trace on A* is a linear functional $\gamma: A \to F$ such that

$$\gamma(ab) = \gamma(ba) \qquad\qquad (a, b \in A).$$

If $a = 0$ whenever $a \in A$ and $\gamma(ab) = 0$ for all b in A, then the trace γ is *non-degenerate.*

Thus the characters χ_T of 4.14 are traces.

Proposition*. *Every trace on A is of the form*

$$\gamma = \sum_{\sigma \in \bar{A}} k_\sigma \chi_\sigma \qquad\qquad (k_\sigma \in F). \qquad (13)$$

The trace γ is non-degenerate if and only if $k_\sigma \neq 0$ for all σ in $\bar{A}$.

5.24. Now let γ be a fixed non-degenerate trace on A; let $\{k_\sigma\}$ be the non-zero numbers determining γ by (13); and define a bilinear "inner product" on A:

$$[a, b] = \gamma(ab) \qquad\qquad (a, b \in A).$$

In view of the non-degeneracy of γ, we can use $[\ ,\ ]$ to place A and its adjoint space $A^\#$ in one-to-one correspondence. If $\alpha \in A^\#$, let $\alpha^\sim$ be the corresponding element of A, given by

$$[\alpha^\sim, b] = \alpha(b) \qquad\qquad (b \in A).$$

5.25. Proposition*. *Given $\sigma \in \bar{A}$, let us fix a basis of $X(S)$, where S belongs to the class σ; and let $\{\sigma_{pq}(a)\}$ $(p, q = 1, \ldots, d_\sigma)$ be the matrix of S_a with respect to this basis. Thus for each p, q we have a linear functional $\sigma_{pq}: a \mapsto \sigma_{pq}(a)$ on A. Put $e^{pq} = k_\sigma(\sigma_{qp})^\sim$. Then the $\{e^{pq}\}$ $(p, q = 1, \ldots, d_\sigma)$ form a canonical basis of the minimal two-sided ideal J_σ of A associated with σ as in 5.15. In particular $k_\sigma(\chi_\sigma)^\sim$ $(= \sum_p k_\sigma(\sigma_{pp})^\sim)$ is the unit element of J_σ.*

Note: If M is an algebra isomorphic with $M(n, F)$, by a *canonical basis* of M we mean a basis $\{u^{pq}\}(p, q = 1, \ldots, n)$ of M such that

$$u^{pq}u^{rs} = \delta_{qr}u^{ps}.$$

5.26. Corollary* (Orthogonality Relations). *Keeping the notation of 5.25, we have*

$$[(\sigma_{pq})^\sim, (\tau_{rs})^\sim] = k_\sigma^{-1}\delta_{\sigma\tau}\delta_{qr}\delta_{ps}, \qquad [(\chi_\sigma)^\sim, (\chi_\tau)^\sim] = d_\sigma k_\sigma^{-1}\delta_{\sigma\tau}$$

$(\sigma, \tau \in \bar{A}; p, q = 1, \ldots, d_\sigma; r, s = 1, \ldots, d_\tau).$

5.27. Corollary*.   *Let S and T be any finite-dimensional representations of
A. Given $\sigma \in \bar{A}$, let μ_σ and v_σ be the multiplicities of σ in S and T respectively.
Then*

$$[(\chi_S)^\sim, (\chi_T)^\sim] = \sum_{\sigma \in \bar{A}} d_\sigma k_\sigma^{-1} \mu_\sigma v_\sigma, \tag{14}$$

$$[(\chi_T)^\sim, (\chi_\sigma)^\sim] = k_\sigma^{-1} d_\sigma v_\sigma \qquad (\sigma \in \bar{A}). \tag{15}$$

5.28.   Notice that, if R is the regular representation of A, then χ_R is a
non-degenerate trace. In fact, by 5.20,

$$\chi_R = \sum_{\sigma \in \bar{A}} d_\sigma \chi_\sigma. \tag{16}$$

Suppose that the γ fixed in 5.24 was in fact $m\chi_R$, where m is some non-zero
constant in F. Then $k_\sigma = m d_\sigma$ by (16); so (14), for example, becomes:

$$m[(\chi_S)^\sim, (\chi_T)^\sim] = \sum_{\sigma \in \bar{A}} \mu_\sigma v_\sigma. \tag{17}$$

Now clearly the T of 5.27 is irreducible if and only if $\sum_\sigma v_\sigma^2 = 1$. Hence, by
(17) (with $S = T$), *a necessary and sufficient condition for the irreducibility of
T is that*

$$m[(\chi_T)^\sim, (\chi_T)^\sim] = 1. \tag{18}$$

6. Applications to Finite Groups

6.1.   Perhaps the most beautiful field of application of the preceding
generalities is the theory of representations of finite groups, of which we shall
now give a brief introductory account.

Let G be a fixed finite group of order n, with unit e. The fundamental fact
about the representations of G is Maschke's Theorem:

Maschke's Theorem.   *Assume that the characteristic of the ground field F
does not divide the order n of G. Then every representation T of G is completely
reducible.*

Proof.   By 3.12 it is enough to take an arbitary T-stable subspace Y of $X(T)$,
and find a T-stable subspace complementary to Y.

Choose any linear subspace W of $X(T)$ complementary to Y in $X(T)$; and
let P be the idempotent linear operator on $X(T)$ whose range is W and which
annihilates Y. If $x \in G$ we define a linear operator S_x on W:

$$S_x = (P \circ T_x)|W.$$

Since Y is T-stable, one verifies that S is a representation of G on W (equivalent to the quotient of T on $X(T)/Y$).

Since the characteristic of F does not divide n, n^{-1} exists in F; and we can define a linear map $Q: W \to X(T)$ by:

$$Q(\xi) = n^{-1} \sum_{x \in G} T_x^{-1} S_x \xi \qquad (\xi \in W).$$

Since $S_x\xi \equiv T_x\xi \bmod Y$ for each x in G, it follows that $Q(\xi) \equiv \xi \bmod Y$ for all ξ in W. Consequently $Q(W)$ is another linear subspace complementary to Y.

If $y \in G$ and $\xi \in W$, $T_y Q\xi = n^{-1} \sum_x T_{yx^{-1}} S_x \xi$. Replacing x by xy in this summation we get $T_y Q\xi = n^{-1} \sum_x T_{x^{-1}} S_x S_y \xi = Q S_y \xi$. Thus Q intertwines S and T. It follows that the range of Q—that is, the subspace $Q(W)$ complementary to Y—is T-stable. ∎

Remark. If the characteristic of F does divide n, the conclusion of Maschke's Theorem fails. See for example van der Waerden [2], Part II, p. 186.

6.2. Let $B(G)$ stand for the group algebra of G defined in 3.13. If the characteristic of F does not divide n, Maschke's Theorem, together with the correspondence 3.13 between the representations of G and of $B(G)$, shows that every representation of $B(G)$ is completely reducible. Therefore by 5.11 we have:

Theorem. *If the characteristic of F does not divide the order of G, $B(G)$ is semisimple.*

6.3. In view of the fundamental Theorem 6.2, the results 5.18 to 5.27 can all be applied to the finite group situation. For simplicity we shall assume for the rest of this section that *the ground field F is algebraically closed and of characteristic 0.* For example, F might be $\mathbb{C}$.

In supplying detailed proofs of the results that follow, the reader should constantly bear in mind the correspondence 3.13 between representations of G and $B(G)$.

If $\sigma \in \overline{G}$, we put $d_\sigma = \dim(\sigma)$.

6.4. Proposition. $\overline{G}$ *is a finite set; and*

$$\sum_{\sigma \in \overline{G}} d_\sigma^2 = n = \text{order of } G.$$

Proof. By 5.18 (12). ∎

6.5.   Proposition.   Card($\overline{G}$) *is equal to the number of conjugacy classes of elements of G.*

Proof.   If $a \in B(G)$ and $x \in G$, then $(a\delta_x)(y) = a(yx^{-1})$ and $(\delta_x a)(y) = a(x^{-1}y)$. So, if C is the center of $B(G)$, we have $a \in C \Leftrightarrow a\delta_x = \delta_x a$ for all x in $G \Leftrightarrow a(yx^{-1}) = a(x^{-1}y)$ for all x, y in $G \Leftrightarrow a$ is constant on each conjugacy class in G. It follows that dim(C) equals the number r of conjugacy classes in G. Combining this with 5.21 we obtain card($\overline{G}$) = r.   ∎

Remark.   Although $\overline{G}$ and the set of conjugacy classes of elements of G thus have the same number of elements, it is only rarely that there is a *natural* correspondence between the two sets. The situation is somewhat similar to that of a finite-dimensional vector space V and its adjoint $V^{\#}$. Although V and $V^{\#}$ are isomorphic as linear spaces, there is no *natural* isomorphism between them (in the absence of further structure such as an inner product).

6.6.   Since G is Abelian if and only if $B(G)$ is Abelian, we have by 5.19:

Proposition. G *is Abelian if and only if all its irreducible representations are one-dimensional.*

Remark.   Thus, for an Abelian group G, the last statement of 5.19 shows that the algebra isomorphism class of $B(G)$ depends only on the order n of G. Since two Abelian groups of the same order can very well be non-isomorphic, this shows that, in general, the algebra isomorphism class of $B(G)$ does not determine the group isomorphism class of G.

6.7.   Remark.   In simple cases, Propositions 6.4, 6.5, 6.6 may go far toward determining the dimensions of the irreducible representations of G.

For example, suppose we know that G is a non-Abelian group of order 8. By 6.4 the d_σ provide a decomposition of 8 as a sum of squares of positive integers. There are only three such decompositions:

$$8 = 1^2 + 1^2 + \cdots + 1^2,$$

$$8 = 1^2 + 1^2 + 1^2 + 1^2 + 2^2,$$

$$8 = 2^2 + 2^2.$$

Since G is non-Abelian, the first of these is ruled out by 6.6. The last is ruled out because G always has at least one one-dimensional representation, namely the trivial representation (see 3.8, Remark 2). Therefore the second decomposition is the one provided by the d_σ in this case. We conclude that $\overline{G}$ consists of four one-dimensional representations, one two-dimensional representation, and nothing else.

Another general arithmetic fact about the d_σ is the following result, which we shall not prove here: d_σ *divides n for every* σ *in* $\overline{G}$. For a proof, see Dixmier [21], 15.4.3.

6.8. Proposition. *Every element* σ *of* $\overline{G}$ *occurs in the regular representation R of G. In fact the multiplicity of* σ *in R is equal to* $\dim(\sigma)$.

Proof. This follows from 5.20. ∎

Characters of Group Representations

6.9. Definition. If T is a finite-dimensional representation of G, the *character* of T is the F-valued function $x \mapsto \mathrm{Tr}(T_x)$ on G. It will be denoted by χ_T.

Analogously with 4.14, we observe that

$$\chi_T(xy) = \chi_T(yx) \qquad (x, y \in G);$$

that is, χ_T is constant on conjugacy classes of elements of G. Also, χ_T depends only on the equivalence class of T; and we can write χ_τ for χ_T, where τ is the equivalence class to which T belongs.

6.10. If T is a finite-dimensional representation of G, and T' the corresponding representation of $B(G)$ (by 3.13), the character $\chi_{T'}$ of T' is determined from χ_T by the obvious formula:

$$\chi_{T'}(a) = \sum_{x \in G} a(x)\chi_T(x).$$

Therefore, since T is automatically completely reducible (6.1), we obtain from 4.16:

Proposition. *A finite-dimensional representation T of G is determined to within equivalence by* χ_T.

6.11. For the regular representation R of G we have:

Proposition. $\chi_R(x) = n$ *if* $x = e$; $\chi_R(x) = 0$ *if* $x \neq e$.

Proof. The first statement is obvious. The second is evident when we observe that if $x \neq e$ the matrix of R_x with respect to the basis $\{\delta_y : y \in G\}$ of $B(G)$ has zeros everywhere on the diagonal. ∎

Harmonic Analysis of Group Representations.

6.12.   We shall now apply 5.22–5.28 to the group context.

It will be convenient to use the same letter to denote both a representation of G and also the representation of $B(G)$ corresponding to it by 3.13, and to identify $\overline{G}$ with $(B(G))^{\overline{}}$.

As the γ of 5.24 we shall take:

$$\gamma(a) = a(e) \qquad\qquad (a \in B(G)).$$

Evidently γ is a non-degenerate trace on $B(G)$. In fact we have by 6.11 and 6.8:

$$\gamma = n^{-1}\chi_R = n^{-1}\sum_{\sigma\in\overline{G}} d_\sigma \chi_\sigma.$$

Therefore the k_σ of 5.23 (13) are given by:

$$k_\sigma = n^{-1}d_\sigma. \tag{1}$$

The bilinear form $[\ ,\ ]$ of 5.24 becomes:

$$[a, b] = \sum_{x\in G} a(x)b(x^{-1}) \qquad (a, b \in B(G)). \tag{2}$$

Suppose that $\alpha: G \to F$ is considered as an element of $B(G)^{\#}$:

$$\alpha(a) = \sum_{x\in G} \alpha(x)a(x) \qquad\qquad (a \in B(G)).$$

It then follows from (2) that the element $\tilde{\alpha}$ of $B(G)$ defined in 5.24 is given by

$$\tilde{\alpha}(x) = \alpha(x^{-1}) \qquad\qquad (x \in G). \tag{3}$$

Combining 5.16 and 5.25 with (1) and (3), we have:

Proposition*.   *For each σ in $\overline{G}$, the function*

$$x \mapsto n^{-1} d_\sigma \chi_\sigma(x^{-1}) \qquad\qquad (x \in G)$$

in $B(G)$ is the unit element of the minimal two-sided ideal of $B(G)$ corresponding (via 5.15) with σ.

For any σ in $\overline{G}$, any representation T of G, and any ξ in $X(T)$, the σ-component of ξ with respect to T (see 2.13) is

$$\xi_\sigma = n^{-1} d_\sigma \sum_{x\in G} \chi_\sigma(x^{-1})T_x\xi.$$

6.13.   A special case of 6.12 is the harmonic analysis of the regular representation R of G.

Proposition*. *Let* $\sigma \in \overline{G}$; *and suppose S belongs to the class* σ. *If* $a \in B(G)$, *the* σ-*component* a_σ *of* a *with respect to* R *is given by*:

$$a_\sigma(x) = n^{-1} d_\sigma \operatorname{Tr}(S_{x^{-1}} S_a).$$

6.14. From 5.27 we get:

Proposition*. *Let S and T be any two finite-dimensional representations of G. If* $\sigma \in \overline{G}$, *let* μ_σ *and* v_σ *be the multiplicities of* σ *in S and T respectively. Then*

$$\sum_{\sigma \in G} \mu_\sigma v_\sigma = n^{-1} \sum_{x \in G} \chi_S(x)\chi_T(x^{-1}), \tag{4}$$

$$v_\sigma = n^{-1} \sum_{x \in G} \chi_T(x)\chi_\sigma(x^{-1}) \qquad (\sigma \in \overline{G}). \tag{5}$$

In particular, T is irreducible if and only if

$$\sum_{x \in G} \chi_T(x)\chi_T(x^{-1}) = n. \tag{6}$$

Formula (5) is especially convenient for finding the multiplicities with which the different irreducible representations occur in any given finite-dimensional representation of G.

Some Illustrations

6.15. *Finite Abelian groups.* Let G be a finite Abelian group of order n. By 5.19 G consists of n one-dimensional representations (i.e., homomorphisms of G into $F \setminus \{0\}$).

It is well known that G is essentially a direct product of cyclic groups:

$$G = \prod_{i=1}^{r} \mathbb{Z}_{p_i} \tag{7}$$

(where $\mathbb{Z}_{p_i}$ is the cyclic group of finite order p_i). In terms of (7) it is easy to describe $\overline{G}$ explicitly: For each $i = 1, \ldots, r$ we choose a primitive p_i-th root u_i of unity in F (recall that F is algebraically closed), and a generator x_i of $\mathbb{Z}_{p_i}$. Given integers $m_1, \ldots, m_r$, define

$$\chi_{m_1, \ldots, m_r}(x_1^{q_1}, x_2^{q_2}, \ldots, x_r^{q_r}) = u_1^{m_1 q_1} u_2^{m_2 q_2}, \ldots, u_r^{m_r q_r}$$

(for any integers $q_1, \ldots, q_r$). Then $\chi_{m_1, \ldots, m_r} \in \overline{G}$; and if for each i m_i is allowed to run only over $\{0, 1, 2, \ldots, p_i - 1\}$, the $\chi_{m_1, \ldots, m_r}$ run once over $\overline{G}$.

6.16. *Fourier analysis on finite Abelian groups.* Again let G be a finite Abelian group of order n.

Proposition*.   *Let* $a \in B(G)$; *and define*

$$\hat{a}(\chi) = \sum_{x \in G} a(x)\chi(x) \qquad\qquad (\chi \in \overline{G}). \qquad (8)$$

Then

$$a(x) = n^{-1} \sum_{\chi \in \overline{G}} \hat{a}(\chi)\chi(x^{-1}) \qquad\qquad (x \in G). \qquad (9)$$

This is a special case of 6.13.

The map $\hat{a}: \overline{G} \to F$ defined by (8) is called the *Fourier transform* of a; and (9), which reconstructs a from its Fourier transform, is called the *Fourier inversion formula*. Equations (8) and (9) together constitute *Fourier analysis* on finite Abelian groups.

6.17. *Odd and even functions.*   Here is an extremely simple application of Proposition 6.12 to a special case of 2.23.

Let S be any set and π a non-trivial permutation of S of order 2 (i.e., $\pi^2 = 1 =$ identity). Let X be the family of all F-valued functions on S. Then the equation $(T_\pi f)(s) = f(\pi(s))$ $(f \in X; s \in S)$ defines a representation T of the two-element group $G = \{1, \pi\}$ on X.

Now $\overline{G}$ consists of χ_+ and χ_-, where $\chi_+(\pi) = 1$, $\chi_-(\pi) = -1$. The χ_+-subspace $[\chi_-$-subspace$]$ of X (with respect to T) consists of those f for which

$$f(\pi(s)) = f(s) \quad [f(\pi(s)) = -f(s)] \qquad \text{for all } s \text{ in } S;$$

that is, the χ_+-subspace $[\chi_-$-subspace$]$ is the space of even [odd] functions with respect to π. If f is any function in X, the last statement of Proposition 6.12 asserts that the χ_+-component f_+ and χ_--component f_- of f are given by

$$f_+(s) = \tfrac{1}{2}(f(s) + f(\pi(s))),$$

$$f_-(s) = \tfrac{1}{2}(f(s) - f(\pi(s))).$$

These are the well-known formulae for expressing an arbitrary function as a sum of an odd function and an even function.

6.18. *The dihedral groups.*   Let G be the semidirect product of the cyclic group $N = \mathbb{Z}_p = \{e, u, u^2, \ldots, u^{p-1}\}$ $(p = 3, 4, \ldots)$ and the two-element group $H = \{e, w\}$. We suppose that N and H are embedded as subgroups in G, the action of w by inner automorphism on N being:

$$wu^r w = u^{-r} \qquad\qquad (r = 0, 1, 2, \ldots).$$

G is called the *dihedral group of order 2p*. We shall find all its irreducible representations.

The elements of $\overline{N}$ are of course one-dimensional; and to each χ in $\overline{N}$ there is an "inverse" element $\chi^{-1} : m \mapsto \chi(m^{-1}) = (\chi(m))^{-1}$ of $\overline{N}$.

Given any χ in $\overline{N}$ one easily constructs a two-dimensional representation T^χ of G (in matrix form) as follows:

$$T^\chi_m = \begin{pmatrix} \chi(m) & 0 \\ 0 & \chi(m)^{-1} \end{pmatrix} \qquad (m \in N),$$

$$T^\chi_w = \begin{pmatrix} 0 & 1 \\ 1 & 0 \end{pmatrix}.$$

Notice that $T^\chi | N \cong \chi \oplus \chi^{-1}$. It follows that, if $\chi, \chi' \in \overline{N}$, then $T^{\chi'}$ and T^χ are inequivalent unless either $\chi' = \chi$ or $\chi' = \chi^{-1}$. Further, the matrix T^χ_w implements the equivalence of $T^{\chi^{-1}}$ and T^χ. Therefore

$$T^{\chi'} \cong T^\chi \qquad \text{if and only if either } \chi' = \chi \text{ or } \chi' = \chi^{-1}. \qquad (10)$$

We claim that

$$T^\chi \text{ is irreducible provided } \chi \neq \chi^{-1}. \qquad (11)$$

Indeed: Assume $\chi \neq \chi^{-1}$; and let Y be a non-trivial T^χ-stable subspace. In particular Y is stable under $T^\chi | N \cong \chi \oplus \chi^{-1}$. Since $\chi^{-1} \not\cong \chi$, it follows from 2.17 that Y is the sum of its intersections with the χ- and χ^{-1}-subspaces. But, if $\{e_1, e_2\}$ is the natural basis of F^2, the χ- and χ^{-1}-subspaces for $T^\chi | N$ are Fe_1 and Fe_2 respectively. Therefore Y, being non-trivial, is either Fe_1 or Fe_2. But Y is also stable under T^χ_w, which interchanges Fe_1 and Fe_2; and we have a contradiction, proving the claim.

Choose a primitive p-th root of unity t in F. Then $\overline{N} = \{\chi_0, \chi_1, \ldots, \chi_{p-1}\}$, where

$$\chi_q(u^r) = t^{rq} \qquad (q, r \text{ integers}).$$

We must now consider separately the cases that p is even or odd.

Assume first that p is odd. Then $(\chi_q)^{-1} = \chi_{p-q}$; and $(\chi_q)^{-1} \neq \chi_q$ for all $q = 1, \ldots, p-1$. So there are $\frac{1}{2}(p-1)$ distinct two-element sets $\{\chi_1, \chi_{p-1}\}$, $\{\chi_2, \chi_{p-2}\}, \ldots, \{\chi_{(p-1)/2}, \chi_{(p+1)/2}\}$, each consisting of an element of $\overline{N}$ and its inverse. Consequently, by (10) and (11), the $T^{\chi_1}, T^{\chi_2}, \ldots, T^{\chi_{(p-1)/2}}$ are $\frac{1}{2}(p-1)$ inequivalent irreducible two-dimensional representations of G. In addition, there are two distinct one-dimensional representations of G, the trivial representation and the representation ψ defined by: $\psi(m) = 1$ for $m \in N$; $\psi(w) = -1$. Since

$$\tfrac{1}{2}(p-1) \cdot 2^2 + 1^2 + 1^2 = 2p = \text{order of } G,$$

it follows from 6.4 that these are *all* the irreducible representations of G.

Now assume that p is even. In that case $(\chi_q)^{-1} \neq \chi_q$ provided $q = 1, \ldots, p - 1$ and $q \neq \frac{1}{2}p$. Thus the sets $\{\chi_1, \chi_{p-1}\}, \ldots, \{\chi_{(p-2)/2}, \chi_{(p+2)/2}\}$ are distinct two-element sets; and so by (10) and (11) the $T^{\chi_1}, \ldots, T^{\chi_{(p-2)/2}}$ are $\frac{1}{2}(p - 2)$ inequivalent irreducible two-dimensional representations of G. In addition we now have 4 one-dimensional representations of G. The easiest way to describe these is to observe that $M = \{e, u^2, u^4, \ldots, u^{p-2}\}$ is a normal subgroup of G, and that G/M is an Abelian 4-element group. Thus the four (one-dimensional) elements of $(G/M)^-$ give rise to four distinct one-dimensional elements of $\overline{G}$. Since

$$\tfrac{1}{2}(p - 2) \cdot 2^2 + 4 \cdot 1^2 = 2p = \text{order of } G,$$

it follows from 6.4 that these are *all* the irreducible representations of G. We now have:

Proposition.   *Let G be the dihedral group of order $2p$ $(p = 3, 4, \ldots)$. If p is odd, $\overline{G}$ consists of two one-dimensional and $\frac{1}{2}(p - 1)$ two-dimensional elements. If p is even, $\overline{G}$ consists of four one-dimensional and $\frac{1}{2}(p - 2)$ two-dimensional elements.*

Remark.   The representation theory of the dihedral groups will appear again in XII.1.24 as an example of a systematic general method of studying $\overline{G}$.

7.   The Complex Field and *-Algebras

*The Structure of Certain *-Algebras*

7.1.   In this final section of the chapter we shall take the ground field F to be $\mathbb{C}$. Since $\mathbb{C}$ is algebraically closed and of characteristic 0, every one of the results of this chapter up to now is valid.

Throughout most of this work we shall be interested in *-algebras and in *-representations of these, rather than mere algebras and representations of algebras. So it is worthwhile to ask what information can be extracted from this chapter about *-algebras.

7.2.   It is well known that, for each positive integer n, the total matrix algebra $M(n, \mathbb{C})$ becomes a *-algebra when equipped with the adjoint operation *:

$$(a^*)_{ij} = \overline{a_{ji}} \tag{1}$$

($^-$ being complex conjugation). Equipped with the operation (1), $M(n, \mathbb{C})$ is called the $n \times n$ *total matrix *-algebra*. Notice that

$$a^*a = 0 \Rightarrow a = 0 \qquad\qquad (a \in M(n\ \mathbb{C})). \qquad (2)$$

7.3.   Conversely, we have:

Proposition.   *Let $a \mapsto a'$ be another involution operation on the algebra $M(n, \mathbb{C})$, such that $M(n, \mathbb{C})$, equipped with its usual algebra structure and the operation $'$, is a *-algebra satisfying:*

$$a'a = 0 \Rightarrow a = 0 \qquad\qquad (a \in M(n, \mathbb{C})). \qquad (3)$$

Then there exists a non-singular matrix h in $M(n, \mathbb{C})$ such that

$$ha'h^{-1} = (hah^{-1})^* \qquad\qquad (4)$$

for all a in $M(n, \mathbb{C})$. That is, $M(n, \mathbb{C})$ equipped with the involution $'$ is *-isomorphic to the total matrix *-algebra.

Proof.   Abbreviate $M(n, \mathbb{C})$ to M. We shall regard M as acting on $\mathbb{C}^n$ in the usual way.

By the properties of involutions, $a \mapsto (a')^*$ is an automorphism of the algebra M. Hence by 5.10 there is a non-singular matrix g in M such that $(a')^* = g^*ag^{*-1}$, or

$$a' = g^{-1}a^*g \qquad \text{for all } a \text{ in } M. \qquad (5)$$

Since $a = a'' = g^{-1}g^*ag^{*-1}g$ for all a (by (5)), it follows that $g^{-1}g^*$ commutes with every a in M, whence

$$g^* = \lambda g \qquad \text{for some } \lambda \text{ in } \mathbb{C}. \qquad (6)$$

Let $\| \ \|$ be any norm on M under which $a \mapsto a^*$ is an isometry. Applying $\| \ \|$ to both sides of (6) we deduce that $|\lambda| = 1$. Hence, replacing g by μg, where $\mu^2 = \lambda$, we can assume that $g^* = g$. (Notice that this replacement does not invalidate (5).)

We shall now make use of the spectral resolution of the non-singular Hermitian matrix g. Suppose that some of the eigenvalues of g are positive and some negative. Then there exists a non-zero vector ξ in $\mathbb{C}^n$ such that $(g\xi, \xi) = 0$ ((,) being the ordinary inner product in $\mathbb{C}^n$). Hence the orthogonal projection p of $\mathbb{C}^n$ onto $\mathbb{C}\xi$ satisfies $pgp = 0$, or (by (5)) $p'p = g^{-1}pgp = 0$. This contradicts (3). Hence the eigenvalues of g are either all positive or all negative. This means that there is a non-singular matrix h satisfying $g = \pm h^*h$. Combining this with (5) we get (4).   ∎

7.4.   Theorem. *Let A be a finite-dimensional *-algebra (with involution *) satisfying*

$$a^*a = 0 \Rightarrow a = 0 \qquad\qquad (a \in A). \qquad (7)$$

Then A is *-isomorphic to a finite direct sum of total matrix *-algebras.

Note: If $A_1, \ldots, A_r$ are *-algebras, the algebra direct sum $\sum_i^\oplus A_i$, equipped with the involution $\langle a_1, \ldots, a_r \rangle \mapsto \langle a_1^*, \ldots, a_r^* \rangle$, is called the *direct sum of the *-algebras A_i.*

Proof.   First we claim that the algebra underlying A is semi-simple. Indeed, otherwise there is a non-zero element a of the radical R of A. Thus $a^*a \in R$, so that by 5.4 $(a^*a)^{2^n} = 0$ for some positive integer n. On the other hand, by iterating (7) we find that $(a^*a)^{2^n} \neq 0$ for all positive integers n. This contradiction proves the claim.

Thus it follows from 5.14 that (as an algebra) A is the algebra direct sum of its minimal two-sided ideals. Now let J be a minimal two-sided ideal of the algebra underlying A; then the properties of * imply that J^* is also a minimal two-sided ideal. If $J^* \neq J$, then J^* and J are linearly independent (by 5.14) and so $J^*J \subset J^* \cap J = \{0\}$, whence $a^*a = 0$ for all a in J. This contradicts (7), showing that $J^* = J$. So each minimal two-sided ideal J is stable under *, hence a *-algebra in its own right.

By 5.14 each minimal two-sided ideal J, as an algebra, is isomorphic to a total matrix algebra. Combining this fact with (7) and Proposition 7.3, we conclude that J is *-isomorphic with a total matrix *-algebra. Therefore, since A is the direct sum of its minimal two-sided ideals, the proof is complete.   ■

Remark.   The involution of a *-algebra is often said to be *proper* when it satisfies (7).

Finite-Dimensional *-Representations

7.5.   For the rest of this section we will be discussing a *-algebra A. The algebra underlying A will be called A_0.

Definition.   Let X be a finite-dimensional Hilbert space. A *-*representation of A on X* is a representation T of A_0 on X satisfying the additional condition that $T_{(a^*)} = (T_a)^*$ (adjoint operator) for all a in A.

Thus a *-representation of A on X is a *-homomorphism of A into the *-algebra of all linear operators on X (the involution being the adjoint operation).

As usual we denote by $X(T)$ the Hilbert space on which T acts.

The inner product of the Hilbert space X is denoted by $(\ , \ )_X$, or more simply by $(\ , \ )$.

7.6. All the concepts defined in §1 and §3 for representations apply in particular to *-representations. In addition, we shall need the relation of unitary equivalence between *-representations:

Definition. Two *-representations T' and T of A, acting on (finite-dimensional) Hilbert spaces X' and X respectively, are *unitarily equivalent* if there exists a linear isometry f of X' onto X such that

$$T_a \circ f = f \circ T'_a \qquad (a \in A);$$

that is, if they are equivalent, and the equivalence is implemented by an isometry.

In the future we shall refer to the equivalence defined in 1.6 as *algebraic equivalence*, to distinguish it from the stronger relation of unitary equivalence.

7.7. If $T^1, \ldots, T^r$ are *-representations of A acting on (finite-dimensional) Hilbert spaces $X_1, \ldots, X_r$ respectively, then the direct sum representation $T = \sum_i^{\oplus} T^i$, acting on the Hilbert space direct sum $X = \sum_i^{\oplus} X_i$, is obviously a *-representation. We call it the *Hilbert direct sum* $\sum_i^{\oplus} T^i$.

7.8. The first fundamental fact about *-representations is that they are completely reducible with *orthogonal* irreducible subspaces.

Proposition. *Let T be a finite-dimensional *-representation of A. Then $X(T)$ can be written as a sum of orthogonal T-stable T-irreducible subspaces.*

Proof. If Y is any T-stable subspace, we claim that $Y^{\perp}$ (the orthogonal complement of Y) is also T-stable. Indeed, if $a \in A$, $\xi \in Y$, and $\eta \in Y^{\perp}$, we have $(T_a \eta, \xi) = (\eta, (T_a)^* \xi) = (\eta, T_{a^*} \xi) = 0$ since $T_{a^*} \xi \in Y$. Fixing a and η and letting ξ vary, we conclude from this that $T_a \eta \in Y^{\perp}$. By the arbitrariness of a and η this implies the claim.

In view of this claim, we can continue breaking down $X(T)$ into successively smaller and smaller mutually orthogonal stable subspaces until (by the finite-dimensionality of $X(T)$) we are brought to a stop by the irreducibilty of all the subspaces so obtained. This completes the demonstration. ∎

7.9.   Let $M = M(n, \mathbb{C})$ be the $n \times n$ total matrix *-algebra. If $\mathbb{C}^n$ is made into a Hilbert space by the usual inner product $(\xi, \eta) = \sum_{i=1}^{n} \xi_i \overline{\eta_i}$, the natural action of M on $\mathbb{C}^n$ is an irreducible *-representation of M, which we shall call W.

Proposition.   *To within unitary equivalence, W is the only (finite-dimensional) irreducible *-representation of M.*

The proof is a close imitation of the proof of 5.7. (Notice especially that the $\xi^1, \ldots, \xi^n$ of the proof of 5.7 form an orthonormal set in the present context, provided $\|\xi\| = 1$.) We leave the details to the reader.

7.10.   Theorem.   *Let T and T' be two algebraically equivalent (finite-dimensional) *-representations of A. Then T and T' are unitarily equivalent.*

Proof.   First assume that T and T' are irreducible. Since they are equivalent, they have the same kernel J. By Burnside's Theorem (4.10) A/J is *-isomorphic with a total matrix *-algebra. So the unitary equivalence of T and T' is obtained by applying 7.9 to the corresponding *-representations of A/J.

In the general case, by 7.8 T and T' are Hilbert direct sums of irreducible *-representations $\{T^i\}$ and $\{T'^j\}$ respectively. Since T and T' are algebraically equivalent, Theorem 2.20 enables us to set up a one-to-one correspondence between the $\{T^i\}$ and the $\{T'^j\}$ such that, if T^i and T'^j correspond, they are algebraically equivalent, and hence by the preceding paragraph unitarily equivalent. Thus T and T', being Hilbert direct sums of pairwise unitarily equivalent *-representations, are themselves unitarily equivalent.   ∎

Remark.   There is another proof of this Theorem, more direct and much more capable of generalization, based on the polar decomposition of operators. See Mackey [15].

7.11.   Corollary.   *A finite-dimensional *-algebra A has at most $\dim(A)$ unitary equivalence classes of irreducible *-representations.*

Proof.   By 7.10 and 5.11.   ∎

7.12.   Proposition.   *Suppose that A is a finite-dimensional *-algebra satisfying (7). Then every finite-dimensional representation T of A_0 is algebraically equivalent to some *-representation of A. (In other words, there exists an inner product making $X(T)$ a Hilbert space and T a *-representation.)*

Proof. By 7.4 and 5.11 T is completely reducible. Thus we have only to prove the Proposition under the assumption that T is either a zero representation or irreducible. If it is a zero representation the assertion is obvious. Assume that T is irreducible. By 5.15 T is essentially just the unique irreducible representation of one of the minimal two-sided ideals J of A_0. But by 7.4 J is a total matrix *-algebra; and so by the opening remarks of 7.9 J has an irreducible *-representation, which must be algebraically equivalent to T. ∎

Remark. This Proposition fails if either of the two hypotheses on A (finite-dimensionality and Condition (7)) fails.

Here is an example in which A is finite-dimensional but (7) fails. Let $A = \mathbb{C}^2$, with multiplication $\langle u, v \rangle \langle u', v' \rangle = \langle uu', vv' \rangle$ and involution $\langle u, v \rangle^* = \langle \bar{v}, \bar{u} \rangle$. Putting $a = \langle 1, 0 \rangle$, we have $a^*a = 0$. Let T be a faithful (finite-dimensional) representation of A. If there were an inner product making T a *-representation, we would have $T_a \neq 0$, hence $0 \neq (T_a)^* T_a = T_{a^*a} = 0$, a contradiction.

Examples where A is infinite-dimensional and (7) holds, but the Proposition fails, will be pointed out in 7.21 in the context of infinite groups.

7.13. Assume that A is finite-dimensional and satisfies (7). Combining 7.10 and 7.12 we see that the algebraic equivalence classes of finite-dimensional representations of A_0 are essentially the same objects as the unitary equivalence classes of finite-dimensional *-representations of A. Indeed, every algebraic equivalence class of finite-dimensional representations of A_0 contains one and only one unitary equivalence class of finite-dimensional *-representations of A.

7.14. Corollary. *If A is finite-dimensional and satisfies (7), and if T is a finite-dimensional representation of A_0, then*

$$\chi_T(a^*) = \overline{\chi_T(a)} \qquad\qquad (a \in A).$$

Proof. Combine 7.12 with the observation that $\mathrm{Tr}(g^*) = \overline{\mathrm{Tr}(g)}$ for any g in $M(n, \mathbb{C})$. ∎

7.15. Proposition*. *Let A be a finite-dimensional *-algebra satisfying (7); and adopt the constructions and notation of Proposition 5.25. If in addition S is a *-representation belonging to the class σ, then the canonical basis $\{e^{pq}\}$ $(p, q = 1, \ldots, d_\sigma)$ of J_σ satisfies $(e^{pq})^* = e^{qp}$.*

The Group *-Algebra and Unitary Group Representations

7.16. Let G be any group (finite or infinite), with unit e. In 3.13 we defined the (discrete) group algebra $B(G)$ of G. Now that $F = \mathbb{C}$, we can define an involution * on $B(G)$ as follows:

$$f^*(x) = \overline{f(x^{-1})} \qquad\qquad (x \in G). \qquad (8)$$

One easily verifies that $B(G)$ is a *-algebra under the operation (8); we call it the *(discrete) group *-algebra of G.*

Notice that $(\delta_x)^* = \delta_{x^{-1}}$ $(x \in G)$.

If $0 \neq f \in B(G)$, we have

$$(f^*f)(e) = \sum_{x \in G} |f(x)|^2 \neq 0;$$

and so $B(G)$ *satisfies Condition* (7).

7.17. Definition. Let X be a finite-dimensional Hilbert space. A *unitary representation* of G on X is a representation T of G on X such that T_x is a unitary operator for each x in G.

Let T be a representation of G on the finite-dimensional Hilbert space X. Since $T_{(x^{-1})} = (T_x)^{-1}$, T will be unitary if and only if $T_{x^{-1}} = (T_x)^*$ for each x in G. If T' is the representation of $B(G)$ corresponding to T by 3.13, this implies that T' is a *-representation if and only if T is unitary.

Thus, by 3.13, the unitary representations of G are in natural one-to-one correspondence with the non-degenerate *-representations of $B(G)$.

7.18. Definition. Two unitary representations T and S of G are *unitarily equivalent* if there is a linear isometry of $X(T)$ onto $X(S)$ which intertwines T and S.

Thus T and S are unitarily equivalent if and only if the corresponding *-representations T' and S' of $B(G)$ are unitarily equivalent.

7.19. Suppose now that G is a finite group. Then $B(G)$ is finite-dimensional and satisfies (7); so 7.13 (with the correspondence 7.17) implies the following conclusion:

Every algebraic equivalence class of finite-dimensional representations of G contains exactly one unitary equivalence class of (finite-dimensional) unitary representations of G.

Thus the results of §6 can all be applied to the unitary representation theory of G. The reader can easily make this statement precise.

7.20. From 7.19 we deduce (as in 7.14) that, if T is any finite-dimensional representation of the finite group G,

$$\chi_T(x^{-1}) = \overline{\chi_T(x)} \qquad\qquad (x \in G). \qquad (9)$$

Thus, equations (4), (5) of 6.14 can now be written in the form

$$\sum_{\sigma \in \overline{G}} \mu_\sigma v_\sigma = n^{-1} \sum_{x \in G} \chi_S(x)\overline{\chi_T(x)}, \qquad (10)$$

$$v_\sigma = n^{-1} \sum_{x \in G} \chi_T(x)\overline{\chi_\sigma(x)} \qquad\qquad (\sigma \in \overline{G}). \qquad (11)$$

Equation (6) of 6.14 asserts that T is irreducible if and only if

$$\sum_{x \in G} |\chi_T(x)|^2 = n = \text{order of } G. \qquad (12)$$

7.21. *Remark.* If G is not finite, then in general there will exist finite-dimensional, even one-dimensional, representations which are not equivalent to a unitary representation. For example, if G is the additive group of the integers and $\chi(m) = e^m$ ($m \in G$), χ is a one-dimensional representation but is certainly not unitary. Likewise, the corresponding representation $f \mapsto \sum_{m \in G} f(m)e^m$ of $B(G)$ is not a *-representation.

8. Exercises for Chapter IV

Note: All algebras in these exercises are associative and lie over a commutative field F unless stated otherwise.

1. Is $x \mapsto \begin{pmatrix} \cos x & -\sin x \\ \sin x & \cos x \end{pmatrix}$ a representation of the additive group of reals? Interpret geometrically.

2. Consider the symmetric group S_n and let E be an n-dimensional vector space over a field F. If $\{v_1, v_2, \ldots, v_n\}$ is a basis for E, define $T_\sigma: E \to E$ by $T_\sigma(v_i) = v_{\sigma(i)}$ for $\sigma \in S_n, 1 \le i \le n$, and extend by linearity. Prove that $\sigma \mapsto T_\sigma$ is a representation T of S_n. Is T faithful? Is T irreducible?

3. Let M be a finite collection of non-singular $n \times n$ matrices over a field F which is closed under multiplication. Show that F^n is completely reducible under M.

4. Let G be a cyclic group of prime order p with fixed generator x, and let F be a field of characteristic p. Show that $T = \begin{pmatrix} 1 & 1 \\ 0 & 1 \end{pmatrix}$ satisfies $T^p = I$, and that $x \mapsto T$ defines a representation of G which is not irreducible. Is T completely reducible?

5. List all one-dimensional representations of the symmetric group S_3.

6. Is the left regular representation of a finite group ever irreducible?

7. Let T be a representation of an algebra A with space $X(T)$. Show that if for any two vectors ξ and η in $X(T)$ with $\xi \neq 0$ there exists $a \in A$ such that $T_a \xi = \eta$, then T is irreducible (see 3.7).

8. Give a simple direct proof of Lemma 3.11 and of Corollary 3.12 when the operator set T is finite dimensional.

9. Let G be a cyclic group of order three. Show that the discrete group algebra $B(G)$ over a field F has nonzero divisors of zero (an element a of a ring R is a *divisor of zero* if there exists a non-zero element b in R such that $ab = 0$ or a non-zero element c in R such that $ca = 0$).

10. Suppose that T is a representation of a group G; let H be a normal subgroup of finite index not divisible by the characteristic of the ground field F. Show that if $T|H$ is completely reducible, then T is completely reducible.

11. Let G be a finite group. Show that a linear subspace L of $B(G)$ is a left ideal of $B(G)$ if and only if L is an invariant subspace of the left regular representation.

12. Let E be a finite dimensional linear space over a field F. Show that the algebra of linear operators on E is simple.

13. Let A be the algebra of all complex 3×3 matrices of the form

$$\begin{pmatrix} a_{11} & a_{12} & a_{13} \\ 0 & a_{22} & a_{23} \\ 0 & 0 & a_{33} \end{pmatrix}$$

Determine the radical of A.

14. Let A be an algebra over a field F and I an ideal in A. Show that the algebra $(A/I)_n$ of $n \times n$ matrices over A/I is isomorphic to the algebra A_n/I_n.

15. Prove that if I is a minimal right ideal in an algebra A, then either $I^2 = \{0\}$ or $I = eA$, where e is an idempotent in A.

16. Let I be a two-sided ideal in an algebra A, and f a homomorphism of I onto an algebra B with unit. Prove that if I is contained in the center of A, then there exists a unique homomorphism $g: A \to B$ such that $g|I = f$.

17. Give an example to show that a subrepresentation of a non-degenerate representation may be degenerate. Also, give an example of a quotient representation of a non-degenerate representation which is degenerate (see 3.17).

18. Let A be a complex algebra, and T an irreducible representation of A on a complex infinite-dimensional space such that, for some $a \in A$, T_a is non-zero and has finite-dimensional range. Prove that T is totally irreducible (see 4.5).

19. Prove Proposition 4.13.

20. Give proofs of Propositions 5.18, 5.19, 5.20, and 5.21.

21. Find all irreducible representations, over the complex field, of a cyclic group of order n.

22. Let G be a non-cyclic finite Abelian group. Show that no one-dimensional representation of G is faithful.

23. Prove Proposition 5.23.

24. Prove Proposition 5.25.

25. Let G be a finite group of order n with unit element e. Show that, if $x \in G$,

$$\sum_{\sigma \in \widehat{G}} d_\sigma \chi_\sigma(x) = \begin{cases} n, & \text{if } x = e \\ 0, & \text{if } x \neq e. \end{cases}$$

(All representations are over the complex field.)

26. Let G be a finite group. By 6.5 the elements $\sigma^1, \sigma^2, \ldots, \sigma^m$ of $\widehat{G}$ are equal in number to the conjugacy classes $\xi^1, \xi^2, \ldots, \xi^m$ in G. An $m \times m$ table giving the values of $\chi_{\sigma^i}(\xi^j)$, where i numbers the rows and j the columns, is called the *character table* of G. (Here, of course, $\chi_{\sigma^i}(\xi^j)$ means the constant value of χ_{σ^i} on ξ^j; see 6.9.)

In this problem all representations are over the complex field.

(A) The dihedral group of order 6 (i.e., the symmetric group S_3 on three objects) has 3 irreducible representations I, A, T and 3 conjugacy classes. Show that the character table is

	$\{e\}$	$3 = 1 + 2$	$3 = 3$
I	1	1	1
A	1	-1	1
T	2	0	-1

(Here the conjugacy classes, as with the symmetric group on any number n of objects, are described by the corresponding partitions of n as a sum of positive integers.)

(B) Let G be either the dihedral group of order 8 or the quaternion group of order 8 (the latter group consisting of the quaternions ± 1, $\pm i$, $\pm j$, $\pm k$ under multiplication). Show that the character table of both of these groups is the same (with suitable labelling of the elements of $\overline{G}$ and the conjugacy classes in each), namely:

1	1	1	1	1
1	1	1	-1	-1
1	1	-1	1	-1
1	1	-1	-1	1
2	-2	0	0	0

(C) Let G be the symmetric group S_4 on 4 objects. The five conjugacy classes in G are indexed by the partitions of 4 as a sum of positive integers:

$\xi_1 : 4 = 1 + 1 + 1 + 1$ (the class consisting of the unit e only),

$\xi_2 : 4 = 2 + 1 + 1$,

$\xi_3 : 4 = 2 + 2$,

$\xi_4 : 4 = 3 + 1$,

$\xi_5 : 4 = 4$.

Show that the character table of G is

	ξ_1	ξ_2	ξ_3	ξ_4	ξ_5
I	1	1	1	1	1
A	1	-1	1	1	-1
T	2	0	2	-1	0
U	3	1	-1	0	-1
V	3	-1	-1	0	1

[Hint for (C): Let N be the 4-element normal subgroup of G consisting of the unit e and the three elements of the conjugacy class corresponding to $4 = 2 + 2$. Then $G/N \cong S_3$; so the three elements of $(S_3)^\frown$ can be lifted to G, giving the first three rows of the above table. To get a three dimensional irreducible representation, let S be the representation of G acting on $\mathbb{C}^4$ by: $(S_p a)_i = a_{p^{-1}(i)}$ $(i = 1, 2, 3, 4; p \in G)$. This S is not irreducible, since $\mathbb{C}(1, 1, 1, 1)$ is stable under S; but the three-dimensional subrepresentation complementary to $\mathbb{C}(1, 1, 1, 1)$ is irreducible. Thus we obtain a fourth element of $\overline{G}$. To get the character of the fifth element V of $\overline{G}$, now use Exercise 25.]

(D) Let G be the group A_4 of all *even* permutations of 4 objects. This has four conjugacy classes and hence four irreducible representations. Show that the character table of G (with suitable labelling of the elements of $\overline{G}$ and the conjugacy classes) is:

1	1	1	1
1	1	ω	ω^2
1	1	ω^2	ω
3	-1	0	0

(where ω is a primitive cube root of unity).

27. Prove Proposition 6.12.

28. Show that the number of distinct one-dimensional representations of a finite group must divide the order of the group.

29. If S and T are two representations (over the complex field) of the finite group G, then by $S \otimes T$ (the *inner tensor product* of S and T) we mean the representation of G acting on the tensor product linear space $X(S) \otimes X(T)$, and given by

$$(S \otimes T)_x = S_x \otimes T_x \qquad\qquad (x \in G).$$

(I) If S and T are finite-dimensional, show that

$$\chi_{S \otimes T}(x) = \chi_S(x)\chi_T(x) \qquad\qquad (x \in G).$$

(II) In general $S \otimes T$ will not be irreducible even if S and T are irreducible. However, if the character table of G is known, we can use 6.14 and the preceding (I) to decompose $S \otimes T$ as a direct sum of irreducibles for each S, $T \in \overline{G}$. This gives us the *tensor product multiplication table* for G. For example, for the dihedral group of order 6 (see Exercise 26(A)) we get the simple multiplication table:

$$I \otimes I = I, I \otimes A = A, I \otimes T = T,$$

$$A \otimes A = I, A \otimes T = T,$$

$$T \otimes T = I \oplus A \oplus T.$$

Use Exercise 26(B), (C), (D) above to construct the tensor product multiplication tables for the quaternion group, the dihedral group of order 8, S_4, and A_4.

30. Let r be a positive integer, X an n-dimensional complex linear space, and let

$$X^{\otimes r} = X \otimes X \otimes \cdots \otimes X \quad (r \text{ times}).$$

Let G be the symmetric group S_r of all permutations of $\{1, \ldots, r\}$; and let $Q^{(r;n)}$ be the natural representation of G on $X^{\otimes r}$:

$$Q_p^{(r;n)}(\xi_1 \otimes \xi_2 \otimes \cdots \otimes \xi_r) = \xi_{p^{-1}(1)} \otimes \xi_{p^{-1}(2)} \otimes \cdots \otimes \xi_{p^{-1}(r)} \qquad (p \in G; \xi_i \in X).$$

The harmonic analysis of $Q^{(r;n)}$ is very important in the representation theory of $GL(n, \mathbb{C})$.

(I)   If p belongs to the conjugacy class in S_r corresponding to the partition

$$r = r_1 + r_2 + \cdots + r_s \quad (r_1, \ldots, r_s \text{ positive integers}),$$

   show that

$$\chi_{Q^{(r;n)}}(p) = n^s.$$

(II)   Using (I) and the character tables of Exercise 26, find the irreducible decomposition of $Q^{(r;n)}$ for $r = 2, 3, 4$ and arbitrary n.

   [*Note*: Further exercises on the representation theory of finite groups will appear in Chapter IX, after we have defined induced representations; see Exercises 36 and 37, Chapter IX.]

31. An involution $a \mapsto a^*$ in a (not necessarily finite-dimensional) *-algebra A is said to be *proper* if $a^*a = 0$ implies $a = 0$ (see Remark 7.4). The involution is said to be *quasi-proper* if $a^*a = 0$ implies $aa^* = 0$.

(a)   Show that every proper involution is quasi-proper;

(b)   Give an example of a quasi-proper involution which is not proper;

(c)   Give an example of an involution which is not quasi-proper;

(d)   Show that if the involution is proper, then $ab = 0$ if and only if $a^*ab = 0$;

(e)   Show that if A has a unit e and the involution is proper, then $a^*a = -e$ implies $aa^* = -e$.

32. Prove that a *-algebra with proper involution has no nonzero left or right nilpotent ideals. (An ideal I is called *nilpotent* if $I^n = \{0\}$ for some positive integer n, where I^n denotes the set of all finite sums of products of n elements taken from I).

33. Let A be any *-algebra. A finite-dimensional representation T of A (that is, of the associative algebra underlying A) is said to be *essentially involutory* if there exists an inner product on $X(T)$ making T a *-representation of A. Prove the following:

(I)   Suppose A is finite-dimensional and the involution is proper (see Exercise 31). Then every finite-dimensional representation of A is essentially involutory.

(II)   Let T be any finite-dimensional representation of A. For T to be essentially involutory it is necessary and sufficient that the following two conditions hold: (a) $\text{Ker}(T)$ is closed under *; (b) if $a \in A$ and $T_{a^*a} = 0$ then $T_a = 0$.

(III)   Given an example showing that condition (II) (a) is not sufficient to make T essentially involutory, even if the condition of (I) holds. [Hint to Part (III): Let $n \geq 2$; and let b be an invertible $n \times n$ Hermitian matrix such that $(b^{-1}\xi, \xi) = 0$ for some unit vector $\xi \in \mathbb{C}^n$. Now let A be the *-algebra of all $(n \times n)$-matrix-valued functions f which are continuous on the closed unit disk $\{u : |u| \leq 1\}$ and analytic on the open unit disk and satisfy $f(-i) =$

$b^{-1}f(i)b$; the *-operation in A will be given by $f^*(z) = (f(\bar{z}))^*$. Show that A satisfies the condition of (I), and that the representation $T: f \mapsto f(i)$ of A satisfies (II) (a) but not (II) (b).]

Notes and Remarks

The results presented in this chapter are classical. A special case of the Density Theorem (4.3) (namely the case when $r = 1$) was proved in 1945 by Jacobson [1]. The case for arbitrary r can be found (in somewhat different language) in Bourbaki [7, §4, no. 2, Théorème 1]. Burnside's Theorem (see Corollary 4.10) can be traced back to the 1905 paper of Burnside [1]. The central importance of the concept of a radical (as given in 5.1) was pointed out in the classical paper of Jacobson [3]. A detailed exposition of the radical and related results can be found in Herstein [2] and Jacobson [4]. Wedderburn's Theorem 5.11 was proved in 1908 by Wedderburn [1]. The fundamental result 6.1 due to Maschke [1] was established in 1899. General references which treat the material of the first six sections of this chapter are Albert [1], Bourbaki [7], Burrow [1], Curtis and Reiner [1, 2], Dornhoff [1], Keown [1], Lang [1], Serre [1], and van der Waerden [2].

Abstract *-algebras over the complex field were first introduced in 1943 by Gelfand and Naimark [1] in their work characterizing norm-closed *-subalgebras of bounded linear operators on a Hilbert space by a few simple axioms. The theory presented in §7 is a forerunner to the infinite-dimensional representation theory of Banach *-algebras and C*-algebras given in Chapter VI.

V   Locally Convex Representations and Banach Algebras

In this chapter all linear spaces and algebras are over the complex field.

Our purpose now, roughly speaking, is to extend the algebraic results of Chapter IV to a topological setting. We begin in §1 by redefining the basic concepts of representation theory for locally convex representations, that is, representations acting by continuous operators on a locally convex space. A few of the less profound results of Chapter IV have more or less natural generalizations to the topological context (see §§1, 2); but generally speaking, our progress is disappointing. Hardly any trace of multiplicity theory remains valid (see 1.20); and (see 1.19) there is no known analogue of the Extended Jacobson Density Theorem (IV.4.3), from which so many beautiful results flowed in Chapter IV. Evidently locally convex repesentations form too general a category for most purposes; and in succeeding chapters we shall restrict attention almost entirely to the much narrower category of *-representations of *-algebras. The main purpose of this chapter is to present those results about topological non-involutory representations and algebras which will be needed for the involutory theory in the remainder of the work.

In searching for infinite-dimensional topological versions of the structure theory of finite-dimensional semisimple algebras (see §IV.5), we must of course expect to topologize not only the representation spaces but the algebras themselves. The most useful kind of topologized algebras are Banach algebras. One of the most important results on Banach algebras is Mazur's Theorem (6.12), which states that a Banach *division* algebra is one-dimensional. This of course generalizes the fact that a finite-dimensional division algebra over $\mathbb{C}$ is one-dimensional, which was so fruitful in §§4, 5 of Chapter IV. As we shall show in §7, Mazur's Theorem is the main tool for developing Gelfand's presentation of a semisimple commutative Banach algebra as an algebra of continuous complex functions on a locally compact Hausdorff space—a development which generalizes IV.5.19.

Probably the most striking result of §6 is Gelfand's Theorem (6.17), which relates the norm in a Banach algebra to the spectrum of its elements.

Of the sections of this chapter not yet mentioned, §3 is a brief discussion of the so-called Naimark relation between locally convex representations. This relation has turned out (in view of 3.6) to be of great importance for the classification of (non-unitary) representations of many important groups. §4 and §5 are preparatory to §6. §8 gives the merest hint of the scope of a vast branch of modern functional analysis, the theory of function algebras, and proves one or two useful facts about one special kind of function algebra, namely $\mathscr{C}_0(S)$. §9 contains a proof of the Cohen–Hewitt factorization theorem for Banach algebras with approximate unit.

1. Locally Convex Representations; Fundamental Definitions

1.1. Throughout §1 to §3 A is fixed (complex) algebra.

Definition. A *locally convex representation of A* is a representation T of A (see IV.3.3) whose space $X(T)$ is an LCS and such that $T_a: X(T) \to X(T)$ is continuous for every a in A.

Remark. Though it does not appear explicitly in the notation, the topology of $X(T)$ is to be included as part of the content of the concept of a locally convex representation.

Remark. From now on the term "*algebraic representation*" will be used for a representation in the purely algebraic sense of IV.3.3.

If T is a locally convex representation, the *algebraic representation underlying T* is the algebraic representation that results from forgetting the topology of $X(T)$.

A finite-dimensional linear space X has a unique topology making it an LCS; and all linear operators on X are continuous in that topology. So there is no difference between algebraic and locally convex representations in the finite-dimensional context.

1.2. *Definition.*   Let T be a locally convex representation of A. If $X(T)$ is a Fréchet space (i.e., a complete metrizable LCS), T is called a *Fréchet representation*. If $X(T)$ is a Banach space, T is a *Banach representation*. If A is a *-algebra, $X(T)$ is a Hilbert space, and $T_{(a*)} = (T_a)^*$ for every a in A, then T is a *-*representation* of A.

Remark.   Finite-dimensional *-representations were discussed briefly in Chapter IV, §7.

1.3.   Let T be a locally convex representation of A. If Y is a T-stable subspace of $X(T)$, the subrepresentation $^Y T$ of T acting on Y is a locally convex representation with respect to the relativized topology of Y. If the linear subspace Y is closed, the quotient space $\tilde{X} = X(T)/Y$ is an LCS under the quotient topology; and the quotient $\tilde{T}$ of T on $\tilde{X}$ (defined algebraically in IV.1.4) is a locally convex representation with respect to the quotient topology.

In this connection, notice (from the continuity of the T_a) that the closure of a T-stable linear subspace is T-stable.

1.4. *Definition.*   Let T and S be two locally convex representations of A. A linear map $f: X(T) \to X(S)$ which is continuous and intertwines the algebraic representations underlying T and S is said to *intertwine T and S*, or to be T, S *intertwining*.

If there exists a linear homeomorphism f of $X(T)$ onto $X(S)$ which intertwines T and S, T and S are called *homeomorphically equivalent* (*under f*)—in symbols, $T \cong S$.

1.5. *Definition.*   Let $T^1, \ldots, T^n$ be finitely many locally convex representations of A. If we give to $\sum_{i=1}^{n \oplus} X(T^i)$ the LCS direct sum topology, then the direct sum representation $\sum_{i=1}^{n \oplus} T^i$ (defined algebraically in IV.1.10) becomes a locally convex representation, called the *locally convex direct sum of the T^i*.

1.6. *Definition.*   Let T be a locally convex representation of A. A vector ξ in $X(T)$ is called a *cyclic vector for T* if $\{T_a \xi : a \in A\}$ is dense in $X(T)$.

Remark.   By the Hahn–Banach Theorem, a linear subspace Y of $X(T)$ is dense in $X(T)$ if and only if $\alpha \in X(T)^*$, $\alpha(Y) = \{0\} \Rightarrow \alpha = 0$. Therefore ξ is cyclic for T if and only if there is no non-zero functional α in $X(T)^*$ such that $\alpha(T_a \xi) = 0$ for all a in A.

1.7. *Definition*. A locally convex representation T of A is *non-degenerate* if the following two conditions hold: (i) There is no non-zero vector ξ in $X(T)$ such that $T_a\xi = 0$ for all a in A; and (ii) the linear span of $\{T_a\xi : a \in A, \xi \in X(T)\}$ is dense in $X(T)$. Otherwise T is *degenerate*.

Remark. If A has a unit element 1, T is non-degenerate if and only if T_1 is the identity operator on $X(T)$.

Remark. If T is a non-degenerate locally convex representation of A, then T is still non-degenerate when restricted to the linear span B of $\{ab : a, b \in A\}$ (which is of course a two-sided ideal of A).

1.8. *Definition*. Let T be a locally convex representation of A. We say that T is *irreducible* if (i) $T_a \neq 0$ for some a (that is, T is not a zero representation), and (ii) there is no closed T-stable linear subspace of $X(T)$ other than $\{0\}$ and $X(T)$.

1.9. Proposition. *A locally convex representation T of A is irreducible if and only if $X(T) \neq \{0\}$ and every non-zero vector ξ in $X(T)$ is cyclic for T.*
The proof is similar to that of IV.3.7.

1.10. *Definition*. A locally convex representation T of A is *totally irreducible* if (i) $X(T) \neq \{0\}$, and (ii) for each positive integer n, each $2n$-termed sequence $\xi_1, \ldots, \xi_n, \eta_1, \ldots, \eta_n$ of elements of $X(T)$ such that the $\xi_1, \ldots, \xi_n$ are linearly independent, and each neighborhood U of 0 in $X(T)$, there exists an element a of A such that $T_a\xi_i - \eta_i \in U$ for all $i = 1, \ldots, n$.

Remark. Thus, T is totally irreducible if and only if $X(T) \neq \{0\}$ and, for each positive integer n and each n-termed sequence $\xi_1, \ldots, \xi_n$ of linearly independent vectors in $X(T)$, the vector $\xi_1 \oplus \cdots \oplus \xi_n$ is cyclic for the locally convex direct sum $T \oplus \cdots \oplus T$ (n times).

1.11. Evidently total irreducibility implies irreducibility (by 1.9), and irreducibility implies non-degeneracy.
In view of 1.9 and Remarks 1.6 and 1.10, each of these three properties depends on the topology of $X(T)$ only through $X(T)^*$.

1.12. Let T be a locally convex representation, and T^0 the algebraic representation underlying it. If T^0 is non-degenerate [irreducible, totally irreducible] in the purely algebraic sense of IV.3.17 [IV. 3.6, IV.4.5], then T is non-degenerate [irreducible, totally irreducible] in the topological sense of this section. But the converse is far from true (see 1.14).

It is useful to call T *algebraically non-degenerate* [*algebraically irreducible, algebraically totally irreducible*] if T^0 has these properties in the purely algebraic sense. For additional clarity, we may sometimes refer to the properties defined in 1.7, 1.8, and 1.10 as *topological non-degeneracy, topological irreducibility*, and *topological total irreducibility*.

1.13. Remark. Every linear space X becomes an LCS when equipped with the weak topology generated by $X^{\#}$ (so that $\xi_i \to \xi$ in X if and only if $\alpha(\xi_i) \to \alpha(\xi)$ for all α in $X^{\#}$). We shall denote this LCS by X_ω. Every linear subspace of X is closed in X_ω; and every linear map $f: X \to Y$ (X, Y being linear spaces) is continuous from X_ω to Y_ω. Thus each algebraic representation T of A becomes automatically a locally convex representation, which we shall call T^ω, acting on $X(T)_\omega$.

Let T be an algebraic representation of A. The reader will verify without difficulty that T is non-degenerate [irreducible, totally irreducible] in the sense of Chapter IV if and only if T^ω is non-degenerate [irreducible, totally irreducible] in the sense of the present section. Thus, if we regard T and T^ω as identified, *the purely algebraic definitions of Chapter IV become* (*in case $F = \mathbb{C}$*) *special cases of our present definitions.*

1.14. Remark. Given a locally convex representation T of A, we thus have four notions of "irreducibility," the following implications clearly holding between them:

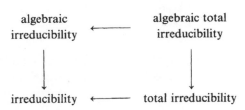

The reader should verify, by giving examples, that no other implications hold except those indicated above. (To see that total irreducibility does not imply algebraic irreducibility, let X be a Hilbert space with a countable infinite orthonormal basis $\{\xi_1, \xi_2, \ldots\}$, and consider the identity representation T of the algebra A consisting of those (bounded) linear operators a on X such that for some n we have (i) $a\xi_i = 0$ for all $i > n$, and (ii) range$(a) \subset \sum_{i=1}^{n} \mathbb{C}\xi_i$.)

1.15. Burnside's Theorem (IV.4.10) says that for finite-dimensional representations all four kinds of "irreducibility" coincide. In 6.20 we shall generalize Burnside's Theorem by proving that every algebraically irreducible *Banach* representation is algebraically totally irreducible.

For *-representations we shall see in VI.25.7 that irreducibility implies total irreducibility.

1.16. Proposition. *Let T be a totally irreducible locally convex representation of A. Then the only continuous linear operators* $g: X(T) \to X(T)$ *which intertwine T with itself are the "scalar" operators* $\lambda 1$ $(\lambda \in \mathbb{C})$. *In particular, if b belongs to the center of A,* T_b *is a scalar operator.*

Proof. Let g be T, T intertwining; and suppose ξ and $g\xi$ are linearly independent. By the total irreducibility of T there is a net $\{a_i\}$ of elements of A such that $T_{a_i}\xi \to \xi$ and $T_{a_i}g\xi \to 0$. But $T_{a_i}g\xi = gT_{a_i}\xi \to g\xi$ by the continuity of g, giving $g\xi = 0$, a contradiction. So ξ and $g\xi$ are linearly dependent. The rest of the proof is a repetition of the argument of IV.4.5.   ∎

1.17. Corollary. *A totally irreducible locally convex representation of an Abelian algebra is one-dimensional.*

1.18. Generalizing IV.3.16, we have:

Proposition. *Let J be a two-sided ideal of A, and T an irreducible locally convex representation of A. Then either* $J \subset \mathrm{Ker}(T)$ *or* $T|J$ *is irreducible.*

The proof is based on 1.9 and closely imitates that of IV.3.16. Details are left to the reader.

Remark. The above Proposition remains true if "irreducible" is replaced by "totally irreducible" in the hypothesis and conclusion. The proof is left as an exercise to the reader.

1.19. *Unsolved Problems.* The program of trying to generalize the algebraic results of Chapter IV to the context of locally convex representations leads to interesting unsolved problems. For example, let us ask the following question: Under what special circumstances is an irreducible locally convex representation necessarily totally irreducible? One or two positive results in this direction were mentioned in 1.15. In view of 1.16, for a closer look at the matter we shall split the question into two parts:

(I) Under what circumstances is it true that, if T is an irreducible locally convex representation of A, the only continuous T, T intertwining operators are the "scalar" operators?

(II) If T is an irreducible locally convex representation such that the only continuous T, T intertwining operators are the "scalar" operators, is T totally irreducible?

As we have seen in Remark IV.4.7, the implication (I) is false in general, even for algebraic representations. It has been conjectured that it holds for *Banach* representations. However we point out that a famous special case of this conjecture was recently shown to be false. Indeed, C. J. Read [1] has exhibited a Banach space X of dimension greater than 1 and a bounded linear operator $T: X \to X$ such that X has no non-trivial closed stable linear subspaces.

The implication (II), if true, would be a topological extension of the Jacobson Density Theorem (see IV.4.6). However, its truth or falsity appears to be unknown.

Just as in Remark IV.4.6, the proof of 1.16 used only the case $n = 2$ of the definition 1.10 of total irreducibility. Thus the following question, weaker than (II), suggests itself:

(III) Is the case $n = 2$ of 1.10 sufficient for total irreducibility?

This also appears to be unknown.

1.20. At this point it would be natural to try to develop a multiplicity theory for locally convex representations, analogous to the algebraic theory of §IV.2. The prospects for such a theory are however discouraging. Indeed, given a homeomorphic equivalence class σ of irreducible locally convex representations of A, one would presumably begin (in imitation of IV.2.11) by defining the σ-subspace $X_\sigma(T)$ of a locally convex representation T as the closure in $X(T)$ of the linear span of $\{Y : Y$ is a closed T-stable subspace of $X(T)$, and $^Y T$ is of class $\sigma\}$. Unfortunately, in contrast with IV.2.12, the $X_\sigma(T)$ as so defined fail in general to be linearly independent. One can construct examples in which $\{0\} \neq X_\tau(T) \subset X_\sigma(T)$, where σ and τ are entirely different homeomorphic equivalence classes.

Here is one such example. Let N be the discrete space of positive integers, and M the one-point compactification of N (∞ being the point at infinity). If S is any set, let $\mathscr{L}(S)$ denote the linear space of all complex functions on S which vanish at all but finitely many points. Define A' to be the algebra of all those linear endomorphisms of $\mathscr{L}(N)$ which annihilate all but finitely many δ_n (where $\delta_n(m) = \delta_{nm}$); and let A be the algebra $\{a + \lambda I : a \in A', \lambda \in \mathbb{C}, I = \text{identity operator on } \mathscr{L}(N)\}$. The identity representation W of A on $\mathscr{L}(N)$

is obviously algebraically irreducible. For each n in N let us form the permutation π_n of N which interchanges r and $n + r$ for each $r = 1, \ldots, n$, and leaves r unmoved if $r > 2n$. Thus π_n induces a bijective linear endomorphism π_n' of $\mathscr{L}(N)$ (of order 2), and $W^{(n)}: a \mapsto \pi_n' a \pi_n'$ is a representation (acting on $\mathscr{L}(N)$) equivalent to W. Also, let ρ be the one-dimensional representation $a' + \lambda I \mapsto \lambda$ of A ($a' \in A'$; $\lambda \in \mathbb{C}$); and $W^{(\infty)}: a \mapsto \rho(a)I$ ($a \in A$) the "multiple" of ρ acting on $\mathscr{L}(N)$. We now form the algebraic direct sum representation

$$V = \sum_{n \in N}{}^{\oplus} W^{(n)} \oplus W^{(\infty)} = \sum_{i \in M}{}^{\oplus} W^{(i)},$$

and identify $X(V)$ with $\mathscr{L}(M \times N)$ in the obvious way. Now let us give to $\mathscr{L}(N)\,[\mathscr{L}(M \times N)]$ the weak locally convex topology generated by the linear functionals of the form

$$f \mapsto \sum_{n \in N} f(n)\phi(n) \qquad \left[f \mapsto \sum_{p \in M \times N} f(p)\phi(p) \right],$$

where $\phi \in \mathscr{L}(N)$ $[\phi \in \mathscr{L}(M \times N)]$. One verifies that, with these topologies, W and V are locally convex representations of A on $\mathscr{L}(N)$ and $\mathscr{L}(M \times N)$ respectively; and that W is homeomorphically equivalent to each $W^{(n)}$. Notice that

$$\delta_{(m,n)} \to \delta_{(\infty,n)} \quad \text{as} \quad m \to \infty$$

for each n in N ($\delta_{(m,n)}$ being the element of $\mathscr{L}(M \times N)$ which is 1 at (m, n) and 0 elsewhere). It follows that the W-subspace of V is all of $X(V)$. At the same time $X(V)$ has a non-zero ρ-subspace (consisting of those f which vanish except on $\{\infty\} \times N$).

1.21.   We end this section with a result similar in flavor to 1.18.

Suppose that T is a locally convex representation of A, and q an idempotent element ($q^2 = q$) of A. Thus $Y = \text{range}(T_q)$ is a closed linear subspace of $X(T)$ which is stable under T_b whenever b belongs to the subalgebra $B = qAq$ of A. Let us denote by $T^{(q)}$ the locally convex representation $b \mapsto T_b | Y$ of B acting on Y.

Proposition.   *With these assumptions, suppose that $T_q \neq 0$. Then, if T is irreducible, $T^{(q)}$ is also irreducible. If T is totally irreducible, $T^{(q)}$ is also totally irreducible.*

Proof. Assume that T is irreducible. By 1.9, given any vectors ξ, η in Y with $\xi \neq 0$, we can find a net $\{a_i\}$ of elements of A such that $T_{a_i}\xi \to \eta$. Then

$$T_{qa_iq}\xi = T_q T_{a_i} T_q \xi$$

$$= T_q T_{a_i} \xi \qquad \text{(since } \xi \in Y\text{)}$$

$$\to T_q \eta = \eta \qquad \text{(since } \eta \in Y\text{)}.$$

By 1.9, the arbitrariness of ξ, η, and the fact that $qa_iq \in B$, this shows that $T^{(q)}$ is irreducible.

The statement about total irreducibility is proved by an exactly similar argument. ∎

2. Extending Locally Convex Representations of Two-Sided Ideals

2.1. We shall see in Remark VI.19.11 that the immediate analogue of IV.3.18 for locally convex representations (even for *-representations) is false. In this short section we shall obtain as strong an analogue of IV.3.18 as possible for locally convex representations. In VI.19.11 we will derive a different analogue of IV.3.18 for *-representations.

2.2. Fix a two-sided ideal J of the algebra A; and let S be a locally convex representation of J. We shall suppose that there is a net $\{e_i\}$ of elements of J such that

(I) $S_{e_i}\xi \to \xi$ in $X(S)$ for all ξ in $X(S)$;

(II) $\{S_{e_i}: i \text{ varying}\}$ is an equicontinuous family of operators on $X(S)$.

Define $A^{(S)}$ to be the set of those elements a of A such that there exists a continuous linear operator T_a on $X(S)$ satisfying

$$T_a S_b = S_{ab} \qquad \text{for all } b \text{ in } J. \tag{1}$$

(T_a is unique in virtue of (I).) Thus $A^{(S)}$ is a subalgebra of A containing J; and $T: a \mapsto T_a$ is a locally convex representation of $A^{(S)}$ on $X(S)$ extending S.

Proposition. *Assume in addition that $X(S)$ is complete. Then an element a of A belongs to $A^{(S)}$ if and only if $\{S_{e_i a}: i \text{ varying}\}$ is equicontinuous.*

Proof. We denote the linear span of $\{S_b\xi: b \in J, \xi \in X(S)\}$ by X_0; by (I) X_0 is dense in $X(S)$.

Let $a \in A$. If $\eta = S_b\xi$ ($b \in J, \xi \in X(S)$), then by (I)

$$S_{ab}\xi = \lim_i S_{e_i(ab)}\xi = \lim_i S_{e_i a}\eta.$$

It follows that the equation

$$T_a \eta = \lim_i S_{e_i a} \eta \qquad\qquad (\eta \in X_0) \qquad (2)$$

defines a linear operator T_a on X_0 satisfying

$$T_a S_b = S_{ab} \qquad\qquad (b \in J). \qquad (3)$$

Since X_0 is dense and $X(S)$ is complete, T_a will extend to a continuous operator on $X(S)$, giving $a \in A^{(S)}$, if and only if T_a is continuous on X_0.

Assume that $\{S_{e_i a} : i \text{ varying}\}$ is equicontinuous. Given a closed $X(S)$-neighborhood W of 0, we can choose another $X(S)$-neighborhood V of 0 such that $S_{e_i a}(V) \subset W$ for all i. So, if $\eta \in V \cap X_0$, (2) gives $T_a \eta = \lim_i S_{e_i a} \eta \in \bar{W} = W$. Thus T_a is continuous on X_0. Conversely, assume that T_a extends to a continuous operator on $X(S)$. Then by (3) $S_b T_a S_{e_i} = S_{bae_i} = S_{ba} S_{e_i}$ for all b in J and all i, implying by (I) that $S_b T_a = S_{ba}$ $(b \in J)$. In particular $S_{e_i a} = S_{e_i} T_a$; whence (II) implies the equicontinuity of $\{S_{e_i a} : i \text{ varying}\}$. ∎

Remark. Keeping the above notation, we can repeat the last part of the proof to obtain an added assertion: *If D is any subset of $A^{(S)}$ such that $\{S_{e_i a} : a \in D, i \text{ varying}\}$ is equicontinuous, then $\{T_a : a \in D\}$ is equicontinuous.*

2.3. As before, let J be a two-sided ideal of A. In this number we shall suppose that the underlying linear space of J is an LCS, in whose topology the multiplication in J is separately continuous. Let $\{e_i\}$ be a net of elements of J such that

$$e_i b \to b \text{ in } J \qquad \text{for all } b \text{ in } J. \qquad (4)$$

Let S be a locally convex representation of J such that 2.2(II) holds, and also the following conditions (III) and (IV):

(III) S is strongly continuous, that is, if $b_i \to b$ in J then $S_{b_i} \xi \to S_b \xi$ for all ξ in $X(S)$.

(IV) The linear span of $\{S_b \xi : b \in J, \xi \in X(S)\}$ is dense in $X(S)$.

We claim that 2.2(I) then holds.

Indeed: By (4) and (III)

$$S_{e_i} \xi \to \xi \qquad \text{for all } \xi \text{ in } L, \qquad (5)$$

where L denotes the linear span of $\{S_b \eta : b \in J, \eta \in X(S)\}$. By (IV) L is dense in $X(S)$. Now let ξ be an arbitrary element of $X(S)$, and U an arbitrary convex open neighborhood of 0 in $X(S)$. The $\{S_{e_i}\}$ are equicontinuous (by 2.2(II)); so we can choose a symmetric neighborhood V of 0 such that $V \subset \frac{1}{3}U$ and

$S_{e_i}V \subset \frac{1}{3}U$ for all i. Since L is dense, there is an element ξ' of $L \cap (\xi + V)$; and by (5) $S_{e_i}\xi' \in \xi' + V$ for all large enough i. Therefore

$$S_{e_i}\xi = S_{e_i}(\xi - \xi') + S_{e_i}\xi' \in S_{e_i}V + \xi' + V \subset \tfrac{1}{3}U + \xi + V + V \subset \xi + U$$

for all large enough i. Thus $S_{e_i}\xi \to \xi$; and the claim is proved.

Now define $A^{(S)}$ and T as in 2.2. In our present context T has the following continuity property:

Proposition. *Let $\{a_j\}$ be a net of elements of $A^{(S)}$ and a an element of $A^{(S)}$ such that (i) $a_j b \to ab$ in J for all b in J, and (ii) $\{T_{a_j}:j\ varying\}$ is equicontinuous. Then $T_{a_j}\xi \to T_a\xi$ for all ξ in $X(S)$.*

Proof. By (ii) and (IV) we have only to verify that $T_{a_j}\xi \to T_a\xi$ when ξ is of the form $S_b\eta$ ($b \in J$; $\eta \in X(S)$). But this follows from (1), (i), and (III). ∎

2.4. Here is an important special case of the above discussion:

Theorem. *Suppose that A is a normed algebra, J is a two-sided ideal of A, and $\{e_i\}$ is a left approximate unit of J (that is, $\{\|e_i\|:i\ varying\}$ is bounded, and $e_i b \to b$ for all b in J). Let S be a locally convex representation of J on a complete LCS $X(S)$, such that (i) the linear span of $\{S_b\xi : b \in J, \xi \in X(S)\}$ is dense in $X(S)$, and (ii) $\{S_b : b \in J, \|b\| \leq 1\}$ is equicontinuous. Then*

(I) *There is a unique locally convex representation T of A on $X(S)$ which extends S.*

(II) *The family $\{T_a : a \in A, \|a\| \leq 1\}$ is equicontinuous.*

(III) *If $\{a_j\}$ is a bounded net of elements of A, $a \in A$, and $a_j b \to ab$ in J for all b in J, then $T_{a_j}\xi \to T_a\xi$ for all ξ in $X(S)$.*

Proof. We note that 2.2(II) holds; and hypothesis (ii) implies 2.3(III). So by 2.3, 2.2(I) holds. Therefore by Proposition 2.2 and (ii) $A^{(S)} = A$. So (I) is proved.

The equicontinuity of $\{T_a : \|a\| \leq 1\}$ now follows from (ii) and Remark 2.2. Finally, (III) follows from (II) and Proposition 2.3. ∎

3. The Naimark Relation

3.1. For the successful classification of, say, the irreducible locally convex representations of a given algebra A, the relation of homeomorphic equivalence (see 1.4) turns out to be far too strong. Whether or not two locally convex representations are homeomorphically equivalent depends heavily on the detailed topological structure of the spaces in which they act. We should like to find a weaker equivalence relation between locally convex representations T which, roughly speaking, would disregard the detailed topological structure of $X(T)$ and concentrate mainly on the formulae defining the T_a. No completely satisfactory such relation has been found. However, in this section we shall present the so-called Naimark relation between locally convex representations, which in certain narrow but important contexts becomes an equivalence relation giving rise to a perfectly satisfactory classification theory (see 3.6). The contexts where the Naimark relation is important lie outside the sphere of *-representations; so our discussion of it will be brief.

3.2. Definition. Let S and T be two locally convex representations of the algebra A. We say that S and T are *Naimark-related* if there exist a dense S-stable linear subspace Y of $X(S)$, a dense T-stable linear subspace Z of $X(T)$, and a linear bijection $F: Y \to Z$ such that (i) $FS_a\xi = T_aF\xi$ ($a \in A$, $\xi \in Y$), and (ii) F is *closed* in the sense that the graph of F is closed as a subset of $X(S) \times X(T)$.

3.3. Obviously, homeomorphic equivalence implies Naimark-relatedness.

If the S and T of 3.2 are finite-dimensional, the Naimark relation is the same as (algebraic) equivalence.

Suppose that S and T are algebraically irreducible Fréchet representations. If they are Naimark-related, the Y and Z of 3.2 must be all of $X(S)$ and $X(T)$ respectively; and by the Closed Graph Theorem the F of 3.2 is a homeomorphism. So, for such representations, the Naimark relation coincides with homeomorphic equivalence.

We shall show in VI.13.16 that for *-representations the Naimark relation is the same as unitary equivalence.

3.4. In general, the Naimark relation between locally convex representations is evidently reflexive and symmetric, but *it is not transitive, hence not an equivalence relation*. This is its great disadvantage. In VI.13.17 we shall give an explicit example of the failure of transitivity of the Naimark relation. To conclude this section we shall mention briefly one narrow but important non-involutory context in which the Naimark relation *is* transitive. For proofs the reader is referred to the literature (see Fell [10]).

3.5. Fix a locally convex representation T of A. We shall denote by $J(T)$ the two-sided ideal of A consisting of those a such that T_a is of finite rank.

Definition. The locally convex representation T is said to be *locally finite-dimensional* if $T|J(T)$ is non-degenerate (in the sense of 1.7).

3.6. Proposition*. *If R, S, T are three locally finite-dimensional locally convex representations of A, and if R and S are Naimark-related and S and T are Naimark-related, then R and T are Naimark-related.*

In other words, on the set of locally finite-dimensional locally convex representations of A the Naimark relation is an equivalence relation.

3.7. Here is another property which locally finite-dimensional representations possess (in common with finite-dimensional representations and *-representations):

Proposition*. *A locally finite-dimensional locally convex representation of A is irreducible if and only if it is totally irreducible.*

3.8. In the irreducible locally finite-dimensional context there is a simple criterion for Naimark-relatedness.

Proposition*. *Two irreducible locally finite-dimensional locally convex representations S and T of A are Naimark-related if and only if* $\operatorname{Ker}(S) = \operatorname{Ker}(T)$.

For finite-dimensional representations this result is an immediate corollary of IV.4.8.

4. Elementary Remarks on Normed Algebras; Examples

4.1. For the rest of this chapter we shall be dealing with normed algebras and their locally convex representations. Normed algebras and Banach algebras have been defined in Chapter I.

4.2. Let A, $\| \ \|$ be a normed algebra; and let B, $\| \ \|$ be the Banach space completion of the underlying normed linear space of A. The inequality

$$\|ab\| \leq \|a\| \ \|b\| \qquad\qquad (a, b \in A) \qquad (1)$$

evidently permits us to extend the multiplication on A to an (associative) multiplication on B making the latter a Banach algebra; and this extension is unique. With the extended multiplication, B, $\| \ \|$ is called the *Banach algebra completion of A, $\| \ \|$.*

4.3. *Example.* Let X be a normed linear space, and $\mathcal{O}(X)$, as usual, the algebra (under composition) of all bounded linear operators on X. Under the *operator norm*

$$\|a\| = \sup\{\|a\xi\| : \xi \in X, \ \|\xi\| \leq 1\} \qquad (a \in \mathcal{O}(X)),$$

$\mathcal{O}(X)$ is a normed algebra. In fact, if X is complete (i.e., a Banach space), the reader will easily verify that $\mathcal{O}(X)$ is complete (hence a Banach algebra).

4.4. Let A, $\| \ \|$ be a normed algebra with a unit $\mathbb{1}$. Inequality (1) implies that $\|\mathbb{1}\| \geq 1$. Now it can perfectly well happen that $\|\mathbb{1}\| > 1$; (indeed, any normed algebra remains a normed algebra when its norm is multiplied by a constant greater than 1). However, even if $\|\mathbb{1}\| > 1$, it is always possible to introduce into A a new norm $\| \ \|'$, equivalent to $\| \ \|$ (i.e., giving rise to the same topology), under which A continues to be a normed algebra, and which satisfies $\|\mathbb{1}\|' = 1$. To see this, let λ_a be the operator $b \mapsto ab$ on A of left multiplication by a ($a \in A$); and define $\|a\|'$ to be the operator norm of λ_a. Evidently A, $\| \ \|'$ is a normed algebra (see 4.3), and $\|\mathbb{1}\|' = 1$. We have

$$\|a\| \geq \|a\|' \qquad \qquad \text{(by (1))}$$
$$= \|\lambda_a\| \geq \|\mathbb{1}\|^{-1}\|a \cdot \mathbb{1}\|$$
$$= \|\mathbb{1}\|^{-1}\|a\|,$$

whence the norms $\| \ \|$ and $\| \ \|'$ are equivalent.

4.5. As another application of 4.3 we notice:

Proposition. *Any finite-dimensional algebra A becomes a Banach algebra when equipped with a suitable norm.*

Proof. Choose a faithful finite-dimensional representation T of A (see Remark IV.5.3); introduce a norm into $X(T)$; and define $\|a\|$ ($a \in A$) to be the operator norm of T_a on $X(T)$. By 4.3 A becomes a (finite-dimensional) normed algebra, hence a Banach algebra. ∎

Remark. As a variant of this proof, we use Remark IV.5.3 to observe that A is a subalgebra of some total matrix algebra $M(n, \mathbb{C})$. But $M(n, \mathbb{C})$ is a Banach algebra under many different norms, for example under

$$\|a\| = \sum_{i, j=1}^{n} |a_{ij}|.$$

4.6. *Example.* Let S be any set, and $\mathscr{B}(S)$ the algebra of all bounded complex-valued functions on S (addition, scalar multiplication, and multiplication being defined pointwise on S). If $f \in \mathscr{B}(S)$, the (finite) quantity

$$\|f\|_\infty = \sup\{|f(s)| : s \in S\}$$

is called the *supremum norm* of f; and it is easy to see that $\mathscr{B}(S)$, $\|\ \|_\infty$ is a commutative Banach algebra.

If S is a topological space, the space $\mathscr{B}_c(S)$ of all continuous bounded complex-valued functions on S is a norm-closed subalgebra of $\mathscr{B}(S)$, hence a Banach algebra in its own right.

Suppose in addition that S is a non-compact locally compact Hausdorff space. The space $\mathscr{C}_0(S)$ of all those continuous complex-valued functions on S which vanish at infinity is a norm-closed subalgebra of $\mathscr{B}_c(S)$, and hence a commutative Banach algebra in its own right under the supremum norm.

4.7. *Example.* Let μ be a measure on a δ-ring of subsets of a set S (see II.1.2), and $\mathscr{L}_\infty(\mu)$, $\|\ \|_\infty$ the Banach space of μ-essentially bounded locally μ-measurable functions defined in II.7.10. It is clear that $\mathscr{L}_\infty(\mu)$, $\|\ \|_\infty$ is a Banach algebra under the operation of pointwise multiplication.

4.8. *Example.* The following example illustrates, as we shall see later (8.4), a very important phenomenon in the theory of commutative Banach algebras.

Let D be the unit disk $\{z \in \mathbb{C} : |z| \le 1\}$ in the complex plane; and let A be the family of all those complex-valued functions on D which are continuous on D and analytic on the interior $\{z \in \mathbb{C} : |z| < 1\}$ of D. Clearly A is a normed subalgebra of $\mathscr{B}_c(D)$ (with the supremum norm). In fact A is complete, hence a Banach algebra, under the supremum norm. This follows from the well-known fact (proved by the use of Cauchy's Integral Formula) that the limit of a uniformly convergent sequence of analytic functions is analytic.

4.9. *Examples.* Let G be a locally compact group. We have seen in III.10.15 that the Banach space $\mathscr{M}_r(G)$ of all bounded regular complex Borel measures on G (with the total variation norm) is a Banach algebra under convolution—the so-called measure algebra of G.

If λ stands for a left Haar measure on G, the Banach space $\mathscr{L}_1(\lambda)$ of all λ-summable complex functions on G becomes a Banach algebra under the operation of convolution, called the $\mathscr{L}_1$ group algebra of G (see III.11.9).

4.10.   *Adjunction of a Unit.* Let A for the moment be any algebra without a unit element. There is a natural way of embedding A in an algebra A_1 with unit: We define A_1 to be the linear space direct sum $A \oplus \mathbb{C}$ of A and the one-dimensional space $\mathbb{C}$; and introduce into A_1 the multiplication:

$$\langle a, \lambda \rangle \langle b, \mu \rangle = \langle ab + \mu a + \lambda b, \lambda \mu \rangle \tag{2}$$

$(a, b \in A; \lambda, \mu \in \mathbb{C})$; this makes A_1 an algebra with unit $\mathbf{1} = \langle 0, 1 \rangle$. We refer to A_1 as the algebra *obtained from A by adjoining a unit (or the unitization of A)*. One identifies a with $\langle a, 0 \rangle$ $(a \in A)$, so that $A \subset A_1$ and $\langle a, \lambda \rangle = a + \lambda\mathbf{1}$. Note that the unitization of A can be formed whether or not A has a unit.

Suppose now that the above A is a normed algebra. Then A_1 can be made into a normed algebra by means of the norm

$$\|a + \lambda\mathbf{1}\| = \|a\| + |\lambda| \qquad (a \in A; \lambda \in \mathbb{C}). \tag{3}$$

If A is complete (i.e., a Banach algebra) then so is A_1.

4.11.   *Quotient Normed Algebras.* Let A be a normed algebra, and I a norm-closed two-sided ideal of A. It is well known that the quotient A/I is a normed linear space under the norm

$$\|a + I\| = \inf\{\|a + b\| : b \in I\} \qquad (a \in A). \tag{4}$$

In fact, with the quotient algebra structure, A/I is a normed algebra. Indeed, we have only to prove the inequality (1). If $\alpha = a + I$ and $\beta = b + I$ $(a, b \in A)$, then

$$\|\alpha\beta\| = \inf\{\|ab + w\| : w \in I\}$$

$$\leq \inf\{\|(a + u)(b + v)\| : u, v \in I\}$$

$$\leq \inf\{\|a + u\| \, \|b + v\| : u, v \in I\} \leq \|\alpha\| \, \|\beta\|.$$

So (1) holds. We call A/I the *quotient normed algebra (modulo I)*.

If A is a Banach algebra, then A/I is well known to be complete, hence a Banach algebra.

4.12. In normed algebras which do not have a unit, the following weaker topological notion of an "approximate unit" is often important.

Definition. A *left [right] approximate unit* of a normed algebra A is a net $\{e_i\}$ of elements of A such that (i) for some positive number k, $\|e_i\| \leq k$ for all i, and (ii) $\lim_i (e_i a) = a$ $[\lim_i (a e_i) = a]$ for all a in A (in the norm-topology of A).

A net which is both a left and a right approximate unit of A is called simply an *approximate unit of A*.

Remark. In the case that A is the group algebra of a locally compact group G, one must distinguish approximate units of A from the more restrictive concept of an approximate unit of G (as defined in III.11.17).

4.13. Example. If G is a locally compact group, its $\mathscr{L}_1$ group algebra $\mathscr{L}_1(\lambda)$ will not have a unit element unless G is discrete (see III.11.21). But $\mathscr{L}_1(\lambda)$ always has an approximate unit by III.11.19.

4.14. We shall end this section with some remarks on locally convex representations of normed algebras.

Let A be a normed algebra, and I a closed left ideal of A. In IV.3.9 we defined the so-called natural representation R of A on A/I:

$$R_a(b + I) = ab + I \qquad\qquad (a, b \in A).$$

Since I is closed, A/I becomes a normed linear space (complete if A is complete) under the quotient norm (4). Notice that R is then a locally convex representation of A, that is, the R_a are continuous; in fact,

$$\|R_a\| \le \|a\| \tag{5}$$

$(a \in A; \|R_a\|$ is the operator norm of R_a on A/I). To prove (5), we observe that for all b in A and c in I $\|ab + I\| \le \|ab + ac\| \le \|a\| \|b + c\|$; and the infimum of the last expression, as c runs over I, is $\|a\| \|b + I\|$.

4.15. If A has a unit $\mathcal{1}$, a locally convex representation T of A is non-degenerate if and only if $T_{\mathcal{1}}$ is the identity operator. For approximate units we have an analogous statement.

Proposition. *Let T be a locally convex representation of a normed algebra A such that $\{T_a: \|a\| \le 1\}$ is an equicontinuous family of operators on $X(T)$. Assume that A has a left approximate unit $\{e_i\}$. Then T is non-degenerate if and only if*

$$T_{e_i}\xi \xrightarrow[i]{} \xi \tag{6}$$

for all ξ in $X(T)$.

Proof. Clearly (6) implies non-degeneracy. Assume conversely that T is non-degenerate. The equicontinuity hypothesis on T implies that $a \mapsto T_a\xi$ is continuous on A to $X(T)$ for each ξ in $X(T)$. So, since $e_i a \to a$, $T_{e_i}(T_a\eta) \to T_a\eta$ whenever $a \in A$ and $\eta \in X(T)$. That is, $T_{e_i}\xi \to \xi$ whenever ξ belongs to the linear span L of $\{T_a\eta : a \in A, \eta \in X(T)\}$. Now the non-degeneracy of T says that L is dense in $X(T)$. Also, since the $\|e_i\|$ are bounded in i, the $\{T_{e_i}\}$ form an equicontinuous family. Applying the argument of 2.3 to the last three facts, we conclude that $T_{e_i}\xi \to \xi$ for all ξ in $X(T)$. ∎

5. The Spectrum

5.1. Definition. Let A be an algebra with unit $\mathbb{1}$. If $a \in A$, the *spectrum of a in A*, in symbols $\mathrm{Sp}_A(a)$, is defined as the set of all those complex numbers λ such that $a - \lambda\mathbb{1}$ has no inverse in A.

If A is an algebra *without* unit, the *spectrum* $\mathrm{Sp}_A(a)$ *of* an element a of A is defined to be the same as $\mathrm{Sp}_{A_1}(a)$, where A_1 is the algebra obtained by adjoining a unit to A (see 4.10).

5.2. Remarks. Let A be any algebra.

(I)   If A has no unit element, $0 \in \mathrm{Sp}_A(a)$ for every a in A. If A has a unit $\mathbb{1}$, $\mathrm{Sp}_A(\mathbb{1}) = \{1\}$.

(II) $\mathrm{Sp}_A(\lambda a) = \lambda\,\mathrm{Sp}_A(a)$ $(a \in A; \lambda \in \mathbb{C})$

(unless $\lambda = 0$ and $\mathrm{Sp}_A(a)$ is void).

(III) If A has a unit $\mathbb{1}$, $a \in A$, and $\lambda \in \mathbb{C}$, then $\mathrm{Sp}_A(a + \lambda\mathbb{1}) = \mathrm{Sp}_A(a) + \lambda$.

(IV)   If A and B are algebras with units $\mathbb{1}$ and $\mathbb{1}'$ respectively, and if $\phi: A \to B$ is an algebra homomorphism with $\phi(\mathbb{1}) = \mathbb{1}'$, then

$$\mathrm{Sp}_B(\phi(a)) \subset \mathrm{Sp}_A(a) \qquad\qquad (a \in A). \qquad (1)$$

In particular this holds if A is a subalgebra of B containing $\mathbb{1}'$ and ϕ is the identity injection.

In general the inclusion (1) is proper, as we shall see in 5.3(IV). For an important context in which equality holds in (1) see 5.4. See also 6.16.

5.3. Examples.

(I)   Let A be the algebra (under the pointwise linear operations and multiplication) of all continuous complex functions on some topological space S. This algebra has a unit $\mathbb{1}$; and a function f in A has an inverse in A if and only if $0 \notin \mathrm{range}(f)$. Thus:

$$\mathrm{Sp}_A(f) = \mathrm{range}(f) \qquad\qquad (f \in A).$$

(II)   Let A be the algebra of all continuous *bounded* complex functions on the topological space S. Then an element f of A has an inverse in A if and only if 0 is not in the closure $\overline{\mathrm{range}(f)}$ of $\mathrm{range}(f)$. Thus:

$$\mathrm{Sp}_A(f) = \overline{\mathrm{range}(f)} \qquad\qquad (f \in A).$$

(III) Let A be the total matrix algebra $M(n, \mathbb{C})$ of all $n \times n$ complex matrices. A matrix a in A has an inverse if and only if there is no non-zero vector x in $\mathbb{C}^n$ such that $ax = 0$. Thus $\text{Sp}_A(a)$ is just the set of all eigenvalues of a, that is, the set of those complex numbers λ such that $ax = \lambda x$ for some non-zero vector x in $\mathbb{C}^n$.

(IV) Here is an example of inequality in (1). Let B be the algebra (under pointwise operations) of all continuous complex functions on $[0, 1]$, A the subalgebra of all polynomial functions on $[0, 1]$ (with complex coefficients), and ϕ the identity injection. If $f \in A$, then $\text{Sp}_B(f) = \text{range}(f)$ (by (I)), while $\text{Sp}_A(f) = \mathbb{C}$ unless f is a constant function.

5.4. Here is a situation in which equality holds in (1).

Proposition. *If B is any algebra with unit $\mathbb{1}$, A is any maximal commutative subalgebra of B, and $a \in A$, then*

$$\text{Sp}_A(a) = \text{Sp}_B(a).$$

Proof. Observe that $\mathbb{1} \in A$. If $\lambda \notin \text{Sp}_B(a)$ then $(a - \lambda\mathbb{1})^{-1}$ exists in B, and commutes with all b in A since $a - \lambda\mathbb{1}$ does. It follows from the maximality of A that $(a - \lambda\mathbb{1})^{-1} \in A$, whence $\lambda \notin \text{Sp}_A(a)$. Consequently $\text{Sp}_A(a) \subset \text{Sp}_B(a)$. Combining this with (1) we obtain the Proposition.   ■

5.5. The following Proposition will be helpful at a critical juncture in Chapter VI.

Proposition. *If a, b are elements of an algebra A, the symmetric difference of $\text{Sp}_A(ab)$ and $\text{Sp}_A(ba)$ consists at most of the number 0; that is, $\text{Sp}_A(ab) \cup \{0\} = \text{Sp}_A(ba) \cup \{0\}$.*

Proof. We may as well suppose that A has a unit $\mathbb{1}$. It is evidently enough to show that if $u = (\mathbb{1} - ab)^{-1}$ exists, then so does $(\mathbb{1} - ba)^{-1}$. We have:

$$(\mathbb{1} + bua)(\mathbb{1} - ba) = \mathbb{1} + b[u(\mathbb{1} - ab) - \mathbb{1}]a = \mathbb{1},$$

$$(\mathbb{1} - ba)(\mathbb{1} + bua) = \mathbb{1} + b[(\mathbb{1} - ab)u - \mathbb{1}]a = \mathbb{1},$$

whence $(\mathbb{1} - ba)^{-1} = \mathbb{1} + bua$.   ■

Remark. It is easy to construct examples of linear operators a and b on infinite-dimensional linear spaces such that ab has an inverse but ba does not. In this situation 0 belongs to the spectrum of ba but not to that of ab. This shows that $\text{Sp}_A(ba)$ and $\text{Sp}_A(ab)$ may differ as regards the number 0.

5.6. Proposition.   *Let A be an algebra with unit $\mathbf{1}$, and P a polynomial in one variable with complex coefficients. Then, if $a \in A$,*

$$\mathrm{Sp}_A(P(a)) = P(\mathrm{Sp}_A(a)) \qquad (= \{P(\lambda): \lambda \in \mathrm{Sp}_A(a)\}).$$

Proof.   Let λ_0 be any complex number, and factor $P(\lambda) - \lambda_0$ into linear factors:

$$P(\lambda) - \lambda_0 = k(\lambda - r_1)(\lambda - r_2)\ldots(\lambda - r_n) \qquad (k \neq 0).$$

Then

$$P(a) - \lambda_0\mathbf{1} = k(a - r_1\mathbf{1})(a - r_2\mathbf{1})\ldots(a - r_n\mathbf{1}).$$

Thus $\lambda_0 \in \mathrm{Sp}_A(P(a))$ if and only if for some i $(a - r_i\mathbf{1})^{-1}$ does not exist, that is, if and only if some r_i belongs to $\mathrm{Sp}_A(a)$. But the $r_1,\ldots,r_n$ are exactly the solutions of $P(\lambda) = \lambda_0$. Thus $\lambda_0 \in \mathrm{Sp}_A(P(a))$ if and only if $\lambda_0 \in P(\mathrm{Sp}_A(a))$.   ∎

5.7.   For handling algebras without a unit the notion of an adverse is convenient. Let A be any algebra.

Definition.   If $a, b \in A$, put

$$a \circ b = a + b - ab.$$

The operation $\circ$ is associative and has 0 as a unit. This is easily seen from the fact that, if A has a unit $\mathbf{1}$, we have

$$\mathbf{1} - (a \circ b) = (\mathbf{1} - a)(\mathbf{1} - b). \tag{2}$$

We shall say that b is a *left adverse* [*right adverse, adverse*] of a if $b \circ a = 0$ [$a \circ b = 0$, $a \circ b = b \circ a = 0$].

If a has a left adverse b and a right adverse c, then, as with inverses, we have $b = c$, and b is the adverse of a. The adverse of a, if it exists, is thus unique.

Remark.   The term "adverse" is often referred to in the literature as "quasi-inverse" (see, for example, Rickart [2]).

By (2), in the presence of a unit, a has a left adverse [right adverse, adverse] if and only if $\mathbf{1} - a$ has a left inverse [right inverse, inverse]. Notice also that, if A has no unit and A_1 is the result of adjoining one, and if b is a left or right adverse in A_1 of an element a of A, then b is automatically in A. Thus, whether A has a unit or not, the spectrum of an element of A can be described as follows:

5.8. Proposition. *Let a be an element of the algebra A. Then a non-zero complex number λ belongs to $\mathrm{Sp}_A(a)$ if and only if $\lambda^{-1}a$ has no adverse in A. Further, $0 \notin \mathrm{Sp}_A(a)$ if and only if A has a unit and a^{-1} exists in A.*

5.9. The existence of inverses and adverses is related to the existence of proper ideals containing a given element. Let A be an algebra.

Definition. Let I be a right [left] ideal of A. An element u of A is a *left* [*right*] *identity modulo I* if

$$ua - a \in I \qquad [au - a \in I]$$

for all a in A.

Note that such an element u cannot be in I unless $I = A$.

Definition. A right [left] ideal I of A is *right-regular* [*left-regular*] (or simply *regular* if no ambiguity can arise) if there exists in A a left [right] identity modulo I.

Remark. The term "regular ideal" is often referred to in the literature as "modular ideal" (see, for example, Rickart [2]).

If A has a unit, every left or right ideal of A is obviously regular.

Note that a two-sided ideal of A might be left-regular but not right-regular, or vice versa. Indeed, the algebra A of all complex matrices of the form $\begin{pmatrix} a & 0 \\ b & 0 \end{pmatrix}$ has a right unit but no left unit. So the zero ideal is left-regular in A but not right-regular.

5.10. Let A be an algebra. An ideal I of A is of course *proper* if $I \neq A$. By a *maximal* ideal we mean one which is maximal among proper ideals.

Proposition. *Every proper left-regular left ideal [right-regular right ideal] I of A is contained in some maximal left [right] ideal J of A.*

Note. J is of course automatically regular, since a right (or left) identity modulo I is the same modulo J.

Proof. Consider the case that I is a left ideal; and let u be a right identity modulo I. By Zorn's Lemma there is a maximal family $\mathscr{F}$ of proper left ideals containing I which is simply ordered by inclusion. We claim that the union J

of $\mathscr{F}$ is proper. Indeed, if $u \in J$, then $u \in K$ for some K in $\mathscr{F}$, in which case $K = A$ by a remark in 5.9, contradicting the properness of K. So $u \notin J$, and $J \neq A$. Thus J is a maximal left ideal containing I.

The same argument works of course if I is a right ideal. ∎

Remark. A proper left ideal of A which is not left-regular need not be contained in any maximal left ideal of A (see Exercise 37).

5.11. Proposition. *An element a of an algebra A with unit 1 has a left [right] inverse if and only if a does not belong to any maximal left [right] ideal of A.*

Proof. If a has a left inverse b, then any left ideal I containing a contains $ba = 1$, and hence coincides with A; so a does not belong to any proper left ideal. If a has no left inverse, then $\{ba : b \in A\}$ is a *proper* left ideal containing a, and so by 5.10 there is a maximal left ideal containing a.

Similarly for right inverses. ∎

5.12. Proposition. *Let A be any algebra. If $a \in A$, then a has a left [right] adverse if and only if a is not a right [left] identity modulo any regular maximal left [right] ideal of A.*

Proof. Suppose that a is a left identity modulo a regular maximal right ideal I. If b is a right adverse of a, then $a = ab - b \in I$, contradicting the properness of I. The same is true with "left" and "right" reversed. So the "only if" part is proved.

Suppose that a has no right adverse. Then $J = \{ab - b : b \in A\}$ is a right ideal and is proper; indeed, $a \notin J$ (since a has no right adverse). Evidently a is a left identity modulo J. Extending J to a maximal right ideal by 5.10, and observing that the same argument holds with "left" and "right" interchanged, we obtain the "if" part of the Proposition. ∎

5.13. Let I be a left ideal of the algebra A, and R the natural representation of A on A/I. It was pointed out in IV.3.9 that I is a maximal (proper) left ideal if and only if R is irreducible as an operator set. If in addition I is regular then R is irreducible as a representation. In fact we have:

Proposition. *A subset I of the algebra A is a regular maximal left ideal of A if and only if there exist an irreducible representation T of A and a non-zero vector ξ in X(T) such that*

$$I = \{a \in A : T_a \xi = 0\}. \tag{3}$$

If this is the case then T is equivalent to the natural representation R of A on A/I.

Proof. Suppose that (3) holds, T being irreducible and ξ a non-zero vector in $X(T)$. Then the correspondence $a + I \mapsto T_a \xi$ $(a \in A)$ is clearly an equivalence between R and T (proving the last statement of the Proposition). So by IV.3.9 I is a maximal left ideal. If u is any element of A such that $T_u \xi = \xi$, then for all a in A $T_{au-a} \xi = T_a T_u \xi - T_a \xi = 0$, or $au - a \in I$; so I is regular.

Conversely, let I be a regular maximal left ideal, with right identity u. Then the natural representation R of A on A/I is irreducible as an operator set by IV.3.9; and, for any a in A, we have $R_a(u + I) = 0 \Leftrightarrow au \in I \Leftrightarrow a \in I$. It follows that R is not the zero representation, and hence is irreducible; and also that (3) holds with $T = R$, $\xi = u + I$. ∎

5.14. Corollary. *Every irreducible representation of an algebra A is equivalent to the natural representation of A on A/I for some regular maximal left ideal I of A.*

5.15. Remark. The most obvious examples of non-regular maximal left ideals are the linear subspaces I of an algebra A which are of codimension 1 in A and satisfy $A^2 \subset I$. For these, the natural representation R of A on A/I is of course a one-dimensional zero representation. But there exist non-regular maximal left ideals I which are not of this type (see 5.16). For these, R is irreducible, and provides a counter-example to the natural conjecture that a left ideal I of A is regular maximal if and only if the natural representation of A on A/I is irreducible.

5.16. Proposition*. *Let X be an infinite-dimensional Hilbert space, and Y a non-closed linear subspace of X of codimension 1. Let A be the algebra of all bounded linear operators on X of finite rank; and put*

$$I = \{a \in A : \text{range}(a) \subset Y\}.$$

Then I is a non-regular maximal right ideal of A.

6. Spectra in Banach Algebras, Mazur's Theorem, Gelfand's Theorem

6.1. *In this section A is a fixed Banach algebra.*

6.2. Proposition. *If A has a unit 1, then $(1 - a)^{-1}$ exists whenever $a \in A$ and $\|a\| < 1$. Further, the map $a \mapsto (1 - a)^{-1}$ is continuous on $\{a \in A: \|a\| < 1\}$.*

Proof. The geometric series

$$1 + a + a^2 + a^3 + \cdots \tag{1}$$

converges uniformly absolutely on $\{a: \|a\| \le r\}$ for each $0 < r < 1$. Its sum is clearly $(1 - a)^{-1}$. ∎

6.3. Corollary. *Whether A has a unit or not, every element a of A satisfying $\|a\| < 1$ has an adverse a' given by:*

$$a' = -a - a^2 - a^3 - a^4 - \cdots . \tag{2}$$

The map $a \mapsto a'$ is continuous on $\{a: \|a\| < 1\}$.

We notice from (2) that

$$\|a'\| \le (1 - \|a\|)^{-1} \|a\| \qquad \text{for } \|a\| < 1. \tag{3}$$

6.4. Proposition. *Assume that A has a unit 1. Then $\{a \in A: a^{-1} \text{ exists in } A\}$ is an open subset of A; and $a \mapsto a^{-1}$ is continuous on this set.*

Proof. If a^{-1} exists and $\|b\| < \|a^{-1}\|^{-1}$, then by 6.2 $(a + b)^{-1} = [a(1 + a^{-1}b)]^{-1}$ exists. Thus $\{a \in A: a^{-1} \text{ exists}\}$ is open. Also, by 6.2 $b \mapsto (1 + a^{-1}b)^{-1}$ is continuous at $b = 0$; hence $b \mapsto (a + b)^{-1} = (1 + a^{-1}b)^{-1}a^{-1}$ is continuous at $b = 0$. ∎

6.5. Corollary. *For each a in A, $\mathrm{Sp}_A(a)$ is a closed bounded subset of $\mathbb{C}$. In fact $|\lambda| \le \|a\|$ for all λ in $\mathrm{Sp}_A(a)$.*

Proof. Adjoining a unit, we may assume from the beginning that A has a unit 1.

By 6.4, $\{\lambda \in \mathbb{C}: (a - \lambda 1)^{-1} \text{ exists}\}$ is open. Hence its complement $\mathrm{Sp}_A(a)$ is closed.

If $|\lambda| > \|a\|$, then by 6.2 $(1 - \lambda^{-1}a)^{-1}$ exists, whence $(a - \lambda 1)^{-1}$ exists; so $\lambda \notin \mathrm{Sp}_A(a)$. This proves the last statement of the Corollary. In particular $\mathrm{Sp}_A(a)$ is a bounded set. ∎

6.6. Proposition.   *A regular maximal left (or right) ideal of A is norm-closed.*

Proof.   Assume that I is a proper regular left ideal of A, and that u is a right identity modulo I. We claim that

$$\|u - a\| \geq 1 \qquad \text{for all } a \text{ in } I. \tag{4}$$

Indeed: If $a \in I$ and $\|u - a\| < 1$, then by 6.3 $u - a$ has an adverse b. Thus $(u - a) + b - b(u - a) = 0$, whence

$$u = a - ba + (bu - b) \in I.$$

But by a remark in 5.9 this contradicts the properness of I. Thus (4) holds.

It follows from (4) that the closure $\bar{I}$ of I cannot contain u, and hence is also a proper regular left ideal. Thus, since I is maximal, it coincides with its closure.   ∎

Remark.   A maximal left ideal I of A which is *not* regular need not be closed. The simplest example is obtained by taking A to be an infinite-dimensional Banach space, considering A as a Banach algebra with trivial multiplication $(ab = 0$ for all $a, b)$, and taking I to be any dense linear subspace of A with codimension 1.

6.7.   The following Proposition is very important inasmuch as it enables us to apply analytic function theory to Banach algebras.

Proposition.   *Suppose that A has a unit 1; and let a be any element of A. For each continuous linear functional E on A, the function f given by:*

$$f(\lambda) = E((a - \lambda 1)^{-1})$$

is analytic on the complement of $\mathrm{Sp}_A(a)$, *including the point ∞ at infinity, and satisfies* $f(\infty) = 0$.

Remark.   Since $\mathrm{Sp}_A(a)$ is bounded by 6.5, its complement contains a neighborhood of ∞.

Proof.   If a^{-1} exists, then by 6.4 $(a - \lambda 1)^{-1}$ exists for small non-zero complex λ, and

$$\lambda^{-1}[(a - \lambda 1)^{-1} - a^{-1}] = \lambda^{-1}[(1 - \lambda a^{-1})^{-1} - 1]a^{-1}. \tag{5}$$

Now from the power series expansion (1) we see that

$$\lim_{\lambda \to 0} \lambda^{-1}[(1 - \lambda a^{-1})^{-1} - 1] = a^{-1}.$$

Hence by (5)

$$\lim_{\lambda \to 0} \lambda^{-1}[(a - \lambda \mathbf{1})^{-1} - a^{-1}] = a^{-2}. \tag{6}$$

If μ is any number in $\mathbb{C} \setminus \mathrm{Sp}_A(a)$, we can replace a by $a - \mu\mathbf{1}$ and λ by $\lambda - \mu$ in (6), getting

$$\lim_{\lambda \to \mu} (\lambda - \mu)^{-1}[(a - \lambda\mathbf{1})^{-1} - (a - \mu\mathbf{1})^{-1}] = (a - \mu\mathbf{1})^{-2}. \tag{7}$$

It follows from (7) that, if E is any continuous linear functional on A, the function $f: \lambda \mapsto E((a - \lambda\mathbf{1})^{-1})$ is analytic on $\mathbb{C} \setminus \mathrm{Sp}_A(a)$.

It remains to show that f is analytic at ∞ and satisfies $f(\infty) = 0$. For small λ we have by (1)

$$f(\lambda^{-1}) = -\lambda E((\mathbf{1} - \lambda a)^{-1})$$

$$= -\lambda[E(\mathbf{1}) + \lambda E(a) + \lambda^2 E(a^2) + \ldots].$$

Since $|E(a^n)| \le \| E \| \, \| a \|^n$, the right side is analytic around 0 and has the value 0 when $\lambda = 0$. ∎

6.8.   Now let us drop the assumption that A has a unit. Proposition 6.7 can then be stated in terms of adverses. If a is an element of A, 5.8 shows that λa has an adverse $(\lambda a)'$ for all λ in the open set

$$W_a = \{\lambda \in \mathbb{C} : \lambda = 0 \text{ or } \lambda^{-1} \notin \mathrm{Sp}_A(a)\}.$$

Proposition. *If $a \in A$ and E is a continuous linear functional on A, then $\lambda \mapsto E((\lambda a)')$ is analytic on W_a.*

Proof.   Adjoining a unit $\mathbf{1}$ if necessary, we recall that, for $\lambda \ne 0$,

$$(\lambda a)' = \mathbf{1} + \lambda^{-1}(a - \lambda^{-1}\mathbf{1})^{-1}.$$

Hence for $\lambda \ne 0$ the Proposition follows from 6.7. For $\lambda = 0$ we can apply the power series expansion (2). ∎

6.9.   The following Theorem is very important.

Theorem.   *Let a be an element of a normed algebra B. Then $\mathrm{Sp}_B(a)$ is non-void. Thus, if B is a Banach algebra, $\mathrm{Sp}_B(a)$ is a closed non-void bounded subset of $\mathbb{C}$.*

Proof.   Adjoining a unit if necessary, we may assume that B has a unit. If B_c is the Banach algebra completion of B, we have $\mathrm{Sp}_{B_c}(a) \subset \mathrm{Sp}_B(a)$ by 5.2(IV); so it is sufficient to prove that $\mathrm{Sp}_{B_c}(a)$ is non-void. We may therefore assume that B is complete.

Suppose that $\mathrm{Sp}_B(a)$ is void. For each continuous linear functional E on A, 6.7 then asserts that $\lambda \mapsto E((a - \lambda\mathbb{1})^{-1})$ is an entire function on $\mathbb{C}$ which vanishes at ∞. By Liouville's Theorem this implies that

$$E((a - \lambda\mathbb{1})^{-1}) = 0 \qquad (8)$$

for all complex λ and all E. But by the Hahn–Banach Theorem, to every non-zero element b of B there is a continuous linear functional E on B such that $E(b) \neq 0$. Hence we conclude from (8) that $(a - \lambda\mathbb{1})^{-1} = 0$ (for all complex λ). But this is impossible. So $\mathrm{Sp}_B(a)$ is non-void.

The last statement of the Theorem now follows from 6.5. ∎

6.10. Remark. By 5.3(III), Theorem 6.9 applied to the total matrix algebra $M(n, \mathbb{C})$ says that every $n \times n$ complex matrix a has an "eigenvector," that is, a non-zero vector x in $\mathbb{C}^n$ such that $ax = \lambda x$ for some complex number λ. This well known fact also follows immediately from the Fundamental Theorem of Algebra applied to the characteristic equation of a. It is therefore worth recalling that Liouville's Theorem, which played the crucial role in the proof of 6.9, also supplies one of the classical proofs of the Fundamental Theorem of Algebra. Indeed, if p is a non-constant polynomial in one complex indeterminate, and if $p(z) \neq 0$ for all z in $\mathbb{C}$, then $z \mapsto (p(z))^{-1}$ is an entire function on $\mathbb{C}$ which vanishes at ∞. Hence by Liouville's Theorem $(p(z))^{-1} \equiv 0$, an impossibility.

6.11. Here is an interesting application of 6.9 which has a bearing on the Heisenberg commutation relations of quantum mechanics.

Corollary. *Let B be a normed algebra with a unit $\mathbb{1}$. There cannot exist two elements a, b of B satisfying $ab - ba = \mathbb{1}$.*

Proof. Replacing B by its Banach algebra completion, we may as well assume that B is a Banach algebra. Suppose that $ab - ba = \mathbb{1}$, and put $S = \mathrm{Sp}_B(ba)$. From 5.2(III) and 5.5 we conclude that the symmetric difference of S and $S + 1$ is contained in $\{0\}$. But this contradicts the fact (6.9) that S is non-void and bounded. ∎

Remark. There are of course (unnormed) algebras containing elements a, b satisfying $ab - ba = \mathbb{1}$. For example, in the algebra of operators on functions of one variable x, take a to be d/dx and b to be multiplication by x.

6.12. One of the most fundamental consequences of 6.9 is Mazur's Theorem on normed division algebras.

A *normed division algebra* is of course a normed algebra B which has a unit 1, and in which every non-zero element a has an inverse a^{-1}.

Theorem (Mazur). *A normed division algebra B is one-dimensional, that is, $B \cong \mathbb{C}$.*

Proof. If $a \in B$, then by 6.9 $(a - \lambda 1)^{-1}$ fails to exist for some complex λ. By the definition of a division algebra this implies that $a - \lambda 1 = 0$, or $a = \lambda 1$. So $B = \mathbb{C}1$. ∎

6.13. *Remark*. It is essential for 6.9 and 6.12 that the algebras should be normed. Take for example the division algebra B of all complex rational functions

$$\frac{a_0 x^n + a_1 x^{n-1} + \cdots + a_n}{b_0 x^m + b_1 x^{m-1} + \cdots + b_m}$$

in one indeterminate x. The only elements of B which have non-void spectrum are the complex multiples of the unit. This, incidentally, shows by 6.9 that there is no norm which makes B into a normed algebra.

6.14. At the end of this section (6.20 and 6.21) we will present some important consequences of 6.12 for irreducible representations on Banach spaces.

6.15. Here is another interesting consequence of the application of analytic function theory to Banach algebras. We recall from 6.5 that $\mathrm{Sp}_A(a)$ is a closed bounded set.

Proposition. *Suppose that the Banach algebra A has a unit 1, and that B is a norm-closed subalgebra of A containing 1. If $a \in B$ and W is the unbounded connected component of $\mathbb{C} \setminus \mathrm{Sp}_A(a)$, we have*

$$\mathrm{Sp}_A(a) \subset \mathrm{Sp}_B(a) \subset \mathbb{C} \setminus W. \tag{9}$$

Proof. The first inclusion (9) was observed in 5.2(IV).

To prove the second, we take an arbitrary continuous linear functional E on A which vanishes on B. By 6.7 the function $f(\lambda) = E((a - \lambda 1)^{-1})$ is analytic on $\mathbb{C} \setminus \mathrm{Sp}_A(a)$. On the other hand, by 6.5 (applied to B) $(a - \lambda 1)^{-1} \in B$ for all large enough λ; so $f(\lambda) = 0$ for all large λ. By analytic continuation we deduce that f vanishes on W. By the Hahn–Banach Theorem every element of $A \setminus B$ is separated from 0 by some continuous linear functional E vanishing on B. The last two facts imply that $(a - \lambda\textsf{1})^{-1} \in B$ for all λ in W. It follows that $\text{Sp}_B(a) \subset \mathbb{C} \setminus W$. ∎

6.16. Corollary. *Let A, B, and a be as in 6.15. Then, if $\mathbb{C} \setminus \text{Sp}_A(a)$ is connected, we have*

$$\text{Sp}_B(a) = \text{Sp}_A(a).$$

This will be the case, for example, if $\text{Sp}_A(a) \subset \mathbb{R}$.

6.17. We come now to a remarkable theorem of Gelfand.

If a is any element of A, we define the *spectral radius* $\|a\|_{\text{sp}}$ of a to be $\sup\{|\lambda| : \lambda \in \text{Sp}_A(a)\}$. By 6.9 $\|a\|_{\text{sp}}$ is non-negative and finite.

Theorem (Gelfand). *For any element a of the Banach algebra A we have*

$$\|a\|_{\text{sp}} = \lim_{n \to \infty} \|a^n\|^{1/n}. \tag{10}$$

Proof. By 5.6 (applied after adjoining a unit if necessary) we have $\|a^n\|_{\text{sp}} = (\|a\|_{\text{sp}})^n$ for each positive integer n. By 6.5, $\|a^n\|_{\text{sp}} \leq \|a^n\|$. Combining these two facts we get $\|a^n\|^{1/n} \geq \|a\|_{\text{sp}}$ $(n = 1, 2, \ldots)$, whence

$$\liminf_{n \to \infty} \|a^n\|^{1/n} \geq \|a\|_{\text{sp}}. \tag{11}$$

Now let E be any continuous linear functional on A. By 6.8 the function $f : \lambda \mapsto E((\lambda a)')$ is analytic for $|\lambda| < \|a\|_{\text{sp}}^{-1}$, and so can be expanded in a power series about 0, whose coefficients are obtained by substituting λa for a in 6.3(2):

$$f(\lambda) = -\lambda E(a) - \lambda^2 E(a^2) - \lambda^3 E(a^3) - \cdots. \tag{12}$$

Since f is analytic in $\{\lambda : |\lambda| < \|a\|_{\text{sp}}^{-1}\}$, the right side of (12) must converge for $|\lambda| < \|a\|_{\text{sp}}^{-1}$. Consequently, if we fix a complex λ satisfying $|\lambda| < \|a\|_{\text{sp}}^{-1}$, the set of numbers

$$\{E(\lambda^n a^n) : n = 1, 2, \ldots\}$$

will be bounded. By the arbitrariness of E and the Uniform Boundedness Principle, this says that $\{\lambda^n a^n : n = 1, 2, \ldots\}$ is bounded:

$$\|\lambda^n a^n\| \leq k \qquad (n = 1, 2, \ldots).$$

(Of course, k depends on λ). This implies that

$$\|a^n\|^{1/n} \leq k^{1/n} |\lambda|^{-1} \qquad (n = 1, 2, \ldots). \tag{13}$$

Letting $n \to \infty$ in (13) we obtain

$$\limsup_{n \to \infty} \| a^n \|^{1/n} \leq |\lambda|^{-1}. \tag{14}$$

Since (14) is true for every λ in $\mathbb{C}$ satisfying $|\lambda| < \| a \|_{sp}^{-1}$, we get

$$\limsup_{n \to \infty} \| a^n \|^{1/n} \leq \| a \|_{sp}. \tag{15}$$

Combining (11) and (15) we obtain (10). ∎

6.18.   Corollary. *If B is a norm-closed subalgebra of A, and $a \in B$, the spectral radii of a with respect to A and B are the same.*

Proof.   Apply (10) in A and B.   ∎

This Corollary could also be deduced from 6.15.

6.19.   Corollary. *If B is a normed algebra and a is an element of B such that* $\mathrm{Sp}_B(a) = \{0\}$, *then* $\lim_{n \to \infty} \| a^n \|^{1/n} = 0$.

Proof.   Apply 6.17 to the completion of B.   ∎

Remark.   An element a of A which satisfies $\lim_{n \to \infty} \| a^n \|^{1/n} = 0$ is called a *generalized nilpotent element.*

We leave it as an exercise for the reader to show that a generalized nilpotent element of a finite-dimensional algebra is nilpotent.

Remark.   An example in Rickart [2], Appendix A.1.1, shows that the set of all generalized nilpotent elements of a Banach algebra A need not be norm-closed in A. This shows that in general the numerical function $a \mapsto \| a \|_{sp}$ will not be continuous on A (though by Exercise 49 it must be at least upper semi-continuous). If however A is a *commutative* Banach algebra, then $a \mapsto \| a \|_{sp}$ will be continuous on A (see Exercise 50).

6.20.   Mazur's Theorem enables us to prove an interesting generalization of Burnside's Theorem (see IV.4.10).

Proposition.   *Let B be any algebra, and T an algebraically irreducible locally convex representation of B on a normed linear space X(T). Then T is algebraically totally irreducible.*

Proof.   Let $\mathscr{D}$ be the commuting division algebra of T (see IV.4.2). By IV.4.6 it is enough to show that $\mathscr{D}$ is one-dimensional; and hence by 6.12 it is enough to show that $\mathscr{D}$ is a normed division algebra under a suitable norm.

We shall denote by $\| \ \|$ the norm on $X(T)$, and also the operator norm of bounded linear operators on $X(T)$. Fix a nonzero vector ξ in $X(T)$; and, if $D \in \mathcal{D}$, set

$$\|D\|_1 = \inf\{\|T_a\| : a \in B, \ T_a\xi = D\xi\}. \tag{16}$$

Since T is algebraically irreducible, the set on the right side of (16) is non-void, and $0 \le \|D\|_1 < \infty$. Note that $\|D\|_1 \ge \|\xi\|^{-1}\|D\xi\|$, so that $\|D\|_1 = 0 \Leftrightarrow D\xi = 0 \Leftrightarrow D = 0$. It is evident that

$$\|\lambda D\|_1 = |\lambda| \|D\|_1 \qquad (D \in \mathcal{D}; \lambda \in \mathbb{C}). \tag{17}$$

Let D, E be in $\mathcal{D}$. We have

$$\|D + E\|_1 \le \inf\{\|T_{a+b}\| : a, b \in B, \ T_a\xi = D\xi, \ T_b\xi = E\xi\}$$

$$\le \inf\{\|T_a\| + \|T_b\| : a, b \in B, \ T_a\xi = D\xi, \ T_b\xi = E\xi\}$$

$$= \|D\|_1 + \|E\|_1. \tag{18}$$

Further, if $a, b \in B$, $T_a\xi = D\xi$, and $T_b\xi = E\xi$, then $T_{ba}\xi = T_bD\xi = DT_b\xi = DE\xi$; so

$$\|DE\|_1 \le \|D\|_1\|E\|_1 \tag{19}$$

by a calculation similar to (18).

By (17), (18), and (19), $\mathcal{D}$ is a normed division algebra under $\| \ \|_1$. As we have seen this completes the proof. ∎

6.21. Corollary. *Let A be a Banach algebra, and T any algebraically irreducible algebraic representation of A. Then T is algebraically totally irreducible.*

Proof. By 5.14 T can be identified with the natural representation of A on A/I, where I is some regular maximal left ideal of A. By 6.6 I is norm-closed. So by 4.14 T is a Banach representation of A; and the conclusion follows from 6.20. ∎

6.22. Corollary. *An algebraically irreducible representation T of a commutative Banach algebra A is one-dimensional.*

Proof. By 6.21, T is algebraically totally irreducible. Now apply IV.4.7. ∎

6.23. Notice the following useful fact which emerged from the proof of 6.21 (in view of 4.14(5)):

Proposition.   *If T is any algebraically irreducible algebraic representation of a Banach algebra A, then X(T) can be normed so that it becomes a Banach space, and so that*

$$\| T_a \| \leq \| a \| \qquad \text{for all a in A.}$$

6.24.   Since any two norms on a finite-dimensional space are equivalent, 6.23 implies the following Corollary:

Corollary.   *Every finite-dimensional irreducible representation T of a Banach algebra A is continuous (i.e., there exists a positive constant k such that $\| T_a \| \leq k \| a \|$ for $a \in A$, $\| T_a \|$ being the operator norm of T_a with respect to any fixed norm on X(T)).*

Remark.   It follows immediately from this Corollary that a finite-dimensional completely reducible representation of a Banach algebra A is automatically continuous. However, a finite-dimensional representation T of A which is not completely reducible need not be continuous. For example, let A have trivial multiplication, take a non-continuous linear functional ϕ on A, and set

$$T_a = \begin{pmatrix} 0 & 0 \\ \phi(a) & 0 \end{pmatrix}.$$

7.   Commutative Banach Algebras

7.1.   *In this section A is a fixed commutative Banach algebra.*

By a *complex homomorphism* of A we shall mean a non-zero one-dimensional representation of A, that is, a non-zero linear functional $\phi: A \rightarrow \mathbb{C}$ satisfying $\phi(ab) = \phi(a)\phi(b)$ $(a, b \in A)$.

Definition.   The space of all complex homomorphisms of A will be denoted here by $A^\dagger$. We always equip $A^\dagger$ with the topology of pointwise convergence on A; this is called the *Gelfand topology* of $A^\dagger$. Equipped with this topology, $A^\dagger$ is referred to as the *Gelfand space* of A.

If A has a unit $\mathbf{1}$, then of course

$$\phi(\mathbf{1}) = 1 \qquad \text{for all } \phi \text{ in } A^\dagger. \tag{1}$$

7.2.   By 6.21 and IV.4.7, every algebraically irreducible (algebraic) representation of A is one-dimensional. So $A^\dagger$ as defined in 7.1 coincides with the $\overline{A}$ defined in IV.3.8.

7.3. Proposition. *Every element ϕ of $A^{\dagger}$ is continuous; in fact, $\|\phi\| \leq 1$.*

Proof. This follows immediately from 6.23. ∎

7.4. Proposition. *The map $\phi \mapsto \mathrm{Ker}(\phi)$ $(\phi \in A^{\dagger})$ is a bijection of $A^{\dagger}$ onto the set of all regular maximal ideals of A.*

Proof. Let ϕ be in $A^{\dagger}$. Then $\mathrm{Ker}(\phi)$ is of codimension 1 and so is a maximal ideal of A. Any element u of A for which $\phi(u) = 1$ is an identity modulo $\mathrm{Ker}(\phi)$; so $\mathrm{Ker}(\phi)$ is regular. Conversely, if u is any identity modulo $\mathrm{Ker}(\phi)$, then $\phi(u)\phi(a) - \phi(a) = \phi(ua - a) = 0$ for all a in A; so that $\phi(u) = 1$. This shows that ϕ is determined by $\mathrm{Ker}(\phi)$.

It remains only to show that every regular maximal ideal I of A is of the form $\mathrm{Ker}(\phi)$ for some ϕ in $A^{\dagger}$. Now by 5.13 the natural representation of A on A/I is irreducible, hence by 7.2 one-dimensional. It follows that $A/I \cong \mathbb{C}$; and so the quotient homomorphism $A \to A/I$ can be identified with an element ϕ of $A^{\dagger}$. Evidently $I = \mathrm{Ker}(\phi)$. ∎

Remark. In the crucial last paragraph of the above proof we appealed to 7.2, which in turn depended on 6.20 and 6.21 and therefore on the version IV.4.6 of the Jacobson Density Theorem. We can easily avoid this appeal to the Jacobson Theorem as follows:

Let I be a regular maximal ideal of A, and u an identity modulo I. By 6.6 I is closed; hence A/I is a Banach algebra. Let $\pi: A \to A/I$ be the quotient homomorphism. Then $\pi(u)$ is the unit element of A/I. If $a \in A \setminus I$, $\pi^{-1}(\pi(Aa))$ is an ideal of A properly containing I, and so equal to A; it follows that $\pi(a)$ has an inverse in A/I. Therefore A/I is a normed division algebra, and thus by 6.12 one-dimensional. Consequently, as before, π can be identified with an element of $A^{\dagger}$, with kernel I.

7.5. Proposition. *The Gelfand space $A^{\dagger}$ of A is a locally compact Hausdorff space. If A has a unit $A^{\dagger}$ is compact.*

Proof. Let S be the unit ball $\{\psi \in A^{*}: \|\psi\| \leq 1\}$ of A^{*}. It is well known that S is compact in the weak* topology (i.e., the topology of pointwise convergence on A). By 7.3 $A^{\dagger} \subset S$. If ϕ belongs to the weak* closure $\mathrm{cl}(A^{\dagger})$ of $A^{\dagger}$ in S, an easy argument with limits shows that $\phi(ab) = \phi(a)\phi(b)$ $(a, b \in A)$; so either $\phi \in A^{\dagger}$ or $\phi = 0$. Thus we have two cases:

Case I: $0 \notin \mathrm{cl}(A^{\dagger})$. Then by the preceding paragraph $A^{\dagger}$ is weakly* closed in S, hence weakly* compact.

Case II: $0 \in \mathrm{cl}(A^\dagger)$. Then $\mathrm{cl}(A^\dagger) = A^\dagger \cup \{0\}$, and $A^\dagger$ is obtained by removing one point from the weakly* compact space $\mathrm{cl}(A^\dagger)$. So $A^\dagger$ is locally compact.

If A has a unit then by (1) Case II cannot arise. So $A^\dagger$ must be compact.   ∎

7.6. Proposition. *If A is separable, then $A^\dagger$ satisfies the second axiom of countability.*

Proof. It is well known that the unit ball of the adjoint space of a separable Banach space is second-countable in the weak* topology. Hence, if A is separable, the S of the proof of 7.5 is second-countable; and hence so is its subspace $A^\dagger$.   ∎

7.7. Definition. If $a \in A$, let us denote by $\hat{a}$ the complex-valued function on $A^\dagger$ given by:

$$\hat{a}(\phi) = \phi(a) \qquad\qquad (\phi \in A^\dagger).$$

The function $\hat{a}$ is called the *Gelfand transform* of a. It is clear from the definition of $A^\dagger$ that $\hat{a}$ is continuous for each a in A, and that

$$(a + b)\hat{} = \hat{a} + \hat{b},$$

$$(\lambda a)\hat{} = \lambda \hat{a},$$

$$(ab)\hat{} = \hat{a}\hat{b}$$

$(a, b \in A; \lambda \in \mathbb{C})$. Since $\|\phi\| \leq 1$ for $\phi \in A^\dagger$ (7.3), we have

$$|\hat{a}(\phi)| \leq \|a\| \qquad\qquad (a \in A; \phi \in A^\dagger). \qquad (2)$$

7.8. Proposition. *If $A^\dagger$ is not compact, each $\hat{a}$ $(a \in A)$ vanishes at infinity on $A^\dagger$.*

Proof. If $A^\dagger$ is not compact, the proof of 7.5 shows that the one-point compactification of $A^\dagger$ can be identified topologically with $\mathrm{cl}(A^\dagger) = A^\dagger \cup \{0\}$ (the latter carrying the weak* topology). Each function $\hat{a}$ extends to a continuous function $\phi \mapsto \phi(a)$ on $\mathrm{cl}(A^\dagger)$; and the value of this function at 0 (the "point at infinity" for $A^\dagger$) is 0.   ∎

7.9. From 7.7 and 7.8 we obtain (referring to 4.6 for the meaning of $\mathscr{B}_c(A^\dagger)$, $\mathscr{C}_0(A^\dagger)$):

Proposition. *The Gelfand transform map $a \mapsto \hat{a}$ is a norm-decreasing algebra-homomorphism of A into $\mathscr{B}_c(A^\dagger)$, in fact into $\mathscr{C}_0(A^\dagger)$ if $A^\dagger$ is not compact.*

7.10. Definition. The kernel of the Gelfand transform map $a \mapsto \hat{a}$ $(a \in A)$ is referred to as the *radical* of A. It is of course a closed ideal of A. If the radical of A is $\{0\}$—that is, if the elements of $A^\dagger$ distinguish points of A—then A is called *semisimple*.

By 7.4 the radical of A is the intersection of all the regular maximal ideals of A.

By 7.2, these definitions are special cases of the general algebraic definitions IV.5.1.

7.11. Notice from 7.9 that A is semisimple if and only if, as an algebra, it is algebraically isomorphic with a subalgebra of $\mathscr{C}_0(S)$ for some locally compact Hausdorff space S.

7.12. Remark. Proposition 7.5 suggests the question of whether the existence of a unit is *necessary* for the compactness of $A^\dagger$. This is obviously not so in general. Indeed, if A is finite-dimensional, it need not have a unit, but $A^\dagger$ will always be finite, hence compact. However, if A is finite-dimensional *and semisimple*, then by IV.5.12 it has a unit. Analogously, it has been shown by Šilov that any *semisimple* commutative Banach algebra whose Gelfand space is compact must have a unit (see Rickart [2], Theorem 3.6.6). In other words, an arbitrary commutative Banach algebra A has compact Gelfand space if and only if A modulo its radical has a unit.

7.13. What is the connection between $\hat{a}$ and $\mathrm{Sp}_A(a)$?

Proposition. *Let a be an element of A. If A has a unit, $\mathrm{Sp}_A(a) = \mathrm{range}(\hat{a})$. If A has no unit, $\mathrm{Sp}_A(a) = \mathrm{range}(\hat{a}) \cup \{0\}$.*

Proof. We begin by observing (see the proof of 7.4) that, if $\phi \in A^\dagger$ and $b \in A$,

$$b \text{ is an identity modulo } \mathrm{Ker}(\phi) \Leftrightarrow \phi(b) = 1. \tag{3}$$

Suppose $0 \neq \lambda \in \mathbb{C}$. By 5.8, $\lambda \in \mathrm{Sp}_A(a)$ if and only if $\lambda^{-1}a$ has no adverse in A; by 5.12, this is the case if and only if $\lambda^{-1}a$ is an identity modulo some regular maximal ideal of A; and by (3) and 7.4 this happens if and only if $\lambda = \phi(a)$ for some ϕ in $A^\dagger$. This shows that $\mathrm{range}(\hat{a})$ and $\mathrm{Sp}_A(a)$ differ at most by the number 0.

If A has no unit, then 0 is in $\mathrm{Sp}_A(a)$; so by the preceding paragraph $\mathrm{Sp}_A(a) = \mathrm{range}(\hat{a}) \cup \{0\}$.

Let A have a unit. By 5.11 and 7.4 $0 \in \mathrm{Sp}_A(a)$ if and only if $a \in \mathrm{Ker}(\phi)$ for some ϕ in $A^\dagger$, that is, if and only if $0 \in \mathrm{range}(\hat{a})$. Combining this with the first paragraph we get $\mathrm{Sp}_A(a) = \mathrm{range}(\hat{a})$. ■

7.14. Corollary. *If A has a unit and $a \in A$, then a^{-1} exists if and only if $0 \notin \text{range}(\hat{a})$.*

7.15. Similarly, even if A has no unit, we have:

Corollary. *An element a of A has an adverse in A if and only if $1 \notin \text{range}(\hat{a})$.*

Proof. See the first paragraph of the proof of 7.13. ∎

7.16. Proposition. *If $a \in A$, then*

$$\|\hat{a}\|_{\infty} = \|a\|_{\text{sp}} = \lim_{n \to \infty} \|a^n\|^{1/n}.$$

Proof. Combine 7.13 with 6.17. ∎

7.17. Corollary. *The radical of A consists of those elements a of A for which $\lim_{n \to \infty} \|a^n\|^{1/n} = 0$.*

Thus the radical of A consists of the generalized nilpotent elements of A (see 6.19). For the finite-dimensional special case of this see Remark IV.5.4.

7.18. *The Gelfand Space and Adjunction of a Unit.* Suppose that A has no unit, and that A_1 is the commutative Banach algebra obtained by adjoining a unit $\mathbb{1}$ to A.

Each ϕ in $A^\dagger$ extends to a unique element ϕ_1 of $(A_1)^\dagger$ (on setting $\phi_1(\mathbb{1}) = 1$). Clearly $\{\phi_1 : \phi \in A^\dagger\}$ contains all the elements of $(A_1)^\dagger$ which do not vanish on A, that is, all except the complex homomorphism $\tau : a + \lambda \mathbb{1} \mapsto \lambda$ ($a \in A$; $\lambda \in \mathbb{C}$). Further, $\phi \mapsto \phi_1$ is a homeomorphism with respect to the Gelfand topologies of $A^\dagger$ and $(A_1)^\dagger$. So we have:

Proposition. *Suppose that A has no unit. Then the map $\phi \mapsto \phi_1$ defined above is a homeomorphism of $A^\dagger$ onto $(A_1)^\dagger \setminus \{\tau\}$.*

Thus, if $A^\dagger$ is not already compact, $(A_1)^\dagger$ can be identified with the one-point compactification of $A^\dagger$ (τ being the point at infinity).

8. Function Algebras and $\mathscr{C}_0(S)$

8.1. Let S be a fixed locally compact Hausdorff space. As in 4.6, $\mathscr{C}_0(S)$ denotes the Banach algebra of all continuous complex functions on S vanishing at infinity, with the supremum norm. (If S happens to be compact,

$\mathscr{C}_0(S)$ will mean just the algebra $\mathscr{C}(S)$ of all continuous complex functions on S.)

8.2. Suppose that A is a subalgebra of $\mathscr{C}_0(S)$ which is itself a (commutative) Banach algebra under some norm which may or may not coincide with the supremum norm of $\mathscr{C}_0(S)$. We shall assume in addition that:

(i) A vanishes nowhere on S (that is, for each s in S we have $f(s) \neq 0$ for some f in A);

(ii) A separates points of S (that is, if $s, t \in S$ and $s \neq t$, we have $f(s) \neq f(t)$ for some f in A).

Such a Banach algebra A is called a *function algebra on S.*

What is the relation between S and $A^{\dagger}$? Evidently each s in S gives rise to an element ϕ_s of $A^{\dagger}$, namely the "evaluation homomorphism":

$$\phi_s(f) = f(s) \qquad\qquad (f \in A).$$

We have:

Proposition. *The map $\phi : s \mapsto \phi_s$ is a homeomorphism of S onto a closed subset of $A^{\dagger}$.*

Proof. It follows from hypotheses (i) and (ii) that ϕ is one-to-one and into $A^{\dagger}$. It is evidently continuous.

Suppose that $\{s_i\}$ is a net of elements of S and α is an element of $A^{\dagger}$ such that

$$\phi_{s_i} \to \alpha \text{ in } A^{\dagger}. \qquad\qquad (1)$$

Then we claim that $\alpha = \phi_t$ for some t in S, and that $s_i \to t$. Indeed: Passing to a subnet of $\{s_i\}$ we may assume that $\{s_i\}$ converges to some t in the one-point compactification S_0 of S. But then by (1) $\alpha(f) = f(t)$ for all f in A. Since $\alpha \neq 0$, t cannot be the point at infinity; thus $t \in S$, and $\alpha = \phi_t$. The latter statement determines t uniquely (by (ii)). Thus every subnet of $\{s_i\}$ has a subnet converging to this t. It follows that $s_i \to t$; and the claim is proved.

This claim implies that ϕ^{-1} is continuous and that $\phi(S)$ is closed in $A^{\dagger}$. So the proof is finished. ∎

8.3. Remark. Every function algebra is semisimple. Conversely, every semisimple commutative Banach algebra A is isometrically isomorphic (under the Gelfand transform map) with a function algebra on $A^{\dagger}$. (See 7.11.)

8.4.   *Example.* We shall give here an example, very important for the theory of function algebras, showing that the homeomorphism ϕ of 8.2 need not be *onto* $A^\dagger$.

Let D and A be as in 4.8; and as usual let $\mathbb{E}$ be the boundary $\{z \in \mathbb{C} : |z| = 1\}$ of D. By the Maximum Modulus Principle of analytic function theory, the restriction map $\rho : f \mapsto f | \mathbb{E}$ is an isometric isomorphism of A into $\mathscr{C}(\mathbb{E})$; and so the range of ρ, which we shall call B, is a norm-closed subalgebra of $\mathscr{C}(\mathbb{E})$, hence a function algebra on $\mathbb{E}$ under the supremum norm. Now every point z of D gives rise to an element ψ_z of $B^\dagger$:

$$\psi_z(g) = (\rho^{-1}(g))(z) \qquad\qquad (g \in B).$$

The map $\psi : z \mapsto \psi_z$ of D into $B^\dagger$ is of course one-to-one. So, if $|z| < 1$, ψ_z is an element of $B^\dagger$ differing from all evaluations at points of $\mathbb{E}$. Thus, in this case, the ϕ of 8.2 does not map $\mathbb{E}$ onto $B^\dagger$.

One can show that in this example $B^\dagger \cong D$. More precisely:

Proposition*:   *The map ψ is a homeomorphism of D onto $B^\dagger$.*

Remark. This example suggests that function algebras might form an appropriate setting for an abstract generalization of analytic function theory. Indeed, suppose that A is any function algebra on a locally compact Hausdorff space S such that the ϕ of 8.2 maps S onto a proper subset of $A^\dagger$. Will it be the case that the Gelfand transforms $\hat{a}$ $(a \in A)$ show properties analogous to analyticity on $A^\dagger \setminus \phi(S)$? This question is too vague as it stands. Nevertheless, there are profound problems in the theory of functions of several complex variables which amount to the study of the relationship between S and $A^\dagger$ in special cases of the context of 8.2. See Wermer [1].

8.5.   *Remark.* Let A be a function algebra on S; and let $s_1, \ldots, s_n$ be a finite sequence of pairwise distinct points of S. By 8.2 the n one-dimensional representations $\phi_{s_1}, \ldots, \phi_{s_n}$ of A are non-zero and pairwise distinct. So IV.4.8 applied to the ϕ_{s_i} shows that, given any complex numbers $\lambda_1, \ldots, \lambda_n$ we can find an element a of A such that

$$a(s_i) = \lambda_i \qquad \text{for all } i = 1, \ldots, n.$$

That is, given any f in $\mathscr{C}_0(S)$ we can find an element a of A coinciding with f on any preassigned finite set F of points of S.

The finiteness of F is essential to the last statement. The stronger conjecture that A is necessarily dense in $\mathscr{C}_0(S)$ is far from true. This is illustrated in Example 8.4, where A and B fail to be dense in $\mathscr{C}(D)$ and $\mathscr{C}(\mathbb{E})$ respectively.

Notice that, if in addition we suppose that A is closed under complex conjugation, then the Stone–Weierstrass Theorem (Appendix A) asserts that A is dense in $\mathscr{C}_0(S)$.

The Function Algebra $\mathscr{C}_0(S)$

8.6. Let S be a locally compact Hausdorff space, fixed for the rest of the section. The most obvious function algebra on S is $\mathscr{C}_0(S)$ itself. We shall show that when $A = \mathscr{C}_0(S)$ the ϕ of 8.2 is *onto* $A^\dagger$; that is, $(\mathscr{C}_0(S))^\dagger$ can be identified, both setwise and topologically, with S.

As in 8.2 each s in S gives rise to the complex homomorphism $\phi_s: f \mapsto f(s)$ of $\mathscr{C}_0(S)$. Let I_s denote the kernel of ϕ_s, that is, the regular maximal ideal of $\mathscr{C}_0(S)$ consisting of all f which vanish at s.

Proposition. *If J is an ideal of $\mathscr{C}_0(S)$ such that $J \not\subset I_s$ for all s in S, then J contains all functions in $\mathscr{C}_0(S)$ which have compact support.*

Proof. Let g be a function in $\mathscr{C}_0(S)$ vanishing outside the compact subset D of S. For each s in D there is by hypothesis a function f_s in J such that $f_s(s) \neq 0$, and hence $f_s(t) \neq 0$ for all t in some open neighborhood U_s of s. Since D is compact, we can choose finitely many $s_1, \ldots, s_n$ in D such that $D \subset U_{s_1} \cup \cdots \cup U_{s_n}$. Then $h = f_{s_1}\bar{f}_{s_1} + \cdots + f_{s_n}\bar{f}_{s_n}$ belongs to J and is positive everywhere on D. Therefore the function p defined by

$$p(x) = \begin{cases} h(x)^{-1}g(x) & \text{if } h(x) \neq 0 \\ 0 & \text{if } h(x) = 0 \end{cases}$$

belongs to $\mathscr{C}_0(S)$, and

$$g = ph \in J. \qquad \blacksquare$$

8.7. Corollary. *Every proper closed ideal of $\mathscr{C}_0(S)$ is contained in some I_s $(s \in S)$. If S is compact, every proper ideal of $\mathscr{C}(S)$ (whether closed or not) is contained in some I_s.*

Proof. This follows from 8.6 when we recall that the set of all functions in $\mathscr{C}_0(S)$ with compact support is dense in $\mathscr{C}_0(S)$, and is equal to $\mathscr{C}(S)$ if S is compact. $\blacksquare$

8.8 Corollary. $S \cong (\mathscr{C}_0(S))^\dagger$. *That is, the map ϕ of 8.2 is a homeomorphism of S onto $(\mathscr{C}_0(S))^\dagger$. Every regular maximal ideal of $\mathscr{C}_0(S)$ is of the form I_s for some s in S.*

Proof. Since regular maximal ideals are closed (6.6), the second statement follows from 8.7. The first statement then follows from the second statement and 7.4 and 8.2. ∎

Remark. When we identify S with $\mathscr{C}_0(S)^\dagger$, each element of $\mathscr{C}_0(S)$ is its own Gelfand transform.

8.9. We shall now classify *all closed ideals of* $\mathscr{C}_0(S)$.

Proposition. *Every closed ideal I of $\mathscr{C}_0(S)$ is the intersection of the set of all regular maximal ideals which contain it. That is,*

$$I = \{f \in \mathscr{C}_0(S) : f(s) = 0 \text{ for all } s \text{ in } Q\}$$

for some closed subset Q of S.

Proof. Let us define $Q = \{s \in S : f(s) = 0 \text{ for all } f \text{ in } I\}$; and put $J = \{f \in \mathscr{C}_0(S) : f(s) = 0 \text{ for all } s \text{ in } Q\}$. We shall show that $I = J$.

Since Q is closed, $S \setminus Q$ is open, hence locally compact; and the restriction map $\rho : f \mapsto f|(S \setminus Q)$ is easily seen to be a linear isometry of J onto $\mathscr{C}_0(S \setminus Q)$. Evidently $I \subset J$; so $\rho(I)$ is a closed ideal of $\mathscr{C}_0(S \setminus Q)$. If $\rho(I) \neq \mathscr{C}_0(S \setminus Q)$, then by 8.7 (applied to $\mathscr{C}_0(S \setminus Q)$) there is an element s of $S \setminus Q$ such that $f(s) = 0$ for all f in I, contradicting the definition of Q. So $\rho(I) = \mathscr{C}_0(S \setminus Q)$, whence $I = J$. ∎

Remark. Let I and Q be as in the above Proposition. From Tietze's Extension Theorem we easily deduce that the map sending $f + I$ into $f|Q$ ($f \in \mathscr{C}_0(S)$) is an isometric isomorphism of the quotient Banach algebra $\mathscr{C}_0(S)/I$ onto $\mathscr{C}_0(Q)$.

8.10. The following result will be useful later on.

Proposition. *Let $\| \ \|_\infty$ be the usual supremum norm on $\mathscr{C}_0(S)$, and $\| \ \|$ any other norm on $\mathscr{C}_0(S)$ which makes the latter a (not necessarily complete) normed algebra (under the usual pointwise operations). Then*

$$\|f\| \geq \|f\|_\infty \qquad \text{for all } f \text{ in } \mathscr{C}_0(S).$$

Proof. Let A be the commutative Banach algebra obtained by completing $\mathscr{C}_0(S)$ with respect to $\| \ \|$. Then by 8.8 $A^\dagger$ can be identified with the set W of those s in S such that $\phi_s : f \mapsto f(s)$ is continuous with respect to $\| \ \|$.

We claim that W is dense in S. Indeed, suppose it were not. Then there would exist an element g of $\mathscr{C}_0(S)$ taking the value 1 on some non-void open set U but not taking the value 1 anywhere on W. From 7.15 it follows that g has an adverse h in A:

$$g + h - gh = 0. \tag{2}$$

Now let f be a non-zero function in $\mathscr{C}_0(S)$ which vanishes outside U; thus $gf = f$. Multiplying (2) by f, we get $f + hf - hf = 0$, or $f = 0$, a contradiction. So W is dense in S.

It follows that for any f in $\mathscr{C}_0(S)$

$$\| f \| \geq \sup_{s \in W} | f(s) | \qquad \text{(by 7.9)}$$

$$= \sup_{s \in S} | f(s) | \qquad \text{(since } W \text{ is dense in } S\text{)}$$

$$= \| f \|_\infty . \qquad \blacksquare$$

8.11. In conclusion we mention a function algebra of a quite different sort from $\mathscr{C}_0(S)$ or the example of 8.4.

Let n be a positive integer, and let D_n stand for the algebra, under pointwise multiplication, of all continuous complex functions on $[0, 1]$ which are n times continuously differentiable on $[0, 1]$.

Proposition*. *D_n is a Banach algebra under the norm*

$$\| f \| = \sum_{k=0}^{n} (k!)^{-1} \sup\{| f^{(k)}(t) |: 0 \leq t \leq 1\}$$

($f^{(k)}$ denoting the k-th derivative of f). Further, $(D_n)^\dagger \cong [0, 1]$; that is, the ϕ of 8.2 maps $[0, 1]$ onto $(D_n)^\dagger$.

To prove the last statement one may repeat the argument of 8.6.

Thus D_n is a function algebra on $[0, 1]$ which resembles $\mathscr{C}([0, 1])$ inasmuch as its Gelfand space is just $[0, 1]$. Other algebraic properties, however, distinguish D_n sharply from $\mathscr{C}([0, 1])$. To see this, let us say that a proper closed ideal I of a commutative Banach algebra A is *primary* if $I \subset J$ for one and only one regular maximal ideal J of A. It follows from 8.8 that a primary ideal of $\mathscr{C}_0(S)$ is necessarily maximal. However, D_n has non-maximal primary ideals. Indeed, for each t in $[0, 1]$, the $n + 1$ sets

$$J_r^{(t)} = \{ f \in D_n : f(t) = f'(t) = \cdots = f^{(r)}(t) = 0 \}$$

($r = 0, 1, \ldots, n$) form a nest of distinct closed primary ideals of D_n contained in the maximal ideal $J_0^{(t)}$. (As a matter of fact, the $J_r^{(t)}$ are the *only* primary ideals contained in $J_0^{(t)}$.)

Thus, Proposition 8.8 is not true for D_n. Whitney [1] has proved the following modification of 8.8 for D_n: *Every closed ideal of D_n is the intersection of the primary ideals of D_n which contain it.*

From this example the reader may be tempted to conjecture that, for any commutative Banach algebra A, every closed ideal of A is the intersection of the primary ideals of A which contain it. This is false. If A is the $\mathcal{L}_1$ group algebra of a locally compact Abelian group, it can be shown that: (i) Every primary ideal of A is maximal; (ii) in general it is not possible to represent an arbitrary closed ideal of A as an intersection of regular maximal ideals. (See for example Bourbaki [12], Chap. 2, §3, no. 1).

8.12. *Functions with Absolutely Convergent Fourier Series.* Let A be the linear space of all those complex continous functions f on $\mathbb{R}$ which can be expressed as the sum of an absolutely convergent series of the form

$$f(t) = \sum_{n \in \mathbb{Z}} c_n e^{int} \qquad (t \in \mathbb{R}), \qquad (3)$$

where the c_n are complex numbers satisfying

$$\sum_{n \in \mathbb{Z}} |c_n| < \infty.$$

Such functions are said to *have absolutely convergent Fourier series.* If f and the c_n are related by (3) one verifies that

$$c_n = (2\pi)^{-1} \int_{-\pi}^{\pi} f(t) e^{-int} \, dt.$$

Thus the c_n are determined by f; and the equation

$$\|f\| = \sum_{n \in \mathbb{Z}} |c_n|$$

defines a norm on A, making A a Banach space isometrically isomorphic with the Banach space of all summable complex functions on $\mathbb{Z}$. If f and g are functions in A corresponding via (3) with the complex coefficients $\{c_n : n \in \mathbb{Z}\}$ and $\{d_n : n \in \mathbb{Z}\}$ respectively, the pointwise product function fg is given by the absolutely convergent series

$$f(t) \cdot g(t) = \sum_{n \in \mathbb{Z}} b_n e^{int},$$

where

$$b_n = \sum_{m \in \mathbb{Z}} c_m d_{n-m'} \qquad (4)$$

so that

$$\sum_{n \in \mathbb{Z}} |b_n| \leq \sum_{n,m} |c_m| \, \|d_{n-m}\| = \|f\| \cdot \|g\|.$$

Thus, under pointwise multiplication and the norm $\|\ \|$, A is a commutative Banach algebra with unit. In view of (4), A is in fact isometrically isomorphic with the $\mathscr{L}_1$ group algebra of the additive group $\mathbb{Z}$.

Since the functions in A are periodic with period 2π, one can regard A as a function algebra on the circle group $\mathbb{E}$.

What are the elements of $A^\dagger$? We claim that they are exactly the evaluations at points of $\mathbb{R}$. Indeed: Let $\phi \in A^\dagger$; and let g be the element $t \mapsto e^{it}$ of A. Then both g and g^{-1} have norm 1 in A; so, by 7.3, $|\phi(g)| \leq 1$ and $|\phi(g)|^{-1} = |\phi(g^{-1})| \leq 1$. It follows that $|\phi(g)| = 1$; that is, there is a number u in $\mathbb{R}$ such that $\phi(g) = g(u)$. Thus $\phi(h) = h(u)$ whenever h is in the subalgebra of A generated by g and g^{-1}. Since this subalgebra is evidently dense in A, we have $\phi(h) = h(u)$ for all h in A, and the claim is proved.

Thus, as in the case of $\mathscr{C}_0(S)$ and the D_n of 8.11, the Gelfand space of A consists just of the evaluations of the functions in A at the points of their domain.

Combining this with Corollary 7.14 we obtain the following famous theorem of Norbert Wiener (for which the theory of Banach algebras thus appears to be the natural setting):

Wiener's Theorem. *If $f : \mathbb{R} \to \mathbb{C}$ is a continuous function with absolutely convergent Fourier series, and if $f(t) \neq 0$ for all real t, then f^{-1} has absolutely convergent Fourier series.*

9. Factorization in Banach Algebras

9.1. The possibility of decomposing a given algebraic expression into a product of two (or more) factors enjoying special properties has proved useful in algebra and analysis. In 9.2 we prove a factorization theorem which states that if A is a Banach algebra with left approximate unit, then for $c \in A$ there are elements a, b in A such that $c = ab$, where b belongs to the closed left ideal generated by c, and b is arbitrarily close to c. Actually a slight, but very useful, generalization of this result will be treated.

Recall that the Banach algebra A has a *left approximate unit* if there is a net $\{e_\alpha\}$ in A and a positive real constant M such that $\|e_\alpha\| \leq M$ for all α, and $e_\alpha a \to a$ for all $a \in A$. Equivalently, A has a left approximate unit if there is a positive real constant M such that for each finite subset $\{a_1, \ldots, a_n\}$ of A and each $\varepsilon > 0$ there exists $e \in A$ with $\|ea_i - a_i\| < \varepsilon$, $i = 1, \ldots, n$, and $\|e\| \leq M$.

9.2. Theorem. (Cohen–Hewitt).   *Let A be a Banach algebra having a left approximate unit $\{e_\alpha\}$; and let T be a Banach representation of A, acting on a Banach space X, such that for some constant K we have $\|T_a\| \leq K\|a\|$ for all a in A. Then $\{T_a x : a \in A, x \in X\}$ is a closed linear subspace of X.*

Proof.   If A has a unit element, the theorem is trivial. So let us assume that A has no unit, adjoin 1 to get the Banach algebra A_1, and extend T to a representation of A_1 by letting T_1 be the identity operator.

For brevity we will write ax instead of $T_a x$ throughout this proof, and up to the end of 9.4.

Now the closed linear span X_0 of the set $\{ax : a \in A, \ x \in X\}$ is $\{x \in X : \lim_\alpha e_\alpha x = x\}$. Given $x \in X_0$, following the original argument of Cohen, we shall define a sequence $\{e_n\}$ of elements in A with $\|e_n\| \leq M$ so that, if $\beta = M^{-1}$, then

$$b_n = [(1 + \beta)1 - \beta e_n]^{-1} \ldots [(1 + \beta)1 - \beta e_1]^{-1}$$

$$= (1 + \beta)^{-n}1 + a_n \qquad\qquad (a_n \in A) \qquad (1)$$

converges in A_1 to a limit $b \in A$ and $y_n = b_n^{-1} x$ converges in $X_0 \subset X$ to an element y.

Observe that since $\|e_n\| \leq M$ then $1 - \beta(1 + \beta)^{-1} e_n$ is invertible, so that $(1 + \beta)1 - \beta e_n$ is also. By the series expansion for inverses (see 6.2) it follows that

$$[(1 + \beta)1 - \beta e_n]^{-1} = (1 + \beta)^{-1}1 + a$$

for some $a \in A$. Hence the relation (1) defines the elements b_n and a_n in the spaces A_1 and A, respectively. Then we have $x = \lim_{n \to \infty} b_n y_n = by$.

We turn now to the problem of defining the sequence $\{e_n\}$. This is done as follows: Set $a_0 = 0$, $b_0 = 1$ and choose $\{e_n\}_{n=1}^{\infty}$ inductively so that for all $n \geq 0$

$$\|e_{n+1}\| \leq M,$$

$$\|e_{n+1} a_n - a_n\| < M(1 + \beta)^{-n-1},$$

$$\|x - e_{n+1} x\| < \|b_n^{-1}\|^{-1} M(1 + \beta)^{-n-1}.$$

Since $b_n - a_n \to 0$, to prove that $\{b_n\}$ is convergent, it suffices to show that $\{a_n\}$ is Cauchy. Now

$$a_{n+1} = b_{n+1} - (1 + \beta)^{-n-1}\mathbf{1}$$

$$= [(1 + \beta)\mathbf{1} - \beta e_{n+1}]^{-1}[(1 + \beta)^{-n}\mathbf{1} + a_n] - (1 + \beta)^{-n-1}\mathbf{1},$$

so that

$$\|a_{n+1} - a_n\|$$

$$= \|[(1 + \beta)\mathbf{1} - \beta e_{n+1}]^{-1}[\beta(1 + \beta)^{-n-1}e_{n+1} + \beta e_{n+1}a_n - \beta a_n]\|.$$

Expanding $[\mathbf{1} - (1 + \beta)e_{n+1}]^{-1}$ in power series as in 6.2 and using $\|e_{n+1}\| \leq M$, we obtain $\|[(1 + \beta)\mathbf{1} - \beta e_{n+1}]^{-1}\| \leq M$. Since

$$\|\beta(1 + \beta)^{-n-1}e_{n+1}\| \leq (1 + \beta)^{-n-1}$$

and

$$\beta\|e_{n+1}a_n - a_n\| \leq (1 + \beta)^{-n-1},$$

it follows that $\|a_{n+1} - a_n\| \leq 2M(1 + \beta)^{-n-1}$. Hence for $m > n$

$$\|a_n - a_m\| \leq \|a_n - a_{n+1}\| + \|a_{n+1} - a_{n+2}\| + \cdots + \|a_{m-1} - a_m\|$$

$$\leq 2M(1 + \beta)^{-n-1}(1 + (1 + \beta)^{-1} + \cdots + (1 + \beta)^{n-m+1})$$

$$\leq 2M(M + 1)(1 + \beta)^{-n-1} \to 0 \qquad \text{as } n \to \infty,$$

and so $\{a_n\}$ is Cauchy in A.

Furthermore, $y_{n+1} = b_n^{-1}[(1 + \beta)\mathbf{1} - \beta e_{n+1}]x$, so that

$$\|y_{n+1} - y_n\| = \|b_n^{-1}(\beta x - \beta e_{n+1}x\|$$

$$\leq \beta\|b_n^{-1}\| \cdot \|x - e_{n+1}x\|$$

$$\leq (1 + \beta)^{-n-1}$$

and hence, as with the a_n's, the sequence $\{y_n\}$ is Cauchy, as required. ∎

9.3. Corollary. *Let S be a compact subset of the X_0 of 9.2. Then there exists an element a of A and a continuous map $g: S \to X_0$ such that*

$$x = ag(x) \qquad \text{for all } x \in S.$$

Proof. Let B denote the Banach space of continuous functions from S into X_0 under the natural linear operations and norm $\|f\| = \sup\{\|f(s)\| : a \in A\}$; and consider the Banach representation T of A on B given by $(T_a f)(s) = af(s)$ for $a \in A$, $s \in S$, $f \in B$. If $\{e_\alpha\}$ is a left approximate unit in A, then $e_\alpha x \to x$

for each $x \in X_0$, and since $\{e_\alpha\}$ is bounded, $e_\alpha x \to x$ uniformly on compact subsets of X_0. If $f \in B$, then $f(S)$ is compact; so $e_\alpha f(s) \to f(s)$ uniformly on S. Hence, $B_0 = B$. Let $i \in B$ be the identity, i.e., $i(s) = s$ for all $s \in S$. Then by 9.2 there exists $a \in A$ and $g \in B$ such that $i = ag$ and, in particular, $S = ag(S)$. Setting $T = g(S)$, the corollary follows.   ∎

9.4.   Remark. If $\{x_n\}$ is a sequence in X_0 which converges to 0, then Corollary 9.3 can be applied with $S = \{0, x_1, x_2, \ldots\}$. Let g be as in the proof of 9.3 and set $y_n = g(x_n)$. Then $ay_n = x_n$, $ag(0) = 0$, and $y_n \to g(0)$. Set $z_n = y_n - g(0)$. Then $az_n = x_n$ and $z_n \to 0$. This factoring of null sequences has several useful applications.

9.5.   We recall that a Banach algebra A *has a (two-sided) approximate unit* if there is a net $\{e_\alpha\}$ of elements of A, and a constant M, such that $\|e_\alpha\| \le M$ for all α, and $\lim e_\alpha a = a = \lim a e_\alpha$ for all $a \in A$.

Corollary.   *Let A be a Banach algebra with a (two-sided) approximate unit. Then, for any norm-compact subset S of A, there are elements a, b of A and continuous functions $g : S \to A$ and $h : S \to A$ such that*

$$x = ag(x) = h(x)b \qquad \text{for all } x \in S.$$

Proof.   Apply 9.3 to the regular representation of A by left multiplication on A, and to the regular representation of the "reverse" algebra of A by right multiplication on A.   ∎

10.   Exercises for Chapter V

1. Give an example of a finite-dimensional non-degenerate representation of an algebra A which has a degenerate subrepresentation. [Note: This cannot happen in the involutory context; see VI.9.9.]

2. Given a locally convex representation T of an algebra A, show that the implications described in 1.14 between the four notions of irreducibility hold, and construct suitable examples to show that no other implications hold.

3. Let T be a topologically irreducible locally convex representation of an algebra A. If Y is any finite-dimensional subspace, show that there is an element a of A such that $T_a | Y$ is one-to-one on Y. If T is algebraically irreducible, show that a can even be chosen so that T_a is the identity operator on Y.

4. Provide a detailed proof of Proposition 1.18.

5. Let J be a two-sided ideal of an algebra A, and T a totally irreducible locally convex representation of A. Prove that either $J \subset \mathrm{Ker}(T)$ or $T|J$ is totally irreducible.

6. Let S and T be two topologically irreducible locally convex representations of an algebra A. Show that S and T are Naimark-related if and only if there exist non-zero elements $\xi \in X(S)$, $\eta \in X(T)$, $\alpha \in X(S)^*$, and $\beta \in X(T)^*$ such that

$$\alpha(S_a\xi) = \beta(T_a\eta) \quad \text{for all} \quad a \in A.$$

7. Let A be the algebra of all complex polynomial functions of one variable. If $p \in A$, let

$$\|p\|_1 = \sup\{|p(t)|: 0 \le t \le 1\},$$

$$\|p\|_2 = \sum_{i=0}^{n} |a_i|, \text{ where } p(t) = \sum_{i=0}^{n} a_i t^i \quad (t \in \mathbb{R}).$$

(i) Show that $A, \| \ \|_1$ and $A, \| \ \|_2$ are both normed algebras.

(ii) What are the completions of A with respect to these two norms?

8. Let A be the algebra of all complex matrices of the form $\begin{pmatrix} \alpha & \beta \\ 0 & 0 \end{pmatrix}$ with the usual linear operations and matrix product. Find the algebra A_1 obtained from A by adjoining a unit.

9. Let A be a normed algebra with unit 1. Let $x \in A$ and set $\alpha_n = \|x^{2^n} - 1\|$. Prove that if $\limsup \alpha_n < 1$, then for some n, $x^{2^n} = 1$. Show in addition if every $\alpha_k < 2$, then $x = 1$.

10. Show that a normed algebra A has a left approximate unit (bounded by k) if and only if for every finite subset $\{a_1, \ldots, a_n\}$ of elements of A and every $\varepsilon > 0$ there exists e in A (with $\|e\| \le k$) such that $\|a_i - ea_i\| < \varepsilon$ for $i = 1, \ldots, n$.

11. A left approximate unit $\{e_i\}_{i \in I}$ in a normed algebra A is said to be *sequential* if I is the set of positive integers with the usual order. Show that if A is separable and has a left approximate unit, then A has a sequential left approximate unit bounded by the same constant.

12. Let S be a non-compact locally compact Hausdorff space. Show that the Banach algebra $\mathscr{C}_0(S)$ of continuous complex-valued functions on S which vanish at infinity has an approximate unit by explicitly constructing one.

13. Let A denote the Banach algebra of complex-valued functions on $[0, 1]$ with continuous first derivatives satisfying $f(0) = 0$, where the norm of an element f is defined by

$$\|f\| = \sup_{0 \le t \le 1} |f(t)| + \sup_{0 \le t \le 1} |f'(t)|.$$

Show that A does not admit an approximate unit.

14. Let A be a normed algebra with a right unit. Prove that if A has a left approximate unit, then A has a two-sided unit.

15. Let A be a normed algebra and I a closed two-sided ideal in A. Prove that if both I and A/I have left approximate units, then A has a left approximate unit. (It is not known if this result remains true if the approximate units are not required to be bounded.) [See Doran and Wichmann [2, p. 43].]

16. An approximate unit $\{e_i\}$ in a normed algebra A is called *central* if $e_i a = a e_i$ for all a in A and all i. Show that the $\mathscr{L}_1$ group algebra of a compact group G has a central approximate unit (with bound 1).

17. Give an example of a Banach algebra such that each element is a sum of products but which also contains an element which is not a product.

18. Let A be a Banach algebra such that, for every Banach algebra B, every homomorphism of B into A is continuous. Prove that A contains no non-zero element a with $a^2 = 0$.

19. Prove that if A is an algebra and I is a two-sided ideal of A, then for each a in A, $\mathrm{Sp}_{A/I}(a + I) \subset \bigcap_{b \in a + I} \mathrm{Sp}_A(b)$.

20. Let A be an algebra with unit $\mathbf{1}$. Prove the "resolvent equation" $x(\lambda) - x(\mu) = (\lambda - \mu)x(\lambda)x(\mu)$ for all $\lambda, \mu \notin \mathrm{Sp}_A(x)$, where $x \in A$ and $x(\lambda) = (x - \lambda\mathbf{1})^{-1}$.

21. Give an example of a Banach algebra A with unit and elements a, b in A such that $\mathrm{Sp}_A(ab) \neq \mathrm{Sp}_A(ba)$ (see 5.5).

22. Show that every non-void closed bounded subset of the complex plane is the spectrum of some element in a suitable Banach algebra.

23. Let A be a Banach algebra with unit. If a is a boundary point of the set of non-invertible elements of A, is 0 a boundary point of the spectrum of a?

24. Let A be a Banach algebra with unit. Let T be any component of the group of invertible elements of A. Prove that the union of the spectra of all elements of T is equal to the set of all nonzero complex numbers.

25. Let A be an algebra with unit $\mathbf{1}$. Show that if a is a nilpotent element of A (i.e., $a^n = 0$ for some positive integer n), then the element $\mathbf{1} + a$ is invertible in A.

26. Let A be an algebra with unit $\mathbf{1}$.
(a) Prove that if $a, b \in A$ and at least two elements of the set $\{a, b, ab, ba\}$ are invertible in A, then all of them are invertible.
(b) Prove that if A is finite-dimensional and ab is invertible, then both a and b are invertible.
(c) Show that (b) is false if the assumption of finite-dimensionality is dropped.

27. Give an example of a normed algebra A with unit $\mathbf{1}$ and a singular element a in A satisfying $\|a - \mathbf{1}\| < 1$.

28. Let A be a Banach algebra with unit $\mathbf{1}$. Given an element a in A, define the *exponential function* of a by $\exp(a) = \sum_{k=0}^{\infty} a^k/k!$ (set $a^0 = \mathbf{1}$). Show that the series defining $\exp(x)$ is absolutely convergent and establish the following:
(a) If a and b commute, then $\exp(a + b) = \exp(a)\exp(b)$;
(b) $\exp(-a) = (\exp(a))^{-1}$ for all $a \in A$;
(c) If $a \in A$ and $\|\mathbf{1} - a\| < 1$, there exists $b \in A$ with $\exp(b) = a$;
(d) If A is commutative, $\exp(A) = \{\exp(a): a \in A\}$ is the connected component of $\mathbf{1}$ in the group of invertible elements of A.

29. Let A be a Banach algebra with unit 1. Show that if $\exp(a + b) = \exp(a)\exp(b)$ for all a, b in A, then A is commutative.

30. Let A be an algebra and let a and b be elements of A with adverses a' and b' (see 5.7). Prove that $a + b$ has an adverse if and only if $a'b'$ has an adverse.

31. Prove that if a and b are elements of an algebra A such that $a = ab$ and b has an adverse, then $a = 0$.

32. Prove that if a is an element in an algebra A such that a^n has an adverse for some positive integer n, then a has an adverse. Show that the converse is false.

33. Let A be a Banach algebra, and a an element of A with $\|a\| < 1$. Define the sequence $\{b_n\}$ inductively as follows: $b_0 = 0$, $b_n = -a + 2ab_{n-1} + b_{n-1}^2 - ab_{n-1}^2$ ($n = 1, 2, \ldots$). Show that $\{b_n\}$ converges in norm to the adverse a' of a.

34. Let A be an algebra and I a left-regular two-sided ideal of A. Prove that if J is a left-regular two-sided ideal of A, then $I \cap J$ is a left-regular two-sided ideal of A.

35. Let A be an algebra and A_1 the algebra obtained from A by adjoining a unit. Let I be an arbitrary maximal right ideal of A_1 distinct from A. Prove that $A \cap I$ is a regular maximal right ideal of A, and the mapping $I \to A \cap I$ is a bijective correspondence between the family of all maximal right ideals of A_1 distinct from A and the family of all regular maximal right ideals of A.

36. Prove that if I is a right-regular right ideal of an algebra A and u is a left identity modulo I, then u^n is a left identity modulo I for all positive integers n.

37. Consider the commutative Banach algebra A consisting of all complex sequences a such that $\lim_{n \to \infty} a_n = 0$ (with pointwise linear operations and multiplication, and the supremum norm $\|a\| = \sup\{|a_n|\}_{n=1}^{\infty}$). Let I be the ideal of A consisting of those a such that $a_n = 0$ for all sufficiently large n. Show that I is not contained in any maximal proper ideal of A.

38. Prove Proposition 5.16.

39. Show by an example that a Banach algebra over the real numbers may admit elements with void spectrum. Hence 6.9 fails for real Banach algebras.

40. Give an example of a non-nilpotent generalized nilpotent element in a Banach algebra (see 6.19).

41. Prove that a generalized nilpotent element in a finite-dimensional Banach algebra is nilpotent.

42. Prove that if a and b are commuting elements in a Banach algebra A, then
(a) $\mathrm{Sp}_A(ab) \subset \mathrm{Sp}_A(a)\mathrm{Sp}_A(b)$;
(b) $\mathrm{Sp}_A(a + b) \subset \mathrm{Sp}_A(a) + \mathrm{Sp}_A(b)$.

43. Let A be a Banach algebra and let $r(a) = \|a\|_{\mathrm{sp}}$, the spectral radius of a in A. Show that r has the following properties for a, b in A.
(a) $0 \leq r(a) \leq \|a\|$;
(b) $r(\lambda a) = |\lambda| r(a)$, $\lambda \in \mathbb{C}$;
(c) $r(ab) = r(ba)$;
(d) $r(a^k) = r(a)^k$, $k = 1, 2, 3, \ldots$;
(e) If $ab = ba$, $r(ab) \leq r(a)r(b)$;
(f) If $ab = ba$, $r(a + b) \leq r(a) + r(b)$.

44. Let A and B be commutative Banach algebras with B semisimple. Prove that if $f: A \to B$ is a homomorphism, then f is continuous. [Hint: Use 7.3 and the Closed Graph Theorem.]

45. Let A be a commutative Banach algebra with unit. Show that an element a in A has an inverse in A if and only if the Gelfand transform $\hat{a}$ has an inverse in $\mathscr{C}(A^{\dagger})$, and that $(a^{-1})\hat{} = 1/\hat{a}$.

46. Give an example to show that 7.3 may fail if the algebra A is not complete.

47. Give an example of a commutative Banach algebra A and a complex homomorphism ϕ on A such that $\|\phi\| < 1$.

48. Give an example of a commutative Banach algebra A with unit such that

(a) The Gelfand transform map $a \mapsto \hat{a}$ is not injective;

(b) The Gelfand transform map $a \mapsto \hat{a}$ is not surjective.

49. Let A be a Banach algebra, a an element of A, and U an open subset of $\mathbb{C}$ with $\mathrm{Sp}_A(a) \subset U$. Show that there exists $\delta > 0$ such that $\mathrm{Sp}_A(b) \subset U$ whenever $\|b - a\| < \delta$. (See Rickart [2], Theorem 1.6.16.)

50. Let $\mathscr{S}$ be the metric space of all non-void compact subsets of $\mathbb{C}$, with the metric

$$d(A, B) = \max\{\sup\{\mathrm{dist}(x, B): x \in A\}, \sup\{\mathrm{dist}(x, A): x \in B\}\}.$$

(Here, of course, $\mathrm{dist}(x, B) = \inf\{|x - y|: y \in B\}$.) Prove that, if A is a commutative Banach algebra, then $a \mapsto \mathrm{Sp}_A(a)$ is continuous on A to $\mathscr{S}$. (See Rickart [2], Theorem 1.6.17.) It follows in particular that $a \mapsto \|a\|_{\mathrm{sp}}$ is continuous on A whenever A is a commutative Banach algebra.

51. Let A be a Banach algebra with unit 1. Show that an element a in A such that $\|a\|_{\mathrm{sp}} = 0$ is not invertible.

52. Let $\{\lambda_n\}$ be a fixed sequence of nonzero complex numbers with the following properties:

(a) $|\lambda_n| \geq |\lambda_{n+1}|$, $n = 1, 2, 3, \ldots$;

(b) $\lambda_n \to 0$;

(c) $n^{\varepsilon} |\lambda_{n+1}|^{n+1}/|\lambda_n|^n \to \infty$ for any fixed $\varepsilon > 0$;

(such sequences exist, e.g., $\lambda_n = 1/\log(n + 1)$). Let A denote the set of all formal power series $a(t) = \sum_{n=1}^{\infty} c_n t^n$ with complex coefficients such that $\sum_{n=1}^{\infty} |c_n| |\lambda_n|^n < \infty$.

(i) Show that if A is furnished with pointwise operations and norm $\|a\| = \sum_{n=1}^{\infty} |c_n| |\lambda_n|^n$, then A is a commutative Banach algebra without identity whose Gelfand space $A^{\dagger}$ is void.

(ii) A *derivation* on an algebra A is a linear operator D of A into itself such that $D(ab) = a(Db) + (Da)b$ for all a, b in A. Show that the algebra in (i) admits no nonzero continuous derivations.

53. Let E denote the Banach space $\mathscr{C}([0, 1])$ of continuous complex-valued functions on $[0, 1]$ with supremum norm. Define a linear operator $T: E \to E$ by

$$(Tf)(s) = \int_0^s f(t)\,dt,$$

for $f \in E$ and $s \in [0, 1]$.

(a) Show that T is continuous and determine $\|T\|$.

(b) Show that for $n = 2, 3, \ldots$

$$(T^n f)(s) = \int_0^s \frac{(s-t)^{n-1}}{(n-1)!} f(t)\,dt.$$

(c) Determine $\|T\|_{sp}$ in $\mathcal{O}(E)$.

(d) Show that the commutative subalgebra A of $\mathcal{O}(E)$ generated by T and the identity operator has a Gelfand space $A^\dagger$ consisting of one point.

54. Let A be a Banach algebra with unit and let K denote the closed two-sided ideal in A generated by the set $\{ab - ba: a, b \in A\}$ of commutators. Show that A/K is a commutative Banach algebra. Moreover, show that if I is a closed two-sided ideal of A such that A/I is commutative, then $K \subset I$.

55. Let A be a commutative Banach algebra with unit 1. Prove that if A is separable, then the Gelfand space $A^\dagger$ is metrizable.

56. Let A denote the (noncommutative) Banach algebra of all 2×2 complex matrices $x = (x_{ij})$ with $x_{12} = 0$ and with norm $\|x\| = \max\{\sum_{j=1}^2 |x_{ij}|: i = 1, 2\}$.

(a) Find all complex homomorphisms on A. How many are there?

(b) Determine the radical of A.

57. An element a in a Banach algebra A is called a *topological divisor of zero* if there is a sequence $\{b_n\}$ in A, with $\|b_n\| = 1$, such that $ab_n \to 0$ and $b_n a \to 0$ as $n \to \infty$. Show that every boundary point a of the group of invertible elements of A is a topological divisor of zero.

58. Let A be a commutative normed algebra with unit 1, and B its completion. Let M and N be maximal ideals in A and B respectively. If the bar denotes closure in B, prove that:

(a) $\overline{N \cap A} = N$.

(b) $\overline{M}$ is a maximal ideal in B.

59. Prove Proposition 8.4.

60. Give an example of a commutative Banach algebra A in which all maximal ideals are principal (an ideal in A is *principal* if it is generated by a single element). Is such an algebra semisimple?

61. Give an example of a commutative Banach algebra in which each maximal ideal is the closure of a principal ideal.

62. Prove Proposition 8.11.

63. Show that, if S is a locally compact Hausdorff space which is not compact, then every maximal proper ideal of $\mathscr{C}_0(S)$ is regular.

64. Let A be a function algebra on a locally compact Hausdorff space S; and consider the following two conditions:

(a) $S \cong A^\dagger$ (i.e., the map ϕ of 8.2 is onto $A^\dagger$);

(b) A is closed under complex conjugation.

Give examples showing that neither of these conditions implies the other.

65. Let S be a fixed compact subset of the complex plane; and let us define A as the family of all those complex functions on S which can be approached, uniformly on S,

by the restrictions to S of entire (or equivalently polynomial) functions on $\mathbb{C}$. Thus A (with the supremum norm on S) is a function algebra on S.

   (a) Prove that for each $\phi \in A^\dagger$ there is a unique point z_ϕ (not necessarily in S) such that $\phi(p|S) = p(z_\phi)$ for every polynomial p on $\mathbb{C}$.

   (b) Show that the map $\phi \mapsto z_\phi$ is a homeomorphism of $A^\dagger$ onto the compact subset $W = \{z_\phi : \phi \in A^\dagger\}$ of $\mathbb{C}$. So $A^\dagger$ can identified (both setwise and topologically) with W.

   (c) If V is the unbounded connected component of $\mathbb{C} \setminus S$, show that $S \subset W \subset \mathbb{C} \setminus V$. [As an example of this situation, see Example 8.4.]

   66. Let A be as in Example 8.4, considered as a function algebra on $\mathbb{E}$. Show that A is a *maximal* proper uniformly closed function algebra on $\mathbb{E}$.

   67. Let S be a compact Hausdorff space; and V a closed subset of S. Let I be the closed ideal of $\mathscr{C}(S)$ consisting of those f in $\mathscr{C}(S)$ such that $f(x) = 0$ for all x in V. Show that $\mathscr{C}(S)/I$ can be canonically identified with $\mathscr{C}(V)$.

   68. If A is a Banach algebra, let us define $B(A)$ as the closed linear span of all products ab where $a, b \in A$.

It is obvious that if A has an approximate unit then $B(A) = A$.

Give an example of a Banach algebra A such that $B(A) = A$ and yet A has no approximate unit. [Hint: Let D_1 be as in 8.11; and take A to be the closed subalgebra of D_1 consisting of those f such that $f(0) = f'(0) = 0$.]

Notes and Remarks

Representations of groups and algebras on locally convex spaces (in particular on Banach spaces) have been considered by many authors. For example, see Berezin [1], Bonsall and Duncan [1], Bourbaki [12], Fell [9, 10], Godement [9], Kirillov [6], de Leeuw and Glicksberg [1], Litvinov [1], Naimark [6], and G. Warner [1].

The notion of Naimark-relatedness was introduced in 1958 by Naimark [6] in his study of the Banach representations of the Lorentz group. The basic properties of this relation were established by Fell in [10]. The book of G. Warner [1] contains a survey of some of these properties. Further results on Naimark-relatedness have been obtained by Barnes [1].

The notion of an abstract normed algebra was introduced by Nagumo [1] in 1936 under the name "linear metric ring." During the late 1930's the term "normed ring" was introduced by the Soviet mathematicians. The present term "Banach algebra" was used for the first time in 1945 by Ambrose [2] in his work on generalizing the $\mathscr{L}_2$ algebra of a compact group.

Approximate units appeared in the literature as early as 1927 (see Peter and Weyl [1]), and have been an important tool since. A comprehensive treatment of them is given in the book by Doran and Wichmann [2].

The first writers to use the circle operation (defined in 5.7) were Perlis [1] and Jacobson [2] (although they worked with $x + y + xy$ instead of $x + y - xy$). The term "adverse" is due to Segal [3], while "quasi-inverse" was used by Perlis and Jacobson. Regular ideals were introduced into the literature by Segal [3] who also developed their basic properties; Jacobson coined the term "modular ideal."

Mazur [1] announced Theorem 6.12 in 1938 for algebras over the field of real numbers, making no mention of the complex case. For real scalars Mazur's Theorem states that a normed division algebra is isomorphic to either the reals, the complexes, or the quaternions. The complex version, as given in 6.12, can be obtained from the real version in a standard way since every complex normed algebra is also a real normed algebra, and the possibilities of the reals and the quaternions are easily eliminated in the complex case.

The details of Mazur's proof of 6.12 were too lengthy to be included in his announcement, and it was Gelfand [3] who furnished the first published proof of the complex version. Although the algebras considered by Gelfand were commutative, his proof, which is essentially the one we have given in §6, did not make use of this fact. It is perhaps of some historical interest to note that Mazur's original unpublished proof can be found in the 1973 book on Banach algebras by Żelazko [1]. For additional information on Mazur's Theorem and related results we refer the reader to Belfi and Doran [1].

The fundamental result 6.17, which generalized a result of Beurling [1] for the group algebra of the real line, is also due to Gelfand [3] as are the beautiful results in §7 on commutative Banach algebras and their maximal ideal theory. The results on function algebras in §8 are well-known and can be found in one form or another, for example, in Gelfand, Raikov, and Šilov [2], Naimark [8], and Rickart [2]. Theorem 8.12 was originally published in 1932 by Wiener [1], using classical methods of real analysis. The short Banach algebra-theoretic proof given in §8 was observed by Gelfand [3]; this single result was responsible for drawing considerable attention to the young field of Banach algebras.

The Factorization Theorem 9.2 was established by Cohen [1] in 1959 for Banach algebras with approximate unit; and was generalized to the setting of Banach modules by Hewitt [1] in 1964. The result in Remark 9.4 was first observed by Varopoulos [1] who used it to prove automatic continuity of positive linear functionals on Banach *-algebras with approximate unit (see 18.15, Chapter VI). The monograph by Doran and Wichmann [2] contains further results on factorization in Banach algebras and Banach modules.

VI C^*-Algebras and Their *-Representations

The theory of the *-representations of *-algebras is incomparably deeper and
stronger than what can be obtained (at present at least) in the more general
non-involutory context of locally convex (or even Banach) representations of
algebras. For the rest of this work we shall be mostly concerned with
*-algebras and their *-representations. In this involutory context, with the help
of topological and measure-theoretic machinery, we shall find some striking
"continuous analogues" of certain of the algebraic results of Chapter IV. One
example of this is the Glimm projection-valued measure, to be developed in
the next chapter, which is a beautiful analogue of the σ-subspace theory of
§IV.2. This chapter will lay the foundation for all such developments.

After a few elementary remarks in §1, §2 introduces the important property
of symmetry of a *-algebra; this says that elements of the form a^*a have
positivity properties. In §3 we define the *-algebras of greatest interest,
namely the C^*-algebras. The Gelfand–Naimark Theorem (22.12) asserts that,
up to isomorphism, these are just the norm-closed *-algebras of bounded
operators on Hilbert space; and it is quite remarkable that the objects
characterized "geometrically" in this way admit so elegant an abstract
characterization as the Definition 3.1.

Before proceeding with the theory of general C^*-algebras one must settle
the structure of commutative C^*-algebras. This is done in §4; it is shown that

the most general commutative C^*-algebra is just $\mathscr{C}_0(S)$, S being a locally compact Hausdorff space. In the succeeding §§5–8 some of the implications of the commutative theory for general C^*-algebras are developed. They are based on the evident fact that any normal element of a C^*-algebra A can be placed inside a commutative C^*-subalgebra of A. In particular, in §7 we prove the crucial fact that every C^*-algebra is symmetric. In §8 we find that every C^*-algebra has, if not a unit, at least an approximate unit, and deduce from this several technically important facts about ideals and homomorphisms of C^*-algebras.

In §9 we begin the study of *-representations. In §10, with any Banach *-algebra A we associate a certain C^*-algebra A_c, the C^*-completion of A, whose *-representation theory coincides with that of A. If A is commutative then so is A_c. From this, combining the structure theory of commutative C^*-algebras (§4) with the "projection-valued analogue of the Riesz Theorem" (§II.12), one obtains a description of any *-representation of a commutative Banach *-algebra A in terms of a projection-valued measure on $\hat{A}$ (Stone's Theorem 10.10).

§11 contains the modern approach to the spectral theory of bounded normal operators, based on Stone's Theorem; and §12 generalizes this to unbounded normal operators. Using this we obtain in §13 the so-called Mackey's form of Schur's Lemma, which is basic to the multiplicity theory of *-representations. In §14 we use this to obtain the multiplicity theory of discretely decomposable *-representations, analogous to the purely algebraic theory of §IV.2. The beautiful multiplicity theory of general (not discretely decomposable) *-representations, based on the theory of von Neumann algebras, is unfortunately beyond the scope of this work.

Perhaps the most important special non-commutative C^*-algebras are the algebras of compact operators on Hilbert space. These are studied in §15. It is shown that their spectral theory is "discrete," and that their *-representation theory is quite analogous to that of finite-dimensional total matrix algebras. §16 is a historical digression, pointing out the relevance of the spectral theory of compact operators to the Sturm–Liouville theory of the classical differential equations of vibration (from which, indeed, the theory of operators on Hilbert space received its first impulse). In §17 we define inductive limits of C^*-algebras; and obtain as special cases of this construction not only the algebras of compact operators but also the so-called Glimm algebras, which are important both in quantum theory and as examples of non-Type I C^*-algebras.

§18 begins the study of positive functionals. It is shown in §§19, 20 that they are vital tools for the *-representation theory of general non-commutative

*-algebras. Some idea of their importance can be gauged from the following considerations: Let A be a commutative C^*-algebra with unit. By §4 this is essentially just $\mathscr{C}(S)$, where S is a compact Hausdorff space; and so by the Riesz Theorem a positive linear functional p on A is just a regular Borel measure on S. From such a measure we can build the Hilbert space $\mathscr{L}_2(p)$; and $A \cong \mathscr{C}(S)$ has a natural *-representation T on $\mathscr{L}_2(p)$ by multiplication of functions $((T_f\phi)(s) = f(s)\phi(s)$ for $f \in A$, $\phi \in \mathscr{L}_2(p))$. This T will be irreducible if and only if p is concentrated at a single point s_0 of S (so that $T: f \mapsto f(s_0)$ is one-dimensional); and this amounts to saying that p is "indecomposable," in the sense that p cannot be written as a sum of two measures q_1 and q_2 on S unless q_1 and q_2 are both multiples of p.

Now suppose A is any (non-commutative) C^*-algebra, and p a positive linear functional on A. It turns out to be fruitful to press the analogy of the commutative case and to think of p as a "non-commutative measure"; and then to try to generalize to such p as much ordinary measure theory as possible. In §19 we give the Gelfand–Naimark–Segal construction of the non-commutative analogue of the Hilbert space $\mathscr{L}_2(p)$ and of the *-representation T of A generated by p; and in §20 we show that, even in the non-commutative case, the irreducibility of T is equivalent to the indecomposability of p.

In ordinary measure theory it is important to deal not only with bounded measures on a compact space but also with unbounded measures. Analogously we are often interested in positive functionals in the general C^*-algebra context which are "unbounded," and thus defined only on a dense *-subalgebra B of a C^*-algebra A. §21 deals with such "unbounded" positive functionals in the commutative context only. Our main purpose here is to prove the abstract Plancherel Theorem (which will take on a more familiar appearance in Chapter X, in the context of locally compact Abelian groups).

In §22 we at last prove the Gelfand–Naimark Theorem. Positive functionals, and their relation to *-representations, are of the essence here. Two other tools are crucial: First, the structure of commutative C^*-algebras, which provides us with nontrivial positive functionals p on an arbitrary commutative C^*-subalgebra B of A; and, secondly, the Krein Extension Theorem, which permits us to extend p to a positive functional on all of A. Putting these results together we obtain enough *-representations of A to separate points of A. Applying in addition the Krein–Milman Theorem, one even obtains enough *irreducible* *-representations to separate points of A.

In §23 certain interesting results on the compact operator algebras, notably Rosenberg's Theorem, are obtained by means of the Extension Theorem of §22.

Perhaps the most unfortunate shortcoming of this work is the absence of the theory of von Neumann algebras. In preparation for Chapter XII, in §24 we at least prove the fundamental von Neumann Double Commuter Theorem, discuss primary (factor) representations, and define what it means for the *-representation theory of a *-algebra to be of Type I.

§25 centers around Kadison's Irreducibility Theorem. From this result and its corollaries it follows that there is essentially no difference between the space of unitary equivalence classes of irreducible *-representations of a C^*-algebra A and the space of algebraic equivalence classes of algebraically irreducible algebraic representations of A. Perhaps even more important is the conclusion that every norm-closed left ideal of a C^*-algebra is the intersection of the regular maximal left ideals that contain it (25.12(III)). This was proved in V.8.9 for commutative C^*-algebras; and we observed in V.8.11 that it is false for general Banach *-algebras (even commutative ones).

§25 is not needed for the rest of the work, and may be omitted if the reader wishes.

1. *-Algebras; Elementary Remarks and Examples

1.1.   Let A be a fixed *-algebra.

An element a of A is *Hermitian* (or *self-adjoint*) if $a^* = a$, and *normal* if $a^*a = aa^*$. Of course self-adjointness implies normality.

Every element a of A can be written in one and only one way in the form

$$a = b + ic, \tag{1}$$

where b and c are self-adjoint. Indeed, (1) holds if we take b and c to be the self-adjoint elements

$$b = \tfrac{1}{2}(a + a^*), \qquad c = -\tfrac{1}{2}i(a - a^*); \tag{2}$$

and, conversely, (1) implies that $a^* = b - ic$, which combined with (1) gives (2).

By a *subalgebra* or an *ideal* of A we mean a subalgebra or ideal of the associative algebra underlying A. A subalgebra of A which is closed under the involution operation is a *-*subalgebra* of A. A left or right ideal I of A which is closed under involution is automatically a two-sided ideal; indeed, if I is a left ideal, $a \in A$, and $b \in I$, then $ba = (a^*b^*)^* \in I$. Such a two-sided ideal is called a *-*ideal* of A.

Observe that an element a of A is normal if and only if a belongs to some commutative *-subalgebra of A.

If A has a unit $\mathbb{1}$, then $\mathbb{1}$ is self-adjoint. An element u of A is then *unitary* if $u^{-1} = u^*$, that is, $u^*u = uu^* = \mathbb{1}$. Unitary elements are of course normal.

1.2.   Proposition.   *A maximal commutative *-subalgebra B of A is a maximal commutative subalgebra of A.*

Proof.   Let a be an element of A which commutes with all of B; we must show that $a \in B$. Since B is a *-subalgebra, $a^*b = (b^*a)^* = (ab^*)^* = ba^*$ for all b in B. Thus a^*, and likewise the Hermitian elements $c = \frac{1}{2}(a + a^*)$ and $d = -\frac{1}{2}i(a - a^*)$, commute with all of B. It follows that the subalgebra C of A generated by B and c is commutative and closed under involution, and hence equal to B by the maximality of the latter. Therefore $c \in B$. Similarly $d \in B$. Consequently $a = c + id \in B$.   ∎

1.3.   Adjunction of a Unit.   If the *-algebra A has no unit, the result A_1 of adjoining a unit $\mathbb{1}$ to A (see V.4.10) becomes a *-algebra when we set

$$(a + \lambda \mathbb{1})^* = a^* + \bar{\lambda} \mathbb{1} \qquad\qquad (a \in A; \lambda \in \mathbb{C}).$$

If A is a normed *-algebra (or Banach *-algebra), then so is A_1 when it is equipped with the above *-operation and the norm

$$\|a + \lambda \mathbb{1}\| = \|a\| + |\lambda|.$$

1.4.   If A has a unit and a is invertible in A, then evidently $(a^*)^{-1} = (a^{-1})^*$. Thus, whether A has a unit or not, we have:

Proposition.   *For any element a of A,*

$$\mathrm{Sp}(a^*) = \overline{\mathrm{Sp}_A(a)}.$$

1.5.   Quotient *-Algebras.   Let I be an *-ideal of A. Then A/I is a *-algebra under the quotient algebra structure and the involution

$$(a + I)^* = a^* + I \qquad\qquad (a \in A).$$

If A is a normed [Banach] *-algebra, and in addition I is norm-closed, then A/I, with the quotient norm of V.4.11, is a normed [Banach] *-algebra.

1.6.   Example.   Let X be a Hilbert space. We saw in V.4.3 that $\mathcal{O}(X)$ is a Banach algebra under the operator norm. If we equip $\mathcal{O}(X)$ with the adjoint operation *, defined by the requirement that

$$(a\xi, \eta) = (\xi, a^*\eta) \qquad\qquad (a \in \mathcal{O}(X); \xi, \eta \in X),$$

then $\mathcal{O}(X)$ becomes a Banach *-algebra.

1.7.   Examples.   Let S be any set. The commutative Banach algebra $\mathcal{B}(S)$ (defined in V.4.6) becomes a Banach *-algebra when involution is taken to be the operation of complex conjugation. The same of course is true of the norm-closed subalgebras $\mathscr{B}_c(S)$ and $\mathscr{C}_0(S)$ of $\mathscr{B}(S)$ mentioned in V.4.6, and also of the Banach algebra $\mathscr{L}_\infty(\mu)$ of V.4.7.

1.8.   *Example.* Let A be the Banach algebra of continuous complex functions analytic on the interior of the unit disk D, defined in V.4.8. Evidently A is not closed under complex conjugation. It is however closed under the involution * given by

$$f^*(z) = \overline{f(\bar{z})} \qquad\qquad (z \in D).$$

With this involution A becomes a Banach *-algebra.

1.9.   *Examples.* Let G be a locally compact group. We have seen in III.10.15 that the measure algebra $\mathscr{M}_r(G)$ of G is a Banach *-algebra. Likewise the $\mathscr{L}_1$ group algebra of G, defined in III.11.9, is a Banach *-algebra.

1.10.   *Example.* Let A be an infinite-dimensional Banach space, and let E be a Hamel basis for A, chosen so that $\|e\| = 1$ for each e in E. Let $\{e_n : n = 1, 2, \ldots\}$ be a sequence of distinct elements of E and define

$$e_{2n-1}^* = n e_{2n}, \; e_{2n}^* = \frac{1}{n} e_{2n-1} \qquad\qquad (n = 1, 2, \ldots).$$

For all other elements of E, let $e^* = e$, and then extend the mapping $e \mapsto e^*$ to all of A by conjugate linearity; that is,

$$(\lambda_1 d_1 + \cdots + \lambda_k d_k)^* = \bar{\lambda}_1 d_1^* + \cdots + \bar{\lambda}_k d_k^* \qquad (\lambda_i \in \mathbb{C}, d_i \in E, i = 1, 2, \ldots, k)$$

Finally, we make A into a Banach algebra by introducing the trivial multiplication $ab = 0$ for a, b in A. Then A is an example of a Banach algebra with *discontinuous* involution. By adjoining an identity to A, we obtain a Banach algebra with discontinuous involution in which the multiplication is not trivial.

2.   Symmetric *-Algebras

2.1. There is an important property which a *-algebra may possess, called symmetry.

Definition. A *-algebra A is *symmetric* if, for every element a of A, $-a^*a$ has an adverse (V.5.7) in A.

If A has a unit element $\mathbf{1}$, this amounts to saying that $(\mathbf{1} + a^*a)^{-1}$ exists for every a in A.

2.2. Proposition. *If A is a symmetric *-algebra and a is a Hermitian element of A, then $\mathrm{Sp}_A(a) \subset \mathbb{R}$.*

Proof. Adjoin a unit $\mathbf{1}$ to A if there is not one already. By hypothesis, the element

$$a^*a + \mathbf{1} = a^2 + \mathbf{1} = (a + i\mathbf{1})(a - i\mathbf{1})$$

has an inverse. It follows that $a - i\mathbf{1}$ has an inverse, and hence that

$$i \notin \mathrm{Sp}_A(a). \tag{1}$$

Suppose now that some non-real complex number λ belongs to $\mathrm{Sp}_A(a)$. Since λ is not real, λ and λ^2 are linearly independent over the reals; so there are real numbers r and s such that

$$r\lambda + s\lambda^2 = i. \tag{2}$$

From (2) and V.5.6 we conclude that $u = ra + sa^2$ has i in its spectrum. But u is Hermitian (since r and s are real); and (1) was valid for any Hermitian element of A, in particular for u. So we have a contradiction. ∎

2.3. Proposition. *A *-algebra A is symmetric if and only if, for every a in A, $\mathrm{Sp}_A(a^*a)$ consists entirely of non-negative real numbers.*

Proof. If $\mathrm{Sp}_A(a^*a)$ does not contain -1, then $-a^*a$ has an adverse (V.5.8). So the "if" part holds. Conversely, assume that A is symmetric, and that $a \in A$. For any positive number r, the hypothesis asserts that $-r(a^*a) = -(r^{1/2}a)^*(r^{1/2}a)$ has an adverse, and hence (V.5.8) that $-r^{-1}$ is not in the spectrum of a^*a. This and 2.2 imply that $\mathrm{Sp}_A(a^*a)$ is real and non-negative, proving the "only if" part. ∎

2.4. Proposition. *Let A be a symmetric *-algebra without unit. Then the result A_1 of adjoining a unit $\mathbf{1}$ to A (as in 1.3) is symmetric.*

Proof. Let $b = a + \lambda\mathbf{1}$ $(a \in A; \lambda \in \mathbb{C})$; and set $\mu = 1 + |\lambda|^2$, $c = \bar{\lambda}a + \lambda a^* + a^*a$, $d = -\mu^{-1}c$. Thus

$$\mathbf{1} - d = \mu^{-1}(\mathbf{1} + b^*b). \tag{3}$$

We now take the polynomial $p(t) = 1 + (\mu - 1)t^2 - \mu t^3$, and notice, since $d^* = d$, that

$$p(d) = \mathbf{1} + (bd)^*bd. \tag{4}$$

Now assume that $(\mathbf{1} + b^*b)^{-1}$ does not exist in A_1. By (3) this means that $1 \in \mathrm{Sp}(d)$, whence by V.5.6

$$0 = p(1) \in \mathrm{Sp}(p(d)).$$

By (4) this implies that

$$(1 + (bd)^*bd)^{-1} \text{ does not exist.} \tag{5}$$

Now A is an ideal of A_1, and $d \in A$; so $bd \in A$. Hence (5) contradicts the symmetry of A. Thus $(1 + b^*b)^{-1}$ must exist. Since b was an arbitrary element of A_1, the latter is symmetric. ∎

Remark. Let A be a *-algebra and I a *-ideal of A. Then A is symmetric if and only if both I and A/I are symmetric (see J. Wichmann [3]). Since A_1/A is isomorphic to $\mathbb{C}$ (a symmetric *-algebra), this result generalizes the preceding proposition.

2.5. Here is an interesting criterion for the symmetry of a commutative Banach *-algebra.

Proposition. *A commutative Banach *-algebra A is symmetric if and only if*

$$\phi(a^*) = \overline{\phi(a)} \tag{6}$$

for all a in A and ϕ in $A^\dagger$.

Note. Here $A^\dagger$ is the Gelfand space of the commutative Banach algebra underlying A; see V.7.1.

Proof. If (6) holds for all ϕ and a, then $\phi(a^*a) = \overline{\phi(a)}\phi(a) \geq 0$ for all ϕ and a; and it follows from V.7.13 that $\mathrm{Sp}(a^*a)$ is real and non-negative for every a in A. So A is symmetric.

Conversely, let A be symmetric. Assume that $\phi \in A^\dagger$, $b = b^* \in A$, and that $\phi(b)$ is not real. By the same argument as in the proof of 2.2 we obtain a Hermitian element c of A (of the form $rb + sb^2$, where r and s are real) such that $\phi(c) = i$. It follows that $\phi(c^*c) = (\phi(c))^2 = -1$. Thus by V.7.15 $-c^*c$ has no adverse, contradicting the symmetry of A. We have proved:

$$\phi(b) \text{ is real whenever } \phi \in A^\dagger \text{ and } b^* = b \in A. \tag{7}$$

Now let a be an element of A, and write $a = b + ic$, where b, c are Hermitian (as in 1.1). For any ϕ in $A^\dagger$ we have by (7)

$$\phi(a^*) = \phi(b - ic) = \phi(b) - i\phi(c)$$
$$= (\phi(b) + i\phi(c))^- = \overline{\phi(a)};$$

and this is (6). ∎

2.6. Are the Banach *-algebras mentioned in Examples 1.6–1.9 symmetric?

The Banach *-algebras of 1.6 and 1.7 are in fact C^*-algebras (to be defined in §3); and we shall see in 7.11 that all C^*-algebras are symmetric.

The commutative Banach *-algebra A of Example 1.8 is *not* symmetric. Indeed, if u is a non-real point of the unit disk D, the homomorphism $\phi: f \mapsto f(u)$ (evaluation at u) is an element of $A^\dagger$ which does not satisfy (6) of Proposition 2.5.

Let G be a locally compact group. Is the $\mathscr{L}_1$ group algebra of G symmetric? If G is either Abelian or compact, the answer is "yes" (see X.1.3 for Abelian G and either Anusiak [1] or Van Dijk [1] for compact G); but in general it is "no" (see for example §29 of Naimark [8]). The characterization of those G whose $\mathscr{L}_1$ group algebras are symmetric is an interesting open problem (see Aupetit [1, p. 131]).

As regards the measure algebra $\mathscr{M}_r(G)$ of G, it has been shown by Williamson [1] that if G is a non-discrete locally compact Abelian group, $\mathscr{M}_r(G)$ is never symmetric. See also Hewitt and Kakutani [1].

3. C*-Algebras

3.1. For us, the most important Banach *-algebras are the C^*-algebras.

Definition. A C^*-*algebra* is a Banach *-algebra A whose norm $\| \ \|$ satisfies the identity:

$$\|a^*a\| = \|a\|^2 \qquad \text{for all } a \text{ in } A. \tag{1}$$

3.2. Remark. The notion of a C^*-algebra can be defined by apparently weaker conditions. For example, one can define a C^*-algebra as a Banach algebra which is also a *-algebra satisfying

$$\|a\|^2 \leq \|a^*a\| \qquad \text{for all } a \text{ in } A. \tag{2}$$

Indeed: (2) gives $\|a\|^2 \leq \|a^*a\| \leq \|a^*\| \|a\|$, whence, if $\|a\| \neq 0$, $\|a\| \leq \|a^*\|$. Reversing the roles of a and a^*, we obtain $\|a^*\| \leq \|a\|$. Therefore $\|a^*\| = \|a\|$; and A is a Banach *-algebra. So it follows from (2) that $\|a^2\| \leq \|a^*a\| \leq \|a^*\| \|a\| = \|a\|^2$, giving (1).

In 1973 Araki and Elliott [1] proved the following two results:

Proposition*. *Let A be a Banach space which is also a *-algebra satisfying* $\|a^*a\| = \|a\|^2$ *for all a in A. Then A is a C^*-algebra.*

Proposition*.   *Let A be a Banach space which is also a *-algebra; and suppose that*

(i) $a \mapsto a^*$ *is continuous on A, and*
(ii) $\|a^*a\| = \|a^*\| \|a\|$ *for all a in A.*

Then A is a C*-algebra.

Recently Magyar and Sebestyén [1] have shown that continuity of the involution can be dropped from the second of these propositions. We refer the reader to their paper and the book of Doran and Belfi [1] for more information concerning various characterizations of C*-algebras. A brief account of the history of the axiomatics of C*-algebras is given in the bibliographical notes at the end of the chapter.

3.3. It is sometimes useful to refer to a normed *-algebra whose norm satisfies (1) as a *pre-C*-algebra*. The completion of a pre-C*-algebra is a C*-algebra.

3.4. If a C*-algebra has a unit $\mathbb{1}$, then by (1) $\|\mathbb{1}\|^2 = \|\mathbb{1}^*\mathbb{1}\| = \|\mathbb{1}\|$; hence $\|\mathbb{1}\| = 1$.

3.5.   *Examples.*   The Banach *-algebra $\mathcal{O}(X)$ of 1.6 is a C*-algebra; for it is well known that (1) holds for the operator norm of bounded linear operators on a Hilbert space. Thus, any norm-closed *-subalgebra of $\mathcal{O}(X)$ is a C*-algebra. Conversely, we shall show in 22.12 that every C*-algebra is isometrically *-isomorphic with a norm-closed *-subalgebra of $\mathcal{O}(X)$ for some Hilbert space X. This famous theorem of Gelfand and Naimark is the raison d'être of the above definition of C*-algebras.

It is often convenient to refer to a norm-closed *-subalgebra A of $\mathcal{O}(X)$ as a *concrete C*-algebra*. This concept includes the Hilbert space X on which A acts.

Again, the $\mathcal{B}(S)$ of Example 1.7 is a commutative C*-algebra. In 4.3 we shall show that every commutative C*-algebra A is isometrically *-isomorphic with $\mathcal{C}_0(S)$ for some locally compact Hausdorff space S.

As for the $\mathcal{L}_1$ group algebra of a locally compact group, it is not hard to prove:

Proposition*.   *The $\mathcal{L}_1$ group algebra of a locally compact group G cannot be a C*-algebra unless G has only one element.*

3.6. Here is a very important fact relating norms and spectra in a C*-algebra. (Recall that $\| \ \|_{\text{sp}}$ is the spectral radius defined in V.6.17.)

Proposition. *If a is a normal element of a C*-algebra A, then*

$$\|a\|_{sp} = \|a\|. \tag{3}$$

In particular, if a is Hermitian (3) holds.

Proof. Since $aa^* = a^*a$ we have $\|a^2\|^2 = \|(a^2)^*a^2\| = \|(a^*a)a^*a\| = \|a^*a\|^2 = \|a\|^4$. Hence $\|a^2\| = \|a\|^2$. Iterating we obtain

$$\|a^{(2^n)}\| = \|a\|^{2^n} \qquad \text{for all } n = 1, 2, \ldots \tag{4}$$

On the other hand, by V.6.17,

$$\lim_{n \to \infty} \|a^{(2^n)}\|^{2^{-n}} = \|a\|_{sp}. \tag{5}$$

Combining (4) and (5) we get (3). ∎

3.7. This leads to a very significant result.

Theorem. *Any *-homomorphism F of a Banach *-algebra B into a C*-algebra A is continuous; in fact, $\|F\| \leq 1$.*

Proof. Let $a \in B$. If $0 \neq \lambda \in \mathbb{C}$ and $\lambda \notin \mathrm{Sp}_B(a)$, then $\lambda^{-1}a$ has an adverse in B (V.5.8); it follows, on applying F, that $\lambda^{-1}F(a)$ has an adverse in A, so that $\lambda \notin \mathrm{Sp}_A(F(a))$ (by V.5.8). Thus $\mathrm{Sp}_A(F(a)) \subset \mathrm{Sp}_B(a) \cup \{0\}$, whence

$$\|F(a)\|_{sp} \leq \|a\|_{sp}. \tag{6}$$

Now let b be any element of B, and put $a = b^*b$. Since $F(a)$ is Hermitian,

$$\|F(b)\|^2 = \|F(b)^*F(b)\| \qquad \text{(by (1))}$$

$$= \|F(a)\| = \|F(a)\|_{sp} \qquad \text{(by 3.6)}$$

$$\leq \|a\|_{sp} \qquad \text{(by (6))}$$

$$\leq \|a\| \qquad \text{(by V.6.5)}$$

$$\leq \|b^*\| \, \|b\| = \|b\|^2.$$

So $\|F(b)\| \leq \|b\|$; and the theorem is proved. ∎

3.8. Corollary. *Every *-representation T (see V.1.2) of a Banach *-algebra is continuous. In fact $\|T\| \leq 1$.*

Proof. Apply 3.7, recalling from 3.5 that $\mathcal{O}(X)$ is a C^*-algebra for each Hilbert space X. ∎

3.9. Corollary. *If two C*-algebras A and B are *-isomorphic under a *-isomorphism F, then F is an isometry.*

Thus the norm in a C*-algebra is determined purely by its algebraic structure.

This corollary will be strengthened in 8.8.

Adjoining a Unit to a C*-Algebra

3.10. Let A be a C*-algebra without a unit, and A_1 the *-algebra (without norm) obtained by adjoining a unit 1 to A as in 1.3. Is A_1 a C*-algebra? More precisely, can A_1 be made a C*-algebra by introducing into it a suitable norm $\| \ \|_1$? The answer is "yes," as the following proposition shows. The norm $\| \ \|_1$ is unique in view of 3.9, and different from (though equivalent to) the norm introduced into A_1 in 1.3.

Proposition. *Let A and A_1 be as above. For each a in A and λ in $\mathbb{C}$, define*

$$\|a + \lambda 1\|_1 = \sup\{\|ab + \lambda b\| : \|b\| \le 1, b \in A\} \tag{7}$$

(the norm of left multiplication by $a + \lambda 1$ in A). Then A_1 is a C*-algebra under the norm $\| \ \|_1$. Further, $\| \ \|_1$ coincides with $\| \ \|$ on A; and is equivalent (in the sense of giving rise to the same topology of A_1) to the norm of A_1 introduced in 1.3.

Proof. First we claim that $\|a\|_1 = \|a\|$ for $a \in A$. Indeed, the inequality $\|ab\| \le \|a\| \|b\|$ shows that $\|a\|_1 \le \|a\|$. On the other hand $\|a\|^2 = \|aa^*\| \le \|a\|_1 \|a^*\| = \|a\|_1 \|a\|$; so $\|a\| \le \|a\|_1$. This proves the claim.

Notice that, since A has no unit, it cannot even have a left unit u. For then u^* would be a right unit and $u = uu^* = u^*$; so u would be a unit, contrary to the hypothesis.

Now suppose that $a \in A$, $\lambda \in \mathbb{C}$, and $\|a + \lambda 1\|_1 = 0$. If $\lambda \ne 0$, this implies that $ab + \lambda b = 0$, or $(-\lambda^{-1}a)b = b$, for all b in A; whence $-\lambda^{-1}a$ is a left unit of A, contradicting the preceding paragraph. So $\lambda = 0$, and $\|a\| = 0$, that is, $a = 0$. Thus $\|a + \lambda 1\|_1 = 0$ implies $a + \lambda 1 = 0$. In view of the definition of $\| \ \|_1$ as the norm of a left-multiplication operator, it follows that $A_1, \| \ \|_1$ is a normed algebra.

Next we shall show that

$$\|b\|_1^2 \le \|b^*b\|_1 \qquad \text{for all } b \text{ in } A_1. \tag{8}$$

For this, let $0 < \gamma < 1$; and choose a in A so that $\|a\| = 1$ and $\gamma\|b\|_1 \le \|ba\|$. Then

$$\gamma^2 \|b\|_1^2 \le \|ba\|^2 = \|(ba)^*(ba)\|$$

$$= \|a^*b^*ba\|$$

$$\le \|b^*b\|_1 \qquad\qquad \text{(since } \|a^*\| = \|a\| = 1\text{)}.$$

By the arbitrariness of γ, this proves (8).

Thus $A_1, \|\ \|_1$ is a normed algebra and also a *-algebra satisfying (8). By the argument of 3.2 this shows that $A_1, \|\ \|_1$ is a normed *-algebra satisfying the identity (1). Its completeness results from Remark 3.11 (which follows this proof), since A_1 has a complete subspace A of codimension 1. The last statement of the proposition is also a consequence of 3.11. So the proof is finished. ∎

3.11.   Remark.   In the preceding proof we have made use of the following general fact about normed linear spaces:

If a normed linear space X has a complete linear subspace Y of finite codimension n in X, then X is complete, and X is naturally isomorphic (as an LCS) with $Y \oplus \mathbb{C}^n$.

The proof of this is quite easy, and proceeds by induction in n. To make the inductive step we may as well assume $n = 1$. Fix $u \in X \setminus Y$; and suppose $\{x_n + \lambda_n u\}$ is a Cauchy sequence $(x_n \in Y; \lambda_n \in \mathbb{C})$. If $\lambda_n \to \infty$, the boundedness of $\{x_n + \lambda_n u\}$ shows that $\lambda_n^{-1}(x_n + \lambda_n u) \to 0$, or $-\lambda_n^{-1} x_n \to u$; since Y is complete, hence closed, this implies $u \in Y$, a contradiction. So we may pass to a subsequence and assume $\lambda_n \to \lambda \in \mathbb{C}$. Then $\{x_n\}$ must be Cauchy, and so converges in Y to some x. Consequently $x_n + \lambda_n u \to x + \lambda u$; and X is complete. The last statement is an easy consequence of the same argument.

3.12.   Remark.   Suppose that B is a C^*-algebra with unit $\mathbb{1}$; and that A is a norm-closed *-subalgebra of B with $\mathbb{1} \notin A$. Then by 3.11 $A_1 = A + \mathbb{C}\mathbb{1}$ is also a norm-closed *-subalgebra of B, and so by 3.9 coincides (if A has no unit) with the A_1 constructed in 3.10.

Direct Sums of C*-Algebras

3.13. For each i in some index set I let A_i be a C^*-algebra; and let A be the family of all those functions a on I such that

(i) $a_i \in A_i$ for each i,
(ii) $\{\|a_i\| : i \in I\}$ is bounded.

If we equip A with the obvious pointwise operations $((a + b)_i = a_i + b_i$; $(\lambda a)_i = \lambda a_i$; $(ab)_i = a_i b_i$; $(a^*)_i = (a_i)^*)$ and the supremum norm $\|a\| = \sup\{\|a_i\| : i \in I\}$, then clearly A becomes a C^*-algebra. It is called the C^*-*direct product* of the A_i.

The closed *-subalgebra B of A consisting of those a in A which "vanish at infinity" (that is, for each $\varepsilon > 0$ we can find a finite subset F of I such that $\|a_i\| < \varepsilon$ for $i \notin F$) is called the C^*-*direct sum* of the A_i. (This is of course the same as the C_0 direct sum of C^*-algebras as defined in Chapter I, §5).

Finite-Dimensional C*-Algebras

3.14. Proposition. *Every finite-dimensional C*-algebra A is isometrically *-isomorphic with a C*-direct sum of finitely many total matrix *-algebras $\mathcal{O}(X)$ (X being a finite-dimensional Hilbert space).*

Proof. The identity (1) implies that $a^*a = 0 \Rightarrow a = 0$. Hence by IV.7.4 A is *-isomorphic with a C^*-direct sum of finitely many such $\mathcal{O}(X)$. By 3.9 the *-isomorphism is an isometry. ∎

4. Commutative C^*-Algebras

4.1. In this section we shall completely determine the structure of a commutative C^*-algebra. This will be indispensable for two reasons. First, we will be able to deduce from it a surprising number of facts about non-commutative C^*-algebras. Secondly, it will lead to a complete analysis of the *-representations of all commutative Banach *-algebras (10.10).

4.2. Let A be a commutative Banach *-algebra. As usual (V.7.1) $A^\dagger$ is the Gelfand space of the commutative Banach algebra underlying A.

Definition. We denote by $\hat{A}$ the closed subset of $A^\dagger$ consisting of those ϕ in $A^\dagger$ such that

$$\phi(a^*) = \overline{\phi(a)} \qquad \text{for all } a \text{ in } A.$$

An element of $A^\dagger$ is called *symmetric* if it lies in $\hat{A}$.

By 2.5, A is symmetric if and only if $\hat{A} = A^\dagger$.

4.3. Theorem. *Let A be a commutative C*-algebra. Then $\hat{A} = A^\dagger$. Furthermore, the Gelfand transform map $a \mapsto \hat{a}$ (see V.7.7) is an isometric *-isomorphism of A onto $\mathscr{C}_0(\hat{A})$.*

Proof. We shall first show that:

$$\text{If } a^* = a \in A \text{ and } \phi \in A^\dagger, \quad \text{then } \phi(a) \text{ is real.} \tag{1}$$

Let $a^* = a \in A$ and $\phi \in A^\dagger$. If A has no unit, ϕ extends to a complex homomorphism of the C*-algebra (see 3.10) obtained by adjoining a unit to A. Thus it will be sufficient to prove (1) under the assumption that A has a unit $\mathbf{1}$.

Let $\phi(a) = r + is$, where r and s are real; and let $b = a + it\mathbf{1}$, where t is real. Since $\|\phi\| \leq 1$ (V.7.3), we have:

$$r^2 + s^2 + 2st + t^2 = |r + i(s + t)|^2$$
$$= |\phi(b)|^2 \leq \|b\|^2 = \|b^*b\|$$
$$= \|a^2 + t^2\mathbf{1}\| \leq \|a^2\| + t^2.$$

Cancelling t^2 from the first and last expressions, we get

$$r^2 + s^2 + 2st \leq \|a^2\|.$$

This holds for all real t—an impossibility unless $s = 0$. So (1) is proved.

We now drop the assumption that A has a unit.

If $\phi \in A^\dagger$ and $a = b + ic \in A$ (b, c being Hermitian), then by (1) $\phi(a^*) = \phi(b) - i\phi(c) = (\phi(b) + i\phi(c))^- = \overline{\phi(a)}$; so the first statement of the theorem is proved. Thus the Gelfand transform map is a *-homomorphism of A into $\mathscr{C}_0(A^\dagger)$.

Next we will show that $a \mapsto \hat{a}$ is an isometry. If a is Hermitian, we have

$$\|a\| = \|a\|_{sp} \qquad\qquad \text{(by 3.6)}$$
$$= \|\hat{a}\|_\infty \qquad\qquad \text{(by V.7.13).}$$

Consequently, if b is an arbitrary element of A,

$$\|b\|^2 = \|b^*b\| = \|(b^*b)\hat{\,}\|_\infty = \|(b^*)\hat{\,}\,\hat{b}\|_\infty$$
$$= \|\bar{\hat{b}}\hat{b}\|_\infty \qquad\qquad (\text{since } (b^*)\hat{\,} = \bar{\hat{b}})$$
$$= \|\hat{b}\|_\infty^2.$$

So $a \mapsto \hat{a}$ is an isometric *-isomorphism of A onto some *-subalgebra B of $\mathscr{C}_0(\hat{A})$. Since A is complete, B must be norm-closed. Also, B separates points of $\hat{A}$ and does not vanish anywhere on $\hat{A}$. Consequently the Stone-Weierstrass Theorem (A8) asserts that $B = \mathscr{C}_0(\hat{A})$. The theorem is now completely proved.   ∎

4.4. Corollary.   *A commutative C*-algebra is semisimple and symmetric.*

Proof. By 4.3 and 2.5.   ∎

4.5. Remark.   In view of the first statement of 4.3, we shall usually denote the Gelfand space of a commutative C^*-algebra A by $\hat{A}$ instead of $A^\dagger$.

The correspondence $a \mapsto \hat{a}$ between elements of A and functions on $\hat{A}$ is often called the *functional representation* of the commutative C^*-algebra A.

Notice that an element a of A is Hermitian [unitary] if and only if the values of $\hat{a}$ are all real [of absolute value 1].

4.6. Remark.   By V.8.8, every locally compact Hausdorff space S can serve as the Gelfand space of a commutative C^*-algebra, namely, $\mathscr{C}_0(S)$. Thus Theorem 4.3 sets up a one-to-one correspondence between the *-isomorphism classes of commutative C^*-algebras and the homeomorphism classes of locally compact Hausdorff spaces. Under this correspondence the commutative C^*-algebras with unit go into the compact Hausdorff spaces.

5.   Spectra in Subalgebras of C^*-Algebras

5.1. Lemma.   *Let A be a commutative C*-algebra with unit $\mathbb{1}$, and a and b two elements of A such that*

$$\phi(b) = \psi(b)) \qquad \text{whenever } \phi, \psi \in \hat{A} \text{ and } \phi(a) = \psi(a).$$

Then b belongs to the norm-closed *-subalgebra of A generated by a and $\mathbb{1}$.

Proof.   Use the functional representation (4.5) of A; and apply the Stone-Weierstrass Theorem in the form A7.   ∎

5.2. Proposition.   *Let A be any C*-algebra with unit $\mathbb{1}$, and B a norm-closed *-subalgebra of A containing $\mathbb{1}$. Then $\mathrm{Sp}_A(a) = \mathrm{Sp}_B(a)$ for every a in B.*

Proof.   We shall show that, if $a \in B$ and a^{-1} exists in A, then $a^{-1} \in B$.

To do this, first assume that a is Hermitian. In that case a^{-1} is also Hermitian, and since a^{-1} commutes with a we can embed both a and a^{-1} in a commutative norm-closed *-subalgebra C of A containing 1. Since $\phi(a)\phi(a^{-1}) = 1$, $\phi(a^{-1})$ is determined by $\phi(a)$ for each ϕ in $\hat{C}$; and we conclude by 5.1 that a^{-1} lies in the norm-closed *-subalgebra generated by a and 1. Since a and 1 are in B, this asserts that $a^{-1} \in B$.

Now drop the assumption that a is Hermitian. Let $b = a^*a$. Since a^{-1} exists in A, so does b^{-1}. But b is Hermitian; so by the preceding paragraph $b^{-1} \in B$. Hence the equality $1 = (b^{-1}a^*)a$ shows that $a^{-1} = b^{-1}a^* \in B$.

We have now established the first statement of the proof. Applied to $a - \lambda 1$ instead of a, it shows that

$$\mathrm{Sp}_B(a) \subset \mathrm{Sp}_A(a) \qquad \text{for all } a \text{ in } B.$$

Since $\mathrm{Sp}_A(a) \subset \mathrm{Sp}_B(a)$ trivially (V.5.2(IV)), the proof is complete. ∎

5.3. We can now remove the hypothesis about units in 5.2. In that case $\mathrm{Sp}_A(a)$ and $\mathrm{Sp}_B(a)$ must coincide except possibly for the number 0.

Proposition. *Let B be any norm-closed *-subalgebra of a C^*-algebra A. Then*

$$\mathrm{Sp}_A(a) \cup \{0\} = \mathrm{Sp}_B(a) \cup \{0\} \qquad \text{for all } a \text{ in } B.$$

Proof. Fix $a \in B$. We first assume that A has a unit 1 and $1 \notin B$; and define $B' = B + \mathbb{C}1$. Two cases are possible: Either B has no unit, in which case B' is the *-algebra obtained by adjoining one, and $\mathrm{Sp}_{B'}(a) = \mathrm{Sp}_B(a)$ by definition; or B has its own unit $e \neq 1$, in which case one checks easily that $\mathrm{Sp}_{B'}(a) = \mathrm{Sp}_B(a) \cup \{0\}$. By 3.12, B' is closed in A; hence $\mathrm{Sp}_A(a) = \mathrm{Sp}_{B'}(a)$ by 5.2. So the proposition is proved for this case.

Next, assume that A has no unit; and adjoin 1 to get A_1. By definition $\mathrm{Sp}_{A_1}(a) = \mathrm{Sp}_A(a)$; so for this case the proposition follows from the preceding paragraph.

This together with 5.2 disposes of all the cases. ∎

5.4. If the B of 5.3 is commutative, and if $a \mapsto \hat{a}$ is its Gelfand transform map, it follows from 5.2, 5.3, and V.7.13 that for all a in B

$$\mathrm{Sp}_A(a) \cup \{0\} = \mathrm{range}(\hat{a}) \cup \{0\}. \tag{1}$$

and

$$\mathrm{Sp}_A(a) = \mathrm{range}(\hat{a}) \qquad \text{if } A \text{ and } B \text{ have a common unit.} \tag{2}$$

Now, a is a Hermitian element of B if and only if $\mathrm{range}(\hat{a}) \subset \mathbb{R}$. Hence (1) implies:

5.5. Proposition. *A normal element a of a C^*-algebra A is Hermitian if and only if $\mathrm{Sp}_A(a) \subset \mathbb{R}$.*

Remark. The word "normal" cannot of course be omitted from this proposition. For example, a nilpotent matrix a has spectrum $\{0\}$ in the $n \times n$ total matrix *-algebra, but certainly need not be Hermitian.

5.6. Likewise, if A has a unit which is also in B, then by 4.5 a is unitary if and only if range($\hat{a}$) $\subset \{\lambda : |\lambda| = 1\}$. So we have:

Proposition. *A normal element a of a C^*-algebra A with unit is unitary if and only if* $\mathrm{Sp}_A(a) \subset \{\lambda \in \mathbb{C} : |\lambda| = 1\}$.

6. The Functional Calculus in C^*-Algebras

6.1. It turns out that, because of the structure of commutative C^*-algebras, any continuous function on $\mathbb{C}$ to $\mathbb{C}$ can be made to act on the normal elements of a C^*-algebra A to produce other normal elements of A.

Let us denote by $\mathscr{F}$ the family of all continuous functions $f : \mathbb{C} \to \mathbb{C}$, considered as a *-algebra under the usual pointwise operations (complex conjugation being the involution operation). Let e denote the unit of $\mathscr{F}$ (so that $e(z) \equiv 1$), and w the identity function $z \mapsto z$ ($z \in \mathbb{C}$).

Proposition. *Let A be a C^*-algebra with unit $\mathbb{1}$, and a a fixed normal element of A. Then there exists a unique *-homomorphism $\Phi : \mathscr{F} \to A$ with the following properties:*

  (i) $\Phi(e) = \mathbb{1}$;
  (ii) $\Phi(w) = a$;
  (iii) *If $f_i \to f$ in $\mathscr{F}$ uniformly on compact sets, then $\Phi(f_i) \to \Phi(f)$ in A.*

Proof. We first observe the uniqueness of Φ. Indeed, if p is a complex polynomial in z and $\bar{z}$, then (i), (ii), and the *-homomorphism property of Φ show that $\Phi(p) = p(a, a^*)$. The uniqueness of Φ now follows from this and (iii), together with the Stone–Weierstrass Theorem A8, which asserts that any f in $\mathscr{F}$ can be approximated uniformly on compact sets by polynomials in z and $\bar{z}$.

Since a is normal it belongs to some norm-closed commutative *-subalgebra of A containing $\mathbb{1}$. So, to prove the existence of Φ, we may as well suppose that A itself is commutative. In fact, by Theorem 4.3, we may suppose that $A = \mathscr{C}(S)$, S being a compact Hausdorff space. But then the definition

$$\Phi(f) = f \circ a \qquad\qquad (f \in \mathscr{F}) \qquad (1)$$

has all the properties required of Φ. ∎

6.2. *Definition.* Continuing the notation of 6.1, we shall in the future denote the element $\phi(f)$ by $f(a)$.

By (1), $f(a)$ is evaluated by embedding the (normal) element a in a commutative C^*-subalgebra B containing 1, and composing f with the Gelfand transform of a. The uniqueness assertion of 6.1 means that the result is independent of the particular B.

Of course $f(a)$ is again normal, and commutes with a and a^*.

6.3. *Definition.* In 6.2 we supposed that A had a unit. Now let A be a C^*-algebra without unit, and adjoin a unit 1 to get the C^*-algebra A_1 (as in 3.10). If a is a normal element of A and $f \in \mathscr{F}$, then $f(a)$ (evaluated in A_1) belongs to A if and only if $f(0) = 0$. (Indeed, this is evident for polynomials in z and $\bar{z}$; to prove it for arbitrary f in $\mathscr{F}$, we express f as a uniform-on-compacta limit of polynomials p satisfying $p(0) = f(0)$.) Thus, if $f \in \mathscr{F}$ and $f(0) = 0$, then $f(a)$ still has a well-defined meaning and lies in A for any normal element a of A.

The above element $f(a)$ is called *the result of applying the function f to a.* This operation is known as the *functional calculus* on C^*-algebras.

6.4. Here are some further properties of the functional calculus.

Proposition. *Let A be a C^*-algebra, and let $f, g \in \mathscr{F}$. If A does not have a unit, we assume that $f(0) = g(0) = 0$. Let a be a normal element of A. Then:*

(i) $\|f(a)\| = \sup\{|f(z)| : z \in \mathrm{Sp}_A(a)\}$.
(ii) $g(f(a)) = (g \circ f)(a)$.
(iii) *If f and g coincide on $\mathrm{Sp}_A(a)$, then $f(a) = g(a)$.*

Proof. Apply (1) and 5.4(2). ∎

Remark. It follows from (iii) that $f(a)$ has meaning for complex-valued functions f which are defined and continuous merely on $\mathrm{Sp}_A(a)$ (and vanish at 0 if A has no unit).

6.5. Proposition. *Let A and A' be two C^*-algebras and $\phi: A \to A'$ a $*$-homomorphism; and suppose that $f \in \mathscr{F}$. We shall assume that either $f(0) = 0$ or A and A' have unit elements 1 and $1'$ respectively satisfying $\phi(1) = 1'$. Then $\phi(f(a)) = f(\phi(a))$ for every normal element a of A.*

Proof. Certainly this is true if f is a polynomial in z and $\bar{z}$. For arbitrary f we prove it by approximating f uniformly on compact sets by polynomials, and then using 6.1(iii) and the continuity of ϕ (3.7). ∎

6.6. We shall frequently make important applications of the functional calculus on C^*-algebras. Here is an application to the continuity properties of the spectrum of a variable normal element of A. We leave the details of the proof as an exercise to the reader.

Proposition*. *Let A be a C^*-algebra with unit, and denote by A_n the set of all normal elements of A. If $a_i \to a$ in A_n, then:*

  (I) *For each open subset V of $\mathbb{C}$ containing $\mathrm{Sp}_A(a)$, we have $\mathrm{Sp}_A(a_i) \subset V$ for all large enough i.*

  (II) *Every open subset V of $\mathbb{C}$ which intersects $\mathrm{Sp}_A(a)$ also intersects $\mathrm{Sp}_A(a_i)$ for all large enough i.*

The first step in proving (II) is to show that $a \mapsto f(a)$ is continuous on A_n to A_n for each f in $\mathscr{F}$. Now, given an open set V with $\lambda \in V \cap \mathrm{Sp}_A(a)$, we choose a function f in $\mathscr{F}$ vanishing outside V and satisfying $f(\lambda) \neq 0$. Then $f(a) \neq 0$; so $f(a_i) \neq 0$ for all large enough i; and the latter implies that $V \cap \mathrm{Sp}_A(a_i) \neq \emptyset$ for all large enough i.

Remark. Conditions (I) and (II) together assert that $a \mapsto \mathrm{Sp}_A(a)$ is continuous on A_n with respect to a certain natural topology of the space of closed bounded subsets of $\mathbb{C}$.

Remark. One might conjecture that (I) and (II) would hold as $a_i \to a$ in any Banach algebra A whatever. This is true for (I); and it is true for (II) if A is commutative. But (II) fails in general. See Remark, V.6.19.

7. Positive Elements and Symmetry of C^*-Algebras

7.1. The most significant result of this section will be the fact (7.11) that every C^*-algebra is symmetric. In connection with this we will develop the very important notion of positivity in a C^*-algebra.

7.2. Let A be a fixed C^*-algebra.

Definition. An element a of A is *positive* (*in A*) if (i) a is Hermitian, and (ii) $\mathrm{Sp}_A(a)$ consists entirely of non-negative real numbers.

The set of all positive elements of A will be denoted by A_+.

Remark. In view of 5.5, we could replace the word "Hermitian" in the preceding definition by "normal."

7.3. Proposition. *If B is a norm-closed *-subalgebra of a C*-algebra A, an element of B is positive in B if and only if it is positive in A.*

Proof. By 5.3. ∎

7.4. Proposition. *Let a be a Hermitian element of a C*-algebra A; let B be a norm-closed commutative *-subalgebra of A containing a; and let $b \mapsto \hat{b}$ be the Gelfand transform map for B. Then $a \in A_+$ if and only if $\hat{a}(\phi) \geq 0$ for all ϕ in $\hat{B}$.*

Proof. By 5.4(1). ∎

Since any Hermitian element of A can be embedded in such an algebra B, we thus obtain:

7.5. Corollary. *An element a of a C*-algebra A is in A_+ if and only if $a = b^2$ for some Hermitian b in A.*

7.6. Corollary. *If A is a C*-algebra, and a and $-a$ are both in A_+, then $a = 0$.*

7.7. Corollary. *Let a be a Hermitian element of a C*-algebra A with unit $\mathbb{1}$; and let $r \geq \|a\|$. Then $a \in A_+$ if and only if $\|r\mathbb{1} - a\| \leq r$.*

Proof. $\|r\mathbb{1} - a\| \leq r$ if and only if $|r - \hat{a}(\phi)| \leq r$ for all ϕ in $\hat{B}$ (B being as in 7.4). Since $\hat{a}(\phi) \leq r$, this is equivalent to saying that $\hat{a}(\phi) \geq 0$ for all ϕ. Now apply 7.4. ∎

7.8. Corollary. *A_+ is norm-closed in A.*

7.9. Proposition. *If A is a C*-algebra, and a and b are in A_+, then $a + b \in A_+$.*

Proof. By 3.10 we may as well assume that A has a unit $\mathbb{1}$. Then by 7.7

$$\| \|a\|\mathbb{1} - a\| \leq \|a\|,$$

$$\| \|b\|\mathbb{1} - b\| \leq \|b\|.$$

Adding these and using the triangle inequality, we get

$$\|(\|a\| + \|b\|)\mathbb{1} - (a + b)\| \leq \|a\| + \|b\|. \tag{1}$$

Now $a + b$ is Hermitian, and $\|a + b\| \leq \|a\| + \|b\|$. So we can apply 7.7 with a and r replaced by $a + b$ and $\|a\| + \|b\|$, and conclude from (1) that $a + b \in A_+$. ∎

7.10. Corollary. *If $a_1, \ldots, a_n$ are positive elements of a C*-algebra A, and $a_1 + \cdots + a_n = 0$, then $a_i = 0$ for each i.*

Proof. $a_2 + \cdots + a_n$ is positive by 7.9. Therefore both a_1 and $-a_1 = a_2 + \cdots + a_n$ are positive. So $a_1 = 0$ by 7.6. Similarly for each a_i.   ∎

7.11. Theorem. *Every C*-algebra A is symmetric.*

Note. By 2.3 this amounts to saying that a^*a is positive for every a in A.

Proof. Take an element a of A; let $b = a^*a$; and embed the Hermitian element b in a norm-closed commutative *-subalgebra B of A. By 7.4 and the above Note we have only to show that $\hat{b} \geq 0$ everywhere on $\hat{B}$ ($\hat{b}$ being the Gelfand transform on B). Assume then that $\hat{b}(\phi) < 0$ for some ϕ. By 4.3 there is a non-zero Hermitian element c of B such that $\hat{c}$ vanishes except where $\hat{b}$ is negative, that is,

$$c^2 b \neq 0, \qquad (c^2 b)\hat{\ } \leq 0. \tag{2}$$

Setting $d = ac$, we have $d^*d = c^2 b$. So, by (2) and 7.4,

$$0 \neq -d^*d \in A_+. \tag{3}$$

Now $-dd^*$ is Hermitian. Hence by (3) and V.5.5,

$$-dd^* \in A_+. \tag{4}$$

Now setting $d = u + iv$ (where u, v are Hermitian) and multiplying out $dd^* + d^*d$, we get

$$(-d^*d) + (-dd^*) + 2u^2 + 2v^2 = 0. \tag{5}$$

By (3), (4), and 7.5, all four terms of (5) are positive. So by 7.10 they are all 0, contradicting (3). Therefore $\hat{b} \geq 0$; and the proof is complete.   ∎

7.12. Remark. The proof of 7.11 was not discovered until many years after the first introduction of C^*-algebras by Gelfand and Naimark. In their original definition, symmetry was included as one of the postulates for a C^*-algebra (see the Notes and Remarks).

7.13. Corollary. *If A is a C*-algebra, the positive elements of A are precisely those of the form a^*a $(a \in A)$.*

Proof. By 7.5 and 7.11.   ∎

7.14. **Corollary.** *If A is a C*-algebra, $a \in A$, and $b \in A_+$, then $a^*ba \in A_+$.*

Proof. By 7.13 $b = c^*c$ for some c. Thus $a^*ba = a^*c^*ca = (ca)^*ca$; and this is positive by 7.13. ∎

7.15. ***n*-th Roots in *C**-Algebras.** Here is an important application of the functional calculus to positive elements.

Proposition. *Let A be a C*-algebra, and let $a \in A_+$. Then for each positive integer n there is a unique $b \in A_+$ such that $b^n = a$.*

We refer to this b as $a^{1/n}$, the *positive n-th root of a.*

Proof. Let $f(t) = t^{1/n}$ for $0 \leq t \in \mathbb{R}$. Since $\mathrm{Sp}_A(a) \subset \{t : t \geq 0\}$, $f(a)$ is well defined (by Remark 6.4); and $(f(a))^n = a$ by 6.4. Since $f \geq 0$, $f(a) \in A_+$. So the existence of b has been proved.

Now let b be any element of A_+ with $b^n = a$. Thus, if $g(t) = t^n$ $(t \in \mathbb{R})$, we have $g(b) = a$, and $f \circ g$ is the identity function on $\{t : t \geq 0\}$. Therefore, since $\mathrm{Sp}_A(b)$ is nonnegative,

$$b = (f \circ g)(b) \qquad \text{(by 6.4(iii))}$$

$$= f(g(b)) \qquad \text{(by 6.4(ii))}$$

$$= f(a).$$

So b is unique. ∎

7.16. As another application of the functional calculus we have:

Proposition. *For every Hermitian element a of the C*-algebra A, there are unique positive elements a_+ and a_- of A such that*

$$a = a_+ - a_-, \qquad a_+a_- = a_-a_+ = 0.$$

We have $\|a_+\| \leq \|a\|$, $\|a_-\| \leq \|a\|$.

Proof. Let $f : \mathbb{R} \to \mathbb{R}$ and $g : \mathbb{R} \to \mathbb{R}$ be the continuous functions given by:

$$f(t) = \begin{cases} t & \text{for } t \geq 0, \\ 0 & \text{for } t \leq 0, \end{cases}$$

$$g(t) = \begin{cases} 0 & \text{for } t \geq 0, \\ -t & \text{for } t \leq 0. \end{cases}$$

By 5.5 and Remark 6.4 $f(a)$ and $g(a)$ have meaning, and by 7.2 are positive. Since $f(t) - g(t) \equiv t$ and $f(t)g(t) \equiv 0$, we have by 6.1 $a = f(a) - g(a)$ and $f(a)g(a) = 0$. Putting $a_+ = f(a)$ and $a_- = g(a)$ we see that the required elements exist, and that $\|a_\pm\| \le \|a\|$.

To prove their uniqueness, let b and c be positive elements of A satisfying $a = b - c$, $bc = 0$. Since $cb = (bc)^* = 0^* = 0 = bc$, we can embed both b and c in a commutative C^*-subalgebra B of A. Identifying the elements of B with their functional representations, we conclude from 7.4 that $b(\phi) \ge 0$ and $c(\phi) \ge 0$ for all ϕ in $\hat{B}$. Also $b(\phi)c(\phi) = (bc)(\phi) = 0$. Thus it is evident that $f(a) = f(b - c) = b$, $g(a) = g(b - c) = c$; that is, $b = a_+$ and $c = a_-$. So a_+ and a_- are unique.   ∎

Definition. a_+ and a_- are called the *positive* and *negative parts* respectively of the Hermitian element a.

7.17.   Combining 7.16 and 1.1(1), we obtain:

Corollary.   *Every element a of the C^*-algebra A is of the form $a_1 - a_2 + i(a_3 - a_4)$, where $a_i \in A_+$ and $\|a_i\| \le \|a\|$ $(i = 1, 2, 3, 4)$.*

The Order Relation Defined by Positivity

7.18.   Fix a C^*-algebra A. Obviously $tA_+ \subset A_+$ if $0 \le t \in \mathbb{R}$. Hence by 7.6 and 7.9 A_+ is a *proper cone* in A; that is, it is closed under addition and multiplication by non-negative scalars, and cannot contain both a and $-a$ unless $a = 0$. We now define an order relation by means of A_+.

Definition.   If $a, b \in A$, we shall say that $a \le b$ (or $b \ge a$) *in A if and only if* $b - a \in A_+$.

The properties of A_+ mentioned above easily imply the following:

(i) $a \le a$.
(ii) $a \le b \le c \Rightarrow a \le c$ (transitivity).
(iii) $a \le b \le a \Rightarrow a = b$ (antisymmetry).
(iv) $a \le b, 0 \le \lambda \in \mathbb{R} \Rightarrow \lambda a \le \lambda b$.
(v) $a \le b, c \in A \Rightarrow a + c \le b + c$.
(vi) $a \ge 0$ in $A \Leftrightarrow a \in A_+$.

If A is commutative, and is identified via its functional representation with $\mathscr{C}_0(\hat{A})$, then by 7.4 the ordering $\le$ in A is just the pointwise ordering of functions: $a \le b \Leftrightarrow a(\phi) \le b(\phi)$ for all ϕ in $\hat{A}$.

Let B be a norm-closed *-subalgebra of the C^*-algebra A. Proposition 7.3 shows that the ordering of elements of B with respect to B is the same as with respect to A.

7.19. Proposition. *If a, b, c are elements of a C^*-algebra A, then*

$$b \leq c \Rightarrow a^*ba \leq a^*ca.$$

Proof. By 7.14. ∎

7.20. Proposition. *If a, b are elements of a C^*-algebra A and $0 \leq a \leq b$, then $\|a\| \leq \|b\|$.*

Proof. We can assume without loss of generality that A has a unit $\mathbb{1}$. The functional representation of b (in the closed *-subalgebra generated by b and $\mathbb{1}$) shows that $b \leq \|b\|\mathbb{1}$. Consequently $0 \leq a \leq \|b\|\mathbb{1}$. From this, looking at a in *its* functional representation, we conclude that $\|a\| \leq \|b\|$. ∎

7.21. Proposition. *Let A be a C^*-algebra with unit $\mathbb{1}$; and let a, b be elements of A_+ such that $a \leq b$. Then, if a^{-1} and b^{-1} exist, we have $b^{-1} \leq a^{-1}$.*

Proof. If $a = \mathbb{1}$ this is obvious from the functional representation of b. In the general case 7.19 gives $\mathbb{1} = a^{-1/2}aa^{-1/2} \leq a^{-1/2}ba^{-1/2}$ (see 7.15), whence by the preceding sentence $\mathbb{1} \geq a^{1/2}b^{-1}a^{1/2}$. Again using 7.19, we get $a^{-1} = a^{-1/2}\mathbb{1}a^{-1/2} \geq a^{-1/2}(a^{1/2}b^{-1}a^{1/2})a^{-1/2} = b^{-1}$. ∎

7.22. For later use we prove here the following technical result:

Proposition. *Let A be a C^*-algebra with unit $\mathbb{1}$. If $0 \leq a \leq b$ in A, then*

$$\|a^{1/2}(b + \varepsilon\mathbb{1})^{-1/2}\| \leq 1$$

for every positive number ε.

Proof. Let $\varepsilon > 0$. Since $a + \varepsilon\mathbb{1} \leq b + \varepsilon\mathbb{1}$, we have $(b + \varepsilon\mathbb{1})^{-1} \leq (a + \varepsilon\mathbb{1})^{-1}$ by 7.21. So by 7.19

$$a^{1/2}(b + \varepsilon\mathbb{1})^{-1}a^{1/2} \leq a^{1/2}(a + \varepsilon\mathbb{1})^{-1}a^{1/2} = a(a + \varepsilon\mathbb{1})^{-1}. \tag{6}$$

Now since $t(t + \varepsilon)^{-1} \leq 1$ whenever $0 \leq t \in \mathbb{R}$, the functional representation of a shows that $\|a(a + \varepsilon\mathbb{1})^{-1}\| \leq 1$. Hence by (6) and 7.20

$$\|a^{1/2}(b + \varepsilon\mathbb{1})^{-1}a^{1/2}\| \leq 1. \tag{7}$$

Putting $c = a^{1/2}(b + \varepsilon 1)^{-1/2}$, we have $cc^* = a^{1/2}(b + \varepsilon 1)^{-1}a^{1/2}$, and hence $\|c\|^2 = \|cc^*\| \leq 1$ by (7). This completes the proof.   ∎

7.23.   Remark.   Let A be a C^*-algebra, and A_h the family of all Hermitian elements of A, considered as an ordered set under $\leq$.

Suppose that A is commutative. Then 4.3 and 7.18 allow us to identify A_h with the space of all real-valued functions in $\mathscr{C}_0(\hat{A})$, with the pointwise ordering relation. Thus any two elements f and g of A_h have a least upper bound $f \vee g$ and a greatest lower bound $f \wedge g$ in A_h.

However, if A is not commutative, least upper bounds and greatest lower bounds of pairs of elements of A_h will not in general exist. Here is a simple example:

Consider the C^*-algebra A of all 2×2 complex matrices (with the adjoint operation, of course, as the involution); and put $u = \begin{pmatrix} 1 & 0 \\ 0 & -1 \end{pmatrix}$. We claim that the two elements 0 and u have no least upper bound in A_h.

To prove this, we begin by noting that the element $\begin{pmatrix} \alpha & \beta \\ \bar{\beta} & \gamma \end{pmatrix}$ (α, γ real) of A_h is in A_+ if and only if $\alpha \geq 0$, $\gamma \geq 0$, and $|\beta|^2 \leq \alpha\gamma$. Suppose now that $r = \begin{pmatrix} \rho & \sigma \\ \bar{\sigma} & \tau \end{pmatrix}$ (ρ, τ real) is the least upper bound of 0 and u in A_h. Since $v = \begin{pmatrix} 1 & 0 \\ 0 & 0 \end{pmatrix}$ is an upper bound of 0 and u, we must have $0 \leq r, u \leq r, r \leq v$. These conditions, together with the above criterion for positivity, imply that $r = v$. On the other hand, for each $\varepsilon > 0$ the matrix

$$\begin{pmatrix} 1 + \varepsilon & (\varepsilon(1 + \varepsilon))^{1/2} \\ (\varepsilon(1 + \varepsilon))^{1/2} & \varepsilon \end{pmatrix}$$

majorizes both 0 and u, but not v.

7.24.   We come now to the "polar decomposition" of an invertible element of a C^*-algebra.

Theorem.   *Let A be a C^*-algebra with unit 1, and a an element of A which is invertible (i.e., a^{-1} exists). Then there is a unique unitary element u of A and a unique positive element p of A such that $a = up$. In fact $p = (a^*a)^{1/2}$ (see 7.13, 7.15).*

Proof.   Since a is invertible, so is a^*a; and hence by the functional calculus so is the positive element $p = (a^*a)^{1/2}$. Defining $u = ap^{-1}$, we have $u^*u = p^{-1}a^*ap^{-1} = p^{-1}p^2p^{-1} = 1$; so u^* is a left inverse of u. Since $upa^{-1} = 1$, u

also has a right inverse. So the left inverse u^* is also the right inverse and hence the inverse of u. So u is unitary; and $a = up$ is a factorization of a of the required kind.

If $a = vq$ is another such factorization (with v unitary and q positive), then $a^*a = qv^*vq = q^2$; so $q = (a^*a)^{1/2} = p$ by 7.15. Then $v = aq^{-1} = ap^{-1} = u$. So the factorization of the given kind is unique. ■

Remark. This should be compared with the "unbounded polar decomposition" 13.5, 13.9.

The Closed Convex Hull of the Unitary Elements in a C^*-Algebra

7.25. Theorem. *Let A be a C^*-algebra with unit 1. Then the closed unit ball of A is the closed convex hull of the group U of unitary elements of A.*

Proof. Let B denote the open unit ball of A and $\bar{H}(U)$ the closed convex hull of U. It suffices to show that $B \subset \bar{H}(U)$. So let us take $a \in B$. To show that $a \in \bar{H}(U)$, we claim that it is enough to show that for every $u \in U$ the element $b = (a + u)/2$ is in $\bar{H}(U)$. For then $U \subset 2\bar{H}(U) - a$, which is closed and convex, so $\bar{H}(U) \subset 2\bar{H}(U) - a$, $(a + \bar{H}(U))/2 \subset \bar{H}(U)$; if $a_0 = u$, and a sequence $\{a_n\}$ is defined iteratively by $a_{n+1} = (a + a_n)/2$ for $n = 1, 2, \ldots$, then $a_n \in \bar{H}(U)$ and $a_n \to a$.

However, the element b can be rewritten as $b = (au^{-1} + 1)u/2$, and since $\|bu^{-1} - 1\| < 1$ it follows that b is invertible (see V.6.1). Hence by 7.24 $b = v|b|$, where v is unitary and $|b| = (b^*b)^{1/2}$. Since $|b| \le 1$, the functional calculus shows that $|b| = (x + x^*)/2$, where $x = |b| + i(1 - |b|^2)^{1/2}$ is also unitary. ■

7.26. Remark. The result in 7.25 is due to Russo and Dye [1]; the clever proof given here, which substantially simplifies and shortens the original one, is due to Gardner [3]. Theorem 7.25 is known to hold in much greater generality (e.g., see Doran and Belfi [1]). We remark that Gardner's proof has led to interesting studies of convex combinations of unitary operators (see Kadison and Pedersen [1] and C. L. Olsen and Pedersen [1]).

8. Approximate Units in a C^*-Algebra; Applications to Ideals and Quotients

8.1. We observed in V.4.13 that a Banach algebra, even if it has no unit, may have a so-called approximate unit. This notion, though weaker than that of a unit element, is strong enough to serve the purposes of the latter in many topological contexts. It will turn out (8.4) that a C^*-algebra always has an approximate unit. Let us recall the definiton.

8.2. Definition. Let A be a normed algebra. By an *approximate unit* of A we mean a net $\{e_i\}$ $(i \in I)$ of elements of A such that (i) $\{\|e_i\|: i \in I\}$ is bounded, and (ii) $\|e_i a - a\| \to 0$ and $\|ae_i - a\| \to 0$ for all a in A.

A unit of A is obviously an approximate unit in a natural sense.

Note.   This definition should be distinguished from the Definition III.11.17 of an approximate unit on a locally compact group G. An approximate unit in the latter sense is certainly an approximate unit of the group algebra in the present sense but the converse is far from true.

8.3. Lemma. *Let A be a C^*-algebra and J a right ideal of A. Then there exists a net $\{c_\sigma\}$ of elements of J with the following properties:* (i) $\|c_\sigma\| \leq 1$ *for all σ;* (ii) $c_\sigma \in A_+$ *for all σ;* (iii) $\sigma \prec \tau \Rightarrow c_\sigma \leq c_\tau$; (iv) $\|c_\sigma a - a\| \underset{\sigma}{\to} 0$ *for all a in the norm-closure $\bar{J}$ of J.*

Proof.   Let $\mathscr{S}$ be the directed set (ordered by inclusion) of all finite non-void subsets of J. Let σ be a set in $\mathscr{S}$ having n elements. We set

$$b_\sigma = \sum_{a \in \sigma} aa^*, \qquad c_\sigma = b_\sigma(n^{-1}\mathbf{1} + b_\sigma)^{-1}.$$

(Here c_σ is calculated in the C^*-algebra with unit $\mathbf{1}$ adjoined if necessary; but it clearly lies in A.) Since $b_\sigma \geq 0$ and $0 \leq t(n^{-1} + t)^{-1} \leq 1$ for $0 \leq t \in \mathbb{R}$, the functional representation of b_σ shows that $0 \leq c_\sigma \leq \mathbf{1}$, hence $\|c_\sigma\| \leq 1$. Also

$$\sum_{a \in \sigma} [(c_\sigma - \mathbf{1})a][(c_\sigma - \mathbf{1})a]^* = (c_\sigma - \mathbf{1})b_\sigma(c_\sigma - \mathbf{1})$$

$$= n^{-2}b_\sigma(n^{-1}\mathbf{1} + b_\sigma)^{-2};$$

and $0 \leq n^{-2}t(n^{-1} + t)^{-2} \leq (4n)^{-1}$ for $0 \leq t \in \mathbb{R}$. So again the functional representation of b_σ shows that $\|\sum_{a \in \sigma} [(c_\sigma - \mathbf{1})a][(c_\sigma - \mathbf{1})a]^*\| \leq (4n)^{-1}$; whence by 7.20

$$\|(c_\sigma - \mathbf{1})a\|^2 = \|[(c_\sigma - \mathbf{1})a][(c_\sigma - \mathbf{1})a]^*\| \leq (4n)^{-1}$$

for all a in σ. Since $n \to \infty$ as σ increases, this gives

$$\|c_\sigma a - a\| \underset{\sigma}{\to} 0 \tag{1}$$

for all a in J. Since $\|c_\sigma\| \leq 1$, (1) holds in fact for all a in $\bar{J}$. Thus the c_σ satisfy (i), (ii), and (iv) of the proposition. It is also evident that b_σ and c_σ belong to J.

It remains only to show that $\sigma \subset \tau \Rightarrow c_\sigma \leq c_\tau$. Assume that σ and τ are sets in $\mathscr{S}$ containing n and p elements respectively, with $\sigma \subset \tau$ (so that $n \leq p$). Then $b_\sigma \leq b_\tau$; so by 7.21

$$(n^{-1}\mathbb{1} + b_\sigma)^{-1} \geq (n^{-1}\mathbb{1} + b_\tau)^{-1}. \tag{2}$$

(Here we adjoin $\mathbb{1}$ if necessary.) Further, since $p \geq n$, we have $p^{-1}(p^{-1} + t)^{-1} \leq n^{-1}(n^{-1} + t)^{-1}$ for all real $t \geq 0$; and hence by the functional representation of b_τ,

$$p^{-1}(p^{-1}\mathbb{1} + b_\tau)^{-1} \leq n^{-1}(n^{-1}\mathbb{1} + b_\tau)^{-1}. \tag{3}$$

Combining (2) and (3) we get

$$c_\sigma = \mathbb{1} - n^{-1}(n^{-1}\mathbb{1} + b_\sigma)^{-1} \leq \mathbb{1} - n^{-1}(n^{-1}\mathbb{1} + b_\tau)^{-1}$$

$$\leq \mathbb{1} - p^{-1}(p^{-1}\mathbb{1} + b_\tau)^{-1} = c_\tau;$$

and the proof is finished. ∎

8.4. If in 8.3 $\bar{J}$ is closed under *, we can apply * to condition (iv) and conclude that $\|ac_\sigma - a\| \to 0$ for all a in $\bar{J}$. Thus we have:

Theorem. *In any C*-algebra A there exists an approximate unit $\{e_i\}$ with the further properties:* (i) $\|e_i\| \leq 1$ *for all i;* (ii) $e_i \in A_+$ *for all i;* (iii) $i \prec j \Rightarrow e_i \leq e_j$.
In fact, if J is any dense left or right ideal of A, we can require in addition that $e_i \in J$ for all i.

8.5. Proposition. *Let J be a norm-closed right [left] ideal of a C*-algebra A. If $a \in A$, then $a \in J$ if and only if $aa^* \in J$ $[a^*a \in J]$.*

Proof. Suppose J is a norm-closed right ideal; and let $\{c_\sigma\}$ be the net of elements of $J \cap A_+$ constructed in 8.3.

The "only if" part is obvious. To prove the "if" part, assume $aa^* \in J$. Then

$$(c_\sigma a - a)(c_\sigma a - a)^* = (c_\sigma aa^* - aa^*)c_\sigma - (c_\sigma aa^* - aa^*) \to 0 \quad \text{(by 8.3(i), (iv)).}$$

Therefore $c_\sigma a - a \to 0$. Since $c_\sigma a \in J$ and J is closed, this gives $a \in J$; and the proof is complete for right ideals.

For a left ideal J, the required result is deduced by applying the preceding paragraph to the right ideal J^*. ∎

8.6.   Corollary.   *Let I be a norm-closed two-sided ideal of the C*-algebra A. Then I is a *-ideal (hence a C*-algebra in its own right). Further, if J is a norm-closed left [right] ideal of I, then J is a left [right] ideal of A.*

Proof.   Let $a \in I$. Since I is a left ideal, $a^*a = a^*(a^*)^* \in I$. Since I is a right ideal, we can apply 8.5 to this, getting $a^* \in I$. So I is a *-ideal.

Let J be a norm-closed left ideal of I; and suppose $a \in J$, $b \in A$. Since I is *-closed and $a \in I$, we have $a^*b^*b \in I$ and so $(ba)^*ba = (a^*b^*b)a \in J$. Since $ba \in I$, this implies $ba \in J$ by 8.5 (applied in I). So J is a left ideal of A.

If J is a right ideal of I, we apply the preceding paragraph to J^*.   ∎

8.7.   Let I be a norm-closed (two-sided) *-ideal of a C^*-algebra A. The quotient A/I, with the quotient norm $\| \ \|$, is certainly a Banach *-algebra (1.5). In fact, as one would hope, we have:

Proposition.   *A/I is a C*-algebra.*

Proof.   For $a \in A$, let $\tilde{a}$ be the image of a in A/I. It is enough by 3.2 to show that $\|\tilde{a}\|^2 \leq \|\tilde{a}^*\tilde{a}\|$.

Since I is a C^*-algebra in its own right, by 8.4 it has an approximate unit $\{e_i\}$ with the properties 8.4(i), (ii), (iii). We claim that

$$\|\tilde{a}\| = \lim_i \|a - ae_i\| \qquad\qquad (a \in A). \qquad (4)$$

Indeed: By 8.4(i), (ii) $\|1 - e_i\| \leq 1$. So, if $a \in A$ and $b \in I$,

$$\|a + b\| \geq \limsup_i \|(a + b)(1 - e_i)\|$$

$$= \limsup_i \|a - ae_i + b - be_i\|$$

$$= \limsup_i \|a - ae_i\| \qquad \left(\text{since } \lim_i \|b - be_i\| = 0\right)$$

$$\geq \liminf_i \|a - ae_i\|$$

$$\geq \inf\{\|a + b\| : b \in I\} = \|\tilde{a}\|.$$

From this and the arbitrariness of b we get (4).

Again let $a \in A$, $b \in I$. By (4),

$$\|\tilde{a}\|^2 = \lim_i \|a - ae_i\|^2 = \lim_i \|(a - ae_i)^*(a - ae_i)\|$$

$$= \lim_i \|(a - ae_i)^*(a - ae_i) + (b - e_ib)(1 - e_i)\| \qquad \text{(since } e_ib \to b\text{)}$$

$$= \lim_i \|(1 - e_i)(a^*a + b)(1 - e_i)\|$$

$$\leq \|a^*a + b\| \qquad \qquad \text{(since } \|1 - e_i\| \leq 1\text{)}.$$

By the arbitrariness of b, this gives $\|\tilde{a}\|^2 \leq \|(a^*a)^{\sim}\| = \|\tilde{a}^*\tilde{a}\|$. $\blacksquare$

8.8. The following proposition strengthens 3.9 and, when combined with 8.7, gives rise to the important proposition 8.10.

Proposition. *Let F be a *-isomorphism of a C*-algebra A into a C*-algebra B. Then F is an isometry; in particular its range is norm-closed.*

Proof. Let D be a norm-closed commutative *-subalgebra of A. By 4.3 D is isometrically *-isomorphic with $\mathscr{C}_0(\hat{D})$. On the other hand, since F is an isomorphism, D is a normed algebra under the norm $\| \ \|'$ given by $\|a\|' = \|F(a)\|$ $(a \in D)$. Combining these two facts with V.8.10 we deduce that

$$\|F(a)\| \geq \|a\| \qquad \qquad (a \in D). \qquad (5)$$

Since any Hermitian element of A can be embedded in such a D, (5) holds for all Hermitian a in A. Now, for any b in A, the element b^*b is Hermitian, and so by (5)

$$\|F(b)\|^2 = \|F(b)^*F(b)\| = \|F(b^*b)\|$$

$$\geq \|b^*b\| = \|b\|^2;$$

thus (5) holds for all a in A. Combining this with 3.7, we find that F is an isometry. $\blacksquare$

8.9. *Remark.* The proof of the preceding proposition shows in fact that *any *-isomorphism F of a C*-algebra A into a normed *-algebra B satisfies $\|F(a)\| \geq \|a\|$ for all a in A.*

8.10. Proposition. *If F is any *-homomorphism of a C*-algebra A into a C*-algebra B, then the range of F is norm-closed in B.*

Proof. The *-ideal Ker(F) of A is norm-closed because of the continuity of F (3.7); so by 8.7 $A/\text{Ker}(F)$ is a C*-algebra. On the other hand F induces a *-isomorphism $F': A/\text{Ker}(F) \to B$. Thus by 8.8 range($F'$), which is the same as range (F), is norm-closed.   ∎

8.11.   Corollary.   *If I is a norm-closed two-sided ideal of a C*-algebra A, and B is a norm-closed *-subalgebra of A, then $I + B$ is norm-closed.*

Proof.   Apply 8.10 to the *-homomorphism $a \mapsto a + I$ of B into A/I.   ∎

8.12.   Proposition.   *Let $\mathscr{I}$ be any family of norm-closed two-sided ideals of a C*-algebra A; and put $J = \bigcap \mathscr{I}$. Then*

$$\|a + J\|_{A/J} = \sup\{\|a + I\|_{A/I} : I \in \mathscr{I}\}$$

for all a in A.

Proof.   For each I in $\mathscr{I}$ let $F_I: A/J \to A/I$ be the quotient *-homomorphism. Then the map F sending each α in A/J into the function $I \mapsto F_I(\alpha)$ is a *-isomorphism of the C*-algebra A/J (see 8.7) into the C*-direct product of the $\{A/I\}$ ($I \in \mathscr{I}$). Therefore F is an isometry by 8.8; and this implies the required conclusion.   ∎

8.13.   *Characterization of C*-Direct Sums.*   Let A be the C*-direct sum $\sum_{i \in I}^{\oplus} A_i$ of C*-algebras A_i (see 3.13). For each i put $I_i = \{a \in A : a_j = 0$ for all $j \neq i\}$. Then I_i is a closed two-sided ideal of A *-isomorphic with A_i and satisfying: (i) $I_i \cap I_j = \{0\}$ for $i \neq j$; (ii) $\sum_{i \in I} I_i$ is dense in A.

Conversely, we shall show that the above properties (i), (ii) characterize C*-direct sums.

Proposition.   *Let A be a C*-algebra, and $\mathscr{I}$ a collection of closed two-sided ideals of A such that (i) $I \cap J = \{0\}$ if $I, J \in \mathscr{I}$ and $I \neq J$, and (ii) $\sum \mathscr{I}$ is dense in A. Then the C*-direct sum $A' = \sum_{I \in \mathscr{I}}^{\oplus} I$ is *-isomorphic with A under a *-isomorphism Φ which sends a into $\sum_{I \in \mathscr{I}} a_I$ whenever $a \in A'$ and $a_I = 0$ for all but finitely many I in $\mathscr{I}$.*

Proof.   Our first step is to notice that, if K, L are closed two-sided ideals of A,

$$KL = \{0\} \Leftrightarrow K \cap L = \{0\}.$$

Indeed: Evidently $KL \subset K \cap L$, so the implication $\Leftarrow$ holds. Conversely, if $KL = \{0\}$ and $a \in K \cap L$, then $a^*a \in KL$, whence $a^*a = 0$ or $a = 0$. This proves the implication $\Rightarrow$.

Next we claim that the elements of $\mathscr{I}$ are linearly independent in A. Indeed, if $I_1, \ldots, I_{n+1}$ are distinct elements of $\mathscr{I}$, hypothesis (i) gives $(I_1 + \cdots + I_n)I_{n+1} = \sum_{j=1}^{n} I_j I_{n+1} = \{0\}$. By the preceding paragraph this implies $(I_1 + \cdots + I_n) \cap I_{n+1} = \{0\}$, proving the claim.

Now let A_0' be the dense *-algebra of A' consisting of those a for which $a_I = 0$ for all but finitely many I; and define $\Phi_0 : A_0' \to A$ as follows;

$$\Phi_0(a) = \sum_{I \in \mathscr{I}} a_I \qquad\qquad (a \in A_0').$$

In view of (i), Φ_0 is a *-homomorphism. For any finite subset $\mathscr{F}$ of $\mathscr{I}$, the preceding claim shows that $\Phi_0 | \sum_{I \in \mathscr{F}}^{\oplus} I$ is one-to-one, hence by 8.8 (since $\sum_{I \in \mathscr{F}}^{\oplus} I$ is a C*-algebra) an isometry. Thus Φ_0 itself is an isometry on A_0', and so extends to an (isometric) *-isomorphism Φ of A' into A. By hypothesis (ii) $\Phi_0(A_0')$ is dense in A; so range$(\Phi) = A$. ■

8.14. The following technical fact will be useful in a later chapter.

Proposition. *Let X be a Hilbert space, and $\mathscr{S}$ a family of closed linear subspaces of X such that the linear span of $\mathscr{S}$ is dense in X. Further, let a be a bounded linear operator on X such that each Y in $\mathscr{S}$ is stable under both a and a^*. Then*

$$\|a\| = \sup\{\|a|Y\| : Y \in \mathscr{S}\}.$$

Proof. Let A be the set of all those b in $\mathcal{O}(X)$ such that each Y in $\mathscr{S}$ is stable under both b and b^*. Evidently A is a norm-closed *-subalgebra of $\mathcal{O}(X)$, and hence a C*-algebra. For each b in A let $F(b)$ be the collection $\{b|Y\}$, indexed by the elements Y of $\mathscr{S}$. Thus F is a *-homomorphism of A into the C*-direct product of the $\mathcal{O}(Y)$ $(Y \in \mathscr{S})$. Since the linear span of $\mathscr{S}$ is dense, F is a *-isomorphism; so by 8.8 it is an isometry. This applied to a gives the required conclusion. ■

8.15. Here is another interesting consequence of 8.3.

Proposition. *Let A be a C*-algebra with a unit $\mathbf{1}$; let I be a closed left or right ideal of A; and let $a \in A_+$. Then $a \in I$ if and only if, for each $\varepsilon > 0$, there is an element u of $I \cap A_+$ such that $a \leq u + \varepsilon\mathbf{1}$.*

Proof. The "only if" part is obvious. To prove the "if" part, it is enough to handle the case of a right ideal I (since $I^* \cap A_+ = I \cap A_+$); so we shall assume that I is a right ideal, and that a satisfies the given condition.

Set $b = a^{1/2}$. Let $\{c_\sigma\}$ be a net of elements of I with the properties of 8.3. We claim that $\lim_\sigma c_\sigma b = b$.

Indeed, given $\varepsilon > 0$, choose u in $I \cap A_+$ so that $a \leq u + \varepsilon 1$. Then

$$(b - c_\sigma b)(b - c_\sigma b)^* = (1 - c_\sigma)a(1 - c_\sigma)$$

$$\leq (1 - c_\sigma)u(1 - c_\sigma) + \varepsilon(1 - c_\sigma)^2 \qquad \text{(by 7.17)}.$$

Therefore, by 7.20

$$\|b - c_\sigma b\|^2 \leq \|1 - c_\sigma\| \|u - c_\sigma u\| + \varepsilon \|1 - c_\sigma\|^2. \tag{6}$$

Now, since $u \in I$, $\lim_\sigma \|u - c_\sigma u\| = 0$ by 8.3. Therefore, by (6), noting that $\|1 - c_\sigma\| \leq 1$, we have $\|b - c_\sigma b\|^2 < 2\varepsilon$ for all large enough σ. So the claim is proved.

Since $c_\sigma \in I$ and I is closed, it follows from the above claim that $b \in I$. Consequently $a = b^2 \in I$. ∎

8.16. Corollary. *If I is a norm-closed left or right ideal of any C*-algebra A, and if $0 \leq a \leq b \in I$, then $a \in I$.*

8.17. The following proposition will be useful in Chapter XI.

Proposition. *Let A be a C*-algebra, and P a norm-closed subset of A_+ with the following properties: (i) P is a cone (i.e., it is closed under addition and scalar multiplication by non-negative reals), and (ii) if $a \in A$ and $b \in P$ then $a^*ba \in P$. Then (I) the linear span I of P in A is a self-adjoint closed two-sided ideal of A, and (II) $I_+ = P$.*

Proof. First we assert that

$$\text{if } b \in A_+ \quad \text{and} \quad b \leq a \in P, \qquad \text{then } b \in P. \tag{7}$$

To prove this, adjoin a unit 1 if necessary and denote $a + \varepsilon 1$ $(0 < \varepsilon \in \mathbb{R})$ by a_ε. Since $b^{1/2}a_\varepsilon^{-1/2} \in A$, hypothesis (ii) says that

$$c_\varepsilon = b^{1/2}a_\varepsilon^{-1/2}aa_\varepsilon^{-1/2}b^{1/2} \in P. \tag{8}$$

But

$$\|c_\varepsilon - b\| = \|b^{1/2}a_\varepsilon^{-1/2}(a - a_\varepsilon)a_\varepsilon^{-1/2}b^{1/2}\|$$

$$\leq \|b^{1/2}a_\varepsilon^{-1/2}\|^2 \|a - a_\varepsilon\|. \tag{9}$$

Since $a_\varepsilon \to a$ as $\varepsilon \to 0$, and since $\|b^{1/2}a_\varepsilon^{-1/2}\| \leq 1$ by 7.22, (9) implies that $c_\varepsilon \to b$ as $\varepsilon \to 0$. This, (8), and the closedness of P give (7).

Put $J = \{a \in A : a^*a \in P\}$. We claim that

$$J \text{ is a right ideal of } A. \tag{10}$$

Indeed, first we show that J is closed under addition. Let $a, b \in J$. By 7.13 $0 \leq (a \pm b)^*(a \pm b) = a^*a + b^*b \pm a^*b \pm b^*a$. Taking the minus signs we get $a^*b + b^*a \leq a^*a + b^*b$. So, taking the plus signs, we conclude

$$(a + b)^*(a + b) \leq 2a^*a + 2b^*b. \tag{11}$$

Since $a, b \in J$, hypothesis (i) gives $2a^*a + 2b^*b \in P$. So (7) and (11) imply that $a + b \in J$. Now, since J is closed under addition, it is clearly a linear subspace of A. If $a \in J$ and $b \in A$, we have $(ab)^*(ab) = b^*(a^*a)b \in P$ by (ii); so $ab \in J$; and J is a right ideal of A.

Next, define K as the linear span of $\{b^*a : a, b \in J\}$. We shall show that

$$K = I \ (= \text{linear span of } P). \tag{12}$$

Indeed, $a \in P \Rightarrow a^{1/2} \in J \Rightarrow a \in K$; so $I \subset K$. Conversely, let $b, a \in J$. By the polarization identity b^*a is a linear combination of elements of the form c^*c where c is a linear combination of b and a. Since by (10) J is linear, such c are in J, and so $c^*c \in P$. It follows that $b^*a \in I$. Consequently $K \subset I$. Combining this with $I \subset K$ we get (12).

Obviously I is self-adjoint. By (10) K is a right ideal of A. So by (12) I is a self-adjoint right ideal and hence a two-sided ideal of A. Thus the norm-closure B of I is a closed two-sided ideal of A.

We shall now show that

$$I \cap A_+ = P. \tag{13}$$

Take $a \in I \cap A_+$. Since $a \in I$ and P is a cone, we can write

$$a = a_1 - a_2 + ia_3 - ia_4,$$

where the a_i are in P. Since a is Hermitian, $ia_3 - ia_4 = 0$; so $a = a_1 - a_2$. Thus we have $a \leq a_1 \in P$, whence $a \in P$ by (7). We have shown that $I \cap A_+ \subset P$. The reverse inclusion being trivial, (13) holds.

Now let a be any element of B_+. Since I is a dense ideal of B, 8.4 gives us an approximate unit $\{e_i\}$ of B consisting of elements of $I \cap A_+$. By (13) this means that the e_i are all in P. Since $a^{1/2} \in B$ we have $a^{1/2}e_i \underset{i}{\to} a^{1/2}$, hence

$$a^{1/2}e_i^2 a^{1/2} = a^{1/2}e_i(a^{1/2}e_i)^* \underset{i}{\to} a. \tag{14}$$

By hypothesis (ii), $a^{1/2}e_i^2 a^{1/2} = (a^{1/2}e_i^{1/2})e_i(a^{1/2}e_i^{1/2})^* \in P$. So by (14) $a \in P$. We have now shown that $B_+ \subset P$. Since $P \subset B_+$ (see 7.3), it follows that

$$B_+ = P. \tag{15}$$

By 7.17 any C^*-algebra is the linear span of the set of its positive elements. This means in view of (15) that

$$I = B. \tag{16}$$

Consequently I was closed all the time, and conclusion (I) is proved. Finally, (15) and (16) give conclusion (II).   ∎

8.18.   Here is a result suggestive of the Stone–Weierstrass Theorem.

Proposition.   *Let I be an index set (with at least two elements); for each i in I let A_i be a C^*-algebra; and form the C^*-direct sum $A = \sum_{i \in I}^{\oplus} A_i$ (see 3.13). Suppose that B is a closed *-subalgebra of A with the following property: For any two distinct indices i, j in I and any elements α of A_i and β of A_j, there is an element c of B such that $c_i = \alpha$ and $c_j = \beta$. Then $B = A$.*

Proof.   Fix $i_0 \in I$. Multiplying together finitely many elements of B each of which has a prescribed value at i_0 and vanishes at some prescribed $i \neq i_0$, we conclude that, given any α in A_{i_0} and any finite subset F of $I \setminus \{i_0\}$, there is an element c^F of B satisfying $c_{i_0}^F = \alpha$, $c_i^F = 0$ for $i \in F$. By the functional calculus we may assume that $\{\|c^F\| : F \text{ varying}\}$ is bounded. Thus, for any b in B, $bc^F \to d$ as F increases, where $d_i = 0$ for $i \neq i_0$ and $d_{i0} = b_{i_0}\alpha$. Such d therefore belong to B. From this it is easy to complete the argument that $B = A$.   ∎

Remark.   If each $A_i = \mathbb{C}$, then the above proposition is a special case of the Stone–Weierstrass Theorem A8.

As we have seen in §4, the subject of the Stone–Weierstrass Theorem A8, namely $\mathscr{C}_0(S)$ (S being locally compact and Hausdorff), is just the general commutative C^*-algebra. Thus it is natural to ask for a generalization of the Stone–Weierstrass Theorem to arbitrary non-commutative C^*-algebras. Such a generalization was given by Glimm [2]; see also Dixmier [21], §11. One version of Glimm's generalization, which patently contains the above proposition as a special case, is the "Stone–Weierstrass–Glimm Theorem" 1.4 of Fell [3]. Further results on the Stone–Weierstrass Theorem have been obtained by Akemann [1], Akemann and Anderson [1], Anderson and Bunce [1], Elliott [1], Ringrose [1], and Sakai [1].

9.   Elementary Remarks on *-Representations

9.1.   It is time now to begin the study of *-representations.

Throughout this section we fix an arbitrary *-algebra A.

Let X be a pre-Hilbert space, with inner product (,). By a *pre-*-representation of A acting in X* we mean a representation T, whose space is X, of the underlying algebra of A (in the purely algebraic sense of IV.3.3), such that

$$(T_a \xi, \eta) = (\xi, T_{(a^*)}\eta) \qquad \text{for all } a \in A \text{ and } \xi, \eta \in X.$$

The space X is of course denoted by $X(T)$. If in addition X is a Hilbert space (i.e., complete) and the operators T_a ($a \in A$) are all continuous on X, then (as in V.1.2) T is a **-representation of A on X*.

Thus a *-representation of A on a Hilbert space X is a *-homomorphism of A into the C^*-algebra $\mathcal{O}(X)$.

Pre-*-representations will not figure prominently in this work; and the definitions that follow will be given in the context of *-representations.

9.2.   *Definition.*   Two *-representations S and T of A are *unitarily equivalent* (in symbols $S \cong T$) if they are homeomorphically equivalent (V.1.4) under an isometry f, that is, if there is a linear surjective isometry $f: X(S) \to X(T)$ which intertwines S and T.

Obviously unitary equivalence is an equivalence relation in the ordinary sense. We shall see in 13.16 that two *-representations S and T are unitarily equivalent if and only if they are homeomorphically equivalent.

9.3.   If S and T are *-representations of A and $f: X(S) \to X(T)$ is S, T intertwining, notice that the adjoint map $f^*: X(T) \to X(S)$ (defined by $(\xi, f^*(\eta)) = (f(\xi), \eta)$, $\xi \in X(S)$, $\eta \in X(T)$) is T, S intertwining. Indeed, if $\xi \in X(S)$, $\eta \in X(T)$, and $a \in A$, we have $(\xi, f^*(T_a\eta)) = (T_{a^*}f(\xi), \eta) = (f(S_{a^*}\xi), \eta) = (\xi, S_a f^*(\eta))$. Fixing a, and letting ξ and η vary, we conclude that $f^* \circ T_a = S_a \circ f^*$.

In particular, the set B of all T, T intertwining operators is a *-subalgebra of $\mathcal{O}(X(T))$. This B is called the *commuting algebra* of T, and will be very important to us throughout this work.

9.4.   *-Representations have a simple but vital property which is largely responsible for the depth to which *-representation theory can be carried, as contrasted with the theory of general locally convex representations. It is this:

Proposition. *If T is a *-representation of A, and Y is a T-stable linear subspace of $X(T)$, then the orthogonal complement $Y^\perp$ of Y (in $X(T)$) is also T-stable.*

Proof. If $\xi \in Y^\perp$, $a \in A$, and $\eta \in Y$, then $(T_a \xi, \eta) = (\xi, T_{a^*} \eta) = 0$ (since $T_{a^*} \eta \in Y$). Fixing ξ and a and letting η vary, we conclude that $T_a \xi \in Y^\perp$. So, by the arbitrariness of a and ξ, $Y^\perp$ is T-stable. ∎

9.5. Corollary. *Let T be a *-representation of A, and p a projection on $X(T)$. Put $Y = \text{range}(p)$. Then Y is T-stable if and only if p belongs to the commuting algebra of T.*

Proof. Assume that p belongs to the commuting algebra of T. If $\xi \in Y$ and $a \in A$, then $p(T_a \xi) = T_a(p\xi) = T_a \xi$; so $T_a \xi \in Y$. Thus Y is T-stable.

Conversely, assume that Y is T-stable; and let $a \in A$, $\xi \in X(T)$. We can write $\xi = \xi_1 + \xi_2$, where $\xi_1 = p\xi \in Y$ and $\xi_2 \in Y^\perp$. Since by 9.4 Y and $Y^\perp$ are both T-stable, we have $T_a \xi = T_a \xi_1 + T_a \xi_2$, where $T_a \xi_1 \in Y$ and $T_a \xi_2 \in Y^\perp$. Therefore $p(T_a \xi) = T_a \xi_1 = T_a(p\xi)$; whence $p \circ T_a = T_a \circ p$ by the arbitrariness of ξ. So p lies in the commuting algebra of T. ∎

9.6. Let T be a *-representation of A. As an example of two orthogonal complementary stable subspaces, let us define:

$$N(T) = \{\xi \in X(T): T_a \xi = 0 \text{ for all } a \text{ in } A\},$$

$$R(T) = \text{the closed linear span of } \{T_a \xi : a \in A, \xi \in X(T)\}.$$

It is evident that $N(T)$ and $R(T)$ are closed stable linear subspaces of $X(T)$.

Proposition. $N(T) = R(T)^\perp$.

Proof. If $\xi \in X(T)$, then $\xi \in R(T)^\perp \Leftrightarrow (\xi, T_a \eta) = 0$ for all a in A and η in $X(T) \Leftrightarrow (T_{a^*} \xi, \eta) = 0$ for all a and $\eta \Leftrightarrow T_a \xi = 0$ for all $a \Leftrightarrow \xi \in N(T)$. ∎

Definition. $N(T)$ is called the *null space of T*; $R(T)$ is called the *essential space of T*.

9.7. For a *-representation T of A, Conditions (i) and (ii) of the Definition V.1.7 of the non-degeneracy of T are equivalent to the assertions $N(T) = \{0\}$ and $R(T) = X(T)$ respectively. By 9.6 these two assertions are equivalent. So either of them alone is equivalent to the non-degeneracy of T.

9.8. Notice that the subrepresentation $^{R(T)}T$ of T is non-degenerate in view of 9.7. It is called the *non-degenerate part of T*.

9.9. In contrast to the general locally convex situation (see IV.3.17), we have:

Proposition. *If T is a non-degenerate *-representation of A, and Y is a closed T-stable subspace of $X(T)$, then $^Y T$ is non-degenerate.*

Proof. Clearly $N(T) = \{0\}$ implies $N(^Y T) = \{0\}$; and the latter implies by 9.7 that $^Y T$ is non-degenerate. ∎

9.10. Corollary. *If T is a non-degenerate *-representation of A and $\xi \in X(T)$, then ξ belongs to the closure of $\{T_a \xi : a \in A\}$.*

Proof. Let Y be the closure of $\{\lambda \xi + T_a \xi : \lambda \in \mathbb{C}, \; a \in A\}$. Clearly Y is T-stable; so by 9.9 $^Y T$ is non-degenerate, whence $R(^Y T) = Y$. But $R(^Y T)$ is the closure of the linear span of $\{T_b \eta : b \in A, \; \eta \in Y\}$, which equals the closure of $\{T_a \xi : a \in A\}$. Since $\xi \in Y$, this implies the required conclusion. ∎

9.11. Corollary. *Let T be a *-representation of A. A vector ξ in $X(T)$ is cyclic for T (see V.1.6) if and only if (i) T is non-degenerate, and (ii) the smallest closed T-stable linear subspace of $X(T)$ containing ξ is $X(T)$.*

Proof. By 9.7 and 9.10. ∎

9.12. Proposition. *If A is a separable Banach *-algebra and T is a cyclic *-representation of A, then $X(T)$ is separable.*

Proof. Let ξ be a cyclic vector for T; and let C be a dense countable subset of A. By 3.8 T is continuous; so $\{T_a \xi : a \in C\}$ is dense in $\{T_a \xi : a \in A\}$. Since ξ is cyclic, the latter set is dense in $X(T)$. Thus $\{T_a \xi : a \in C\}$ is a countable dense subset of $X(T)$. ∎

Hilbert Direct Sums of *-Representations

9.13. Suppose that for each i in an index set I we are given a *-representation T^i of A with the property that

$$\{\|T_a^i\| : i \in I\} \text{ is bounded for each } a \in A. \tag{1}$$

Form the Hilbert space direct sum $\mathscr{X} = \sum_{i \in I}^{\oplus} X(T^i)$; and for each a in A let T_a be the element of $\mathcal{O}(\mathscr{X})$ coinciding on $X(T^i)$ with T_a^i (for each i in I). Such an operator T_a exists in view of (1). Then $T : a \mapsto T_a$ is a *-representation of A on $\mathscr{X}$.

Definition. T is called the (*Hilbert*) *direct sum of the* T^i, and is denoted by $\sum_{i \in I}^{\oplus} T^i$.

Notice that (1) holds automatically if A is a Banach *-algebra (by 3.8), or if I is finite.

In the case of only finitely many indices $i = 1, 2, \ldots, n$, as usual we often write $T^1 \oplus T^2 \oplus \cdots \oplus T^n$ instead of $\sum_{i=1}^{\oplus n} T^i$.

Obviously the formation of direct sums preserves unitary equivalence.

If T is a *-representation of A, and $X(T) = \sum_{i \in I}^{\oplus} Y_i$ is a Hilbert direct sum decomposition of $X(T)$ into closed T-stable subspaces Y_i, then

$$T \cong \sum_{i \in I}^{\oplus Y_i} T.$$

9.14. By 9.6 and 9.8 every *-representation T of A is the direct sum of the zero *-representation $^{N(T)}T$ and the non-degenerate *-representation $^{R(T)}T$.

9.15. If $T = \sum_{i \in I}^{\oplus} T^i$ is the Hilbert direct sum of *-representations T^i of A, clearly T is non-degenerate if and only if T^i is non-degenerate for each i in I.

9.16. Proposition. *Every non-degenerate *-representation T of A is unitarily equivalent to a Hilbert direct sum of cyclic *-representations of A.*

Proof. By Zorn's lemma there is a maximal family $\{Y_i : i \in I\}$ of non-zero closed pairwise orthogonal T-stable subspaces of $X(T)$ such that each ^{Y_i}T is cyclic. We have only to show that $Y = \sum_i Y_i$ is dense in $X(T)$. Suppose it is not; and choose a non-zero vector ξ in $Y^\perp$. By 9.10 the subrepresentation of T acting on the closure Z of $\{T_a \xi : a \in A\}$ has ξ as a cyclic vector. So we can adjoin Z to the $\{Y_i\}$, contradicting the maximality of $\{Y_i\}$. ∎

10. The C*-Completion of a Banach *-Algebra; Stone's Theorem

10.1. One of the principal long-range objectives of representation theory is the classification of all *-representations of a given Banach *-algebra A. For this reason it is interesting to find that every such A gives rise to a C*-algebra A_c (called the C*-completion of A) such that the *-representations of A and those of A_c are in natural one-to-one correspondence. This shows that the *-representation theory of C*-algebras is no less general than that of arbitrary Banach *-algebras.

The C*-completion of A may be thought of as a "photograph" of A using "light" which registers only such properties of A as appear in the structure of its *-representations. Since C^*-algebras are vastly simpler in structure than arbitrary Banach *-algebras, such a "photograph" often results in a great gain in simplicity. (See for example 10.7.)

Throughout this section we fix a Banach *-algebra A.

10.2.   Definition.   The *reducing ideal* of A is defined as the intersection I of the kernels of all *-representations of A. (The reducing ideal is often called the **-radical* of A.)

If $I = \{0\}$ (i.e., if the points of A are separated by its *-representations) we say that A is *reduced* (or **-semisimple*).

Since *-representations of A are continuous (3.8), I is a norm-closed *-ideal of A, and the quotient Banach *-algebra A/I is automatically reduced.

10.3.   Assume for the moment that A is reduced; and for each a in A set

$$\|a\|_c = \sup_T \|T_a\|,$$

where T runs over all *-representations of A. By 3.8 $\|a\|_c$ is finite; in fact

$$\|a\|_c \leq \|a\|. \tag{1}$$

Since A is reduced, $a \neq 0 \Rightarrow \|a\|_c \neq 0$. Evidently A is a pre-C^*-algebra (3.3) under $\| \ \|_c$. We call $\| \ \|_c$ the C^*-*norm* of A.

10.4.   Definition.   Quite generally, if I is the reducing ideal of A, the C^*-algebra obtained by completing A/I with respect to its C^*-norm is called the C^*-*completion* of A, and is denoted by A_c.

10.5.   Proposition.   *Let I be the reducing ideal of A, and $\pi: a \mapsto a + I$ the natural *-homomorphism of A into A_c. Then the map $T \mapsto T' = T \circ \pi$ is a one-to-one correspondence between the set of all *-representations T of A_c and the set of all *-representations T' of A. This correspondence preserves closed stable subspaces (that is, a closed subspace Y of $X(T)$ is T-stable if and only if it is T'-stable); in particular, T is irreducible if and only if T' is. Similarly, it preserves non-degeneracy, (bounded) intertwining operators, unitary equivalence, and Hilbert direct sums.*

Proof.   Since $\pi(A)$ is dense in A_c, T is determined by $T \circ \pi$; so the correspondence is one-to-one.

Any *-representation T' of A vanishes on I by definition, and so gives rise to a *-representation T'' of A/I. Since by definition $\|T''\| \leq 1$ with respect to the C^*-norm of A/I, T'' extends to a *-representation T of A_c; and we have $T' = T \circ \pi$. So the correspondence is surjective.

The remaining statements are almost evident.   ■

Warning.   If T' is a *faithful* *-representation of A, corresponding via the above proposition with a *-representation T of A_c, there is no reason why T should be faithful.

10.6.   Proposition.   *A C*-algebra A is reduced if and only if it is isometrically *-isomorphic with a norm-closed *-subalgebra of $\mathcal{O}(X)$ for some Hilbert space X.*

Proof.   The "if" part is obvious. Assume then that A is reduced; and let $\mathcal{T}$ be a collection of *-representations of A which distinguishes points of A. The result follows on applying 8.8 to the (faithful) Hilbert direct sum $T = \sum_{S \in \mathcal{T}}^{\oplus} S.$ ■

Remark.   Later in this Chapter (22.11) we shall see that every C^*-algebra is reduced. By the above Proposition this will immediately lead to the Gelfand–Naimark Theorem referred to in 3.5.

The Commutative Case

10.7.   We shall now describe the C^*-completion of a commutative Banach *-algebra in terms of its Gelfand transform.

Suppose that A is commutative. As in 4.2, $\hat{A}$ is the subset of symmetric elements of the Gelfand space $A^\dagger$; and $a \mapsto \hat{a}$ is the Gelfand transform map. Let $\pi: A \to A_c$ be as in 10.5.

Proposition.   *The reducing ideal I of A consists of those a in A such that $\hat{a}(\phi) = 0$ for all ϕ in $\hat{A}$. The equation*

$$F(\pi(a)) = \hat{a}|\hat{A} \qquad\qquad (a \in A) \qquad (2)$$

*determines an isometric *-isomorphism F of A_c onto $\mathscr{C}_0(\hat{A})$.*

Proof.   Each element ϕ of $\hat{A}$ is a one-dimensional *-representation of A. It follows that

$$\sup\{|\hat{a}(\phi)|: \phi \in \hat{A}\} \leq \|\pi(a)\|_c \qquad\qquad (a \in A). \qquad (3)$$

On the other hand, take any *-representation T of A; and let B be the norm-closure of range(T). Thus B is a commutative C^*-algebra. If a is any element of A, it follows on applying 4.3 to B that there is an element ψ of $\hat{B}$ satisfying $|\psi(T_a)| = \|T_a\|$. Since $\psi': a \mapsto \psi(T_a)$ belongs to $\hat{A}$, this implies that

$$\|T_a\| = |\psi'(a)| \leq \sup\{|\hat{a}(\phi)|: \phi \in \hat{A}\}.$$

By the arbitrariness of T we conclude that

$$\| \pi(a) \|_c \leq \sup\{|\hat{a}(\phi)| : \phi \in \hat{A}\}. \tag{4}$$

By (3) and (4),

$$\| \pi(a) \|_c = \sup\{|\hat{a}(\phi)| : \phi \in \hat{A}\} \qquad (a \in A). \tag{5}$$

The first statement of the proposition follows immediately from (5). Further, in view of (5), equation (2) defines an isometric *-isomorphism F of A_c into $\mathscr{C}_0(\hat{A})$. The range of F is a *-subalgebra of $\mathscr{C}_0(\hat{A})$ which automatically separates points of $\hat{A}$ and does not vanish at any point of $\hat{A}$. This implies by the Stone-Weierstrass Theorem A8 that range(F) is dense in $\mathscr{C}_0(\hat{A})$, hence equal to $\mathscr{C}_0(\hat{A})$. ∎

Remark. Thus we can identify A_c with $\mathscr{C}_0(\hat{A})$. Since $\mathscr{C}_0(\hat{A})$ has rather simple properties, as compared with those of an arbitrary commutative Banach *-algebra, this illustrates the point made in 10.1.

10.8. Corollary. *The radical of a commutative Banach *-algebra A is contained in its reducing ideal. In particular, if A is reduced it is semisimple. In a symmetric commutative Banach *-algebra, the reducing ideal and the radical are the same.*

10.9. *Examples.* Let D be the unit disk $\{z \in \mathbb{C} : |z| \leq 1\}$; and let B be the Banach *-algebra whose underlying Banach algebra coincides with $\mathscr{C}(D)$, the involution in B being given by:

$$f^*(z) = \overline{f(\bar{z})} \qquad (f \in B; z \in D).$$

By V.8.8 $B^\dagger \cong D$. But clearly $\hat{B} \cong D \cap \mathbb{R}$. Though semisimple, B is neither symmetric nor reduced.

Now let A be the Banach *-algebra of 1.8. This is a norm-closed *-subalgebra of the above B. Again $A^\dagger \cong D$ (see V.8.4) and $\hat{A} = D \cap \mathbb{R}$. Since an analytic function on the disk is uniquely determined by its restriction to $D \cap \mathbb{R}$, A is reduced (though it is not symmetric). By 10.7 $A_c = \mathscr{C}(D \cap \mathbb{R}) = \mathscr{C}([-1, 1])$.

Next, let E be the norm-closed *-ideal of A consisting of those f such that $f(i) = f(-i) = 0$. It is easy to check that $E^\dagger \cong D \setminus \{i, -i\}$ and $\hat{E} \cong D \cap \mathbb{R} = [-1, 1]$. Thus by 10.7 $E_c \cong \mathscr{C}([-1, 1])$; and E is an example of a reduced commutative Banach *-algebra without unit element whose C*-completion has a unit element.

Stone's Theorem

10.10. Putting together 10.5, 10.7, and II.12.8, we obtain the following very important description of the *-representations of a commutative Banach *-algebra in terms of projection-valued measures.

Theorem (Stone). *Let A be a commutative Banach *-algebra. If X is a Hilbert space and P is a regular X-projection-valued Borel measure (see II.11.9) on $\hat{A}$, the equation*

$$T_a = \int_{\hat{A}} \hat{a}(\phi)dP\phi \qquad\qquad (a \in A) \qquad (6)$$

defines a non-degenerate *-representation T of A on X. Conversely, every non-degenerate *-representation T of A on a Hilbert space X determines a unique regular X-projection-valued Borel measure P on $\hat{A}$ such that (6) holds.

Note. The right side of (6) is of course a spectral integral, as defined in II.11.7.

Proof. Let P be a regular X-projection-valued Borel measure on $\hat{A}$. By II.12.9, $T': f \mapsto \int f\, dP$ ($f \in \mathscr{C}_0(\hat{A})$) is a non-degenerate *-representation of $\mathscr{C}_0(\hat{A})$ on X. Since by 10.7 the map $a \mapsto \hat{a}|\hat{A}$ is a *-homomorphism of A onto a dense *-subalgebra of $\mathscr{C}_0(\hat{A})$, the composed map $T: a \mapsto T'_{(\hat{a}|\hat{A})} = \int_{\hat{A}} \hat{a}\, dP$ is a non-degenerate *-representation of A.

Conversely, let T be a non-degenerate *-representation of A on X. Combining 10.5 and 10.7, we obtain a (non-degenerate) *-representation T' of $\mathscr{C}_0(\hat{A})$ on X such that

$$T_a = T'_{(\hat{a}|\hat{A})} \qquad \text{for all } a \text{ in } A. \qquad (7)$$

By II.12.8 there is a regular X-projection-valued Borel measure P on $\hat{A}$ satisfying

$$T'_f = \int f\, dP \qquad \text{for all } f \text{ in } \mathscr{C}_0(\hat{A}). \qquad (8)$$

Combining (7) and (8), we get (6).

Let Q be another regular X-projection-valued Borel measure which is also related to the given T by (6). Thus

$$\int \hat{a}\, dQ = T_a = \int \hat{a}\, dP \qquad (9)$$

for all a in A. Since by 10.7 $\{\hat{a}|\hat{A}:a\in A\}$ is dense in $\mathscr{C}_0(\hat{A})$ (with respect to the supremum norm), it follows from (9) and II.11.8(V) that $\int f\,dQ = \int f\,dP$ for all f in $\mathscr{C}_0(\hat{A})$. By the uniqueness assertion of II.12.8 this implies that $Q = P$.

Thus the P satisfying (6) exists and is unique; and the proof is complete. ∎

Definition. The above P is referred to as the *spectral measure* of the non-degenerate *-representation T of A.

10.11. The next two propositions supplement Theorem 10.10, and will be useful in the discussion that follows.

Proposition. *Let P be the spectral measure of the non-degenerate *-representation T of a commutative Banach *-algebra A. If $\phi \in \hat{A}$, then*

$$\text{range}(P(\{\phi\})) = \{\xi \in X(T): T_a\xi = \phi(a)\xi \text{ for all } a \text{ in } A\}.$$

Proof. Relation (6), and the definition of the spectral integral, show that $\xi \in \text{range } P(\{\phi\}) \Rightarrow T_a\xi = \phi(a)\xi$ for all a.

Conversely, suppose that $T_a\xi = \phi(a)\xi$ for all a in A. If T' is the *-representation of $A_c(\cong \mathscr{C}_0(\hat{A}))$ corresponding to T, it follows by continuity that

$$T'_f\xi = f(\phi)\xi \qquad \text{for all } f \text{ in } \mathscr{C}_0(\hat{A}). \tag{10}$$

Now assume that $\xi \notin \text{range}(P(\{\phi\}))$; and denote by μ the regular Borel measure $W \mapsto \|P(W)\xi\|^2$ on $\hat{A}$. Thus μ is not carried by $\{\phi\}$; and so by regularity we can find a compact subset C of $\hat{A}$ such that $\phi \notin C$, $\mu(C) > 0$. Choose an element f of $\mathscr{C}_0(\hat{A})$ which is positive everywhere on C and vanishes at ϕ. By (10) $T'_f\xi = 0$. But by II.11.8(VII) $\|T'_f\xi\|^2 = \int |f(\phi)|^2\,d\mu\phi \geq \int_C |f(\phi)|^2\,d\mu\phi > 0$; and we have a contradiction. So $\xi \in \text{range } P(\{\phi\})$; and the proof is finished. ∎

10.12. Proposition. *Let A be a commutative Banach *-algebra, and μ any regular Borel measure on $\hat{A}$. The equation*

$$(T_af)(\phi) = \hat{a}(\phi)f(\phi) \qquad (a\in A; f\in \mathscr{L}_2(\mu); \phi\in\hat{A})$$

*defines a non-degenerate *-representation T of A on $\mathscr{L}_2(\mu)$; and the spectral measure P of T is given by*

$$P(W)f = \text{Ch}_W f \ (f\in\mathscr{L}_2(\mu); W \text{ a Borel subset of } \hat{A}).$$

10.13. Let A be a commutative Banach *-algebra for which $\hat{A}$ happens to be a *discrete* space. Given a non-degenerate *-representation T of A, with spectral measure P, let us define $Y_\phi = $ range $P(\{\phi\})$ for each ϕ in $\hat{A}$. Since $\hat{A}$ is discrete and P is regular,

$$X(T) = \sum_{\phi \in \hat{A}}^{\oplus} Y_\phi \qquad \text{(Hilbert direct sum).} \qquad (11)$$

By 10.11,

$$T_a \xi = \phi(a)\xi \qquad \text{for } a \in A, \ \phi \in \hat{A}, \ \xi \in Y_\phi. \qquad (12)$$

Together, (11) and (12) assert that T is a Hilbert direct sum of one-dimensional *-representations. Notice that Y_ϕ is the ϕ-subspace of $X(T)$ in the strict algebraic sense of IV.2.11. We have shown that every *-representation of A is "completely reducible," not in the strict algebraic sense of IV.2.1 (which would require that $X(T) = \sum_{\phi \in \hat{A}} Y_\phi$ algebraically), but in the analogous *-representational sense expressed by (10) and (11). We shall make a further study of "*-representational complete reducibility" in §14.

Let us now drop the assumption that $\hat{A}$ is discrete. Then in general there will exist a non-zero (numerical) regular Borel measure μ on $\hat{A}$ assigning zero measure to every one-element set; and from this by 10.12 one can construct a non-degenerate *-representation T of A whose spectral measure assigns zero measure to every one-element set. By 10.11 this implies that T has no one-dimensional subrepresentations, and hence, as we shall see in 14.3, no closed irreducible T-stable subspaces at all. Thus T is as far as possible from being completely reducible (in the *-representational sense).

Nevertheless, for an arbitrary (non-degenerate) *-representation T of A, the spectral measure P of T provides a very appropriate generalization of the purely algebraic notion, encountered in IV.2.12, of decomposing a completely reducible representation into its σ-subspaces. Thus, if W is a Borel subset of $\hat{A}$, range$(P(W))$ can be regarded as the "W-subspace" of $X(T)$ (wth respect to T). In the special case where $\hat{A}$ is discrete, the "W-subspace" would be simply the Hilbert direct sum of the ϕ-subspaces for which $\phi \in W$. In the general case, as we have seen, no such description is possible. However, roughly speaking, if W is a "small" neighborhood of a point ϕ in $\hat{A}$, we can say that the vectors ξ in range$(P(W))$ will be those for which $T_a \xi$ is "nearly" equal to $\phi(a)\xi$ $(a \in A)$.

Remark. As we have seen above, a *-representation may have no irreducible stable subspaces. At the same time, by 9.4, every closed stable subspace has a stable orthogonal complement. Contrast this with the purely algebraic situation described in IV.3.12.

Remark. In IV.2.22, by combining the harmonic analysis into σ-subspaces with the notion of multiplicity, we showed that, in the purely algebraic context, a completely reducible representation T is determined to within equivalence by a "multiplicity function" which assigns to each class $σ$ of irreducible representations a cardinal number, the multiplicity of $σ$ in T. It turns out that there is a beautiful generalization of this to the present setting. Indeed, a *-representation T of a commutative Banach *-algebra A is described to within unitary equivalence by assigning to each cardinal number $α$ a certain measure theoretic object on $\hat{A}$ which generalizes the notion of "the set of those ϕ in $\hat{A}$ which have multiplicity $α$ in T." Unfortunately this involutory multiplicity theory is beyond the scope of this work.

10.14. *Definition.* Let A be a commutative Banach *-algebra. If T is a non-degenerate *-representation of A and P its spectral measure, we refer to the closed support of P (see II.11.12) as the *spectrum of* T.

Thus the spectrum of T is a closed subset of $\hat{A}$.

Combining (6) with II.11.8(VII), we conclude:

Proposition. *The kernel of a non-degenerate *-representation T of A consists of those a in A such that $\hat{a}$ vanishes everywhere on the spectrum of T.*

Suppose in addition that A is a (commutative) C^*-algebra. Then T is faithful if and only if its spectrum is all of $\hat{A}$.

11. The Spectral Theory of Bounded Normal Operators

11.1. The preceding section has a beautiful application to the spectral theory of normal operators on Hilbert space.

Fix a Hilbert space X. As usual $\mathcal{O}(X)$ is the C^*-algebra of all bounded linear operators on X. If $b \in \mathcal{O}(X)$, we shall abbreviate $\mathrm{Sp}_{\mathcal{O}(X)}(b)$ (V.5.1) to $\mathrm{Sp}(b)$.

Spectral Theorem for Normal Operators. *Let a be a bounded normal operator on X. Then there exists a unique X-projection-valued Borel measure P on the complex plane $\mathbb{C}$ with the following properties* (i) *and* (ii):

(i)   *The closed support D (II.11.12) of P is bounded in $\mathbb{C}$.*

(ii) $a = \int_{\mathbb{C}} \lambda \, dP\lambda.$

This P has the further property that

(iii) $D = \mathrm{Sp}(a).$

Remark. Notice that the spectral integral in (ii) exists since by (i) the integrand λ is bounded except on a P-null set.

Proof. Since a is normal, the norm-closed *-subalgebra A of $\mathcal{O}(X)$ generated by a and the identity operator $\mathbb{1}$ is commutative. Each element ϕ of $\hat{A}$, since $\phi(a^*) = \overline{\phi(a)}$, is determined everywhere on A by the value of $\phi(a)$. Thus the map

$$\Gamma : \phi \mapsto \phi(a) \qquad\qquad (\phi \in \hat{A}) \qquad (1)$$

is continuous and one-to-one, hence a homeomorphism, from the compact space $\hat{A}$ onto the compact subset $\mathrm{Sp}(a)$ of $\mathbb{C}$ (see V.7.13). We shall therefore identify $\hat{A}$ with $\mathrm{Sp}(a)$. With this identification, the Gelfand transform of a becomes the function $\lambda \mapsto \hat{a}(\Gamma^{-1}(\lambda)) = \lambda$, that is, the identity function on $\mathrm{Sp}(a)$. Thus, by 10.10, the identity representation of A in X gives rise to an X-projection-valued Borel measure P on $\mathrm{Sp}(a)$ such that

$$a = \int_{\mathrm{Sp}(a)} \lambda \, dP\lambda. \qquad (2)$$

Since the identity representation is faithful, 10.14 shows that the closed support D of P is equal to $\mathrm{Sp}(a)$.

Extending P to the Borel σ-field of $\mathbb{C}$ by setting $P(W) = P(W \cap \mathrm{Sp}(a))$, we thus obtain a Borel projection-valued measure on $\mathbb{C}$ satisfying properties (i), (ii), and (iii).

To show that P is uniquely determined by properties (i) and (ii), let P' be another Borel X-projection-valued measure on $\mathbb{C}$ satisfying (i) and (ii); and let E be a compact subset of $\mathbb{C}$ containing the supports of both P and P'. Consider the two *-representations $T : f \mapsto \int_E f(\lambda) dP\lambda$ and $T' : f \mapsto \int_E f(\lambda) dP'\lambda$ of $\mathscr{C}(E)$. If w and e are the identity function $\lambda \mapsto \lambda$ and the constant function $\lambda \mapsto 1$ on E respectively, condition (ii) (satisfied by both P and P') says that $T_w = a = T'_w$; also $T_e = \mathbb{1} = T'_e$. Since by the Stone-Weierstrass Theorem w and e generate $\mathscr{C}(E)$ (as a C^*-algebra), it follows that $T' = T$. Hence by the uniqueness clause in 10.10 $P' = P$. ∎

Definition. The P of the above theorem is called the *spectral measure* of the normal operator a.

11.2. Combining 11.1(iii) with 5.5, 5.6, and 7.3, we notice:

Proposition. *Let a be a bounded normal operator on a Hilbert space X, and P its spectral measure. Then a is Hermitian [positive, unitary] if and only if P is carried by $\mathbb{R}$ [$\{t \in \mathbb{R} : t \geq 0\}$, $\{\lambda \in \mathbb{C} : |\lambda| = 1\}$].*

11.3. Proposition. *A bounded normal operator a on a Hilbert space X lies in the norm-closure (in $\mathcal{O}(X)$) of the linear span of the range of its spectral measure.*

Proof. This follows from 11.1(ii) together with the definition of a spectral integral (II.11.7). ∎

Remark. The definition of the spectral integral enables us to state this proposition a little more precisely as follows:

Let a be a bounded normal operator on X, and ε any positive number. Then there exist pairwise orthogonal projections $p_1, \ldots, p_n$ (n depending of course on a and ε) such that $\sum_{i=1}^{n} p_i = 1$, and complex numbers $\lambda_1, \ldots, \lambda_n$, satisfying $\| a - \sum_{i=1}^{n} \lambda_i p_i \| < \varepsilon$.

If X is finite-dimensional, it is well known from the theory of matrices that a normal operator a on X is actually of the form $\sum_{i=1}^{n} \lambda_i p_i$ (where the λ_i are complex numbers and the p_i are orthogonal projections). Thus the above statement shows clearly the manner in which 11.1 generalizes the classical spectral theory of finite-dimensional normal operators.

11.4. Proposition 11.3 says that a normal operator a can be approximated in the norm-topology by linear combinations of projections $P(W)$ belonging to the range of its spectral measure P. We shall now show that, conversely, any such projection $P(W)$ can be approximated by polynomials in a and a^*—but only in the strong topology.

Proposition. *Let a be a bounded normal operator on X, with spectral measure P. Then every projection in range(P) is the limit, in the strong topology, of a net of polynomials in a and a^*.*

Proof. Let $\xi_1, \ldots, \xi_n$ be any vectors in X, and ε any positive number. Let μ_i be the Borel measure $W \mapsto \| P(W)\xi_i \|^2$ on $\mathrm{Sp}(a)$; and set $\mu = \sum_{i=1}^{n} \mu_i$. Now take any Borel subset V of $\mathrm{Sp}(a)$. Since μ is regular (II.8.8), by the Stone–Weierstrass Theorem and the density of continuous functions in $\mathcal{L}_2(\mu)$ (II.15.9) we can choose a complex polynomial p in two indeterminates such that

$$\int_{\mathrm{Sp}(a)} |\mathrm{Ch}_V(\lambda) - p(\lambda, \bar{\lambda})|^2 \, d\mu\lambda < \varepsilon.$$

But, by II.11.8, this implies that

$$\| P(V)\xi_i - p(a, a^*)\xi_i \|^2 < \varepsilon \qquad \text{for all } i.$$

This completes the proof. ∎

The Point Spectrum and Approximate Point Spectrum

11.5.   The presence of vectors in X gives rise to an interesting classification of the elements of $\mathrm{Sp}(a)$ when $a \in \mathcal{O}(X)$.

Definition.   Let a be in $\mathcal{O}(X)$. We define the *point spectrum* $\mathrm{PSp}(a)$ *of* a to consist of all λ in $\mathbb{C}$ such that there exists a non-zero vector ξ in X satisfying $a\xi = \lambda\xi$. Such a ξ is called an *eigenvector for a and* λ.

The following proposition is an easy corollary of 10.11 (applied to the A of the proof of 11.1):

Proposition.   *Let a be a bounded normal operator on X and P its spectral measure. A complex number λ belongs to $\mathrm{PSp}(a)$ if and only if $P(\{\lambda\}) \neq 0$. In fact, for all λ in $\mathbb{C}$,*

$$\mathrm{range}(P(\{\lambda\})) = \{\xi \in X : a\xi = \lambda\xi\}. \tag{3}$$

11.6.   *Definition.*   Let a be in $\mathcal{O}(X)$. By the *approximate point spectrum* $\mathrm{APSp}(a)$ *of* a we mean the set of those λ in $\mathbb{C}$ such that, for each $\varepsilon > 0$, there is a vector ξ in X satisfying $\| \xi \| = 1$, $\| a\xi - \lambda\xi \| < \varepsilon$.
Clearly

$$\mathrm{PSp}(a) \subset \mathrm{APSp}(a) \subset \mathrm{Sp}(a). \tag{4}$$

In general both of these inclusions are proper. As regards the second one, suppose that X is infinite-dimensional and that u is a linear isometry of X onto a (closed) proper linear subspace of X. Then it is easily seen that $0 \in \mathrm{Sp}(u)$ but $0 \notin \mathrm{APSp}(u)$. As regards the first, see Example 11.12.

If X is finite-dimensional, then $\mathrm{Sp}(a) = \mathrm{PSp}(a) = \mathrm{APSp}(a)$.

11.7.   If a is normal, the second inclusion (4) is an equality.

Proposition.   *If a is a bounded normal operator on X, then $\mathrm{Sp}(a) = \mathrm{APSp}(a)$.*

Proof. Let $\lambda_0 \in \mathrm{Sp}(a)$. By (4) it is enough to show that $\lambda_0 \in \mathrm{APSp}(a)$.

Let P be the spectral measure of a. By 11.1(iii) λ_0 belongs to the closed support of P; so, if D is the disk about λ_0 of radius $\varepsilon > 0$, we have $P(D) \neq 0$. Choosing a vector ξ of norm 1 in range$(P(D))$, we have by II.11.8

$$\| a\xi - \lambda_0 \xi \|^2 = \int_{\mathbb{C}} |\lambda - \lambda_0|^2 \, d\| P\xi \|^2 \lambda, \tag{5}$$

where $\| P\xi \|^2(W) = \| P(W)\xi \|^2$. Since $\xi \in \mathrm{range}(P(D))$ and $\| P\xi \|^2$ has total mass 1, the integral in (5) is majorized by ε^2. Thus

$$\| a\xi - \lambda_0 \xi \| \leq \varepsilon,$$

and the arbitrariness of ε shows that $\lambda_0 \in \mathrm{APSp}(a)$. ∎

11.8. Corollary. *If a is a bounded normal operator on X, we have*

$$\| a \|_{\mathrm{sp}} = \sup\{|(a\xi, \xi)| : \xi \in X, \| \xi \| = 1\} = \| a \|. \tag{6}$$

Proof. Here $\| a \|_{\mathrm{sp}}$ is the spectral norm of V.6.17. It was proved in 3.6 that

$$\| a \|_{\mathrm{sp}} = \| a \|. \tag{7}$$

Now (whether a is normal or not) it is evident from the definition of $\mathrm{APSp}(a)$ that

$$\sup\{|\lambda| : \lambda \in \mathrm{APSp}(a)\} \leq k_a \leq \| a \|, \tag{8}$$

where we have written $k_a = \sup\{|(a\xi, \xi)| : \xi \in X, \| \xi \| = 1\}$. By 11.7 the first term of (8) is just $\| a \|_{\mathrm{sp}}$. So (6) follows from (7) and (8). ∎

11.9. Remark. For any a in $\mathcal{O}(X)$ it is a fact that the *boundary* of $\mathrm{Sp}(a)$ is contained in $\mathrm{APSp}(a)$ (see for example Problem 63 of Halmos [3]). It follows that the first term of (8) is always equal to $\| a \|_{\mathrm{sp}}$, giving

$$\| a \|_{\mathrm{sp}} \leq \sup\{|(a\xi, \xi)| : \xi \in X, \| \xi \| = 1\} \leq \| a \|, \tag{9}$$

for *all* a in $\mathcal{O}(X)$. In general both the inequalities in (9) are proper. For example, if $X = \mathbb{C}^2$, $a = \begin{pmatrix} 0 & 0 \\ 1 & 0 \end{pmatrix}$, we have $\| a \|_{\mathrm{sp}} = 0$, $\sup\{|(a\xi, \xi)| : \| \xi \| = 1\} = \frac{1}{2}$, $\| a \| = 1$.

11.10. As another corollary, we show that, for operators, the abstract definition 7.2 of positivity coincides with the usual operator definition:

Corollary. *If $a \in \mathcal{O}(X)$, the following two conditions are equivalent:* (i) *a is positive in $\mathcal{O}(X)$ in the sense of 7.2;* (ii) *$(a\xi, \xi) \geq 0$ for all ξ in X.*

Proof.   Assume (i). By 7.5 $a = b^2$ for some Hermitian b in $\mathcal{O}(X)$. Hence, if $\xi \in X$,

$$(a\xi, \xi) = (b^2\xi, \xi) = (b\xi, b\xi) \geq 0;$$

so (ii) holds.

Conversely, assume (ii). In particular $(a\xi, \xi) = (\xi, a\xi)$; so by the polarization identity $(a\xi, \eta) = (\xi, a\eta)$ for all ξ, η in X. Thus a is Hermitian. Now we claim that $\mathrm{APSp}(a) \subset \{\lambda \in \mathbb{R} : \lambda \geq 0\}$. Indeed: Let $\lambda \in \mathrm{APSp}(a)$. Then, given $\varepsilon > 0$, we have $\|\xi\| = 1$, $\|a\xi - \lambda\xi\| < \varepsilon$ for some ξ, whence $|(a\xi, \xi) - \lambda| = |(a\xi - \lambda\xi, \xi)| < \varepsilon$. Since $(a\xi, \xi) \geq 0$ and ε was arbitrary, this implies that $\lambda \in \mathbb{R}$, $\lambda \geq 0$. So the claim is proved. Since a is Hermitian, 11.7 applied to this claim gives $\mathrm{Sp}(a) \subset \{\lambda \in \mathbb{R} : \lambda \geq 0\}$, whence (i) holds. ∎

11.11.   Example.   Take an orthonormal basis $\{\xi_i : i \in I\}$ of X; let $\{\lambda_i : i \in I\}$ be a bounded family of complex numbers indexed by I; and let a be the element of $\mathcal{O}(X)$ determined by

$$a\xi_i = \lambda_i \xi_i \qquad\qquad (i \in I).$$

Notice that a is normal; in fact

$$a^*\xi_i = \bar{\lambda}_i \xi_i \qquad\qquad (i \in I).$$

Let P be the spectral measure of a. One easily deduces from Proposition 11.5 that, for each Borel subset W of $\mathbb{C}$, range$(P(W))$ is the closure of the linear span of $\{\xi_i : \lambda_i \in W\}$. Thus, by 11.5, $\mathrm{PSp}(a) = \{\lambda_i : i \in I\}$. By 11.1, $\mathrm{Sp}(a)$ is the closed support of P, which is the closure of $\{\lambda_i : i \in I\}$.

11.12.   Example.   Let ρ stand for Lebesgue measure on $[0, 1]$; and let a be the bounded linear operator on $X = \mathcal{L}_2(\rho)$ which multiplies each function by the identity function:

$$(a(f))(t) = tf(t) \qquad\qquad (f \in X; 0 \leq t \leq 1).$$

The spectrum of a is $[0, 1]$; and one verifies from 10.12 that the spectral measure P of a is given by

$$P(W)(f) = \mathrm{Ch}_W f$$

$(f \in X; W$ a Borel subset of $[0, 1])$. The point spectrum of a is void.

11.13.   Remark.   Let X be infinite-dimensional; and suppose the numbers λ_i in 11.11 were chosen so that the closure of $\{\lambda_i : i \in I\}$ is equal to $[0, 1]$. Then the spectrum of the a in 11.11 is the same as that of the a in 11.12. This shows the impossibility of finding a condition on the set $\text{Sp}(a)$ which will be necessary and sufficient for $P\text{Sp}(a)$ to be non-trivial. There are, however, sufficient conditions, as the following proposition shows:

Proposition*. *Let a be a bounded normal operator on X, and assume that every non-void subset S of $\text{Sp}(a)$ has an element which is isolated relative to S. Then X has an orthonormal basis consisting of eigenvectors for a.*

A Commutation Theorem

11.14. It will be important to set down in greater detail just how a normal operator a on a Hilbert space X determines its spectral measure P.

Proposition. *Let λ_0 be a complex number, k a positive number, and W the disk $\{\lambda \in \mathbb{C} : |\lambda - \lambda_0| \leq k\}$. Then, for each ξ in X, the following three conditions are equivalent:*

(i) $\xi \in \text{range}(P(W))$;

(ii) $\|(a - \lambda_0 1)^n \xi\| \leq k^n \|\xi\|$ *for all positive integers n;*

(iii) *the set of all numbers $k^{-n} \|(a - \lambda_0 1)^n \xi\|$, as n runs over all positive integers, is bounded.*

Proof. By 11.1(ii) and II.11.8,

$$(a - \lambda_0 1)^n = \int_{\mathbb{C}} (\lambda - \lambda_0)^n \, dP\lambda,$$

$$\|(a - \lambda_0 1)^n \xi\|^2 = \int_{\mathbb{C}} |\lambda - \lambda_0|^{2n} \, d\|P\xi\|^2 \lambda \qquad (\xi \in X), \qquad (10)$$

where $\|P\xi\|^2$ is the Borel measure on $\mathbb{C}$ whose value at V is $\|P(V)\xi\|^2$.

Now assume (i). Then the measure $\|P\xi\|^2$ is carried by W; so (10) gives $\|(a - \lambda_0 1)^n \xi\|^2 \leq k^{2n} \|P\xi\|^2(W) = k^{2n} \|\xi\|^2$. Thus (ii) holds.

Obviously (ii) $\Rightarrow$ (iii). We shall complete the proof by showing that (iii) $\Rightarrow$ (i).

Assume that (i) fails. Then there is a number $r > k$ such that $P(V)\xi \neq 0$, where $V = \{\lambda \in \mathbb{C} : |\lambda - \lambda_0| \geq r\}$. Hence by (10)

$$\|(a - \lambda_0 1)^n \xi\|^2 \geq \int_V |\lambda - \lambda_0|^{2n} \, d\|P\xi\|^2 \lambda \geq r^{2n} \|P(V)\xi\|^2,$$

whence

$$k^{-2n} \|(a - \lambda_0 1)^n \xi\|^2 \geq (k^{-1} r)^{2n} \|P(V)\xi\|^2. \qquad (11)$$

Since $r > k$ and $P(V)\xi \neq 0$, the right side of (11) is unbounded in n. Hence, by (11), condition (iii) fails. Consequently (iii) $\Rightarrow$ (i). ∎

11.15.   Remark.   A projection-valued Borel measure P on $\mathbb{C}$ is determined when its values on all disks are known. Indeed: Every open subset U of $\mathbb{C}$ is a countable union of disks; therefore the knowledge of P on disks gives us $P(U)$ for all open sets U. Also, P is necessarily regular by II.11.10. Therefore the knowledge of P on open sets determines $P(W)$ for every Borel subset W of $\mathbb{C}$.

In view of this, Proposition 11.14 provides us with an independent proof of the uniqueness assertion in 11.1.

11.16.   Theorem.   *Let a be a normal element of $\mathcal{O}(X)$, with spectral measure P; and let b be any element of $\mathcal{O}(X)$ such that $ba = ab$. Then $bP(W) = P(W)b$ for every Borel subset W of $\mathbb{C}$. In particular $ba^* = a^*b$.*

Proof.   Let $\lambda_0 \in \mathbb{C}, k > 0, W = \{\lambda \in \mathbb{C}: |\lambda - \lambda_0| \leq k\}$. If $\xi \in \text{range}(P(W))$, then by 11.14(iii) the set $\{k^{-n}\|(a - \lambda_0 \mathbb{1})^n \xi\|: n = 1, 2, \ldots\}$ is bounded. Now

$$k^{-n}\|(a - \lambda_0 \mathbb{1})^n b\xi\| = k^{-n}\|b(a - \lambda_0 \mathbb{1})^n \xi\|$$

$$\leq k^{-n}\|b\|\|(a - \lambda_0 \mathbb{1})^n \xi\|;$$

so the numbers $k^{-n}\|(a - \lambda_0 \mathbb{1})^n b\xi\|$ $(n = 1, 2, \ldots)$ also form a bounded set. Consequently, by 11.14, $b\xi \in \text{range}(P(W))$.

We have thus shown that range$(P(W))$ is stable under b for every closed disk W. Since every open subset of $\mathbb{C}$ is a countable union of closed disks, and since P is regular (II.11.10), range$(P(W))$ is stable under b for every Borel subset W of $\mathbb{C}$. So (by the argument of 9.5) $bP(W) = P(W)b$ for every Borel set W.

The fact that $ba^* = a^*b$ now follows from the preceding paragraph and 11.3. ∎

11.17.   Corollary.   *Let a and b be as in 11.16. Then $bf(a) = f(a)b$ for every continuous function $f: \mathbb{C} \to \mathbb{C}$.*

Proof.   Let $\{p_\nu\}$ be a net of polynomials in two variables such that $p_\nu(\lambda, \bar{\lambda}) \to f(\lambda)$ uniformly on compact sets; such a net exists by the Stone–Weierstrass Theorem. By 6.1 $p_\nu(a, a^*) \to f(a)$ in the operator norm topology. But $bp_\nu(a, a^*) = p_\nu(a, a^*)b$ by 11.16. Hence $bf(a) = f(a)b$. ∎

12. The Spectral Theory of Unbounded Normal Operators

12.1. Theorem 11.1 has a beautiful and straightforward generalization to *unbounded* normal operators on X.

As before, X is a fixed Hilbert space. In this and the next section the terms "transformation" and "operator" are understood in the "unbounded" sense of Appendix B. Normal operators are defined in B22.

Spectral Theorem for Unbounded Normal Operators. *Let E be a normal operator in X. There exists a unique X-projection-valued Borel measure P on the complex plane such that*

$$E = \int_{\mathbb{C}} \lambda \, dP\lambda. \tag{1}$$

Note. The right side of (1) is the spectral integral defined for unbounded integrands in II.11.15.

Remark. By the last remark of II.11.15, the right side of (1) cannot be a bounded operator unless $\lambda \mapsto \lambda$ is P-essentially bounded, that is, unless P has compact support. So, if E is a bounded operator, the present theorem reduces to 11.1.

Proof of Existence. We shall first prove the existence of P. Put

$$B = (\mathbf{1} + E^*E)^{-1}, \qquad C = EB$$

($\mathbf{1}$ being the identity operator on X). By B11 and B31 B and C are bounded operators, B being positive and C normal; and $BC = CB$. Hence B and C can be embedded in some commutative norm-closed *-subalgebra $\mathscr{A}$ of $\mathcal{O}(X)$ containing $\mathbf{1}$. By Stone's Theorem 10.10 there is a regular X-projection-valued Borel measure Q on $\hat{\mathscr{A}}$ such that

$$D = \int \hat{D}(\phi) dQ\phi$$

for all D in $\mathscr{A}$ ($\hat{D}$ being the Gelfand transform of D). By II.11.8(VII)

$$(D\xi, \xi) = \int \hat{D}(\phi) d\|Q\xi\|^2 \phi \qquad (D \in \mathscr{A}; \, \xi \in X), \tag{2}$$

where $\|Q\xi\|^2(W) = \|Q(W)\xi\|^2$.

Now we recall from B11(14) that $(B\xi, \xi) = 0 \Rightarrow \xi = 0$. From this and (2) (and the positivity of B), we deduce that

$$\hat{B}(\phi) > 0 \qquad \text{for } Q\text{-almost all } \phi \text{ in } \hat{\mathscr{A}}. \tag{3}$$

In view of this, the equation

$$f(\phi) = (\hat{B}(\phi))^{-1}\hat{C}(\phi) \qquad (4)$$

defines f as a complex Borel function Q-almost everywhere on $\hat{\mathscr{A}}$. Thus by
II.11.15, II.11.16,

$$E' = \int_{\hat{\mathscr{A}}} f \, dQ \qquad (5)$$

is a well-defined normal operator in X.

We claim that

$$E' = E. \qquad (6)$$

Indeed: By (3) we can choose a sequence $\{S_n\}$ of pairwise disjoint Borel
subsets of $\hat{\mathscr{A}}$ such that (i) $\hat{\mathscr{A}} \setminus \bigcup_n S_n$ is Q-null, and (ii) $\hat{B}$ is bounded away
from 0 on each S_n. It follows that, for each n, the operator $B_n = \int_{S_n} \hat{B} \, dQ$
on $X_n = $ range $Q(S_n)$ has a bounded inverse $B_n^{-1} = \int_{S_n} \hat{B}^{-1} \, dQ$; and by (4)
and (5),

$$E' \text{ coincides on } X_n \text{ with } CB_n^{-1}. \qquad (7)$$

On the other hand, if $\xi \in X_n$, we have, by the definition of C, $E\xi = EBB_n^{-1}\xi = CB_n^{-1}\xi$; so E also coincides on X_n with CB_n^{-1}. Combining this fact with (7)
and .B28 we obtain (6).

Now let P be the X-projection-valued Borel measure on $\mathbb{C}$ given by
$P(W) = Q(f^{-1}(W))$. Applying II.11.23 to (5) and (6) we obtain

$$E = \int_{\mathbb{C}} \lambda \, dP\lambda.$$

This proves the existence of the required P.

The uniqueness of P will follow from the next proposition.

12.2.   Proposition.   *Let P be any X-projection-valued Borel measure on $\mathbb{C}$,
and put*

$$E = \int_{\mathbb{C}} \lambda \, dP\lambda.$$

*Take a complex number λ_0 and a positive number k; and let W be the disk
$\{\lambda \in \mathbb{C} : |\lambda - \lambda_0| \le k\}$. Then for each ξ in X the following three conditions are
equivalent:*

(i) $\xi \in$ range $P(W)$;

(ii)   *for all* $n = 1, 2, \ldots,$ $\xi \in \text{domain}((E - \lambda_0 \mathbf{1})^n)$ *and* $\|(E - \lambda_0 \mathbf{1})^n \xi\| \le k^n \|\xi\|$;

(iii) $\xi \in \mathrm{domain}((E - \lambda_0 \mathbf{1})^n)$ *for all* $n = 1, 2, \ldots,$ *and* $\{k^{-n} \|(E - \lambda_0 \mathbf{1})^n \xi\|:$ $n = 1, 2, \ldots\}$ *is bounded.*

Proof. In view of II.11.18 and II.11.21, this is obtained by an argument similar to that of 11.14. ∎

12.3. Corollary. *If P_1 and P_2 are two X-projection-valued measures on* $\mathbb{C}$ *satisfying*

$$\int_{\mathbb{C}} \lambda \, dP_1 \lambda = \int_{\mathbb{C}} \lambda \, dP_2 \lambda,$$

then $P_1 = P_2$.

Proof. This follows from 12.2 by the argument of 11.15. ∎

12.4. Corollary 12.3 is of course the uniqueness statement in 12.1. Thus the proof of Theorem 12.1 is now complete. ∎

Definition. As in 11.1, the P of 12.1 is called the *spectral measure* of the normal operator E.

As we mentioned in Remark 12.1, E is bounded if and only if its spectral measure has compact support.

12.5. As another corollary of 12.2, we have, as in 11.5:

Corollary. *If P is the spectral measure of the normal operator E in X, and if $\lambda_0 \in \mathbb{C}$, then*

$$\mathrm{range}\, P(\{\lambda_0\}) = \{\xi \in \mathrm{domain}(E): E\xi = \lambda_0 \xi\}.$$

In particular, E is one-to-one if and only if $P(\{0\}) = 0$.

Proof. Apply 12.2 to a sequence of disks W with centers at λ_0 and with radii decreasing to 0. ∎

Functions on Normal Operators

12.6. Definition. Let E be a normal operator in the Hilbert space X, and $f: \mathbb{C} \to \mathbb{C}$ any Borel function. We define the normal operator $f(E)$ as follows:

$$f(E) = \int_{\mathbb{C}} f(\lambda) dP\lambda,$$

where P stands for the spectral measure of T.

12.7.   Suppose in 12.6 that $g: \mathbb{C} \to \mathbb{C}$ is another Borel function. By II.11.16, $f(E) = g(E)$ if and only if $f(\lambda) = g(\lambda)$ for P-almost all λ.

Thus $f(E)$ is well defined even if the Borel function f is defined only P-almost everywhere.

12.8.   If $c(\lambda) = \bar{\lambda}$ (complex conjugate), then by II.11.21(IV) $c(E) = E^*$.

If W is a Borel subset of $\mathbb{C}$, then $\mathrm{Ch}_W(E) = P(W)$.

12.9.   If the f of 12.6 is a polynomial function of degree n:

$$f(\lambda) = a_0 \lambda^n + a_1 \lambda^{n-1} + \cdots + a_n$$

$(a_i \in \mathbb{C}; a_0 \neq 0)$, then $f(E)$ as defined in 12.6 coincides with $a_0 E^n + a_1 E^{n-1} + \cdots + a_n \mathbf{1}$ as defined by successive applications of the addition, scalar multiplication, and composition of B9. In fact the domain of $f(E)$ is just domain(E^n).

This follows easily from II.11.21, II.11.22.

12.10.   If the normal operator E is one-to-one, then by 12.5 the function $\rho: \lambda \mapsto \lambda^{-1}$ is defined P-almost everywhere on $\mathbb{C}$; and by II.11.21(15)

$$\rho(E) = E^{-1}.$$

In particular, E^{-1} is normal.

12.11.   Let P' be the spectral measure of the $f(E)$ of 12.6. By II.11.23

$$P'(W) = P(f^{-1}(W))$$

for all Borel subsets W of $\mathbb{C}$.

12.12.   Proposition.   *Let Q be an X-projection-valued measure on a σ-field S of subsets of some set S. Let $f: S \to \mathbb{C}$ be Q-measurable, and let $\phi: \mathbb{C} \to \mathbb{C}$ be a Borel function. If we put $E = \int_S f\, dQ$, then*

$$\phi(E) = \int_S \phi(f(s))\, dQs. \tag{8}$$

Proof.   Let P and P' be the spectral measures of E and $\phi(E)$. By II.11.23, if W is any Borel subset of $\mathbb{C}$, we have $P(W) = Q(f^{-1}(W))$, $P'(W) = Q((\phi \circ f)^{-1}(W))$. But $(\phi \circ f)^{-1}(W) = f^{-1}(\phi^{-1}(W))$; so

$$P'(W) = P(\phi^{-1}(W)).$$

By 12.11 this gives (8).   ∎

12.13.   Corollary.   *Suppose in 12.6 that* $g: \mathbb{C} \to \mathbb{C}$ *is another Borel function. Then*

$$g(f(E)) = (g \circ f)(E).$$

Commutation Theorem

12.14.   As a generalization of 11.16 we have:

Theorem.   *Let E be a normal operator in X, with spectral measure P; and let B be any element of* $\mathcal{O}(X)$ *such that BcommE (B12). Then* $BP(W) = P(W)B$ *for every Borel subset W of* $\mathbb{C}$. *More generally, Bcommf(E) for every Borel function* $f: \mathbb{C} \to \mathbb{C}$.

Proof.   The fact that B commutes with all $P(W)$ follows from 12.2 just as in the argument of 11.16. The last statement is now obtained from II.11.19.   ∎

Spectra of Unbounded Normal Operators

12.15.   Definition.   As with bounded operators, we shall define the *spectrum* Sp(E) of an operator E in $\mathcal{U}(X)$ (see B4) as $\{\lambda \in \mathbb{C} : (E - \lambda \mathbb{1})^{-1}$ is not a bounded operator on $X\}$.

12.16.   Just as in 11.1, 11.7, we have:

Proposition.   *Let E be a normal operator in X, with spectral measure P. For any complex number* λ_0 *the following three conditions are equivalent:*

   (i) $\lambda_0 \in \mathrm{Sp}(E)$;
   (ii) λ_0 *belongs to the closed support of P;*
   (iii) *for each* $\varepsilon > 0$ *there is a vector* ξ *in* domain(E) *such that* $\|\xi\| = 1$ *and* $\|E\xi - \lambda_0 \xi\| < \varepsilon$.

Proof.   Assume that (ii) fails. Then $\lambda \mapsto (\lambda - \lambda_0)^{-1}$ is P-essentially bounded; so $\int (\lambda - \lambda_0)^{-1} dP\lambda$ is a bounded operator. On the other hand,

$$(E - \lambda_0 \mathbb{1})^{-1} = \int (\lambda - \lambda_0)^{-1} dP\lambda$$

by 12.10. Thus $(E - \lambda_0 \mathbb{1})^{-1}$ is bounded, and (i) fails. We have shown that (i) $\Rightarrow$ (ii).

The same argument as in 11.7 shows that (ii) $\Rightarrow$ (iii). Finally, it is quite evident that (iii) $\Rightarrow$ (i).   ∎

12.17. Thus the spectrum of a normal operator E coincides with the closed support of its spectral measure P. The spectrum of E is a closed set and is non-void (if $X \neq \{0\}$); it is bounded if and only if E is bounded.

Since $\mathbb{C} \setminus \mathrm{Sp}(E)$ is P-null, 12.7 tells us that $f(E)$ is well-defined for any Borel function $f : \mathrm{Sp}(E) \to \mathbb{C}$.

12.18. Proposition. *An operator E in X is self-adjoint if and only if E is normal and $\mathrm{Sp}(E) \subset \mathbb{R}$.*

Proof. Since by B25 a self-adjoint operator is always normal, we may assume to begin with that E is normal.

Let P be the spectral measure of E. By II.11.21(IV) and 12.7, $E^* = E \Leftrightarrow \bar{\lambda} = \lambda$ for P-almost all $\lambda \Leftrightarrow P(\mathbb{C} \setminus \mathbb{R}) = 0 \Leftrightarrow \mathrm{Sp}(E) \subset \mathbb{R}$ (the last step by 12.17).   ∎

Positive Operators

12.19. Definition. An operator E in X will be called *positive* if it is normal and

$$(E\xi, \xi) \geq 0 \qquad \text{for all } \xi \text{ in domain}(E).$$

This coincides with the usual definition if E is bounded (see 11.10).

12.20. Proposition. *A normal operator E in X is positive if and only if $\mathrm{Sp}(E) \subset \{\lambda \in \mathbb{R} : \lambda \geq 0\}$. In particular a positive operator is self-adjoint.*

Proof. If $\mathrm{Sp}(E) \subset \{\lambda \in \mathbb{R} : \lambda \geq 0\}$, then the spectral measure P of E is carried by $\{\lambda : \lambda \geq 0\}$ in view of 12.17, and so the positivity of E follows from II.11.18(12).

Conversely, assume that E is positive. If $\lambda_0 \in \mathrm{Sp}(E)$, λ_0 satisfies 12.16(iii). Then the same argument that we used in 11.10 to conclude that $\mathrm{APSp}(a) \subset \{\lambda : \lambda \geq 0\}$ shows here also that $\lambda_0 \geq 0$. This proves the required equivalence.

The last statement now follows from 12.18.   ∎

12.21. Corollary. *Let E be a normal operator in X, with spectral measure P; and let $f : \mathbb{C} \to \mathbb{C}$ be Borel. Then $f(E)$ is self-adjoint [positive] if and only if $f(\lambda)$ is real [real and non-negative] for P-almost all λ.*

Proof. By 12.11, 12.18, and 12.20. ∎

12.22. Proposition. *Let Y be another Hilbert space, and E a transformation in $\mathcal{U}(X, Y)$. Then E^*E is a positive operator in X.*

Proof. E^*E is self-adjoint by Appendix B30. If $\xi \in \text{domain}(E^*E)$, then $(E^*E\xi, \xi) = (E\xi, E\xi) \geq 0$. ∎

n-th Roots of Positive Operators

12.23. Fix a positive integer n; and let $f(t) = t^{1/n}$ (positive n-th root) for $t \geq 0$.

Proposition. *Let E be any positive operator in X. Then $f(E)$ is the unique positive operator D in X satisfying $D^n = E$.*

Proof. $f(E)$ is well-defined and positive by 12.17, 12.20, and 12.21. By 12.9

$$(f(E))^n = f^n(E) = E.$$

Let D be any positive operator satisfying $D^n = E$. Then

$$f(E) = f(D^n) = D$$

by 12.9 and 12.13. ∎

The operator $f(E)$ is called the *positive n-th root of E*, and is denoted by $E^{1/n}$.

12.24. Corollary. *An operator in X is positive if and only if it is of the form E^*E, where $E \in \mathcal{U}(X)$.*

Proof. By 12.22 and 12.23 (applied with $n = 2$). ∎

13. Polar Decomposition of Operators; Mackey's Form of Schur's Lemma

13.1. We fix two Hilbert spaces X and Y.

To begin with, we must introduce the notion of a partial isometry.

Lemma. *Let $U \in \mathcal{O}(X, Y)$, and put $N = \text{Ker}(U)$. Then the following conditions are equivalent:*

(i) $U|N^\perp$ *is an isometry of $N^\perp$ into Y;*

(ii) $UU^*U = U$;

(iii) UU^* *is a projection in $\mathcal{O}(Y)$;*

(iv) U^*U *is a projection in $\mathcal{O}(X)$.*

Proof.   Multiplying (ii) on the right by U^* we obtain (iii). If (iii) holds, then

$$(UU^*U - U)(UU^*U - U)^* = (UU^*)^3 - 2(UU^*)^2 + UU^* = 0,$$

so that $UU^*U - U = 0$, whence (ii) holds. Thus (ii) $\Leftrightarrow$ (iii). Taking adjoints in (ii) we get $U^*UU^* = U^*$; so, replacing U by U^* in the preceding argument, we obtain (ii) $\Leftrightarrow$ (iv). Therefore (ii) $\Leftrightarrow$ (iii) $\Leftrightarrow$ (iv).

Assume (ii), (iii), (iv). Clearly $U\xi = 0 \Leftrightarrow U^*U\xi = 0$; therefore $N^\perp$ is the range of the projection U^*U. Thus, if $\xi \in N^\perp$, we have $U^*U\xi = \xi$ or $\|\xi\|^2 = (U^*U\xi, \xi) = \|U\xi\|^2$. Hence (i) holds.

Conversely, assume (i); and let $\xi \in N^\perp$. If $\eta \in N$, $(U^*U\xi, \eta) = (U\xi, U\eta) = 0 = (\xi, \eta)$. If $\eta \in N^\perp$, $(U^*U\xi, \eta) = (U\xi, U\eta) = (\xi, \eta)$ by (i). So $(U^*U\xi, \eta) = (\xi, \eta)$ for every η in X; that is, $U^*U\xi = \xi$. This means that U^*U is the identity map on $N^\perp$ and annihilates N, and so is the projection onto $N^\perp$. We have now shown that (i) $\Rightarrow$ (iv).

Thus (i) is equivalent to the other conditions.   ∎

13.2.   Definition.   An element U of $\mathcal{O}(X, Y)$ satisfying conditions (i)–(iv) of 13.1 is called a *partial isometry* of X into Y. The subspace $N^\perp = \text{range}(U^*U)$ is called the *initial space* of U, and the range of U, i.e., the image of $N^\perp$ under the isometry $U|N^\perp$, is called the *final space* of U.

One verifies the following further easy facts:

(I) U^*U and UU^* are the projections of X and Y onto the initial and final space of U respectively.

(II) U^* is a partial isometry of Y into X whose initial [final] space is the final [initial] space of U.

(III) U^* and U are inverses of each other when each is restricted to its initial space.

13.3.   We remind the reader that in this section the terms "transformation" and "operator" are understood in the "unbounded" sense of Appendix B.

The following easy observation relates partial isometries with metric equivalence (see B20).

Proposition.   *Let Z be a third Hilbert space; and let E_1 and E_2 be two metrically equivalent transformations (in the sense of B2) from X to Y and from X to Z respectively. Then there is a unique partial isometry U from Y to Z having the following two properties:*

(i) $E_2 = UE_1$;

(ii)   *the initial space of U is the closure of the range of E_1.*

U has the further properties:

(iii) $E_1 = U^*E_2$;
(iv) *the final space of U is the closure of the range of E_2.*

Proof. U is determined by the relation $UE_1\xi = E_2\xi$ $(\xi \in \text{domain}(E_i))$. Details are left to the reader. ∎

13.4. Proposition. *Let E be any transformation in $\mathcal{U}(X, Y)$. There is a unique positive operator P in X which is metrically equivalent with E. In fact $P = (E^*E)^{1/2}$.*

Proof. Define P to be the positive operator $(E^*E)^{1/2}$ (see 12.22); thus $P^2 = E^*E$. Let P_1 and E_1 denote the restriction of P and E respectively to domain(P^2) ($=$domain(E^*E)). If $\xi \in \text{domain}(P^2)$,

$$\|P_1\xi\|^2 = \|P\xi\|^2 = (P^2\xi, \xi)$$

$$= (E^*E\xi, \xi) = \|E\xi\|^2 = \|E_1\xi\|;$$

so P_1 and E_1 are metrically equivalent. By Theorem B11 the closures of P_1 and E_1 are P and E respectively. Hence, by B20, P and E are metrically equivalent.

It remains to show that any other positive operator P' in X which is metrically equivalent to E (and hence also to P) must be equal to P. To do this, let U be the partial isometry given to us by 13.3, whose initial space is the closure of range(P) and which satisfies $P' = UP$. Since P and P' are self-adjoint, this implies by B9(viii) that $P' = PU^*$. Now since the range of the projection U^*U contains that of P, we have $U^*UP = P$. Putting these facts together we find $P'^2 = PU^*UP = P^2$. By the uniqueness of square roots (12.23) this implies $P' = P$. ∎

13.5. Theorem. *Every transformation E in $\mathcal{U}(X, Y)$ is of the form*

$$E = UP, \tag{1}$$

where:

(i)   *P is a positive operator in X,*
(ii) *U is a partial isometry from X to Y,*
(iii) *the initial space of U is the closure of range(P).*

Furthermore, U and P are uniquely determined by (1), (i), (ii), (iii); and we have:

$$P = (E^*E)^{1/2}. \tag{2}$$

Proof. This follows very easily from 13.3 and 13.4, if we notice that the conditions on U and P imply the metric equivalence of P and E. ∎

Definition. Equation (1) is called the *polar decomposition* of the transformation E.

Remark. The polar decomposition of a transformation is a vast generalization of the polar representation of a complex number z:

$$z = ur \qquad (0 \le r \in \mathbb{R}; u \in \mathbb{C}, |u| = 1).$$

13.6. In the polar decomposition (1), notice that P and E are metrically equivalent, and that the final space of U is the closure of range(E). We claim that the initial space of U is the orthogonal complement (in X) of the closed subspace $\mathrm{Ker}(E) = \{\xi \in \mathrm{domain}(E) : E\xi = 0\}$. Indeed, this follows from B10(1) (applied to P) if we remember that $\mathrm{Ker}(E) = \mathrm{Ker}(P)$.

13.7. Keeping the notation of 13.5, notice that the polar decomposition of E^* is

$$E^* = U^*P', \qquad\qquad\qquad (3)$$

where P' is the positive operator UPU^* on Y.

13.8. Definition. Let us say that a transformation E in $\mathscr{U}(X, Y)$ is *non-singular* if $\mathrm{Ker}(E) = \{0\}$ (i.e., E is one-to-one) and range(E) is dense in Y.

By B10(2) the condition that range(E) is dense can be replaced by the condition that E^* is one-to-one, and so can be omitted if E is a normal operator (since then E and E^* are metrically equivalent).

13.9. Proposition. *If the E of 13.5 is non-singular, then, in the polar decomposition (1), P is non-singular and U is a linear isometry of X onto Y.*

Mackey's Form of Schur's Lemma

13.10. We shall now be interested in *-representations of a fixed *-algebra A.

The notion of an intertwining transformation is a trivial extension of the definition in IV.1.6:

Definition. Let S, T be *-representations of A and E a transformation of $X(S)$ into $X(T)$. We say that E *intertwines S and T* (or is S, T *intertwining*) if

(i) domain(E) and range(E) are S-stable and T-stable respectively,
(ii) $ES_a\xi = T_aE\xi$ for all a in A and ξ in domain(E).

Conditions (i) and (ii) are clearly equivalent to asserting that E, considered as a linear subspace of $X(S) \oplus X(T)$, is stable under the direct sum representation $S \oplus T$.

13.11. It is clear that sums, scalar multiples, and compositions of intertwining transformations (in the sense of B8) are also intertwining.

Furthermore, if E is S, T intertwining as in 13.10 and if in addition E^* is a transformation, then E^* is T, S intertwining. Indeed, assume that $\langle \eta', \xi' \rangle \in E^*$. Then, for all a in A and ξ in domain(E),

$$(E\xi, T_a\eta') = (T_{a^*}E\xi, \eta')$$

$$= (ES_{a^*}\xi, \eta') \qquad \text{(since } E \text{ intertwines)}$$

$$= (S_{a^*}\xi, \xi') \qquad \text{(since } \langle \eta', \xi' \rangle \in E^*)$$

$$= (\xi, S_a\xi').$$

This says that $\langle T_a\eta', S_a\xi' \rangle \in E^*$, and hence that E^* is $(T \oplus S)$-stable.

13.12. If S is a *-representation of A and E is an operator in $X(S)$, then evidently E is S, S intertwining if and only if S_a comm E (in the sense of B12) for all a in A.

13.13. Proposition. *Let S and T be two *-representations of A, and E an S, T intertwining transformation in $\mathcal{U}(X(S), X(T))$. If*

$$E = UP$$

is the polar decomposition of E (as in 13.5), then P is S, S intertwining and U is S, T intertwining.

Proof. It follows from 13.11 that E^*E is S, S intertwining, that is (13.12), that

$$S_a \text{ comm } E^*E \qquad\qquad (a \in A). \qquad (4)$$

Now E^*E is a positive operator (by 12.22). Hence (4) and 12.14 give S_a comm$(E^*E)^{1/2} = P$. So P is S, S intertwining.

To prove that U is S, T intertwining it is enough to verify the equation

$$US_a\xi = T_aU\xi \qquad\qquad (5)$$

for the cases $U\xi = 0$ and $\xi \in \text{range}(P)$ (since these two kinds of ξ together span a dense subspace of $X(S)$). Now $U\xi = 0 \Leftrightarrow E\xi = 0$ (see 13.6); so the validity of (5) if $U\xi = 0$ follows from the S-stability of $\{\xi : E\xi = 0\}$. If

$\xi = P\eta$, then $US_a\xi = US_aP\eta = UPS_a\eta$ (since P is S, S intertwining) $= ES_a\eta$ $= T_aE\eta$ (since E is S, T intertwining) $= T_aUP\eta = T_aU\xi$; so (5) holds for $\xi \in \mathrm{range}(P)$.   ■

13.14.   Mackey's Form of Schur's Lemma.   *Let S and T be two *-representations of A, and E an S, T intertwining transformation in $\mathcal{U}(X(S), X(T))$. Put $N = \mathrm{Ker}(E)$, $R = $ closure of range(E). Then $N^\perp$ and R are stable under S and T respectively; and the subrepresentations of S and T acting on $N^\perp$ and R respectively are unitarily equivalent.*

Proof.   The stability of N and R is easily verified as in IV.1.8. Thus by 9.4 $N^\perp$ is S-stable. Let

$$E = UP$$

be the polar decomposition of E. From 13.6 we see that U sends $N^\perp$ isometrically onto R; and by 13.13 U is S, T intertwining. This gives the required conclusion.   ■

Remark.   The formal similarity between this theorem and Schur's Lemma IV.1.8 is quite evident.

13.15.   Corollary.   *Let S and T be two *-representations of A, S being irreducible; and suppose there exists a non-zero S, T intertwining transformation E in $\mathcal{U}(X(S), X(T))$. Then S is unitarily equivalent to some subrepresentation of T. In particular, if T is also irreducible, then S and T are unitarily equivalent.*

Proof.   Since S is irreducible, the proper closed S-stable subspace $\mathrm{Ker}(E)$ of $X(S)$ must be $\{0\}$. Hence by 13.14 S is unitarily equivalent to ^RT, where $R = \overline{\mathrm{range}(E)}$. If T is also irreducible, then R, being a non-zero closed T-stable subspace of $X(T)$, must be all of $X(T)$; and so the second statement holds.   ■

*-Representations and the Naimark Relation

13.16.   Suppose that the E of 13.14 is one-to-one and that range(E) is dense in $X(T)$. The existence of such an E means that S and T are Naimark-related (see V.3.2). On the other hand, by 13.9, U is then a linear isometry of $X(S)$ onto $X(T)$. We therefore have:

Corollary. *For any two *-representations S and T of A, the following three conditions are equivalent:*

(i)   *S and T are unitarily equivalent;*
(ii)   *S and T are homeomorphically equivalent (see* V.1.4);
(iii) *S and T are Naimark-related.*

Proof. We have seen above that (iii) ⇒ (i) in virtue of 13.14. The implications (i) ⇒ (ii) ⇒ (iii) are almost immediate.   ∎

Remark. For finite-dimensional *-representations the above corollary was proved by a different method in IV.7.10.

13.17. Example. We shall conclude this section with an explicit example of the non-transitivity of the Naimark relation (see V.3.4).

Let A be the commutative C^*-algebra $\mathscr{C}([0, 1])$ (with the supremum norm). Choose two Borel measures μ and ν on $[0, 1]$, both of which have $[0, 1]$ as their closed support, but which are not equivalent as measures (i.e., have different null sets). For example, we might take μ to be Lebesgue measure, and ν to be the sum of μ and a point mass at some point of $[0, 1]$.

We shall now define three Banach representations $T^{(1)}$, $T^{(2)}$, $T^{(3)}$ of A. The Banach spaces in which they act will be $\mathscr{L}_2(\mu)$, $\mathscr{L}_2(\nu)$ and A respectively; and the operators are given by the same formula in all cases:

$$(T_f^{(i)}\phi)(t) = f(t)\phi(t)$$

$(i = 1, 2, 3; f \in A; \phi \in X(T^{(i)}); 0 \le t \le 1)$. Now the identity injection $j: A \to \mathscr{L}_2(\mu)$ is continuous (hence closed), one-to-one (since the closed support of μ is $[0, 1]$), maps A onto a dense subset of $\mathscr{L}_2(\mu)$, and intertwines $T^{(3)}$ and $T^{(1)}$. So $T^{(1)}$ and $T^{(3)}$ are Naimark-related. Similarly $T^{(2)}$ and $T^{(3)}$ are Naimark-related. Now notice that $T^{(1)}$ and $T^{(2)}$ are *-representations and are not unitarily equivalent. For, if they were unitarily equivalent, their spectral measures (see 10.10) would have the same null sets; and by 10.12 this would imply that the null sets of μ and ν were the same. So $T^{(1)}$ and $T^{(2)}$ are not unitarily equivalent, and hence by 13.16 not Naimark-related.

Consequently, the three Banach representations $T^{(1)}$, $T^{(2)}$, $T^{(3)}$ provide an example of the non-transitivity of the Naimark-relation.

Remark 1. The above example also shows the indispensability of the hypothesis in 13.14, 13.15, 13.16 that the intertwining transformation be *closed*. Indeed, the identity map $j: A \to A$, regarded as a transformation from $\mathscr{L}_2(\mu)$ to $\mathscr{L}_2(\nu)$, is one-to-one, has dense domain and range, and intertwines

$T^{(1)}$ and $T^{(2)}$; and yet $T^{(1)}$ and $T^{(2)}$, as we have seen, are not unitarily equivalent. The trouble is, of course, that j is not closed.

Remark 2.   The above three Banach representations $T^{(i)}$ are of course far from irreducible. Can examples of the non-transitivity of the Naimark relation be found which involve only *irreducible Banach* representations? The answer to this seems not to be known (though we do know of examples involving only irreducible *Fréchet* representations; see Fell [10], Appendix, Example 4).

14.   A Criterion for Irreducibility; Discrete Multiplicity Theory

14.1.   The Spectral Theorem gives rise to an important criterion for the irreducibility of *-representations.
   We fix a *-algebra A.

Theorem.   *Let T be a non-zero *-representation of A. Then T is irreducible if and only if the commuting algebra B of T (see 9.3) consists of the scalar operators only.*

Proof.   If T has a non-trivial closed stable subspace Y and p is projection onto Y, then (by 9.5) p is a non-scalar operator in B. So the "if" part holds.
   Conversely, suppose that u is a non-scalar operator in B. Since B is a *-algebra, $v = \frac{1}{2}(u + u^*)$ and $w = \frac{1}{2}i(u - u^*)$ are also in B. Now v and w are Hermitian and at least one of them is not a scalar operator (since $u = v - iw$). So we may assume from the outset that u is Hermitian. Let P be its spectral measure. Since u is not a scalar operator, 11.3 implies that some projection p in range(P) is different from 0 and $\mathbb{1}$. By 11.16 $p \in B$; and therefore the non-trivial closed subspace range(p) is T-stable. So T is not irreducible; and the "only if" part is proved.   ∎

14.2.   Corollary.   *Let T be an irreducible *-representation of A. If a belongs to the center C of A, then T_a is a scalar operator:*

$$T_a = \chi(a)\mathbb{1}_{X(T)}.$$

*The complex function χ so obtained is a one-dimensional *-representation of C.*

   We call this χ the *central character of T.*

14.3.   Corollary.   *If A is commutative, every irreducible *-representation T of A is one-dimensional.*

Proof. By 14.2 the T_a are all scalar operators; and so cannot act irreducibly on a space of dimension greater than one. ∎

Remark. This corollary could have been deduced directly from Stone's Theorem 10.10.

14.4. *Remark.* The "if" statement in Theorem 14.1 shows a striking difference between the behaviors of *-representations and representations of non-involutory algebras. Consider for example the (non-involutory) algebra B of all 2×2 triangular matrices $\begin{pmatrix} a & b \\ 0 & c \end{pmatrix}$ $(a, b, c \in \mathbb{C})$. The only 2×2 matrices which commute with all of B are the scalar matrices; and yet the identity representation of B is not irreducible.

14.5. Theorem 14.1 has an important generalization.

Theorem. *Let T be an irreducible *-representation of A; and put $X = X(T)$. Let Y and Z be two other Hilbert spaces; form the Hilbert tensor products $X \otimes Y$ and $X \otimes Z$; and suppose that F is a transformation in $\mathcal{O}(X \otimes Y, X \otimes Z)$ satisfying*

$$F(T_a \otimes 1_Y) = (T_a \otimes 1_Z)F \qquad (1)$$

for all a in A. Then

$$F = 1_X \otimes u$$

for some unique u in $\mathcal{O}(Y, Z)$.

Proof. To begin with we assume that $Y = Z$.

Let $\{\eta_i : i \in I\}$ be an orthonormal basis of Y; and for each i let $e_i : X \to X \otimes Y$ be the isometry sending ξ into $\xi \otimes \eta_i$. Thus e_i^* is the inverse of e_i on $X \otimes \eta_i$, and annihilates $(X \otimes \eta_i)^\perp$. It follows that

$$\zeta = \sum_i (e_i^*\zeta \otimes \eta_i) \qquad (\zeta \in X \otimes Y). \qquad (2)$$

Notice also that

$$(T_a \otimes 1_Y)e_i = e_i T_a \qquad \text{for each } a. \qquad (3)$$

For each i, j, set $F_{ij} = e_j^* F e_i$. By (1) and (3) $T_a F_{ij} = T_a e_j^* F e_i = e_j^*(T_a \otimes 1_Y)F e_i = e_j^* F(T_a \otimes 1_Y)e_i = F_{ij} T_a$ for all a. Hence by 14.1 there are complex numbers λ_{ij} such that

$$F_{ij} = \lambda_{ij} 1_X \qquad \text{for all } i, j \text{ in } I. \qquad (4)$$

Now fix a vector ξ in X. For each j in I we have by (2) and (4)

$$F(\xi \otimes \eta_j) = \sum_i [e_i^* F(\xi \otimes \eta_j) \otimes \eta_i] = \sum_i (F_{ji} \xi \otimes \eta_i)$$

$$= \sum_i (\xi \otimes \lambda_{ji} \eta_i). \tag{5}$$

It follows from (5) that $\xi \otimes Y$ is stable under F, hence that there is an operator u_ξ in $\mathcal{O}(Y)$ satisfying

$$F(\xi \otimes \eta) = \xi \otimes u_\xi \eta \qquad (\eta \in Y). \tag{6}$$

Since the λ_{ji} in (5) do not depend on ξ, we conclude from (5) that u_ξ is independent of ξ. Therefore (6) implies that $F = 1_X \otimes u$, where $u = u_\xi$ for each ξ.

We have proved the existence of the required u in case $Z = Y$. Let us now drop this assumption. Set $W = Y \oplus Z$; and let $p: W \to Y$ be projection onto Y. Thus $1_X \otimes p$ projects $X \otimes W$ onto $X \otimes Y$; and so $\Phi = F(1_X \otimes p)$ is an operator in $\mathcal{O}(X \otimes W)$ commuting with all $T_a \otimes 1_W$. Applying the preceding paragraph to Φ, we obtain

$$\Phi = 1_X \otimes v, \qquad \text{where } v \in \mathcal{O}(W).$$

The reader will easily verify that $v(W) \subset Z$, and that $F = 1_X \otimes u$, where $u = v|Y$.

Thus the existence of u has been established. Its uniqueness is obvious. ∎

Discretely Decomposable *-Representations

14.6. We pointed out in V.1.20 that for arbitrary locally convex representations there is no analogue of the purely algebraic multiplicity theory of §IV.2. For *-representations however, as we shall now see, the theory of §IV.2 has a perfect analogue.

Definition. A *-representation T of A is *discretely decomposable* if it is unitarily equivalent to a Hilbert direct sum of irreducible *-representations —or, equivalently, if we can write $X(T)$ as a Hilbert direct sum $\sum_i^\oplus Y_i$, where each Y_i is a closed T-stable subspace of $X(T)$ on which T acts irreducibly.

Remark. This is the same as the "*-representational complete reducibility" mentioned in 10.13.

If the *-representation T is finite-dimensional, successive application of 9.4 leads to the Hilbert direct sum decomposition of $X(T)$ into minimal, hence irreducible, subspaces. So we have:

Proposition. *Every finite-dimensional ∗-representation T of A is discretely decomposable, hence completely reducible.*

Remark. We pointed out in 10.13 the existence of (infinite-dimensional) non-degenerate ∗-representations which are not discretely decomposable. In this connection notice that, in view of 9.4, there is no obvious analogue of IV.3.12 for ∗-representations.

14.7. In analogy with IV.2.9 we have:

Proposition. *Let T be a ∗-representation of A, $\mathcal{Y}$ the family of all those closed T-stable subspaces of X(T) on which T acts irreducibly, and Z the closure of the linear span of $\mathcal{Y}$. Then T is discretely decomposable if and only if $Z = X(T)$.*

Proof. The "only if" part is obvious.

Assume that $Z = X(T)$. By Zorn's Lemma there is a maximal pairwise orthogonal subfamily $\mathcal{Y}_0$ of $\mathcal{Y}$. Let W be the closure of the linear span of $\mathcal{Y}_0$. We must show that $W = X(T)$. Suppose then that $W \neq X(T)$. Then $W^{\perp} \neq \{0\}$; and the projection p onto $W^{\perp}$ is non-zero. Since $Z = X(T)$, there is an element Y of $\mathcal{Y}$ such that p does not annihilate Y. On the other hand, W is T-stable, hence so is $W^{\perp}$; so p belongs to the commuting algebra of T. Thus $p|Y$ is a (bounded) non-zero intertwining operator for ^{Y}T and $^{W^{\perp}}T$. Since ^{Y}T is irreducible, this implies by 13.15 that ^{Y}T is unitarily equivalent to some subrepresentation ^{R}T of $^{W^{\perp}}T$. In particular $R \in \mathcal{Y}$, and could have been adjoined to $\mathcal{Y}_0$, contradicting the maximality of the latter. Hence $W = X(T)$, showing that T is discretely decomposable. This proves the "if" statement. ∎

14.8. Remark. For any ∗-representation T of A we can define $\mathcal{Y}$ and Z as in Proposition 14.7. Then Z is a closed T-stable subspace, and by 14.7 ^{Z}T is discretely decomposable. In fact Z is evidently the largest closed stable subspace with this property. We refer to ^{Z}T as the *discretely decomposable part of T*.

14.9. Proposition. *A subrepresentation of a discretely decomposable ∗-representation T (acting on a closed T-stable subspace W of X(T)) is discretely decomposable. A Hilbert direct sum of discretely decomposable ∗representations is discretely decomposable.*

Proof. To prove the first statement, define $\mathcal{Y}$ as in 14.7; and let p be the projection onto W. By 13.15, for each Y in $\mathcal{Y}$ either $p(Y) = \{0\}$ or $\overline{p(Y)} \in \mathcal{Y}$. Since by 14.7 $\mathcal{Y}$ spans a dense subspace of $X(T)$, the $\overline{p(Y)}$ $(Y \in \mathcal{Y})$ must span a dense subspace of W. Applying 14.7 again we conclude that ^WT is discretely decomposable.

The second statement of the proposition is obvious.   ∎

14.10.   As in IV.2.11 we now raise the question of whether the direct sum decomposition of a discretely decomposable *-representation into irreducible parts is in some sense unique.

In preparation for the answer (Theorem 14.13) we need a simple conse-quence of 14.5. If T is any *-representation of A and α is a cardinal number, let $\alpha \cdot T$ be the Hilbert direct sum of α copies of T; that is,

$$\alpha \cdot T = \sum_{i \in I}^{\oplus} T,$$

where I is any index set of cardinality α. (Of course $O \cdot T$ is the zero representation on $\{0\}$.)

Lemma.   *Let T be an irreducible *-representation of A, and α and β two cardinal numbers such that $\alpha \cdot T$ and $\beta \cdot T$ are unitarily equivalent. Then $\alpha = \beta$.*

Proof.   Let $X = X(T)$; and let Y and Z be Hilbert spaces of dimension α and β respectively. Then $\alpha \cdot T$ and $\beta \cdot T$ act on the Hilbert tensor products $X \otimes Y$ and $X \otimes Z$, with

$$(\alpha \cdot T)_a = T_a \otimes 1_Y, \qquad (\beta \cdot T)_a = T_a \otimes 1_Z.$$

Applying 14.5 to the isometry F which implements the unitary equivalence of $\alpha \cdot T$ and $\beta \cdot T$, we conclude that $F = 1_X \otimes u$, where $u \in \mathcal{O}(Y, Z)$. It is easy to see that u must be an isometry onto Z. So $\alpha = \dim(Y) = \dim(Z) = \beta$.   ∎

14.11.   Let us denote by $\mathscr{I}$ the family of all unitary equivalence classes of irreducible *-representations of A.

In analogy with IV.2.11 we make the following definition:

Definition.   If T is a *-representation of A and σ is any element of $\mathscr{I}$, we define the *(closed)* σ-*subspace* $X_\sigma(T)$ of $X(T)$ as the closure of the linear span of the set of those closed T-stable subspaces Y of $X(T)$ for which ^YT is of class σ.

14.12.   Proposition.   *Let T be a *-representation of A. If σ, τ are distinct elements of $\mathscr{I}$, then $X_\sigma(T) \perp X_\tau(T)$. If T is discretely decomposable,*

$$X(T) = \sum_{\sigma \in \mathscr{I}}^{\oplus} X_\sigma(T) \qquad (Hilbert\ direct\ sum). \qquad (7)$$

Proof. Let σ, τ be distinct elements of $\mathscr{I}$. To prove the first statement it is enough to take closed T-stable subspaces Y and Z such that $^Y T$ and $^Z T$ are of classes σ and τ respectively, and show that $Y \perp Z$. Now the projection p onto Z is T, T intertwining, and hence $p|Y$ intertwines $^Y T$ and $^Z T$. If $p|Y$ were non-zero, then by 13.15 we would have $^Y T \cong {^Z T}$, contradicting $\sigma \neq \tau$. So $p|Y = 0$, whence $Y \perp Z$. This proves the first statement.

It follows that the right side of (7) always exists. If T is discretely decomposable it must equal $X(T)$. ∎

14.13.   Theorem.   *Let T be a discretely decomposable *-representation of A:*

$$T = \sum_{i \in I}{}^{\oplus} T^i \tag{8}$$

(T^i being irreducible for each i; we identify $X(T^i)$ with a closed subspace of $X(T)$). For each σ in $\mathscr{I}$ set $I_\sigma = \{i \in I : T^i \text{ is of class } \sigma\}$. Then

(i) $X_\sigma(T) = \sum_{i \in I_\sigma}^{\oplus} X(T^i)$ *(Hilbert direct sum);*
(ii)   card(I_σ) *depends only on T and σ, not on the particular decomposition* (8).

Proof.   The right side of (i) is contained in $X_\sigma(T)$ by the definition of the latter. So, since by 14.12 the different $X_\sigma(T)$ are orthogonal, (8) implies (i).

In view of (i), (ii) follows from the application of Lemma 14.10 to the subrepresentation of T acting on $X_\sigma(T)$. ∎

14.14.   Definition.   The cardinal number card(I_σ) occurring in 14.13 is called the *(Hilbert) multiplicity $m_\sigma(T)$ of σ in T*. If $m_\sigma(T) \neq 0$, we say that σ *occurs in T*.

Evidently, for any discretely decomposable *-representation T,

$$T \cong \sum_{\sigma \in \mathscr{I}}{}^{\oplus} m_\sigma(T) \cdot S^\sigma \qquad \text{(Hilbert direct sum))}$$

(where S^σ is some representative of the class σ).

Remark.   The discussion in IV.2.13 now applies, with obvious small variations, to the question of the uniqueness of the decomposition (8).

14.15.   In analogy with IV.2.16 we have:

Proposition.   *Let T be a *-representation of A. Then $b(X_\sigma(T)) \subset X_\sigma(T)$ for all σ in $\mathscr{I}$ and all b in the commuting algebra of T.*

Proof.   Let σ be in $\mathscr{I}$ and b be in the commuting algebra of T; and let q be the projection onto $X_\sigma(T)^\perp$. Since $X_\sigma(T)$ is T-stable, q belongs to the commuting algebra of T; hence so does qb. Now take a closed T-stable subspace Y of $X_\sigma(T)$ on which T acts irreducibly (of class σ). If qb does not annihilate Y, then by 13.15 T has an irreducible subrepresentation of class σ acting on a subspace of $X_\sigma(T)^\perp$—an impossibility. So qb annihilates all such Y, and hence annihilates the closure $X_\sigma(T)$ of their linear span. Thus $qb\xi = 0$, whence $b\xi \in X_\sigma(T)$, for all ξ in $X_\sigma(T)$.   ∎

15. Compact Operators and Hilbert–Schmidt Operators

15.1.   In Stone's Theorem (10.10) we classified the *-representations of commutative Banach *-algebras. In this section we shall take up a very special kind of non-commutative C^*-algebra, the C^*-algebra of compact operators, and completely classify its *-representations. The results will be in complete analogy with the earlier results on total matrix algebras (see IV.5.7).

15.2.   The reader will recall that a subset A of a metric space S, d is *precompact* if for each $\varepsilon > 0$ there is a finite subset F of A such that every x in A satisfies $d(x, y) < \varepsilon$ for some y in F. If S, d is complete, a subset A of S is precompact if and only if its closure is compact.

Now let X and Y be two fixed Banach spaces.

A bounded linear map a on X to Y is *compact* if the a-image of the unit ball $X_1 = \{\xi \in X : \|\xi\| \le 1\}$ of X is precompact in Y.

Equivalently a is compact if and only if $a(X_1)$ has compact closure in Y. This implies that $a(W)$ has compact closure for every norm-bounded subset W of X.

Let $\mathcal{O}_c(X, Y)$ and $\mathcal{O}_F(X, Y)$ denote the space of all compact elements of $\mathcal{O}(X, Y)$, and the space of all maps in $\mathcal{O}(X, Y)$ of finite rank, respectively. Evidently

$$\mathcal{O}_F(X, Y) \subset \mathcal{O}_c(X, Y). \tag{1}$$

As usual we abbreviate $\mathcal{O}_c(X, X)$ and $\mathcal{O}_F(X, X)$ to $\mathcal{O}_c(X)$ and $\mathcal{O}_F(X)$.

15.3.   Proposition.   *If $a, b \in \mathcal{O}_c(X, Y)$, $c \in \mathcal{O}(X)$, $d \in \mathcal{O}(Y)$, and $\lambda \in \mathbb{C}$, then*

$$\lambda a, \qquad a + b, \qquad da \quad and \quad ac$$

are in $\mathcal{O}_c(X, Y)$. In particular, $\mathcal{O}_c(X)$ is a two-sided ideal of $\mathcal{O}(X)$.

The proof is left as an easy exercise for the reader.

15.4.   It is a well known fact that the unit ball $\{\xi \in X : \|\xi\| \leq 1\}$ of the Banach space X cannot be norm-compact if X is infinite-dimensional. From this we deduce easily that an operator a in $\mathcal{O}(X, Y)$ cannot be compact if there exists a positive number k and an infinite-dimensional subspace Z of X such that $\|a\xi\| \geq k\|\xi\|$ for all ξ in Z. From this in turn we see that an idempotent operator a in $\mathcal{O}(X)$ is compact if and only if a is of finite rank.

15.5.   Proposition. $\mathcal{O}_c(X, Y)$ *is closed in* $\mathcal{O}(X, Y)$ *in the operator norm.*

Proof.   Let a belong to the norm-closure of $\mathcal{O}_c(X, Y)$; let X_1 be the unit ball of X; and let $\varepsilon > 0$. Choose $b \in \mathcal{O}_c(X, Y)$ so that $\|a - b\| < \frac{1}{3}\varepsilon$. We can then find a finite subset F of X_1 such that every ξ in X_1 satisfies $\|b\xi - b\eta\| < \frac{1}{3}\varepsilon$ for some η in F. For this ξ and η we have $\|a\xi - a\eta\| \leq \|(a - b)\xi\| + \|b\xi - b\eta\| + \|(a - b)\eta\| < 2\|a - b\| + \frac{1}{3}\varepsilon < \varepsilon$. By the arbitrariness of ξ this implies that a is compact.   ∎

15.6.   Corollary. $\mathcal{O}_c(X)$ *is a norm-closed two-sided ideal of* $\mathcal{O}(X)$. *If* X *is a Hilbert space,* $\mathcal{O}_c(X)$ *is also closed under the adjoint operation.*

Proof.   The second statement follows from the first by 8.6. It also follows easily and directly from the next proposition.

Remark.   Thus, if X is a Hilbert space, $\mathcal{O}_c(X)$ is a C*-algebra in its own right. Notice that $\mathcal{O}_c(X)$ does not have a unit element unless X is finite-dimensional.

15.7.   It follows from (1) and 15.5 that $\mathcal{O}_c(X, Y) \supset \overline{\mathcal{O}_F(X, Y)}$. If Y is a Hilbert space, we shall show that this is an equality.

Proposition.   *If* Y *is a Hilbert space,* $\mathcal{O}_c(X, Y) = \overline{\mathcal{O}_F(X, Y)}$.

Proof.   We need only show that

$$\mathcal{O}_c(X, Y) \subset \overline{\mathcal{O}_F(X, Y)}. \tag{2}$$

Take an operator a in $\mathcal{O}_c(X, Y)$; and let $\varepsilon > 0$. By the compactness of a there is a finite subset F of the unit ball X_1 of X such that for every ξ in X_1 we have

$$\|a\xi - a\eta\| < \varepsilon \tag{3}$$

for some η in F. Let Z be the (finite-dimensional closed) subspace of Y spanned by $\{a\eta : \eta \in F\}$; and let p be projection onto Z. For each ξ in X_1, $pa\xi$

is the element of Z nearest to $a\xi$; but by (3) there is an element of Z nearer to $a\xi$ than ε. Therefore $\|pa\xi - a\xi\| < \varepsilon$ for all ξ in X_1, that is, $\|pa - a\| \leq \varepsilon$. Since pa has finite rank and ε is arbitrary, this says that a belongs to $\overline{\mathcal{O}_F(X, Y)}$. We have thus proved (2). ∎

Remark. Does the above proposition hold for arbitrary Banach spaces X and Y? For a long time this was a famous unsolved problem. But in 1973 Enflo [1] gave an intricate counter-example showing that the answer is "no".

15.8. The following characterization of compact operators is sometimes useful.

Proposition. *Let X be a Hilbert space. An operator a in $\mathcal{O}(X, Y)$ is compact if and only if the restriction a_1 of a to the unit ball X_1 of X is continuous with respect to the weak topology of X and the norm-topology of Y.*

Proof. Assume $a \in \mathcal{O}_c(X, Y)$; and let W be the norm-closure of $a(X_1)$. Thus W is norm-compact. We now use the general topological fact that if a space carries two topologies, a compact topology $\mathcal{T}$ and a Hausdorff topology $\mathcal{T}'$, and if $\mathcal{T}' \subset \mathcal{T}$, then $\mathcal{T}' = \mathcal{T}$. It follows that the weak and norm topologies of Y coincide on W. Since $a_1 : X_1 \to W$, and since every element of $\mathcal{O}(X, Y)$ is weakly continuous, we see that a_1 is continuous with respect to the weak topology of X and the norm topology of Y.

Conversely, assume the conclusion of the last sentence. Since the unit ball of a Hilbert space is weakly compact, its continuous image $a_1(X_1)$ must be norm-compact. Therefore a is a compact operator. ∎

Remark. From this proposition it follows that, for a compact transformation a whose domain is a Hilbert space the image $a(X_1)$ of the unit ball of X is not only precompact but actually compact in the norm-topology.

The Spectral Theorem for Compact Normal Operators

15.9. Let X be a Hilbert space, $\{\xi_i\}$ $(i \in I)$ an orthonormal basis of X, and a the normal operator on X defined by

$$a\xi_i = \lambda_i \xi_i \qquad\qquad (i \in I), \qquad (4)$$

where the λ_i are in $\mathbb{C}$ (and $\{\lambda_i\}$ is bounded). It follows easily from the definition of compactness and 15.4 that a will be compact if and only if

$$\lim_i \lambda_i = 0 \qquad\qquad\qquad (5)$$

(that is, for each $\varepsilon > 0$ there is a finite subset F of I such that $i \in I \setminus F \Rightarrow |\lambda_i| < \varepsilon$). The next theorem will show that every compact normal operator is of this form.

15.10. Theorem. *Let a be a compact normal operator on a Hilbert space X. Then there is an orthonormal basis $\{\xi_i\}$ of X such that a is described as in (4) by numbers λ_i satisfying (5). The spectrum of a has no limit point other than 0.*

Proof. Let P be the spectral measure of a. Let us set X_0, X_1, and X_n ($n = 2, 3, \ldots$) equal to the ranges of $P(\{0\})$, $P(\{\lambda : |\lambda| \geq 1\})$, and $P(\{\lambda : (n-1)^{-1} > |\lambda| \geq n^{-1}\})$ respectively. Then

$$X = \sum_{n=0}^{\infty} {}^{\oplus} X_n \qquad \text{(Hilbert direct sum).} \qquad (6)$$

If $n \geq 1$ and $\xi \in X_n$ it follows from II.11.8 that

$$\|a\xi\|^2 = \int_{|\lambda| \geq n^{-1}} |\lambda|^2 \, d\|P\xi\|^2 \lambda \geq n^{-2} \|\xi\|^2.$$

By 15.4 this implies that the X_n with $n \geq 1$ are all finite-dimensional. Since each X_n is a-stable, the spectral theorem for normal operators on finite-dimensional Hilbert spaces assures us that for each $n \geq 1$ there is an orthonormal basis of X_n consisting of eigenvectors of a. This last conclusion is also true for X_0, since a is the zero operator on X_0. Therefore by (6) there is an orthonormal basis $\{\xi_i\}$ of X consisting of eigenvectors of a. If $\{\lambda_i\}$ are the numbers satisfying (4), then by 15.9 (5) must hold.

The last statement is now an easy consequence of what we have already proved, together with 11.1(iii). ∎

15.11. Corollary. *Let a be a compact normal operator on a Hilbert space X, with spectral measure P. If W is a Borel subset of $\mathbb{C}$ whose closure does not contain 0, then $P(W)$ lies in the norm-closed *-subalgebra A of $\mathcal{O}_c(X)$ generated by a.*

Proof. Since (by the last statement of 15.10) $W \cap \mathrm{Sp}(a)$ contains only finitely many points, there is a continuous function $f : \mathbb{C} \to \mathbb{C}$ coinciding with Ch_W on $\mathrm{Sp}(a)$ Thus $f(a) = P(W)$. But, since $f(0) = 0$, $f(a) \in A$. ∎

The *-Representations of $\mathcal{O}_c(X)$

15.12. Let X be a Hilbert space. By 15.6 $\mathcal{O}_c(X)$ is a C^*-algebra in its own right. We are now going to investigate the *-representations of $\mathcal{O}_c(X)$. The most obvious such *-representation is the identity representation sending each a in $\mathcal{O}_c(X)$ into itself. This *-representation, which we will call I, acts of course on X, and is easily seen to be irreducible.

Theorem.   *Every non-degenerate *-representation of $\mathcal{O}_c(X)$ is unitarily equivalent to a direct sum of copies of I. In particular, I is (to within unitary equivalence) the only irreducible *-representation of $\mathcal{O}_c(X)$.*

Proof.   Let $\{\xi_i\}$ $(i \in J)$ be an orthonormal basis of X; and pick out one vector of this basis, say $\xi = \xi_0$. For each i let us define q_i to be the element of $\mathcal{O}_F(X)$ which sends ξ into ξ_i and $\xi^\perp$ into $\{0\}$. Notice that q_0 is projection onto $\mathbb{C}\xi$, that q_i^* sends ξ_i into ξ and $\xi_i^\perp$ into $\{0\}$, and that

$$q_i^* q_j = \delta_{ij} q_0. \tag{7}$$

We now set $q_{ij} = q_i q_j^*$, and notice that

$$q_{ij}\xi_k = \delta_{jk}\xi_i, \qquad q_{ij}q_{kl} = \delta_{jk}q_{il}, \qquad q_{ij}^* = q_{ji}. \tag{8}$$

We claim that the linear span D of the q_{ij} is norm-dense in $\mathcal{O}_c(X)$. Indeed: Any element a of $\mathcal{O}_F(X)$ of rank 1 is of the form $\xi \mapsto (\xi, \eta)\zeta$, where η and ζ are vectors in X. Approximating η and ζ by linear combinations of the ξ_i, we find that a can be approximated in norm by linear combinations of operators of the form $\xi \mapsto (\xi, \xi_j)\xi_i$. But the latter is just q_{ij}. Therefore every element a of $\mathcal{O}_F(X)$ of rank 1 lies in the norm-closure $\bar{D}$ of D. Since the linear span of the set of elements of $\mathcal{O}_F(X)$ of rank 1 is just $\mathcal{O}_F(X)$ itself, we get $\mathcal{O}_F(X) \subset \bar{D}$. Therefore by 15.7 $\mathcal{O}_c(X) \subset \bar{D}$; and the claim is proved.

In view of (8) and the above claim, the q_{ij} act like the canonical matrix units of a total matrix algebra, and the rest of the proof will resemble that of IV.5.7.

Take an arbitrary non-degenerate *-representation T of $\mathcal{O}_c(X)$. Since q_0 is a projection, so is T_{q_0}. Let $\{\eta_\alpha\}$ be an orthonormal basis of range(T_{q_0}). Since $(T_{q_i}\eta_\alpha, T_{q_j}\eta_\beta) = (T_{q_j^*q_i}\eta_\alpha, \eta_\beta) = \delta_{ij}(\eta_\alpha, \eta_\beta)$ (by (7)) $= \delta_{ij}\delta_{\alpha\beta}$, it follows that the $T_{q_i}\xi_\alpha$ (i, α both varying) form an orthonormal set in $X(T)$. We claim that the $T_{q_i}\xi_\alpha$ in fact form an orthonormal basis of $X(T)$. To see this observe that, by the non-degeneracy of T and the denseness of D in $\mathcal{O}_c(X)$, the vectors $T_{q_{ij}}\eta$ ($\eta \in X(T)$; i, j varying) span a dense subspace of $X(T)$. In view of the evident relation $q_{ij} = q_i q_0 q_j^*$, we have $T_{q_{ij}}\eta = T_{q_i}T_{q_0}(T_{q_j^*}\eta)$. But $T_{q_0}(T_{q_j^*}\eta)$ belongs to the closed linear span of the η_α; so $T_{q_{ij}}\eta$ belongs to the closed linear span of the $T_{q_i}\eta_\alpha$ (i, α varying). It follows that the $T_{q_i}\eta_\alpha$ span a dense subspace of $X(T)$. Since they are known to be orthonormal, the claim is proved.

For each fixed α let X_α be the closed linear span of the $T_{q_i}\eta_\alpha$ (i varying). In view of the last claim we have

$$X(T) = \sum_\alpha{}^\oplus X_\alpha. \tag{9}$$

Thus, the theorem will be proved if we can show that each X_α is T-stable, and that T restricted to each X_α is unitarily equivalent to I. But this is easy. Fix α, and let U be the linear isometry of X onto X_α satisfying $U\xi_i = T_{q_i}\eta_\alpha$ for each i. For all indices k, r, i in J we then have:

$$UI_{q_{kr}}\xi_i = Uq_{kr}\xi_i = \delta_{ri}U\xi_k \qquad \text{(by (8))}$$

$$= \delta_{ri}T_{q_k}\eta_\alpha = T_{q_{kr}}T_{q_i}\eta_\alpha \qquad \text{(by (8), since } q_i = q_{i0})$$

$$= T_{q_{kr}}U\xi_i.$$

So U intertwines the restrictions of I and T to D. Since we have seen that D is dense in $\mathcal{O}_c(X)$, it follows that U intertwines I and T. This shows that X_α is T-stable, and that T restricted to X_α is equivalent to I, proving the theorem. ∎

15.13. Let us define a C^*-algebra A to be *elementary* if, for some Hilbert space X, A is $*$-isomorphic with $\mathcal{O}_c(X)$. Theorem 15.12 then says that elementary C^*-algebras have the interesting property of possessing, to within unitary equivalence, *exactly one* irreducible $*$-representation. For separable C^*-algebras we shall obtain (in 23.1) a remarkable converse of this, due to Rosenberg: A separable C^*-algebra A which has (up to unitary equivalence) only one irreducible $*$-representation must be elementary. In this connection, notice that $\mathcal{O}_c(X)$ is separable whenever the Hilbert space X is separable; this follows easily from the first claim in the proof of 15.12.

15.14. Corollary. *If X is a Hilbert space, every $*$-automorphism τ of $\mathcal{O}_c(X)$ is realized by a unitary operator, that is, there exists a unitary operator u on X such that $\tau(a) = u^{-1}au$ for all a in $\mathcal{O}_c(X)$.*

Proof. As in 15.12 let I be the identity representation of $\mathcal{O}_c(X)$. Clearly $a \mapsto I_{\tau(a)}$ is an irreducible $*$-representation of $\mathcal{O}_c(X)$, and so by 15.12 is unitarily equivalent to I. That is, there is a unitary operator u on X such that $\tau(a) = u^{-1}au$ for all a in $\mathcal{O}_c(X)$. ∎

15.15. If X is a Hilbert space, notice that $\mathcal{O}_c(X)$ is *topologically simple*, that is, there are no norm-closed two-sided ideals of $\mathcal{O}_c(X)$ except itself and $\{0\}$. Indeed, let J be a norm-closed two-sided ideal of $\mathcal{O}_c(X)$ containing a non-zero element a. Letting ξ_i, q_i, q_{ij} be as in the proof of 15.12, we can choose indices r, s so that $(\xi_r, a\xi_s) \neq 0$. This implies that $q_r^* a q_s$ is a *non-zero* multiple of q_0, hence (since $a \in J$) that $q_0 \in J$. Therefore $q_{ij} = q_i q_0 q_j^* \in J$ for all i, j. By the first claim in the proof of 15.12 this gives $J = \mathcal{O}_c(X)$.

15.16.   The following analogue of Burnside's Theorem (IV.4.10) has very important consequences.

Proposition.   *Let X be a Hilbert space, and A a norm-closed *-subalgebra of $\mathcal{O}_c(X)$ such that the identity representation of A on X is irreducible. Then $A = \mathcal{O}_c(X)$.*

Proof.   The first step is to show that A contains a projection q of rank 1.

Since $A \neq \{0\}$, A contains a non-zero (compact) Hermitian operator a, and so by 15.11 contains some non-zero projection p (belonging to the range of the spectral measure of a). By 15.4 p is of finite rank. By V.1.21 pAp acts irreducibly on the finite-dimensional space $Y = \text{range}(p)$; and so by Burnside's Theorem (IV.4.10) there is a projection q in pAp of rank 1. This completes the first step of the proof.

Let ξ_0 be a unit vector in the range of the above q; embed ξ_0 in an orthonormal basis $\{\xi_i\}$ of X; and define the q_i, q_{ij} as in the proof of 15.12. For each fixed i, the irreducibility of A gives us a sequence $\{a_n\}$ of elements of A such that $a_n \xi_0 \xrightarrow[n]{} \xi_i$; this implies that $\|a_n q - q_i\| \to 0$, whence $q_i \in A$ for every i. Thus $q_{ij} \in A$ for every i, j. But, as we saw in the proof of 15.12, the q_{ij} span a dense subspace of $\mathcal{O}_c(X)$. Consequently A, being norm-closed, contains all of $\mathcal{O}_c(X)$.   ∎

15.17. Corollary.   *Let A be any C*-algebra, and T an irreducible *-representation of A such that $\text{range}(T)$ contains at least one non-zero compact operator on $X(T)$. Then $\text{range}(T) \supset \mathcal{O}_c(X(T))$.*

Proof.   Define $I = \{a \in A : T_a \in \mathcal{O}_c(X(T))\}$. By 15.6 I is a norm-closed (two-sided) *-ideal of A. Since by hypothesis $I \not\subset \text{Ker}(T)$, V.1.18 shows that $T|I$ is irreducible. Combining this with 8.10, we see that $T(I)$ is a norm-closed *-subalgebra of $\mathcal{O}_c(X(T))$ which acts irreducibly on $X(T)$. So by 15.16 $T(I) = \mathcal{O}_c(X(T))$; and the proof is complete.   ∎

15.18. Corollary.   *Let S and T be two irreducible *-representations of a C*-algebra A such that $\text{Ker}(S) = \text{Ker}(T)$; and assume that $\text{range}(S)$ contains some non-zero compact operator. Then S and T are unitarily equivalent.*

Proof. Put $I = \mathrm{Ker}(S) = \mathrm{Ker}(T)$; and let $J = \{a \in A : S_a$ is a compact opera-
tor$\}$. By hypothesis $S(J) \neq \{0\}$; so by 15.17 $S(J) = \mathcal{O}_c(X(S))$. Thus $J/I \cong$
$\mathcal{O}_c(X(S))$, and J/I is elementary. Now J is a closed two-sided ideal of A not
contained in $\mathrm{Ker}(T)$. So by V.1.18 $T|J$ is irreducible; that is, $a + I \mapsto T_a$
$(a \in J)$ is an irreducible *-representation of J/I. By 15.12 the two irreducible
*-representations $a + I \mapsto S_a$ and $a + I \mapsto T_a$ of J/I are unitarily equivalent;
let u be the linear isometry of $X(S)$ onto $X(T)$ implementing this equivalence.
We then have, for any a in A, b in J, and ξ in $X(S)$.

$$uS_a(S_b\xi) = uS_{ab}\xi$$

$$= T_{ab}u\xi \qquad\qquad \text{(since } ab \in J\text{)}$$

$$= T_a(T_b u\xi) = T_a(u(S_b\xi)).$$

Since $\{S_b\xi : b \in J, \xi \in X(S)\} = X(S)$, this implies that $uS_a = T_a u$. Consequently
u implements the unitary equivalence of S and T. ∎

Remark. We have shown that an irreducible *-representation T of a C^*-
algebra is determined by its kernel to within unitary equivalence, provided
range(T) contains some non-zero compact operator. This is far from true,
however, if range(T) contains no non-zero compact operators; see 17.5.

The next proposition is clearly of Stone-Weierstrass type:

Proposition. *Let $\{X_i\}_{i \in I}$ be an indexed collection of Hilbert spaces. Let A be
the C^*-direct sum $\sum_{i \in I}^{\oplus} \mathcal{O}_c(X_i)$; and let B be a norm-closed *-subalgebra of A
such that:*

(i)   *For each $i \in I$ the *-representation $T^i : b \mapsto b_i$ $(b \in B)$ of B is irreducible;*
(ii) *T^i and T^j are unitarily inequivalent whenever $i, j \in I$ and $i \neq j$.*

Then $B = A$.

Proof. Fix elements $i, j \in I$ with $i \neq j$ and an element β of $\mathcal{O}_c(X_i)$. By 8.18 we
have only to show that there is an element b of B such that $b_i = \beta$ and $b_j = 0$.
Now, by 15.15 and 15.16, range$(T^k) = \mathcal{O}_c(X_k)$ for each k, and the latter is
topologically simple. Since (by the preceding corollary) $\mathrm{Ker}(T^i) \neq \mathrm{Ker}(T^j)$,
the topological simplicity of $\mathcal{O}_c(X_j)$ implies that $\mathrm{Ker}(T^j) \not\subset \mathrm{Ker}(T^i)$; and
hence by the topological simplicity of $\mathcal{O}_c(X_i)$ the image of $\mathrm{Ker}(T^j)$ under T^i is
all of $\mathcal{O}(X_i)$. This proves the existence of the required b. ∎

Hilbert–Schmidt Transformations

15.19.   Let X and Y be two Hilbert spaces. Between $\mathcal{O}_F(X, Y)$ and $\mathcal{O}_c(X, Y)$ there lies an important class of transformations known as the Hilbert–Schmidt transformations, which we will now define.

Let $\{\xi_i\}$ and $\{\eta_r\}$ be orthonormal bases of X and Y respectively. If $a \in \mathcal{O}(X, Y)$, we can form the sum $\sum_i \|a\xi_i\|^2$ (whose value may be ∞); and we have

$$\sum_i \|a\xi_i\|^2 = \sum_{i,r} |(a\xi_i, \eta_r)|^2$$

$$= \sum_{i,r} |(\xi_i, a^*\eta_r)|^2$$

$$= \sum_r \|a^*\eta_r\|^2. \tag{10}$$

Since the bases $\{\xi_i\}$ and $\{\eta_r\}$ can be chosen independently of each other, (10) says that the sum $\sum_i \|a\xi_i\|^2$ depends only on a, not on the particular basis $\{\xi_i\}$. If $\sum_i \|a\xi_i\|^2 < \infty$, we shall say that a is a *Hilbert–Schmidt transformation on X to Y*. The set of all such Hilbert–Schmidt transformations is denoted by $\mathcal{O}_H(X, Y)$. As usual, if $X = Y$ we write $\mathcal{O}_H(X)$ for $\mathcal{O}_H(X, X)$.

The reader will easily verify that $\mathcal{O}_H(X, Y)$ is a (not necessarily closed) linear subspace of $\mathcal{O}(X, Y)$. In fact $\mathcal{O}_H(X, Y)$ is a normed linear space, not only under the operator norm $\| \ \|$, but under the norm $\|\| \ \|\|$ defined by

$$\|\|a\|\| = \left(\sum_i \|a\xi_i\|^2 \right)^{1/2} \qquad (a \in \mathcal{O}_H(X, Y)). \tag{11}$$

$\|\|a\|\|$ is called the *Hilbert–Schmidt norm* of a.

Since any unit vector in X is a member of some orthonormal basis, one checks that

$$\|a\| \leq \|\|a\|\| \qquad \text{for } a \in \mathcal{O}_H(X, Y). \tag{12}$$

If $a \in \mathcal{O}_F(X, Y)$, we can choose the basis $\{\xi_i\}$ of X so that $a\xi_i = 0$ for all but finitely many i. It follows that

$$\mathcal{O}_F(X, Y) \subset \mathcal{O}_H(X, Y). \tag{13}$$

Furthermore, given any a in $\mathcal{O}_H(X, Y)$, and any finite subset D of the index set $\{i\}$, there is an element b of $\mathcal{O}_F(X, Y)$ such that $b\xi_i = a\xi_i$ for $i \in D$ and $b\xi_i = 0$ for $i \notin D$. If D is large enough, $\|\|b - a\|\|$ will be arbitrarily small. Therefore

$$\mathcal{O}_F(X, Y) \text{ is } \|\| \ \|\| - \text{dense in } \mathcal{O}_H(X, Y). \tag{14}$$

Since $\mathcal{O}_c(X, Y)$ is a norm-closed subspace of $\mathcal{O}(X, Y)$ containing $\mathcal{O}_F(X, Y)$, it follows from (12) and (14) that

$$\mathcal{O}_H(X, Y) \subset \mathcal{O}_c(X, Y). \tag{15}$$

Observe that $\mathcal{O}_H(X, Y)$ is in fact a Hilbert space under the norm $\|\| \ \|\|$. Indeed: As before we fix an orthonormal basis $\{\xi_i : i \in I\}$ of X and set $\mathcal{Y} = \sum_{i \in I}^{\oplus} Y$ (Hilbert direct sum). Then $\Phi : a \mapsto \Phi(a) = \sum_{i \in I}^{\oplus} a\xi_i$ is a linear isometry of $\mathcal{O}_H(X, Y)$ into $\mathcal{Y}$. We shall show that range(Φ) = $\mathcal{Y}$. Let $\zeta \in \mathcal{Y}$. For any ξ in X Schwarz's Inequality gives $\sum_i |(\xi, \xi_i)| \|\zeta_i\| \le \|\xi\| \|\zeta\|$; therefore $\sum_i (\xi, \xi_i)\zeta_i$ converges absolutely in Y to an element of norm no greater than $\|\xi\| \|\zeta\|$; and $a : \xi \mapsto \sum_i (\xi, \xi_i)\zeta_i$ is a bounded linear transformation of X into Y. Since $a\xi_i = \zeta_i$, we have $\sum_i \|a\xi_i\|^2 < \infty$, hence $a \in \mathcal{O}_H(X, Y)$ and $\Phi(a) = \zeta$. Thus range(Φ) = $\mathcal{Y}$. It follows that $\mathcal{O}_H(X, Y)$, with the norm $\|\| \ \|\|$, is linearly isometric with $\mathcal{Y}$; and so is itself a Hilbert space. Its inner product $[\ , \ ]$ is given by

$$[a, b] = \sum_i (a\xi_i, b\xi_i) \qquad (a, b \in \mathcal{O}_H(X, Y)). \tag{16}$$

15.20. Now take a single Hilbert space X. It follows from (10) that if $a \in \mathcal{O}_H(X)$, then

$$a^* \in \mathcal{O}_H(X), \qquad \|a^*\| = \|a\|. \tag{17}$$

Furthermore, if $a \in \mathcal{O}_H(X)$, $b \in \mathcal{O}(X)$, and $\{\xi_i\}$ is an orthonormal basis of X, then

$$\sum_i \|ba\xi_i\|^2 \le \|b\|^2 \sum_i \|a\xi_i\|^2,$$

whence

$$ba \in \mathcal{O}_H(X), \qquad \|ba\| \le \|b\| \|a\|. \tag{18}$$

Also, by (17) and (18),

$$ab \in \mathcal{O}_H(X), \qquad \|ab\| \le \|b\| \|a\|. \tag{19}$$

Thus $\mathcal{O}_H(X)$ is a two-sided self-adjoint ideal of $\mathcal{O}(X)$.

From (12), (17), and (18) it follows that $\mathcal{O}_H(X)$, $\|\| \ \|\|$ is not only a Hilbert space but a Banach *-algebra.

Integral Operators

15.21. Let μ and ν be regular Borel measures on the locally compact Hausdorff spaces S and T respectively. As in II.9.5 we form their regular product $\mu \times \nu$ on $S \times T$.

Proposition.   *Let A belong to $\mathscr{L}_2(\mu \times v)$. The equation*

$$(\alpha(f))(s) = \int_T A(s,t)f(t)dvt \qquad (20)$$

($f \in \mathscr{L}_2(v)$; $s \in S$) defines α as a Hilbert–Schmidt transformation from $\mathscr{L}_2(v)$ to $\mathscr{L}_2(\mu)$, whose Hilbert–Schmidt norm is equal to $\|A\|_{\mathscr{L}_2(\mu \times v)}$.

  Every Hilbert–Schmidt transformation from $\mathscr{L}_2(v)$ to $\mathscr{L}_2(\mu)$ is obtained as in (20) from some A in $\mathscr{L}_2(\mu \times v)$.

Proof.   Let f be in $\mathscr{L}_2(v)$ and let W be any μ-measurable subset of S of finite μ-measure. Then both A and $\langle s,t \rangle \mapsto \mathrm{Ch}_W(s)f(t)$ are in $\mathscr{L}_2(\mu \times v)$; so their product is $(\mu \times v)$-summable (see II.2.6). From this and Fubini's Theorem (II.9.8(III)) we conclude that the function $\alpha(f)$ defined by (20) is at least locally μ-summable on S. In fact, by Schwarz's Inequality, for μ-almost all s

$$|(\alpha(f))(s)|^2 \le \|f\|^2 \int_T |A(s,t)|^2 \, dvt.$$

Integrating both sides of this with respect to s and using Fubini's Theorem again, we find that $\alpha(f) \in \mathscr{L}_2(\mu)$, $\|\alpha(f)\| \le \|f\| \|A\|$. It follows that $\alpha \in \mathcal{O}(\mathscr{L}_2(v), \mathscr{L}_2(\mu))$.

  To see that α is a Hilbert–Schmidt transformation, choose orthonormal bases $\{g_i : i \in I\}$ and $\{f_j : j \in J\}$ of $\mathscr{L}_2(\mu)$ and $\mathscr{L}_2(v)$ respectively. By II.15.17 the functions $h_{ij} : \langle s,t \rangle \mapsto g_i(s)\overline{f_j(t)}$ $(i \in I, j \in J)$ form an orthonormal basis of $\mathscr{L}_2(\mu \times v)$. Now

$$(\alpha(f_j), g_i) = \int_S \int_T A(s,t)f_j(t)\overline{g_i(s)}dvt \, d\mu s$$

$$= \int_{S \times T} A(s,t)\overline{h_{ij}(s,t)}d(\mu \times v)\langle s,t \rangle \qquad \text{(by Fubini's Theorem)}$$

$$= (A, h_{ij})_{\mathscr{L}_2(\mu \times v)};$$

and therefore

$$\sum_j \|\alpha(f_j)\|^2 = \sum_{j,i} |(\alpha(f_j), g_i)|^2$$

$$= \sum_{j,i} |(A, h_{ij})|^2$$

$$= \|A\|^2$$

(since $\{h_{ij}\}$ is an orthonormal basis of $\mathscr{L}_2(\mu \times v)$). It follows by 15.19 that α is a Hilbert–Schmidt transformation with $\|\alpha\| = \|A\|$.

Thus the map $A \mapsto \alpha$ defined by (20) is a linear isometry of $\mathscr{L}_2(\mu \times v)$ onto a closed subspace of $\mathcal{O}_H(\mathscr{L}_2(v), \mathscr{L}_2(\mu))$. To see that the map is surjective, it is enough by (14) to see that its range contains every continuous transformation α of rank 1. But such an α is of the form $f \mapsto (f, g)h$, where $g \in \mathscr{L}_2(v)$, $h \in \mathscr{L}_2(\mu)$; and so is defined by (20) when we take $A(s, t) = h(s)\overline{g(t)}$. ∎

Remarks. Transformations α expressed in the form (20) are called *integral operators*; and A is called the *integral kernel* of α.

If S and T are discrete spaces and μ and v are "counting measures" (giving measure 1 to each one-element set), the above proposition asserts what was already implicit in the discussion of 15.18—that $\mathcal{O}_H(X, Y)$ is in natural correspondence with the set of all complex "matrices" $\{\alpha_{ir}\}$ such that $\sum_{i,r}|\alpha_{ir}|^2 < \infty$ ($\{i\}$ and $\{r\}$ being the index set of an orthonormal basis of X and Y respectively).

16. The Sturm–Liouville Theory

16.1. Compact and Hilbert–Schmidt operators were the first kinds of operators on abstract Hilbert space to be studied in depth; and they in turn had their historical origin in the Sturm–Liouville theory of the differential equations arising in physical problems of vibration and heat conduction. In this section, which is a historical digression from our main theme (and never needed hereafter), we give a brief account of the beautiful relation between Hilbert–Schmidt operators and the Sturm–Liouville theory.

16.2. Let a be a positive number, and q a continuous function on $[0, a]$ to the non-negative reals. We consider the problem of finding a function $u = u(x, t)$ $(0 \leq x \leq a; \ 0 \leq t)$ which (a) satisfies the partial differential equation

$$\frac{\partial^2 u}{\partial t^2} = \frac{\partial^2 u}{\partial x^2} - q(x)u, \tag{1}$$

(b) satisfies the boundary conditions

$$u(0, t) = u(a, t) = 0 \qquad \text{for all } t, \tag{2}$$

and (c) gives to u and $\partial u/\partial t$ prescribed values for $t = 0$. Physically, this is the problem of determining the vibrations of a homogeneous string which is stretched between two fixed points and is also subject to a "restoring force" at each point x, of strength $q(x)$, tending to restore the string to its equilibrium position.

The classical procedure for solving this problem, initiated by Fourier in the analogous problem of heat conduction, is the method of separation of variables (see for example Courant and Hilbert [1], Vol. I, Chap. V, §3). This consists in looking for non-trivial solutions of conditions (1) and (2) which have the form of a product of a function of space and a function of time:

$$u(x, t) = v(x)w(t). \qquad (3)$$

Substituting (3) in (1) and (2) we find that, for some real number λ, v must satisfy the ordinary differential equation

$$v''(x) + (\lambda - q(x))v(x) = 0 \qquad (0 \leq x \leq a); \qquad (4)$$

also it must satisfy the boundary conditions

$$v(0) = v(a) = 0. \qquad (5)$$

Now the Sturm–Liouville theory begins with conditions (4) and (5). It considers two main problems: First, what are the numbers λ for which there exist non-trivial functions v satisfying (4) and (5)? (Physically, these numbers λ are the squares of the "natural frequencies" of the vibrating system.) Secondly, if we take all such functions v, obtained from all possible λ, are there "enough" of them? More precisely, do they span a dense subspace of the $\mathscr{L}_2$ space of Lebesgue measure on $[0, a]$? (Only if the answer to this question is "yes" will we be able to use linear combinations of the functions (3) to satisfy the original condition (c), as well as (a) and (b).)

We now state and prove the principal theorem of the Sturm–Liouville theory, which in its parts (I) and (II) answers the two main questions above.

16.3. As before, a is a positive constant; and q is a continuous function on $[0, a]$ to the non-negative reals.

For each real number λ let S_λ denote the real-linear space of all twice continuously differentiable real functions v on the closed interval $[0, a]$ which satisfy (4) and (5). (By the derivatives at the end-points 0 and a we mean of course the right-hand and left-hand derivatives at those points.) Define $\Lambda = \{\lambda \in \mathbb{R} : S_\lambda \neq \{0\}\}$; and let $X = \mathscr{L}_2(m)$ be the Hilbert space of all complex functions on $[0, a]$ which are square-integrable with respect to Lebesgue measure m on $[0, a]$.

Theorem (Sturm–Liouville).

(I) Λ *is countably infinite, and is of the form* $\{\lambda_n : n = 1, 2, \ldots\}$, *where*

$$0 < \lambda_1 < \lambda_2 < \lambda_3 < \ldots.$$

and $\sum_{n=1}^{\infty} \lambda_n^{-2} < \infty$ (*in particular* $\lim_{n\to\infty} \lambda_n = \infty$).

(II) *For each λ in Λ, S_λ is exactly one-dimensional. If for each λ in Λ we choose an element v_λ of S_λ of norm 1 in X, then $\{v_\lambda : \lambda \in \Lambda\}$ is an orthonormal basis of X.*

Proof. We shall have to presuppose the following basic fact about the solutions of second-order linear differential equations on the interval $[0, a]$: Let p_1, p_2, p_3 be continuous real-valued functions on $[0, a]$; and fix a number b in $[0, a]$. For each pair r, s of real numbers, there is a unique twice continuously differentiable real function v on $[0, a]$ such that:

$$v''(x) + p_1(x)v'(x) + p_2(x)v(x) = p_3(x) \tag{6}$$

$(0 \le x \le a)$, and

$$v(b) = r, \qquad v'(b) = s. \tag{7}$$

Using this, we shall first show that S_λ is at most one-dimensional. Fix $\lambda \in \mathbb{R}$; and let v_1 and v_2 be two non-zero functions in S_λ. If $v_1'(0) = 0$, then (since $v_1(0) = 0$) the preceding paragraph applied to (4) says that v_1 coincides with the zero solution of (4); that is, $v_1 \equiv 0$, contrary to hypothesis. So $v_1'(0) \ne 0$; and we have $kv_1'(0) = v_2'(0)$ for some real constant k. Then the preceding paragraph applied to the solutions kv_1 and v_2 of (4) shows that

$$kv_1 = v_2. \tag{8}$$

Since v_1 and v_2 were arbitrary non-zero elements of S_λ, (8) shows that S_λ is at most one-dimensional.

We shall now prove that

$$S_\lambda = \{0\} \qquad \text{for all } \lambda \le 0. \tag{9}$$

Assume that $0 \ne v \in S_\lambda$, where $\lambda \le 0$. By the argument leading to (8), $v'(0) \ne 0$; so, multiplying v by -1 if necessary, we may suppose that

$$v'(0) > 0. \tag{10}$$

Now the set $W = \{x \in [0, a] : 0 < x, v(x) = 0\}$ is non-void (since $a \in W$); so we can define $c = glb(W)$. By (10) and the continuity of v,

$$c > 0, v(c) = 0, v(x) > 0 \qquad \text{for } 0 < x < c. \tag{11}$$

Rolle's Theorem applied to (11) gives

$$v'(c_1) = 0 \qquad \text{for some } 0 < c_1 < c; \tag{12}$$

and the Mean Value Theorem applied to (10) and (12) shows that

$$v''(c_2) < 0 \qquad \text{for some } 0 < c_2 < c_1. \tag{13}$$

But by (4)

$$v''(c_2) = (-\lambda + q(c_2))v(c_2) \tag{14}$$

and the right side of (14) is non-negative (since $\lambda \leq 0$, $q(c_2) \geq 0$, and $v(c_2) > 0$ by (11)). This contradicts (13). So (9) has been proved.

Next we introduce the so-called Green's function G, which converts the "differential" problem (4), (5) into an "integral" problem.

By the first paragraph of this proof, there are non-zero twice continuously differentiable real functions w_1, w_2 on $[0, a]$ satisfying

$$w_i''(x) - q(x)w_i(x) = 0 \qquad (0 \leq x \leq a). \tag{15}$$

$$w_1(0) = 0, \qquad w_2(a) = 0. \tag{16}$$

By the argument leading to (8),

$$w_1'(0) \neq 0. \tag{17}$$

If $w_2(0) = 0$, then by (15) and (16) we would have $0 \neq w_2 \in S_0$, contradicting (9). So

$$w_2(0) \neq 0. \tag{18}$$

Using (15) we get

$$\frac{d}{dx}(w_1 w_2' - w_2 w_1') = w_1 w_2'' - w_2 w_1''$$

$$= w_1 q w_2 - w_2 q w_1 = 0.$$

So $w_1 w_2' - w_2 w_1'$ is constant; in fact

$$(w_1 w_2' - w_2 w_1')(x) = (w_1 w_2' - w_2 w_1')(0)$$

$$= -w_2(0)w_1'(0) \neq 0 \qquad \text{(by (16), (17), (18))}.$$

Thus we can multiply either w_1 or w_2 by a suitable real constant, and assume henceforth that

$$w_1(x)w_2'(x) - w_2(x)w_1'(x) = 1 \tag{19}$$

for all $0 \leq x \leq a$.

We now define the *Green's function* $G: [0, a] \times [0, a] \to \mathbb{R}$ as follows:

$$G(x, \xi) = \begin{cases} w_1(x)w_2(\xi) & \text{if } 0 \leq x \leq \xi \leq a, \\ w_2(x)w_1(\xi) & \text{if } 0 \leq \xi \leq x \leq a. \end{cases}$$

Evidently G is continuous; and

$$G(x, \xi) = G(\xi, x) \qquad\qquad (x, \xi \in [0, a]). \qquad (20)$$

Now the main fact about G is the following claim: Given any continuous function $f: [0, a] \to \mathbb{R}$, there is a unique twice continuously differentiable real function v on $[0, a]$ satisfying

$$v''(x) - q(x)v(x) = f(x) \qquad (0 \le x \le a) \qquad (21)$$

and

$$v(0) = v(a) = 0; \qquad (22)$$

and this v is given by the following formula:

$$v(x) = \int_0^a G(x, \xi)f(\xi)d\xi \qquad (0 \le x \le a). \qquad (23)$$

To prove this, we first observe that a function v satisfying (21) and (22) exists. Indeed: By the first paragraph of the proof there exists a twice continuously differentiable function v^* on $[0, a]$ such that $v^*(a) = 0$ and $v^{*''} - qv^* = f$ on $[0, a]$. By (18) we have $v^*(0) + kw_2(0) = 0$ for some constant k. Therefore $v = v^* + kw_2$ satisfies (21) and (22). To prove the claim, it remains only to show that this function v must satisfy (23).

By (21) and the definition of G we have for $0 \le x \le a$:

$$\int_0^a G(x, \xi)f(\xi)d\xi = \int_0^a G(x, \xi)(v''(\xi) - q(\xi)v(\xi))d\xi$$

$$= w_2(x) \int_0^x w_1(\xi)(v''(\xi) - q(\xi)v(\xi))d\xi$$

$$+ w_1(x) \int_x^a w_2(\xi)(v''(\xi) - q(\xi)v(\xi))d\xi. \qquad (24)$$

Integrating by parts, we get:

$$\int_0^x w_1 v'' \, d\xi = (w_1 v' - vw_1') \Big|_0^x + \int_0^x w_1'' v \, d\xi$$

$$= w_1(x)v'(x) - v(x)w_1'(x) + \int_0^x w_1'' v \, d\xi \qquad (25)$$

(since $w_1(0) = v(0) = 0$); and

$$\int_x^a w_2 v'' \, d\xi = (w_2 v' - v w_2') \bigg|_x^a + \int_x^a w_2'' v \, d\xi$$

$$= v(x) w_2'(x) - w_2(x) v'(x) + \int_x^a w_2'' v \, d\xi \qquad (26)$$

(since $w_2(a) = v(a) = 0$). Combining (15), (24), (25), (26), we have

$$\int_0^a G(x, \xi) f(\xi) d\xi = v(x)[w_1(x) w_2'(x) - w_2(x) w_1'(x)]$$

$$= v(x) \qquad \qquad \text{(by (19),}$$

proving (23).

Recalling 15.21, let us now denote by K the integral operator on X whose integral kernel is G:

$$(Kf)(x) = \int_0^a G(x, \xi) f(\xi) d\xi \qquad (f \in X; \ \ 0 \le x \le a).$$

Since G is continuous, an easy argument by uniform continuity (which we leave to the reader) shows that the functions in the range of K are continuous. It is evident from (20) that K is Hermitian.

Now if $0 \ne \lambda \in \mathbb{R}$ we have

$$S_\lambda = \{v \in X : K(v) = -\lambda^{-1} v\}. \qquad (27)$$

Indeed: Suppose that $v \in X$, $K(v) = -\lambda^{-1} v$. Then $v \in \text{range}(K)$, so v is continuous; and (23) holds when v and f are replaced by $-\lambda^{-1} v$ and v. It follows from the previous claim that the same is true of (21) and (22); that is,

$$v'' + (\lambda - q)v = 0, \qquad v(0) = v(a) = 0.$$

Consequently $v \in S_\lambda$. The same argument in reverse shows that $v \in S_\lambda \Rightarrow K(v) = -\lambda^{-1} v$. So (27) is proved.

Observe that

$$\text{range}(K) \text{ is dense in } X. \qquad (28)$$

To see this, take any twice continuously differentiable real function v on $[0, a]$ satisfying $v(0) = v(a) = 0$. Then $f = v'' - qv$ is continuous on $[0, a]$; and by the above claim $K(f) = v$, whence $v \in \text{range}(K)$. It is easy to see that the set of all such v is dense in X. Therefore (28) holds.

Since K is a Hilbert–Schmidt operator (see 15.21), hence compact (15.19), and is also Hermitian, we can apply to it the Spectral Theorem 15.10. We thus obtain an orthonormal basis $\{v_n\}(n = 1, 2, ...)$ of X and a sequence of real numbers μ_n $(n = 1, 2, ...)$ such that

$$K(v_n) = \mu_n v_n \qquad (n = 1, 2, ...). \qquad (29)$$

Since K is Hilbert–Schmidt, the μ_n satisfy

$$\sum_{n=1}^{\infty} \mu_n^2 < \infty. \qquad (30)$$

If $\mu_n = 0$ for some n, then $v_n \perp$ range(K), contradicting (28); so the μ_n are all non-zero.

Recalling 11.11, we see from (29) and (27) that

$$\Lambda = \{-\mu_n^{-1} : n = 1, 2, ...\}. \qquad (31)$$

This and (30), together with the fact (9) that the elements of Λ are positive, proves (I) of the theorem. We have already shown that S_λ is one-dimensional for $\lambda \in \Lambda$. The rest of statement (II) of the theorem now follows from the fact that the $\{v_n\}$ form a basis of X.   ∎

16.4. Example. The best known special case of 16.3 is of course that in which $q(x) \equiv 0$. The reader will verify that in that case

$$\Lambda = \{n^2 \pi^2 a^{-2} : n = 1, 2, ...\}.$$

For $\lambda = n^2\pi^2 a^{-2}$, S_λ consists of all multiples of the sine function $x \mapsto \sin(n\pi x a^{-1})$.

16.5. Remark. Properly speaking, what we have presented is a special case of the Sturm–Liouville theory. To generalize it, one could begin by replacing (4) by a second-order linear homogeneous equation of less restricted form. However, it can be shown that, under rather general conditions, such an equation can be reduced to the form (4) by suitable changes of variable. Another avenue for generalizing the problem is to replace the boundary conditions (5) by more general homogeneous linear boundary conditions:

$$k_1 v(0) + k_2 v'(0) = 0, \qquad h_1 v(a) + h_2 v'(a) = 0$$

(where k_1, k_2 are not both 0, and h_1, h_2 are not both 0). Here again, both the results and their proof are essentially similar to 16.3. We will not pursue the matter further.

17.   Inductive Limits of C^*-Algebras

17.1.   In this section we will describe a construction which leads to an interesting class of topologically simple C^*-algebras other than $\mathcal{O}_c(X)$ (see 15.15).

17.2.   Let I be a set directed by a relation $\prec$. For each i in I let A_i be a C^*-algebra; and for each pair i, j of elements of I with $i \prec j$ let $F_{ji}: A_i \to A_j$ be a *-homomorphism such that

$$F_{ii} = \text{identity}; \qquad F_{kj}F_{ji} = F_{ki} \qquad (i, j, k \in I; i \prec j \prec k). \qquad (1)$$

Such a system $\{A_i\}, \{F_{ji}\}$ $(i, j \in I; i \prec j)$ will be called a *directed system of C^*-algebras*. By 3.7 the F_{ji} are norm-decreasing.

Let P be the set $\{\langle i, a \rangle : i \in I; a \in A_i\}$; and introduce into P an equivalence relation $\sim$ as follows:

$$\langle i, a \rangle \sim \langle j, b \rangle \Leftrightarrow \lim_k \| F_{ki}(a) - F_{kj}(b) \| = 0.$$

Let $\langle i, a \rangle^\sim$ be the $\sim$-class containing $\langle i, a \rangle$, and $\mathcal{A}' = P/\sim$ the family of all such classes. One verifies that $\mathcal{A}'$ becomes a pre-C^*-algebra under the operations and norm

$$\left. \begin{array}{c} \lambda\langle i, a \rangle^\sim = \langle i, \lambda a \rangle^\sim \qquad (\lambda \in \mathbb{C}), \\[4pt] \langle i, a \rangle^\sim + \langle j, b \rangle^\sim = \langle k, F_{ki}(a) + F_{kj}(b) \rangle^\sim, \\[4pt] \langle i, a \rangle^\sim \cdot \langle j, b \rangle^\sim = \langle k, F_{ki}(a)F_{kj}(b) \rangle^\sim, \\[4pt] \langle i, a \rangle^{\sim *} = \langle i, a^* \rangle^\sim, \end{array} \right\} \qquad (2)$$

$$\|\langle i, a \rangle^\sim\| = \lim_k \| F_{ki}(a) \| \qquad (3)$$

$(i, j \in I; a \in A_i; b \in A_j; i \prec k; j \prec k)$. (It is easy to see that the definitions (2) are independent of k. The limit in (3) exists since the $\| F_{ki}(a) \|$ are non-negative and decreasing in k.) The completion of $\mathcal{A}'$ is a C^*-algebra $\mathcal{A}$ which is referred to as the *inductive limit of the system* $\{A_i\}, \{F_{ji}\}$.

Notice that $\mathcal{A}$ is essentially unaffected if the net $\{A_i\}$ is replaced by a subnet, with a corresponding replacement of the F_{ji}.

For each i, the map $F_i: a \mapsto \langle i, a \rangle^\sim$ is a norm-decreasing *-homomorhism of A_i into $\mathcal{A}$, and

$$F_j F_{ji} = F_i \qquad \text{whenever} \quad i \prec j. \qquad (4)$$

The F_i are the *canonical *-homomorphisms* of A_i into $\mathcal{A}$.

17.3. Suppose in 17.2 that the F_{ji} are all one-to-one (hence by 8.8 are isometries); then we will speak of the directed system $\{A_i\}$, $\{F_{ji}\}$ as *isometric*. In that case of course $\|\langle i, a\rangle^\sim\| = \|a\|$; and the $F_i: A_i \to \mathscr{A}$ are also isometric.

Proposition. *Suppose that the directed system of C^*-algebras $\{A_i\}$, $\{F_{ji}\}$ is isometric, and that each A_i is topologically simple. Then the inductive limit $\mathscr{A}$ is topologically simple.*

Proof. Suppose J is a proper closed two sided ideal of $\mathscr{A}$, so that $\pi: \mathscr{A} \to \mathscr{A}/J$ is a non-zero *-homomorphism of $\mathscr{A}$ into the C^*-algebra $\mathscr{A}/J$ (8.7). It is enough to show that π is an isometry, hence one-to-one; for then $J = \mathrm{Ker}(\pi) = \{0\}$.

Let $\pi_i = \pi \circ F_i$. Since $\bigcup_i F_i(A_i)$ is dense in $\mathscr{A}$, some π_i must be non-zero. Fix i_0 so that $\pi_{i_0} \neq 0$. Observing that $\pi_{i_0} = \pi F_{i_0} = \pi F_i F_{ii_0} = \pi_i F_{ii_0}$ for $i_0 \prec i$, we conclude that $\pi_i \neq 0$ for all $i \succ i_0$. Since π_i is a *-homomorphism and A_i is topologically simple, this implies that π_i is one-to-one and hence (8.8) an isometry for $i \succ i_0$. It follows that $\pi|F_i(A_i)$ is an isometry for $i \succ i_0$. Since $\bigcup_{i \succ i_0} F_i(A_i)$ is dense in $\mathscr{A}$, π must be an isometry. ∎

17.4. As an example of the preceding construction, let S be any non-void set and I the directed set (under inclusion) of all non-void finite subsets of S. If $F \in I$, let M_F be the $F \times F$ total matrix *-algebra; and if $F, G \in I$ and $F \subset G$, let $\Phi_{GF}: M_F \to M_G$ be given by

$$(\Phi_{GF}(a))_{rs} = \begin{cases} a_{rs} & \text{if } r, s \in F, \\ 0 & \text{if either } r \notin F \text{ or } s \notin F \end{cases}$$

$(a \in M_F;\ r,\ s \in G)$. Then $\{M_F\}$, $\{\Phi_{GF}\}$ is an isometric directed system of C^*-algebras. Let $\mathscr{M}$ be its inductive limit. Since each M_F is simple, $\mathscr{M}$ is topologically simple by 17.3. As a matter of fact, the reader will verify without difficulty that $\mathscr{M}$ is just the C^*-algebra $\mathcal{O}_c(X)$ of all compact operators on a Hilbert space X whose Hilbert dimension is $\mathrm{card}(S)$.

17.5. Here is a more novel example, in which the directed system again consists of total matrix *-algebras, but the connecting maps F_{ji} are of a different nature from those of 17.4.

Let $\{X_n\}$ $(n = 1, 2, \ldots)$ be a sequence of Hilbert spaces of finite dimension $d_n > 1$; and let $A_n = \mathcal{O}(X_1) \otimes \cdots \otimes \mathcal{O}(X_n) \cong \mathcal{O}(X_1 \otimes X_2 \otimes \cdots \otimes X_n)$. If $m < n$, we take $F_{nm}: A_m \to A_n$ to be the *-isomorphism

$$a \mapsto a \otimes 1_{nm} \qquad\qquad (a \in A_m),$$

where, 1_{nm} stands for the identity operator on $X_{m+1} \otimes \cdots \otimes X_n$. Thus $\{A_n\}$, $\{F_{nm}\}$ $(n, m = 1, 2, \ldots; m < n)$ is an isometric directed system of C*-algebras. Let $\mathscr{A}$ be its inductive limit. By 17.3 $\mathscr{A}$ is topologically simple. It is called the $(d_1, d_2, \ldots)$ *Glimm algebra*.

Though sharing with $\mathcal{O}_c(X)$ the property of topological simplicity, $\mathscr{A}$ is a quite different kind of C*-algebra from $\mathcal{O}_c(X)$. For one thing $\mathscr{A}$ has a unit element (the image in $\mathscr{A}$ of the unit element of each A_n); whereas $\mathcal{O}_c(X)$ has no unit if X is infinite-dimensional. A more striking difference lies in the fact that, unlike $\mathcal{O}_c(X)$ (see 15.12), $\mathscr{A}$ has uncountably many unitarily inequivalent irreducible *-representations. This will be shown in XII.9.14.

Even from Rosenberg's Theorem (23.1) it will follow that $\mathscr{A}$ must have at least two unitarily inequivalent irreducible *-representations S and T. Since $\mathscr{A}$ is topologically simple, S and T have the same kernel $\{0\}$.

In the special case that each X_n is two-dimensional, the $\mathscr{A}$ constructed above is called the *CAR algebra*. ("CAR" stands for "canonical anticommutation relations," and refers to the application of this algebra in quantum theory.)

18. Positive Functionals

18.1. Having disposed of the *-representations of commutative Banach *-algebras (§10) and elementary C*-algebras (§15), we will now take up one of the most important tools in the theory of *-representations of arbitrary *-algebras, namely positive functionals.

To begin with, let A be an arbitrary *-algebra.

18.2. *Definition*. A linear functional f on A is

 (i) *real* if $f(a^*) = \overline{f(a)}$ for all a in A,
 (ii) *positive* if $f(a^*a) \geq 0$ for all a in A.

It follows immediately from 1.1(1) that f is real if and only if $f(a)$ is real for every Hermitian element a of A. Further, by the argument leading to 1.1(1), every linear functional f on A can be written in one and only one way in the form

$$f = f_1 + if_2,$$

where f_1 and f_2 are real.

18.3. *Remark*. In general a positive functional need not be real. Indeed, suppose A is a *-algebra with trivial multiplication ($ab = 0$ for all a, b). Then every linear functional on A is positive, but not every one is real.

However, a positive functional is "real" on elements of the form ab, as the next proposition shows.

18.4. Proposition. *If p is a positive linear functional on A, then*

$$p(b^*a) = \overline{p(a^*b)} \tag{1}$$

and

$$|p(b^*a)|^2 \le p(a^*a)p(b^*b) \tag{2}$$

for all a, b in A.

Proof. The map $\langle a, b \rangle \mapsto p(b^*a)$ is conjugate-bilinear and positive on $A \times A$. Now apply Schwarz's Inequality and the polarization identity. ∎

18.5. Corollary. *If A has a unit element, every positive linear functional on A is real.*

18.6. By A_1 we shall mean A if A already has a unit; otherwise A_1 will denote the *-algebra obtained by adjoining a unit $\mathbb{1}$ to A (see 1.3).

Definition. A positive linear functional p on A is *extendable* if it can be extended to a positive linear functional on A_1.

If A has a unit, a positive linear functional on A is trivially extendable.

By 18.5 an extendable positive linear functional on A is real. But the converse is false; see 18.8.

18.7. Proposition. *A positive linear functional p on A is extendable if and only if*

(i) *p is real,*
(ii) *there is a non-negative number C such that $|p(a)|^2 \le Cp(a^*a)$ for all a in A.*

Proof. Let p be extendable to a positive linear functional p' on A_1. By (2) applied to p' (with $b = \mathbb{1}$), (ii) holds with $C = p'(\mathbb{1})$. Also p is real by 18.5. So the "only if" part holds.

Conversely, assume (i) and (ii). We may as well suppose A has no unit. Extend p to a linear functional p' on A_1 by setting $p'(\mathbb{1}) = C$. For any a in A and $\lambda \in \mathbb{C}$,

$$p'((a + \lambda\mathbb{1})^*(a + \lambda\mathbb{1})) = p'(|\lambda|^2\mathbb{1} + \bar{\lambda}a + \lambda a^* + a^*a)$$

$$= |\lambda|^2 C + 2\operatorname{Re}(\bar{\lambda}p(a)) + p(a^*a)$$

$$\ge |\lambda|^2 C - 2|\lambda||p(a)| + p(a^*a). \tag{3}$$

Now by (ii) the discriminant $4|p(a)|^2 - 4Cp(a^*a)$ of the right side of (3) is non-positive. It follows that the right side of (3) is non-negative, whence by (3) the functional p' is positive on A_1. So p is extendable, and the "if" part holds. ∎

Remark. We have shown that, if (i) and (ii) hold, and A has no unit, the positive extension of p to A_1 can be made by taking $p(\1) = C$.

Remark. Gil de Lamadrid [1] and Sebestyén [2] have recently shown, independently, that (i) implies (ii) if A is a Banach *-algebra.

18.8.   Example.   Let G be a locally compact group with unit e; and define a linear functional p on the convolution *-algebra $\mathscr{L}(G)$ (see §III.11) by:

$$p(f) = f(e) \qquad\qquad (f \in \mathscr{L}(G)). \qquad (4)$$

The reader will easily check that p is positive and real. But suppose that G is not discrete (so that $\mathscr{L}(G)$ has no unit). Then p is not extendable. For condition 18.7(ii) would then assert that $|f(e)|^2 \le C\|f\|_2^2$ for all f in $\mathscr{L}(G)$; whereas it is clear from the non-discreteness of G that we can find an element f of $\mathscr{L}(G)$ having arbitrarily large $f(e)$ and arbitrarily small $\|f\|_2$.

The positive functional (4) turns out to be extremely important for harmonic analysis on groups.

18.9.   Proposition.   *Let B be a (two-sided) *-ideal of the *-algebra A, b any fixed element of B, and p any positive linear functional on B. Then $q: a \mapsto p(b^*ab)$ is an extendable positive linear functional on A; and, if A has no unit, the positive linear extension of q to A_1 (see 18.6) can be made by setting $q(\1) = p(b^*b)$.*

Proof.   Let A_1 be as in 18.6. Then b is a two-sided *-ideal of A_1 as well as of A; so the equation

$$q_1(a) = p(b^*ab)$$

defines a linear functional q_1 on A_1 which extends q. Since

$$q_1(a^*a) = p((ab)^*ab) \ge 0,$$

q_1 is positive. Evidently $q_1(\1) = p(b^*b)$. ∎

Positive Functionals on Normed *-Algebras

18.10. For the rest of this section we suppose that A is a *normed* *-algebra. We recall from 8.2 the definition of an approximate unit.

18.11. Proposition. *If A has an approximate unit $\{e_i\}$, then any continuous positive linear functional p on A is extendable.*

Proof. For each a in A, we have $ae_i \to a$; so, by the continuity of p and 18.4,

$$p(a^*) = \lim_i p(e_i^* a^*) = \lim_i \overline{p(ae_i)} = \overline{p(a)}.$$

Thus p is real. Again using continuity and 18.4,

$$|p(a)|^2 = \lim_i |p(e_i^* a)|^2 \le \sup_i p(e_i^* e_i) p(a^* a). \tag{5}$$

If k is an upper bound of $\{\|e_i\| : i \text{ varying}\}$ (see 8.2), the right side of (5) is majorized by $k^2 \|p\| p(a^* a)$. Hence by 18.7 p is extendable. ∎

Remark. Taking Remark 18.7 into account, we have shown that, if A has no unit, the positive extension to A_1 of the p of the above proposition can be made by setting $p(1) = k^2 \|p\|$, where k is any upper bound of $\{\|e_i\| : i \text{ varying}\}$.

18.12. Theorem. *A positive linear functional p on a Banach *-algebra A with unit 1 is continuous. In fact $\|p\| \le p(1)$. If $\|1\| = 1$, then $\|p\| = p(1)$.*

Proof. The binomial series

$$(1 - \lambda)^{1/2} = 1 - \frac{\lambda}{2} - \frac{\lambda^2}{8} + \cdots \tag{6}$$

converges absolutely for $|\lambda| < 1$. Hence, replacing 1 by 1 and λ by a Hermitian element a of A with $\|a\| < 1$, we see that

$$1 - \frac{a}{2} - \frac{a^2}{8} + \cdots \tag{7}$$

converges absolutely to a Hermitian element b of A. Multiplying the series (7) by itself termwise (see the Remark following this proof) and using (6), we find

$$b^* b = b^2 = 1 - a. \tag{8}$$

This gives $p(1 - a) = p(b^* b) \ge 0$, or $p(a) \le p(1)$. Likewise $-p(a) = p(-a) \le p(1)$. Thus

$$|p(a)| \le p(1) \qquad \text{whenever } a = a^*, \|a\| < 1. \tag{9}$$

For arbitrary Hermitian a, we apply (9) to $(\|a\| + \varepsilon)^{-1}a$, where ε is an arbitrarily small positive number, getting

$$|p(a)| \leq \|a\|p(\mathbf{1}) \qquad \text{for all Hermitian } a. \tag{10}$$

Now for any b in A we can apply (10) to the Hermitian element b^*b, getting by 18.4

$$|p(b)|^2 \leq p(b^*b)p(\mathbf{1}) \leq \|b^*b\|p(\mathbf{1})^2 \leq \|b\|^2 p(\mathbf{1})^2.$$

Therefore

$$\|p\| \leq p(\mathbf{1}), \tag{11}$$

and p is continuous.

If $\|\mathbf{1}\| = 1$, then $\|p\| \geq p(\mathbf{1})$, and equality holds in (11).   ∎

Remark. At one stage in the above proof we assumed the validity of termwise multiplication of two absolutely convergent series in a Banach algebra. We leave to the reader the task of making and verifying the precise statement of this validity.

18.13.   Corollary.   *An extendable positive linear functional on a Banach *-algebra is continuous.*

Proof.   Apply 18.12 to the Banach *-algebra A_1.   ∎

Combining this with 18.11, we see that *a positive linear functional on a Banach *-algebra with an approximate unit is continuous if and only if it is extendable.*

18.14.   Corollary.   *Let A be a Banach *-algebra, B a (not necessarily closed) *-ideal of A, and p any positive linear functional on B. Then*

$$|p(b^*ab)| \leq \|a\|p(b^*b) ,$$

for all a in A and b in B.

Proof.   Combine 18.12 with 18.9.   ∎

18.15.   For Banach *-algebras with approximate unit the situation is especially pleasant.

Theorem.   *If A is a Banach *-algebra with an approximate unit, then every positive linear functional p on A is continuous and extendable.*

Proof. For a fixed element $a \in A$ the linear functional $x \mapsto p(a^*xa)$ is positive and extendable by 18.9, and therefore continuous by 18.13. So, since

$$4axb = \sum_{k=0}^{3} i^k(a + i^k b^*)x(a + i^k b^*)^* \qquad \text{(polarization identity)},$$

the linear functional $_aP_b\colon x \mapsto p(axb)$ is continuous on A for each $a, b \in A$. Now let $\{x_n\}$ be any sequence in A such that $x_n \to 0$. Applying V.9.4 twice, to A and its "reverse" Banach algebra, we find elements $a, b \in A$ and a sequence $\{y_n\}$ in A such that $y_n \to 0$ and $x_n = ay_nb$ for all n. The continuity of $_aP_b$ now gives $p(x_n) = {_aP_b}(y_n) \to 0$. Therefore p is continuous. ∎

Since every C^*-algebra has an approximate unit (see 8.4) we obtain:

Corollary. *If A is a C^*-algebra, every positive linear functional on A is continuous and extendable.*

Remark. The automatic continuity of a positive linear functional on a Banach algebra with involution and approximate unit can be established even when the involution is not assumed to be continuous. The argument (which will not be given here) depends on a "square-root" lemma due to Ford [1]. We refer the reader to Bonsall and Duncan [1] or Doran and Belfi [1] for more information.

18.16. We conclude this section with an interesting and important application of 18.12. It states that any pre-*-representation of a Banach *-algebra is essentially a *-representation.

Proposition. *Every pre-*-representation T of a Banach *-algebra A satisfies*

$$\|T_a\xi\| \le \|a\|\|\xi\| \qquad (a \in A; \xi \in X(T)). \qquad (13)$$

*Thus each T_a extends to a bounded linear operator T'_a on the completion X' of $X(T)$, and $T'\colon a \mapsto T'_a$ is a *-representation of A.*

Proof. If A has no unit we can adjoin one to get a Banach *-algebra A_1, and then extend T to a pre-*-representation of A_1 by sending the unit of A_1 into the identity operator on $X(T)$. So we may as well assume from the beginning that A has a unit $\mathbf{1}$.

Take a vector ξ in $X(T)$, and set

$$p(a) = (T_a\xi, \xi) \qquad (a \in A).$$

Then $p(a^*a) = (T_{a^*a}\xi, \xi) = (T_a\xi, T_a\xi) \geq 0$; so p is a positive linear functional on A. By 18.12 this implies that for all a in A

$$\|T_a\xi\|^2 = (T_{a^*a}\xi, \xi) = p(a^*a)$$

$$\leq \|a^*a\|p(\mathbf{1}) = \|a^*a\|\,\|T_{\mathbf{1}}\xi\|^2$$

$$\leq \|a\|^2\|\xi\|^2$$

(since $T_{\mathbf{1}}\xi \perp (\xi - T_{\mathbf{1}}\xi)$ and so $\|T_{\mathbf{1}}\xi\| \leq \|\xi\|$). Thus (13) holds.   ∎

19.   Positive Functionals and *-Representations

19.1.   The great virtue of positive functionals is that (under certain general hypotheses) one can construct *-representations from them. This construction is the topic of this section.

We fix a *-algebra A.

19.2.   To motivate what is to come, suppose that T is a *-representation of A, and $\xi \in X(T)$. Then we claim that the function p given by

$$p(a) = (T_a\xi, \xi) \qquad\qquad (a \in A) \qquad (1)$$

is an extendable positive linear functional on A. Indeed: $p(a^*a) = (T_{a^*a}\xi, \xi) = \|T_a\xi\|^2 \geq 0$; so p is positive. If A has no unit, T can always be extended to a *-representation T' of the *-algebra A_1 (obtained by adjoining a unit $\mathbf{1}$ to A) by setting $T'_{\mathbf{1}}$ equal to the identity operator on $X(T)$. The functional $p': a \mapsto (T'_a\xi, \xi)$ $(a \in A_1)$ is then positive on A_1 and extends p. So p is extendable.

In this section we shall be interested in the converse question: Given an extendable positive linear functional p on A, is p derived from some *-representation T and some vector ξ in $X(T)$, as in (1)? We shall find that if A is a *Banach* *-algebra the answer is always "yes" (see 19.5 and 19.6).

19.3.   Fix any positive functional p on the *-algebra A. Even without any further hypotheses on p or A we can at least construct a pre-*-representation of A (see 9.1).

Let I denote $\{a \in A : p(a^*a) = 0\}$. We assert that

$$I = \{a \in A : p(ba) = 0 \text{ for all } b \text{ in } A\}. \qquad (2)$$

Indeed: Clearly I contains the right side of (2). If $a \in I$, then 18.4 gives $|p(ba)|^2 \leq p(bb^*)p(a^*a) = 0$ for all b. So (2) holds.

It follows from (2) that I is a left ideal of A. We call it the *null ideal* of p.

Let X stand for the quotient space A/I, and $\rho: A \to X$ for the quotient map. Notice that X is a pre-Hilbert space under the inner product $(\ ,\ )$ given by

$$(\rho(a), \rho(b)) = p(b^*a). \tag{3}$$

Indeed, the legitimacy of the definition (3) follows from (2) and the fact (18.4) that $p(b^*a) = \overline{p(a^*b)}$. Evidently $(\ ,\ )$ is linear in the first variable and conjugate-linear in the second. Its positivity follows from that of p. Finally, if $(\rho(a), \rho(a)) = 0$, then $a \in I$, so $\rho(a) = 0$. Thus X is a pre-Hilbert space.

Let T' be the natural representation of A on $X = A/I$ (see IV.3.9):

$$T'_a(\rho(b)) = \rho(ab).$$

If $a, b, c \in A$, we have

$$(T'_a\rho(b), \rho(c)) = p(c^*ab) = p((a^*c)^*b) = (\rho(b), T'_{a^*}\rho(c)).$$

Thus

$$(T'_a\xi, \eta) = (\xi, T'_{a^*}\eta) \qquad (a \in A; \xi, \eta \in X). \tag{4}$$

So T' is a pre-*-representation of A acting on X.

Now let X_c be the Hilbert space completion of $X, (\ ,\ )$. In order to be able to extend each T'_a to a bounded linear operator T_a on X_c, and thus to obtain a *-representation T of A on X_c, we must know that each T'_a is continuous on X with respect to $(\ ,\ )$. Clearly this will be the case if and only if the following Condition (R) holds:

Definition. We say that the positive linear functional p on A satisfies *Condition (R)* if, for each a in A, there is a non-negative constant C_a (depending on a) such that

$$p(b^*a^*ab) \le C_a p(b^*b) \qquad \text{for all } b \text{ in } A. \tag{5}$$

Definition. If p satisfies Condition (R), the *-representation T of A on X_c, defined above, is said to be *generated by p*.

Remark 1. In Chapter XI we shall encounter a generalization of this construction of T from p, very important in the theory of induced group representations.

Remark 2. If $X = A/I$ is finite-dimensional, p automatically satisfies Condition (R) and generates a *-representation of A.

19.4.   *Remark*.   Later on (19.16) we will give examples showing that neither of the two properties, extendability and Condition (R), implies the other. Notice what happens if p satisfies Condition (R) but is not extendable. In that case p generates a *-representation T of A; but by 19.2 there is no vector ξ in $X(T)$ such that p is given by (1). On the other hand, if both Condition (R) and extendability hold, we shall see in 19.6 that there does exist a ξ in $X(T)$ satisfying (1).

19.5.   It is an interesting and important fact that, if A is a Banach *-algebra, Condition (R) always holds.

Proposition.   *If A is a Banach *-algebra, Condition (R) holds for every positive linear functional p on A.*

Proof.   Applying 18.14 with $B = A$ and a replaced by a^*a, we obtain (5), with $C_a = \|a^*a\|$.   ∎

Remark.   Thus every positive linear functional on a Banach *-algebra A generates a *-representation T of A as in 19.3.

19.6.   We are now ready to answer the converse question raised in 19.2.

Proposition.   *Suppose that p is a positive linear functional on A which satisfies Condition (R) and is extendable. Then the *-representation T of A generated by p is cyclic, and there is a cyclic vector ξ in $X(T)$ such that p is given by (1).*

Proof.   Let A_1 be A if A has a unit, otherwise the *-algebra obtained by adjoining a unit $\mathbb{1}$. Let p' be a positive extension of p to A_1. If I and I' are the null ideals of p and p' respectively, we have $I = I' \cap A$. Thus $X = A/I$ may be regarded as a linear subspace of $X' = A_1/I'$; and the inner products (3) for X and X' coincide on X. Hence the completion X_c of X is a closed linear subspace of the completion X'_c of X'.

Let $\rho: A_1 \to X'$ be the quotient map; and let us define ξ to be the orthogonal projection of $\rho(\mathbb{1})$ into X_c. We shall show that ξ has the properties required in the proposition.

Since X is dense in X_c, there is a sequence $\{b_n\}$ of elements of A such that

$$\rho(b_n) \to \xi \quad \text{in} \quad X_c. \tag{6}$$

Then, for any a in A,

$$p(a) = (\rho(\mathbf{1}), \rho(a^*)) = (\xi, \rho(a^*)) \qquad \text{(since } \rho(a^*) \in X_c)$$

$$= \lim_n (\rho(b_n), \rho(a^*)) = \lim_n p(ab_n)$$

$$= \lim_n (\rho(ab_n), \rho(\mathbf{1})) = \lim_n (\rho(ab_n), \xi) \qquad \text{(since } \rho(ab_n) \in X_c)$$

$$= \lim_n (T_a \rho(b_n), \xi) = (T_a \xi, \xi).$$

Thus (1) holds.

The preceding equalities have shown that

$$p(a) = \lim_n p(ab_n) \qquad\qquad (a \in A). \qquad (7)$$

Taking adjoints and using the reality of p, we also have

$$p(a) = \lim_n p(b_n^* a) \qquad\qquad (a \in A). \qquad (8)$$

Further, by (1) and (6), we have for $a \in A$

$$p(b_n^* a^* a b_n) = (T_{a^* a} \rho(b_n), \rho(b_n))$$

$$\underset{n}{\to} (T_{a^* a} \xi, \xi) = p(a^* a). \qquad (9)$$

Combining (7), (8), and (9), we get for every a in A

$$\| \rho(ab_n - a) \|^2 = p((ab_n - a)^*(ab_n - a))$$

$$= p(b_n^* a^* a b_n) - p(a^* a b_n) - p(b_n^* a^* a) + p(a^* a)$$

$$\to 0 \qquad \text{as } n \to \infty,$$

or

$$T_a \rho(b_n) \to \rho(a) \qquad \text{in } X_c \text{ as } n \to \infty. \qquad (10)$$

But, by (6) and the continuity of T_a on X_c, $T_a(\rho(b_n)) \to T_a \xi$. Combining this with (10) we have

$$T_a \xi = \rho(a) \qquad\qquad (a \in A). \qquad (11)$$

From (11) and the denseness of $\rho(A) (=X)$ in X_c, it follows that ξ is a cyclic vector for T. The proof is now complete. ∎

19.7. *Definition.* By a *cyclic pair* for the *-algebra A we shall mean a pair $\langle T, \xi \rangle$, where T is a *-representation of A and ξ is a cyclic vector for T.

Proposition. *Let p be a positive linear functional on A. Then the following two conditions are equivalent:*

(i) *p is extendable and satisfies Condition (R);*

(ii) *there is a cyclic pair $\langle T, \xi \rangle$ for A such that* (1) *holds.*

Proof. In view of 19.2 and 19.6, all that remains to be proved is that (1) implies Condition (R). But from (1) we get $p(b^*a^*ab) = (T_{b^*a^*ab}\xi, \xi) = (T_{a^*a}T_b\xi, T_b\xi) \leq \| T_{a^*a} \| \, \| T_b\xi \|^2 = \| T_{a^*a} \| p(b^*b)$, proving Condition (R). ∎

Corollary. *Let A be a Banach *-algebra with an approximate unit. Then every positive linear functional on A is obtained by* (1) *from some cyclic pair $\langle T, \xi \rangle$.*

Proof. By 18.13, 18.15 and 19.5. ∎

19.8. To what extent is the cyclic pair $\langle T, \xi \rangle$ of 19.7(ii) uniquely determined by p?

We shall say that two cyclic pairs $\langle T, \xi \rangle$ and $\langle T', \xi' \rangle$ for A are *equivalent* if T and T' are unitarily equivalent under a linear isometry Φ such that $\Phi(\xi) = \xi'$.

The next proposition asserts that the p of 19.7 uniquely determines the corresponding cyclic pair up to equivalence.

Proposition. *Two cyclic pairs $\langle T, \xi \rangle$ and $\langle T', \xi' \rangle$ for the *-algebra A are equivalent if and only if*

$$(T_a\xi, \xi) = (T'_a\xi', \xi') \qquad \text{for all a in A.} \tag{12}$$

Proof. The necessity of (12) is obvious. Conversely, assume that (12) holds. Then, for any a in A,

$$\| T_a\xi \|^2 = (T_{a^*a}\xi, \xi) = (T'_{a^*a}\xi', \xi') = \| T'_a\xi \|^2.$$

It follows that the equation

$$\Phi T_a\xi = T'_a\xi' \qquad\qquad (a \in A) \tag{13}$$

defines a linear isometry Φ of $T(A)\xi$ onto $T'(A)\xi'$. Since ξ and ξ' are cyclic vectors, these spaces are dense in $X(T)$ and $X(T')$ respectively; so Φ extends to an isometry of $X(T)$ onto $X(T')$. For all a, b in A, (13) gives $\Phi T_b(T_a\xi) = \Phi(T_{ba}\xi) = T'_{ba}\xi' = T'_b\Phi(T_a\xi)$; so that

$$\Phi T_b = T'_b\Phi. \tag{14}$$

Also, by (13) and (14), $T'_a(\Phi\xi - \xi') = 0$ for all a in A. Since T' is non-degenerate, this gives

$$\Phi\xi = \xi'. \tag{15}$$

Now (14) and (15) assert that the cyclic pairs $\langle T, \xi \rangle$ and $\langle T', \xi' \rangle$ are equivalent. ∎

Remark. In the above proof of the sufficiency of (12), all we really needed was the validity of (12) for elements a of the form b^*b ($b \in A$). This verifies the following remark which will be useful later:

If p and q are two extendable positive linear functionals on A satisfying Condition (R), and if $p(b^*b) = q(b^*b)$ for all b in A, then $p = q$.

19.9. We summarize the results of this section:

Theorem. *Let A be a *-algebra. There is a one-to-one correspondence between the family of all equivalence classes of cyclic pairs $\langle T, \xi \rangle$ for A and the family of all extendable positive linear functionals p on A which satisfy Condition (R). The correspondence is given (in the direction $\langle T, \xi \rangle \mapsto p$) by:*

$$p(a) = (T_a \xi, \xi) \qquad\qquad (a \in A).$$

If A is a Banach *-algebra with an approximate unit, then every positive linear functional is continuous, extendable, and satisfies condition (R).

Miscellaneous Consequences

19.10. Here is an application of 19.8. Let e be a self-adjoint idempotent element of the *-algebra A. If T is a *-representation of A, then $T^{(e)}: b \mapsto T_b | \mathrm{range}(T_e)$ ($b \in B$) is a *-representation of the *-subalgebra $B = eAe$ of A, acting on $\mathrm{range}(T_e)$ (see V.1.21). In fact, by V.1.21, if T is irreducible and $T_e \neq 0$, then $T^{(e)}$ is irreducible.

Proposition. *In the above context suppose that S and T are two irreducible *-representations of A with $S_e \neq 0$, $T_e \neq 0$. Then S and T are unitarily equivalent if and only if $S^{(e)}$ and $T^{(e)}$ are unitarily equivalent.*

Proof. The "only if" part is obvious.

Assume that $S^{(e)} \cong T^{(e)}$. Then there are non-zero vectors ξ, η in $X(S^{(e)})$ and $X(T^{(e)})$ respectively such that

$$(S_b^{(e)} \xi, \xi) = (T_b^{(e)} \eta, \eta) \qquad\qquad (b \in B). \qquad (16)$$

Let a be any element of A. Since $eae \in B$, $S_e \xi = \xi$, and $T_e \eta = \eta$, (16) gives

$$(S_a \xi, \xi) = (S_{eae}^{(e)} \xi, \xi) = (T_{eae}^{(e)} \eta, \eta)$$

$$= (T_a \eta, \eta). \qquad (17)$$

Since S and T are irreducible ξ and η are cyclic vectors. So, by 19.8, (17) implies that $S \cong T$. ∎

19.11. We shall show that a *-representation of a *-ideal of a Banach *-algebra A can be extended to all of A.

Proposition. *Let A be a Banach *-algebra, B a (not necessarily closed) *-ideal of A, and T a non-degenerate *-representation of B. Then T can be extended in one and only one way to a *-representation T' of A (acting in $X(T)$). In particular T is norm-continuous.*

Proof. We shall first assume that T is cyclic, with cyclic vector ξ. Let $p(b) = (T_b\xi, \xi)$ $(b \in B)$. By 18.14 we have

$$\|T_{ab}\xi\|^2 = p(b^*a^*ab)$$

$$\leq \|a\|^2 p(b^*b) = \|a\|^2 \|T_b\xi\|^2 \tag{18}$$

for each a in A, b in B. For each a in A, the inequality (18) (together with the fact that $\{T_b\xi : b \in B\}$ is dense since ξ is a cyclic vector) enables us to define a linear operator T'_a on $X(T)$ satisfying

$$T'_a T_b \xi = T_{ab}\xi \qquad\qquad (b \in B), \tag{19}$$

$$\|T'_a\| \leq \|a\|. \tag{20}$$

Given a in A and c in B, we have by (19)

$$T'_a T_c(T_b\xi) = T'_a T_{cb}\xi = T_{acb}\xi = T_{ac}(T_b\xi)$$

for any b in B, showing that

$$T'_a T_c = T_{ac} \qquad\qquad (a \in A; c \in B). \tag{21}$$

Also, if $a \in A$, (19) gives, for all b, c in B

$$(T'_a T_b\xi, T_c\xi) = (T_{ab}\xi, T_c\xi) = (\xi, T_{(ab)^*c}\xi)$$

$$= (\xi, T_{b^*}T_{a^*c}\xi) = (T_b\xi, T'_{a^*}T_c\xi);$$

from which, by the cyclicity of ξ, we obtain

$$(T'_a)^* = T'_{a^*}. \tag{22}$$

Now (21) and (22) imply that T' is a *-representation of A extending T. This proves the existence of T' if T is cyclic.

Now drop the assumption that T is cyclic. By 9.16 $T \cong \sum_{i \in I}^{\oplus} T^i$, where each T^i is a cyclic *-representation of B. Extending each T^i to a *-representation

$T^{i'}$ of A by the preceding paragraph, and noting (20), we form the direct sum $T' = \sum_{i \in I}^{\oplus} T^{i'}$. This is evidently an extension of T to A.

Since T is non-degenerate, the equation (21), which any extension T' of T must satisfy, uniquely determines T'. So T' is unique. ∎

Remark. Proposition 19.11 may not hold if A is merely a *-algebra. As an example, let A be the *-algebra of all (possibly unbounded) complex sequences with pointwise operations; and let B be the *-ideal of A consisting of those sequences a which vanish eventually. Let X be the Hilbert space of all complex square-summable sequences, and S the non-degenerate *-representation of B on X by multiplication operators. Then S *cannot* be extended to a *-representation of A acting on X. (Indeed: Suppose it can. Let $a \in A$ be the sequence given by $a_n = n$ $(n = 1, 2, 3, \ldots)$. Fix a positive integer k; and let $\xi \in X, \xi_n = \delta_{kn}$, and $b \in B, b_n = \delta_{kn}$. Then $S_b \xi = \xi, ab = kb$; so $S_a \xi = S_a S_b \xi = S_{ab} \xi = k S_b \xi = k\xi$. Since k is arbitrary, S_a must be unbounded.)

19.12. Proposition. *Let A be a Banach *-algebra having an approximate unit $\{e_i\}$ such that $\|e_i\| \le 1$ for all i; and let p be a positive linear functional on A, corresponding via 19.9 with the cyclic pair $\langle T, \xi \rangle$. Then:*

(I) $\|p\| = \|\xi\|^2 = \lim_i p(e_i) = \lim_i p(e_i^* e_i)$.

(II) *If A has no unit, A_1 is the *-algebra obtained by adjoining a unit to A, and $\| \ \|_1$ is any Banach *-algebra norm on A_1 which extends $\| \ \|$, then p can be positively extended to A_1, $\| \ \|_1$ without increase of norm.*

Proof.

(I) By V.4.15 (and 3.8) $T_{e_i}\eta \to \eta$ for all η in $X(T)$. It follows that $p(e_i) = (T_{e_i}\xi, \xi) \underset{i}{\to} \|\xi\|^2$ and $p(e_i^* e_i) = \|T_{e_i}\xi\|^2 \underset{i}{\to} \|\xi\|^2$. Since $\|e_i\| \le 1$, $|p(e_i)| \le \|p\|$ for all i. Consequently $\|\xi\|^2 \le \|p\|$. On the other hand $|p(a)| = |(T_a\xi, \xi)| \le \|T_a\| \|\xi\|^2 \le \|a\| \|\xi\|^2$ (3.8); so $\|p\| \le \|\xi\|^2$. Thus (I) holds.

(II) Extend T to a *-representation T' of A_1 by setting $T'_{\mathbb{1}}$ equal to the identity operator; and let p' be the positive extension $a \mapsto (T'_a\xi, \xi)$ of p. Then by 18.12 and (I) $\|p'\| \le p'(\mathbb{1}) = \|\xi\|^2 = \|p\| \le \|p'\|$. ∎

Remark. By 8.4, any C^*-algebra A has an approximate unit $\{e_i\}$ with $\|e_i\| \le 1$, and hence the above proposition is applicable to it.

19.13. Corollary. *Let A be as in 19.12. If $p_1, \ldots, p_n$ are positive linear functionals on A, then $\|p_1 + \cdots + p_n\| = \|p_1\| + \cdots + \|p_n\|$.*

19.14.   Proposition.   *Let A be a Banach *-algebra, and I its reducing ideal (see 10.2); and let E stand for the set of all extendable positive linear functionals on A. Then the following three conditions on an element a of A are equivalent:*

(i) $a \in I$;

(ii) $p(a^*a) = 0$ *for all p in E*;

(iii) $p(a) = 0$ *for all p in E*.

Suppose in addition that A has an approximate unit $\{e_i\}$; *and let* $\| \ \|_c$ *be the C*-norm of A/I (see 10.3). Then, for* $a \in A$,

$$\|a + I\|_c = \sup\{p(a^*a): p \in E, \lim_i p(e_i) = 1\} \tag{23}$$

Proof.   The first part follows very easily from 19.9 (and 19.5).

As for the second part, if p corresponds as in 19.9 with the cyclic pair $\langle T, \xi \rangle$, the condition $\lim p(e_i) = 1$ means that $\|\xi\| = 1$ (since $T_{e_i} \to 1$ strongly as in the proof of 19.12). From this (23) follows.   ∎

19.15.   Corollary.   *A Banach *-algebra A is reduced (see 10.2) if and only if, for each non-zero a in A, there is an extendable positive linear functional p on A such that* $p(a^*a) \neq 0$.

19.16.   Examples.   We end the section with three examples. The first two will show that neither of the two properties of extendability and condition (R) implies the other.

(I)   Let G and p be as in Example 18.8. We saw in 18.8 that p is positive but not extendable. Now, if $f, g \in \mathscr{L}(G)$, we have

$$p(g^* * f^* * f * g) = (g^* * f^* * f * g)(e)$$

$$= \|f * g\|_2^2 \leq \|f\|_1^2 \|g\|_2^2 \qquad \text{(see III.11.14(16))}$$

$$= \|f\|_1^2 p(g^* * g).$$

Thus p satisfies Condition (R).

(II)   Let A be the (commutative) *-algebra of all polynomials with complex coefficients in one real variable, multiplication being defined pointwise and the *-operation being complex conjugation. The equation

$$p(f) = \int_0^\infty f(x)e^{-x}\, dx \qquad (f \in A)$$

clearly defines a positive linear functional p on A (extendable since A has a unit). We claim that p does not satisfy Condition (R).

Indeed, let n be a positive integer, and set $a(x) = x$, $b(x) = x^n$ ($x \in \mathbb{R}$). Integrating by parts, we check that

$$p(b^*a^*ab) = (2n + 1)(2n + 2)p(b^*b).$$

It follows that for this a there can be no constant C_a satisfying the definition of condition (R).

(III) If p satisfies Condition (R) but is not extendable, the *-representation which it generates may fail to be cyclic or even non-degenerate. Here is a very simple example. Put $A = \mathbb{C}^2$ (as a linear space) with multiplication and involution:

$$\langle u, v \rangle \langle u', v' \rangle = \langle 0, uu' \rangle,$$

$$\langle u, v \rangle^* = \langle \bar{u}, \bar{v} \rangle.$$

Then A is a *-algebra, and the positive linear functional $p \colon \langle u, v \rangle \mapsto v$ on A generates the one-dimensional zero representation of A.

20. Indecomposable Positive Functionals and Irreducible *-Representations

20.1. In the correspondence established in Theorem 19.9, how are the properties of the cyclic pair $\langle T, \xi \rangle$ correlated with properties of p? For example, what property of p is correlated with *irreducibility* of T? The main object of this section is to answer this question.

As before we fix a *-algebra A. By $\mathscr{P}(A)$ we denote the set of all extendable positive linear functionals on A satisfying Condition (R). Thus, if A were a Banach *-algebra with an approximate unit, $\mathscr{P}(A)$ would be (see 19.9) just the set of all positive linear functionals on A.

Clearly $\mathscr{P}(A)$ is a cone; that is, it is closed under addition and multiplication by non-negative scalars.

20.2. Definition. If p, q are in $\mathscr{P}(A)$, we say that p is *subordinate to* q (in symbols, $p \prec q$) if there is a complex number λ such that

$$\lambda q - p \text{ is a positive functional on } A. \tag{1}$$

Observe that the λ of (1), if it exists, may always be chosen non-negative. One verifies that the only element of $\mathscr{P}(A)$ subordinate to 0 is 0.

20.3. Theorem. *Let p be an element of $\mathscr{P}(A)$ corresponding via 19.9 with a cyclic pair $\langle T, \xi \rangle$ for A (so that 19.2(1) holds). Then there is a one-to-one correspondence $q \leftrightarrow B$ between the set of all q in $\mathscr{P}(A)$ which are subordinate to p, and the set of all positive bounded linear operators B on $X(T)$ such that*

$$BT_a = T_a B \qquad \text{for all } a \text{ in } A. \tag{2}$$

The correspondence (in the direction $B \mapsto q$) is given by

$$q(a) = (T_a B \xi, \xi) \qquad\qquad (a \in A). \tag{3}$$

Proof. Let B be a positive bounded linear operator on $X(T)$ satisfying (2). By 11.16 the positive square root $B^{1/2}$ also commutes with all T_a. So, if q is defined by (3),

$$q(a) = (T_a B^{1/2} B^{1/2} \xi, \xi) = (T_a B^{1/2} \xi, B^{1/2} \xi);$$

and hence $q \in \mathscr{P}(A)$. Putting $\lambda = \|B\|$, we have $(B\eta, \eta) \le \lambda(\eta, \eta)$ for all η; so, if $a \in A$, $q(a^*a) = (T_{a^*a} B \xi, \xi) = (BT_a \xi, T_a \xi) \le \lambda(T_a \xi, T_a \xi) = \lambda p(a^*a)$; therefore $\lambda p - q$ is positive. Thus $q \prec p$.

We observe that q uniquely determines B. Indeed, $q(b^*a) = (T_{b^*a} B \xi, \xi) = (BT_a \xi, T_b \xi)$; and the $T_a \xi$ and $T_b \xi$ each run over a dense subset of $X(T)$. So q determines $(B\eta, \eta')$ for all η and η' in $X(T)$, and hence B itself.

It remains only to show that every q in $\mathscr{P}(A)$ subordinate to p is given by (3) for some positive operator on $X(T)$ satisfying (2).

So let q be an element of $\mathscr{P}(A)$ and λ a non-negative number satisfying

$$0 \le q(a^*a) \le \lambda p(a^*a) \qquad\qquad (a \in A). \tag{4}$$

Let $X' = \{T_a \xi : a \in A\}$. By (4) and Schwarz's Inequality, the equation

$$\beta(T_a \xi, T_b \xi) = q(b^*a) \qquad\qquad (a, b \in A) \tag{5}$$

defines a conjugate-bilinear functional β on $X' \times X'$; and

$$\beta(\eta, \eta) \ge 0 \qquad\qquad (\eta \in X'). \tag{6}$$

Further, by Schwarz's Inequality and (4),

$$|\beta(T_a \xi, T_b \xi)|^2 \le q(a^*a)q(b^*b)$$

$$\le \lambda^2 p(a^*a)p(b^*b)$$

$$= \lambda^2 \|T_a \xi\|^2 \|T_b \xi\|^2.$$

Thus β extends to a continuous conjugate-bilinear functional on $X(T) \times X(T)$, and so gives rise to a bounded linear operator B on $X(T)$ satisfying

$$(BT_a \xi, T_b \xi) = \beta(T_a \xi, T_b \xi) = q(b^*a) \tag{7}$$

$(a, b \in A)$. By (6) B is positive. If $a, b, c \in A$ and $\eta = T_a \xi$, $\eta' = T_b \xi$,

$$(BT_c \eta, \eta') = (BT_{ca} \xi, T_b \xi) = q(b^* ca) \qquad \text{(by (7))}$$

$$= q((c^* b)^* a) = (BT_a \xi, T_{c*b} \xi)$$

$$= (T_c B \eta, \eta').$$

Since η and η' run over a dense subset of $X(T)$, this implies that B satisfies (2).

To complete the proof we must show that q is related to B by (3). Define

$$q'(a) = (T_a B \xi, \xi) \qquad (a \in A).$$

We pointed out at the beginning of the proof that $q' \in \mathscr{P}(A)$. By (7) and (2)

$$q(b^* b) = (BT_b \xi, T_b \xi) = (T_{b*b} B \xi, \xi) = q'(b^* b)$$

for all b in A. So by the remark at the end of 19.8 $q' = q$; and the proof is complete. ∎

20.4.   Definition.   A non-zero element p of $\mathscr{P}(A)$ is *indecomposable* (or *pure*) if the only elements of $\mathscr{P}(A)$ which are subordinate to p are the multiples λp of p $(0 \le \lambda \in \mathbb{R})$.

The following theorem now answers the question raised at the beginning of this section.

Theorem.   *Let p be a non-zero element of $\mathscr{P}(A)$, and $\langle T, \xi \rangle$ the cyclic pair corresponding with p by 19.9. Then T is irreducible if and only if p is indecomposable.*

Proof.   In the correspondence $q \leftrightarrow B$ of 20.3 p goes into the identity operator $\mathbb{1}$ on $X(T)$. By 14.1 T is irreducible if and only if the $\lambda\mathbb{1}$ $(0 \le \lambda \in \mathbb{R})$ are the only positive operators commuting with all T_a. By 20.3 this amounts to saying that the λp $(\lambda \ge 0)$ are the only elements of $\mathscr{P}(A)$ subordinate to p, that is, that p is indecomposable. ∎

20.5.   If the *-algebra A has no unit, and A_1 is the *-algebra obtained by adjoining a unit to A, it follows from 20.4 that every indecomposable element of $\mathscr{P}(A)$ extends to an indecomposable element of $\mathscr{P}(A_1)$. Conversely, if p is an indecomposable element of $\mathscr{P}(A_1)$, then either p vanishes on A or $p|A$ is an indecomposable element of $\mathscr{P}(A)$. The reader should supply the details of the argument.

20.6.   Proposition.   *If A is Abelian, the indecomposable elements of $\mathscr{P}(A)$ are just the positive multiples of the (non-zero) one-dimensional *-representations of A.*

Proof. This follows from 20.4 and 14.3. ∎

Ergodic Positive Functionals

20.7. The notion of indecomposability has an interesting generalization
—namely ergodicity—to the context where A is acted upon by a group.

As before let A be a *-algebra; and let G be a (discrete) group, with unit e,
acting by *-automorphisms on A (that is, A is a left G-space such that, for
each x in G, $a \mapsto xa$ is a *-automorphism of A).

A linear functional p on A is *G-invariant* if $p(xa) = p(a)$ for all x in G and a
in A.

Proposition*. *Let* $\langle T, \xi \rangle$ *be a non-zero cyclic pair for* A *such that the
corresponding element* $p : a \mapsto (T_a \xi, \xi)$ *of* $\mathscr{P}(A)$ *is G-invariant.*

(I)   *There is a unique unitary representation* U *of* G *acting on* $X(T)$ *and
satisfying*

$$T_{xa} = U_x T_a U_x^* \qquad (x \in G; a \in A), \qquad (8)$$

$$U_x \xi = \xi \qquad\qquad\qquad (x \in G). \qquad (9)$$

(II)   *If* $q \in \mathscr{P}(A)$ *and* q *is subordinate to* p, *and if* B *is the positive linear
operator on* $X(T)$ *corresponding to* q *as in 20.3, then* q *is G-invariant if
and only if* $BU_x = U_x B$ *for all* x *in* G.

(III) $X(T)$ *is irreducible under the combined action of* T *and* U *if and only if
the only G-invariant elements of* $\mathscr{P}(A)$ *which are subordinate to* p *are
the non-negative multiples of* p.

Definition. If the conditions of (III) of the above proposition hold, p is said
to be *ergodic* (*with respect to the given action of* G *on* A).

Remark. If $G = \{e\}$, then of course ergodicity is just indecomposability.

20.8. The term "ergodic" points to an interesting connection with physics.
We shall close this section with a brief discussion of the physical significance
of positive linear functionals and of their indecomposability and ergodicity.

Let S be some physical system. From the standpoint of classical (i.e., pre-
quantum) physics, associated with S is a certain set M called the phase space

of S, the points of which are the possible complete descriptions of the configuration of S (including the positions, velocities, etc., of all its parts) at a given instant. We emphasize that each point of M represents *complete* knowledge of the instantaneous configuration of S.

More generally, we must take into consideration the possibility of our having only partial knowledge of the configuration of S. Such partial knowledge is described by a probability measure μ on M; the number $\mu(W)$ (W being a suitable subset of M) is the probability that the exact configuration of the system, if it were known, would be found to lie in W. Complete knowledge occurs only when μ is concentrated at a single point of M.

It is reasonable that the phase space M of S should carry a topology; let us assume that M is in fact a locally compact Hausdorff space. Then $A = \mathscr{C}_0(M)$ is a commutative C^*-algebra; and according to the Riesz Theorem (II.8.12) the probability measures on M (i.e., the regular Borel measures of total mass 1) are just the positive linear functionals on A of norm 1. By 20.6 (see also the introduction to this chapter) those probability measures which represent complete knowledge—that is, which are concentrated at a single point—are abstractly characterized among all probability measures by the property of *indecomposability*.

A physical observable (relating to S), considered operationally, is simply a procedure for performing a measurement on S. The result of the measurement is a real number which depends on the configuration of the system. Thus a physical observable can be identified with a (presumably continuous) real-valued function on M. In particular the real-valued (i.e., self-adjoint) elements of A are physical observables. Given partial knowledge of the state of the system, described by a probability measure μ on M, the *expected value* $E_\mu(a)$ of an observable a is the integral $\int_M a\, d\mu$ (if this exists); and the *dispersion* of a (i.e., its expected deviation from $E_\mu(a)$) is the square root of the quantity $E_\mu((a - E_\mu(a))^2) = E_\mu(a^2) - (E_\mu(a))^2$. In a state of complete knowledge, when μ is concentrated at a point of M, each physical observable has zero dispersion; its value is then known with certainty.

So much for the classical description of a physical system. The change from the classical to the quantum-mechanical framework consists in essence in allowing the A of the above discussion to be *non-commutative*. Instead of describing S in terms of a locally compact Hausdorff phase space M (or equivalently, a *commutative* C^*-algebra $A = \mathscr{C}_0(M)$), we begin with an arbitrary (in general non-commutative) C^*-algebra A. The self-adjoint elements of A are now thought of as the physical observables of S. The positive linear functionals p of norm 1 on A (which generalize the probability measures on M in the classical context) are thought of as descriptions of possible states of partial knowledge of the configuration of S. For brevity we refer to such p as *states of A*. If p is a state and a is a physical observable, $p(a)$ is the expected value of a in the state p; and, as in the classical case, $p(a^2) - (p(a))^2$ is the square of the dispersion of a. If $p(a^2) - (p(a))^2 = 0$, we say that a has a precise value, namely $p(a)$, in the state p.

Again generalizing the commutative situation, we shall refer to states of A which are indecomposable as *pure states*. In the classical context, the pure states were those probability measures which were concentrated at a single point of M. They represented "maximal knowledge" of S—a description of S which is not a probabilistic combination of two, more exact, descriptions. In the non-commutative (quantum-mechanical) context, the pure states again represent maximally precise knowledge of S—knowledge which cannot be represented as a probabilistic combination of two, more exact, descriptions. But there is this difference: *In the non-commutative situation a pure state will not in general give zero dispersion to all observables.* That is, even when our knowledge of the configuration of the system is as accurate as is theoretically possible, some physical observables will not have precisely defined values. This fundamental consequence of the non-commutativity of A mirrors the famous Uncertainty Principle of quantum mechanics, according to which there is no physical state of a particle in which *both* its position *and* momentum have precisely defined values.

Let us be a little more specific. Let p be a pure state of A, and $\langle T, \xi \rangle$ the corresponding cyclic pair for A (so that T is irreducible). The reader will verify that a physical observable a (i.e., a self-adjoint element of A) has zero dispersion in the state p if and only if $T_a \xi = p(a)\xi$. Thus p will give zero dispersion to *all* physical observables if and only if $\mathbb{C}\xi$ is T-stable; and, since T is irreducible, this is only possible if T is one-dimensional. So the only pure states of A which give zero dispersion to all physical observables are the one-dimensional *-representations of A. On the other hand, if A is non-commutative it follows from 22.14 that it has irreducible *-representations which are of dimension greater than one. Hence it has pure states which do not give zero dispersion to all physical observables.

Let us now try to give a physical interpretation of 20.7 (at least when G is the additive group $\mathbb{R}$ of the reals).

Return for the moment to the classical description of S, starting from the phase space M; and let us think of S as undergoing change or motion under the guidance of the laws of nature. If we assume that S is isolated from all changeable influences outside itself, then by the principle of causality each real number t generates a permutation ω_t of M characterized as follows: If the configuration of S at a certain instant is $m \in M$, then t units of time later the configuration of S will be $\omega_t(m)$. Simple physical considerations dictate that

$$\omega_{t_1} \circ \omega_{t_2} = \omega_{t_1 + t_2}, \qquad \omega_0 = \text{identity permutation},$$

$$\langle t, m \rangle \mapsto \omega_t(m) \qquad \text{is continuous.}$$

Thus M becomes a topological $\mathbb{R}$-space; and $\mathbb{R}$ also acts on $A = \mathscr{C}_0(M)$ by *-automorphisms:

$$(ta)(m) = a(\omega_{-t}(m)) \qquad (a \in A; t \in \mathbb{R}; m \in M).$$

More generally, let us suppose that S is a quantum-mechanical system, so that A is some non-commutative C^*-algebra. The time-development of S is again described by specifying an action of $\mathbb{R}$ on A by *-automorphisms (see the beginning of 20.7): If $t \in \mathbb{R}$ and a is a physical observable, ta is that observable whose expected value is the same as the expected value of a would have been t units of time earlier.

Let such an action of $\mathbb{R}$ on A be given. What is the physical meaning of the $\mathbb{R}$-invariance of a state p (see 20.7)? It means that p is an *equilibrium state* in the sense that the expected values of the physical observables do not change with time.

It is a fundamental assumption of statistical mechanics that the "macroscopic state" of a large thermodynamical system in equilibrium, with billions of molecules moving randomly, is nothing but an equilibrium state of this sort. The detailed motions of the individual molecules—that is, the "microscopic" or pure state of the system—are unknown; only the expected values of the observables are calculable. Furthermore, the macroscopic state of a large physical system in equilibrium is not an arbitrary $\mathbb{R}$-invariant state p, but a "maximally precise" one, in which the observables are known as precisely as possible subject to the requirement of $\mathbb{R}$-invariance; that is, p cannot be written as a probabilistic combination (i.e., a convex combination) of two distinct $\mathbb{R}$-invariant states. In other words, the macroscopic state of a thermodynamical system S in equilibrium is an ergodic state of the corresponding C^*-algebra A (with respect to the "time-development" of A), in the sense of 20.7.

It is a vital part of one's assumptions about the macroscopic systems of thermodynamics that the ergodic $\mathbb{R}$-invariant states are relatively few in number. Though the pure states of the system may be hugely complicated, its ergodic time-invariant states should be indexed by only a few parameters describing its macroscopic properties, such as pressure, temperature, etc. This of course is excellent motivation for the study of ergodic $\mathbb{R}$-invariant states. We shall return briefly to the subject in XII.8.24.

21.   Positive Functionals on Commutative *-Algebras; the Generalized Bochner and Plancherel Theorems

21.1.   In §19 we related positive functionals to *-representations. In §10 we classified the *-representations of a commutative Banach *-algebra in terms of projection-valued measures. Putting these two steps together, we obtain the generalized Bochner Theorem, classifying the extendable positive functionals on a commutative Banach *-algebra A in terms of measures on $\hat{A}$.

21.2.   Fix a commutative Banach *-algebra A. Let $\hat{A}$ be the symmetric part of $A^\dagger$, as in 4.2; and let $a \mapsto \hat{a}$ be the Gelfand transform map of A.

Generalized Bochner Theorem.   *For each bounded regular (non-negative) Borel measure μ on $\hat{A}$, the equation*

$$p(a) = \int_{\hat{A}} \hat{a}(\phi)d\mu\phi \qquad\qquad (a \in A) \qquad (1)$$

defines an extendable positive linear functional p on A. Conversely, to every extendable positive linear functional p on A there is a unique bounded regular (non-negative) Borel measure μ on $\hat{A}$ such that (1) holds.

Proof.   Given μ, it is clear that the p defined by (1) is positive. If the total variation norm of μ is k, Schwarz's Inequality gives $|p(a)|^2 \leq k \int_{\hat{A}} |\hat{a}(\phi)|^2 \, d\mu\phi = kp(a^*a)$. So p is extendable.

In the proof of 10.7 we observed that $\{\hat{a}|\hat{A} : a \in A\}$ is dense in $\mathscr{C}_0(\hat{A})$ in the supremum norm. Therefore by II.8.11 the p of (1) uniquely determines μ.

Let p be an extendable positive linear functional on A. It remains only to show that p is related to some μ as in (1). To do this, we use 19.9 (see 19.5) to obtain a cyclic pair $\langle T, \xi \rangle$ for A satisfying

$$p(a) = (T_a\xi, \xi) \qquad\qquad (a \in A). \qquad (2)$$

Let P be the spectral measure of T (10.10). Defining $\mu(W) = (P(W)\xi, \xi)$ (for Borel subsets W of $\hat{A}$), we see from (2) and II.11.8(VII) that

$$p(a) = \int_{\hat{A}} \hat{a}(\phi)d\mu\phi \qquad\qquad (a \in A).$$

Since μ is a bounded regular Borel measure, the proof is complete.   ∎

21.3. Now many of the important positive linear functionals occurring in harmonic analysis are defined not on a Banach *-algebra as in 21.2, but only on a dense *-subalgebra of a Banach *-algebra. For example, if G is a non-discrete locally compact group, the important functional $f \mapsto f(e)$ (evaluation at the unit element; see 18.8) does not make sense for arbitrary functions f in the $\mathscr{L}_1$ group algebra of G, but only for continuous functions. Thus it would be desirable to generalize Theorem 21.2 to such functionals. This is what we are going to do next. The result (Theorem 21.4) will differ from 21.2 mainly in that the measure μ on $\hat{A}$ need no longer be bounded. In the case of the functional $f \mapsto f(e)$ mentioned above, we shall see in Chapter X that 21.4 (or rather 21.6) becomes the classical Plancherel formula.

21.4. Fix a commutative Banach *-algebra A. For simplicity (though this is not essential) we shall assume that A is symmetric, that is, $A^\dagger = \hat{A}$ (see 2.5). Let B be a dense *-subalgebra of A; and let p be a positive linear functional on B satisfying the following two conditions:

(i) For some constant $k \geq 0$ we have

$$p(b^*a^*ab) \leq k\|a\|^2 p(b^*b) \qquad (a, b \in B). \qquad (3)$$

(In particular p satisfies Condition (R) on B.)

(ii) For each a in B and each $\varepsilon > 0$, there are elements b, c of B such that

$$p((a - bc)^*(a - bc)) < \varepsilon. \qquad (4)$$

Theorem. *Let A, B, p be as above. Then there is a unique (not necessarily bounded) regular (non-negative) Borel measure μ on $\hat{A}$ such that, for all a and b in B,*

$$\hat{a}\hat{b} \quad \text{is } \mu\text{-summable and} \quad p(ab) = \int_{\hat{A}} \hat{a}(\phi)\hat{b}(\phi)d\mu\phi \qquad (5)$$

Further, for fixed a, b in B, the functional $x \mapsto p(axb)$ is continuous on B in the A-norm, and so extends to a continuous linear functional q on A; and we have

$$q(x) = \int \hat{a}(\phi)\hat{b}(\phi)\hat{x}(\phi)d\mu\phi \qquad (6)$$

for all x in A.

Proof. Let I be the null ideal $\{b \in B : p(b^*b) = 0\}$ of p, $\rho : B \to X = B/I$ the quotient map, and X_c the completion of X with respect to the inner product $(\rho(a), \rho(b)) = p(b^*a)$. In view of condition (i) and the denseness of B in A, the

*-representation of B on X_c generated by p extends to a *-representation T of A on X_c. By condition (ii) T is non-degenerate.

Thus by 10.10 T gives rise to a regular X_c-projection-valued Borel measure P on $\hat{A}$ satisfying

$$T_a = \int_{\hat{A}} \hat{a}(\phi) dP\phi \qquad\qquad (a \in A). \qquad (7)$$

If $b \in B$, let us denote by π_b the bounded regular Borel measure on $\hat{A}$ given by

$$\pi_b(W) = (P(W)\rho(b), \rho(b)) \qquad\qquad (8)$$

(for Borel subsets W of $\hat{A}$). We claim that

$$|\hat{a}(\phi)|^2 \, d\pi_b\phi = |\hat{b}(\phi)|^2 \, d\pi_a\phi \qquad\qquad (9)$$

for all a, b in B. Indeed: Both sides of (9) are bounded regular Borel measures; call then v_1 and v_2 respectively. As we saw in the proof of 10.7, $\{\hat{c} : c \in B\}$ is dense in $\mathscr{C}_0(\hat{A})$ in the supremum norm. Hence to verify that $v_1 = v_2$ it is enough to show that

$$\int \hat{c}(\phi) dv_1\phi = \int \hat{c}(\phi) dv_2\phi \qquad\qquad (10)$$

for all c in B. But, if $c \in B$, we have by (7) and II.7.6

$$\int \hat{c}(\phi) dv_1\phi = \int (caa^*)\hat{\ }(\phi) d\pi_b\phi$$

$$= (T_{caa^*}\rho(b), \rho(b)) = p(b^*caa^*b).$$

Similarly, interchanging a and b, we get

$$\int \hat{c}(\phi) dv_2\phi = p(a^*cbb^*a).$$

By the commutativity of A, the last two results imply (10), and hence (9).

In view of (9), the measure $|\hat{a}(\phi)|^{-2} \, d\pi_a\phi$ is independent of the particular element a of B, at least on any open set where $\hat{a}$ never vanishes. We wish now to define μ as that measure on $\hat{A}$ which, for each a in B, coincides with $|\hat{a}(\phi)|^{-2} \, d\pi_a\phi$ on $\{\phi : \hat{a}(\phi) \neq 0\}$.

To do this rigorously, we first observe that, given any compact subset C of $\hat{A}$, there exists $b \in B$ such that

$$\hat{b}(\phi) \neq 0 \qquad \text{for all } \phi \text{ in } C. \qquad (11)$$

Indeed: Cover C with finitely many open sets $\{U_i\}$ such that for each i we have an element b_i in B satisfying $\hat{b}_i(\phi) \neq 0$ for all ϕ in U_i. Then $b = \sum_i b_i^* b_i$ will satisfy (11). Consequently, given any $f \in \mathcal{L}(\hat{A})$, we can choose $b \in B$ such that $\hat{b}$ does not vanish anywhere on the compact support C of f. With this b we define

$$\mu(f) = \int_C f(\phi) |\hat{b}(\phi)|^{-2} \, d\pi_b \phi. \tag{12}$$

Now the right side of (12) is independent of b. Indeed, if a is another element of B such that $\hat{a}$ never vanishes on C, (9) and II.7.6 give

$$\int_C f(\phi)|\hat{b}(\phi)|^{-2} \, d\pi_b \phi = \int_C f(\phi)|\hat{b}(\phi)|^{-2}|\hat{a}(\phi)|^{-2}|\hat{a}(\phi)|^2 \, d\pi_b \phi$$

$$= \int_C f(\phi)|\hat{b}(\phi)|^{-2}|\hat{a}(\phi)|^{-2}|\hat{b}(\phi)|^2 \, d\pi_a \phi$$

$$= \int_C f(\phi)|\hat{a}(\phi)|^{-2} \, d\pi_a \phi.$$

So (12) is independent of b. It follows that μ, as defined by (12), is an integral on $\mathcal{L}(\hat{A})$, and so by II..8.12 gives rise to a (non-negative) regular Borel measure on $\hat{A}$ which we also call μ,

For any a in B we have

$$d\pi_a \phi = |\hat{a}(\phi)|^2 \, d\mu\phi. \tag{13}$$

Indeed: Let f be any element of $\mathcal{L}(\hat{A})$, and b an element of B such that $\hat{b}$ never vanishes on the compact support of f. Then (12) and (9) give

$$\int f(\phi)|\hat{a}(\phi)|^2 \, d\mu\phi = \int f(\phi)|\hat{a}(\phi)|^2|\hat{b}(\phi)|^{-2} \, d\pi_b \phi$$

$$= \int f(\phi) d\pi_a \phi.$$

So (13) holds.

Since π_a is a bounded measure, it follows from (13) that $|\hat{a}|^2 = \widehat{(a^*a)}$ is μ-summable on $\hat{A}$ for all a in B; in fact

$$p(a^*a) = \|\rho(a)\|^2 = \pi_a(\hat{A})$$

$$= \int \widehat{(a^*a)}(\phi) d\mu\phi. \tag{14}$$

By the polarization identity every product bc ($b, c \in B$) is a linear combination of terms of the form a^*a ($a \in B$). Hence (14) implies (5).

To prove (6) let us fix two elements a, b of B. Since the *-representation T is norm-continuous, the map $x \mapsto p(axb) = (T_x \rho(b), \rho(a^*))$ ($x \in B$) is continuous in the A-norm, and so extends to a continuous linear functional q on A. If $x \in A$ and $x_n \to x$, $x_n \in B$, we have by (5)

$$p(ax_nb) = \int \hat{a}(\phi)\hat{x}_n(\phi)\hat{b}(\phi)d\mu\phi. \tag{15}$$

Now let $n \to \infty$ in (15). The left side approaches $q(x) = p(axb)$. From the summability of $\hat{a}\hat{b}$ and the fact that $\hat{x}_n \to \hat{x}$ uniformly on $\hat{A}$, we see that the right side of (15) approaches $\int \hat{a}(\phi)\hat{x}(\phi)\hat{b}(\phi)d\mu\phi$. Thus in the limit we obtain (6).

It remains only to show that the regular Borel measure μ satisfying (5) is unique. Let μ' be another regular Borel measure on $\hat{A}$ satisfying (5). Then in particular

$$\int \hat{a}(\phi)|\hat{b}(\phi)|^2 \, d\mu\phi = \int \hat{a}(\phi)|\hat{b}(\phi)|^2 \, d\mu'\phi \tag{16}$$

for $a, b \in B$. Now consider the b in (16) as fixed. By (5) $|\hat{b}(\phi)|^2 \, d\mu\phi$ and $|\hat{b}(\phi)|^2 \, d\mu'\phi$ are bounded; and we have already pointed out that $\{\hat{a}: a \in B\}$ is dense in $\mathscr{C}_0(\hat{A})$. So it follows from (16) that

$$|\hat{b}(\phi)|^2 \, d\mu\phi = |\hat{b}(\phi)|^2 \, d\mu'\phi$$

for every b in B. Hence, by an evident argument based on (11), we conclude that $\mu' = \mu$; and the proof is complete. ■

Definition.   It is reasonable to refer to the above μ as the *Gelfand transform* of p on $\hat{A}$.

21.5. Remark.   Hypothesis (ii) of 21.4 was used only to show the non-degeneracy of the *-representation T generated by p. But it is worthwhile observing that Theorem 21.4 becomes false if (ii) is omitted. Here is a simple example. Let A be the Banach *-algebra of all continuous complex functions on $[0, 1]$ which vanish at 0 (with pointwise multiplication and complex conjugation, and the supremum norm). Let B be the dense *-subalgebra of all functions in A which are twice continuously differentiable on the closed interval $[0, 1]$; and put

$$p(f) = f''(0) \qquad\qquad (f \in B).$$

Notice from V.8.8 that $A \cong]0, 1]$. We leave it as an exercise for the reader to verify that p is positive on B, and satisfies 21.4(i) but not 21.4(ii); and that the conclusion of Theorem 21.4 is false for p.

21.6. We can deduce from 21.4 a simple and elegant description of the *-representation T.

Let A, B, and p be as in 21.4. Suppose that T is the *-representation of A generated by p as in the first paragraph of the proof of 21.4; and let $\rho: B \to X_c = X(T)$ be the quotient map defined there. Let us denote by μ the Gelfand transform of p on $\hat{A}$.

Generalized Plancherel Theorem.   *The equation*

$$\Phi(\rho(a)) = \hat{a} \qquad\qquad (a \in B) \qquad (17)$$

*determines a linear isometry Φ of $X(T)$ onto $\mathscr{L}_2(\mu)$. The *-representation T is unitarily equivalent under Φ with the *-representation W of A on $\mathscr{L}_2(\mu)$ given by*

$$(W_a f)(\phi) = \hat{a}(\phi)f(\phi) \qquad\qquad (18)$$

$(a \in A; f \in \mathscr{L}_2(\mu); \phi \in \hat{A})$.

Proof.   It follows from 21.4(5) (and the denseness of $\rho(B)$ in $X(T)$) that (17) does determine a linear isometry of $X(T)$ into $\mathscr{L}_2(\mu)$. Certainly the W defined by (18) is a *-representation of A, and it follows from (17) that $W_a \Phi = \Phi T_a$ for $a \in B$, and hence by continuity for all a in A. So it remains only to verify that Φ is *onto* $\mathscr{L}_2(\mu)$.

To see this, it is enough to fix a function $f \in \mathscr{L}(\hat{A})$, and to approximate f in the $\mathscr{L}_2(\mu)$ norm by elements of range(Φ).

By (11) we can choose an element b of B such that $\hat{b}$ never vanishes on the compact support of f. Then the function g given by

$$g(\phi) = \begin{cases} (\hat{b}(\phi))^{-1}f(\phi) & \text{if } \hat{b}(\phi) \neq 0 \\ 0 & \text{if } \hat{b}(\phi) = 0 \end{cases}$$

is in $\mathscr{L}(\hat{A})$. So, since $\{\hat{c} : c \in B\}$ is dense in $\mathscr{C}_0(\hat{A})$, there is a sequence $\{c_n\}$ of elements of B such that $\hat{c}_n \to g$ uniformly on $\hat{A}$. But then, since $\hat{b} \in \mathscr{L}_2(\mu)$, it is clear that

$$\hat{b}\hat{c}_n \to \hat{b}g = f \quad \text{in } \mathscr{L}_2(\mu).$$

Since $\hat{b}\hat{c}_n \in \text{range}(\Phi)$, we have approximated f in the desired way.   ∎

21.7.   *Remark*.   Keep the notation of 21.6; and let P be the spectral measure of T. By Theorem 21.6 the image of P under Φ is the spectral measure Q of W. By 10.12 Q is the natural $\mathscr{L}_2(\mu)$-projection-valued measure on $\hat{A}$ which sends each Borel subset S of $\hat{A}$ into multiplication by Ch_S. In particular, as regards sets in $\mathscr{S}(\hat{A})$, the measure-theoretic equivalence class of P is the same as that of μ.

Thus Theorem 21.6 also provides us with an elegant description of the spectral measure of T.

22.   The Existence of Positive Functionals and *-Representations of C*-Algebras; the Gelfand–Naimark Theorem

22.1.   In 10.6 we raised but left unanswered a very important question: Is every C^*-algebra reduced, that is, are there enough *-representations of a C^*-algebra A to separate points of A? In this section we shall show that the answer is "yes." This, as we pointed out in 10.6, will immediately give us the famous Gelfand–Naimark Representation Theorem for C^*-algebras. Even more, we shall prove that the points of a C^*-algebra A are separated by the *irreducible* *-representations of A. This will imply that the same is true for any reduced Banach *-algebra A. From this we shall see in Chapter VIII that any locally compact group G has enough irreducible unitary representations to separate points of G.

The tools for doing this are five: 1) The fact that commutative C^*-algebras are reduced (§4); 2) the symmetry of C^*-algebras (§7) and its consequences for the ordering relation 7.18; 3) the correspondence between *-representations and positive functionals (§19); 4) the Krein Extension Theorem for positive functionals; 5) the Krein–Milman Theorem on extreme points of compact convex sets (see for example Kelley and Namioka [1], p. 131). The last two of these fall properly into the domain of the theory of linear topological spaces. However, we will prove the Krein Extension Theorem for positive functionals, since it is perhaps less well known than the Krein–Milman Theorem.

22.2.   A real linear space V is *partially ordered* by a relation $\leq$ on V if

(i) $\leq$ is reflexive and transitive,
(ii) $x \leq y \Rightarrow x + z \leq y + z$ for $x, y, z \in V$,
(iii) $x \leq y \Rightarrow \lambda x \leq \lambda y$ for $x, y \in V$, $0 \leq \lambda \in \mathbb{R}$.

Krein Extension Theorem. *Let V be a real linear space partially ordered by a relation $\leq$. Let W be a real linear subspace of V with the following property:*

$$\text{For each } x \text{ in } V, \text{ there are elements } y \quad\quad (1)$$
$$\text{and } z \text{ in } W \text{ such that } y \leq x \leq z.$$

Further, let p be a real-valued linear functional on W satisfying $p(x) \geq 0$ whenever $0 \leq x \in W$. Then p can be extended to a real linear functional q on V satisfying $q(x) \geq 0$ whenever $0 \leq x \in V$.

Proof. By Zorn's Lemma it is enough to extend p to a linear functional q on the subspace W_0 of V generated by W and one non-zero vector $x_0 \in V \setminus W$, in such a way that $0 \leq x \in W_0 \Rightarrow q(x) \geq 0$.

Put $a = \inf\{p(v) : x_0 \leq v \in W\}$, $b = \sup\{p(u) : x_0 \geq u \in W\}$. By condition (1) this infimum and supremum are being taken over non-void sets and are therefore finite; and $a \geq b$. Choose a number c with $a \geq c \geq b$; then

$$p(u) \leq c \leq p(v) \quad\quad \text{whenever } u, v \in W \text{ and } u \leq x_0 \leq v. \quad\quad (2)$$

Now extend p to a linear functional q on W_0 by setting $q(x_0) = c$; and suppose that $x + \lambda x_0 \geq 0$ ($x \in W$; $\lambda \in \mathbb{R}$). We must show

$$q(x + \lambda x_0) \geq 0. \quad\quad (3)$$

If $\lambda = 0$, then $x \geq 0$, and so $q(x + \lambda x_0) = p(x) \geq 0$.

Assume $\lambda > 0$. Then $x + \lambda x_0 \geq 0$ gives $x_0 \geq -\lambda^{-1} x \in W$; so by (2) $p(-\lambda^{-1} x) \leq c$, or $q(x + \lambda x_0) \geq 0$.

Assume $\lambda < 0$. Then $x + \lambda x_0 \geq 0$ gives $x_0 \leq -\lambda^{-1} x \in W$; so by (2) $c \leq p(-\lambda^{-1} x)$, or $q(x + \lambda x_0) \geq 0$.

Thus in all cases (3) holds, and the theorem is proved. ∎

22.3. Next we shall set the stage for the application of the Krein–Milman Theorem.

Proposition. *Let A be a Banach $*$-algebra with unit $\mathbb{1}$; and let $\mathcal{N}(A)$ be the set of all positive linear functionals p on A for which $p(\mathbb{1}) = 1$. Then: (I) $\mathcal{N}(A)$ is compact in the topology of pointwise convergence on A; (II) $\mathcal{N}(A)$ is convex (as a subset of the conjugate space A^* of A), and its extreme points are precisely the elements of $\mathcal{N}(A)$ which are indecomposable (in the sense of 20.4).*

Proof. By 18.12 $\mathcal{N}(A)$ is contained in the unit ball A_1^* of A^*. Since positivity is preserved on passing to pointwise limits, $\mathcal{N}(A)$ is a pointwise-closed subset of A_1^*. But A_1^* is compact in the pointwise convergence topology. Therefore (I) holds.

It is obvious that $\mathcal{N}(A)$ is convex. If $r = \lambda p + (1 - \lambda)q$, where $p, q, r \in \mathcal{N}(A)$ and $0 < \lambda < 1$, then p and q are subordinate to r (see 20.2); so, if r is indecomposable, p and q are multiples of r and hence equal to r. Thus indecomposable elements of $\mathcal{N}(A)$ are extreme in $\mathcal{N}(A)$. Conversely, let r be an extreme point of $\mathcal{N}(A)$, and p a positive linear functional subordinate to r. Then $p + q = \lambda r$ for some $\lambda > 0$ and some positive linear functional q. Assuming without loss of generality that $p \neq 0 \neq q$, writing the last equation in the form

$$r = (\lambda^{-1}p(\tfrac{1}{2}))p(\tfrac{1}{2})^{-1}p + (\lambda^{-1}q(\tfrac{1}{2}))q(\tfrac{1}{2})^{-1}q,$$

and using the extreme property of r, we see that $p = p(\tfrac{1}{2})r$. So r is indecomposable.   ∎

22.4.   Notice that if A has no unit, the following modification of 22.3 holds:

Proposition.   *Let A be a Banach *-algebra with an approximate unit $\{e_i\}$ satisfying $\|e_i\| \leq 1$; and let $\mathcal{N}'(A)$ be the set of all (continuous) positive linear functionals p on A for which $\|p\| \leq 1$. Then:*

(I)   *$\mathcal{N}'(A)$ is compact in the pointwise convergence topology;*
(II)   *$\mathcal{N}'(A)$ is convex, and the set of its extreme points is $E \cup \{0\}$, where E is the set of all indecomposable elements of $\mathcal{N}'(A)$ of norm 1.*

The proof is similar to that of 22.3, and is left to the reader. (At one step we need 19.13.)

22.5.   We now prove the fundamental extension theorem for positive functionals on C^*-algebras, which will imply in particular that all C^*-algebras are reduced. Recall that all positive linear functionals on a C^*-algebra A are continuous, extendable, and satisfy Condition (R) (see 18.15 and 19.5); that is, they belong to $\mathcal{P}(A)$ (20.1).

Theorem.   *Let A be any C^*-algebra, B a closed *-subalgebra of A, and p a positive linear functional on B. Then p can be extended to a positive linear functional q on A. If p is indecomposable (on B), the positive extension q can be taken to be indecomposable (on A).*

Proof.   We shall first prove the theorem under the assumption that A has a unit $\mathbf{1}$ with $\mathbf{1} \in B$.

Let $A_h = \{a \in A : a^* = a\}$, $B_h = A_h \cap B$. Then A_h is a partially ordered real linear space (see 22.2) under the relation $\leq$ introduced in 7.18. In view of 7.3, the restriction of $\leq$ to B_h is just the partial ordering of B_h deduced as in 7.18 from the notion of positivity in B. Therefore $p' = p|B_h$ is a real-valued functional on B_h satisfying $p'(a) \geq 0$ whenever $0 \leq a \in B_h$. Now since $1 \in B_h$ and $-\|a\| \cdot 1 \leq a \leq \|a\| \cdot 1$ for all a in A_h, B_h satisfies 22.2(1) as a subspace of A_h. Therefore by the Krein Extension Theorem 22.2, p' extends to a real linear functional q' on A_h such that $0 \leq a \in A_h \Rightarrow q'(a) \geq 0$. The equation

$$q(a + ib) = q'(a) + iq'(b) \qquad\qquad (a, b \in A_h)$$

now defines a positive complex-linear extension q of p to all of A.

Now assume that p was indecomposable; we wish to show that p can be extended to an indecomposable positive linear functional on A. Without loss of generality suppose $p(1) = 1$; and let Q be the set of all positive linear functionals on A which extend p. Clearly Q is a pointwise-closed convex subset of $\mathcal{N}(A)$ (see 22.3). Thus, since $\mathcal{N}(A)$ is compact (22.3), Q is compact. By the preceding paragraph Q is non-void. So the Krein–Milman Theorem assures us that Q has some extreme point q. To show that q is indecomposable, it is enough by 22.3 to show that q is an extreme point not merely of Q but of $\mathcal{N}(A)$. To do this, assume that

$$q = \lambda q_1 + (1 - \lambda)q_2 \qquad (q_i \in \mathcal{N}(A), 0 < \lambda < 1). \qquad (4)$$

Restricting both sides of this equation to B, recalling that p is indecomposable, and applying 22.3 to $\mathcal{N}(B)$, we conclude that $q_1|B = q_2|B = p$. Hence q_1 and q_2 are in Q; and the relation (4), combined with the fact that q is extreme in Q, gives $q_1 = q_2 = q$. Therefore q is extreme in $\mathcal{N}(A)$, and so indecomposable.

We have proved the theorem in case A has a unit 1 and $1 \in B$. Now drop this assumption.

Let A_1 stand for A if A has a unit 1, otherwise the result of adjoining a unit 1 to A. Let B_1 stand for B if 1 (the unit of A_1) is in B; otherwise $B_1 = B + \mathbb{C}1$. We have seen in 3.12 that B_1 is a closed *-subalgebra of A_1. Now extend p to a positive functional p^0 on B_1. By the preceding part of the proof, p^0 extends to a positive functional q^0 on A_1, whose restriction q to A is the required positive extension of p to A. If p was indecomposable, then by 20.5 p^0 could be taken indecomposable, and hence, by the preceding part of this proof, q^0 could be likewise chosen indecomposable. Since q^0 extends the non-zero functional p, $q = q^0|A$ is non-zero and hence indecomposable by 20.5. This completes the proof. ■

22.6. We now translate 22.5 into a statement about extensions of representations. Let A be a C^*-algebra and B a closed *-subalgebra of A. A *-representation T of A is said to be an *extension* of a *-representation S of B

if there exists a closed subspace Y of $X(T)$ which is stable under $T|B$, such that the subrepresentation of $T|B$ acting on Y is unitarily equivalent to S.

Theorem.   *Let A, B be as above. Any non-degenerate [cyclic, irreducible] *-representation of B has a non-degenerate [cyclic, irreducible] extension to A.*

Proof.   We shall first prove this for the cyclic case. Let S be a cyclic *-representation of B, with cyclic vector ξ, and put $p(b) = (S_b\xi, \xi)$. By 22.5 p extends to a positive linear functional q on A; and by 19.7 there is a cyclic pair $\langle T, \eta \rangle$ for A such that

$$q(a) = (T_a\eta, \eta) \qquad\qquad (a \in A).$$

We now define Y to be the closure of $\{T_b\eta : b \in B\}$ in $X(T)$. This is clearly stable under $T|B$. Let η' be the projection of η on Y, and put $\eta'' = \eta - \eta' \in Y^\perp$. If $b \in B$, then $T_b\eta = T_b\eta' + T_b\eta''$. Since $T_b\eta'' \in Y^\perp$ and $T_b\eta'' = T_b\eta - T_b\eta' \in Y$, we get $T_b\eta'' = 0$. It follows that $T_b\eta = T_b\eta'$ for all b in B; so η' is a cyclic vector for the subrepresentation S' of $T|B$ on Y. Further, if $b \in B$,

$$(S_b'\eta', \eta') = (T_b\eta, \eta) = q(b)$$

$$= p(b) = (S_b\xi, \xi).$$

From this and 19.8 we conclude that $S' \cong S$. Thus T is a cyclic extension of S to A.

In the above argument suppose that S had been irreducible. Then by 20.4 p would be indecomposable. So by 22.5 q could have been chosen indecomposable, in which case by 20.4 T would be irreducible.

Finally, let S be any non-degenerate *-representation of B. By 9.16 $S \cong \sum_{i \in I}^\oplus S^i$, where each S^i is cyclic. By the preceding part of the proof each S^i has a cyclic extension T^i to A. Clearly $T = \sum_{i \in I}^\oplus T^i$ is a non-degenerate extension of S to A.   ∎

22.7.   If B is a two-sided ideal of A, the extension in 22.6 can be made *without enlarging the space of the representation.*

Corollary.   *Let B be a closed two-sided ideal of the C*-algebra A. Any non-degenerate *-representation S of B can be extended to a (non-degenerate) *-representation T of A acting in the same space as S.*

This is a special case of 19.11. It can also easily be obtained directly from 22.6.

22.8.   *Remark.* Although 22.7 is true for arbitrary Banach *-algebras (by 19.11), this is not so for 22.6. Here is a simple example.

Let A be the (Banach) *-algebra whose underlying linear space is $\mathbb{C}^2$, with multiplication and involution given by:

$$\langle u, v \rangle \langle u', v' \rangle = \langle uu', vv' \rangle,$$

$$\langle u, v \rangle^* = \langle \bar{v}, \bar{u} \rangle.$$

Since $\langle 0, 1 \rangle^* \langle 0, 1 \rangle = \langle 0, 0 \rangle$ and $\langle 1, 0 \rangle^* \langle 1, 0 \rangle = \langle 0, 0 \rangle$, the only *-representations of A are the zero representations. On the other hand, the *-subalgebra $B = \{\langle u, u \rangle : u \in \mathbb{C}\}$ of A has the non-zero one-dimensional *-representation $\langle u, u \rangle \mapsto u$. The latter therefore has no extension to A.

22.9. As another immediate special case of 22.6 we have:

Corollary. *Let B be a norm-closed commutative *-subalgebra of a C^*-algebra A. For any ϕ in $\hat{B}$, there exists an irreducible *-representation T of A and a non-zero vector ξ in $X(T)$ such that*

$$T_b \xi = \phi(b)\xi \qquad \text{for all } b \text{ in } B.$$

22.10.   Corollary. *Let a be any element of a non-zero C^*-algebra A. There exists an irreducible *-representation T of A and a unit vector ξ in $X(T)$ such that $\|T_a \xi\| = \|a\|$.*

Proof. Since $b = a^*a$ is Hermitian, there is a norm-closed commutative *-subalgebra B of A with $b \in B$. By §4 there is an element ϕ of $\hat{B}$ such that $|\phi(b)| = \|b\|$. By 22.9 we can find an irreducible *-representation T of A and a unit vector ξ in $X(T)$ such that $T_c \xi = \phi(c)\xi$ for all c in B; in particular $T_b \xi = \phi(b)\xi$. Therefore

$$\|T_a \xi\|^2 = (T_b \xi, \xi) = \phi(b) = \|b\| = \|a\|^2. \qquad \blacksquare$$

Remark. The conclusion of this corollary implies of course that $\|T_a\| = \|a\|$.

22.11.   Corollary. *Every C^*-algebra is reduced.*

22.12. As we saw in 10.6, 22.11 leads immediately to the following famous result:

Gelfand–Naimark Theorem. *Every C^*-algebra A is isometrically *-isomorphic with some norm-closed *-subalgebra of $\mathcal{O}(X)$ for some Hilbert space X.*

22.13. By IV.7.4 a finite-dimensional C^*-algebra can be realized as a $*$-subalgebra of $\mathcal{O}(X)$ for some finite-dimensional Hilbert space X. We have a similar supplement to 22.12 for separable C^*-algebras.

Proposition. *A separable C^*-algebra A is isometrically $*$-isomorphic with a norm-closed $*$-subalgebra of $\mathcal{O}(X)$ for some separable Hilbert space X.*

Proof. By 22.12 we may as well assume that A is a closed $*$-subalgebra of $\mathcal{O}(Y)$ for some (not necessarily separable) Hilbert space Y. Choose a countable dense subset C of A. There clearly exists a countable set D of unit vectors in Y such that

$$\|a\| = \sup\{\|a\xi\| : \xi \in D\} \tag{5}$$

for every a in C. It then follows from the denseness of C that (5) holds for every a in A.

Let X be the smallest A-stable closed linear subspace of Y containing D. By the argument of 9.12 X is separable. By (5) $a \mapsto a|X$ is an isometry on A; so A is isometrically $*$-isomorphic with the closed $*$-subalgebra $\{a|X : a \in A\}$ of $\mathcal{O}(X)$. ∎

22.14.   Proposition. *Let A be a reduced Banach $*$-algebra and $\| \ \|_c$ its C^*-norm (10.3). For $a \in A$ we have*

$$\|a\|_c = \sup\{\|T_a\| : T \text{ is an irreducible } *\text{-representation of } A\}.$$

In particular, the irreducible $*$-representations of A separate the points of A.

Proof. By 22.10 (applied to the C^*-completion of A) and 10.5. ∎

22.15. As another corollary of the Gelfand–Naimark Theorem we have the following useful criterion for positivity.

Proposition. *An element a of a C^*-algebra A is positive if and only if $p(a) \geq 0$ for all positive linear functionals p on A.*

Proof. The "only if" part is obvious. To prove the converse, we may assume by 22.12 that A is a concrete C^*-algebra of operators in a Hilbert space X. For each ξ in X, $b \mapsto (b\xi, \xi)$ is a positive linear functional on A, and so $(a\xi, \xi) \geq 0$. Hence a is positive by 11.10 and 7.3. ∎

23. Application of Extension Techniques to the Algebra of Compact Operators

23.1. As an interesting application of the Extension Theorem 22.6 we shall prove the following theorem of Rosenberg:

Theorem. *A separable C*-algebra A which has (to within unitary equivalence) only one irreducible *-representation is elementary (i.e., is *-isomorphic with $\mathcal{O}_c(X)$ for some Hilbert space X).*

Proof. The first step is to show that such an A is topologically simple (see 15.15). Let I be a proper closed two-sided ideal of A. Then A/I is a non-zero C^*-algebra (8.7), and so by 22.10 has an irreducible *-representation S. Thus $T: a \mapsto S_{a+I}$ is an irreducible *-representation of A—the only one. It follows that any non-zero element of I, being in $\mathrm{Ker}(T)$, could not be distinguished from 0 by irreducible *-representations of A. So by 22.10 $I = \{0\}$; and A is topologically simple.

In particular, the unique irreducible *-representation T of A has zero kernel, i.e., is faithful.

Take a (norm-closed) maximal commutative *-subalgebra B of A. We claim that the Gelfand space $\hat{B}$ of B is countable. Indeed, by 22.9, to each ϕ in $\hat{B}$ we can associate a unit vector ξ_ϕ in $X(T)$ such that

$$T_b \xi_\phi = \phi(b)\xi_\phi \qquad \text{for all } b \text{ in } B. \tag{1}$$

If ϕ, ψ are distinct elements of $\hat{B}$, we can find a Hermitian element b of B such that $\phi(b) \neq \psi(b)$; and this implies by (1) that $\phi(b)(\xi_\phi, \xi_\psi) = (T_b \xi_\phi, \xi_\psi) = (\xi_\phi, T_b \xi_\psi) = \psi(b)(\xi_\phi, \xi_\psi)$, whence $(\xi_\phi, \xi_\psi) = 0$. Thus if $\hat{B}$ were uncountable, $X(T)$ would have uncountably many orthogonal unit vectors, contradicting the fact that A is separable and hence (9.12) $X(T)$ is separable. Thus $\hat{B}$ is countable.

By the Baire Category Theorem, the countable locally compact Hausdorff space $\hat{B}$ has at least one isolated point ϕ_0. Since $B \cong \mathscr{C}_0(\hat{B})$, there is a projection p in B such that $\phi_0(p) = 1$ and $\phi(p) = 0$ for $\phi \neq \phi_0$. Every element b of B is of the form

$$b = \lambda p + c, \tag{2}$$

where $\lambda \in \mathbb{C}$, $c \in B$, and $cp = 0$.

Let a be any element of A. If b is any element of B, written in the form (2), we have (since $cp = pc = 0$)

$$b(pap) = \lambda(pap) + cpap$$

$$= \lambda(pap)$$

$$= \lambda(pap) + papc = (pap)b.$$

Thus pap commutes with all elements of B, and so by the maximality of B (see 1.2) belongs to B. By the arbitrariness of a, this implies that pAp is a commuting set.

Now $T_p \neq 0$, since T is faithful. So by V.1.21 the *-representation $a \mapsto T_a|\text{range}(T_p)$ of pAp is irreducible. By 14.3 and the commutativity of pAp this means that T_p is of rank 1.

We have shown that range(T) contains some non-zero compact operator. By 15.17 it follows that

$$\mathscr{O}_c(X(T)) \subset \text{range}(T). \tag{3}$$

Now $\mathscr{O}_c(X(T))$ is a closed two-sided ideal of $\mathscr{O}(X(T))$; hence its inverse image under T is a closed two-sided ideal I of A. By (3) $I \neq \{0\}$. Hence, by the topological simplicity of A, $I = A$. Therefore (3) becomes

$$\text{range}(T) = \mathscr{O}_c(X(T)). \tag{4}$$

Since T is faithful, (4) together with 3.9 shows that A is isometrically *-isomorphic with $\mathscr{O}_c(X(T))$, and so is elementary. This completes the proof. ∎

Remark. Does Rosenberg's Theorem hold without the hypothesis of the separability of A? This is at present an unsolved problem, of great importance for the theory of the *-representations of non-separable C^*-algebras.

23.2   We shall now prove a proposition which has considerable interest of its own, and also leads to the structure theory of norm-closed *-subalgebras of $\mathscr{O}_c(X)$.

Proposition. *Let T be any non-degenerate *-representation of a *-algebra such that* range(T) $\subset \mathscr{O}_c(X(T))$. *Then T is discretely decomposable* (14.6).

Proof.   Put $X = X(T)$. By hypothesis the norm-closure B of range(T) is a *-subalgebra of $\mathscr{O}_c(X)$. By 22.10 B has an irreducible *-representation S; and by 22.6 S has an extension to an irreducible *-representation V of $\mathscr{O}_c(X)$. But, in view of 15.12, V must be the identity *-representation of $\mathcal{O}_c(X)$. This implies that there exists a closed subspace Y of X, stable under B (hence under T), on which B (hence also T) acts irreducibly.

Now let $\{Y_i\}$ be a maximal pairwise orthogonal family of closed T-stable T-irreducible subspaces of X. If $X = \sum_i^\oplus Y_i$ (Hilbert direct sum), the proof is finished. Otherwise, we apply the preceding paragraph to the subrepresentation of T acting on $Z = (\sum_i^\oplus Y_i)^\perp$, obtaining a T-irreducible closed subspace of Z and thus contradicting the maximality of $\{Y_i\}$. ∎

23.3. Theorem. *For any C*-algebra A, the following two conditions are equivalent:* (I) A *is *-isomorphic with a norm-closed *-subalgebra of an elementary C*-algebra;* (II) A *is *-isomorphic with a C*-direct sum of elementary C*-algebras.*

Proof. The implication (II) ⇒ (I) is very easy. To see the converse, take a norm-closed (non-degenerate) *-subalgebra B of $\mathcal{O}_c(X)$, and by 23.2 write $X = \sum_{i\in I}^\oplus Y_i$, where Y_i is irreducible under B for each i in I. Choose a maximal subset J of I such that the subrepresentations S^j of the identity representation of B, acting on the different Y_j for $j\in J$, are pairwise inequivalent. Thus $\Phi: b \mapsto \sum_{j\in J}^\oplus S_b^j$ is a *-isomorphism of B into the C^*-direct sum of the $\mathcal{O}_c(Y_j)$ ($j\in J$). Now apply Proposition 15.18 to range(Φ). ∎

This theorem generalizes 3.14.

Definition. A C^*-algebra satisfying conditions (I) and (II) of the above theorem will be said to be *of compact type*.

23.4. It is easy to deduce from 23.3 just what a norm-closed *-subalgebra of $\mathcal{O}_c(X)$ looks like "geometrically."

Let I be an index set; and for each i in I let X_i and Y_i be two non-zero Hilbert spaces, Y_i being finite-dimensional. We form the Hilbert direct sum $Z = \sum_{i\in I}^\oplus (X_i \otimes Y_i)$; and denote by B the concrete C^*-algebra of all operators on Z of the form $\sum_{i\in I}^\oplus (\beta_i \otimes 1_i)$, where (i) for each i, $\beta_i \in \mathcal{O}_c(X_i)$ and 1_i is the identity operator on Y_i, and (ii) $\lim_i \|\beta_i\| = 0$. Then evidently $B \subset \mathcal{O}_c(Z)$. A concrete C^*-algebra of compact type constructed in this way is said to be *canonical*.

Proposition. *Every concrete C*-algebra A of compact type, acting non-degenerately on its Hilbert space X, is unitarily equivalent to some canonical concrete C*-algebra B of compact type. (That is, if Y is the Hilbert space of B, there is a linear isometry u of X onto Y such that $A = \{u^{-1}bu : b\in B\}$.)*

24.   Von Neumann Algebras and *-Algebras with Type I Representation Theory

24.1.   This section introduces, all too briefly, the fundamental notions of a von Neumann algebra and of a *-algebra having a Type I representation theory.

Let X be a fixed Hilbert space. If $M \subset \mathcal{O}(X)$, the *commuting algebra* of M (see 9.3) is the subalgebra $\{a \in \mathcal{O}(X): ab = ba$ for all b in $M\}$ of $\mathcal{O}(X)$. *In this section we shall denote the commuting algebra of M by M'.* Evidently the algebra M' contains the identity operator 1_X and is closed in the weak operator topology. If M is closed under the adjoint operation, then so is M'.

By M'' we mean of course $(M')'$. Notice that

$$M \subset M''. \tag{1}$$

24.2.   Von Neumann Double Commuter Theorem.   *Let A be a *-subalgebra of $\mathcal{O}(X)$ acting non-degenerately on X (that is, its identity representation on X is non-degenerate). Then the following three objects are equal: (i) A'', (ii) the weak operator closure of A, and (iii) the strong operator closure of A.*

Proof.   Since A'' is weakly closed we have (i) $\supset$ (ii) $\supset$ (iii). So by (1) it is enough to prove that A is dense in A'' in the strong operator topology.

To do this we shall first show that

$$b\xi \text{ is in the closure of } A\xi \tag{2}$$

whenever $\xi \in X$ and $b \in A''$. Let p be projection onto the closure Z of $A\xi$. Then Z is A-stable and so $p \in A'$. Thus $bp = pb$ (since $b \in A''$); and hence Z is stable under b. Since $\xi \in Z$ by 9.10, we therefore have $b\xi \in Z$, and (2) is proved.

Now let Y be any Hilbert space, and form the Hilbert tensor product $\mathscr{X} = X \otimes Y$. For each a in $\mathcal{O}(X)$ let $\tilde{a}$ denote the operator $a \otimes 1_Y$ in $\mathcal{O}(\mathscr{X})$. Thus $\tilde{A} = \{\tilde{a}: a \in A\}$ is a *-algebra of operators acting non-degenerately on $\mathscr{X}$. We now claim that

$$\text{if } b \in A'', \text{ then } \tilde{b} \in (\tilde{A})''. \tag{3}$$

To see this, let g be any operator in $\tilde{A}'$. We introduce a basis $\{\eta_i\}$ of Y and maps e_i just as in the proof of 14.5, and write $g_{ij} = e_j^* g e_i \in \mathcal{O}(X)$. By the same calculation as in 14.5 we conclude that each g_{ij} is in A'. Now let $b \in A''$; then b

commutes with the g_{ij}. So, for any i and any ξ in X, we have (using 14.5(2), 14.5(3)):

$$\tilde{b}ge_i\xi = \tilde{b}\left[\sum_j e_j^* ge_i\xi \otimes \eta_j\right]$$

$$= \sum_j (bg_{ij}\xi \otimes \eta_j)$$

$$= \sum_j (g_{ij}b\xi \otimes \eta_j)$$

$$= ge_ib\xi = g\tilde{b}e_i\xi.$$

Since the $e_i\xi$ (i, ξ varying) span a dense subspace of X, this implies that $\tilde{b}g = g\tilde{b}$. By the arbitrariness of g this gives $\tilde{b} \in (\tilde{A})''$, proving (3).

We are now ready to show that A is strongly dense in A''. Let $b \in A''$; take any finite sequence of vectors $\xi_1, \ldots, \xi_n$ in X; let Y be an n-dimensional Hilbert space with orthonormal basis $\eta_1, \ldots, \eta_n$; and form the operators $\tilde{a}$ ($a \in \mathcal{O}(X)$) in $\mathcal{X} = X \otimes Y$ as in the preceding paragraph. By (3) $\tilde{b} \in (\tilde{A})''$. Hence we can apply (2) with A, b, and ξ replaced by $\tilde{A}$, $\tilde{b}$, and $\zeta = \sum_{i=1}^n (\xi_i \otimes \eta_i)$, and obtain a sequence $\{a_n\}$ of elements of A such that $\tilde{a}_n\zeta \to \tilde{b}\zeta$, that is,

$$a_n\xi_i \to b\xi_i \qquad \text{for each } i = 1, \ldots, n.$$

This says that A is strongly dense in A'', proving the theorem. ∎

24.3.   Remark.   The equality of (ii) and (iii) in Theorem 24.2 is a fact of much more general validity. Indeed, we shall see in 25.2 that the weak operator topology of $\mathcal{O}(X)$ is just the weak topology generated by the set of all strongly continuous linear functionals on $\mathcal{O}(X)$. Thus, by a well-known theorem on locally convex linear spaces (see for example Kelley and Namioka [1], Theorem 17.1) the weak and strong operator closures of M will coincide for any convex subset M of $\mathcal{O}(X)$, in particular for any linear subspace M of $\mathcal{O}(X)$.

However, even if X is finite-dimensional and M is a (non-self-adjoint) subalgebra of $\mathcal{O}(X)$ containing 1_X, the equation $M'' = M$ is usually false. For example, if $\dim(X) = 2$ and M consists of all triangular matrices $\begin{pmatrix} a & b \\ 0 & c \end{pmatrix}$ ($a, b, c \in \mathbb{C}$), then $M'' = \mathcal{O}(X)$ (see 14.3).

24.4.   Definition.   A *-subalgebra A of $\mathcal{O}(X)$ with the property that $A'' = A$ is called a *von Neumann algebra* of operators on the Hilbert space X.

In view of the Double Commuter Theorem there are several equivalent definitions of a von Neumann algebra. For example, a *-subalgebra A of $\mathcal{O}(X)$ is a von Neumann algebra if and only if it satisfies any one of the following conditions:

(I) A is weakly closed and $1_X \in A$;

(II) A is strongly closed and $1_X \in A$;

(III) A is weakly closed and acts non-degenerately on X;

(IV) A is strongly closed and acts non-degenerately on X.

The equivalence of (I)–(IV) with the definition of a von Neumann algebra is an immediate consequence of 24.2.

24.5.   The commuting algebra A' of any *-subalgebra A of $\mathcal{O}(X)$ is a von Neumann algebra (by 24.4(I)). Thus the von Neumann algebras on X occur in pairs, A being paired with its commuting algebra A'—the iteration of this process returning us to $A'' = A$. Notice that the von Neumann algebras A and A' have the same center Z:

$$Z = A \cap A'. \tag{4}$$

24.6.   Definition.   If $M \subset \mathcal{O}(X)$, the smallest von Neumann algebra A of operators on X containing M (which exists by 24.4(I)) is called the *von Neumann algebra generated by M*.

If M is a *-subalgebra of $\mathcal{O}(X)$ acting non-degenerately on X, by 24.4(III)(IV) A is just the weak (or strong) operator closure of M.

24.7.   The Double Commuter Theorem has a very important implication for *-algebras of operators which act irreducibly on a Hilbert space.

Corollary.   *If T is an irreducible *-representation of a *-algebra A, then* range(T) *is strongly dense in* $\mathcal{O}(X(T))$; *that is to say, T is totally irreducible in the sense of* V.1.10.

Proof.   By 14.1 (range(T))' consists of the scalar operators only. Therefore (range(T))'' = $\mathcal{O}(X(T))$. Now apply 24.2.   ∎

24.8.   We shall now introduce the very important idea of a primary representation of Type I. This concept will be used in Chapter XII; but its full development lies outside the scope of this work.

Definition. A von Neumann algebra A on X is a *factor* if its center Z consists of the scalar operators only (i.e., $Z = \mathbb{C}\mathbf{1}_X$).

By (4) A is a factor if and only if its commuting algebra A' is a factor.

Definition. A non-degenerate *-representation T of a *-algebra A is *primary* (or a *factor representation*) if the von Neumann algebra generated by range(T) is a factor—or, equivalently, if the commuting algebra of T is a factor.

24.9. Now $\mathcal{O}(X)$ is a von Neumann algebra on X whose center is $\mathbb{C}\mathbf{1}_X$. (The latter statement follows from 14.1, and is easily checked directly.) Thus $\mathcal{O}(X)$, and likewise any von Neumann algebra *-isomorphic with it, is a factor. We give a name to factors of this special kind.

Definition. A von Neumann algebra B on X is called a *factor of Type I* if there exists a Hilbert space Y such that B is *-isomorphic with $\mathcal{O}(Y)$.

A non-degenerate *-representation T of a *-algebra A is called a *primary* (or *factor*) *representation of Type I* if the commuting algebra of T is a factor of Type I.

Notice especially that the property of being primary of Type I depends only on the *-isomorphism type of the commuting algebra of the *-representation.

24.10. Here is a different characterization of Type I primary representations.

Proposition. *A non-degenerate *-representation T of the *-algebra A is a primary representation of Type I if and only if there is an irreducible *-representation S of A such that T is unitarily equivalent to a Hilbert direct sum of copies of S (i.e., $T \cong \sum_{i \in I}^{\oplus} S$ for some index set I).*

Proof. If T is a direct sum of copies of an irreducible *-representation S, we have (to within unitary equivalence):

$$X(T) = X(S) \otimes Y,$$

$$T_a = S_a \otimes \mathbf{1}_Y \qquad\qquad (a \in A),$$

for some Hilbert space Y. So, by 14.5, the commuting algebra of T is $\{\mathbf{1}_{X(S)} \otimes b : b \in \mathcal{O}(Y)\}$, which is evidently *-isomorphic with $\mathcal{O}(Y)$. Thus T is primary of Type I.

Conversely, assume that T is primary of Type I. Thus there is a Hilbert space Y and a *-isomorphism Φ of $\mathcal{O}(Y)$ onto the commuting algebra B of T.

We choose an orthonormal basis $\{\eta_i\}$ ($i \in I$) of Y, and denote by q_i the projection of Y onto $\mathbb{C}\eta_i$. Since Φ is a *-isomorphism and q_i is a minimal non-zero projection in $\mathcal{O}(Y)$, the image $p_i = \Phi(q_i)$ of q_i must be a minimal non-zero projection in the commuting algebra B. This implies that $Z_i = \text{range}(p_i)$ is stable and irreducible under T for each i in I. The Z_i are pairwise orthogonal. If their sum were not dense in $X(T)$, there would exist a non-zero projection r in B such that $rp_i = 0$ for all i, implying that $\Phi^{-1}(r)q_i = 0$ for all i; but the latter is impossible unless $\Phi^{-1}(r) = 0$, a contradiction. So $\sum_i Z_i$ is dense in $X(T)$. Hence, if S^i stands for the irreducible subrepresentation of T acting on Z_i, we have

$$T = \sum_{i \in I}^{\oplus} S^i \qquad \text{(Hilbert direct sum).} \qquad (5)$$

Now, if i, j are distinct elements of I, the element u of $\mathcal{O}(Y)$ which sends η_i into η_j and annihilates $\eta_i^{\perp}$ satisfies $u^*u = q_i$, $uu^* = q_j$. So $v = \Phi(u)$ belongs to B and satisfies $v^*v = p_i$, $vv^* = p_j$. Thus v is a partial isometry (see 13.2), mapping Z_i isometrically onto Z_j and intertwining S^i and S^j. Hence $S^i \cong S^j$ for all i, j in I. This together with (5) asserts that T is a Hilbert direct sum of copies of a single irreducible *-representation of A. ∎

In particular, of course, an irreducible *-representation is primary of Type I.

24.11. Corollary.   *If the von Neumann algebra B is a factor of Type I, then its commuting algebra B' is also a factor of Type I.*

24.12. Corollary.   *A primary *-representation T of a *-algebra A is of Type I if and only if it is discretely decomposable (14.6).*

Proof.   By 24.10 the Type I property implies discrete decomposability. Conversely, assume that T is discretely decomposable; and let σ be a unitary equivalence class of irreducible *-representations of A which occurs in T. Thus the projection p of $X(T)$ onto $X_\sigma(T)$ is non-zero and belongs to the commuting algebra B of T. In fact, by 14.15, p is in the center of B. Since T is primary, this implies that $p = \mathbf{1}$, i.e., $X_\sigma(T) = X(T)$. So T is a Hilbert direct sum of copies of σ, hence of Type I. ∎

24.13.   We come now to an extremely important definition.

Definition.   A *-algebra A is said to have a *Type I *-representation theory* if every primary *-representation of A is of Type I.

Remark. Much profound work has been done in the last thirty-five years on the classification of C^*-algebras and locally compact groups according to whether their *-representation theories are or are not of Type I. It turns out that the distinction between the two alternatives is very striking and incisive. If a separable C^*-algebra A has a Type I *-representation theory, then the space of its irreducible *-representations is in a certain precise sense "smooth" and therefore "classifiable." In addition it turns out that, for each irreducible *-representation T of such an A, the range of T contains some non-zero compact operator; thus (see 15.18) the irreducible *-representations of A are determined by their kernels. If on the other hand A does not have a Type I representation theory then none of these pleasant properties hold.

The Glimm algebras of 17.5 are examples of C^*-algebras which do *not* have a Type I *-representation theory (though we shall not prove this fact). We end this section with two situations in which it is easy to prove that the Type I property does hold.

24.14. Proposition. *A commutative *-algebra A has a Type I *-representation theory.*

Proof. Let T be a non-degenerate primary *-representation of A on $X = X(T)$. It must be shown that T is of Type I.

Denote by A^0 the norm-closure of range(T), and by T^0 the identity *-representation of A^0. Evidently the commuting algebras of T and T^0 are the same. So it is enough to prove that T^0 is of Type I. Thus, since A^0 is a commutative C^*-algebra, we may and shall assume from the beginning that A is a commutative C^*-algebra.

Let P be the spectral measure (10.10) of T. From the regularity of P it is easy to verify that range(P) is contained in the center of the von Neumann algebra generated by T. (In fact, we shall verify a more general statement in VII.8.13). Since T is primary it follows that range(P) consists of 0 and 1_X only. Applying II.11.13 we conclude that P is concentrated at one point ϕ on $\hat{A}$ (i.e., $P(W) = \mathrm{Ch}_W(\phi)1_X$ for each Borel subset W of $\hat{A}$). By the definition of the spectral measure this implies that $T_a = \phi(a)1_X$ ($a \in A$); in other words T is a direct sum of copies of ϕ. So by 24.10 T is of Type I. ∎

24.15. Proposition*. *A C^*-algebra A of compact type (see 23.3) has a Type I *-representation theory.*

Using 23.3 and 15.12, one shows that every non-degenerate *-representation of A is discretely decomposable. Now apply 24.12.

24.16.   Remark.   In VII.9.11 we shall obtain a very general sufficient condition for a (non-commutative) C*-algebra to have a Type I *-representation theory.

25.   Kadison's Irreducibility Theorem and Related Properties of C*-Algebras

25.1.   It turns out that every irreducible *-representation of a C*-algebra A is automatically algebraically irreducible. In fact, there is no essential difference between the space of unitary equivalence classes of irreducible *-representations of A and the space of algebraic equivalence classes of algebraically irreducible (algebraic) representations of A. To establish these surprising facts, we must first prove the Kaplansky Density Theorem.

25.2.   Lemma.   *Let X be a Hilbert space. Every strongly continuous linear functional ϕ on $\mathcal{O}(X)$ belongs to the linear span of the set of all functionals on $\mathcal{O}(X)$ of the form $a \mapsto (a\xi, \eta)$ (where $\xi, \eta \in X$).*

Proof.   By the strong continuity of ϕ we can find finitely many vectors $\xi_1, \ldots, \xi_n$ in X such that, for any a in $\mathcal{O}(X)$,

$$\|a\xi_i\| \leq 1 \quad \text{for all } i = 1, \ldots, n \Rightarrow |\phi(a)| \leq 1.$$

It follows that, if $\{a_v\}$ is any net in $\mathcal{O}(X)$ such that $a_v \xi_i \underset{v}{\to} 0$ for each i, then $\phi(a_v) \underset{v}{\to} 0$. In particular,

$$a\xi_i = 0 \quad \text{for each } i \Rightarrow \phi(a) = 0. \tag{1}$$

Now let $T : \mathcal{O}(X) \mapsto \mathscr{X} = \sum_{i=1}^{n \oplus} X$ be defined by $(T(a))_i = a\xi_i$. By (1) ϕ vanishes on the kernel of T; and, by the continuity statement preceding (1), the functional ϕ' on $T(\mathcal{O}(X))$ deduced from ϕ is continuous in the norm of $\mathscr{X}$. Therefore we can find an element η of $\mathscr{X}$ such that

$$\phi(a) = (T(a), \eta)_{\mathscr{X}}$$

$$= \sum_{i=1}^{n} (a\xi_i, \eta_i) \qquad (a \in A).$$

This proves the Lemma.   ∎

25.3.   Kaplansky Density Theorem.   *Let X be a Hilbert space and A a *-subalgebra of $\mathcal{O}(X)$. Let b be an operator in $\mathcal{O}(X)$ belonging to the weak operator closure of A; and suppose $\|b\| \leq 1$. Then b belongs to the strong* *operator closure of* $A_1 = \{a \in A : \|a\| \le 1\}$. *If in addition* $b^* = b$, *then b belongs to the strong operator closure of* $A_1 \cap \{a : a^* = a\}$.

Proof. We may, and shall, assume without cost that A is norm-closed, hence a C^*-algebra.

Suppose first that $b^* = b$. By hypothesis some net $\{a_i\}$ of elements of A converges weakly to b. Since the adjoint operation is weakly continuous, this implies that $\frac{1}{2}(a_i + a_i^*) \to b$ weakly; so b lies in the weak closure of $M = \{a \in A : a^* = a\}$. Now by 25.2 the strongly continuous and the weakly continuous linear functionals on $\mathcal{O}(X)$ are the same. So by a well-known theorem on convex sets (see for example Kelley and Namioka [1], Theorem 17.1) the weak and strong closures of the convex set M are the same. Thus b lies in the strong closure of M.

We shall now make vigorous use of the functional calculus on Hermitian elements of A (see §6). Let $g(r) = 2r(1 + r^2)^{-1}$ (r real). Then $g|[-1, 1]$ is strictly monotone increasing with range $[-1, 1]$. So $h = (g|[-1, 1])^{-1}$ is continuous on $[-1, 1]$, and, since $\|b\| \le 1$, we can apply h to b. Denote $h(b)$ by b'. Thus b' belongs to the norm-closed algebra generated by b, hence to the weak closure of A, hence (like b) to the strong closure of M. Note also that

$$g(b') = g(h(b)) = b. \tag{2}$$

We now choose a net $\{b_v'\}$ of elements of M converging strongly to b'; and set $g(b_v') = b_v$. Thus $b_v \in M$; in fact, since $|g(r)| \le 1$ for all real r, we have $\|b_v\| \le 1$, that is, $b_v \in A_1 \cap \{a : a^* = a\}$. We claim that $b_v \to b$ strongly. This will prove the statement of the theorem for Hermitian elements.

To prove the claim, observe from (2) that $b = 2b'(1 + b'^2)^{-1}$ and $b_v = 2b_v'(1 + b_v'^2)^{-1}$; so

$$\begin{aligned} \tfrac{1}{2}(b_v - b) &= (1 + b_v'^2)^{-1}[b_v'(1 + b'^2) - (1 + b_v'^2)b'](1 + b'^2)^{-1} \\ &= (1 + b_v'^2)^{-1}(b_v' - b')(1 + b'^2)^{-1} \\ &\quad + (1 + b_v'^2)^{-1}b_v'(b' - b_v')b'(1 + b'^2)^{-1} \\ &= (1 + b_v'^2)^{-1}(b_v' - b')(1 + b'^2)^{-1} + \tfrac{1}{4}b_v(b' - b_v')b. \tag{3} \end{aligned}$$

Since $\|(1 + b_v'^2)^{-1}\| \le 1$, $\|b_v\| \le 1$ for all v, and $b_v' \to b'$ strongly, (3) implies that $b_v \to b$ strongly. So the claim is proved.

We now drop the assumption that $b^* = b$; and consider the C^*-algebra A^0 of all bounded linear operators on $Y = X \oplus X$ having matrices of the form $\begin{pmatrix} a_{11} & a_{12} \\ a_{21} & a_{22} \end{pmatrix}$, where $a_{ij} \in A$. Since $\|b\| \le 1$ and b is in the weak closure of A, the matrix $b^0 = \begin{pmatrix} 0 & b \\ b^* & 0 \end{pmatrix}$ is Hermitian, satisfies $\|b^0\| \leq 1$, and belongs to the weak closure of A^0. Hence by the preceding part of the proof b^0 is the strong limit of a net of elements

$$\left\{ \begin{pmatrix} a^v_{11} & a^v_{12} \\ a^v_{21} & a^v_{22} \end{pmatrix} \right\}$$

of the unit ball of A^0. From this it follows that $a^v_{12} \in A$, $\|a^v_{12}\| \leq 1$ and $a^v_{12} \to b$ strongly. This completes the proof. ∎

25.4.   Lemma.   *Let r be a non-negative number, X a Hilbert space, $\xi_1, \ldots, \xi_n$ a finite n-termed orthonormal sequence of vectors in X, and $\eta_1, \ldots, \eta_n$ an n-termed sequence of vectors in X with $\|\eta_i\| \leq r$ for all i. Then there exists an operator b in $\mathcal{O}(X)$ such that (i) $b\xi_i = \eta_i$ for all $i = 1, \ldots, n$, and (ii) $\|b\| \leq r(2n)^{1/2}$. If in addition*

$$(\eta_i, \xi_j) = (\xi_i, \eta_j) \qquad \text{for all } i, j, \tag{4}$$

the above b can be chosen to be Hermitian.

Proof.   Since we are concerned with only a finite-dimensional subspace of X, we may as well assume X is of finite dimension m, and enlarge $\{\xi_i\}$ to an orthonormal basis

$$\xi_1, \ldots, \xi_n, \ldots, \xi_m$$

of X. Define b as the operator on X whose matrix β with respect to this basis is as follows: (a) $\beta_{ij} = (\eta_j, \xi_i)$ for $1 \leq i \leq m$, $1 \leq j \leq n$; (b) if (4) holds, then $\beta_{ij} = (\xi_j, \eta_i)$ for all $1 \leq i \leq n$, $1 \leq j \leq m$; (c) $\beta_{ij} = 0$ for those i, j not covered by (a), (b). Thus $b\xi_i = \eta_i$ for all i, b is Hermitian if (4) holds, and

$$\|b\|^2 \leq \text{Trace}(b^*b) = \sum_{i,j=1}^{n} |\beta_{ij}|^2$$

$$\leq 2 \sum_{i=1}^{n} \|\eta_i\|^2 \leq 2nr^2. \qquad \blacksquare$$

25.5.   Before proving Kadison's Irreducibility Theorem we want to remind the reader of the following crucial result on general Banach spaces.

Let X and Y be two real or two complex Banach spaces; and if $0 < r \in \mathbb{R}$, put $X_r = \{x \in X : \|x\| < r\}$, $Y_r = \{y \in Y : \|y\| < r\}$.

Proposition. *Let $E: X \to Y$ be a bounded linear map. If r, s are positive real numbers such that Y_s is contained in the closure of $E(X_r)$, then $Y_s \subset E(X_r)$.*

The proof of this is contained, for example, in the proof of Theorem 3, p. 36 of Day [1].

25.6. Kadison Irreducibility Theorem. *Let A be a C^*-algebra and T an irreducible $*$-representation of A on X. Let $\xi_1, \ldots, \xi_n$ be a finite n-termed orthonormal sequence of vectors in X, and $\eta_1, \ldots, \eta_n$ an n-termed sequence of vectors in X with $\|\eta_i\| < r$. Then there is an element b of A such that $T_b \xi_i = \eta_i$ for each $i = 1, \ldots, n$ and $\|b\| < r(2n)^{1/2}$. If in addition (4) holds, b can be taken to be Hermitian.*

Proof. In view of 8.10 we may replace A by $T(A)$, and assume that A is a norm-closed $*$-subalgebra of $\mathcal{O}(X)$ acting irreducibly on X.

By 24.7 A is strongly dense in $\mathcal{O}(X)$. So by 25.3, for each positive r the ball $A_r = \{a \in A : \|a\| < r\}$ is strongly dense in the ball $\mathcal{O}_r(X) = \{b \in \mathcal{O}(X) : \|b\| < r\}$. Let $\mathcal{X} = \sum_{i=1}^{n \oplus} X$, considered as a Banach space with norm $\|\zeta\| = \sup\{\|\zeta_i\| : i = 1, 2, \ldots, n\}$; and define $\Phi: A \to \mathcal{X}$ as the continuous linear map given by

$$\Phi(a) = \langle a\xi_1, a\xi_2, \ldots, a\xi_n \rangle.$$

By 25.4 and the denseness of A_r in $\mathcal{O}_r(X)$, we conclude that the closure of $\Phi(A_{r(2n)^{1/2}})$ contains $\mathcal{X}_r$. Therefore by 25.5 $\mathcal{X}_r \subset \Phi(A_{r(2n)^{1/2}})$. Since $\langle \eta_1, \ldots, \eta_n \rangle \in \mathcal{X}_r$, this proves the first statement of the theorem.

Now assume in addition that (4) holds. let A^h be the real Banach space of all Hermitian elements of A, and $\mathcal{X}^h$ the real Banach subspace of $\mathcal{X}$ consisting of those ζ which satisfy $(\zeta_i, \xi_j) = (\xi_i, \zeta_j)$ for all i, j. In particular $\langle \eta_1, \ldots, \eta_n \rangle \in \mathcal{X}_r^h$. Noticing that $\Phi(A_h) \subset \mathcal{X}^h$, we apply the same argument as in the preceding paragraph (using the statement about Hermitian elements in 25.3) to show that $\mathcal{X}_r^h \subset \Phi(A_{r(2n)^{1/2}}^h)$. This completes the proof. ∎

25.7. From this we draw the surprising conclusion:

Corollary. *An irreducible $*$-representation of a C^*-algebra is algebraically totally irreducible.*

Thus, for $*$-representations of C^*-algebras, all the four notions of irreducibility given in V.1.14 coincide.

Positive Functionals and Regular Maximal Left Ideals

25.8. We shall conclude this chapter by setting up a one-to-one correspondence between the indecomposable positive functionals p satisfying $\|p\| = 1$ and the regular maximal left ideals of a C^*-algebra A. From this the following interesting fact will emerge: Every algebraically irreducible (purely algebraic) representation of A is equivalent (in the purely algebraic sense) to some irreducible *-representation of A.

25.9. Let p be a positive linear functional on a C^*-algebra A. We recall from 19.3 that the null ideal

$$I_p = \{a \in A : p(a^*a) = 0\}$$

is a left ideal of A, and that A/I_p is a pre-Hilbert space under the inner product

$$(a + I_p, b + I_p) = p(b^*a). \tag{5}$$

Proposition.   *Suppose that the positive linear functional p on A is indecomposable. Then the space A/I_p is complete with respect to the inner product (5). The (irreducible) *-representation of A generated by p is unitarily equivalent to the natural representation S of A on A/I_p.*

Proof.   Let $\langle T, \xi \rangle$ be the cyclic pair corresponding by 19.9 to p; T is thus irreducible (20.4). The map

$$a + I_p \mapsto T_a \xi \tag{6}$$

is a linear isometry of A/I_p (with the inner product (5)) onto $T(A)\xi$. But by 25.7 $T(A)\xi = X(T)$. So A/I_p is complete. The map (6) sets up the required unitary equivalence between S and T.   ∎

25.10. The next lemma will show that an indecomposable positive functional on a C^*-algebra is essentially determined by its null ideal.

Lemma.   *Let p be a positive linear functional on a C^*-algebra A, and I its null ideal. Then $I + I^* \subset \text{Ker}(p)$. If p is indecomposable, then $I + I^* = \text{Ker}(p)$.*

Proof.   Since p is automatically extendable (18.15), $p(a^*a) = 0$ implies $p(a) = 0$; so $I \subset \text{Ker}(p)$. Also $\text{Ker}(p)$ is self-adjoint (since $p(a^*) = \overline{p(a)}$); so $I + I^* \subset \text{Ker}(p)$; and the first statement is proved.

Now assume p is indecomposable; let $\langle T, \xi \rangle$ be the corresponding cyclic pair (T being irreducible); and let $b \in \text{Ker}(p)$. Putting $T_b \xi = \eta$, we have

$(\eta, \xi) = p(b) = 0$. By 25.6, therefore, there exists a Hermitian element a of A such that

$$T_a \xi = 0, \ T_a \eta = \eta.$$

This implies that a and $b - ab$ are both in I. Hence

$$b^* = (b - ab)^* + b^* a \in I^* + I,$$

so that $b \in I^* + I$. We have shown that $\mathrm{Ker}(p) \subset I^* + I$. ∎

25.11. The next proposition considerably strengthens the Gelfand–Naimark Theorem. Indeed, the latter is implied by the special case $I = \{0\}$.

Proposition. *Let A be a C*-algebra, and I any norm-closed left ideal of A. Then*

$$I = \bigcap \{I_p : p \text{ is an indecomposable positive linear}$$

$$\text{functional on } A \text{ such that } I \subset I_p\}. \tag{7}$$

Proof. We may as well assume that A has a unit 1. Otherwise we adjoin 1 to get A_1, and remember (from 20.5) that the restrictions to A of indecomposable positive functionals on A_1 are either 0 or indecomposable.

Define $J = \bigcap \{I_p : p \text{ is a positive linear functional on } A \text{ such that } I \subset I_p\}$. We will first show that $J = I$.

Fix an element a in J, and let $\varepsilon > 0$. Let S be the set of all positive linear functionals p on A such that $P(1) = 1$ and $p(a^*a) \geq \varepsilon$. Since S is closed in the weak topology (i.e., the topology of pointwise convergence on A), and contained in the (weakly compact) unit ball of A^*, S is weakly compact. Now since $a \in J$, $p \in S \Rightarrow I \not\subset I_p$. Hence to each p in S there is an element b_p of I with $p(b_p^* b_p) > 0$; and by continuity we can find a weak neighborhood U_p of p such that $q(b_p^* b_p) > 0$ for all q in U_p. Hence by the compactness of S there exists a finite weakly open covering $U_1, \ldots, U_n$ of S, and elements $b_1, \ldots, b_n$ of I, such that

$$q(b_i^* b_i) > 0 \qquad \text{for each } i \text{ and each } q \text{ in } U_i.$$

Therefore, if we put $b = \sum_{i=1}^n b_i^* b_i$, we have

$$0 \leq b \in I, \tag{8}$$

$$q(b) > 0 \qquad \text{for all } q \text{ in } S. \tag{9}$$

In view of (9) we can multiply b by a large positive number and assume that

$$q(b) > q(a^*a) \qquad \text{for all } q \text{ in } S.$$

By the definition of S it then follows that

$$q(b - a^*a + \varepsilon 1) \geq 0 \tag{10}$$

for all positive linear functionals q satisfying $q(1) = 1$, and hence for all positive linear functionals q on A. From (10) and 22.15 we get $b - a^*a + \varepsilon 1 \geq 0$, or

$$a^*a \leq b + \varepsilon 1. \tag{11}$$

Since $0 \leq b \in I$ and ε was arbitrary, we deduce from (11) and 8.15 that $a^*a \in I$. This and 8.5 imply that $a \in I$.

By the arbitrariness of a we have shown that $J \subset I$. Obviously $I \subset J$. Therefore $J = I$.

Next, let W be the set of all positive linear functionals p on A such that $p(1) = 1$ and $I \subset I_p$. We deduce in the usual manner that W is convex and weakly compact. Hence by the Krein–Milman Theorem W is the weak closure of the convex hull of the set W_e of all extreme points of W. From this it follows easily that $\bigcap \{I_p : p \in W\} = \bigcap \{I_p : p \in W_e\}$. Now we have proved above that $I = \bigcap \{I_p : p \in W\}$. Therefore

$$I = \bigcap \{I_p : p \in W_e\}. \tag{12}$$

In view of (12), the proposition will be proved if we show that the elements of W_e are indecomposable.

Let $p \in W_e$, and suppose that $p = \lambda p_1 + (1 - \lambda)p_2$, where p_1, p_2 are positive linear functionals with $p_i(1) = 1$ and $0 < \lambda < 1$. Clearly $I \subset I_p \subset I_{p_i}$; so $p_i \in W$ $(i = 1, 2)$. Since p is extreme in W, this gives $p_1 = p_2 = p$. Thus p is indecomposable by 22.3. This completes the proof. ∎

25.12.   Theorem.   *Let A be any C*-algebra.*

(I)   *If p is any positive linear functional on A, the null ideal I_p of p is a regular maximal left ideal of A if and only if p is indecomposable.*

(II)   *The map $p \mapsto I_p$ is a one-to-one correspondence between the set of all indecomposable positive linear functionals p of norm 1 on A and the set of all regular maximal left ideals I of A.*

(III)   *Every closed left ideal J of A is the intersection of the set of all regular maximal left ideals of A which contain J.*

Proof.

(a) Let p be an indecomposable positive linear functional on A, and $\langle T, \xi \rangle$ the corresponding cyclic pair. By 20.4 and 25.7 T is algebraically irreducible. Since $I_p = \{a \in A : T_a \xi = 0\}$, I_p is a regular maximal left ideal by V.5.13.

(b) Let I be any regular maximal left ideal of A. By V.6.6 I is norm-closed. Hence by 25.11 $I \subset I_p$ for some indecomposable positive linear functional p. Since I_p must be proper, the maximality of I asserts that $I = I_p$.

(c) Suppose that I and p are as in (b), and that $I = I_q$, where q is another positive linear functional on A. By 25.10

$$\operatorname{Ker}(p) = I^* + I \subset \operatorname{Ker}(q).$$

Since the kernels of p and q are both of codimension 1, this implies that p and q have the same kernels, and so are proportional. Thus q is also indecomposable.

(d) Now (a) and (c) together prove (I). As for (II), the map $p \mapsto I_p$ has regular maximal left ideals for its values by (a), is one-to-one by (c), and has the right range by (b). Part (III) follows from 25.11 together with Part (I). ∎

Remark. Part (III) of the above theorem was proved for the commutative case in V.8.9. In V.8.11 we noticed that it is far from true for general Banach *-algebras.

25.13. Corollary. *Let A be a C^*-algebra.*

(I) *Every algebraically irreducible (algebraic) representation T of the associative algebra underlying A is algebraically equivalent to some irreducible *-representation of A.*

(II) *If two irreducible *-representations S and T of A are algebraically equivalent (as representations of the algebra underlying A), then they are unitarily equivalent.*

Proof.

(I) Suppose $0 \neq \xi \in X(T)$, $I = \{a \in A : T_a \xi = 0\}$. By V.5.13, I is a regular maximal left ideal, and T is equivalent to the natural representation W of A on A/I. Now by 25.12, $I = I_p$ for some indecomposable positive linear functional p on A; and by 25.9 W is equivalent to the irreducible *-representation S of A generated by p. So T is equivalent to S; and (I) is proved.

(II) Suppose that ξ and η are non-zero vectors in $X(S)$ and $X(T)$ respectively which correspond under an algebraic equivalence between S and T. Then

$$\{a \in A : S_a \xi = 0\} = \{a \in A : T_a \eta = 0\}. \tag{13}$$

Put $p(a) = (S_a\xi, \xi)$, $q(a) = (T_a\eta, \eta)$ $(a \in A)$. By (13) $I_p = I_q$. Hence, by 25.10, since p and q are indecomposable they are proportional to each other. Thus, replacing η by a suitable scalar multiple η' of η, we have

$$(S_a\xi, \xi) = (T_a\eta', \eta') \qquad \text{for all } a \text{ in } A.$$

By 19.8 this implies $S \cong T$. ∎

25.14. *Remark.* 25.13 and 25.7 together make the surprising assertion that, if A is a C^*-algebra, the family of all unitary equivalence classes of irreducible *-representations of A is essentially identical with the family of all algebraic equivalence classes of algebraically irreducible (algebraic) representations of A.

26. Exercises for Chapter VI

1. Let A and B be *-algebras and $f: A \to B$ a *-homomorphism. Prove that:
   (a) $A/\text{Ker}(f)$ is *-isomorphic to $f(A)$.
   (b) If J is any *-ideal of B, then $A/f^{-1}(J)$ is *-isomorphic to $f(A)/J$.
   (c) If I is a *-ideal of A and $\text{Ker}(f) \subset I$, then A/I is *-isomorphic to $f(A)/f(I)$. In particular, if I and J are *-ideals of A with $J \subset I$, then I/J is a *-ideal of A/J and $(A/J)/(I/J)$ is *-isomorphic to A/I.

2. Prove that if B is a *-subalgebra and J is a *-ideal of a *-algebra A, then $B/(B \cap J)$ is *-isomorphic to $(B + J)/J$.

3. Let I be a *-ideal of a *-algebra A and suppose that B_1 and B_2 are *-subalgebras of A which contain I. Prove that if $\pi: A \to A/I$ is the canonical homomorphism, then:
   (a) $B_1 \subset B_2$ if and only $\pi(B_1) \subset \pi(B_2)$.
   (b) $\pi(B_1 \cap B_2) = \pi(B_1) \cap \pi(B_2)$.

4. Let A be a *-algebra with unit; and let $M_n(A)$ denote the *-algebra of $n \times n$ matrices over A with the usual matrix operations and involution $(a_{ij})^* = (a^*)_{ji}$. Prove that:
   (a) If I is a *-ideal of A, then $M_n(I)$ is a *-ideal of $M_n(A)$.
   (b) Every *-ideal of $M_n(A)$ is of the form $M_n(I)$ for some *-ideal I of A.

5. Let A be a *-algebra with proper involution, i.e., if $x^*x = 0$ then $x = 0$. Prove that:
   (a) If $x \in A$ and $(x^*x)^n = 0$ for some positive integer n, then $x = 0$.
   (b) If $x^*x = xx^*$ and $x^n = 0$ for some positive integer n, then $x = 0$.

6. A *-algebra A is said to be *Hermitian* if each Hermitian element in A has real spectrum. Prove that:
   (a) the algebra A is Hermitian if and only if every element of the form $-h^2$, where $h^* = h$, has an adverse.

(b) By Proposition 2.2 every symmetric *-algebra is Hermitian. Give an example to show that the converse is false (this is a bit tricky, see Wichmann [1]).

7. Prove that every maximal commutative *-subalgebra of a symmetric *-algebra A is symmetric.

8. Let A be a *-algebra and I a *-ideal of A. Prove that A is symmetric if and only if both I and A/I are symmetric. (See Wichmann [3].)

9. Prove that the product $\prod_{i \in I} A_i$ of a family $\{A_i\}_{i \in I}$ of *-algebras with the natural involution $(a_i)^* = (a^*)_i$ is symmetric if and only if each A_i is symmetric.

10. Let $x \mapsto x^*$ and $x \mapsto x'$ be two involutions on an algebra A. Prove that the map $x \mapsto x^{\#}$, where $x^{\#} = x'^*{}'$, is an involution on A which is symmetric if and only if $x \mapsto x^*$ is symmetric.

11. Let A be a commutative Banach *-algebra with radical R. Let I be a closed *-ideal in A such that $I \subset R$. Prove that if A/I is symmetric, then A is symmetric.

12. Let A be a Banach *-algebra. The involution is *regular* if $a = a^*$ and $\|a\|_{sp} = 0$ imply $a = 0$. The involution is *proper* if $a^*a = 0$ implies $a = 0$. Prove that if A is a symmetric Banach *-algebra with regular involution, then * is proper.

13. Let I be a proper two-sided ideal in a symmetric Banach *-algebra A with unit. Prove that:

(a) $I + I^*$ is a proper two-sided ideal of A.

(b) Each maximal two-sided ideal of A is a *-ideal of A.

14. Let A be a symmetric Banach *-algebra with unit 1. Prove that if I is any proper left ideal of A, then the closure of $I + I^*$ does not contain 1.

15. Let A be a symmetric Banach *-algebra. Prove that if a is an idempotent element in A with $(a^*a)^n = 0$ for some positive integer n, then $a = 0$.

16. Show that every finite-dimensional *-algebra can be made into a Banach *-algebra (but not in general into a C^*-algebra!) by introducing a suitable norm.

17. Show that every Banach algebra B can be isometrically embedded as a closed two-sided ideal of a Banach *-algebra A. Further, if B has a unit so does A. (Note: there exist Banach algebras which do not admit an involution. See Civin and Yood [1].)

18. Give an example to show that Proposition 3.6 may fail if a is not assumed to be normal.

19. Let A be a pre-C^*-algebra. An element a in A is *idempotent* if $a^2 = a$; an idempotent element a such that $a^* = a$ is a *projection*. Prove that:

(a) $\|a\| = \sup\{\|ab\| : b \in A, \|b\| \le 1\}$ for all a in A.

(b) Any nonzero projection in A has norm 1.

(c) If x is a normal idempotent, then x is a projection.

20. Prove that the commutative C^*-algebra $\mathscr{C}_0(S)$ of continuous complex-valued functions on a locally compact Hausdorff space S has a unit if and only if S is compact.

21. A compact Hausdorff space S is called a *Stonean space* if the closure of every open set is open. Prove that if S is a Stonean space, then every element of the C^*-algebra $\mathscr{C}(S)$ can be uniformly approximated by finite linear combinations of projections, i.e., Hermitian idempotents.

22. Let A be a commutative C^*-algebra which is generated by its projections. Let $a \in A$ and $\varepsilon > 0$. Prove that A contains a projection p which is a multiple of a and satisfies $\|a - pa\| < \varepsilon$.

23. Let A be a commutative C^*-algebra which is generated by a single element a (thus, A is the closure of the set of all polynomials in a and a^* without constant term). Prove that A is isometrically *-isomorphic to the C^*-algebra of continuous complex-valued functions on the spectrum $\mathrm{Sp}_A(a)$ which vanish at zero.

24. Let S be a compact Hausdorff space and T a linear operator on the C^*-algebra $\mathscr{C}(S)$. Prove that if $T(1) = 1$ and $\|T\| = 1$, then $Tf \geq 0$ whenever $f \geq 0$.

25. Let A be a C^*-algebra with unit 1 and A_n the set of all normal elements of A. Let F be the space of all continuous complex-valued functions on $\mathbb{C}$, with the topology of uniform convergence on compact sets.

Prove that the map

$$\langle f, a \rangle \mapsto f(a) \quad \text{(see 6.2)}$$

is continuous on $F \times A_n$ to A_n.

26. Let S be a metric space and $\mathscr{B}_c(S)$ the C^*-algebra of bounded continuous complex-valued functions on S. Prove that if $\mathscr{B}_c(S)$ is separable, then S is compact.

27. Show that the $\mathscr{L}_1$ group algebra of a locally compact group G cannot be a C^*-algebra unless G has only one element (see 3.5).

28. Suppose A is a Banach *-algebra under a norm $\| \ \|$; and suppose that A, as a *-algebra, is also the underlying *-algebra of a C^*-algebra with norm $\| \ \|_c$. Show that $\| \ \|$ and $\| \ \|_c$ are equivalent (i.e., give to A the same norm topology).

29. We define a Banach *-algebra A, $\| \ \|$ to be C^*-*renormable* if its underlying *-algebra A becomes a C^*-algebra under some norm $\| \ \|_c$ (see the preceding exercise). Show that the $\mathscr{L}_1$ group algebra of the additive group of integers is *not* C^*-renormable.

30. Let A be a C^*-algebra with unit. Prove:
(a) If $a \in A$ is Hermitian, then $\mathrm{Sp}_A(a^2) \subset [0, \|a\|^2]$.
(b) If $a \in A_+$, then $a\|a\| \geq a^2$.

31. Let A be a C^*-algebra with unit 1. Prove that if $0 \leq b \leq a$ and $\lambda > 0$, then $(a + \lambda 1)^{-1} \leq (b + \lambda 1)^{-1}$.

32. Let A be a C^*-algebra. Prove that if $a, b, x \in A$ and $0 \leq a \leq b$, then $\|a^{1/2}x\| \leq \|b^{1/2}x\|$.

33. Let A be a C^*-algebra and $a \in A$ Hermitian. Prove that if $f \in \mathscr{C}(\mathrm{Sp}_A(a))$, then $f(a)$ is a positive element of A if and only if $f(\lambda) \geq 0$ for all $\lambda \in \mathrm{Sp}_A(a)$.

34. Let A be a C^*-algebra with unit. Prove that if elements a, b, c, d and x in A satisfy $a \geq b \geq 0$, $c \geq d \geq 0$ and $axc = 0$, then $bxd = 0$.

35. Let A be a C^*-algebra. Prove that if $x, y \in A_+$, then $\|x + y\| \geq \|x\|$.

36. An approximate unit $\{e_i\}_{i \in I}$ in a normed algebra is said to be *sequential* if the index set I is the set of positive integers with their usual order. Show that if S is a locally compact Hausdorff space, then the C^*-algebra $\mathscr{C}_0(S)$ has a sequential approximate unit if and only if S is σ-compact.

37. Let A be a separable C^*-algebra and $\mathscr{I}$ a family of norm-closed two-sided ideals of A whose intersection is $\{0\}$. Show that there is a *countable* subset $\mathscr{I}_0$ of $\mathscr{I}$ whose intersection is $\{0\}$ (see 8.12).

38. Prove that if A is a C^*-algebra such that each closed left ideal in A is a two-sided ideal, then A is commutative. [Hint: Show that each maximal left ideal I in A is also a maximal right ideal; then show each non-zero element in A/I is invertible. Use ideas from 7.4, 7.13 of Chapter V to obtain the result.]

39. Show that a C^*-algebra A has a unit if and only if every maximal commutative *-subalgebra of A has a unit.

40. Show that the decomposition in 9.16 of a non-degenerate *-representation T of a *-algebra A into a direct sum of cyclic *-representations is not unique.

41. Give an elementary proof that a normal operator a on a finite-dimensional Hilbert space X has an orthonormal basis of eigenvectors.

42. Show that if a bounded normal operator a on a Hilbert space X satisfies $a^n = 1$ (1 is the identity operator on X; n is some positive integer), then a is unitary.

43. Let p be a complex polynomial all of whose roots are real. Show that if a is a bounded normal operator on a Hilbert space X such that $p(a) = 0$, then a is Hermitian.

44. Let a be a bounded normal operator on a Hilbert space X, and let A be the norm-closed subalgebra of $\mathcal{O}(X)$ generated by a and the identity operator 1. We have seen that

$$\mathrm{Sp}_{\mathcal{O}(X)}(a) \subset \mathrm{Sp}_A(a). \qquad \qquad \ldots (\dagger)$$

Give an example in which the inclusion ($\dagger$) is proper. [Note: This does not contradict 5.2, since A is not in general a *-subalgebra of $\mathcal{O}(X)$.]

45. Let a be a bounded normal operator on a Hilbert space X. Show that the following three conditions are equivalent:

(i) $\mathrm{Sp}_{\mathcal{O}(X)}(a)$ is a finite set.

(ii) a is a finite linear combination of pairwise commuting projections.

(iii) $p(a) = 0$ for some non-constant polynomial p (with complex coefficients).

46. Show in Example 11.12 that:

(a) The operator a is Hermitian.

(b) $\mathrm{Sp}(a) = [0, 1]$;

(c) The spectral measure of a is given by $P(W)(f) = \mathrm{Ch}_W f$ ($f \in X$, W a Borel subset of $[0, 1]$).

(d) The point spectrum of a is void.

47. Let a be a bounded normal operator on the Hilbert space X, with spectral measure P. A vector ξ in X will be called *cyclic for* a if X itself is the smallest closed linear subspace of X which contains ξ and is stable under a and a^*.

Let ξ be cyclic for a and let μ be the (finite) Borel measure on $\mathrm{Sp}(a)$ given by $\mu(W) = \|P(W)\xi\|^2$ (for Borel subsets W of $\mathrm{Sp}(a)$).

Show that the equation

$$F(\mathrm{Ch}_W) = P(W)\xi \qquad (W \text{ a Borel subset of } \mathbb{C})$$

defines a unique linear surjective isometry $F: \mathscr{L}_2(\mu) \to X$.

If b belongs to the closed *-subalgebra A of $\mathcal{O}(X)$ generated by a and 1, and $f \in \mathscr{L}_2(\mu)$, show that

$$F(\hat{b}f) = b(F(f)).$$

(Here $\hat{b}$ is the Gelfand transform of b in A; recall that $\hat{A} \cong \mathscr{C}(\mathrm{Sp}(a))$.) [Note: This Exercise is the germ of the multiplicity theory of normal operators.]

48. Show that if E is a (not necessarily bounded) normal operator on a Hilbert space X, and B is any bounded operator on X such that B comm E (see Appendix B), then B commutes with all of the projections in the range of the spectral measure of E. In particular B comm E^*.

49. Prove that if f and g are Borel functions from $\mathbb{C}$ into $\mathbb{C}$, and if E is a bounded normal operator on a Hilbert space X with spectral measure Q, then $f(E) = g(E)$ if and only if $f \equiv g$ Q-almost everywhere (see 12.12).

50. Let E be a bounded normal operator on a Hilbert space X. Prove that if f and g are Borel functions from $\mathbb{C}$ into $\mathbb{C}$ which coincide on $\mathrm{Sp}(E)$, then $f(E) = g(E)$. (Hence, to define $f(E)$, we need only specify the values of f on $\mathrm{Sp}(E)$.)

51. Let a and b be bounded normal operators on Hilbert spaces X and Y respectively. Then $a \otimes b$ is a bounded normal operator on the tensor product $X \otimes Y$. Express the spectral measure of $a \otimes b$ in terms of those of a and b.

52. (Spectral Convergence Theorem). Let $\{a_i\}$ be a net of bounded normal operators on a Hilbert space X, and a a normal operator on X, such that:

(i) For some constant k we have $\|a_i\| \le k$ for all i;

(ii) $a_i \to a$ strongly (see §I.5).

Let P_i and P denote the spectral measures of a_i and a respectively. Prove that if W is any Borel subset of $\mathbb{C}$ such that the boundary of W has P-measure 0, we have

$$P_i(W) \to P(W)$$

in the strong operator topology.

53. Let X, Y, Y' be Hilbert spaces, and A a *-subalgebra of $\mathcal{O}(X)$ such that the identity *-representation of A on X is irreducible. Furthermore let $c \in \mathcal{O}(X \otimes Y, X \otimes Y')$ satisfying:

$$c(a \otimes 1_Y) = (a \otimes 1_{Y'})c \quad \text{for all} \quad a \in A.$$

(Here $X \otimes Y$ and $X \otimes Y'$ are the Hilbert tensor products, and 1_Y and $1_{Y'}$ are the identity operators on Y and Y'.)

Prove that

$$c = 1_X \otimes b \quad \text{for some} \quad b \in \mathcal{O}(Y, Y').$$

54. Suppose E and F are two (not necessarily bounded) normal operators on a Hilbert space X, with spectral measures P and Q respectively. Let us say that E comm F (E and F *commute*) if $P(U)$ and $Q(V)$ commute for all Borel subsets U, V of $\mathbb{C}$.

(I)   Show that if one of E or F is bounded, then this definition coincides with that of B12.

(II) Show that if E comm$F$, then there is one and only one normal operator G on X such that $E + F \subset G$, and that this G commutes with both E and F.

(III) Assume that E comm F and that E and F are both positive. Show that $E + F$ is positive.

(IV) Assume that E and F are positive and commute. We shall say that $E \leq F$ if $(E\xi, \xi) \leq (F\xi, \xi)$ for all ξ in domain$(E) \cap$ domain (F). Show that, if $E \leq F$, there is a unique positive operator H commuting with both E and F such that $E + H = F$.

55. It was shown in 7.24 that if a is an invertible element in a C^*-algebra A with unit, then $a = up$ for unique elements u and p in A, where u is unitary and p is positive. Prove that the mappings $a \mapsto u$ and $a \mapsto p$ in this decomposition are norm continuous.

56. Let a be a bounded operator on a Hilbert space X and set $\mathrm{Re}(a) = (a + a^*)/2$, $|a| = (a^*a)^{1/2}$. Prove that if $|a| \leq \mathrm{Re}(a)$, then $a \geq 0$.

57. Suppose that p and u are bounded operators on a Hilbert space X; and assume that p is positive. Prove that if $p \leq \mathrm{Re}(up)$, then $p \leq upu^*$. Moreover, show that if $p \leq \mathrm{Re}(up)$ and $p = upu^*$, then $up = p$.

58. Show that if a is an $n \times n$ matrix such that $|a| \leq |\mathrm{Re}(a)|$, then $a = a^*$.

59. Show that if $a = up$ is the polar decomposition of a bounded operator a on a Hilbert space X, then u and p commute with b and b^*, where b is any operator on X which commutes with a and a^*.

60. Let $a_1 = u_1 p_1$ and $a_2 = u_2 p_2$ be the respective polar decompositions of bounded operators a_1 and a_2 on a Hilbert space X. Prove that the following conditions are equivalent:

(a) a_1 commutes with a_2 and a_2^*;

(b) u_1^*, u_1 and p_1 commute with u_2^*, u_2 and p_2;

(c)   The following five conditions are satisfied: (1) $p_1 p_2 = p_2 p_1$, (2) $u_1 p_2 = p_2 u_1$, (3) $p_1 u_2 = u_2 p_1$, (4) $u_1 u_2 = u_2 u_1$, and (5) $u_1^* u_2 = u_2 u_1^*$.

61. Let $a_1 = u_1 p_1$ and $a_2 = u_2 p_2$ be the respective polar decompositions of bounded operators a_1 and a_2 on a Hilbert space X. Prove that if a_1 commutes with a_2 and a_2^*, then $a_1 a_2 = (u_1 u_2)(p_1 p_2)$ is the polar decomposition of $a_1 a_2$.

62. Let $a = up$ be the polar decomposition of a bounded operator a on a Hilbert space X. Prove that a is normal if and only if u commutes with p and u is unitary on $[\mathrm{Ker}(a)]^\perp$.

63. Let a be any bounded normal operator on a Hilbert space X, with polar decomposition $a = up$. Show that, if b is any bounded linear operator on X commuting with a, and if $b = vq$ is the polar decomposition of b, then the four operators u, p, v, q commute pairwise.

64. Give an example of a C^*-algebra and elements $a, b \in A$ such that $a \geq 0$, $a \neq 0$, $b = b^*$, $-a < b < a$, but $|b| \leq a$ is not true.

65. Let A be a C^*-algebra with unit 1. Prove that:

(a)   If $x \in A$ satisfies $(1 - x^*x)A(1 - xx^*) = \{0\}$, then x is a partial isometry.

(b) The unit 1 is an extreme point of the unit ball B of A.

(c) If B_+ is the set of positive elements in the unit ball B of A, then the extreme points of B_+ consist precisely of all projections in A.

66. Prove that if A is a C^*-algebra and a, b are Hermitian elements of A, then $\|ab\| = \|ba\|$. Does this remain true if a and b are merely assumed to be normal?

67. Let A be a C^*-algebra and a, b Hermitian elements of A such that $\|b\| \le \|a\| \le 1$. Prove that $\|a + b\| \le 1 + 2\|ab\|$.

68. (a) Show that the product of two normal elements in a C^*-algebra need not be normal.

(b) Prove that if a, b, and ab are normal matrices, then ba is normal.

69. Show that if a sequence of commuting normal operators a_n on a Hilbert space X converge strongly to an operator a (i.e., $a_n \xi \to a\xi$ for all ξ in X), then a is normal. Show that the assumption that the operators commute cannot be dropped, even if the a_n are unitary.

70. Let X be a Hilbert space and a, b, and c bounded operators on X satisfying $ca = bc$. Prove that if both a and b are Hermitian and if c is injective and has dense range, then a and b are unitarily equivalent.

71. Let A be a C^*-algebra with identity 1. Let I be a norm-closed two-sided ideal of A and $\pi: A \to A/I$ the canonical homomorphism. Prove that if $b \in A/I$ is Hermitian and satisfies $0 \le b \le 1$, then there exists a Hermitian element $a \in A$ such that $\pi(a) = b$ and $0 \le a \le 1$.

72. Let X be a Hilbert space and f a continuous complex-valued function on the unit circle $\{\lambda \in \mathbb{C} : |\lambda| = 1\}$. Prove that for each $\varepsilon > 0$ there exists $\delta > 0$ such that $\|f(u) - f(v)\| < \varepsilon$ whenever u and v are unitary operators on X satisfying $\|u - v\| < \delta$.

73. Let X be a Hilbert space and $a, u \in \mathcal{O}(X)$. Prove that if u is unitary, then there is a unitary operator v in the C^*-algebra A generated by a and the identity operator 1, such that $\|u - v\| \le 2\|u - a\|$.

74. Let X be a Hilbert space and a_0, a_1 normal operators in $\mathcal{O}(X)$. Prove that if a_1 has finite spectrum, then there exists a normal operator a_2 in the C^*-algebra generated by a_1 and 1 such that $\mathrm{Sp}(a_2) \subset \mathrm{Sp}(a_0)$ and $\|a_0 - a_2\| \le 2\|a_0 - a_1\|$.

75. Let A be a C^*-algebra. Prove that if J is a norm-closed two-sided ideal in A, $a \in J$, $b \in J_+$, $\|b\| \le 1$ and $aa^* \le b^4$, then $a = bc$ for some $c \in J$ with $\|c\| \le 1$.

76. Show that if X is a separable Hilbert space, then the C^*-algebra $\mathcal{O}_c(X)$ of compact operators is separable (see 15.13).

77. Let a be a (bounded) normal operator on a Hilbert space X, with spectral measure P. Show that, for every Borel subset W of $\mathbb{C}$, $P(W)$ is a strong operator limit of polynomials in a and a^* (See §I.5 for the definition of the strong operator topology.)

78. Let a be a compact normal operator on the Hilbert space X, P its spectra measure, and $0 \ne \lambda \in \mathbb{C}$. Show that $P(\{\lambda\})$ belongs to the *norm-closed* *-subalgebra of $\mathcal{O}(X)$ generated by a.

79. Let X, Y be Hilbert spaces, a an element of $\mathcal{O}(X, Y)$, and Z an infinite-dimensional linear suspace of X. Show that, if there exists a positive number k such that $\|a\xi\| \geq k\|\xi\|$ for all $\xi \in Z$, then a is not a compact operator.

In particular, an idempotent operator in $\mathcal{O}(X)$ is compact if and only if its range is finite-dimensional.

80. Give an example of a Banach *-algebra with unit which has no nonzero positive functionals. Can you find an example without a unit?

81. (a) Give an example of an (incomplete) normed *-algebra A with unit and a discontinuous positive functional on A. (Thus, completeness is essential in 18.12.)

(b) Give an example of a Banach *-algebra A which does not have an approximate unit and a discontinuous positive functional on A. (Therefore, the existence of a unit or approximate unit is essential in 18.12 and 18.15.)

82. (Generalized Uncertainty Principle). Let A be a *-algebra with unit 1, and let p be a positive linear functional on A. For each self-adjoint element a in A define

$$\delta(p, a) = \sqrt{p((a - p(a)1)^2)}.$$

Show that, for any two self-adjoint elements a, b of A,

$$\delta(p, a)\delta(p, b) \geq \frac{1}{2} |p(ab - ba)|.$$

83. Let A be a *-algebra with unit 1 and p a positive functional on A such that $p(1) = 1$ and $p(a^2) = p(a)^2$ for all Hermitian a in A. Prove that:

(a) $\mathrm{Ker}(p)$ is a *-ideal of A;

(b) $p(ab) = p(a)p(b)$ for all a, b in A.

84. Let A be a symmetric *-algebra with unit 1 and suppose that f is a linear functional on A which is multiplicative on each subalgebra of A generated by a Hermitian element and 1. Prove that f is multiplicative on A.

85. Let A and B be C^*-algebras with units. We shall call a linear mapping $\phi: A \to B$ which maps the unit of A into the unit of B unital. If $\phi(a^*) = \phi(a)^*$ for all a in A, ϕ is called self-adjoint. If ϕ is unital, self-adjoint, and satisfies $\phi(a^2) = \phi(a)^2$ for all Hermitian a in A, then ϕ is said to be a Jordan homomorphism. Show that if ϕ is a unital, self-adjoint linear mapping of A into B, then the following are equivalent:

(a) ϕ is a Jordan homomorphism;

(b) $\phi(ab + ba) = \phi(a)\phi(b) + \phi(b)\phi(a)$ for all a, b in A;

(c) $\phi(a^n) = \phi(a)^n$ for all a in A and $n = 1, 2, 3, \ldots$.

86. Prove that if A and B are C^*-algebras with unit and $\phi: A \to B$ is a Jordan homomorphism, then:

(a) ϕ maps positive elements in A to positive elements in B.

(b) ϕ is bounded with norm $\|\phi\| = 1$.

(c) $\mathrm{Ker}(\phi)$ is a norm-closed *-ideal of A.

87. Construct an example of a Jordan homomorphism from a C^*-algebra A into a C^*-algebra B which is not a *-homomorphism.

88. Let A be a *-algebra and p a positive functional on A. Let $I_p = \{x \in A : p(x^*x) = 0\}$ and set $P = \bigcap I_p$ where the intersection is taken over all positive functionals on A. The algebra A is said to be *P-commutative* if $ab - ba \in P$ for every a, b in A. Show that:

(a) P is a two-sided ideal in A.

(b) If A is P-commutative, then the algebra A_1 obtained from A by adjoining a unit is P-commutative.

(c) If A is P-commutative and ϕ is a *-homomorphism of A onto a *-algebra B, then B is P-commutative. In particular, if I is a *-ideal of A, then A/I is P-commutative.

(d) If A is a Banach *-algebra with an approximate unit, then the ideal P coincides with the reducing ideal (10.2) of A.

(e) Give an example of a non-commutative P-commutative Banach *-algebra.

89. Give an example of a *-algebra with no irreducible *-representations.

90. Show that a nonzero *-representation of a *-algebra A on a Hilbert space X is irreducible if and only if each nonzero vector in X is cyclic.

91. Let A be a C^*-algebra. An element a in A is *strictly positive* if $p(a) > 0$ for each nonzero positive functional p on A. Prove that if $a \in A$ is strictly positive and T is a non-degenerate *-representation of A on a Hilbert space X, then the subspace $\{T_a \xi : \xi \in X\}$ is dense in X.

92. Show that if A is a C^*-algebra with a sequential approximate unit $\{x_i\}_{i=1}^{\infty}$ bounded by 1, then A contains a strictly positive element.

93. Show that every separable C^*-algebra contains a strictly positive element.

94. Give an example of a C^*-algebra which does not have a sequential approximate unit.

95. Let A be a *-algebra and A_1 the algebra obtained from A by adjoining an identity. Prove that if p is an indecomposable element of $\mathscr{P}(A_1)$, then either p vanishes on A or $p|A$ is an indecomposable element of $\mathscr{P}(A)$ (see 20.5).

96. Supply a proof of Proposition 20.7.

97. Show that the linear functional p defined on the dense *-subalgebra B of the Banach *-algebra A in 21.5 has the properties described.

98. Prove Proposition 22.4.

99. Prove that if p is a positive functional of norm one on a C^*-algebra A and T is the corresponding *-representation, then T is injective if and only if $\{a \in A : p(x^*x) = 0\} = \{0\}$.

100. Let A and B be C^*-algebras with units and let $\phi : A \to B$ be a homomorphism which preserves the units. Prove that if $\|\phi(a)\| \le \|a\|$ for all a in A, then ϕ is a *-homomorphism.

101. Prove that a C^*-algebra A of compact type has a Type I *-representation theory (see 24.15).

102. Let A be a C^*-algebra, B a norm-closed *-subalgebra of A, I a norm-closed left ideal of B, and J the smallest norm-closed left ideal of A containing I. Show that $J \cap B = I$.

Notes and Remarks

The study of C^*-algebras originated in 1943 with the work of Gelfand and Naimark [1], who called them *-rings. They defined a *-ring as a Banach algebra A with involution * and unit 1 satisfying the three conditions:

(1) $\|a^*a\| = \|a^*\| \cdot \|a\|$,

(2) $\|a^*\| = \|a\|$, and

(3) $1 + a^*a$ is invertible for each a in A.

They immediately asked in a footnote to their definition if conditions (2) and (3) could be deleted—apparently recognizing that they were of a different character than condition (1) and were needed primarily because of their method of proof.

In order to clarify the role of condition (3), Banach algebras satisfying $\|x^*x\| = \|x\|^2$ were called B^*-algebras in 1946 by Rickart [1]. The term "C^*-algebra" was introduced in 1947 by Segal [4] to describe those B^*-algebras which are norm-closed *-subalgebras of bounded linear operators on a Hilbert space (the "C" stood for closed in the norm topology and the * for self-adjoint subalgebra). Algebras with involution which satisfy (3) were called symmetric. With these definitions Gelfand and Naimark [1] proved that every symmetric B^*-algebra is isometrically *-isomorphic to a C^*-algebra. The problems of whether or not *all* B^*-algebras are symmetric and whether or not condition (1) implies condition (2) were left open. To discuss the solution of these problems it is convenient to consider the commutative and non-commutative cases separately.

Commutative algebras: In their paper Gelfand and Naimark first established that every commutative Banach algebra A with unit and involution satisfying (1) is isometrically *-isomorphic to $\mathscr{C}(S)$ for some compact Hausdorff space S. Under the assumption of commutativity they were able to show quite easily that condition (1) implies (2). Then, by utilizing a delicate argument depending on the notion of Šilov boundary (see Naimark [8, p. 231]) they proved that A is symmetric. Therefore, in the commutative case Gelfand and Naimark were able to show that conditions (2) and (3) follow from condition (1). Simpler proofs of the symmetry for commutative A were published by Arens [1] in 1946 (the proof of symmetry in 4.3 is Arens' proof), and by Fukamiya [2] in 1952. We remark that the result obtained by Gelfand and Naimark in the commutative case complemented earlier work of Stone [7].

Non-commutative algebras: The 1952 paper of Fukamiya [2] implicitly contained the key result (see Proposition 7.9) needed to eliminate condition

(3) for non-commutative algebras. Independently and nearly simultaneously 7.9 was discovered by Kelley and Vaught [1]; their proof is given in §7. The non-trivial observation that 7.9 was the key to eliminating condition (3) is due to Kaplansky. His clever argument was recorded in J. A. Schatz's review of Fukamiya's paper; see the proof of 7.11 for details.

In marked contrast to the commutative case, the redundancy of condition (2) for non-commutative algebras did not follow easily; in fact, the question remained open until 1960 when a solution for algebras with unit was published by Glimm and Kadison [1]. Their proof was based on a deep "n-fold transitivity" theorem for unitary operators in an irreducible C^*-algebra. In 1967 Vowden [1] utilized approximate units to extend the result to algebras which may not have a unit. By using Theorem 7.24 one can by-pass the Kadison–Glimm transitivity theorem and substantially simplify the theory; for details see Doran and Belfi [1].

The fact that every B^*-algebra is isometrically *-isomorphic to a C^*-algebra (of operators on a Hilbert space) is proved in 22.12. Because of this result there is no longer any distinction between B^*-algebras and C^*-algebras (as defined by Segal). Present day terminology favors the term "C^*-algebra" over the term "B^*-algebra," and most authors (see, for example, Bratteli and Robinson [1], Dixmier [24], Doran and Belfi [1], Goodearl [1], Kadison and Ringrose [1], Pedersen [1], and Takesaki [7]) define C^*-algebras via the abstract axioms as we have done (see 3.1).

We remark that the early editions of Naimark's book [8] referred to C^*-algebras as completely regular algebras; however, this terminology has been changed to C^*-algebra in [8] to conform to present day usage, with completely regular still being reserved for what we have called a "pre-C^*-algebra" (3.3). Finally, we mention that C^*-algebras are called "algèbres stellaires" in Bourbaki [12].

The elementary material in §1 is standard and appears in the earliest writings on algebras with involution; see for example, Gelfand and Naimark [1], Naimark [2, 8], and Rickart [1, 2]. The example of a discontinuous involution in 1.10 is due to F. Bonsall and first appeared in a 1966 paper of J. Duncan (see Bonsall and Duncan [1, p. 194]). For additional examples, see Aupetit [1, pp. 129–130].

Most of the results in §2 on symmetric *-algebras were established in the late 1940's by Gelfand, Raikov, and Šilov [1], [2], and Raikov [1]. For Proposition 2.4 see Doran [1]; a somewhat different proof of 2.4 is given in Wichmann [1]. Further results on the purely algebraic theory of symmetric *-algebras have been established by Wichmann [2], [3] (sample result: the

*-algebra of $n \times n$ matrices over a symmetric *-algebra is symmetric). Proposition 2.5 can be found, for example, in Naimark [8, p. 215] where many additional results on symmetric *-algebras are given. Rickart [2] also contains an excellent treatment of symmetric *-algebras up to 1960. For recent accounts of the subject see Aupetit [1] and Doran and Belfi [1].

Proposition 3.7 is due to Rickart [1], [2, p. 188], where it is shown more generally that every homomorphism of an arbitrary Banach algebra (no involution assumed) into a C^*-algebra is automatically continuous (however the nice estimate $\|F\| \leq 1$ may fail in this generality). The result 3.10 on adjoining a unit to a C^*-algebra was proved in 1946 by Arens [2] for commutative algebras, and by Yood in the general case (see Rickart [2, p. 186]). Berberian [2, p. 328] points out that the result was also considered by Kaplansky in 1952.

For treatments of finite-dimensional C^*-algebras which roughly parallel what has been given in 3.14, see Goodearl [1, pp. 8-10] and Takesaki [6, pp. 50-54].

Theorem 4.3 is the Gelfand–Naimark Theorem for commutative C^*-algebras which we discussed above. The results on spectra in subalgebras of C^*-algebras given in §5 are due to Rickart [1]. The continuous functional calculus for C^*-algebras described in §6 originated in 1943 with the work of Gelfand and Naimark [1]; it is a standard part of the theory and has been treated by many authors (see, for example, Dixmier [24, pp. 12-13], Kadison and Ringrose [1, pp. 239-244], Rickart [2, pp. 241-242], and Takesaki [6, pp. 19-20]).

The results in §7 on positive elements (in particular Theorem 7.9) are due independently to Fukamiya [2] and Kelley and Vaught [1]. Theorem 7.11 was proved by Kaplansky as we pointed out in the opening paragraphs above. The order relation induced by the positive elements in a C^*-algebra was studied by Sherman [1].

Theorem 7.24 on the closed convex hull of the unitary elements in a C^*-algebra is due to Russo and Dye [1]; the proof we have given is due to Gardner [1].

The existence of approximate units in C^*-algebras was shown in 1947 by Segal [4]; we have followed the treatment given by Dixmier [24, p. 18]. For additional results on approximate units in C^*-algebras see the monograph by Doran and Wichmann [2].

The result 8.7 on quotient C^*-algebras was proved in 1949 by Segal [5] and, independently, by Kaplansky [4]; the proof we have given is from Dixmier [24, p. 22], and is due to F. Combes. Proposition 8.8 was shown by

Gelfand and Naimark [1, 4]; our proof is essentially the one given by Kaplansky [4]. The applications of approximate units and quotients in §8 depend on standard techniques (e.g., see Dixmier [24], Pedersen [5], Rickart [2], and Segal [5]). Proposition 8.15 is proved in Dixmier [24, p. 54].

The elementary results in §9 on *-representations originated with Gelfand and Naimark [1]. Basic references are Dixmier [24], Naimark [8], Pedersen [5], Rickart [2], and Takesaki [6]. The reducing ideal (10.2) was introduced in 1948 by Naimark [2]; our definition is due to Kelley and Vaught [1], who called it the *-radical. Naimark [8, pp. 262-264] and Rickart [2, pp. 210-230] give additional results on the reducing ideal.

For the C*-completion of a Banach *-algebra see Dixmier [24, pp. 47-49], Fell [2], and Naimark [8, p. 264]. Dixmier refers to the C*-completion as the "enveloping C*-algebra" while Naimark refers to the norm in the C*-completion as the "minimal regular norm."

The original version of Stone's Theorem (10.10) was proved for one parameter unitary groups of the real line (see Stone [1], [2]). The theorem was generalized to locally compact Abelian groups, independently and simultaneously, by Ambrose [1] and Godement [1]. Naimark [1] proved a similar but not identical result slightly earlier. Various treatments of the theorem are given, for example, in Barut and Raczka [1, p. 160], Conway [1, p. 334], Hewitt and Ross [2, p. 296], Kadison and Ringrose [1, p. 367], Loomis [2, p. 147], and Sugura [1, p. 132]. Hewitt and Ross [2, pp. 326-327] give further historical comments and references to the literature.

The results in §§11, 12 on spectral theory of bounded and unbounded normal operators have a long and distinguished history. The spectral theorem for bounded Hermitian operators goes back to the work of Hilbert [1] in 1906. Proofs of this result were also given by F. Riesz [1, 2, 3], and contributions were made by many others. The modern form of spectral theory of operators in Hilbert space was ushered in with the fundamental work of von Neumann [1, 2] in 1929 and Stone [1, 2] during the period from 1930 to 1932. Stone's book [2], which is primarily concerned with unbounded operators, gives a thorough treatment of spectral theory in separable Hilbert spaces. The book of Riesz and Sz.-Nagy [2] gives a concise account of both the bounded and unbounded cases for arbitrary (non-separable) Hilbert spaces.

The first treatment of the spectral theorem for bounded normal operators from the point of view of C*-algebras was given in 1950 by Dunford [1], although weakly closed algebras had been considered earlier by Yosida, and similar results had been obtained by Stone for Abelian rings of operators (see

Dunford and Schwartz [2, p. 927]). The paper of Fell and Kelley [1] in 1952 is also relevant, and is the point of view adopted in Kadison and Ringrose [1]. The most extensive treatment of spectral theory of bounded and unbounded operators on a Hilbert space presently available is given in Dunford and Schwartz [2], where the reader will find excellent historical notes and extensive references to the literature. The theory is also treated, for example, in Conway [1], Gaal [1], Halmos [2], Hille and Phillips [1], Naimark [8], and Rudin [2]. For an interesting historical survey of spectral theory, see Steen [1].

Spectral theory has been used in the approximation of Hilbert space operators by Apostol, Foias, and Voiculescu [1, 2], Apostol, Fialkow, Herrero, and Voiculescu [1], Apostol and Voiculescu [1], and Voiculescu [1].

The polar decomposition of operators described in §13 is due to von Neumann [4]. The reader may wish to consult Dunford and Schwartz [2, pp. 1245-1274], Riesz and Sz.-Nagy [2, §110], and Stone [2, pp. 329-333] for more information. The results on Mackey's form of Schur's Lemma in §13 are due to Mackey [15, p. 15]. Additional information is given in Gaal [1, pp. 161-168]. The Naimark relation in 13.6 was discussed in V.3.2; see Naimark [6], Fell [10], and G. Warner [1].

The irreducibility criterion in §14 can be traced to Gelfand and Naimark [4], and Naimark [2]. Discretely decomposable representations were introduced by Mackey [15, p. 41]; they have also been treated in G. Warner [1, p. 248]. Such representations play a significant role in the general theory.

The earliest results on compact operators are implicit in the papers of Volterra and Fredholm on integral equations. The notion of "compact operator" is due to Hilbert [1] who expressed the definition in terms of linear equations via bilinear forms on l_2. F. Riesz [3], in 1918, was the first to adopt the abstract "operator-theoretic" point of view. He formulated the so-called "Fredholm alternative" (see Douglas [1, p. 131]) and gave the first proof of the spectral theorem for compact Hermitian operators.

The *-representations of the C^*-algebra of compact operators (see 15.12) were determined by Naimark [2, 3], [8, pp. 298-300]. For the results in 15.13 to 15.18, see Kaplansky [4, 6], Naimark [2], and Dixmier [24, pp. 94-98].

Hilbert–Schmidt operators were studied in 1921 by Carleman [1]; Hille and Tamarkin [1], and Smithies [1] also made significant contributions to the theory. Many additional results on compact and Hilbert–Schmidt operators as well as further references and historical notes are given in Dunford and Schwartz [2, pp. 1009-1062] to which we refer the reader.

The Sturm-Liouville theory described in §16 originated in the 1830's. For historical accounts of this beautiful theory see, for example, Dieudonné [3] and Dunford and Schwartz [2, pp. 1581–1628].

Inductive limits (also called direct limits) of normed algebras were first studied in detail by S. Warner [1]. The important class of topologically simple C*-algebras described as inductive limits in 17.5 (the Glimm algebras) were introduced and subjected to a penetrating analysis by Glimm [1] in 1960. In 1972, Bratteli [1] introduced a new class of C*-algebras, called AF-algebras, as inductive limits of a sequence of finite-dimensional C*-algebras. We wish to make a few remarks about these algebras. First, the term "AF-algebra" stands for "approximately finite-dimensional C*-algebra" and is justified by the following characterization obtained by Bratteli:

THEOREM: A separable C*-algebra is an AF-algebra if and only if for every finite subset $\{a_1, \ldots, a_n\}$ of A and every $\varepsilon > 0$ there exists a finite-dimensional C*-subalgebra B of A and elements $b_1, \ldots, b_n$ of B such that $\|b_j - a_j\| < \varepsilon$ for $j = 1, \ldots, n$.

AF-algebras are important because of their many diverse properties, and because they are relatively easy to investigate. As a result, they have provided many useful examples (and counter-examples!). Bratteli showed that AF-algebras can be studied by means of diagrams (now called Bratteli diagrams). These diagrams determine the limiting AF-algebras up to isomorphism; moreover, the ideal structure of the limiting algebra can be read off from the diagram. A serious drawback of the diagrams, however, is that quite different diagrams can give rise to isomorphic AF-algebras. This difficulty was overcome by Elliott [4] in 1976 when he introduced a new invariant called the "range of the dimension." Roughly speaking, this invariant consists of equivalence classes of idempotents in the algebra, where e and f are defined to be equivalent if there are elements a and b with $ab = e, ba = f$. Armed with this invariant it is possible to associate an ordered group $K_0(A)$ with the AF-algebra A and show that A is "almost determined" by $K_0(A)$ up to isomorphism. In fact, Elliott was able to prove that two AF-algebras A and B with units 1_A and 1_B, respectively, are isomorphic as C*-algebras if and only if $K_0(A)$ and $K_0(B)$ are isomorphic as ordered groups under an isomorphism which takes the equivalence class of 1_A onto the equivalence class of 1_B. Since $K_0(A)$ can be explicitly calculated in many situations, this result is extremely useful. A carefully written introductory account of AF-algebras which discusses this material is given in the book by Goodearl [1]; further results and additional references can be found in the small book by Effros [3].

The now classical results on positive linear functionals and *-representations described in §§18–22 originated with the fundamental work of Gelfand and Naimark [1] and Segal [4]. These results have been treated at length by many authors (see, for example, Berberian [2], Bonsall and Duncan [1], Bratteli and Robinson [1], Dixmier [24], Doran and Belfi [1], Hewitt and Ross [1], Kadison and Ringrose [1], Naimark [8], Rickart [2], and Takesaki [6]). Theorem 18.15 on automatic continuity of positive functionals on a Banach *-algebra with approximate unit is due to Varopoulos [1]. Additional results on automatic continuity of linear mappings are given in Sinclair [1].

The basic construction described in §19 (see 19.3) is known as the *Gelfand–Naimark–Segal* (or GNS) *Construction*. This construction, which generalizes the classical formation of the $\mathscr{L}_2$-Hilbert space associated with a bounded measure on a locally compact Hausdorff space, appeared first in Gelfand and Naimark [1], and in refined form in Segal [4]. It has been used in the theory of operator algebras and harmonic analysis in many different contexts and underlies much of our later work on induced representations.

Theorem 21.2 is a vast generalization of a theorem originally proved by Bochner [1] in 1932 for the real line. As stated the theorem is the result of many hands. For an extensive treatment of the theorem and an excellent review of its history and main contributors, see Hewitt and Ross [2, pp. 291–326].

The original version of the Plancherel Theorem was proved in 1910 by Plancherel [1]. For general locally compact Abelian groups the theorem is due to Weil [1]. Hewitt and Ross [2, p. 226] treat this case and give detailed historical notes on its development. The theorem, as presented in 21.6, is inspired by an abstract formulation of Plancherel's Theorem, in a very general setting, due to Godement [7]. Godement's version of the theorem is given in Loomis [2, p. 99]; however, we should warn the reader that the presentation there contains an error (see Hewitt and Ross [2, p. 251]).

§22 contains the famous Gelfand–Naimark Theorem (22.12) which was originally established in 1943 by Gelfand and Naimark [1] and discussed at the beginning of these notes. For other treatments and additional information, see Berberian [2, p. 265], Conway [1, p. 259], Dixmier [24, pp. 44–46], Doran and Belfi [1, p. 53], Naimark [8, p. 318], Rickart [2, p. 244], and Takesaki [6, p. 42]. Results on extending *-representations can be found in Dixmier [24, pp. 58–60], Fell and Thoma [1], and Naimark [8, pp. 311–312]. The beautiful result given in 23.1 was originally established in 1953 by Rosenberg [1].

The subject of von Neumann algebras originated with a paper published in 1929 by von Neumann [2] in which he proves the celebrated "Double

Commuter Theorem" (24.2). The systematic study of von Neumann algebras began in 1936 with the monumental series of papers by Murray and von Neumann [1, 2, 3], and von Neumann [7, 8]. These authors referred to von Neumann algebras as "rings of operators." Later on, other mathematicians called them W*-algebras (by analogy with C*-algebras, "W" standing for "weak"). Finally, in the original 1957 edition of his definitive monograph Dixmier [25] introduced the name, "von Neumann algebra." For an excellent overview of the vast literature on von Neumann algebras we recommend the two survey articles by Kadison [14, 15]. The books by Dixmier [25], Kadison and Ringrose [1, 2], Sakai [1], and Stratila and Zsidó [1], contain detailed treatments of the theory. The modular theory of operator algebras is treated extensively in Stratila [1].

The notion of "Type I representation" was introduced by Mackey [6, 10, 15], who also adapted the relevant parts of the theory of von Neumann algebras to representation theory. Further results on Type I representations and algebras can be found in Dixmier [24, pp. 121–127], Gaal [1, pp. 198–211], and Mackey [10, 12, 15, 21].

The Kaplansky Density Theorem (25.3) was established in 1951 by Kaplansky [9]; it is an indispensable tool in the theory of operator algebras. Theorem 25.6 was proved by Kadison [9] in 1957; the result in 25.7 came as somewhat of a surprise, as it was generally believed that such a strong result could not be true. See Kadison and Ringrose [1, pp. 329–332] and Pedersen [5, p. 24 and p. 85] for these results. References for the "spectral synthesis" results in 25.8 to 25.13 are Dixmier [24, pp. 54–58] and Pedersen [5, pp. 84–89].

VII The Topology of the Space of *-Representations

In the structure theory of commutative C^*-algebras A (§VI.4), the crucial role is played by the space $\hat{A}$ of all complex *-homomorphisms of A. When this space is topologized with the topology of pointwise convergence, the elements a of A give rise to continuous functions $\hat{a}$ on $\hat{A}$, and A becomes *-isomorphic with a C^*-algebra $B = \{\hat{a} : a \in A\}$ of continuous complex functions on $\hat{A}$. As a matter of fact $\hat{A}$ is locally compact; and the Stone-Weierstrass Theorem shows that B is equal to $\mathscr{C}_0(\hat{A})$. From this structure theory we showed in VI.10.10 that the *-representations of A could be classified in terms of regular projection-valued measures on the space $\hat{A}$.

All this has a striking generalization to non-commutative C^*-algebras. The space of complex *-homomorphisms of a commutative C^*-algebra becomes the space $\hat{A}$ of all unitary equivalence classes of irreducible *-representations of an arbitrary C^*-algebra A. Each element a of A gives rise to a function $\hat{a} : T \mapsto T_a$ on $\hat{A}$ (generalizing the Gelfand transform in the commutative case). One would now like to say that these functions $\hat{a}$ are continuous in some sense. Since the spaces range(T) in which their values lie vary in general from point to point of $\hat{A}$, it is not at all clear what this should mean. But a first requisite for giving meaning to the continuity of the $\hat{a}$ must be a topology for $\hat{A}$. Such a topology will be studied in this chapter. It will turn out (§6) that, although $\hat{A}$ as a topological space need not have any separation properties, it

539

shares with the commutative case the important property of local compactness and the Baire property.

   One would then hope, as a next step, to give meaning to the "continuity" of
the transforms $\hat{a}$, and to prove a "non-commutative Stone–Weierstrass
Theorem" describing $\{\hat{a} : a \in A\}$ as a collection of continuous cross-sections of
an appropriate bundle over $\hat{A}$. A program approximately realizing this hope
is carried out in Fell [3]. It is based on the "non-commutative Stone–
Weierstrass Theorem" of Glimm [2]. But in this work we shall not pursue
this program further. Much more important for our later chapters is the
non-commutative generalization of Stone's Theorem VI.10.10, to be
obtained in §9.

   Here is a brief outline of the contents of the sections of this chapter.

   In §§1,2 we give the definition and elementary properties of the so-called
regional topology of the space of all (not merely irreducible) *-representa-
tions of an arbitrary *-algebra A, and of the closely associated notion of weak
containment. In §3 we discuss the regional topology on the structure space $\hat{A}$.
It turns out that $\hat{A}$ has another natural topology, the hull-kernel topology,
obtained from the associated space $\text{Prim}(A)$ of all primitive ideals (kernels of
irreducible *-representations) of A. In general the regional and hull-kernel
topologies of $\hat{A}$ may differ; but we shall show in §5 that for C^*-algebras the
two coincide.

   §4 discusses an important class of *-subalgebras B of a *-algebra A, the so-
called hereditary subalgebras. These include two-sided *-ideals and also
subalgebras of the form a^*Aa $(a \in A)$. They have the property that there is a
natural homeomorphism of $\{T \in \hat{A} : B \not\subset \text{Ker}(T)\}$ into $\hat{B}$, which in favorable
cases is a surjection onto $\hat{B}$.

   The main result of §5 is the equality of the regional and hull-kernel
topologies of the structure space $\hat{A}$ if A is a C^*-algebra. Related to this is the
important notion (5.16) of the spectrum of a *-representation T of A; this is
the collection of all those S in $\hat{A}$ which belong to the regional closure of $\{T\}$.

   In §6 we prove the facts, mentioned earlier, that the structure space of a
C^*-algebra is locally compact and has the Baire property.

   §7 shows that any finite T_0 space is homeomorphic to the structure space $\hat{A}$
of some C^*-algebra A, thus graphically illustrating the way in which
separation properties may fail in $\hat{A}$.

   §8 is devoted to the study of bundles of C^*-algebras and the C^*-algebras
formed by their cross-sections. By a bundle of C^*-algebras we mean, roughly
speaking, a Banach bundle $\mathscr{B}$ each of whose fibers is a C^*-algebra; and we
observe that the C_0 cross sectional space $\mathscr{C}_0(\mathscr{B})$ is then a C^*-algebra in a
natural way. (The reader should not confuse these bundles of C^*-algebras

with the quite different objects called C^*-algebraic bundles, to be introduced in Chapter VIII.) We describe the structure space of $\mathscr{C}_0(\mathscr{B})$ in terms of the base space of $\mathscr{B}$ and the structure spaces of its fibers. Of particular importance in Chapters VIII and XII will be the bundle $\mathscr{B}$ of *elementary* C^*-algebras naturally constructed from an arbitrary Hilbert bundle $\mathscr{H}$: Here the base spaces of $\mathscr{B}$ and $\mathscr{H}$ are the same; and the fiber B_x of $\mathscr{B}$ is just the C^*-algebra of all compact operators on the fiber H_x of $\mathscr{H}$.

In §9 we show that every non-degenerate *-representation T of a C^*-algebra A gives rise in a canonical manner to a regular $X(T)$-projection-valued measure P on the Borel σ-field of $\hat{A}$. We call P the spectral measure of T. The construction of this spectral measure is due to Glimm [5], although the proof presented here is quite different from Glimm's. If A is Abelian it reduces to the spectral measure of VI.10.10. The spectrum of T, defined in §5, is just the closed support of P. If T is discretely decomposable, P assigns to each Borel subset W of $\hat{A}$ the projection onto the "W-subspace" of T, that is, the closure of the sum of the σ-subspaces with $\sigma \in W$ (see §VI.14). Even if T is not discretely decomposable, range $P(W)$ can still be regarded as the "generalized W-subspace", and P becomes the continuous generalization of the harmonic analysis whose discrete version we encountered in §IV.2 and §VI.14.

The spectral measure of a *-representation T is a weakened form of the direct integral decomposition of T. The theory of direct integral decompositions for *-algebras and groups has been worked out by many authors, beginning with von Neumann [8] (see for example Mackey [7], Godement [7], Dixmier [21]). Unfortunately, the full theory of direct integral decompositions of representations requires some assumption of separability. The theory of the spectral measure as presented in §9, on the other hand, requires no separability assumptions, and is strong enough to supply just that much of direct integral decomposition theory which is needed for the generalized Mackey analysis presented in Chapter XII.

From the point of view of §9 the "minimal *-representations" of A are those whose spectral measures are concentrated at a single point of $\mathrm{Prim}(A)$. §10 studies these briefly. Under certain very general conditions they include all primary *-representations (VI.24.9), and possibly others as well.

1. The Definition and Elementary Properties of the Regional Topology

1.1. Throughout this chapter A is a fixed *-algebra.

In this and subsequent chapters it will become abundantly clear that one needs to topologize the space of *-representations of A. What does it mean to say that one *-representation is "near" to another? Very roughly our answer will be as follows: A *-representation S of A is "near" to a *-representation T if the action of T can be "almost duplicated" by S as regards *finite* subsets of A and *finite-dimensional* subspaces of $X(T)$.

1.2. To be precise, let us denote by $\mathscr{T}(A)$, or $\mathscr{T}$ for short, the family of all (concrete) *-representations of A. (Properly speaking, we ought to restrict $\mathscr{T}$ in some way in order that it should turn out to be a set rather than merely a class in the sense of von Neumann set theory. In applications, however, we shall always be dealing only with well-defined subsets of $\mathscr{T}$, and so no difficulty will arise from this source.)

1.3. Consider an element T of $\mathscr{T}$; and let ε be a positive number, $\xi_1, \ldots, \xi_p$ a finite sequence of vectors in $X(T)$, and F a finite non-void subset of A. Then we shall define

$$U = U(T; \varepsilon; \xi_1, \ldots, \xi_p; F) \tag{1}$$

to be the family of all those T' in $\mathscr{T}$ such that there exist vectors $\xi'_1 \ldots, \xi'_p$ in $X(T')$ satisfying:

$$|(\xi'_i, \xi'_j) - (\xi_i, \xi_j)| < \varepsilon, \tag{2}$$

$$|(T'_a \xi'_i, \xi'_j) - (T_a \xi_i, \xi_j)| < \varepsilon \tag{3}$$

for all a in F and all $i, j = 1, \ldots, p$. The collection of all such families $U(T; \varepsilon; \{\xi_i\}; F)$, with T fixed and ε, p, $\{\xi_i\}$, F varying, will be denoted by $\mathscr{U}(T)$.

Now the following properties of $\mathscr{U}(T)$ are either obvious or easily verified:

(I) $T \in U$ for every U in $\mathscr{U}(T)$.

(II) If U and V are in $\mathscr{U}(T)$, there is a family W in $\mathscr{U}(T)$ such that $W \subset U \cap V$.

(III) If $U \in \mathscr{U}(T)$, there is a family V in $\mathscr{U}(T)$ such that, for each S in V, we have $W \subset U$ for some W in $\mathscr{U}(S)$.

These three properties assure us that there is a unique topology for $\mathscr{T}$ relative to which, for each T in $\mathscr{T}$, $\mathscr{U}(T)$ is a basis of neighborhoods of T. To see this, we define a subset E of $\mathscr{T}$ to be *open* if, for every T in E, we have $U \subset E$ for some U in $\mathscr{U}(T)$. Evidently this notion of openness defines a topology. It remains to show that every family U in $\mathscr{U}(T)$ (where $T \in \mathscr{T}$) contains an open neighborhood of T. Putting $W = \{S \in U : V \subset U \text{ for some } V$ in $\mathscr{U}(S)\}$, we see from (III) that W is open. Since evidently $T \in W \subset U$, the existence of the required topology is proved. Its uniqueness is obvious.

Definition. The topology of $\mathcal{T}$ constructed above, in which $\mathcal{U}(T)$ is a basis of neighborhoods of T for each T in $\mathcal{T}$, will be called the *regional topology (of *-representations of A*).

For brevity, a basis of neighborhoods of T in the regional topology will be called simply a *regional neighborhood basis* of T; sets which are closed in the regional topology will be called *regionally closed*; and so forth.

Remark. In using the term "regional," we are thinking of finite-dimensional subspaces of $X(T)$ as "regions". The definition of the sets U of (1) expresses the fact that the behavior of T on each such region can be approximately duplicated by the behavior of each element of U on an appropriate region of its own.

Notice that any subrepresentation of a *-representation T of A belongs to the regional closure of $\{T\}$.

1.4. The following proposition asserts that there are regional neighborhood bases which are smaller and easier to work with than $\mathcal{U}(T)$.

Proposition. Let T be a *-representation of A; and let Z be any subset of $X(T)$ such that the smallest closed T-stable subspace of $X(T)$ containing Z is all of $X(T)$. Then the set of all $U(T; \varepsilon; \xi_1, \ldots, \xi_p; F)$, where $\varepsilon > 0$, F is a finite subset of A, and $\xi_1, \ldots, \xi_p$ is a finite sequence of vectors in Z, is a regional neighborhood basis of T.

Proof. To begin with, the reader will observe that this is true if Z is a dense subset X' of $X(T)$.

Next, suppose that Z is a subset X'' of $X(T)$ whose linear span X' is dense in $X(T)$. If $\xi_1, \ldots, \xi_p$ is a given finite sequence of vectors in X', we represent each ξ_i as a finite linear combination of vectors $\eta_1, \ldots, \eta_q$ in X'':

$$\xi_i = \sum_{j=1}^{q} \lambda_{ij} \eta_j \qquad (\lambda_{ij} \in \mathbb{C}; i = 1, \ldots, p).$$

For any $\varepsilon > 0$ and any finite subset F of A, putting

$$m = \max\{|\lambda_{ij}| : i = 1, \ldots, p; j = 1, \ldots, q\}$$

and $\delta = \varepsilon q^{-2} m^{-2}$, we see by an easy calculation that

$$U(T; \delta; \eta_1, \ldots, \eta_q; F) \subset U(T; \varepsilon; \xi_1, \ldots, \xi_p; F).$$

This combined with the first sentence of the proof gives the required conclusion in case $Z = X''$.

Finally, let Z be any subset of $X(T)$ with the property assumed in the proposition. Then $X'' = Z \cup \{T_b \xi : b \in A, \xi \in Z\}$ has dense linear span in $X(T)$. If η_1, η_2 are any two vectors in X'' and $a \in A$, one verifies that (η_1, η_2) and $(T_a \eta_1, \eta_2)$ are each of the form either (ξ_1, ξ_2) or $(T_b \xi_1, \xi_2)$, where $\xi_1, \xi_2 \in Z$ and $b \in A$. It follows that for any $\varepsilon > 0$, any finite sequence $\eta_1, \ldots, \eta_q$ of vectors in X'', and any finite subset F of A, there is a finite sequence $\xi_1, \ldots, \xi_r$ of vectors in Z and a finite subset G of A such that

$$U(T; \varepsilon; \xi_1, \ldots, \xi_r; G) \subset U(T; \varepsilon; \eta_1, \ldots, \eta_q; F).$$

Combining this with the preceding paragraph, we obtain the conclusion of the proposition. ■

1.5. If T is cyclic, with cyclic vector ξ, we can take $Z = \{\xi\}$ in 1.4, and conclude that the family of all $U(T; \varepsilon; \xi; F)$, where $\varepsilon > 0$ and F runs over all finite subsets of A, is a regional neighborhood basis of T.

1.6. Here are two easy and useful corollaries of 1.4.

Corollary. *Suppose that $T \in \mathcal{T}$ and $\mathcal{S} \subset \mathcal{T}$, and that $\mathcal{Y}$ is an upward directed family of closed T-stable subspaces of $X(T)$ such that* (i) $\bigcup \mathcal{Y}$ *is dense in $X(T)$, and* (ii) $^Y T$ *belongs to the regional closure of $\mathcal{S}$ for every Y in $\mathcal{Y}$. Then T belongs to the regional closure of $\mathcal{S}$.*

1.7. The next corollary expresses the continuity of the Hilbert direct sum operation on *-representations.

Corollary. *Let I be an index set, and $\mathcal{J}$ the family of those $T = \{T^i\}$ in the Cartesian product $\prod_{i \in I} \mathcal{T}$ such that the Hilbert direct sum $\sum^{\oplus} T = \sum_{i \in I}^{\oplus} T^i$ exists (i.e., $\{\|T_a^i\| : i \in I\}$ is bounded for each a). Then the map $T \mapsto \sum^{\oplus} T$ is continuous on $\mathcal{J}$ to $\mathcal{T}$ with respect to the I-fold Cartesian product of the regional topology with itself (for $\mathcal{J}$) and the regional topology (for $\mathcal{T}$).*

1.8. It is often useful to formulate the definition of the regional topology in terms of convergence. The following statements are easy consequences of Proposition 1.4 and Definition 1.3.

Let T be an element of $\mathcal{T}$; and let Z be a subset of $X(T)$ with the property assumed in 1.4. Suppose now that $\{T^\nu\}$ is a net of elements of $\mathcal{T}$ with the following property: For any finite sequence $\xi_1, \ldots, \xi_p$ of vectors in Z, we can find vectors $\xi_1^\nu, \ldots, \xi_p^\nu$ in $X(T^\nu)$ for each ν such that

$$(\xi_i^\nu, \xi_j^\nu) \underset{\nu}{\to} (\xi_i, \xi_j) \tag{4}$$

and

$$(T_a^{\nu} \zeta_i^{\nu}, \xi_j^{\nu}) \underset{\nu}{\rightarrow} (T_a \zeta_i, \xi_j) \tag{5}$$

for each $i, j = 1, \ldots, p$ and each a in A. Then $T^{\nu} \underset{\nu}{\rightarrow} T$ *in the regional topology.*

Conversely, suppose that $\{T^{\nu}\}$ is a net converging to T in $\mathcal{T}$ in the regional topology. Then every subnet of $\{T^{\nu}\}$ has a subnet $\{T'^{\mu}\}$ with the following property: For every ξ in $X(T)$ and every μ there is a vector ξ_{μ} in $X(T'^{\mu})$ such that

$$(\xi_{\mu}, \eta_{\mu}) \underset{\mu}{\rightarrow} (\xi, \eta) \tag{6}$$

and

$$(T_a'^{\mu} \xi_{\mu}, \eta_{\mu}) \underset{\mu}{\rightarrow} (T_a \xi, \eta) \tag{7}$$

for all ξ, η in $X(T)$ and all a in A.

1.9. Our next proposition will assert that the definition of the regional topology is unaltered if (2) is replaced by:

$$(\xi_i', \xi_j') = (\xi_i, \xi_j) \qquad \text{for all } i, j = 1, \ldots, p. \tag{8}$$

Before proving this we make an observation. Given any positive number ε and any positive integer n, there exists a positive number $\gamma = \gamma(n, \varepsilon)$ with the following property: If $\xi_1, \ldots, \xi_n$ is an n-termed sequence of vectors in a Hilbert space X satisfying $|(\xi_i, \xi_j) - \delta_{ij}| < \gamma$ for all $i, j = 1, \ldots, n$, then we can find an $n \times n$ complex matrix α such that (i) the vectors $\eta_i = \sum_{j=1}^{n} \alpha_{ij} \xi_j$ $(i = 1, \ldots, n)$ form an orthonormal set, and (ii) $\|\alpha - \mathbf{1}\| < \varepsilon$. (Here $\mathbf{1}$ is the $n \times n$ unit matrix, and $\|\alpha - \mathbf{1}\|$ means, say, the Hilbert space operator norm.)

This observation is easily proved by induction in n, applied to the Gram-Schmidt orthogonalization process. We leave it to the reader.

1.10. If $T, \varepsilon, \xi_1, \ldots, \xi_p, F$ are as in 1.3, let us define

$$U' = U'(T; \varepsilon; \xi_1, \ldots, \xi_p; F) \tag{9}$$

as the family of all those T' in $\mathcal{T}$ such that there exist vectors $\xi_1', \ldots, \xi_p'$ in $X(T')$ satisfying (8) and (3). Since (8) $\Rightarrow$ (2),

$$U'(T; \varepsilon; \xi_1, \ldots, \xi_p; F) \subset U(T; \varepsilon; \xi_1, \ldots, \xi_p; F). \tag{10}$$

Proposition. *Each set (9) is a regional neighborhood of T.*

Proof. Let $T, \varepsilon, \xi_1, \ldots, \xi_p, F$ be as in 1.3. Choose a finite orthonormal sequence of vectors $\eta_1, \ldots, \eta_q$ whose linear span contains all the ξ_i. By the argument of the second paragraph of the proof of 1.4 we can find $\varepsilon' > 0$ such that

$$U'(T; \varepsilon'; \eta_1, \ldots, \eta_q; F) \subset U'(T; \varepsilon; \xi_1, \ldots, \xi_p; F). \tag{11}$$

Thus it is sufficient to show that the left side of (11) is a regional neighborhood of T; that is, we may and shall assume at the outset that the $\xi_1, \ldots, \xi_p$ are orthonormal.

Choose a positive number σ satisfying

$$\sigma(\|T_a\| + p^2\sigma)(2 + \sigma) + p^2\sigma < \varepsilon \tag{12}$$

for all a in F; and then choose a positive number δ such that

$$\delta < \sigma, \delta < \gamma(p, \sigma) \qquad \text{(see 1.9).} \tag{13}$$

With this δ, we claim that

$$U(T; \delta; \xi_1, \ldots, \xi_p; F) \subset U'(T; \varepsilon; \xi_1, \ldots, \xi_p; F). \tag{14}$$

Since the left side of (14) is a regional neighborhood of T, (14) will imply that the right side of (14) is also a regional neighborhood of T, completing the proof of the proposition.

To prove (14), we take an arbitrary element T' of the left side of (14). Thus there are vectors $\xi_1', \ldots, \xi_p'$ in $X(T')$ satisfying

$$|(\xi_i', \xi_j') - \delta_{ij}| < \delta \tag{15}$$

(since the ξ_i are orthonormal) and

$$|(T_a' \xi_i', \xi_j') - (T_a \xi_i, \xi_j)| < \delta \tag{16}$$

for all $i, j = 1, \ldots, p$ and a in F. By (15), (13), and 1.9 we can find an $n \times n$ matrix α such that the

$$\eta_i = \sum_{j=1}^{p} \alpha_{ij} \xi_j' \qquad (i = 1, \ldots, p) \tag{17}$$

are orthonormal in $X(T')$, and

$$\|\alpha - 1\| < \sigma. \tag{18}$$

(Here and in the future the norm $\| \ \|$ of matrices is the Hilbert space operator norm.) Now fix $a \in F$; and let π, π', ρ be the $p \times p$ matrices given by

$$\pi_{ij} = (T_a \xi_i, \xi_j), \ \pi_{ij}' = (T_a' \xi_i', \xi_j'),$$

$$\rho_{ij} = (T_a' \eta_i, \eta_j).$$

From (17) we see that

$$\rho = \alpha\pi'\alpha^*,$$

whence by (18)

$$\|\rho - \pi'\| = \|(\alpha - 1)\pi'\alpha^* + \pi'(\alpha - 1)^*\|$$

$$\leq \|\alpha - 1\| \, \|\pi'\|(\|\alpha\| + 1)$$

$$\leq \sigma(2 + \sigma)\|\pi'\|. \tag{19}$$

It follows from (16) and (13) that

$$\|\pi' - \pi\| < p^2\delta < p^2\sigma. \tag{20}$$

So

$$\|\pi'\| < \|\pi\| + p^2\sigma \leq \|T_a\| + p^2\sigma. \tag{21}$$

Combining (19), (20), and (21), we get

$$\|\rho - \pi\| \leq \|\rho - \pi'\| + \|\pi' - \pi\|$$

$$< \sigma(2 + \sigma)(\|T_a\| + p^2\sigma) + p^2\sigma$$

$$< \varepsilon.$$

From this follows

$$|(T'_a\eta_i, \eta_j) - (T_a\xi_i, \xi_j)| < \varepsilon \tag{22}$$

for all i, j and all a in F. Since the η_i are orthonormal, (22) implies that T' belongs to the right side of (14). But T' was an arbitrary element of the left side of (14); so (14) is proved.   ∎

1.11.   From 1.10 (especially 1.10(10)) we obtain:

Corollary. *Proposition   1.4   and   also   1.5   remain   true   when $U(T; \varepsilon; \xi_1, \dots, \xi_p; F)$ and $U(T; \varepsilon; \xi; F)$ are replaced by $U'(T; \varepsilon; \xi_1, \dots, \xi_p; F)$ and $U'(T; \varepsilon; \xi; F)$ respectively.*

1.8 remains true when (4) is replaced by:

$$(\xi_i^\nu, \xi_j^\nu) = (\xi_i, \xi_j) \qquad \text{(for all } i, j, \nu),$$

and (6) is replaced by:

$$(\xi_\mu, \eta_\mu) = (\xi, \eta) \qquad \text{(for all } \xi, \eta, \mu).$$

1.12.   For cyclic representations the above corollary, as applied to 1.8, takes on an especially simple form.

Proposition. *Let T be a cyclic *-representation of A, with cyclic vector ξ. A net $\{T^\nu\}$ of *-representations of A converges regionally to T if and only if every subnet of $\{T^\nu\}$ has a subnet $\{T'^\mu\}$ such that for each μ we can find a vector ξ^μ in $X(T'^\mu)$ with the properties:*

$$\|\xi^\mu\| = \|\xi\| \qquad \text{for all } \mu,$$

$$(T_a'^\mu \xi^\mu, \xi^\mu) \underset{\mu}{\to} (T_a \xi, \xi) \qquad \text{for all } a \text{ in } A.$$

1.13. Let $\mathcal{T}_1$ be the family of all *-homomorphisms of A into $\mathbb{C}$. Each element of $\mathcal{T}_1$ can be regarded as a one-dimensional *-representation of A on $\mathbb{C}$; so the regional topology of $\mathcal{T}$ can be relativized to $\mathcal{T}_1$.

Corollary. *The (relativized) regional topology of $\mathcal{T}_1$ coincides with the topology of pointwise convergence on A.*

This follows easily from 1.12.

1.14. Proposition. *For each a in A, the real function $T \mapsto \|T_a\|$ is lower semi-continuous on $\mathcal{T}$.*

Proof. Given T in $\mathcal{T}$ and $\varepsilon > 0$, choose unit vectors ξ and η in $X(T)$ so that

$$|(T_a \xi, \eta)| > \|T_a\| - \tfrac{1}{2}\varepsilon. \tag{23}$$

By 1.11 there is a regional neighborhood U of T such that, for every T' in U, there are unit vectors ξ', η' in $X(T')$ satisfying

$$|(T_a' \xi', \eta') - (T_a \xi, \eta)| < \tfrac{1}{2}\varepsilon. \tag{24}$$

Combining (23) and (24) we obtain

$$|(T_a' \xi', \eta')| > \|T_a\| - \varepsilon,$$

showing that $\|T_a'\| > \|T_a\| - \varepsilon$ for every T' in U. This completes the proof. ■

Remark. The function $T \mapsto \|T_a\|$ is in general far from continuous on $\mathcal{T}$. See for example Remark 8.21.

Remark. The results 1.12, 1.13, and 1.14 can be obtained without much difficulty directly from the Definition 1.3, without going through the machinery of 1.10.

1.15. Corollary. *If T belongs to the regional closure of a subset $\mathcal{S}$ of $\mathcal{T}$, then*

$$\|T_a\| \le \sup\{\|S_a\| : S \in \mathcal{S}\} \qquad \text{for all } a \text{ in } A.$$

Remark. This result will be strengthened in XI.8.20.

1.16. The next proposition describes the regional topology in terms of *-representations which act on subspaces of one and the same big Hilbert space.

Proposition*. *Assume that A is a Banach *-algebra. Let T be a nondegenerate *-representation of A, and $\mathscr{S}$ a family of *-representations of A. Then the following two conditions are equivalent:*

(I) *T belongs to the regional closure of $\mathscr{S}$.*
(II) *It is possible to embed $X(T)$ as a closed subspace of a larger Hilbert space Y, and to find a net $\{T^i\}$ of *-representations of A such that (i) for each i $X(T^i)$ is a closed subspace of Y, (ii) for each i T^i is unitarily equivalent to some element of $\mathscr{S}$, and (iii) we have*

$$T_a'^i \xi \to T_a \xi \qquad in\ Y$$

for each ξ in $X(T)$ and a in A. (Here $T_a'^i$ is the operator on Y which coincides with T_a^i on $X(T^i)$ and annihilates $X(T^i)^{\perp}$.)

Sketch of proof. Let P_i be projection onto $X(T^i)$. Since A is a Banach *-algebra, $\{T_a^i: i\ \text{varying}\}$ is norm-bounded for each a. Hence, to prove that $(II) \Rightarrow (I)$, it is enough (in view of (iii)) to show that $P_i \xi \to \xi$ for each ξ in $X(T)$. This follows from (iii) and the non-degeneracy of T.

Assume (I). Let Θ be a fixed orthonormal basis of $X(T)$. Embed $X(T)$ in a Hilbert space Y whose dimension is at least as large as that of each $X(T^i)$. Now, given a positive number ε and finite subsets F and G of A and Θ respectively, hypothesis (I) enables us (in view of 1.10) to construct a *-representation $S = S^{F,G,\varepsilon}$ of A such that (a) $X(S) \subset Y$, (b) S is equivalent to an element of $\mathscr{S}$, and (c)

$$|(S_a \xi, \eta) - (T_a \xi, \eta)| < \varepsilon$$

for all ξ, η in G and all a in F. Thus the $\{S^{F,G,\varepsilon}: F, G, \varepsilon\ \text{varying}\}$ form a net, which we abbreviate to $\{T^i\}$, satisfying II(i), (ii) and

$$(T_a^i \xi, \eta) \underset{i}{\to} (T_a \xi, \eta)$$

for all a in A and ξ, η in $X(T)$. Since $\|T_a^i \xi - T_a \xi\|^2 = (T_{a^* a}^i \xi, \xi) - (T_a^i \xi, T_a \xi) - (T_a \xi, T_a^i \xi) + (T_{a^* a} \xi, \xi)$, this implies II(iii). ∎

Dense Subalgebras

1.17.   Suppose that B is a *-subalgebra of A. Then it is evident that the restriction mapping $\Phi: T \mapsto T|B$ of $\mathcal{T}(A)$ into $\mathcal{T}(B)$ is continuous with respect to the regional topologies. If B is dense, we shall show that Φ is a homeomorphism.

Proposition.   *Suppose that A is a Banach *-algebra and B is a dense *-subalgebra of A. Then the restriction map $\Phi: T \mapsto T|B$ is a homeomorphism with respect to the regional topologies of $\mathcal{T}(A)$ and $\mathcal{T}(B)$.*

Proof.   By VI.3.8

$$\|T_a\| \leq \|a\| \qquad (T \in \mathcal{T}(A); a \in A). \qquad (25)$$

Thus it is clear that Φ is one-to-one.

Let $T \in \mathcal{T}(A)$, and consider the regional neighborhood $U = U(T; \varepsilon; \xi_1, \ldots, \xi_p; F)$ of T, where $F = \{a_1, \ldots, a_n\}$ and $\|\xi_i\| \leq 1$ for all i. For each $i = 1, \ldots, n$ let b_i be an element of B such that $\|a_i - b_i\| < \frac{1}{3}\varepsilon$; and put $G = \{b_1, \ldots, b_n\}$. We shall show that, if T' is any element of $\mathcal{T}(A)$ such that $T'|B \in V = U'(T|B; \frac{1}{3}\varepsilon; \xi_1, \ldots, \xi_p; G)$, then $T' \in U$. By 1.10 and the arbitrariness of U this will establish the continuity of Φ^{-1}, proving the proposition.

Let $T' \in \mathcal{T}(A)$, $T'|B \in V$. Thus there are vectors $\xi'_1, \ldots, \xi'_p$ in $X(T')$ such that

$$(\xi'_i, \xi'_j) = (\xi_i, \xi_j) \qquad \text{(all } i, j), \qquad (26)$$

$$|(T'_{b_k}\xi'_i, \xi'_j) - (T_{b_k}\xi_i, \xi_j)| < \frac{1}{3}\varepsilon \qquad \text{(all } i, j, k). \qquad (27)$$

In particular, (26) gives $\|\xi'_i\| = \|\xi_i\| \leq 1$; so by (27) and (25):

$$|(T'_{a_k}\xi'_i, \xi'_j) - (T_{a_k}\xi_i, \xi_j)| \leq |(T'_{b_k}\xi'_i, \xi'_j) - (T_{b_k}\xi_i, \xi_j)| + |(T'_{a_k - b_k}\xi'_i, \xi'_j)|$$

$$+ |(T_{a_k - b_k}\xi_i, \xi_j)|$$

$$< \tfrac{1}{3}\varepsilon + \tfrac{1}{3}\varepsilon + \tfrac{1}{3}\varepsilon = \varepsilon$$

for all $i, j = 1, \ldots, p$ and $k = 1, \ldots, n$. So $T' \in U$.   ∎

1.18.   Corollary.   *Let A be a Banach *-algebra and A_c its C*-completion (VI.10.4). The correspondence $T \mapsto T'$ of VI.10.5 between $\mathcal{T}(A_c)$ and $\mathcal{T}(A)$ is a homeomorphism with respect to the regional topologies.*

Thus the representation theory of C*-algebras is no less general than that of arbitrary Banach *-algebras even as regards the regional topology (see VI.10.1).

Quotient Algebras

1.19. If $a \in A$, then we claim that

$$\mathscr{S} = \{S \in \mathscr{T} : a \in \text{Ker}(S)\}$$

is regionally closed. Indeed, suppose T belongs to the regional closure of $\mathscr{S}$. Fix a vector ξ in $X(T)$ and a number $\varepsilon > 0$. Then $U(T; \varepsilon; \xi; \{a^*a\})$ contains an element S of $\mathscr{S}$; and for some ξ' in $X(S)$ we have

$$|(S_{a^*a}\xi', \xi') - (T_{a^*a}\xi, \xi)| < \varepsilon. \tag{28}$$

Now $(S_{a^*a}\xi', \xi') = 0$ since $a \in \text{Ker}(S)$; so by (28)

$$\|T_a\xi\|^2 = (T_{a^*a}\xi, \xi) < \varepsilon.$$

By the arbitrariness of ξ and ε this implies $a \in \text{Ker}(T)$, or $T \in \mathscr{S}$. So the claim is proved.

It follows of course that, for any subset B of A, $\{T \in \mathscr{T} : B \subset \text{Ker}(T)\}$ is regionally closed.

1.20. Proposition. *Let I be a *-ideal of A, and $\pi : A \to A/I$ the quotient *-homomorphism; and put $\mathscr{T}_0 = \{T \in \mathscr{T}(A) : I \subset \text{Ker}(T)\}$. Then $\mathscr{T}_0$ is regionally closed. The composition map $T \mapsto T \circ \pi$ is a homeomorphism of $\mathscr{T}(A/I)$ onto $\mathscr{T}_0$ with respect to the regional topologies.*

Proof. The first statement was proved in 1.19. The fact that $T \mapsto T \circ \pi$ is a homeomorphism follows from the evident relation

$$U(T; \varepsilon; \xi_1, \dots, \xi_p; \pi(F)) \circ \pi = U(T \circ \pi; \varepsilon; \xi_1, \dots, \xi_p; F) \cap \mathscr{T}_0. \quad \blacksquare$$

Weak Containment

1.21. Closely related to the regional topology is the rather important notion of weak containment.

Given any subfamily $\mathscr{S}$ of $\mathscr{T}(A)$, let us denote by $\mathscr{S}_f$ the family of all finite Hilbert direct sums $S^{(1)} \oplus \cdots \oplus S^{(p)}$, where $S^{(i)} \in \mathscr{S}$ for each i (repetitions of course being allowed).

Definition. A *-representation T of A is *weakly contained in* a subfamily $\mathscr{S}$ of $\mathscr{T}(A)$ if T belongs to the regional closure of $\mathscr{S}_f$.

If T is in the regional closure of $\mathscr{S}$, then obviously T is weakly contained in $\mathscr{S}$. But the converse is false; for, if T is finite-dimensional, $T \oplus T$ is weakly contained in $\{T\}$, though it is not in the regional closure of $\{T\}$.

1.22. The main difference between weak containment and the regional closure is expressed in the following proposition.

Proposition. *A Hilbert direct sum $\sum_i^{\oplus} T^i$ of *-representations of A is weakly contained in a subfamily $\mathscr{S}$ of $\mathscr{T}(A)$ if and only if each T^i is weakly contained in $\mathscr{S}$.*

Proof. The "only if" is evident. The "if" part is true for finite direct sums by 1.7 (since $\mathscr{S}_f$ is closed under the formation of finite direct sums). It then follows for arbitrary direct sums by 1.6. ∎

1.23. Let A be a Banach *-algebra and B a dense *-subalgebra of A. It follows from 1.16 that a *-representation T of A is weakly contained in a subfamily $\mathscr{S}$ of $\mathscr{T}(A)$ if and only if $T|B$ is weakly contained in $\mathscr{S}|B \, (= \{S|B : S \in \mathscr{S}\})$.

Thus (see 1.18) the relation of weak containment for *-representations of a Banach *-algebra A is exactly the same as for the corresponding *-representations of the C^*-completion A_c.

1.24.   Definition. If $\mathscr{S}_1$ and $\mathscr{S}_2$ are two subfamilies of $\mathscr{T}(A)$, we say that $\mathscr{S}_1$ is *weakly contained in* $\mathscr{S}_2$ if every T in $\mathscr{S}_1$ is weakly contained in $\mathscr{S}_2$, that is, if $\mathscr{S}_1$ is contained in the regional closure $(\mathscr{S}_2)_f^-$ of $(\mathscr{S}_2)_f$.

By 1.7, $\mathscr{S}_1 \subset (\mathscr{S}_2)_f^- \Rightarrow (\mathscr{S}_1)_f \subset (\mathscr{S}_2)_f^-$. Therefore, if $\mathscr{S}_1$ is weakly contained in $\mathscr{S}_2$ and $\mathscr{S}_2$ is weakly contained in $\mathscr{S}_3$, $\mathscr{S}_1$ must be weakly contained in $\mathscr{S}_3$.

Definition. Two subfamilies $\mathscr{S}_1$ and $\mathscr{S}_2$ of $\mathscr{T}(A)$ are *weakly equivalent* if each is weakly contained in the other.

By the preceding paragraph this is an equivalence relation in the ordinary sense.

2.   The Regional Topology and Separation Properties

2.1. Two unitarily equivalent *-representations of A are obviously not distinguished from each other by the regional topology of $\mathscr{T}(A)$. Thus, if we wish, we can regard the regional topology as topologizing the space of all unitary equivalence classes of *-representations of A.

2.2.   Definition. Two *-representations S and T of A are *regionally equivalent* if they are not distinguished from each other by the regional topology, that is, if S and T belong to the regional closure of $\{T\}$ and $\{S\}$ respectively.

Unitary equivalence implies regional equivalence, but the converse is false in general. For example, if T is any *-representation of A in a non-zero separable space, and α and β are distinct infinite cardinal numbers, then $\alpha \cdot T$ and $\beta \cdot T$ (see VI.14.10) are regionally equivalent but unitarily inequivalent. This is easily verified. More interesting examples will appear in Remark 5.12.

Thus, the regional topology of the space of all unitary equivalence classes of *-representations of A is not a T_0 space.

By 1.18, two regionally equivalent *-representations S and T of A have the same kernel. Though the converse is false in general, it is true for certain special classes of *-representations (see 2.6, 5.12).

2.3. Proposition. *If S and T are regionally equivalent finite-dimensional *-representations of A, they are unitarily equivalent.*

Proof. Since S belongs to the closure of $\{T\}$, 1.4 (or 1.11) implies that $\dim(S) \leq \dim(T)$. Likewise $\dim(T) \leq \dim(S)$. So $\dim(S) = \dim(T)$.

Choose an orthonormal basis $\xi_1, \ldots, \xi_n$ of $X(S)$. Since S is in $\{T\}^-$, 1.8 and 1.11 assert the existence of a net $\{\langle \eta_1^\nu, \ldots, \eta_n^\nu \rangle\}$ of orthonormal bases of $X(T)$ such that

$$(T_a \eta_i^\nu, \eta_j^\nu) \to (S_a \xi_i, \xi_j) \tag{1}$$

for all i and j and all a in A. Since $X(T)$ is finite-dimensional we can pass to a subnet and assume that $\eta_i^\nu \to \eta_i$ for each $i = 1, \ldots, n$, where $\eta_1, \ldots, \eta_n$ is some orthonormal basis of $X(T)$. By (1) this implies that

$$(T_a \eta_i, \eta_j) = (S_a \xi_i, \xi_j)$$

for all i, j, and a. Thus the linear isometry sending ξ_i into η_i (for each i) sets up the unitary equivalence of S and T. ∎

2.4. For the moment let $\mathscr{R}$ be the space of regional equivalence classes of *-representations of A, with the regional topology. By definition $\mathscr{R}$ is a T_0 space. But it is not a T_1 space. Indeed: Let T be a finite-dimensional *-representation of A and S a subrepresentation of T on a proper subspace of $X(T)$. By Remark 1.3 S is in the regional closure of $\{T\}$; but S and T are not regionally equivalent in virtue of 2.3.

2.5. Notice that if we confine ourselves to *-representations of fixed finite dimension, the space of regional (i.e., unitary) equivalence classes is not only T_1 but Hausdorff.

Proposition*. *Let* n *be a positive integer, and* $\mathcal{R}_n$ *the space of unitary equivalence classes of *-representations of* A *of dimension* n. *Then* $\mathcal{R}_n$ *is Hausdorff under the regional topology.*

To prove this, let X be a fixed n-dimensional Hilbert space, and $\mathcal{T}_n$ the space of all *-representations of A acting on X, topologized with the topology of pointwise convergence on A. The compact group $U(X)$ of unitary operators on X acts continuously as a topological transformation group on $\mathcal{T}_n$:

$$(uT)_a = u T_a u^{-1} \qquad (a \in A; u \in U(X); T \in \mathcal{T}_n).$$

One now shows that $\mathcal{R}_n$ is homeomorphic with the (Hausdorff) quotient space $\mathcal{T}_n/U(X)$.

2.6. Let us consider *-representations of commutative C^*-algebras. For those which are canonically constructed from measures as in VI.10.12, we shall now see that regional equivalence amounts to having the same kernel.

If A, μ, and T are as in VI.10.12, let us write T^μ for T.

Proposition. *Let* A *be a commutative* C^*-*algebra, and* μ *and* ν *two regular Borel measures on* $\hat{A}$. *The following three conditions are equivalent*: (i) μ *and* ν *have the same closed supports*; (ii) $\mathrm{Ker}(T^\mu) = \mathrm{Ker}(T^\nu)$; (iii) T^μ *and* T^ν *are regionally equivalent.*

Proof. (i) $\Leftrightarrow$ (ii) by VI.10.12 and VI.10.14; and (iii) $\Rightarrow$ (ii) by 2.2.

Assume (i); and let S be the common closed support of μ and ν. We shall show that T^μ belongs to the regional closure of $\{T^\nu\}$. Let $g_1, \dots, g_p$ be elements of $\mathcal{L}(\hat{A})$; let F be a finite subset of A; and let $\varepsilon > 0$. Let C, D be compact subsets of $\hat{A}$ such that C contains the closed supports of all the g_i, and the interior of D contains C. By the theory of partitions of unity we can find a finite sequence $w_1, \dots, w_n$ of non-negative-valued elements of $\mathcal{L}(\hat{A})$ such that: (a) $v = \sum_{k=1}^{n} w_k \equiv 1$ on $S \cap C$, $v \le 1$ everywhere, and v vanishes outside D; (b) $S \cap \{\phi \in \hat{A} : w_k(\phi) > 0\}$ is non-void for each k (so that $\int w_k \, d\mu > 0$, $\int w_k \, d\nu > 0$); and (c) the functions $g_i \bar{g}_j$ and $\hat{a} g_i \bar{g}_j$ $(i, j = 1, \dots, p;$

$a \in F$) all have oscillation less than ε on each set $\{\phi \in \hat{A} : w_k(\phi) > 0\}$ ($k = 1, \ldots, n$). Now put

$$t_k = \left(\int w_k \, d\mu \right) \left(\int w_k \, dv \right)^{-1},$$

$$\rho = \left(\sum_{k=1}^{n} t_k w_k \right)^{1/2},$$

$$h_i = \rho g_i \qquad\qquad (i = 1, \ldots, p).$$

Considering the g_i and h_i as elements of $\mathscr{L}_2(\mu)$ and $\mathscr{L}_2(v)$ respectively, we have (see property (a) of the w_k):

$$(g_i, g_j)_{\mathscr{L}_2(\mu)} = \sum_k \int g_i \overline{g_j} w_k \, d\mu,$$

$$(h_i, h_j)_{\mathscr{L}_2(v)} = \sum_k \int g_i \overline{g_j} t_k w_k \, dv.$$

Now fix $i, j = 1, \ldots, p$ and $k = 1, \ldots, n$; and choose a value γ of $g_i \overline{g_j}$ on $\{\phi \in \hat{A} : w_k(\phi) > 0\}$. By condition (c) on w_k,

$$\left| \int g_i \overline{g_j} w_k \, d\mu - \gamma \int w_k \, d\mu \right| \le \varepsilon \int w_k \, d\mu,$$

$$\left| \int g_i \overline{g_j} t_k w_k \, dv - \gamma \int w_k \, d\mu \right| = \left| t_k \int g_i \overline{g_j} w_k \, dv - \gamma t_k \int w_k \, dv \right|$$

$$\le \varepsilon t_k \int w_k \, dv = \varepsilon \int w_k \, d\mu.$$

Therefore

$$\left| \int g_i \overline{g_j} w_k \, d\mu - \int g_i \overline{g_j} t_k w_k \, dv \right| \le 2\varepsilon \int w_k \, d\mu.$$

Summing over k, we get from condition (a)

$$|(h_i, h_j)_{\mathscr{L}_2(v)} - (g_i, g_j)_{\mathscr{L}_2(\mu)}| \le 2\varepsilon \int v \, d\mu \le 2\varepsilon\mu(D). \qquad (2)$$

Similarly

$$|(\hat{a}h_i, h_j)_{\mathscr{L}_2(v)} - (\hat{a}g_i, g_j)_{\mathscr{L}_2(\mu)}| \le 2\varepsilon\mu(D) \qquad (3)$$

for all a in F and $i, j = 1, \ldots, p$.

Since the g_i were chosen from the dense subset $\mathscr{L}(\hat{A})$ of $\mathscr{L}_2(\mu)$, inequalities (2) and (3), together with the arbitrariness of ε and F, show by 1.4 that T^μ is in the regional closure of $\{T^\nu\}$. The same holds on interchanging μ and ν. So T^μ and T^ν are regionally equivalent; and we have shown that (i) $\Rightarrow$ (iii).   ■

3. The Structure Space

3.1. We continue to assume that A is a fixed *-algebra.

Definition. By the *structure space* of A we mean the family of all unitary equivalence classes of irreducible *-representations of A, equipped with the relativized regional topology (see 2.1). *We denote the structure space of A by $\hat{A}$.*

Remark. The structure space $\hat{A}$ is also called the *dual space* of A or the *spectrum* of A in the literature.

We shall usually fail to distinguish notationally between a unitary equivalence class τ belonging to $\hat{A}$ and a *-representation T belonging to τ.

3.2. If A is commutative, the elements of $\hat{A}$ are all one-dimensional by VI.14.3. Thus, if A is a commutative Banach *-algebra the $\hat{A}$ defined above coincides as a set with the $\hat{A}$ of VI.4.2. In that case, by 1.13 the regional and the Gelfand topologies coincide on $\hat{A}$; so the $\hat{A}$ of 3.1 and VI.4.2 coincide as topological spaces.

3.3. If A is a C^*-algebra, then by VI.25.14 $\hat{A}$ essentially coincides as a set with the $\overline{A}$ of IV.3.8.

3.4. ***Remark.*** For non-commutative C^*-algebras A, we shall see later that $\hat{A}$ need not be a T_0 space (Remark 5.12); if it is a T_0 space it need not be T_1 (7.3); and if it is T_1 it need not be Hausdorff (7.9, 7.10).

3.5. Let I be a *-ideal of the *-algebra A, and $\pi: A \to A/I$ the quotient map. By 1.19, the composition map $T \mapsto T \circ \pi$ is a homeomorphism of $(A/I)\hat{\ }$ onto the closed subset $\{S \in \hat{A} : I \subset \operatorname{Ker}(S)\}$ of $\hat{A}$.

3.6. If A is a Banach *-algebra and B is a dense *-subalgebra of A, then by 1.16 the restriction map $T \mapsto T|B$ is a homeomorphism of $\hat{A}$ into (but not necessarily onto) $\hat{B}$.

3.7. Recall from VI.14.2 that each irreducible *-representation T of the *-algebra A gives rise to a one-dimensional *-representation χ_T of the center Z of A, the so-called central character of T.

Proposition. *The map* $T \mapsto \chi_T$ *is continuous from* $\hat{A}$ *to* $\hat{Z}$ *with respect to the regional topologies of* $\hat{A}$ *and* $\hat{Z}$.

Proof. We recall from 1.13 that the regional topology of $\hat{Z}$ is just the topology of pointwise convergence on Z.

Let T be any element of $\hat{A}$. Choose a positive number ε, a finite subset F of Z, and a unit vector ξ in $X(T)$; and put

$$U' = U'(T; \varepsilon; \xi; F) \qquad \text{(see 1.10).}$$

Thus, if $S \in U'$, there is a unit vector ξ' in $X(S)$ such that

$$|\chi_S(a) - \chi_T(a)| = |(S_a \xi', \xi') - (T_a \xi, \xi)| < \varepsilon \qquad \text{for all } a \text{ in } F.$$

Since U' is a regional neighborhood of T (by 1.10), this gives the required conclusion. ∎

The Primitive Ideal Space and the Hull-Kernel Topology

3.8. Definition. A *primitive ideal* of A is a (two-sided) *-ideal of the form $\mathrm{Ker}(T)$, where $T \in \hat{A}$. We denote the set of all primitive ideals of A by $\mathrm{Prim}(A)$; thus

$$\mathrm{Prim}(A) = \{\mathrm{Ker}(T) : T \in \hat{A}\}.$$

3.9. There is an interesting way of topologizing $\mathrm{Prim}(A)$.

Given any subset W of $\mathrm{Prim}(A)$, let us define the *hull-kernel closure* $\bar{W}$ of W to be $\{I \in \mathrm{Prim}(A) : I \supset \bigcap W\}$. We claim that this "closure operation" $W \mapsto \bar{W}$ satisfies the Kuratowski axioms, namely:

(i) $\bar{\emptyset} = \emptyset$,

(ii) $\bar{W} \supset W$,

(iii) $\bar{\bar{W}} = \bar{W}$,

(iv) $(V \cup W)^- = \bar{V} \cup \bar{W}$

(*for all subsets V, W of* $\mathrm{Prim}(A)$).

Indeed, properties (i), (ii) are evident. It is also evident from the definition of $\bar{W}$ that $\bigcap \bar{W} = \bigcap W$, and hence that (iii) holds. To prove (iv) we first observe that $W_1 \subset W_2 \Rightarrow (W_1)^- \subset (W_2)^-$; and this implies

$$\bar{V} \cup \bar{W} \subset (V \cup W)^-. \qquad (1)$$

To obtain the opposite inclusion, take a primitive ideal I not belonging to $\bar{V} \cup \bar{W}$. This means that $I \not\supset \bigcap V$ and $I \not\supset \bigcap W$. Thus, if $T \in \hat{A}$ and $I = \operatorname{Ker}(T)$, we can choose elements a and b of $\bigcap V$ and $\bigcap W$ respectively such that $T_a \neq 0$ and $T_b \neq 0$. Now let ξ be a vector in $X(T)$ such that $T_a \xi \neq 0$. Since T is irreducible, $T_a \xi$ is cyclic for T; so, since $T_b \neq 0$, there is an element c of A such that $T_b(T_c T_a \xi) \neq 0$. This implies that $bca \notin \operatorname{Ker}(T) = I$. But $bca \in \bigcap V \cap \bigcap W = \bigcap(V \cup W)$. Therefore $I \not\supset \bigcap(V \cup W)$, whence $I \notin (V \cup W)^-$. We have now shown that $I \notin \bar{V} \cup \bar{W} \Rightarrow I \notin (V \cup W)^-$; and this gives the inclusion opposite to (1). So (iv) has been proved.

Definition. The topology for $\operatorname{Prim}(A)$ whose closure operation is the hull-kernel closure defined above is called the *hull-kernel topology* of $\operatorname{Prim}(A)$.

The existence (and of course uniqueness) of this topology follows from the Kuratowski axioms (i) − (iv); see Kelley [1], p. 43.

Definition. The space $\operatorname{Prim}(A)$, equipped with the hull-kernel topology, is called the *primitive ideal space* of A.

In the future we always consider $\operatorname{Prim}(A)$ as equipped with the hull-kernel topology.

Notice that $\operatorname{Prim}(A)$ is automatically a T_0 space.

3.10. The surjection $T \mapsto \operatorname{Ker}(T)$ of $\hat{A}$ onto $\operatorname{Prim}(A)$ enables us to pull back the hull-kernel topology of $\operatorname{Prim}(A)$ to a topology on $\hat{A}$, which we call the *hull-kernel topology of $\hat{A}$*. Thus, a subset $\mathscr{S}$ of $\hat{A}$ is open in the hull-kernel topology of $\hat{A}$ if and only if it is of the form $\{T \in \hat{A} : \operatorname{Ker}(T) \in W\}$ for some subset W of $\operatorname{Prim}(A)$ which is open in the hull-kernel topology of $\operatorname{Prim}(A)$.

3.11. Proposition. *The hull-kernel topology of $\hat{A}$ is contained in the regional topology of $\hat{A}$.*

Proof. Take a subset $\mathscr{S}$ of $\hat{A}$ which is closed in the hull-kernel topology. We must show that $\mathscr{S}$ is regionally closed (relative to $\hat{A}$). Suppose that T is in $\hat{A}$ and lies in the regional closure of $\mathscr{S}$. Then by 1.18 $\bigcap\{\operatorname{Ker}(S) : S \in \mathscr{S}\} \subset \operatorname{Ker}(T)$; so T lies in the hull-kernel closure of $\mathscr{S}$, that is, $T \in \mathscr{S}$. Thus $\mathscr{S}$ is regionally closed in $\hat{A}$. ∎

3.12. Example. In general the hull-kernel and regional topologies of $\hat{A}$ are by no means the same. Take for example the commutative Banach *-algebra A of VI.1.8. As we indicated in VI.2.6, $\hat{A}$ is then essentially $[-1, 1]$. The elements of A are analytic on the interior of the unit disk, and so are determined by their values on any non-void open subinterval W of $[-1, 1]$; in particular, if a vanishes on W then $a = 0$. This implies that the hull-kernel closure of any such W is all of $[-1, 1]$. On the other hand, the regional or Gelfand topology of $\hat{A}$ (see 3.2) is just the natural topology of $[-1, 1]$.

3.13. We shall see in 5.11 that C^*-algebras have the striking property that the hull-kernel and regional topologies of their structure spaces are the same. In §8 we shall compute explicitly the structure spaces, along with their topology, for several specific non-commutative C^*-algebras, to illustrate the phenomena which can arise.

4. Restriction of Representations to Hereditary Subalgebras

4.1. Definition. By a *hereditary *-subalgebra* of the *-algebra A we mean a *-subalgebra B of A with the property that, if $a \in A$ and $b, c \in B$, then $bac \in B$.

Evidently any *-ideal of A is a hereditary *-subalgebra of A. For any element a of A, the *-subalgebra a^*Aa of A is a hereditary *-subalgebra.

Remark. Pedersen [5], 1.5.1, defines a closed *-subalgebra B of a C^*-algebra A to be *hereditary in A* if $a \in B$ whenever $0 \le a \le b$, $b \in B$ and $a \in A$. It is not hard to prove that, for C^*-algebras, Pedersen's definition coincides with ours. (Use Pedersen [5], 1.4.5 and 1.5.2.)

4.2. Let B be a hereditary *-subalgebra of A; and denote by $\mathcal{T}_B$ the family of all those T in $\mathcal{T}(A)$ such that the linear span of $\{T_{ab}\xi : a \in A, b \in B, \xi \in X(T)\}$ is dense in $X(T)$. For each T in $\mathcal{T}_B$ let $T^{(B)}$ stand for the non-degenerate part of $T|B$; that is (see VI.9.8), $T^{(B)}$ is the subrepresentation of $T|B$ acting on the closure of the linear span of $\{T_b\xi : b \in B, \xi \in X(T)\}$.

Proposition.

(I) $T \mapsto T^{(B)}$ is one-to-one on $\mathcal{T}_B$ (*as a mapping of unitary equivalence classes*).

(II) $T \mapsto T^{(B)}$ is a homeomorphism on $\mathcal{T}_B$ *with respect to the regional topologies of* $\mathcal{T}_B$ *and* $\mathcal{T}(B)$.

Proof.

(I) Suppose that S, $T \in \mathcal{T}_B$, and that $S^{(B)} \cong T^{(B)}$ under a unitary equivalence $U : X(S^{(B)}) \to X(T^{(B)})$. If a, $a' \in A$, b, $b' \in B$, and ξ, $\eta \in X(S^{(B)})$, we have

$$(T_{ab} U\xi, T_{a'b'} U\eta) = (T_{b'^*a'^*ab} U\xi, U\eta)$$

$$= (S_{b'^*a'^*ab} \xi, \eta) \qquad \text{(since } b'^*a'^*ab \in B)$$

$$= (S_{ab} \xi, S_{a'b'} \eta).$$

It follows that the equation $V(S_{ab}\xi) = T_{ab} U\xi$ ($a \in A$, $b \in B$, $\xi \in X(S^{(B)})$) defines a linear isometry V of the linear span of the $S_{ab}\xi$ onto the linear span of the $T_{ab} U\xi$. By hypothesis the domain and range of V are dense in $X(S)$ and $X(T)$ respectively. So V extends to a linear isometry of $X(S)$ onto $X(T)$, which clearly intertwines S and T.

(II) To prove that $T \mapsto T^{(B)}$ is continuous at $T \in \mathcal{T}_B$, it is sufficient (by 1.4) to take a neighborhood V of $T^{(B)}$ of the form $U(T^{(B)}; \varepsilon; \{T_{b_i}\xi_i\}; F)$ (where $\xi_i \in X(T)$, $b_i \in B$, and F is a finite subset of B), and find a neighborhood W of T such that if $T' \in \mathcal{T}_B \cap W$ then $T'^{(B)} \in V$. We claim that we can take $W = U(T; \varepsilon; \{\xi_i\}; G)$, where G is the (finite) set of all $b_j^* b_i$ and all $b_j^* a b_i$ ($a \in F$). Indeed, suppose T' is in $\mathcal{T}_B$ and belongs to this W. Then there exist $\xi_i' \in X(T')$ such that, putting $\eta_i' = T'_{b_i} \xi_i'$, $\eta_i = T_{b_i} \xi_i$, we have

$$|(\eta_i', \eta_j') - (\eta_i, \eta_j)| = |(T'_{b_j^* b_i} \xi_i', \xi_j') - (T_{b_j^* b_i} \xi_i, \xi_j)| < \varepsilon$$

and

$$|(T_a' \eta_i', \eta_j') - (T_a \eta_i, \eta_j)| = |(T'_{b_j^* a b} \xi_i', \xi_j') - (T_{b_j^* a b} \xi_i, \xi_j)| < \varepsilon$$

for all i, j and a in F. Since the η_i' are in $X(T'^{(B)})$, this proves that $T'^{(B)} \in V$. So $T \mapsto T^{(B)}$ is continuous.

To show that $T^{(B)} \mapsto T$ is continuous at $T^{(B)}$ (where $T \in \mathcal{T}_B$), it is enough (by 1.4) to take a neighborhood $W = U(T; \varepsilon; \{T_{c_i b_i} \xi_i\}; F)$ of T (where $\xi_i \in X(T^{(B)})$, $b_i \in B$, $c_i \in A$, and F is a finite subset of A), and exhibit a neighborhood V of $T^{(B)}$ such that $T'^{(B)} \in V \Rightarrow T' \in W$. We claim that $V = U(T^{(B)}; \varepsilon; \{\xi_i\}; G')$ will have this property provided G' consists of all $b_j^* c_j^* c_i b_i$ and all $b_j^* c_j^* a c_i b_i$ with $a \in F$. (Notice that $G' \subset B$ since B is a hereditary *-subalgebra.) Indeed, let $T'^{(B)} \in V$. Then there exist ξ_i' in $X(T'^{(B)})$ such that, putting $\eta_i' = T'_{c_i b_i} \xi_i'$ and $\eta_i = T_{c_i b_i} \xi_i$, we have:

$$|(\eta_i', \eta_j') - (\eta_i, \eta_j)| = |(T'_{b_j^* c_j^* c_i b_i} \xi_i', \xi_j') - (T_{b_j^* c_j^* c_i bi} \xi_i, \xi_j)| < \varepsilon$$

and

$$|(T'_a\eta'_i, \eta'_j) - (T_a\eta_i, \eta_j)| = |(T'_{b^*_j c^*_j ac_i b_i}\xi'_i, \xi'_j) - (T_{b^*_j c^*_j ac_i b_i}\xi_i, \xi_j)| < \varepsilon$$

for all i, j and a in F. This shows that $T' \in W$. Thus $T^{(B)} \mapsto T$ is continuous. ∎

Remark.   This Proposition will be obtained by another route in XI.7.6.

4.3.   In the context of 4.2, suppose that T is an irreducible element of $\mathcal{T}(A)$. Then $T \in \mathcal{T}_B$ if and only if $B \not\subset \text{Ker}(T)$. In that case we claim that $T^{(B)}$ is also irreducible.

Indeed, let $0 \neq \xi \in X(T^{(B)})$. Since $T^{(B)}$ is certainly non-degenerate, we can choose b in B so that $T_b\xi \neq 0$. By the irreducibility of T, $\{T_{ab}\xi : a \in A\}$ is dense in $X(T)$. So the linear span of $\{T_{cab}\xi : a \in A, c \in B\}$ is dense in $X(T^{(B)})$. But, if $a \in A$ and $b, c \in B$, we have $cab \in B$ and $T_{cab}\xi = T^{(B)}_{cab}\xi$. By the arbitrariness of ξ, $T^{(B)}$ is therefore irreducible.

The above claim generalizes both V.1.18 and V.1.21 in the involutory context.

4.4.   As a consequence of 1.18, 4.2, and 4.3 we have:

Proposition.   *Let B be a hereditary *-subalgebra of A; and set $\hat{A}_B = \{T \in \hat{A} : B \not\subset \text{Ker}(T)\}$. Then $\hat{A}_B$ is regionally open relative to $\hat{A}$; and the map $T \mapsto T^{(B)}$ of 4.2, restricted to $\hat{A}_B$, is a homeomorphism of $\hat{A}_B$ into $\hat{B}$.*

4.5.   As we shall point out in a moment, the image of $\hat{A}_B$ under the map $T \mapsto T^{(B)}$ is not always the whole of $\hat{B}$. So it is interesting to find situations where it *is* the whole of $\hat{B}$.

Proposition.   *Suppose that A is a C*-algebra and that B is a closed hereditary *-subalgebra of A. Then the image of $\hat{A}_B$ under the map $T \mapsto T^{(B)}$ is all of $\hat{B}$.*

Proof.   Let S be any element of $\hat{B}$. By VI.22.6 S can be extended to an element T of $\hat{A}$. Thus S can be taken to be a subrepresentation of $T|B$ acting on the closed subspace Y of $X(T)$. Let us denote by Z the space of $T^{(B)}$. This means that Z is the largest closed $(T|B)$-stable subspace of $X(T)$ on which $T|B$ acts non-degenerately; in particular $Y \subset Z$. On the other hand $T^{(B)}$ is irreducible by 4.3. Consequently $Y = Z$, and $S = T^{(B)}$. ∎

Remark.   If A is merely a Banach *-algebra (even a reduced Banach *-algebra), and B is a closed hereditary *-subalgebra of A, the image of $\hat{A}_B$ under $T \mapsto T^{(B)}$ need *not* be all of $\hat{B}$. The only counter-examples showing this that we know of are the $\mathscr{L}_1$ group algebras A of non-compact semisimple groups G such as $SL(2, \mathbb{R})$; we take B to be of the form eAe, where e is a minimal central idempotent of the group algebra of a maximal compact subgroup of G. Unfortunately, to present such an example in full detail would carry us too far afield.

4.6.   From VI.19.11 we notice another situation in which the map of 4.4 is surjective.

Proposition.   *If A is a Banach *-algebra and B is any *-ideal (closed or not) of A, the image of $\hat{A}_B$ under $T \mapsto T^{(B)}$ is all of $\hat{B}$.*

Thus, under the hypotheses of this proposition, $\hat{A}$ can be written as the disjoint union of the open set $\hat{A}_B$ and the closed set $\hat{A} \setminus \hat{A}_B$; and we can identify $\hat{A}_B$ and $\hat{A} \setminus \hat{A}_B$, both setwise and topologically, with $\hat{B}$ and $(A/B)\hat{\;}$ respectively (see 1.19).

Notice that in this situation $T^{(B)}$ is simply $T|B$.

Remark.   If the A of the above proposition is merely a *-algebra (not a Banach *-algebra), the conclusion of the above proposition is no longer true. For example, let A be the *-algebra (under matrix multiplication and involution) of all infinite complex matrices $a = (a_{nm})$ ($n, m = 1, 2, 3, \ldots$) such that there exists a positive integer k (depending of course on a) for which $|n - m| > k$ implies $a_{nm} = 0$. Let B be the *-ideal of A consisting of those $a \in A$ having only finitely many nonzero entries. Let S be the natural irreducible *-representation of B acting on the Hilbert space X of all complex square-summable sequences. By nearly the same argument as in Remark VI.19.11, S cannot be extended to a *-representation of A acting on X.

4.7.   Let B be any *-algebra without unit, and A ($= B_1$) the *-algebra obtained by adjoining a unit $\mathbb{1}$ to B. Thus B is a (two-sided) *-ideal of A; and obviously any element S of $\hat{B}$ can be extended to an element of $\hat{A}$ by sending $\mathbb{1}$ into the identity operator on $X(S)$. Furthermore, the only element of $\hat{A} \setminus \hat{A}_B$ is the one-dimensional *-representation $\tau : b + \lambda\mathbb{1} \mapsto \lambda$ of A ($b \in B$; $\lambda \in \mathbb{C}$). Thus by 4.4 we have:

Proposition.   *In the above context, the restriction map $T \mapsto T|B$ is a homeomorphism of $\hat{A} \setminus \{\tau\}$ onto $\hat{B}$.*

4.8.   Although 4.6 fails if the word 'Banach' is omitted, the following partial result holds in the purely *-algebraic context (and of course contains 4.7 as a very special case):

Proposition*. *Let B be a *-ideal of the *-algebra A; and let C be the* **-subalgebra of A generated by B and the centralizer $\{a \in A : ab = ba$ for all b in B$\}$ of B in A. Then any element of $\hat{B}$ can be extended to an element of $\hat{C}$ (acting, of course, in the same space).*

This proposition is easy to deduce from Lemma 6 of Fell [10]. It is of considerable importance in the unitary representation theory of Lie groups.

5.   The Regional and Hull-Kernel Topologies on the Structure Space of a C*-Algebra

5.1.   In this section we shall make a deeper investigation of the structure space of a C^*-algebra. One of the principal results will be that the regional and hull-kernel topologies of the structure space of a C^*-algebra are the same.

5.2.   Our program depends heavily on a well-known result from the theory of linear topological spaces, called the Bipolar Theorem, which we shall state without proof.

If X_1 and X_2 are two real linear spaces, a *duality for X_1 and X_2* is a real-valued bilinear form $(\mid) : \langle \xi, \eta \rangle \mapsto (\xi | \eta)$ on $X_1 \times X_2$ which is non-degenerate in the sense that (i) if $\xi \in X_1$ and $(\xi|\eta) = 0$ for all η in X_2 then $\xi = 0$, and (ii) if $\eta \in X_2$ and $(\xi|\eta) = 0$ for all ξ in X_1 then $\eta = 0$. A pair X_1, X_2 of real linear spaces, together with a duality for X_1 nd X_2, is called a *real dual system.*

Now fix a real dual system $X_1, X_2, (\mid)$. By the *weak topology of X_1 [X_2]* we mean the weak topology generated by the family of all linear functionals on X_1 [X_2] of the form $\xi \mapsto (\xi|\eta)$, where $\eta \in X_2$ [$\eta \mapsto (\xi|\eta)$, where $\xi \in X_1$].

If $S \subset X_1$ and $T \subset X_2$, we define the *polar sets S^π and T_π* as follows:

$$S^\pi = \{\eta \in X_2 : (\xi|\eta) \geq -1 \quad \text{for all } \xi \text{ in } S\},$$

$$T_\pi = \{\xi \in X_1 : (\xi|\eta) \geq -1 \quad \text{for all } \eta \text{ in } T\}.$$

Evidently $(S^\pi)_\pi \supset S$, $(T_\pi)^\pi \supset T$.

If $S \subset X_1$, recall that the *convex hull $H(S)$ of S* is the set of all $\sum_{i=1}^{n} \lambda_i \xi_i$, where $n = 1, 2, \ldots$, the ξ_i are in S, and the λ_i are non-negative real numbers satisfying $\sum_{i=1}^{n} \lambda_i = 1$. The *closed convex hull* of S is the weak closure of $H(S)$.

Bipolar Theorem.   *For any subset S of X_1, $(S^\pi)_\pi$ coincides with the closed convex hull of $S \cup \{0\}$ (0 denoting of course the zero element of X_1). Thus $S = (S^\pi)_\pi$ if and only if S is weakly closed, convex, and contains 0.*

An analogous statement obviously holds for subsets of X_2.

For a proof see, for example, Bourbaki [9], Chap. IV, §1, no. 3, Prop. 3.

5.3.   For the rest of this section A is a fixed C^*-algebra. Recall that A_+ is the set of positive elements of A; and that all positive linear functions on A are continuous and extendable (VI.18.15).

5.4.   Proposition.   *Suppose that A has a unit 1; and let P be a the set of all positive linear functionals p on A with $p(1) = 1$. Let Q be a subset of P such that every Hermitian element a of A satisfying $p(a) \geq 0$ for all p in Q is positive. Then the pointwise closure of the convex hull of Q is P.*

Proof.   We consider the real dual system consisting of the real linear space A_h of all Hermitian elements of A, together with the space A_h^* of continuous real linear functionals on A_h (with the obvious duality). We identify Q with a subset of A_h^*, and recall from VI.7.16 the definition of the positive and negative parts $a_\pm$ of a Hermitian element of A. Then, if $a \in A_h$,

$$a \in Q_\pi \Leftrightarrow p(a) \geq -1 \qquad \text{for all } p \text{ in } Q$$

$$\Leftrightarrow p(1 + a) \geq 0 \qquad \text{for all } p \text{ in } Q$$

$$\Leftrightarrow 1 + a \geq 0$$

$$\Leftrightarrow \| a_- \| \leq 1 \qquad \text{(by the functional representation).}$$

Thus the bipolar $(Q_\pi)^\pi$ consists of all p in A_h^* such that $p(a) \geq -1$ whenever $a \in A_h$ and $\| a_- \| \leq 1$, that is, of all positive linear functionals p on A with $\| p \| \leq 1$. Therefore, by Theorem 5.2, $H(Q \cup \{0\})$ is pointwise dense in the set of all positive linear functionals p with $\| p \| \leq 1$, i.e., $p(1) \leq 1$ (see VI.18.12). Evaluating the positive functionals at 1, we see by a simple argument that $H(Q)$ is pointwise dense in P.   ■

Remark.   An example in Kadison [1] shows that, if A has no unit, the natural analogue of the above proposition is false.

5.5   Definition.   If T is a *-representation of A, a positive functional p on A is *associated with* T if for some $\xi \in X(T)$ we have

$$p(a) = (T_a \xi, \xi) \qquad\qquad (a \in A). \qquad (1)$$

The set of all positive functionals associated with T will be called $\Phi(T)$.

If T is non-degenerate and (1) holds, then by VI.8.4 and VI.9.12

$$\| p \| = \| \xi \|^2. \qquad (2)$$

Equivalence Theorem. *Let T be a non-degenerate *-representation and $\mathscr{S}$ a set of non-degenerate *-representations of A. Then the following four conditions are equivalent:*

(I) $\mathscr{S}$ *weakly contains* T.

(II) $\bigcap\{\operatorname{Ker}(S): S \in \mathscr{S}\} \subset \operatorname{Ker}(T)$.

(III) *For every p in $\Phi(T)$ there is a net $\{p_i\}$ of positive functionals, each of which is a finite sum of elements of $\bigcup\{\Phi(S): S \in \mathscr{S}\}$, such that*

$$p_i \underset{i}{\to} p \text{ pointwise on } A \tag{3}$$

and

$$\|p_i\| \underset{i}{\to} \|p\|. \tag{4}$$

(IV) *Every p in $\Phi(T)$ lies in the pointwise closure of the linear span of $\bigcup\{\Phi(S): S \in \mathscr{S}\}$.*

Proof. We first prove that (I) $\Rightarrow$ (III). Assume (I), and let p be as in (1). By the definition of weak containment p can be approximated on any finite subset of A by some q of the form $a \mapsto (S_a \eta, \eta)$, where $S \in \mathscr{S}_f$ (see 1.20) and $\|\eta\|$ approximates $\|\xi\|$. By (2) the latter implies that $\|q\|$ approximates $\|p\|$. Further, since S is of the form $\sum_{j=1}^{n\oplus} S^j$, where $S^j \in \mathscr{S}$, we have $\eta = \sum_{j=1}^{n\oplus} \eta_j$ and $q = \sum_{j=1}^{n} q_j$, where $q_j(a) = (S_a^j \eta_j, \eta_j)$; so q is a sum of elements of $\bigcup\{\Phi(S): S \in \mathscr{S}\}$. Thus (III) holds.

Conversely, we shall show that (III) $\Rightarrow$ (I). Indeed, assume (III); and write $T = \sum_i^{\oplus} T^i$, where T^i is cyclic with cyclic vector ξ_i. By 1.4 it is sufficient to find an element S of $\mathscr{S}_f$ belonging to $U(T; \varepsilon; \{\xi_i: i \in M\}; F)$ (see 1.3), where $\varepsilon > 0$, M is a finite set of indices i, and F is a finite subset of A. By (III) (and (2)), for each i in M we can pick S^i in $\mathscr{S}_f$ and $\eta_i \in X(S^i)$ so that $\|\|\eta_i\| - \|\xi_i\|\| < \varepsilon$ and $|(S_a^i \eta_i, \eta_i) - (T_a^i \xi_i, \xi_i)| < \varepsilon$ for all a in F. Putting $S = \sum_{i \in M}^{\oplus} S^i$, and observing that $(T_a \xi_i, \xi_j) = (S_a \eta_i, \eta_j) = 0$ for $i \neq j$, we conclude that $S \in \mathscr{S}_f \cap U(T; \varepsilon; \{\xi_i: i \in M\}; F)$. So (III) $\Leftrightarrow$ (I).

Obviously (III) $\Rightarrow$ (IV).

Assume (IV); let $a \in \bigcap\{\operatorname{Ker}(S): S \in \mathscr{S}\}$; and take an arbitrary ξ in $X(T)$. By (IV), $(T_a \xi, \xi)$ can be approximated by linear combinations of numbers of the form $(S_a \eta, \eta)$, where $S \in \mathscr{S}$ and $\eta \in X(S)$. But these numbers are all 0. So $(T_a \xi, \xi) = 0$ for all ξ in $X(T)$, whence $T_a = 0$, or $a \in \operatorname{Ker}(T)$. Thus (IV) $\Rightarrow$ (II).

We shall complete the proof by showing that (II) $\Rightarrow$ (III). Assume that (II) holds.

Suppose first that A has a unit $\mathbf{1}$. Let $I = \bigcap\{\operatorname{Ker}(S): S \in \mathscr{S}\}$; put $B = A/I$; and for each S in $\mathscr{S}$ let S' be the representation of B lifted from S. Thus $W = \sum_{S \in \mathscr{S}}^{\oplus} S'$ is a faithful *-representation of B. So by VI.11.10 the set Q of all

p in $\bigcup\{\Phi(S'): S \in \mathscr{S}\}$ such that $p(\dagger) = 1$ satisfies the hypothesis of 5.4 (with B replacing A). Now by (II) T lifts to a *-representation T' of B. Let q be any positive functional on B with $q(\dagger) = 1$ which is associated with T'. By 5.4 q is approximated (pointwise on B) by elements p of the convex hull of Q. Composing q and the p with the quotient map $A \to B$, and recalling that $\|p\| = p(\dagger)$ and $\|q\| = q(\dagger)$, we obtain the statement of (III).

Now suppose that A has no unit, and adjoin $\dagger$ to get the C^*-algebra A_1. Let T^1 and S^1 $(S \in \mathscr{S})$ be the extensions of T and S to A_1 $(T^1_\dagger$ and $S^1_\dagger$ being the identity operators). Notice that $u = a + \lambda\dagger$ $(a \in A; \lambda \in \mathbb{C})$ belongs to $\operatorname{Ker}(T^1)$ if and only if $ub \in \operatorname{Ker}(T)$ for all b in A; and similarly for each S in $\mathscr{S}$. It therefore follows from (II) that

$$\bigcap\{\operatorname{Ker}(S^1): S \in \mathscr{S}\} \subset \operatorname{Ker}(T^1). \tag{5}$$

Now let p be a positive functional associated with T. By (2) p extends without increase of norm to a positive functional p^1 on A_1 associated with T^1. By (5) and the preceding paragraph, $p^1_i \to p^1$ (pointwise on A_1) for some net $\{p^1_i\}$ of elements of the convex hull of $\bigcup\{\Phi(S^1): S \in \mathscr{S}\}$. Thus, putting $p_i = p^1_i|A$, we have $p_i \to p$ pointwise on A; and by (2) $\|p_i\| = p^1_i(\dagger) = \|p\|$. So (III) holds. We have now shown that (II) $\Rightarrow$ (III) under all conditions. ∎

5.6. Suppose that the T of Theorem 5.5 is cyclic, with cyclic vector ξ_0. Then conditions (III) and (IV) of 5.5 need be asserted only for the one positive functional $p_0: a \mapsto (T_a\xi_0, \xi_0)$. More precisely, conditions (I)–(IV) of 5.5 are then equivalent to the following apparently weaker condition:

(IV′) p_0 belongs to the pointwise closure of the linear span of $\bigcup\{\Phi(S): S \in \mathscr{S}\}$.

To see this, by the chain of reasoning in 5.5 we need only verify that (IV′) $\Rightarrow$ (II). But if (IV′) holds and $a \in \bigcap\{\operatorname{Ker}(S): S \in \mathscr{S}\}$, the argument of the implication (IV) $\Rightarrow$ (II) shows that $p_0(bac) = 0$, that is, $0 = (T_{bac}\xi_0, \xi_0) = (T_a(T_c\xi_0), T_{b^*}\xi_0)$, for all b, c in A. Since ξ_0 is cyclic, the $T_c\xi_0$ and $T_{b^*}\xi_0$ run over a dense subset of $X(T)$. So $T_a = 0$; and (II) holds.

5.7. Remark. The equivalence of (II) and (IV) in Theorem 5.5 is valid for any *-algebra. The equivalence of (I) and (III) holds for any Banach *-algebra having an approximate unit $\{e_i\}$ with $\|e_i\| \leq 1$ (and thus ensuring (2); see VI.19.12). It is the equivalence of (I) and (II) which requires the special properties of C^*-algebras. This equivalence fails, for example, for the $\mathscr{L}_1$ group algebra of a non-amenable locally compact group (see Greenleaf [3], §3.5).

5.8. Our next theorem depends upon another important result from the theory of linear topological spaces.

Proposition. *Let K be a compact subset of a locally convex linear topological space X; and assume that the closure L in X of the convex hull of K is compact. Then every extreme point of L lies in K.*

For a proof, see for example Bourbaki [8], Chap. II, §7, Corollary of Prop. 2.

5.9. Theorem. *In Theorem 5.5 suppose that T is irreducible. Then the conditions (I)–(IV) of 5.5 are equivalent to:*

(V) *T lies in the regional closure of $\mathscr{S}$.*

Proof. Take a positive linear functional p associated with T, with $\|p\| = 1$. Let P denote the set of all positive linear functionals q on A with $\|q\| \leq 1$. Since the unit ball of A^* is pointwise compact, P is pointwise compact. Since p is indecomposable (VI.20.4) it is an extreme point of P (by VI.22.4).

Now let $Q = \bigcup \{\Phi(S): S \in \mathscr{S}\} \cap \{q: \|q\| = 1\}$; and assume 5.5(III). Then p is the pointwise limit of a net $\{p_\alpha\}$ of non-negative linear combinations of elements of Q with $\|p_\alpha\| = \|p\| = 1$. If $p_\alpha = \sum_{i=1}^r \lambda_i q_i$ ($\lambda_i \geq 0$, $q_i \in Q$), we have by VI.19.13 $1 = \|p_\alpha\| = \sum_{i=1}^r \lambda_i$; so the p_α belong to the convex hull $H(Q)$ of Q. Thus p is in the pointwise closure $\overline{H(Q)}$ of $H(Q)$.

Since $\overline{H(Q)}$ is closed and contained in the compact set P, it is compact; and since p is extreme in P and belongs to $\overline{H(Q)}$, it is extreme in $\overline{H(Q)}$. Therefore it follows from 5.8 that p belongs to the pointwise closure of Q. Thus by 1.8 5.5(III) $\Rightarrow$ (V).

It follows from the definition of the regional topology that (V) $\Rightarrow$ 5.5(III). So (V) is equivalent to the conditions of 5.5.   ∎

Remark. As in 5.7, for the equivalence of (V) with 5.5(I), (III), it is sufficient that A be a Banach *-algebra with an approximate unit $\{e_i\}$ such that $\|e_i\| \leq 1$.

5.10. Corollary. *If $T \in \hat{A}$ and $\mathscr{S} \subset \hat{A}$, then T is in the regional closure of $\mathscr{S}$ if and only if T is weakly contained in $\mathscr{S}$.*

5.11. The reader will recall that in 3.10 we defined the hull-kernel topology of $\hat{A}$.

Corollary. *The hull-kernel and regional topologies of the structure space $\hat{A}$ of the C*-algebra A are the same.*

Proof. This follows from the equivalence of 5.5(II) and 5.9(V) (provided T is irreducible).   ∎

Remark. As we have pointed out in 3.11, this corollary is not true for arbitrary Banach *-algebras.

5.12. Corollary. *Two elements S and T of $\hat{A}$ are regionally equivalent (2.2) if and only if $\mathrm{Ker}(S) = \mathrm{Ker}(T)$. Thus, $\hat{A}$ is a T_0 space if and only if any two elements of $\hat{A}$ which have the same kernels are unitarily equivalent.*

Proof. Since $\mathrm{Ker}(S) = \mathrm{Ker}(T)$ if and only if S and T are not distinguished by the hull-kernel topology, the corollary follows from 5.11.   ∎

Remark. Suppose that A is topologically simple, so that $\mathrm{Prim}(A)$ has only one element, namely $\{0\}$. Then by the above corollary the elements of $\hat{A}$ are all regionally equivalent to each other. Thus, if $\hat{A}$ has more than one element, as in the case of the Glimm algebras of VI.17.5, $\hat{A}$ will not be a T_0 space.

The Correspondence Between Ideals and Open or Closed Subsets of $\hat{A}$

5.13. Proposition. *The mapping $\Gamma: \mathscr{S} \mapsto I = \cap\{\mathrm{Ker}(S): S \in \mathscr{S}\}$ is a bijection from the set of all closed subsets $\mathscr{S}$ of $\hat{A}$ onto the set of all norm-closed two-sided ideals I of A. Its inverse is the mapping $\Delta: I \mapsto \mathscr{S} = \{T \in \hat{A}: I \subset \mathrm{Ker}(T)\}$.*

Proof. Let I be a closed two-sided ideal of A. Notice that $\Delta(I)$ is closed in $\hat{A}$; indeed, by 5.11 an element T of the closure of $\Delta(I)$ must satisfy $\mathrm{Ker}(T) \supset \cap\{\mathrm{Ker}(S): S \in \Delta(I)\} \supset I$; so $T \in \Delta(I)$. Furthermore, by VI.8.7 and VI.22.11 the elements of $(A/I)\hat{\ }$ distinguish points in A/I. From this it follows that

$$\Gamma(\Delta(I)) = I. \tag{6}$$

For any subset $\mathscr{S}$ of $\hat{A}$, $\Delta(\Gamma(\mathscr{S}))$ is the hull-kernel closure, hence by 5.11 the regional closure, of $\mathscr{S}$ in $\hat{A}$. So if $\mathscr{S}$ is closed in $\hat{A}$ we have

$$\Delta(\Gamma(\mathscr{S})) = \mathscr{S}. \tag{7}$$

By (6) and (7) Γ and Δ are inverses of each other. So Γ is a bijection; and the proposition is proved.   ∎

Remark. This proposition could equally well of course have been stated in terms of the closed subsets of $\mathrm{Prim}(A)$.

5.14.   By 5.13 the map

$$I \mapsto \hat{A} \setminus \Delta(I) \tag{8}$$

is a one-to-one correspondence between the family of all norm-closed two-sided ideals of A and the family of all *open* subsets of $\hat{A}$. Since Δ clearly reverses inclusion, the map (8) preserves inclusion. Notice that $\hat{A} \setminus \Delta(I)$ is the $\hat{A}_I$ of 4.4.

Since (8) preserves inclusion and is a bijection, we have:

$$\hat{A}_{I \cap J} = \hat{A}_I \cap \hat{A}_J, \tag{9}$$

$$\hat{A}_K = \bigcup \{\hat{A}_I : I \in \mathscr{I}\} \tag{10}$$

whenever I and J are closed two-sided ideals of A, $\mathscr{I}$ is a family of closed two-sided ideals of A, and K is the closed linear span of $\bigcup \mathscr{I}$. Notice that $\hat{A}_\emptyset = \{0\}$, $\hat{A}_A = \hat{A}$.

5.15.   Let I be a closed two-sided ideal of A; and define

$$J = \{b \in A : ba = 0 \quad \text{for all } a \text{ in } I\}.$$

It is easy to see that J is a closed two-sided ideal, in fact the largest such object satisfying $JI = \{0\}$ or (in view of the proof of VI.8.13) $J \cap I = \{0\}$. Hence, by 5.14, $\hat{A}_J$ is the largest open subset of $\hat{A}$ which is disjoint from $\hat{A}_I$; in other words,

$$\hat{A}_J = \hat{A} \setminus (\hat{A}_I)^-.$$

In particular, we have:

Proposition.   *If I is a closed two-sided ideal of A, and $\hat{A}_I$ is the open subset of $\hat{A}$ corresponding to I by 5.14, then $\hat{A}_I$ is dense in $\hat{A}$ if and only if $b = 0$ whenever $b \in A$ and $ba = 0$ for all a in I.*

*The Spectrum of a *-Representation*

5.16.   If $\mathscr{S}_1$ and $\mathscr{S}_2$ are any two families of non-degenerate *-representations of A, then by 5.5 $\mathscr{S}_1$ is weakly contained in $\mathscr{S}_2$ if and only if $\bigcap \{\mathrm{Ker}(S) : S \in \mathscr{S}_2\} \subset \bigcap \{\mathrm{Ker}(S) : S \in \mathscr{S}_1\}$. In particular, $\mathscr{S}_1$ and $\mathscr{S}_2$ are weakly equivalent if and only if

$$\bigcap \{\mathrm{Ker}(S) : S \in \mathscr{S}_1\} = \bigcap \{\mathrm{Ker}(S) : S \in \mathscr{S}_2\}.$$

From this and 5.13 we easily deduce:

Proposition.   *If $\mathscr{S}$ is any family of non-degenerate *-representations of A, there is a unique closed subset W of $\hat{A}$ which is weakly equivalent to $\mathscr{S}$.*

Definition. If $\mathscr{S}$ consists of a single (non-degenerate) *-representation S of A, the W of this proposition is often referred to as the *spectrum* of S. Thus the spectrum of S is just the closed subset $\Delta(\mathrm{Ker}(S))$ of $\hat{A}$ which corresponds to $\mathrm{Ker}(S)$ under the correspondence of 5.13.

By 5.10, the spectrum of S is just the set of those T in $\hat{A}$ which are weakly contained in S.

5.17.   Suppose for the moment that A is Abelian. If S is a non-degenerate *-representation of A, with spectral measure P (see VI.10.10), and W is the closed support of P in $\hat{A}$, then $\mathrm{Ker}(S) = \{a \in A : \hat{a}$ vanishes on $W\}$ ($\hat{a}$ being the Gelfand transform of a). From this we conclude that *the spectrum of S is equal to the closed support of P.*

This fact will be generalized to the non-commutative situation in 9.15.

$\hat{A}$ as a Quotient Space

5.18.   Let $E(A)$ be the family of all indecomposable positive linear functions p on A such that $\|p\| = 1$. As usual we give to $E(A)$ the topology of point-wise convergence on A. For each p in $E(A)$ let $T^{(p)}$ be the irreducible *-representation of A generated by p (see VI.20.4). Thus $p \mapsto T^{(p)}$ maps $E(A)$ onto $\hat{A}$.

Proposition.   *The mapping $p \mapsto T^{(p)}$ from $E(A)$ onto $\hat{A}$ is continuous and open.*

Proof.   The continuity follows from 1.8. The openness follows from the original definition of the regional topology.   ∎

Thus $\hat{A}$ carries the quotient topology derived from the surjection $p \mapsto T^{(p)}$ and the pointwise convergence topology of $E(A)$.

5.19.   Corollary.   *If A is separable, $\hat{A}$ and $\mathrm{Prim}(A)$ are second-countable.*

Proof.   It is well known that the unit ball of the adjoint of a separable Banach space is second-countable (in fact, a compact metric space) in the topology of pointwise convergence. So, if A is separable, $E(A)$ is second-countable; and hence by 5.18 the same is true of $\hat{A}$. By 5.11 this implies the second countability of $\mathrm{Prim}(A)$.   ∎

The T_1 Property of $\hat{A}$

5.20. In the separable case Rosenberg's Theorem (VI.23.1) provides an interesting characterization of the T_1 property of $\hat{A}$.

Proposition. *Assume that A is separable. Then for each T in $\hat{A}$ the following two properties are equivalent*: (i) $\{T\}$ *is closed in* $\hat{A}$; (ii) T_a *is a compact operator for every a in A.*

Proof. Assume (ii). Then by VI.15.17 range(T) is the C^*-algebra B of all compact operators on $X(T)$. If S is in the closure of $\{T\}$, then Ker(T) $\subset$ Ker(S); so there is an irreducible *-representation R of B such that $S = R \circ T$. But by VI.15.12 B has to within unitary equivalence only one irreducible *-representation, namely the identity map I. Therefore $R \cong I$; and $S \cong T$. So $\{T\}$ is closed.

Conversely, assume that $\{T\}$ is closed; and let B be the C^*-algebra range(T). It is clear that $R \mapsto R \circ T$ is a bijection of $\hat{B}$ onto $\{S \in \hat{A} : \mathrm{Ker}(T) \subset \mathrm{Ker}(S)\}$. But since $\{T\}$ is closed the latter is a one-element set. So $\hat{B}$ has only one element. This implies by VI.23.1 that B is an elementary C^*-algebra, the range of whose unique irreducible *-representation consists of all compact operators on some Hilbert space. But the identity representation of B is irreducible. So B consists of all compact operators on $X(T)$. ∎

Remark. As the proof shows, the implication (ii) $\Rightarrow$ (i) holds even without the assumption that A is separable.

The Structure Space and C*-Direct Sums

5.21. For each i in some index set I let A_i be a C^*-algebra; and form the C^*-direct sum $A = \sum_i^{\oplus} A_i$ as in VI.3.13. Let $\pi_i : A \to A_i$ be the projection $a \mapsto a_i$. Thus by 1.19, for each i the map $\tau_i : S \mapsto S \circ \pi_i$ is a homeomorphism of $(A_i)\hat{}$ onto a closed subset $(\hat{A})_i$ of $\hat{A}$.

Proposition. *The $(\hat{A})_i$ $(i \in I)$ are pairwise disjoint open-closed subsets of $\hat{A}$, and $\bigcup_{i \in I} (\hat{A})_i = \hat{A}$.*

Proof. For each i let B_i be the closed two-sided ideal of A consisting of those a such that $a_j = 0$ for $j \neq i$. Thus B_i and A_i are *-isomorphic; $B_i \cap B_j = \{0\}$ for $i \neq j$; and $\sum_i B_i$ is dense in A.

Let T be any element of $\hat{A}$. Since $\sum_i B_i$ is dense in A, we have

$$B_i \not\subset \mathrm{Ker}(T) \qquad \text{for some } i. \tag{11}$$

Assume that there are two distinct indices i, j in I satisfying (11). Then by the argument of the proof of (iv) in 3.8 we can find elements $b \in B_i$, $a \in A$, and $c \in B_j$ such that $T_{bac} \neq 0$. But this contradicts the fact that $bac \in B_i \cap B_j = \{0\}$. So there is exactly one i in I such that (11) holds. Thus $T | B_i$ is irreducible; and there is an element S of $(A_i)\hat{\ }$ such that $T = S \circ \pi_i$.

This shows that $\bigcup_i (\hat{A})_i = \hat{A}$. The fact, proved in the last paragraph, that for each T in $\hat{A}$ there is only one i satisfying (11) implies that the $(\hat{A})_i$ are pairwise disjoint. The same fact also implies that $(\hat{A})_i = \{T \in \hat{A} : B_i \not\subset \mathrm{Ker}(T)\}$ for each i. Since the latter set is open by 1.18, $(\hat{A})_i$ is open. ∎

5.22.   We have the following converse of 5.21:

Proposition.   *Let A be any C^*-algebra, and suppose that $\hat{A} = \bigcup \mathscr{W}$, where $\mathscr{W}$ is a family of pairwise disjoint open-closed subsets of $\hat{A}$. For each W in $\mathscr{W}$ let I_W be the closed two-sided ideal of A corresponding with W by the correspondence 5.14(8); that is,*

$$I_W = \bigcap \{\mathrm{Ker}(T) : T \in \hat{A} \setminus W\}.$$

*Then A is *-isomorphic with the C^*-direct sum $\sum^{\oplus}_{W \in \mathscr{W}} I_W$.*

Proof.   By 5.14(9) $I_V \cap I_W = \{0\}$ if $V, W \in \mathscr{W}$ and $V \neq W$; and by 5.14(10) $\sum_{W \in \mathscr{W}} I_W$ is dense in A. The result now follows from VI.8.13. ∎

5.23.   Corollary.   *For any C^*-algebra A the following two conditions are equivalent:* (i) *$\hat{A}$ is topologically connected;* (ii) *A is not *-isomorphic with the C^*-direct sum of two non-zero C^*-algebras.*

5.24.   As the reader may recall from VI.23.3, a C^*-algebra A is *of compact type* if it is *-isomorphic with a C^*-direct sum of elementary C^*-algebras.

Proposition.   *A C^*-algebra A is of compact type if and only if* (i) *$\hat{A}$ is discrete, and* (ii) *T_a is a compact operator for every T in $\hat{A}$ and $a \in A$. If A is separable, then* (i) *by itself is sufficient to imply that A is of compact type.*

Proof.   If A is of compact type, then (i) and (ii) hold in virtue of 5.21 and VI.15.12.

Assume (i). For each T in $\hat{A}$ let I_T be the closed two-sided ideal of A corresponding by 5.14(8) with the open set $\{T\}$. By 5.22 A is *-isomorphic with the C^*-direct sum $\sum^{\oplus}_{T \in \hat{A}} I_T$. Now by 4.6 $(I_T)\hat{\ }$ consists of one element only (namely $T | I_T$). Thus, if we assume condition (ii), it follows from VI.15.17

that each I_T is elementary. If instead of condition (ii) we assume that A (and hence also each I_T) is separable, the elementariness of I_T follows from Rosenberg's Theorem VI.23.1. Hence in either case A is of compact type. ■

6. The Baire Property and Local Compactness of $\hat{A}$

6.1. In this section we shall prove two very important topological properties of the structure space $\hat{A}$ of a C^*-algebra A, namely the Baire property and local compactness.

To prove the Baire property we shall follow the argument of Dixmier [21], 3.4.13.

6.2. Lemma. *Let C be a compact convex subset of a locally convex linear topological space X. Let D be a non-void closed subset of C such that both D and $C \setminus D$ are convex. Then D contains at least one extreme point of C.*

Proof. Suppose that $X = \mathbb{R}^2$ and C is a closed triangle (interior plus boundary) in $\mathbb{R}^2$. In this special case the reader will easily verify that every extreme point of D must lie on the boundary of C.

Now return to the general case, and take an extreme point x of D. (Such exist by the Krein-Milman Theorem.) If x is extreme in C we are finished. If not, take a line L in X such that the segment $C \cap L$ has x as an interior point. Of the two end-points y, z of $C \cap L$, one—say y—must lie in D (since $C \setminus D$ is convex). We claim y is extreme in C. Indeed, suppose it is not; and let M be a line in X such that the segment $C \cap M$ has y in its interior. Let the endpoints of $C \cap M$ be u, v. By the definition of y, the lines L and M are different; and so $u, v,$ and z are the vertices of a (non-degenerate) triangle T in the plane containing L and M. Now $T \subset C$, and x is in the interior of T. Since x is extreme in D, it is extreme in $D \cap T$. But this contradicts the preceding paragraph (applied to T and $D \cap T$). So y is extreme in C, and the lemma is proved. ■

6.3. A (not necessarily Hausdorff) topological space T is said to be a *Baire space* if, whenever $\{U_n\}$ is a sequence of dense open subsets of T, the intersection $\bigcap_{n=1}^{\infty} U_n$ is also dense in T.

It is easy to see that an open subset of a Baire space is a Baire space. A closed subset of a Baire space, however, need not be a Baire space. For example, let T be the uncountable well-ordered set of all countable ordinals;

and equip T with the topology in which the non-void open sets are just the sets $\{\alpha \in T : \alpha \geq \beta\}$, where $\beta \in T$. This is a Baire space. But the closed subset $\{\alpha \in T : \alpha < \beta\}$, where β is any limit ordinal in T, is not a Baire space.

From the Baire Category Theorem we know that any locally compact Hausdorff space is a Baire space, and likewise any complete metric space.

6.4.   Proposition.   *Let C be a compact convex subset of a locally convex real linear topological space X; and let E denote the set of all extreme points of C. Then E is a Baire space (with the relativized topology of X).*

Proof.   If f is a continuous linear functional on X and $r \in \mathbb{R}$, let $U_{f,r} [F_{f,r}]$ be the set of those x in C such that $f(x) < r$ $[f(x) \leq r]$.

Suppose $x \in E$. We claim that the set Γ of those $F_{f,r}$ (f, r varying) such that $x \in U_{f,r}$ forms a basis of neighborhoods of x relative to C. Indeed, by the Hahn–Banach Theorem, $\bigcap \Gamma = \{x\}$. Since the elements of Γ are compact C-neighborhoods of x, the claim will be proved if we show that Γ is downward-directed, that is, the intersection of any two elements of Γ contains a third. So assume that $x \in U_{f_1, r_1} \cap U_{f_2, r_2}$; and put $C_i = C \setminus U_{f_i, r_i}$. Since C_1 and C_2 are compact and convex, the convex hull D of $C_1 \cup C_2$ is compact. Since x is extreme, $x \notin D$. So there is a continuous linear functional f on X and a real number r such that $x \in U_{f,r}$ and $F_{f,r} \cap D = \emptyset$. Thus $F_{f,r} \in \Gamma$ and $F_{f,r} \subset F_{f_i, r_i}$ ($i = 1, 2$), proving the claim.

Now let $\{V_n\}$ be a sequence of dense relatively open subsets of E, and V a non-void relatively open subset of E. The proposition will be proved if we show that $\bigcap_n V_n \cap V \neq \emptyset$. Let W_n and W be open subsets of C such that $W_n \cap E = V_n$, $W \cap E = V$. Since the V_n are dense, we may as well suppose that the W_n are dense in C. Also we may as well suppose that the V_n and W_n are decreasing in n, and (in view of the above claim) that W is of the form U_{f_1, r_1}.

We shall now construct inductively a sequence of pairs $\{\langle f_n, r_n \rangle\}$ such that $U_{f_n, r_n} \cap E \neq \emptyset$ and $F_{f_{n+1}, r_{n+1}} \subset U_{f_n, r_n} \cap W_{n+1}$ for all n. We begin with the f_1, r_1 already before us. Suppose that $\langle f_1, r_1 \rangle, \ldots, \langle f_n, r_n \rangle$ have already been constructed. Since $U_{f_n, r_n} \cap E \neq \emptyset$ and V_{n+1} is dense in E, there is an $x \in U_{f_n, r_n} \cap E \cap W_{n+1}$. Now $U_{f_n, r_n} \cap W_{n+1}$ is a C-neighborhood of x; so, by the above claim, we can find a pair $\langle f_{n+1}, r_{n+1} \rangle$ such that $x \in U_{f_{n+1}, r_{n+1}} \subset F_{f_{n+1}, r_{n+1}} \subset U_{f_n, r_n} \cap W_{n+1}$. This pair has the properties required to complete the inductive construction.

Thus the F_{f_n, r_n} are non-void, compact, and monotone decreasing. So they have non-void compact convex intersection F. We have $F \subset W$ and

$F \subset \bigcap_n W_n$. Notice also that $C \setminus F$ is convex. Thus by 6.2 F contains a point y of E. We have

$$y \in E \cap W \cap \bigcap_n W_n = V \cap \bigcap_n V_n,$$

and the proof is complete.   ∎

6.5. Theorem. *The structure space $\hat{A}$ of a C*-algebra A is a Baire space. Likewise* Prim(A) *is a Baire space.*

Proof. Let C be the family of all positive linear functionals p on A with $\|p\| \leq 1$. Thus C, with the pointwise topology, is a compact convex subset of A^*; and by VI.22.4 the set of extreme points of C is $E(A) \cup \{0\}$ (see 5.19). So by 6.4 $E(A) \cup \{0\}$, and therefore also $E(A)$, is a Baire space in the topology of pointwise convergence on A.

Now let $\{U_n\}$ be a sequence of dense open subsets of $\hat{A}$; and put $V_n = \tau^{-1}(U_n)$, where $\tau: p \mapsto T^{(p)}$ is the mapping 5.18 of $E(A)$ onto $\hat{A}$. By 5.18 τ is continuous and open; therefore each V_n is open and dense in $E(A)$. Since $E(A)$ is a Baire space, $V = \bigcap_n V_n$ is dense in $E(A)$. Hence $\tau(V)$, which is $\bigcap_n U_n$, is dense in $\hat{A}$. This implies that $\hat{A}$ is a Baire space.

By 5.11 Prim(A) is obtained from $\hat{A}$ by identifying points which are not distinguished by the topology of $\hat{A}$. So Prim(A) is also a Baire space.   ∎

6.6 By 3.5 and 5.13 every closed subset of $\hat{A}$ is homeomorphic with the structure space of A/I for some *-ideal I of A, and so by 6.5 is a Baire space. Contrast this with the behavior of general Baire spaces, whose closed subsets, as we have seen in 6.3, need not have the Baire property.

The Local Compactness of $\hat{A}$

6.7. Proposition. *If $a \in A$ and $0 < r \in \mathbb{R}$, then $W = \{T \in \hat{A}: \|T_a\| \geq r\}$ is compact in $\hat{A}$.*

Proof. Let $\{Q_i\}$ be a decreasing net of relatively closed non-void subsets of W. It is enough to show that $\bigcap_i Q_i \neq \emptyset$. For each i, let $I_i = \bigcap\{\text{Ker}(T): T \in Q_i\}$. Since Q_i is non-void we have by VI.3.8

$$\|a + I_i\|_{A/I_i} \geq r \qquad \text{for each } i. \tag{1}$$

Since the I_i are increasing in i, it follows from (1) that

$$\|a + I\|_{A/I} \geq r,$$

where $I = (\bigcup_i I_i)^-$. From this and VI.22.10 we conclude that there exists an element S of $\hat{A}$ such that $I \subset \text{Ker}(S)$ and $\|S_a\| \geq r$. Thus $S \in W$ and $I_i \subset \text{Ker}(S)$ for each i. Since Q_i is closed relatively to W, it follows that $S \in Q_i$ for all i. So $\bigcap_i Q_i \neq \emptyset$. ∎

6.8. A (not necessarily Hausdorff) topological space M is said to be *locally compact* if, for every point t of M, the set of all compact neighborhoods of t is a basis of neighborhoods of t.

If M is Hausdorff, then local compactness in this sense is equivalent to local compactness in the ordinary sense that every point of M has *some* compact neighborhood.

Remark.   For Hausdorff spaces M, compactness implies local compactness and local compactness implies the Baire property. But if M is not Hausdorff both these implications fail. To see that the first implication fails, let N be any topological space which is not locally compact, and let M be its one-point compactification. The failure of the second implication is shown by the example in 6.3.

6.9.   Theorem.   *If A is a C*-algebra, $\hat{A}$ is locally compact. Likewise, $\text{Prim}(A)$ is locally compact.*

Proof.   Let $T \in \hat{A}$, and let U be an open neighborhood of T. Since $\hat{A} \setminus U$ is closed, by 5.11 there is an element a of A such that

$$T_a \neq 0 \quad \text{and} \quad S_a = 0 \qquad \text{for all } S \in \hat{A} \setminus U \tag{2}$$

Choose a positive number r with $r < \|T_a\|$; and put

$$V = \{S \in \hat{A} : \|S_a\| > r\},$$

$$C = \{S \in \hat{A} : \|S_a\| \geq r\}.$$

By the lower semi-continuity of $S \mapsto \|S_a\|$ (1.14), V is open. Hence C is a neighborhood of T, and is compact by 6.7. By (2) $C \subset U$. By the arbitrariness of U this shows that T has a basis of compact neighborhoods. So $\hat{A}$ is locally compact. From this and 5.11 it follows immediately that $\text{Prim}(A)$ is locally compact. ∎

6.10.   Remark.   The question arises of what properties characterize those topological spaces which are homeomorphic to $\hat{A}$ or to $\text{Prim}(A)$ for some C*-algebra A. By 6.5, 6.6, 6.9, such a space must be locally compact, and all its closed subsets must satisfy the Baire property. If it is to be homeomorphic to Prim(A) it must be T_0. Are these properties sufficient? If not, what other properties must it satisfy? We do not know. In this connection notice 7.7.

6.11. Proposition. *If A is a C*-algebra with a unit, then $\hat{A}$ is compact.*

Proof. By 6.7 (taking $a = 1, r = 1$). ∎

Remark. The converse of this is false. Indeed, if X is an infinite-dimensional Hilbert space, $\mathcal{O}_c(X)$ has no unit but its structure space has only one point.

6.12. As we have already observed (see for example the proof of Theorem 6.5) the space C of all positive linear functionals of norm ≤ 1 on a C*-algebra A is a compact convex set in the topology of pointwise convergence, and the set of its extreme points is $E(A) \cup \{0\}$ ($E(A)$ being the family of all indecomposable positive linear functions of norm 1). Furthermore, by 5.18 $\hat{A}$ is a quotient space derived from $E(A)$. Thus, if $E(A)$ were closed (and hence compact) in C, $\hat{A}$ would be compact and we could derive a great deal of other useful information about convergence in $\hat{A}$. Unfortunately, however, the set of extreme points of a compact convex set is not in general closed, in particular $E(A)$ is not in general closed in C. This limits the usefulness of the quotient presentation of $\hat{A}$ in 5.18 (although its usefulness is by no means negligible; see 5.19 and 6.5).

7. C*-Algebras With Finite Stucture Space

7.1. The object of this short section is to show that every *finite* T_0 space can be the structure space of a C*-algebra. In the process we shall explore an interesting construction of C*-algebras which are extensions of given C*-algebras.

As we saw in Remark 5.12, the structure space of a C*-algebra need not be T_0. It appears to be not yet known whether a C*-algebra can have a *finite* non-T_0 structure space. In particular, can there be a (topologically) simple C*-algebra having only a finite number greater than 1 of inequivalent irreducible *-representations?

7.2. We begin with a rather special construction.

Let X and Y be Hilbert spaces, Y being infinite-dimensional; and let A be a closed *-subalgebra of $\mathcal{O}(X)$. Form the Hilbert tensor product $Z = X \otimes Y$; let A' be the closed *-subalgebra $\{a \otimes 1_Y : a \in A\}$ of $\mathcal{O}(Z)$; and set

$$B = A' + \mathcal{O}_c(Z).$$

Since $\mathcal{O}_c(Z)$ is a closed two-sided ideal of $\mathcal{O}(Z)$, it follows from VI.8.11 that B is a closed *-subalgebra of $\mathcal{O}(Z)$, hence a C*-algebra. From VI.15.4 and the infinite-dimensionality of Y one easily deduces that $A' \cap \mathcal{O}_c(Z) = \{0\}$; hence

$$B/\mathcal{O}_c(Z) \cong A' \cong A. \tag{1}$$

Also,

$$b = 0 \quad \text{whenever} \quad b \in B \quad \text{and} \quad bc = 0 \quad \text{for all } c \text{ in } \mathcal{O}_c(Z). \tag{2}$$

By 4.6 and VI.15.12 the open subset of $\hat{B}$ corresponding (by 5.14) to the two-sided ideal $\mathcal{O}_c(Z)$ of B consists of a single point T, namely, the identity representation of B on Z. By (2) and 5.15 $\{T\}$ is dense in $\hat{B}$. Furthermore, by (1) and 4.6, the closed subset $\hat{B} \setminus \{T\}$ of $\hat{B}$ can be identified both setwise and topologically with $\hat{A}$.

Thus, starting from the arbitrary C*-algebra A, we have constructed a new C*-algebra B such that the topological space $\hat{B}$ consists of $\hat{A}$ together with one extra point T not in $\hat{A}$, and has the properties:

(i) The topology of $\hat{B}$ relativized to $\hat{A}$ is that of $\hat{A}$;
(ii) $\{T\}$ is open and dense in $\hat{B}$.

These two properties determine $\hat{B}$ as a topological space when $\hat{A}$ is known.

7.3. As the simplest example of the construction 7.2, consider the case that A is itself elementary. For example, A might be the one-dimensional C*-algebra $\mathbb{C}$ acting on $\mathbb{C}$, in which case $Z = Y$ and $B = \{c + \lambda 1_Y : c \in \mathcal{O}_c(Y), \lambda \in \mathbb{C}\}$. Then $\hat{B}$ consists of two points T and S, the open subsets of $\hat{B}$ being just $\emptyset$, $\hat{B}$, and $\{T\}$.

If we repeat the construction taking the B just derived as the A of 7.2, the new $\hat{B}$ so obtained will consist of *three* points R, T, S, the non-trivial open sets being $\{R\}$ and $\{R, T\}$. Continuing in this way, for any positive integer n we obtain a C*-algebra B_n whose structure space $(B_n)\hat{\ }$ consists of exactly n distinct elements $T^1, T^2, \ldots, T^n$, and has exactly $n + 1$ open subsets

$$\emptyset, \{T^1\}, \{T^1, T^2\}, \{T^1, T^2, T^3\}, \ldots, \{T^1, \ldots, T^{n-1}\}, (B_n)\hat{\ },$$

forming an ordered chain.

For $n \geq 2$, $(B_n)\hat{\ }$ is of course a T_0 space but not T_1.

7.4. By refining the construction of 7.2, one can show that every finite T_0 topological space is the structure space of some C*-algebra. This result is interesting in view of the question raised in 6.10. We shall present it as the conclusion of a series of exercises.

7.5. Proposition*. *Let A be a C^*-algebra, and B a closed *-subalgebra of A with the property that $T|B$ is irreducible for every T in $\hat{A}$. Then:* (I) *Every S in $\hat{B}$ is of the form $T|B$ for some T in $\hat{A}$;* (II) *the topology of $\hat{B}$ is the quotient topology derived from the topology of $\hat{A}$ and the surjection $T \mapsto T|B$ of $\hat{A}$ onto $\hat{B}$.*

Statement (I) follows from VI.22.6.

7.6. Proposition*. *Let A be any C^*-algebra, and Z any closed subset of $\hat{A}$. Then there exists another C^*-algebra C whose structure space $\hat{C}$ is homeomorphic with the topological space W determined as follows:*

  (i) $W = \hat{A} \cup \{T\}$, *where* $T \notin \hat{A}$;
  (ii) $\{T\}$ *is open in W;*
  (iii) *the topology of W relativized to $\hat{A}$ is the topology of $\hat{A}$;*
  (iv) *the W-closure of $\{T\}$ is $\{T\} \cup Z$.*

Sketch of Proof. Let $J = \bigcap\{\mathrm{Ker}(S): S \in Z\}$ and $D = A/J$; and let $\psi: A \to D$ be the quotient *-homomorphism. Let B be the C^*-algebra obtained from the construction of 7.2 when the A of 7.2 is replaced by D. Thus we have a natural surjective *-homomorphism $\phi: B \to D$. Now take C to be the closed *-subalgebra $\{a \oplus b : a \in A,\ b \in B,\ \psi(a) = \phi(b)\}$ of $A \oplus B$; and apply 7.5 to $A \oplus B$ and C. ∎

7.7 Proposition*. *Every finite T_0 topological space W is the structure space of some C^*-algebra.*

Sketch of Proof. We proceed by induction in the number n of elements of W. It is obvious if $n = 1$. If $n > 1$ let U be a minimal non-void open subset of W. By the T_0 property U has only one element T. By the inductive hypothesis there is a C^*-algebra A whose structure space is homeomorphic with $W \setminus \{T\}$. Now apply 7.6, taking Z to be the intersection of $W \setminus \{T\}$ with the closure of $\{T\}$. ∎

8. Bundles of *C**-Algebras

8.1. Roughly speaking, a bundle of C^*-algebras is a Banach bundle $\mathscr{B}$ each of whose fibers is a C^*-algebra. From such an object one constructs a natural C^*-algebra $\mathscr{C}_0(\mathscr{B})$ consisting of all those continuous cross-sections of $\mathscr{B}$ which vanish at infinity. This construction is a fertile source of interesting examples of C^*-algebras. It turns out that the structure space of $\mathscr{C}_0(\mathscr{B})$ has an interesting description in terms of the bundle $\mathscr{B}$ (see 8.8, 8.9).

As we have seen in Chapter VI, every Abelian C^*-algebra is of the form $\mathscr{C}_0(\mathscr{B})$ for some trivial bundle $\mathscr{B}$ with fiber $\mathbb{C}$. While no comparably simple statement is possible for general non-Abelian C^*-algebras, nevertheless bundles of C^*-algebras do play a vital role in the structure theory of general C^*-algebras. Unfortunately we do not have space in this work to elaborate on this role; we can only refer the reader to Fell [3].

The reader should bear in mind that the bundles of C^*-algebras studied in this section are Banach bundles of an entirely different sort from the C^*-algebraic bundles to be introduced in Chapter VIII.

8.2.   Definition.   Let X be a locally compact Hausdorff space. By a *bundle $\mathscr{B}$ of C^*-algebras over X* we mean a Banach bundle $\langle B, \pi \rangle$ over X, together with a multiplication $\cdot$ and involution $*$ in each fibre B_x of $\mathscr{B}$, such that: (i) For each x in X, B_x is a C^*-algebra under the linear operations and norm of $\mathscr{B}$ and the operations $\cdot$ and $*$; (ii) the multiplication $\cdot$ is continuous on $\{\langle a, b \rangle \in B \times B : \pi(a) = \pi(b)\}$ to B; and (iii) the involution $*$ is continuous on B to B.

Remark.   Let $\langle B, \pi \rangle$ be a Banach bundle over X with operations $\cdot$ and $*$ such that (i) holds. Suppose that Γ is a collection of continuous cross-sections of $\langle B, \pi \rangle$ such that for each x in X $\{f(x) : f \in \Gamma\}$ is dense in B_x. Then condition (ii) is equivalent to:

(ii') For each $f, g \in \Gamma$ the cross-section $x \mapsto f(x) \cdot g(x)$ is continuous;

and condition (iii) is equivalent to:

(iii') For each $f \in \Gamma$ the cross-section $x \mapsto f(x)^*$ is continuous.

The proofs of these equivalences are routine applications of II.13.12.

8.3.   Remark.   Let $\mathscr{B}$ be a bundle of C^*-algebras over X; and for each $x \in X$ let D_x be a closed $*$-subalgebra of the fiber B_x such that $\{D_x\}$ is a lower-semicontinuous choice of subspaces; and let $\mathscr{D}$ be the Banach sub-bundle of $\mathscr{B}$ with fibers $\{D_x\}$ (for definitions see Exercise 41, Chapter II). Then, with the operations of multiplication and $*$ restricted from $\mathscr{B}$, $\mathscr{D}$ becomes a bundle of C^*-algebras over X; we call it the *subbundle of C^*-algebras over X with fibers $\{D_x\}$*.

8.4. Fix a bundle $\mathscr{B}$ of C^*-algebras over a locally compact Hausdorff space X. By II.13.19 $\mathscr{B}$ has enough continuous cross-sections.

Let us form the C_0 cross-sectional Banach space $C = \mathscr{C}_0(\langle B, \pi \rangle)$ of the Banach bundle $\langle B, \pi \rangle$ underlying $\mathscr{B}$; and let us define pointwise multiplication and involution in C in the natural way:

$$\left.\begin{aligned} (f \cdot g)(x) &= f(x) \cdot g(x) \\ f^*(x) &= (f(x))^* \end{aligned}\right\} \tag{1}$$

$(f, g \in C : x \in X)$. By 8.2 (ii), (iii) the cross-sections $f \cdot g$ and f^* are continuous; and the properties of norms in Banach *-algebras show that $\|(f \cdot g)(x)\|$ and $\|f^*(x)\|$ go to 0 as $x \to \infty$ in X. So (1) defines elements $f \cdot g$ and f^* of C; and it is easy to verify that C is in fact a C^*-algebra under $\cdot$ and *.

We shall call C the *cross-sectional C^*-algebra of* $\mathscr{B}$, and denote it by $\mathscr{C}_0(\mathscr{B})$.

8.5. It is obvious that any retraction of a bundle $\mathscr{B}$ of C^*-algebras (see II.13.3) is also a bundle of C^*-algebras. In particular, the reduction of $\mathscr{B}$ to a (locally compact) subspace Y of the base space is a bundle of C^*-algebras, which we denote by $\mathscr{B}_Y$.

8.6. Let $\mathscr{B}$ be a bundle of C^*-algebras over a locally compact Hausdorff space X. Let Y be a closed subset of X, and denote by I the closed two-sided ideal of $C = \mathscr{C}_0(\mathscr{B})$ consisting of those f in C such that $f(x) = 0$ for all $x \in Y$.

Notice that I can be identified with $\mathscr{C}_0(\mathscr{B}_{X \setminus Y})$. Also, the mapping $F: f + I \mapsto f | Y$ ($f \in C$) is evidently an isometric *-isomorphism of C/I into $\mathscr{C}_0(\mathscr{B}_Y)$; and by the generalized Tietze Extension Theorem (II.14.8) it is in fact onto $\mathscr{C}_0(\mathscr{B}_Y)$. Thus we can identify C/I with $\mathscr{C}_0(\mathscr{B}_Y)$.

8.7. Let $X, \mathscr{B}$ and C be as in 8.4. The following lemma describes the structure of closed two-sided ideals of C.

Lemma. *Let J be any closed two-sided ideal of C. For each x in X let $J_x = \{f(x): f \in J\}$. Then*

$$J = \{f \in C : f(x) \in J_x \quad for\ all\ x \in X\}. \tag{2}$$

Note. By VI.8.10 (together with the fact that $\mathscr{B}$ has enough continuous cross-sections), each J_x is a norm-closed two-sided ideal of the fiber B_x over x.

Proof. We can assume without loss of generality that X is compact (otherwise replace it by its one-point compactification, assigning the 0-dimensional fiber to the point at infinity).

First we claim that the ideal J is closed under multiplication by functions in $\mathscr{L}(X)$. Indeed, let $f \in J$ and $\lambda \in \mathscr{L}(X)$. Since C is a C^*-algebra it has an approximate unit $\{u_i\}$. Thus $u_i f \to f$, whence $(\lambda u_i)f = \lambda(u_i f) \to \lambda f$, uniformly on X. But $\lambda u_i \in C$, and so $(\lambda u_i)f \in J$ (since J is an ideal). Thus $\lambda f \in J$ (since J is closed).

To prove (2) we must take an element f of C such that $f(x) \in J_x$ for all $x \in X$, and show that $f \in J$. But we have just seen that J is closed under multiplication by elements of $\mathscr{L}(X)$. So the fact that $f \in J$ follows by the argument of the proof of II.14.1.   ∎

The Structure Space of the Cross-Sectional C^*-Algebra

8.8 Let X, $\mathscr{B}$, C be as in 8.4, the fiber of $\mathscr{B}$ over x being called B_x as usual. We shall now investigate the structure space $\hat{C}$ of C.

To begin with, suppose $x \in X$ and $S \in (B_x)\hat{\ }$. Then the equation

$$T_f = S_{f(x)} \qquad\qquad (f \in C) \qquad (3)$$

clearly defines a *-representation T of C. In fact, since $\mathscr{B}$ has enough continuous cross-sections, the ranges of T and S are the same, and so T is irreducible.

Conversely, we have the following important non-commutative version of V.8.8:

Proposition. *Let T be any irreducible *-representation of C. Then there is a unique element x of X and a unique element S of $(B_x)\hat{\ }$ such that T is given by* (3).

Proof. Let $J = \mathrm{Ker}(T)$; and for $x \in X$ let $J_x = \{f(x): f \in J\}$ as in 8.7.

We claim that there cannot be two distinct elements x and y of X such that $J_x \neq B_x$, $J_y \neq B_y$. Indeed, suppose there were such x and y. We can then choose elements ϕ, ψ of $\mathscr{L}(X)$ such that $\phi(x) = \psi(y) = 1$ and $\phi\psi = 0$. Let f' and g' be elements of C such that $f'(x) \notin J_x$ and $g'(y) \notin J_y$; and define $f = \phi f'$, $g = \psi g'$. Since $f(x) = f'(x) \notin J_x$ we have $f \notin J$, and likewise $g \notin J$. Hence T_f and T_g are non-zero operators; and so since T is irreducible, there is an element h of C such that $0 \neq T_f T_h T_g = T_{fhg}$. On the other hand multiplication in C is pointwise, and f and g vanish outside disjoint sets (since ϕ and ψ do); therefore $fhg = 0$. This contradiction proves the claim.

Now if $J_x = B_x$ for all x in X, then $J = C$ by Lemma 8.7, contradicting the properness of J. So by the preceding claim there is exactly one element x of X such that $J_x \neq B_x$. By Lemma 8.7 this implies that

$$\{f \in C: f(x) = 0\} \subset I.$$

This says that T_f depends only on $f(x)$, and hence that there is a (unique) $S \in (B_x)\hat{\ }$ such that (3) holds. $\blacksquare$

8.9 Continuing the notation of 8.8, let

$$Q = \{\langle x, S \rangle : x \in X, S \in (B_x)\hat{\ }\};$$

and for each $\langle x, S \rangle \in Q$ let $T^{x,S}$ denote the element of $\hat{C}$ given by (3). By Proposition 8.8 the mapping

$$\langle x, S \rangle \mapsto T^{x,S} \tag{4}$$

is a bijection of Q onto $\hat{C}$. We must now discuss the *regional topology* of Q, that is, the topology making (4) a homeomorphism with respect to the regional topology of $\hat{C}$.

For this we make a definition (valid for any Banach bundle $\mathcal{B}$):

Definition. From each $x \in X$ let D_x be a linear subspace of B_x. By the *lower-semicontinuous hull* of the $\{D_x\}$ we mean the collection $\{D'_x\}$, where D'_x is the linear subspace of B_x given by

$$D'_x = \{f(x) : f \in C, f(y) \in D_y \quad \text{for all } y \in X\}.$$

The following proposition now describes the regional closure of a subset of Q.

Proposition. *Let* $W \subset Q$; *for each* $x \in X$ *set* $W_x = \{S \in (B_x)\hat{\ } : \langle x, S \rangle \in W\}$; *let* $J_x = \bigcap_{S \in W_x} \mathrm{Ker}(S)$; *and let* $\{J'_x\}$ *be the lower-semicontinuous hull of the* $\{J_x\}$. *Then the regional closure* $\bar{W}$ *of* W *is given by*

$$\bar{W} = \{\langle x, S \rangle \in Q : \mathrm{Ker}(S) \supset J'_x\}. \tag{5}$$

Note: If $W_x = \emptyset$, then of course $J_x = B_x$.

Proof. Assume that $\langle x, S \rangle \in \bar{W}$. If $f \in C$ and $f(y) \in J_y$ for all $y \in X$, then $f \in \mathrm{Ker}(T^{y,R})$ for all $\langle y, R \rangle \in W$, and so by Theorem 5.9 $f \in \mathrm{Ker}(T^{x,S})$, whence $f(x) \in \mathrm{Ker}(S)$. Since $f(x)$ is an arbitrary element of J'_x, we have shown that $J'_x \subset \mathrm{Ker}(S)$. Thus $\bar{W}$ is contained in the right side of (5). The same argument runs in the reverse direction, showing that equality holds in (5). $\blacksquare$

8.10 Corollary. *The projection* $\rho : \langle x, S \rangle \mapsto x$ *of* Q *into* X *is continuous and open.*

Note: Since $\{x \in X : B_x \text{ is 0-dimensional}\}$ is clearly closed, the range of ρ is open in X, so that the openness of ρ means simply that the ρ-image of an open subset of Q is open in X.

Proof. To show that ρ is continuous amounts to showing that $\rho(\bar{W}) \subset \overline{\rho(W)}$, where $W \subset Q$. Suppose then that $\langle x, S \rangle \in \bar{W}$ but $x \notin \overline{\rho(W)}$. We can then find an element f of C such that $S_{f(x)} \neq 0$ and f vanishes on $\rho(W)$. But this implies $\bigcap_{q \in W} \text{Ker}(T^q) \not\subset \text{Ker}(T^{x,S})$, whence by Theorem 5.9 $\langle x, S \rangle \notin \bar{W}$, a contradiction. So $\rho(\bar{W}) \subset \overline{\rho(W)}$, and ρ is continuous.

To prove ρ is open, it is enough (Exercise 40, Chapter II) to take a subset Y of X and an element x of $\bar{Y}$, and show that $\rho^{-1}(x) \subset \overline{\rho^{-1}(Y)}$. Now if the W of Proposition 8.9 is taken to be $\rho^{-1}(Y)$, then the J_y of that Proposition is $\{0\}$ for $y \in Y$, and hence $J'_x = \{0\}$. It follows immediately from Proposition 8.9 that $\rho^{-1}(x) \subset \overline{\rho^{-1}(Y)}$. ∎

8.11.   Corollary.   *For each fixed x in X, the map $S \mapsto T^{x,S}$ is a homeomorphism of $(B_x)\hat{\ }$ into $\hat{C}$.*

Proof.   This follows immediately from Proposition 8.9. ∎

Bundles of C*-Algebras Over Discrete Spaces

8.12.   Let $\mathscr{B}$ be a bundle of C^*-algebras over a *discrete* space X. Then $\mathscr{C}_0(\mathscr{B})$ is just the C^*-direct sum of the fibers B_x $(x \in X)$ in the sense of VI.3.13. In this case Corollaries 8.10 and 8.11 obviously supply the following complete description of the regional topology of the Q of 8.9:

Proposition.   *For each $x \in X$ the subset $Q_x = \{\langle x, S \rangle : S \in (B_x)\hat{\ }\}$ is regionally both open and closed in Q; and the map $S \mapsto \langle x, S \rangle$ $(S \in (B_x)\hat{\ })$ is a homeomorphism of $(B_x)\hat{\ }$ onto Q_x in the regional topologies.*

Trivial Bundles of C*-Algebras

8.13.   Suppose now that the $\mathscr{B}$ of 8.4 is a *trivial bundle of C^*-algebras with fiber A*; that is, A is a fixed C^*-algebra, and the bundle space of $\mathscr{B}$ is $X \times A$ with the Cartesian product topology, the bundle projection π being just $\langle x, a \rangle \mapsto x$. Thus $C = \mathscr{C}_0(X; A)$; and, as in 8.9, $\hat{C}$ can be identified (as a set) with $Q = X \times \hat{A}$ via the bijection $\langle x, S \rangle \mapsto T^{x,S}$. The next proposition asserts that (as we expect!) the regional topology of Q is just the product topology of $X \times \hat{A}$.

Proposition.   *The bijection $\langle x, S \rangle \mapsto T^{x,S}$ of $X \times \hat{A}$ onto $\hat{C}$ is a homeomorphism with respect to the product topology of $X \times \hat{A}$ ($\hat{A}$ carrying its regional topology of course) and the regional topology of $\hat{C}$.*

Proof. (I) Suppose that

$$T^{x_i, S^i} \underset{i}{\to} T^{x, S} \text{ in } \hat{C}. \tag{6}$$

We wish to prove that

$$\langle x_i, S^i \rangle \underset{i}{\to} \langle x, S \rangle \quad \text{in} \quad X \times \hat{A}. \tag{7}$$

Now, by Corollary 8.10, (6) implies that

$$x_i \to x \quad \text{in} \quad X. \tag{8}$$

To prove that $S^i \to S$ in $\hat{A}$, we take an arbitrary element a of $\bigcap_i \text{Ker}(S^i)$; and choose an element f of C such that $f(y) = a$ for all y in some neighborhood of x. By (8) $f \in \text{Ker}(T^{x_i, S^i})$ for all large i; so by (6) $f \in \text{Ker}(T^{x, S})$, whence $a = f(x) \in \text{Ker}(S)$. This proves that $\bigcap_i \text{Ker}(S^i) \subset \text{Ker}(S)$, so that by Theorem 5.9 S belongs to the closure of $\{S^i\}$ in $\hat{A}$. Since the same holds for any subnet of $\{S^i\}$, we have shown that

$$S^i \to S \quad \text{in} \quad \hat{A}. \tag{9}$$

Now (8) and (9) imply (7).

(II) Now assume (7), that is, (8) and (9). We wish to prove (6). Let f be any element of $\bigcap_i \text{Ker}(T^{x_i, S^i})$, so that

$$f(x_i) \in \text{Ker}(S^i) \quad \text{for each} \quad i. \tag{10}$$

Let $\varepsilon > 0$. By the norm-continuity of f there is an index i_0 such that

$$\| f(x_i) - f(x) \| < \varepsilon \quad \text{for all} \quad i \succ i_0.$$

This and (10) combine to give

$$\| S^i_{f(x)} \| < \varepsilon \quad \text{for all} \quad i \succ i_0.$$

Applying VI.8.12 to the last inequality, we have

$$\| f(x) / \bigcap_{i \succ i_0} \text{Ker}(S^i) \| \le \varepsilon.$$

But by (9) $\bigcap_{i \succ i_0} \text{Ker}(S^i) \subset \text{Ker}(S)$; so the last inequality gives

$$\| S_{f(x)} \| = \| f(x) / \text{Ker}(S) \| \le \varepsilon.$$

Since ε was arbitrary this shows that $S_{f(x)} = 0$, or $f \in \text{Ker}(T^{x, S})$. We have thus shown that

$$\bigcap_i \text{Ker}(T^{x_i, S^i}) \subset \text{Ker}(T^{x, S}),$$

and hence (by Theorem 5.9) that $T^{x, S}$ lies in the regional closure of $\{T^{x_1, S_i}\}$. Since the same is true of any subnet, this implies (6).

Now (I) and (II) together give the required homeomorphism.   ∎

Bundles of Elementary C*-Algebras

8.14. A bundle $\mathscr{B}$ of C^*-algebras is called a *bundle of elementary C^*-algebras* if every fiber B_x of $\mathscr{B}$ is a non-zero elementary C^*-algebra (i.e., is *-isomorphic with the C^*-algebra of all compact operators on some non-zero Hilbert space).

Suppose that $\mathscr{B}$ is a bundle of elementary C^*-algebras over a locally compact Hausdorff space X. The structure space of $\mathscr{C}_0(\mathscr{B})$ is then especially simple. Indeed, by VI.15.13 the structure space of each fiber B_x consists of exactly one element which we shall call S^x; so the map

$$x \mapsto \langle x, S^x \rangle \qquad\qquad (x \in X) \qquad (11)$$

is a bijection of X onto the Q of 8.9.

Proposition.   *The map (11) is a homeomorphism of X onto Q (with the regional topology).*

Proof.   By Corollary 8.10.   ∎

Thus, if $\mathscr{B}$ is a bundle of elementary C^*-algebras, the structure space of $\mathscr{C}_0(\mathscr{B})$ can be identified, both setwise and topologically, with the base space X; in particular, the structure space of $\mathscr{C}_0(\mathscr{B})$ is Hausdorff.

Remark.   For separable C^*-algebras, the converse is true: If A is a separable C^*-algebra whose structure space $\hat{A}$ is Hausdorff (hence locally compact Hausdorff; see 6.9), then $A \cong \mathscr{C}_0(\mathscr{B})$ for some bundle $\mathscr{B}$ of elementary C^*-algebras over $\hat{A}$. See Fell [3].

8.15.   The following proposition is an easy and useful generalization of 8.14.

Let $\mathscr{B}$ be a bundle of C^*-algebras over the locally compact Hausdorff space X. Suppose that Y is a closed subset of X such that the fiber B_x at x is an elementary C^*-algebra, having S^x as its unique irreducible *-representation, for all $x \in X \setminus Y$. Replacing $\langle x, S^x \rangle$ by x for $x \in X \setminus Y$, we can identify the Q of 8.9 with the following disjoint union:

$$Q = (X \setminus Y) \cup Q',$$

where $Q' = \{\langle x, T \rangle : x \in Y, T \in (B_x)\hat{\ }\}$.

The regional topology of $Q \cong (\mathscr{C}_0(\mathscr{B}))\hat{\ }$ is now described as follows:

Proposition. *A subset W of Q is regionally closed if and only if the following three conditions hold:*

(i) *$W \cap (X \setminus Y)$ is closed in the relativized topology of $X \setminus Y$;*

(ii) *$W \cap Q'$ is closed in the regional topology of $Q' \cong (\mathscr{C}_0(\mathscr{B}_Y))\hat{\ }$;*

(iii) *$\langle x, T \rangle \in W$ whenever $x \in Y$, $T \in (B_x)\hat{\ }$, and x lies in the X-closure of $W \cap (X \setminus Y)$.*

The proof is an easy consequence of 8.9.

Bundles of Elementary C*-Algebras Derived From Hilbert Bundles.

8.16. Suppose that $\mathscr{H}$ and $\mathscr{K}$ are two Banach bundles over the same base space X. It is reasonable to ask whether there is a natural way to construct a Banach bundle over X whose fiber at each point x of X is the Banach space $\mathcal{O}(H_x, K_x)$ of all bounded linear operators from H_x to K_x (these being the fibers at x of $\mathscr{H}$ and $\mathscr{K}$ respectively). It turns out this is too much to ask (see Exercise 28). However, if $\mathscr{H}$ and $\mathscr{K}$ are *Hilbert* bundles over X, we shall now show that there is indeed a natural Banach bundle over X whose fiber at x is the Banach space $\mathcal{O}_c(H_x, K_x)$ of all *compact* operators from H_x to K_x. In particular, if $\mathscr{H} = \mathscr{K}$ (and all fibers of $\mathscr{H}$ are non-zero), we will obtain a bundle of elementary C*-algebras naturally derived from the Hilbert bundle $\mathscr{H}$.

It should be remarked immediately that by no means every bundle of elementary C*-algebras will arise by this construction from a Hilbert bundle. See Remark 8.20.

8.17. So let $\mathscr{H}$ and $\mathscr{K}$ be two Hilbert bundles (with fibers $\{H_x\}$ and $\{K_x\}$) over the same locally compact base space X. For each $x \in X$ let D_x be the Banach space $\mathcal{O}_x(H_x, K_x)$ of all compact linear operators from H_x to K_x, with the usual Hilbert operator norm. We wish to introduce a topology into the disjoint union $D = \bigcup_{x \in X} D_x$ so that $\langle D, \{D_x\} \rangle$ will be a Banach bundle $\mathscr{D}$ over X.

For this purpose we must introduce cross-sections of $\mathscr{D}$ which are to be continuous. By II.13.19 $\mathscr{H}$ and $\mathscr{K}$ have enough continuous cross-sections; as usual let $\mathscr{C}(\mathscr{H})$ and $\mathscr{C}(\mathscr{K})$ be the spaces of all continuous cross-sections of $\mathscr{H}$ and $\mathscr{K}$ respectively. If $\phi \in \mathscr{C}(\mathscr{H})$ and $\psi \in \mathscr{C}(\mathscr{K})$, we denote by $a_{\phi, \psi}$ the cross-section of $\mathscr{D}$ given by

$$a_{\phi, \psi}(x)(\xi) = (\xi, \phi(x))_{H_x} \psi(x) \qquad (x \in X, \xi \in H_x).$$

Thus, for each x, $a_{\phi, \psi}(x)$ is an element of D_x of rank at most 1.

Proposition.   *There is a unique topology for D making $\mathscr{D}$ a Banach bundle over X, such that the cross-sections $a_{\phi, \psi}$ are continuous for every $\phi \in \mathscr{C}(\mathscr{H})$ and $\psi \in \mathscr{C}(\mathscr{K})$.*

Proof.   Let Γ be the linear span of $\{a_{\phi, \psi} : \phi \in \mathscr{C}(\mathscr{H}), \ \psi \in \mathscr{C}(\mathscr{K})\}$. By II.13.18 we have only to show that (i) $\{\alpha(x) : \alpha \in \Gamma\}$ is dense in D_x for each $x \in X$, and (ii) $x \mapsto \|\alpha(x)\|$ is continuous for each $\alpha \in \Gamma$.

To prove (i) we merely observe that $\{\alpha(x) : \alpha \in \Gamma\}$ is the subspace of D_x consisting of the continuous operators of finite rank, and apply VI.15.7.

To prove (ii), take $\alpha \in \Gamma$ and $u \in X$; we must show that $x \mapsto \|\alpha(x)\|$ is continuous at u. Introduce finite orthonormal sets of vectors $\xi_1, \ldots, \xi_h$ and $\eta_1, \ldots, \eta_k$ in H_u and K_u respectively, with linear spans L and M respectively, such that $\alpha(u)$ annihilates $L^{\perp}$ and $\alpha(u)(L) \subset M$. By the Gram-Schmidt process applied to continuous cross-sections through the ξ_i and η_j, for each x in some neighborhood U of u we can obtain an orthonormal set $\xi_1(x), \ldots, \xi_h(x)$ (with linear span L_x) in H_x and an orthonormal set $\eta_1(x), \ldots, \eta_k(x)$ (with linear span M_x) in K_x, such that $\xi_i(u) = \xi_i$, $\eta_j(u) = \eta_j$, and the $x \mapsto \xi_i(x)$ and $x \mapsto \eta_j(x)$ are continuous cross-sections on U. For each fixed $x \in U$ let $\beta(x)$ be the linear map $H_x \to K_x$ which annihilates $L_x^{\perp}$ and sends L_x into M_x, and whose matrix with respect to the $\xi_i(x)$ and $\eta_i(x)$ coincides with the matrix of $\alpha(u)$ with respect to the ξ_i and η_j. Thus $\beta(u) = \alpha(u)$, and

$$\|\beta(x)\| = \|\alpha(u)\| \qquad \text{for all } x \in U. \tag{12}$$

Now we claim that

$$\lim_{x \to u} \|\alpha(x) - \beta(x)\| = 0. \tag{13}$$

To see this it is enough (by the triangle inequality) to suppose that $\alpha = a_{\phi, \psi}$ (where $\phi \in \mathscr{C}(\mathscr{H})$, $\psi \in \mathscr{C}(\mathscr{K})$, $\phi(u) \in L$, $\psi(u) \in M$). Then by definition $\beta = a_{\phi', \psi'}$, where ϕ' and ψ' are cross-sections of $\mathscr{H}$ and $\mathscr{K}$ continuous on U and $\phi'(u) = \phi(u)$, $\psi'(u) = \psi(u)$. But then

$$\lim_{x \to u} \|\phi(x) - \phi'(x)\| = 0, \quad \lim_{x \to u} \|\psi(x) - \psi'(x)\| = 0; \tag{14}$$

and it is a routine matter to verify from (14) that

$$\lim_{x \to u} \|a_{\phi, \psi}(x) - a_{\phi', \psi'}(x)\| = 0,$$

so that (13) holds for this α and β. This, as we have observed, proves the claim. Now from (12) and (13) we easily derive the continuity of $x \mapsto \|\alpha(x)\|$ at u. So (ii) holds.   ■

Definition. The Banach bundle $\mathscr{D}$ thus constructed will be denoted by $\mathcal{O}_c(\mathscr{H}, \mathscr{K})$. In case $\mathscr{H} = \mathscr{K}$ we write $\mathcal{O}_c(\mathscr{H})$ instead of $\mathcal{O}_c(\mathscr{H}, \mathscr{H})$.

As we expect, $\mathcal{O}_c(\mathscr{H})$ will turn out to be a bundle of (elementary) C*-algebras (provided the fibers are non-zero). To show this we have still to prove that multiplication and involution are continuous in $\mathcal{O}_c(\mathscr{H})$. This will follow from 8.18.

8.18. To prove the continuity of multiplication and involution in $\mathcal{O}_c(\mathscr{H})$ it will be useful to prove the following slightly more general result:

Proposition. *Let $\mathscr{H}$, $\mathscr{K}$ and $\mathscr{L}$ be three Hilbert bundles over the same locally compact base space X. Then (i) the composition map $\langle a, b \rangle \mapsto a \circ b$ is continuous on $\{\langle a, b \rangle \in \mathcal{O}_c(\mathscr{K}, \mathscr{L}) \times \mathcal{O}_c(\mathscr{H}, \mathscr{K}): \pi(a) = \pi(b)\}$ to $\mathcal{O}_c(\mathscr{H}, \mathscr{L})$; and (ii) the involution map $a \mapsto a^*$ is continuous on $\mathcal{O}_c(\mathscr{H}, \mathscr{K})$ to $\mathcal{O}_c(\mathscr{K}, \mathscr{H})$. (Here π stands for the projection map in each bundle.)*

Proof. (i) Let $\rho \in \mathscr{C}(\mathscr{H})$; $\sigma, \sigma' \in \mathscr{C}(\mathscr{K})$; $\tau \in \mathscr{C}(\mathscr{L})$; and let α, β and γ be the continuous cross-sections of $\mathcal{O}_c(\mathscr{K}, \mathscr{L})$, $\mathcal{O}_c(\mathscr{H}, \mathscr{K})$ and $\mathcal{O}_c(\mathscr{H}, \mathscr{L})$ respectively given by

$$\beta(x)(\xi) = (\xi, \rho(x))\sigma(x) \qquad (x \in X, \xi \in H_x),$$

$$\alpha(x)(\eta) = (\eta, \sigma'(x))\tau(x) \qquad (x \in X, \eta \in K_x),$$

$$\gamma(x)(\xi) = (\xi, \rho(x))\tau(x) \qquad (x \in X, \xi \in H_x).$$

We then verify that

$$\alpha(x) \circ \beta(x) = (\sigma(x), \sigma'(x))\gamma(x) \qquad (x \in X). \qquad (15)$$

Since $x \mapsto (\sigma(x), \sigma'(x))$ is continuous on X (by II.13.5), (15) and the continuity of γ assure us that

$$x \mapsto \alpha(x) \circ \beta(x) \qquad \text{is a continuous cross-section of } \mathcal{O}_c(\mathscr{H}, \mathscr{L}) \qquad (16)$$

Now let Γ and Δ be the linear spans of the set of all such cross-sections α and β of $\mathcal{O}_c(\mathscr{K}, \mathscr{L})$ and $\mathcal{O}_c(\mathscr{H}, \mathscr{K})$ respectively. As we pointed out in the proof of 8.17, $\{\alpha(x): \alpha \in \Gamma\}$ is dense in $\mathcal{O}_c(K_x, L_x)$ for each $x \in X$; and similarly for Δ. By (16) and the continuity of the linear operations in a Banach bundle,

$$(16) \text{ is true for every } \alpha \in \Gamma \text{ and } \beta \in \Delta. \qquad (17)$$

In view of (17) and the denseness property of Γ and Δ, the truth of (i) follows by a routine application of II.13.12. (Compare Remark 8.2.)

(ii) The proof of (ii) is based in a similar way on the observation that the β defined at the beginning of this proof satisfies

$$\beta(x)^*(\eta) = (\eta, \sigma(x))\rho(x) \qquad\qquad (x \in X, \eta \in K_x),$$

and therefore that $x \mapsto (\beta(x))^*$ is a continuous cross-section of $\mathcal{O}_c(\mathcal{K}, \mathcal{H})$. ∎

8.19.   Corollary.   *If $\mathcal{H}$ is a Hilbert bundle over a locally compact base space whose fibers are all non-zero, then $\mathcal{O}_c(\mathcal{H})$ is a bundle of elementary C^*-algebras (under the natural operations of composition and involution in each fiber).*

Definition.   We refer to $\mathcal{O}_c(\mathcal{H})$ as the bundle of elementary C^*-algebras *generated by $\mathcal{H}$.*

8.20.   *Remark.*   Not every bundle of elementary C^*-algebras is isomorphic with one generated by a Hilbert bundle.

Indeed: Let $\mathcal{B}$ be a bundle of elementary C^*-algebras over X (with fibers $\{B_x\}$). We shall say that $\mathcal{B}$ *has continuous trace* if for each $x \in X$ and each minimal non-zero projection p in B_x, there is a continuous cross-section α of $\mathcal{B}$ such that (i) $\alpha(x) = p$, and (ii) there is a neighborhood U of x such that $\alpha(y)$ is a minimal non-zero projection in B_y for each y in U.

It is clear that, if $\mathcal{B}$ is to be generated by a Hilbert bundle, it must have continuous trace.

On the other hand there are bundles of elementary C^*-algebras which do not have continuous trace. For example, let $\mathcal{D}$ be the trivial bundle over X whose fiber is the 2×2 total matrix *-algebra M_2. Take a non-isolated point u of X; and let $\mathcal{B}$ be the subbundle of $\mathcal{D}$ with fibers $B_x = M_2$ if $x \neq u$ and $B_u = \{\lambda I : \lambda \in \mathbb{C}\}$ (where I is the 2×2 identity matrix) (see 8.3). Thus $\mathcal{B}$ is a bundle of elementary C^*-algebras. But there is no continuous cross-section α of $\mathcal{B}$ such that $\alpha(u) = I$ and $\alpha(x)$ is a one-dimensional projection in M_2 for each $x \neq u$ in some neighborhood of u. So $\mathcal{B}$ does not have continuous trace, and hence is not isomorphic with any bundle of elementary C^*-algebras generated by a Hilbert bundle.

As a matter of fact it is not even true that every bundle of elementary C^*-algebras having continuous trace is isomorphic to one generated by a Hilbert bundle. See Fell [3].

Specific Examples

8.21 Let H be a Hilbert space, and H_1 and H_2 two non-zero orthogonal closed linear subspaces of H. We shall think of $\mathcal{O}_c(H_1) \oplus \mathcal{O}_c(H_2)$ as the C*-subalgebra of $\mathcal{O}_c(H)$ consisting of those operators a which leave H_1 and H_2 invariant and annihilate $(H_1 + H_2)^\perp$. Now let $\mathcal{D}$ be the trivial bundle of C*-algebras over $[0, 1]$ with fiber $\mathcal{O}_c(H)$; let $\mathcal{B}$ be the subbundle of $\mathcal{D}$ with fibers $B_x = \mathcal{O}_c(H)$ if $0 < x \leq 1$ and $B_0 = \mathcal{O}_c(H_1) \oplus \mathcal{O}_c(H_2)$ (see 8.3); and put $C = \mathcal{C}_0(\mathcal{B})$. Then 8.15 identifies the structure space $\hat{C}$ with $Q =]0, 1] \cup \{O_1, O_2\}$, where: (i) O_1 and O_2 are two distinct objects not belonging to $]0, 1]$; (ii) $]0, 1]$ is open in Q, and the topology of Q relativized to $]0, 1]$ is the ordinary topology of $]0, 1]$; and (iii) for each $i = 1, 2$, the family of sets of the form $\{O_i\} \cup]0, \varepsilon[$, where $0 < \varepsilon < 1$, is a basis of neighborhoods of O_i in Q.

Though a T_1 space, Q is not a Hausdorff space, since a sequence $\{t_n\}$ of points in $]0, 1]$ converging to 0 in $\mathbb{R}$ will converge both to O_1 and to O_2 in Q.

Remark. We have seen in 1.14 that for any C*-algebra A and any $a \in A$, the norm-function of a, namely $T \mapsto \|T_a\|$, is lower semi-continuous on $\hat{A}$. The above C shows that in general the norm-functions are not *continuous*. Indeed, let f be a continuous cross-section of $\mathcal{B}$ such that $f(0) = a_1 \oplus a_2$ ($a_i \in \mathcal{O}_c(H_i)$), where $\|a_1\| < \|a_2\|$; and let ϕ be the norm-function of f with its domain transferred to Q. Then

$$\lim_{t \to 0+} \phi(t) = \lim_{t \to 0+} \|f(t)\| = \|f(0)\| = \max\{\|a_1\|, \|a_2\|\}$$

$$= \|a_2\| \neq \|a_1\| = \phi(O_1);$$

so that ϕ is not continuous at O_1.

As a matter of fact it turns out that, for any C*-algebra A, the condition that the norm-function $T \mapsto \|T_a\|$ be continuous on $\hat{A}$ for every $a \in A$ is equivalent to the condition that $\hat{A}$ be Hausdorff. See Fell [3].

8.22. To motivate the next example, let us make a definition:

Definition. Let A and D be two C*-algebras. We shall say that D is *bundle-approximable by* A if it is possible to find a locally compact Hausdorff space X, a non-isolated point u of X, and a bundle $\mathcal{B}$ of C*-algebras over X with fibers B_x given by (i) $B_u = D$, and (iii) $B_x = A$ for all $x \neq u$ in X.

First of all, it is rather obvious that any C*-subalgebra of A is bundle-approximable by A. Indeed, if D is a C*-subalgebra of A, we can take $\mathcal{E}$ to be the trivial bundle of C*-algebras, with fiber A, over a locally compact

Hausdorff space X having a non-isolated point u, and $\mathcal{B}$ to be the subbundle of $\mathcal{E}$ having fibers D at u and A at all other points.

Now we claim that D may be bundle-approximable by A even if D is not *-isomorphic with any C*-subalgebra of A. To be specific we shall take $A = \mathcal{O}_c(H)$ where H is a separable infinite-dimensional Hilbert space; and we shall show that any separable Abelian C*-algebra D is bundle-approximable by A. This will establish the claim, since an Abelian C*-subalgebra of $\mathcal{O}_c(H)$ must have discrete spectrum (which in general D will not have).

So let D be a separable Abelian C*-algebra. If D is finite-dimensional it is a *-subalgebra of $\mathcal{O}_c(H)$; so we may as well assume that D is infinite-dimensional. Thus by §VI.4 $D = \mathscr{C}_0(M)$, where M is an infinite locally compact Hausdorff space with a countable base of open sets. So there is a sequence $p_1, p_2, p_3, \ldots$ of points of M whose range is dense in M; and since M is infinite the sequence $\{p_i\}$ can be taken to be one-to-one.

Let X be the compact subset $\{0\} \cup \{1/n : n = 1, 2, 3, \ldots\}$ of $\mathbb{R}$. Let H be the separable Hilbert space of all square-summable complex sequences (with the usual inner product $(s, t) = \sum_{i=1}^{\infty} s_i \overline{t_i}$). We propose to define a bundle $\mathcal{B}$ of C*-algebras over X with fibers $B_0 = D$, $B_{1/n} = \mathcal{O}_c(H)$ for all $n = 1, 2, \ldots$. To do this we shall specify a family of cross-sections which are to be continuous. For each $\phi \in D = \mathscr{C}_0(M)$, let γ_ϕ be the $\mathcal{B}$-cross-section defined on X by (i) $\gamma_\phi(0) = \phi$; (ii) if $n = 1, 2, \ldots, \gamma_\phi(1/n)$ is the operator on H of pointwise multiplication by the (bounded) sequence $m_\phi^{(n)}$ given by

$$m_\phi^{(n)}(i) = \begin{cases} \phi(p_i) & \text{if } 1 \leq i \leq n \\ 0 & \text{if } i > n. \end{cases}$$

Since $\{p_i\}$ is dense in M,

$$\left\| \gamma_\phi\left(\frac{1}{n}\right) \right\| = \sup_{i=1}^{n} |\phi(p_i)| \to \sup_{i=1}^{\infty} |\phi(p_i)| = \|\phi\|_0$$

$$= \|\gamma_\phi(0)\| \qquad \text{as } n \to \infty,$$

so that the norm-function $x \mapsto \|\gamma_\phi(x)\|$ is continuous on X. Now let us denote by Γ the linear space of all cross-sections α of the $\{B_x\}$ such that

$$\lim_{n \to \infty} \left\| \alpha\left(\frac{1}{n}\right) - \gamma_{\alpha(0)}\left(\frac{1}{n}\right) \right\| = 0. \tag{18}$$

Since $x \mapsto \|\gamma_{\alpha(0)}(x)\|$ is continuous, (18) shows that $x \mapsto \|\alpha(x)\|$ is continuous for each $\alpha \in \Gamma$. Obviously $\gamma_\phi \in \Gamma$ for each $\phi \in D$; so Γ "fills out" each fiber B_x. Hence by II.13.18 there is a Banach bundle $\mathcal{B}$ over X having the B_x as its fibers and making all the cross-sections in Γ continuous. Since Γ is clearly closed under multiplication and involution, these operations are continuous on $\mathscr{B}$ (see Remark 8.2). So $\mathscr{B}$ is a bundle of C*-algebras.

Thus, *if H is a separable infinite-dimensional Hilbert space, any separable Abelian C*-algebra is bundle-approximable by $\mathscr{O}_c(H)$.*

Remark. By Proposition 8.15, the structure space of $\mathscr{C}_0(\mathscr{B})$ ($\mathscr{B}$ being as above) can be identified with the disjoint union

$$Q = M \cup \left\{ \frac{1}{n} : n = 1, 2, \ldots \right\}$$

topologized as follows: A subset W of Q is closed if and only if (i) $W \cap M$ is closed in M, and (ii) if $W \cap \{1/n : n = 1, 2, \ldots\}$ is infinite then $M \subset W$.

8.23. It is interesting, as yet unsolved, problem to characterize those C*-algebras which are bundle-approximable by a given C*-algebra. Some information bearing on this problem will be found in Lee [2].

9. The Spectral Measure of a *-Representation

9.1. Let A be an arbitrary C*-algebra, and T a non-degenerate *-representation of A. In this section we are going to show that T gives rise to a canonical $X(T)$-projection-valued measure P on the Borel σ-field of $\hat{A}$, generalizing the spectral measure (VI.10.10) of a *-representation of a commutative C*-algebra. This P, like its prototype in the commutative case, is called the spectral measure of T. It can be regarded as a fragment of direct integral decomposition theory; and will be of vital importance for the Mackey-Blattner analysis presented in Chapter XII.

9.2. We begin with a lemma on Boolean algebras. To give the reader unfamiliar with Boolean algebras as little trouble as possible in following the argument, we make the following very redundant definition of a Boolean algebra.

Definition. A *Boolean algebra* is a set B together with two binary operations $\wedge$ and $\vee$ and one unary operation $^-$ on B, satisfying the following postulates for all a, b, c in B:

(1) $\wedge$ and $\vee$ are associative and commutative;
(2) $a \vee a = a, a \wedge a = a$;
(3) $a \vee (b \wedge c) = (a \vee b) \wedge (a \vee c), a \wedge (b \vee c) = (a \wedge b) \vee (a \wedge c)$ (Distributive Laws);

(4) $\bar{\bar{a}} = a$;

(5) $\overline{a \vee b} = \bar{a} \wedge \bar{b}$, $\overline{a \wedge b} = \bar{a} \vee \bar{b}$ (de Morgan's Laws);

(6) there exist (unique) elements θ and 1 in B such that $a \vee \theta = a$, $a \wedge 1 = a$, $a \vee \bar{a} = 1$, and $a \wedge \bar{a} = \theta$ for every a in B. We call θ the *zero element* and 1 the *unit element* of B.

Notice the symmetry of these postulates in $\wedge$ and $\vee$.

These postulates are highly redundant. With some effort they can all be deduced from a much smaller collection.

If $a_1, \ldots, a_n \in B$, we may write $\bigwedge_{i=1}^{n} a_i$ and $\bigvee_{i=1}^{n} a_i$ for $a_1 \wedge \cdots \wedge a_n$ and $a_1 \vee \cdots \vee a_n$ respectively.

9.3. The prime example of a Boolean algebra is provided by the set-theoretic operations. If $\mathscr{S}$ is a field of subsets of a set S, then $\mathscr{S}$ becomes a Boolean algebra on taking $\wedge$, $\vee$, and $^-$ to be intersection, union, and complementation in S respectively. It can be shown (see Stone [4]) that, conversely, every Boolean algebra is isomorphic to some field of sets, when the latter is considered as a Boolean algebra in this way.

9.4. Another example of a Boolean algebra which will be important to us is the following: Let A be any commutative (complex associative) algebra containing a unit 1, and let $B = \{p \in A: p^2 = p\}$. Then B turns out to be a Boolean algebra with respect to the operations $\wedge$, $\vee$, $^-$ defined as follows:

$$p \wedge q = pq, \quad p \vee q = p + q - pq, \quad \bar{p} = 1 - p. \tag{1}$$

The reader will easily verify this. If A is a commutative *-algebra with unit 1, then $B' = \{p \in A: p^2 = p = p^*\}$ is closed under the operations (1) and contains 0 and 1, and so is also a Boolean algebra under the operations (1).

9.5. Let B_1 and B_2 be Boolean algebras (with operations, $\wedge$, $\vee$, and $^-$). The zero elements of both algebras will be called θ, and the unit elements 1. A map $\phi: B_1 \rightarrow B_2$ is a *Boolean homomorphism* if it preserves $\wedge$, $\vee$, and $^-$ (in particular $\phi(\theta) = \theta$ and $\phi(1) = 1$). If $\varepsilon = \pm 1$ and a is in B_1 or B_2, then εa will mean a or $\bar{a}$ according as ε is 1 or -1.

Lemma. *Let S be a subset of B_1 which generates B_1 (that is, the smallest subset of B_1 closed under $\vee$, $\wedge$, and $^-$ and containing S is B_1). Let $f: S \rightarrow B_2$. A necessary and sufficient condition that f extend to a Boolean homomorphism $\phi: B_1 \rightarrow B_2$ is that, for all positive integers n, all $a_1, \ldots, a_n$ in S, and all $\varepsilon_1, \ldots, \varepsilon_n = \pm 1$, we have*

$$\bigwedge_{i=1}^{n} \varepsilon_i a_i = \theta \Rightarrow \bigwedge_{i=1}^{n} \varepsilon_i f(a_i) = \theta. \tag{2}$$

Proof.   The condition is obviously necessary.

Assume that (2) holds. If an element b of B_1 can be written in the form

$$b = \bigvee_{i=1}^{n} \bigwedge_{j=1}^{r_i} \varepsilon_{ij} a_{ij}, \tag{3}$$

and also in the similar form

$$b = \bigvee_{p=1}^{m} \bigwedge_{q=1}^{s_p} \varepsilon'_{pq} a'_{pq} \tag{4}$$

$(a_{ij}, a'_{pq} \in S; \ \varepsilon_{ij}, \varepsilon'_{pq} = \pm 1)$, then we claim that

$$\bigvee_{i=1}^{n} \bigwedge_{j=1}^{r_i} \varepsilon_{ij} f(a_{ij}) = \bigvee_{p=1}^{m} \bigwedge_{q=1}^{s_p} \varepsilon'_{pq} f(a'_{pq}). \tag{5}$$

Indeed, by (4) and the de Morgan and distributive laws for Boolean algebras,

$$\bar{b} = \bigvee_{\lambda} \bigwedge_{p=1}^{m} (-\varepsilon'_{p,\lambda_p}) a'_{p,\lambda_p} \tag{6}$$

(λ running over all m-termed sequences of positive integers with $\lambda_p \le s_p$ for each p). By (3) and (6), for each i and λ

$$\bigwedge_{j=1}^{r_i} \bigwedge_{p=1}^{m} (\varepsilon_{ij} a_{ij}) \wedge (-\varepsilon'_{p,\lambda_p} a'_{p,\lambda_p}) = 0,$$

whence by (2)

$$\bigwedge_{j=1}^{r_i} \bigwedge_{p=1}^{m} \varepsilon_{ij} f(a_{ij}) \wedge (-\varepsilon'_{p,\lambda_p} f(a'_{p,\lambda_p})) = 0. \tag{7}$$

Let c and d be respectively the left and right sides of (5). Replacing the a'_{pq} in (4) and (6) by their images under f, we see that

$$c \wedge \bar{d} = \bigvee_{i=1}^{n} \bigvee_{\lambda} \bigwedge_{j=1}^{r_i} \bigwedge_{p=1}^{m} \varepsilon_{ij} f(a_{ij}) \wedge (-\varepsilon'_{p,\lambda_p} f(a'_{p,\lambda_p}));$$

and by (7) this shows that $c \wedge \bar{d} = 0$. Similarly $d \wedge \bar{c} = 0$. So $c = d$, and (5) is proved.

Now, since S generates B_1, every b in B_1 is of the form (3); and by (5) we can legitimately define $\phi(b)$ to be the left side of (5). Evidently ϕ is the Boolean homomorphism which extends f.   ■

9.6.   Corollary.   *In addition to the hypotheses of Lemma 9.5, assume that θ and $\dagger$ are in S, that S is closed under $\vee$ and $\wedge$, and that*

$$f(\theta) = \theta, f(\dagger) = \dagger, \tag{8}$$

$$f(a \wedge b) = f(a) \wedge f(b), f(a \vee b) = f(a) \vee f(b) \qquad (a, b \in S). \tag{9}$$

Then f extends to a Boolean homomorphism $\phi: B_1 \to B_2$.

Proof.   It must be shown that (2) holds. Assume that

$$\bigwedge_{i=1}^{n} \varepsilon_i a_i = \theta \tag{10}$$

$(a_i \in S; \varepsilon_1, \ldots, \varepsilon_r = 1; \varepsilon_{r+1}, \ldots, \varepsilon_n = -1)$. If we put $b = a_1 \wedge \cdots \wedge a_r$, $c = a_{r+1} \vee \cdots \vee a_n$, we have $b \in S$, $c \in S$, and by (10) $b \wedge \bar{c} = \theta$ or $b \vee c = c$. Applying f to the last equation we get by (9) $f(b) \vee f(c) = f(c)$, so $\theta = f(b) \wedge (f(c))^{-} = \bigwedge_{i=1}^{n} \varepsilon_i f(a_i)$ (by (8) and (9)). Thus (2) holds; and Lemma 9.5 gives the required extension.   ∎

9.7.   Next we prove an extension lemma for projection-valued measures. Let $\mathscr{S}$ be a field of subsets of a set S, and $\mathscr{S}_0$ the σ-field of subsets of S generated by $\mathscr{S}$.

Lemma.   *Let X be a Hilbert space, and P a finitely additive X-projection-valued measure of $\mathscr{S}$ with the further property that $\inf_n P(A_n) = 0$ whenever $\{A_n\}$ is a decreasing sequence of sets in $\mathscr{S}$ such that $\bigcap_n A_n = \emptyset$. Then P extends to a unique (countably additive) X-projection-valued measure on $\mathscr{S}_0$.*

Proof.   It is sufficient to assume that P is cyclic, i.e., that there is a vector ξ such that $\{P(A)\xi: A \in \mathscr{S}\}$ has dense linear span in X. Put $\mu(A) = (P(A)\xi, \xi)$ $(A \in \mathscr{S})$. By the corresponding extension theorem for numerical measures (Halmos [1], Theorem 13A) μ extends to a (countably additive) finite measure on $\mathscr{S}_0$, which we also call μ.

Consider the Hilbert space $\mathscr{L}_2(\mu)$. If $f = \sum_{i=1}^{r} c_i \, \mathrm{Ch}_{A_i} (c_i \in \mathbb{C}; A_i \in \mathscr{S})$, put $F(f) = \sum_{i=1}^{r} c_i P(A_i)\xi$. Since $\{P(A)\xi; A \in \mathscr{S}\}$ and $\{\mathrm{Ch}_A: A \in \mathscr{S}\}$ have dense linear spans in X and $\mathscr{L}_2(\mu)$ respectively (see Halmos [1], Theorem 6B), and since F preserves norms (by the definition of μ), F extends to a linear isometry of $\mathscr{L}_2(\mu)$ onto X. For $A \in \mathscr{S}$ the projection $P(A)$ corresponds under F with multiplication by Ch_A. So P extends to a countably additive X-projection-valued measure on $\mathscr{S}_0$, namely, the image under F of the $\mathscr{L}_2(\mu)$-projection-valued measure carrying each A in $\mathscr{S}_0$ into multiplication by Ch_A.

The uniqueness of the extension is obvious. ∎

9.8. Now take a fixed C^*-algebra A.

Let T be a fixed non-degenerate *-representation of A on a Hilbert space X. We denote range(T) by B. For each closed two-sided ideal I of A, let X_I be the closed linear span of $\{T_a \xi : a \in I, \xi \in X\}$ in X, and $Q(I)$ the projection of X onto X_I. Thus X_I is the essential space of $T|I$ (see VI.9.6).

We observe that $Q(I)$ belongs to the center of the von Neumann algebra generated by B (see VI.24.6). Indeed, X_I is T-stable since I is an ideal; so $Q(I)$ is in the commuting algebra B' of B. Also, if $V \in B'$, we have $VT_a = T_a V$ for $a \in I$, whence it follows immediately that X_I is V-stable. So $Q(I) \in B''$. We have shown that $Q(I) \in B' \cap B''$, and the latter is the center of B'', the von Neumann algebra generated by B (see VI.24.5).

In particular, the $Q(I)$ all commute with each other.

9.9. Proposition. *If $\mathcal{I}$ is any family of closed two-sided ideals of A, and J is the closed linear span of $\mathcal{I}$, then $Q(J) = \sup\{Q(I) : I \in \mathcal{I}\}$.*

Proof. X_J is the closed linear span of $\{T_a \xi : \xi \in X, a \in I, I \in \mathcal{I}\}$, i.e., of $\{X_I : I \in \mathcal{I}\}$. ∎

9.10. Proposition. *If I and J are closed two-sided ideals of A, then $Q(I \cap J) = Q(I)Q(J)$.*

Proof. Suppose $\xi \in X_I \cap X_J$ and $\varepsilon > 0$. Since I and J have approximate units with norm bounded by 1 (VI.8.4), and $T|I$ and $T|J$ act non-degenerately on $X_I \cap X_J$, we can find elements a of I and b of J such that $\|a\| \leq 1$, $\|T_a \xi - \xi\| < \varepsilon$, and $\|T_b \xi - \xi\| < \varepsilon$. Then $\|T_{ab} \xi - \xi\| \leq \|T_a(T_b \xi - \xi)\| + \|T_a \xi - \xi\| < 2\varepsilon$. Since ε is arbitrary and $ab \in I \cap J$, it follows that $\xi \in X_{I \cap J}$. So $X_I \cap X_J \subset X_{I \cap J}$. Obviously $X_{I \cap J} \subset X_I \cap X_J$. Thus $X_{I \cap J} = X_I \cap X_J$, and this implies (since $Q(I)$ and $Q(J)$ commute) that $Q(I \cap J) = Q(I)Q(J)$. ∎

9.11. Now we recall from 5.14 that there is a natural inclusion-preserving bijection F from the family $\mathcal{U}$ of all open subsets of the structure space $\hat{A}$ onto the family of all closed two-sided ideals of A, given by

$$F(U) = \bigcap\{\mathrm{Ker}(S) : S \in \hat{A} \setminus U\} \qquad (U \in \mathcal{U}) \qquad (11)$$

Let $P' = Q \circ F$. Thus P' associates to each U in $\mathcal{U}$ a projection $P'(U) = Q(F(U))$ on X. Notice that $P'(\emptyset) = 0$ and $P'(\hat{A})$ is the identity operator on X. The $P'(U)$ commute pairwise since the $Q(I)$ do.

Proposition. *if $\mathcal{W} \subset \mathcal{U}$, then $P'(\bigcup \mathcal{W}) = \sup\{P'(U): U \in \mathcal{W}\}$. If $U, V \in \mathcal{U}$, then $P'(U \cap V) = P'(U)P'(V)$.*

Proof. This follows from 9.9. and 9.10 together with 5.14(9), (10). ∎

9.12. Theorem. *Let T be a non-degenerate *-representation of a C*-algebra A, acting on Hilbert space $X = X(T)$. There is a unique X-projection-valued measure P on the Borel σ-field $\mathcal{S}_0$ of subsets of $\hat{A}$ which coincides on $\mathcal{U}$ (the family of open subsets of $\hat{A}$) with the P' of 9.11.*

This P is called the *spectral measure of T.*

Proof. Let $\mathcal{S}$ be the field of subsets of $\hat{A}$ generated by $\mathcal{U}$. Thus $\mathcal{S}_0$, the Borel σ-field, is the σ-field generated by $\mathcal{S}$. By 9.11 we can apply 9.6 to the map P' (taking $B_1 = \mathcal{S}$ and B_2 to be the Boolean algebra of all projections in the center of the von Neumann algebra generated by range(T); see 9.3), obtaining a finitely additive X-projection-valued measure P on $\mathcal{S}$ which extends P'.

Now in order to apply 9.7 to obtain a countably additive extension of P to $\mathcal{S}_0$, it is sufficient to prove the condition assumed in 9.7, that is, to take a sequence $\{C'_n\}$ of sets in $\mathcal{S}$ with $\bigcap_n C'_n = \emptyset$ and show that $\inf_n P(C'_n) = 0$. As a matter of fact we shall pass to complements; we shall take a sequence $\{C_n\}$ of sets in $\mathcal{S}$ with $\bigcup_n C_n = \hat{A}$, and shall show that $\sup_n P(C_n) = 1$ (identity projection).

Now any set in $\mathcal{S}$ is a finite union of sets of the form $D \cap E$, where D is open and E is closed. So we may as well assume that $C_n = B_n \setminus D_n$, where B_n and D_n are open and $D_n \subset B_n$.

Assume now that $\sup_n P(C_n) = 1 - R \neq 1$, where R is a nonzero projection. Since R commutes with range(P), we obtain a finitely additive Y-projection-valued measure $\tilde{P}$ on $\mathcal{S}$ by setting

$$Y = \text{range}(R), \quad \tilde{P}(B) = RP(B) \qquad\qquad (B \in \mathcal{S}).$$

Note that $\tilde{P}$ also satisfies the conclusions of Proposition 9.11. In particular, then, there is a unique largest open subset V of $\hat{A}$ such that $\tilde{P}(V) = 0$. Since $R \neq 0$, we have $V \neq \hat{A}$. By the definition of R, $\tilde{P}(C_n) = 0$ for all n. Let W be the closed non-void set $\hat{A} \setminus V$, and put $B'_n = B_n \cap W$, $D'_n = D_n \cap W$. Thus $\tilde{P}(B'_n \setminus D'_n) = \tilde{P}(B_n \setminus D_n) = \tilde{P}(C_n) = 0$ for all n.

Since $\bigcup_n C_n = \hat{A}$, we have $\bigcup_n (B'_n \setminus D'_n) = W$. So by the Baire property of W (see 6.6) we may choose n such that the closure of $B'_n \setminus D'_n$ has non-void interior relative to W. Thus $B'_n \setminus D'_n$ itself has non-void interior relative to W (see the Remark following this proof). So there is an open subset U of $\hat{A}$

such that $\emptyset \neq U \cap W \subset B'_n \setminus D'_n$. But we observed that $\tilde{P}(B'_n D'_n) = 0$. Hence $\tilde{P}(U) = \tilde{P}(U \cap V) + \tilde{P}(U \cap W) \leq \tilde{P}(V) + \tilde{P}(B'_n \setminus D'_n) = 0$, whence $\tilde{P}(U) = 0$. Thus, since U is open, the definition of V implies that $U \subset V$, contradicting the fact that $U \cap W \neq \emptyset$.

Therefore the assumption that $\sup_n P(C_n) \neq 1$ was wrong. As mentioned above, we can thus apply 9.7 to extend P to a countably additive projection-valued measure on $\mathscr{S}_0$, completing the existence part of the proof.

It remains only to note the uniqueness of the extension P. If R is another such extension, the definition of a projection-valued measure shows that $\{D \in \mathscr{S}_0 : P(D) = R(D)\}$ is a σ-subfield of $\mathscr{S}_0$. By hypothesis it must contain $\mathscr{U}$, and so must equal $\mathscr{S}_0$. Thus $P = R$; and the extension is unique. ∎

Remark. In the course of the preceding proof we made use of the following fact: If Z is any topological space, and B, D are open subsets of Z such that the closure of $B \setminus D$ has non-void interior, then $B \setminus D$ itself has non-void interior.

To prove this, suppose $\emptyset \neq U \subset (B \setminus D)^-$, where U is open. Since D is open $(B \setminus D)^- \subset \bar{B} \setminus D$; so $U \subset \bar{B} \setminus D$. In particular $U \cap B \neq \emptyset$. So $U \cap B$ is a non-void open subset of $B \cap (\bar{B} \setminus D) = B \setminus D$.

9.13. If P is the spectral measure of the non-degenerate *-representation T of A, then the range of P is contained in the center Z of the von Neumann algebra generated by range(T).

Indeed, $\{W : P(W) \in Z\}$ is clearly a σ-field of subsets of $\hat{A}$, and by 9.8 contains all open subsets of $\hat{A}$.

9.14. Let P be the spectral measure of T as in 9.12. Let W be a *closed* subset of $\hat{A}$; and put $I = \cap \{\text{Ker}(S) : S \in W\}$. By the definition in 9.11 $P(\hat{A} \setminus W)$ is projection onto the essential space of $T|I$. Therefore $P(W)$ is projection onto the null space of $T|I$, that is, onto $\{\xi \in X(T) : T_a \xi = 0 \text{ for all } a \text{ in } I\}$. Consequently, in view of 5.5, range$P(W)$ can be described as the largest closed T-stable subspace Y of $X(T)$ such that $^Y T$ is weakly contained in W.

9.15. Proposition. *If P is the spectral measure of a non-degenerate *-representation T of A, then the closed support of P (that is, the closed subset $\{S \in \hat{A} : P(V) \neq 0 \text{ for every open neighborhood } V \text{ of } S\}$ of $\hat{A}$) is equal to the spectrum of T (as defined in 5.16).*

Proof. Let W be the spectrum of T. By 5.16 $W = \{S \in \hat{A} : \text{Ker}(S) \supset \text{Ker}(T)\}$. On the other hand, by 9.11 the complement of the closed support C of P is the largest open P-null subset of $\hat{A}$, that is, the open set corresponding via

5.14(8) with Ker(T). This implies that Ker(T) = $\bigcap\{\text{Ker}(S)\colon S \in C\}$, or $C = W$.   ∎

9.16.   The spectral measure behaves as one would expect on passage to subrepresentations and direct sums.

Proposition.   *Let $\{T^i\}$ ($i \in I$) be an indexed collection of nondegenerate *-representations of A; and form their Hilbert direct sum $T = \sum^{\oplus}_{i \in I} T^i$. Then the spectral measure P of T is the direct sum of the spectral measures P^i of the T^i:*

$$P(W) = \sum_{i \in I}^{\oplus} P^i(W)$$

(for any Borel subset W of $\hat{A}$).

The verification of this is left to the reader.

9.17.   It is often convenient to transfer the P of 9.12 to Prim(A) (see 3.9 and 5.11). If $\Gamma\colon \hat{A} \to \text{Prim}(A)$ denotes the surjection $S \mapsto \text{Ker}(S)$, the projection-valued measure $Q\colon W \mapsto P(\Gamma^{-1}(W))$ on the Borel σ-field of Prim(A) is called the *spectral measure of T on* Prim(A).

9.18.   Suppose it happens that Prim(A) is Hausdorff. Then by 6.9 Prim(A) is a locally compact Hausdorff space. In that case the projection-valued measure Q of 9.17 is regular (in the sense of II.11.9). This follows at once on combining Proposition II.11.9 with the fact (9.11) that

$$Q\left(\bigcup_i U_i \right) = \sup_i Q(U_i) \tag{12}$$

for any collection $\{U_i\}$ of open subsets of Prim(A).

In view of this, condition (12) can be regarded as the proper analogue of regularity in the context of non-Hausdorff spaces.

9.19.   In the commutative case, as we expect, the spectral measure of 9.12 is the same as that of VI.10.10.

Proposition.   *If A is commutative and T is a non-degenerate *-representation of A, the spectral measures of T as defined in 9.12 and in VI.10.10 coincide.*

Proof. Let P and P' be the spectral measures of T as defined in VI.10.10 and 9.12 respectively; and let W be a closed subset of $\hat{A}$. If $\xi \in X(T)$, and $\|P\xi\|^2$ denotes the measure $V \mapsto \|P(V)\xi\|^2$, we have:

$$\xi \in \operatorname{range} P(W) \Leftrightarrow \|P\xi\|^2(\hat{A} \setminus W) = 0$$

$$\Leftrightarrow \int \hat{a}(\phi) d\|P\xi\|^2 \phi = 0 \quad \text{whenever } a \in A \text{ and } \hat{a} \text{ vanishes on } W$$

$$\Leftrightarrow T_a \xi = 0 \qquad \text{whenever } a \in A \text{ and } \hat{a} \text{ vanishes on } W$$

(the last equivalence following from VI.10.10(6) and II.11.8(7)). Thus, if $I = \cap\{\operatorname{Ker}(\chi): \chi \in W\}$, $\operatorname{range} P(W)$ consists of those ξ which are annihilated by $T|I$. But, by 9.14, these are just the ξ in range $P'(W)$. So P and P' coincide on closed sets, and hence by II.11.5 everywhere. ∎

9.20. Let C be a closed subset of $\hat{A}$, and I the corresponding closed *-ideal of A:

$$I = \cap\{\operatorname{Ker}(T): T \in C\}.$$

By 5.13 and 1.19 C, as a topological space, can be identified with $(A/I)\hat{\ }$.

Again, let U be an open subset of $\hat{A}$, and J the corresponding closed *-ideal of A:

$$J = \cap\{\operatorname{Ker}(T): T \in (\hat{A} \setminus U)\}.$$

By 5.13 and 4.4 U, as a topological space, can be identified with $\hat{J}$.

Proposition*. *Let T be a non-degenerate *-representation of A, with spectral measure P; and keep the preceding notation.*

(I) *If P is carried by C, then T is lifted from a non-degenerate *-representation T' of A/I (i.e., $T_a = T'_{a+I}$ for $a \in A$); and the spectral measure of T' is just the restriction of P to the family of Borel subsets of C.*

(II) *If P is carried by U, then $T|J$ is non-degenerate; and the spectral measure of $T|J$ is just the restriction of P to the family of Borel subsets of U.*

9.21. Remark. The construction of the spectral measure is not of course confined to C^*-algebras. Suppose that B is any Banach *-algebra, and form its C^*-completion A as in VI.10.4. By VI.10.5 the non-degenerate *-representations of B correspond naturally to those of A; and by 1.17 $\hat{B}$ and $\hat{A}$ are essentially the same. With this identification we define the *spectral* *measure [spectrum]* of a non-degenerate *-representation T of B to be the same as the spectral mesure [spectrum] of the corresponding non-degenerate *-representation of A. Thus 9.15 and 9.19 now hold automatically in the context of Banach *-algebras. By 9.19, 9.15 is a generalization of 5.17.

9.22.   Remark.   If A is commutative, it was shown in VI.10.10 that *every* regular Borel projection-valued measure P on $\hat{A}$ is the spectral measure of some non-degenerate *-representation T of A, T being constructed from P by means of the integral formula VI.10.10(6). In the non-commutative situation, however, the corresponding statement is false in view of simple dimensional considerations. Thus, if A is elementary, so that $\hat{A}$ consists of only one element, an X-projection-valued measure on $\hat{A}$ can only be the spectral measure of a *-representation of A if $\dim(X)$ is a multiple of the dimension of the unique irreducible *-representation of A (see VI.15.12).

Thus we could try, first, to characterize those projection-valued measures P on $\hat{A}$ which are the spectral measures of *-representations T of A, and, secondly, to find an integral formula analogous to VI.10.10(6) by means of which T can be reconstructed from P in the favorable cases. The first of these problems is solved by the multiplicity theory of projection-valued measures. The second is solved (at least for separable C^*-algebras) by the theory of direct integrals of representations. Both of these subjects, multiplicity theory and direct integrals, are beyond the scope of this work.

10.   *-Representations Whose Spectral Measures are Concentrated at a Single Point

10.1.   In this section we shall consider a fixed C^*-algebra A.

A numerical or projection-valued measure P on a σ-field $\mathscr{S}$ of subsets of a set S is said to be *concentrated at* a point x of S if $P(W) = 0$ whenever $W \in \mathscr{S}$ and $x \notin W$. If $\mathscr{S}$ separates the points of S and P is not the zero measure, then of course x, if it exists, is uniquely determined by P.

Definition.   Let I be an element of $\mathrm{Prim}(A)$. A non-degenerate *-representation T of A (with $X(T) \neq \{0\}$) is *concentrated at* I if the spectral measure P of T on $\mathrm{Prim}(A)$ is concentrated at I, or, in other words, if $P(W) = 1$ for every Borel subset W of $\mathrm{Prim}(A)$ containing I.

Since $\mathrm{Prim}(A)$ is a T_0 space, its Borel σ-field separates its points, and so the I, if it exists, is uniquely determined by T.

If $\{I\}$ is a Borel subset of $\mathrm{Prim}(A)$ (which is not always the case), the condition for T to be concentrated at I is simply that $P(\{I\}) = 1$, or $P(\mathrm{Prim}(A) \setminus \{I\}) = 0$.

10.2. The meaning of this concept is clarified by the following assertion:

Proposition. *Let I be an element of $\mathrm{Prim}(A)$, and T a non-degenerate *-representation of A (with $X(T) \neq \{0\}$). Then the following two conditions are equivalent:* (i) *T is concentrated at I;* (ii) *$\mathrm{Ker}(^Y T) = I$ for every non-zero closed T-stable subspace Y of $X(T)$.*

Proof. Let P be the spectral measure of T on $\mathrm{Prim}(A)$.

We assume (ii). If J is any closed two-sided ideal of A with $J \not\subset I$, it follows from (ii) that

$$T|J \text{ is non-degenerate.} \tag{1}$$

Let W be a closed subset of $\mathrm{Prim}(A)$, and put $J = \bigcap W$. If $I \notin W$, then by the definition of the hull-kernel topology $J \not\subset I$, and so by (1) and 9.14 $P(W) = 0$. If on the other hand $I \in W$, then $J \subset I = \mathrm{Ker}(T)$, and so by 9.14 $P(W) = 1$. Thus, for closed subsets W of $\mathrm{Prim}(A)$,

$$P(W) = \begin{cases} 1 & \text{if } I \in W, \\ 0 & \text{if } I \notin W. \end{cases} \tag{2}$$

We claim that (2) implies (i). Indeed, let $\mathscr{F}$ be the family of all those Borel subsets V of $\mathrm{Prim}(A)$ such that *either $I \in V$ and $P(V) = 1$ or $I \notin V$ and $P(V) = 0$.* Evidently $\mathscr{F}$ is a σ-field; and we have seen in (2) that $\mathscr{F}$ contains all closed sets. So all Borel sets are in $\mathscr{F}$, whence (i) holds.

Now assume (i). Since $\bigcap\{I\}^- = I$ and $P(\{I\}^-) = 1$, we have by 9.14 $I \subset \mathrm{Ker}(T)$. If $I \neq \mathrm{Ker}(T)$, then $V = \{J \in \mathrm{Prim}(A): J \supset \mathrm{Ker}(T)\}$ is a closed subset of $\mathrm{Prim}(A)$ not containing I; so $P(V) = 0$, implying by 9.14 that T restricted to $\bigcap V = \mathrm{Ker}(T)$ is non-degenerate, a contradiction. Therefore $I = \mathrm{Ker}(T)$. Now by 9.16 any non-zero subrepresentation T' of T is also concentrated at I, and hence by what we have just proved $I = \mathrm{Ker}(T')$. Hence (ii) holds. ∎

10.3. By 10.2 an irreducible *-representation T of A is obviously concentrated at $\mathrm{Ker}(T)$. In fact, from VI.14.9 (and its proof) combined with 10.2 we see that any direct sum of copies of T is concentrated at $\mathrm{Ker}(T)$. So any primary *-representation of A of Type I is concentrated at some point of $\mathrm{Prim}(A)$ (see VI.24.10).

10.4.   In fact, under very general conditions on A one can prove that *any* primary representation of A is concentrated at some point of $\mathrm{Prim}(A)$. To see this, assume that T is a primary *-representation of A (see VI.24.8), with spectral measure P. It then follows from 9.13 that range(P) consists of 0 and $\mathbb{1}$ only. Thus, in order to be able to prove that every primary representation of A is concentrated at some point of $\mathrm{Prim}(A)$, it is enough to show that $\mathrm{Prim}(A)$ has the following property:

Every projection-valued measure P on the Borel σ-field of $\mathrm{Prim}(A)$ which is regular in the sense of 9.18(12) and whose range is $\{0, \mathbb{1}\}$ is concentrated at some point of $\mathrm{Prim}(A)$.

10.5.   Proposition.   *If A is separable, every primary *-representation of A is concentrated at some point of* $\mathrm{Prim}(A)$.

Proof.   By 5.19 $\mathrm{Prim}(A)$ is second-countable. Thus we can find a countable family C of Borel subsets of $\mathrm{Prim}(A)$ which separates points of $\mathrm{Prim}(A)$ and is closed under complementation. Let P be a projection-valued measure on the Borel σ-field of $\mathrm{Prim}(A)$ whose range is $\{0, \mathbb{1}\}$; and put

$$C_1 = \{V \in C : P(V) = \mathbb{1}\}.$$

Since C_1 is countable, $P(\bigcap C_1) = \mathbb{1}$; in particular $\bigcap C_1 \neq \emptyset$. If $\bigcap C_1$ contained two distinct elements I and J, there would be a set W in C separating I and J; since neither W nor its complement could be in C_1, we would have $P(\mathrm{Prim}(A)) = P(W) + P(\mathrm{Prim}(A) \setminus W) = 0$, a contradiction. So $\bigcap C_1 = \{I\}$ for some I, and $P(\{I\}) = \mathbb{1}$. This establishes the last condition of 10.4.   ∎

10.6.   There is another general condition on A, quite different from separability, under which the last statement of 10.4 holds.

Definition.   A topological space X is called *almost Hausdorff* if, for every non-void closed subset S of X, there is a non-void subset U of S which is relatively open in S and is Hausdorff with its relativized topology.

Proposition.   *If $\mathrm{Prim}(A)$ is almost Hausdorff, every primary *-representation of A is concentrated at some point of* $\mathrm{Prim}(A)$.

Proof.   Let P be a projection-valued measure on the Borel σ-field of $\mathrm{Prim}(A)$ whose range is $\{0, \mathbb{1}\}$ and which satisfies 9.18(12).

Let $\mathscr{S}$ be the family of all *closed* subsets V of $\mathrm{Prim}(A)$ for which $P(V) = \mathbb{1}$; and put $V_0 = \bigcap \mathscr{S}$. By 9.18(12) $V_0 \in \mathscr{S}$, that is, $P(V_0) = \mathbb{1}$. In particular $V_0 \neq \emptyset$.

Notice that, if U is a non-void relatively open subset of V_0, then $V_0 \setminus U$ is a proper closed subset of V_0, and hence $P_0(V \setminus U) = 0$ by the definition of V_0, that is,

$$P(U) = 1. \tag{3}$$

Since $\mathrm{Prim}(A)$ is almost Hausdorff, V_0 has a non-void relatively open Hausdorff subset R. We claim that $R = \{I\}$ for some I in $\mathrm{Prim}(A)$. Indeed: $R \neq \emptyset$ by definition. Assume that I and J are distinct points of R. Since R is Hausdorff, we can put I and J inside disjoint relatively open subsets U and U' of V_0. It then follows by (3) that $P(U) = P(U') = 1$, contradicting the disjointness of U and U'. So there cannot be two distinct points of R, and the claim is proved.

It follows from (3) that $P(R) = 1$, that is (by the above claim), $P(\{I\}) = 1$. This proves the last condition of 10.4. ∎

10.7. Combining 10.2, 10.5, and 10.6, we obtain:

Corollary. *If either A is separable or $\mathrm{Prim}(A)$ is almost Hausdorff, the kernel of every primary *-representation of A is primitive.*

10.8. *Remark.* If A is not separable and $\mathrm{Prim}(A)$ is not almost Hausdorff, it is not known whether the conclusion of Corollary 10.7 holds.

In this connection, notice that there exist locally compact non-Hausdorff spaces for which the last condition of 10.4 fails. Here is an example: Let X be the (uncountable) space of all countable ordinals, topologized so that, if $0 \neq \alpha \in X$, a basis of X-neighborhoods of α is formed by the collection of all

$$\{\beta \in X : \gamma < \beta \leq \alpha\} \cup \{\beta \in X : \delta < \beta\},$$

where $\gamma, \delta \in X$ and $\gamma < \alpha$. If $\alpha = 0$, a basis of neighborhoods of α is formed by the collection of all $\{0\} \cup \{\beta \in X : \delta < \beta\}$, where $\delta \in X$. One verifies that X is T_1, though not Hausdorff (in fact, any two non-void open sets intersect). Further, X is locally compact, and every closed subset of X has the Baire property. Since every countable subset of X has an upper bound in X, one can show easily that, for every Borel subset W of X, either (i) $W \subset \{\beta \in X : \beta \leq \alpha\}$ for some α in X, or (ii) $\{\beta \in X : \beta > \alpha\} \subset W$ for some α in X. If we set $P(W) = 0$ and $P(W) = 1$ for those Borel sets satisfying alternatives (i) and (ii) respectively, P becomes a Borel measure on X which is regular (in the sense of 9.18(12)) and has range $\{0, 1\}$, but is not concentrated at any point.

10.9. *Remark.* A non-degenerate *-representation T of A may be concentrated at a point of $\mathrm{Prim}(A)$ without being primary. For example, let A be one of the Glimm algebras defined in §VI.17. Since A is topologically simple,

Prim(A) consists of just one point (namely $\{0\}$); and so every non-degenerate *-representation of A is trivially concentrated at a point of Prim(A). On the other hand A certainly has non-primary *-representations. Indeed, let S and T be inequivalent irreducible *-representations of A; such exist by VI.17.5. By VI.24.12, their direct sum $S \oplus T$ is not primary.

10.10.   In spite of Remark 10.9, there is one important situation in which the property of being concentrated at a point of Prim(A) does imply primariness.

Proposition.   *Let S be an irreducible *-representation of A such that S_a is a non-zero compact operator for some a in A. Then a non-degenerate *-representation T of A (with $X(T) \neq \{0\}$) is concentrated at $\mathrm{Ker}(S)$ if and only if T is unitarily equivalent to a Hilbert direct sum of copies of S.*

Proof.   The "if" part was observed in 10.3.

Assume that T is concentrated at $\mathrm{Ker}(S)$. By 10.2, every non-zero subrepresentation of T has kernel $\mathrm{Ker}(S)$. Let $I = \{a \in A : S_a \text{ is compact}\}$. Then by hypothesis $I \supsetneqq \mathrm{Ker}(S)$. Combining the last two facts we conclude that $T|I$ is non-degenerate. In fact, since $\mathrm{Ker}(T) = \mathrm{Ker}(S)$, the equation

$$V_{a + \mathrm{Ker}(S)} = T_a \qquad\qquad (a \in I)$$

defines a non-degenerate *-representation V of $I/\mathrm{Ker}(S)$ acting on $X(T)$. Now $I/\mathrm{Ker}(S)$ is elementary by VI.15.17. Hence by VI.15.12 V is a Hilbert direct sum of copies of the unique irreducible *-representation $a + \mathrm{Ker}(S) \mapsto S_a$ of $I/\mathrm{Ker}(S)$. Thus $T|I$ is a Hilbert direct sum of copies of the irreducible *-representation $S|I$. Since any *-representation of A is uniquely determined by its restriction to the two-sided ideal I provided this restriction is non-degenerate, it follows from the last sentence that T is equivalent to a direct sum of copies of S.   ∎

10.11.   Definition.   Let us say that A has a *smooth *-representation theory* if, for every T in $\hat{A}$, there is an element a of A such that T_a is a non-zero compact operator.

By 10.10 (or VI.15.18), a smooth *-representation theory implies that each element of $\hat{A}$ is determined to within unitary equivalence by its kernel, and hence that $\hat{A}$ and Prim(A) essentially coincide as topological spaces.

Theorem.   *Suppose that A has a smooth *-representation theory; and assume in addition that either A is separable or $\hat{A}$ is almost Hausdorff. Then A has a Type I *-representation theory (see VI.24.13).*

Proof. If T is a primary *-representation of A, then by 10.5 or 10.6 it is concentrated at a point of $\mathrm{Prim}(A)$; and hence by 10.10 it is primary of Type I. ∎

Remark. This theorem is a substantial generalization of VI.24.14 and VI.24.15. It shows, for instance, that if $\mathscr{B}$ is a bundle of elementary C^*-algebras (see 8.14) then $\mathscr{C}_0(\mathscr{B})$ has a Type I *-representation theory.

As a matter of fact, profound researches of Glimm and Sakai have shown that the following statement, much stronger than the above theorem, holds: An arbitrary C^*-algebra A has a smooth *-representation theory if and only if it has a Type I *-representation theory. See for example Sakai [1], Theorem 4.6.4.

11. Exercises for Chapter VII

1. Verify that the collection $\mathscr{U}(T)$ satisfies properties (I), (II), and (III) as stated in 1.3.

2. This exercise is concerned with checking a few details in the proof of Proposition 1.4.

   (a) Show, as stated, that the Proposition remains true if Z is a dense subset X' of $X(T)$.

   (b) Check that $U(T; \delta; \eta_1, \ldots, \eta_q; F) \subset U(T; \varepsilon; \xi_1, \ldots, \xi_p; F)$.

   (c) Show that if η_1, η_2 are any two vectors in X'' and $a \in A$, then (η_1, η_2) and $(T_a \eta_1, \eta_2)$ are each of the form either (ξ_1, ξ_2) or $(T_b \xi_1, \xi_2)$, where $\xi_1, \xi_2 \in Z$ and $b \in A$.

3. Prove Corollaries 1.6 and 1.7.

4. Verify the statements in 1.8 reformulating the definition of the regional topology in terms of convergence of nets.

5. Prove the observation made in 1.9.

6. Show that the relativized regional topology of $\mathscr{T}_1$ (the family of all *-homomorphisms of A into $\mathbb{C}$) coincides with the topology of pointwise convergence on A (1.13).

7. Prove 1.12, 1.13, and 1.14 directly from Definition 1.3; that is, do not use 1.10.

8. Fill in the details of the proof sketched of Proposition 1.16.

9. Verify in detail that weak equivalence of two subfamilies $\mathscr{S}_1$ and $\mathscr{S}_2$ of $\mathscr{T}(A)$ is an equivalence relation in the ordinary sense (see 1.24).

10. Let T be a *-representation of a *-algebra in a non-zero separable Hilbert space, and let α and β be distinct infinite cardinal numbers. Prove that $\alpha \cdot T$ and $\beta \cdot T$ are regionally equivalent (2.2) but are not unitarily equivalent.

11. Give the details of the proof of Proposition 2.5.

12. (Partitions of unity) Let S be a normal Hausdorff space and $U_1, \ldots, U_n$ open subsets of S which cover S. Show that there exist continuous functions $h_1, \ldots, h_n$ on S with values in $[0, 1]$ such that $\sum_{k=1}^{n} h_k = 1$ on S and h_k vanishes outside of U_k $(k = 1, \ldots, n)$.

13. Let A be a *-algebra. Show that:

(a) Every primitive ideal (see 3.8) of A is proper.

(b) If I is a primitive ideal of A, and I_1, I_2 are *-ideals of A such that $I_1 I_2 \subset I$, then either $I_1 \subset I$ or $I_2 \subset I$.

(c) If A is a Banach *-algebra, then every primitive ideal of A is norm-closed.

14. Show in detail that the irreducible *-representation S described in Remark 4.6 cannot be extended to a *-representation of the algebra A acting on the given Hilbert space X.

15. Prove Proposition 4.8 from the reference given there.

16. Show that the set J defined in 5.15 is the largest closed two-sided ideal of the C^*-algebra A which satisfies $JI = \{0\}$ or $J \cap I = \{0\}$.

17. Prove that if S is any family of non-degenerate *-representations of a C^*-algebra A, then there is a unique closed subset W of $\hat{A}$ which is weakly equivalent to S (see 5.16).

18. Recall that a topological space T is a *Baire space* if, whenever $\{U_n\}$ is a sequence of dense open subsets of T, the intersection $\bigcap_{n=1}^{\infty} U_n$ is also dense in T (6.3). Show that:

(a) Every non-void open subset of a Baire space is also a Baire space.

(b) T is a Baire space if and only if each non-void open set in T is of second category in T.

(c) A dense G_δ-set in a Baire space is a Baire space.

19. Use the fact that the Hilbert space Y is infinite-dimensional and VI.15.4 to show that $A' \cap \mathcal{O}_c(Z) = \{0\}$ in 7.2.

20. Prove Proposition 7.5.

21. Fill in the details of the sketch of the proof of Proposition 7.6.

22. Complete the proof sketched of Proposition 7.7.

23. Prove the equivalences described in Remark 8.2.

24. Verify that the cross-sectional algebra $\mathscr{C}_0(\mathscr{B})$ of a bundle $\mathscr{B}$ of C^*-algebras (see 8.4) is a C^*-algebra.

25. Show that the identifications described in 8.6 do, in fact, hold.

26. Prove Proposition 8.15.

27. Let X be a locally compact Hausdorff space and u a point of X. Let A be a C^*-algebra, D a C^*-subalgebra of A, $\mathscr{E}$ the trivial bundle of C^*-algebras over X with fiber A, and $\mathscr{B}$ the subbundle of $\mathscr{E}$ with fibers $B_u = D$ and $B_x = A$ for $x \neq u$ (see Exercise 41, Chapter II).

By VII.8.9 the structure space of $\mathscr{C}_0(\mathscr{B})$ can be identified as a set with $Q = \{\langle x, S \rangle : x \in X, S \in \hat{A} \text{ if } x \neq u, S \in \hat{D} \text{ if } x = u\}$. Show that the regional topology of Q is described in terms of convergence of nets as follows:

A net $\{\langle x_i, S^i \rangle\}$ of elements of Q converges regionally to $\langle x, S \rangle \in Q$ if and only if the following three conditions hold:

(i) $x_i \to x$ in X;

(ii)   If $x \neq u$, then $S^i \to S$ in $\hat{A}$;

(iii)   If $x = u$, then $(S^i)^D \to S$ in the regional topology of *-representations of D.

   [Here $(S^i)^D$ means the non-degenerate part of the restriction of S^i to D.]

28.   If $\mathcal{H}$ is a Hilbert bundle over X, we remarked without proof in VII.8.16 that there is no *natural* way to define a Banach bundle $\mathcal{B}$ over X whose fiber at x is the Banach space $\mathcal{O}(H_x)$ of *all* bounded linear operators on H_x. The word "natural" here would imply in particular that any automorphism of $\mathcal{H}$ should give rise to a corresponding automorphism of $\mathcal{B}$. The difficulty in defining $\mathcal{B}$ is suggested by the following problem:

   Let H_0 be an infinite-dimensional Hilbert space; and let $\mathcal{H}$ be the trivial Hilbert bundle over $[0, 1]$ with fiber H_0. By an *automorphism of $\mathcal{H}$* we shall mean a function $t \mapsto u_t$ mapping $[0, 1]$ into the set of all unitary operators on H_0, such that the map

$$\langle t, \xi \rangle \mapsto \langle t, u_t(\xi) \rangle \qquad\qquad (t \in [0, 1], \xi \in H_0)$$

is a homeomorphism of $\mathcal{H}$ onto itself.

   Prove that it is impossible to construct a Banach bundle $\mathcal{B}$ over $[0, 1]$ whose fiber at each point t of $[0, 1]$ is $\mathcal{O}(H_0)$, and which satisfies the two conditions:

(i)   For each $a \in \mathcal{O}(H_0)$, the constant cross-section $t \mapsto a$ of $\mathcal{B}$ is continuous.

(ii)   For each automorphism $t \mapsto u_t$ of $\mathcal{H}$, the corresponding bijection

$$\langle t, a \rangle \mapsto \langle t, u_t a u_t^{-1} \rangle \qquad\qquad (t \in [0, 1], a \in \mathcal{O}(H_0))$$

of $\mathcal{B}$ onto itself is a homeomorphism.

29.   Show that (14) in 8.17 implies that $\lim_{x \to u} \| a_{\phi, \psi}(x) - a_{\phi', \psi'}(x) \| = 0$. Then show that (12) and (13) imply that the map $x \mapsto \| \alpha(x) \|$ is continuous at u.

30.   Verify the details of the proof of Proposition 8.18.

31.   Prove Proposition 9.16.

32.   Prove Proposition 9.20.

33.   Suppose that $\mathcal{B} = \langle B, \{B_x\} \rangle$ is a bundle of C^*-algebras over a locally compact Hausdorff space X. Let u and v be two different points of X; let D be a C^*-algebra; let $\phi : B_u \to D$ and $\psi : B_v \to D$ be surjective *-homomorphisms; and define A to be the C^*-subalgebra $\{ f \in \mathcal{C}_0(\mathcal{B}) : \phi(f(u)) = \psi(f(v)) \}$ of $\mathcal{C}_0(\mathcal{B})$. Show that every irreducible *-representation of $\mathcal{C}_0(\mathcal{B})$ remains irreducible when restricted to A. (Thus Propositions 7.5, 8.8, and 8.9 determine the structure space of A.)

34.   In this Exercise we sketch the construction of a very interesting special bundle $\mathcal{B}$ of C^*-algebras.

   The base space X of $\mathcal{B}$ will be the compact set $\{0\} \cup \{1/n : n = 1, 2, \dots\}$ of real numbers. The fiber $B_{1/n}$ over the point $1/n$ $(n = 1, 2, \dots)$ will be the $n \times n$ total matrix *-algebra M_n; and the fiber B_0 over 0 will be the commutative C^*-algebra $\mathcal{C}(S)$ of all continuous complex functions on the square $S = [-1, 1] \times [-1, 1]$ in the plane. Let B be the disjoint union of the B_x $(x \in X)$.

For each $n = 1, 2, \ldots$, we shall define two Hermitian matrices $a^{(n)}$ and $b^{(n)}$ in M_n as follows:

$$(a^{(n)})_{rs} = \delta_{rs}\cos\left(\frac{r\pi}{n+1}\right) \qquad (\delta_{rs} \text{ being the Kronecker delta}),$$

$$(b^{(n)})_{rs} = \frac{1}{2}(\delta_{r,s+1} + \delta_{r+1,s})$$

$(r, s = 1, \ldots, n)$. Prove that $b^{(n)}$ has the same eigenvalues $\cos(r\pi/(n+1))$ $(r = 1, \ldots, n)$ as $a^{(n)}$ does. (The eigenvector v for $b^{(n)}$ with eigenvalue $\cos(r\pi/(n+1))$ is given by $v_s = \sin(rs\pi/(n+1))$.) Thus $a^{(n)}$ and $b^{(n)}$ are unitarily equivalent.

Furthermore let P be the algebra of all polynomials (with constant terms and complex coefficients) in two non-commuting indeterminates u, v; let Q be the algebra of all polynomials (with constant terms and complex coefficients) in two commuting indeterminates α, β; and let $\phi: P \to Q$ be the natural surjective homomorphism given by $\phi(1) = 1$, $\phi(u) = \alpha$, $\phi(v) = \beta$.

Prove the following:

(i) $\lim_{n \to \infty} \|a^{(n)}b^{(n)} - b^{(n)}a^{(n)}\| = 0$. (Thus $a^{(n)}$ and $b^{(n)}$ "nearly commute" when n is very large.)

(ii) If $p \in P$, then $\lim_{n \to \infty} \|p(a^{(n)}, b^{(n)})\| = \sup\{|\phi(p)(x, y)| : (x, y) \in S\}$.

(iii) There is a unique topology for B making $\mathcal{B} = \langle B, \{B_x\}\rangle$ a bundle of C^*-algebras such that, for each $p \in P$, the following cross-section f_p of $\mathcal{B}$ is continuous:

$$f_p\left(\frac{1}{n}\right) = p(a^{(n)}, b^{(n)}) \qquad\qquad (n = 1, 2, \ldots),$$

$$f_p(0)(x, y) = \phi(p)(x, y) \qquad\qquad ((x, y) \in S).$$

The bundle $\mathcal{B}$ of C^*-algebras is called the *Lee bundle* (see R. Y. Lee [2]). Its interest arises from its relevance to the theory of "split extensions." See D. Voiculescu [2, 3].

Notes and Remarks

The idea of the structure space (also called the dual space or spectrum) of an arbitrary associative algebra was introduced in 1945 by Jacobson [3], who established the basic properties of the hull-kernel topology (also called the Jacobson topology) on the set of primitive ideals. Somewhat earlier (1937) Stone [4] considered a special case of this idea in his investigations of Boolean rings. Essentially the same definition of (hull-kernel) closure in a set of ideals was given in 1941 by Segal [1] in his initial study of the group algebra of a locally compact group. Of course, in the commutative case,

Gelfand [2, 3] had already demonstrated the fundamental importance of the maximal ideal space in the study of commutative Banach algebras.

Godement [4] in 1948, introduced and studied a topology on the irreducible unitary representations of a locally compact group. Kaplansky [6] followed soon after (1951) with a penetrating analysis of the structure spaces of those C^*-algebras whose irreducible *-representations have their range in the compact operators (the so-called CCR-algebras or liminal C^*-algebras). The importance of this study is emphasized by the fact that the group algebras of connected semi-simple Lie groups which have faithful matrix representations belong to this class (see Godement [9]).

The period from 1952 to 1958 saw the appearance of Mackey's fundamental papers [5, 6, 7, 8] which were to have a profound effect on both the theory of group representations and operator algebras.

The present chapter is, to a large extent a rewriting and extension of results in Fell [1, 2, 9] and Dixmier [10, 14]. The regional topology was introduced by Fell [2] and therefore is often referred to in the literature as the Fell topology. Many of the results in the papers of Fell were inspired directly or indirectly by the earlier work of Kaplansky [6] and Mackey [7]. To a lesser extent, the papers by Fell [3, 10] also play a significant role in this chapter.

The monographs of Dixmier [24, pp. 69–93], Pedersen [5, pp. 91–109], and Rickart [2, p. 76 and p. 248] all contain treatments of the structure space (of the three, Dixmier's treatment is the most complete and closest in spirit to what we have given).

Without attempting to be complete, we mention the origin of a few of the main ideas and results of the chapter. The definition and basic properties of the regional topology, as well as the equivalent forms of its definition, are contained in Fell [1, 2]. The notion of weak containment is introduced and studied in Fell [2], where the Equivalence Theorem (5.5) is established. It is also shown in this paper that the hull-kernel and regional topologies coincide for C^*-algebras.

Hereditary subalgebras of C^*-algebras are treated in Pedersen [5, pp. 15–17, 91–96]. This terminology appears to have been introduced by Pedersen [2, p. 63]. Hereditary subalgebras have also been called "order-related" and "facial" in the literature. The term "block subalgebra" was used by Fell [9, pp. 235–237].

The fact that $\hat{A}$ and Prim(A) are Baire spaces (6.5) was proved by Dixmier [10]; this was established earlier for CCR-algebras by Kaplansky [6, Theorem 5.1]. The local compactness of $\hat{A}$ and Prim(A) (see 6.9) was established by Fell [3, p. 244].

C^*-algebras with finite structure spaces have been considered by Behncke and Leptin [1]; however, their point of view is quite different from that taken in §7.

It is a truism among workers in the field of C^*-algebras (see for example the review articles of Mackey [10, 18, 22]) that if a C^*-algebra A has a non-Type I *-representation theory its irreducible *-representations cannot be satisfactorily classified; the structure space $\hat{A}$ is then too intractable. The question then arises: If the *-representation theory of A is not of Type I, are there any other spaces different from but related to $\hat{A}$ which might prove more tractable? Two candidates have been proposed: The first is $\mathrm{Prim}(A)$, which as we have seen, is always at least T_0, and is often a "well-behaved" space even if $\hat{A}$ is not; to classify the points of $\mathrm{Prim}(A)$ is to classify the *kernels* of irreducible *-representations of A. The second is the space A_{nor} of all so-called quasi-equivalence classes of traceable factor representations of A, introduced by H. Halpern and shown by him to have at least a "well-behaved" Borel structure (see Halpern [3]). Pukanszky [4] proved the beautiful result that, if G is a connected locally compact group, $(C^*(G))_{\mathrm{nor}}$ and $\mathrm{Prim}(C^*(G))$ are essentially the same space.

The important results in §9 on the spectral measure of a *-representation originated with Glimm [5]. The "direct integral decomposition" corresponding to Glimm's spectral measure in the separable context was studied by Effros [1].

Further results on the structure spaces of groups and algebras may be found in the following: Auslander and Moore [1], Baggett [2, 3, 4, 5], Baggett and Sund [1], Bernat and Dixmier [1], Bichteler [2, 3], Blattner [6], Boidol [1, 4, 5], Boidol, Leptin, Schürman, and Vahle [1], Bratteli [2], Bratteli and Elliott [1], Britton [1], I. D. Brown [1, 2], Busby [1], Carey and Moran [1, 2], Dague [1], Dauns [1], Dauns and Hofmann [1, 2], Dixmier [12, 13] Effros [1, 2], Effros and Hahn [1], Ernest [2, 3, 11, 12], Gaal [1, pp. 21–26], Gardner [1, 2], Gootman [1, 2, 4], Gootman and Rosenberg [1], Green [1, 2, 3, 4, 5], Halpern [1, 3], Liukkonen [1], Liukkonen and Mosak [1], Mackey [15], Moore and Rosenberg [2], Peters [1], Picardello [1], Poguntke [4], Pukanszky [1, 3], Sauvageot [1, 2], Schochetman [1], Štern [1, 2], G. Warner [1, 2], Williams [1, 2, 3], and Wulfsohn [1, 2, 3].

Appendix A

The Stone–Weierstrass Theorems

A1. One of the most important tools of functional analysis is the Stone–Weierstrass Theorem, which we present in this appendix.

A2. Let S be a compact Hausdorff space; and let A be a set of real-valued continuous functions on S which is closed under $\wedge$ and $\vee$. (Recall that $(f \wedge g)(s) = \min(f(s), g(s))$, $(f \vee g)(s) = \max(f(s), g(s))$.) A real-valued function h on S will be said to be *approximable by A at pairs of points* if for every $\varepsilon > 0$ and every pair s, t of points of S there is a function f in A such that $|h(s) - f(s)| < \varepsilon$ and $|h(t) - f(t)| < \varepsilon$.

Proposition. *Every continuous real-valued function on S which is approximable by A at pairs of points belongs to the uniform closure of A (i.e., the closure of A in $\mathscr{C}_r(S)$ with respect to the supremum norm).*

Proof. Let h be a function in $\mathscr{C}_r(S)$ which is approximable by A at pairs of points; and let $\varepsilon > 0$. For each pair s, t of points of S choose a function $g_{s,t}$ in A satisfying

$$|h(s) - g_{s,t}(s)| < \varepsilon, \; |h(t) - g_{s,t}(t)| < \varepsilon.$$

We now set

$$U_{s,t} = \{r \in S : g_{s,t}(r) < h(r) + \varepsilon\},$$

$$V_{s,t} = \{r \in S : g_{s,t}(r) > h(r) - \varepsilon\}.$$

So $U_{s,t}$ and $V_{s,t}$ are open sets, each containing both s and t. Now fix t. As s varies, the $U_{s,t}$ form an open covering of S. Since S is compact, there are a finite number $s_1, \ldots, s_n$ of points of S such that

$$\bigcup_{i=1}^{n} U_{s_i,t} = S.$$

Therefore, setting

$$g_t = g_{s_1,t} \wedge g_{s_2,t} \wedge \cdots \wedge g_{s_n,t},$$

we have $g_t \in A$, and

$$g_t(r) < h(r) + \varepsilon \qquad \text{for all } r \text{ in } S. \tag{1}$$

Denoting by V_t the open neighborhood $V_{s_1,t} \cap \cdots \cap V_{s_n,t}$ of t, we also have

$$g_t(r) > h(r) - \varepsilon \qquad \text{for all } r \text{ in } V_t. \tag{2}$$

Again we use the compactness of S to find a finite set $t_1, \ldots, t_m$ of points of S satisfying

$$\bigcup_{j=1}^{m} V_{t_j} = S; \tag{3}$$

and set $g = g_{t_1} \vee \cdots \vee g_{t_m}$. Since the g_{t_j} are in A, g is in A. By (1) $g(r) < h(r) + \varepsilon$ for all r in S; and by (2) and (3) $g(r) > h(r) - \varepsilon$ for all r in S. Thus

$$|h(r) - g(r)| < \varepsilon \qquad \text{for all } r \text{ in } S;$$

and we have approximated h arbitrarily closely in the supremum norm by an element of A. ∎

A3. Proposition. *Let S be any set, and A a real algebra (under the pointwise operations) of bounded real-valued functions on S which is complete with respect to the supremum norm $\| \ \|_\infty$. Then A is closed under the operations $\vee$ and $\wedge$.*

Proof. Since $f \vee g = \frac{1}{2}(f + g) + \frac{1}{2}|f - g|$ and $f \wedge g = \frac{1}{2}(f + g) - \frac{1}{2}|f - g|$, it is enough to show that

$$f \in A \Rightarrow |f| \in A. \tag{4}$$

Let $\varepsilon > 0$. If $k > 0$, the power series expansion of $(k + t)^{1/2}$ about $t = 0$ has radius of convergence k. Thus the power series expansion of $(\varepsilon^2 + t)^{1/2}$ about

$t = \frac{1}{2}$ has radius of convergence $\frac{1}{2} + \varepsilon^2$, and so converges uniformly for $0 \le t \le 1$. Therefore we can find a real polynomial p such that

$$|(\varepsilon^2 + t)^{1/2} - p(t)| < \varepsilon \qquad \text{for all} \quad 0 \le t \le 1,$$

and hence

$$|(\varepsilon^2 + t^2)^{1/2} - p(t^2)| < \varepsilon \qquad \text{for } -1 \le t \le 1. \tag{5}$$

Now for any real t,

$$|(\varepsilon^2 + t^2)^{1/2} - |t|| \le \varepsilon. \tag{6}$$

Combining (5) and (6), we have

$$|p(t^2) - |t|| < 2\varepsilon \qquad \text{for } -1 \le t \le 1. \tag{7}$$

Now (7) implies in particular that $|p(0)| < 2\varepsilon$. Thus by (7) the polynomial $q = p - p(0)$, which has zero constant term, satisfies

$$|q(t^2) - |t|| < 4\varepsilon \qquad \text{for } -1 \le t \le 1. \tag{8}$$

Now take any f in A. To prove (4) it is enough to assume that $\|f\|_\infty \le 1$. Since A is an algebra and q has no constant term, $q(f^2): s \mapsto q((f(s))^2)$ belongs to A; and by (8)

$$\|q(f^2) - |f|\|_\infty \le 4\varepsilon.$$

Thus we have approximated $|f|$ arbitrarily closely in the supremum norm by elements of A. Since A is complete, this implies that $|f| \in A$. So (4) is proved, and hence also the Proposition. ∎

A4. A linear space of bounded real or complex-valued functions on a set S is called an *algebra* if it is closed under pointwise multiplication. It is called *uniformly closed* if it is complete with respect to the supremum norm, that is, if it is closed as a subspace of the Banach algebra of all bounded complex functions on S (with the supremum norm).

A5. The following four theorems are different versions of the Stone–Weierstrass Theorem.

Theorem.   *Let S be a compact Hausdorff space, and A a uniformly closed real algebra of real-valued continuous functions on S. Put $N = \{s \in S : f(s) = 0$ for all f in $A\}$. If $s, t \in S \setminus N$, we shall write $s \sim t$ to mean that $f(s) = f(t)$ for all f in A. Then A is equal to the set of those g in $\mathscr{C}_r(S)$ such that*

(i) *$g(s) = 0$ for all s in N,*
(ii) *$g(s) = g(t)$ whenever $s, t \in S \setminus N$ and $s \sim t$.*

Proof.   By A3 A is closed under $\wedge$ and $\vee$. Let g be any function in $\mathscr{C}_r(S)$ satisfying (i) and (ii). We shall show that, given any two points s, t of S, there is an element f of A such that $f(s) = g(s)$, $f(t) = g(t)$. By A2 and the completeness of A, this will prove that $g \in A$.

If either s or t is in N, the existence of such an f is trivial. Assume then that $s \notin N$, $t \notin N$. If $s \sim t$, choose a function f in A not vanishing at s; and multiply f by a suitable constant so that $f(s) = g(s)$; since $s \sim t$ this gives $f(t) = f(s) = g(s) = g(t)$. Finally, assume that $s \notin N$, $t \notin N$, $s \not\sim t$. Then we can find an element h of A such that $0 \neq h(s) \neq h(t) \neq 0$. The equations

$$ah(s) + b(h(s))^2 = g(s),$$

$$ah(t) + b(h(t))^2 = g(t)$$

can then be solved for the real numbers a and b, showing that $f = ah + bh^2$ is an element of A coinciding with g at s and t.

Thus the required f always exists. As we have seen, this implies that $g \in A$.

We have proved that any g in $\mathscr{C}_r(S)$ which satisfies (i) and (ii) beongs to A. The converse is obvious. So the proof of the Theorem is complete.   ■

A6.   Theorem.   *Let S be a locally compact Hausdorff space, and A a uniformly closed subalgebra of $\mathscr{C}_{0r}(S)$. We assume that: (I) A does not vanish anywhere on S (i.e., for each s in S we have $f(s) \neq 0$ for some f in A); (II) A separates points of S (i.e., if s, $t \in S$ and $s \neq t$, we have $f(s) \neq f(t)$ for some f in A). Then $A = \mathscr{C}_{0r}(S)$.*

Proof.   If S is compact this is an immediate special case of A5. If S is not compact, we apply A5 to A considered as an algebra of functions on the one-point compactification of S.   ■

A7.   For algebras of complex-valued functions Theorem A5 takes the following form:

Theorem.   *Let S be a compact Hausdorff space, and A a uniformly closed complex algebra of complex-valued continuous functions on S, satisfying the further condition that $f \in A \Rightarrow \bar{f} \in A$ ($\bar{f}$ being the complex-conjugate of f). Let N and $\sim$ be defined as in Theorem A5. Then A coincides with the set of all g in $\mathscr{C}(S)$ which satisfy (i) and (ii) of A5.*

Proof.   Let A_r stand for the real algebra of all real-valued functions in A. The condition $f \in A \Rightarrow \bar{f} \in A$ implies that a function f in $\mathscr{C}(S)$ belongs to A if and only if its real and imaginary parts belong to A_r. Thus, on applying A5 to the real algebra A_r we obtain the present Theorem.   ■

A8.   As an immediate corollary of A7 we obtain the complex version of A6:

Theorem.   *Let S be a locally compact Hausdorff space, and A a uniformly closed subalgebra of $\mathscr{C}_0(S)$ which is closed under complex conjugation, separates points of S, and does not vanish anywhere on S (see A6). Then $A = \mathscr{C}_0(S)$.*

A9.   Remark.   Theorems A7 and A8 fail if we omit the hypothesis that A is closed under complex conjugation. As an example, take S to be the disk $D = \{z \in \mathbb{C} : |z| \le 1\}$ in the complex plane, and A to be the subalgebra of $\mathscr{C}(D)$ consisting of those f which are analytic in the interior of D. (See V.4.8.)

However, in the special case that S is *finite*, Theorems A7 and A8 hold even without the assumption that A is closed under complex conjugation. See the argument of V.8.5.

A10.   Here is a simple application of A7. Take $S = [0, 1]$; for each n in $\mathbb{Z}$, let $e_n(t) = e^{2\pi i n t}$ $(0 \le t \le 1)$; and let A be the closure of the linear span of $\{e_n : n \in \mathbb{Z}\}$ with respect to the supremum norm. Since $\{e_n : n \in \mathbb{Z}\}$ is closed under multiplication and complex conjugation, the same is true of A. Evidently A vanishes nowhere on $[0, 1]$, and separates every pair of points of $[0, 1]$ except the pair 0, 1. Therefore, by A7,

$$A = \{f \in \mathscr{C}([0, 1]) : f(0) = f(1)\}. \tag{8}$$

A11.   Evidently the right side of (8) is dense in the Hilbert space $\mathscr{L}_2(\lambda)$, where λ stands for Lebesgue measure on $[0, 1]$. Therefore A10 implies that $\{e_n : n \in \mathbb{Z}\}$ is an orthonormal basis of $\mathscr{L}_2(\lambda)$.

Appendix B

Unbounded Operators in Hilbert Space

B1. We shall give a rather rapid introduction to the theory of unbounded linear transformations of Hilbert space, to serve as background for the Spectral Theorem for unbounded normal operators (Chapter VI, §12). Parallel treatments (though leading only to the spectral theory for self-adjoint operators) will be found in Riesz and Sz.-Nagy [1], Chapter VIII, and in Dunford and Schwartz [2], Chapter XII.

B2. We fix two Hilbert spaces X and Y.

A *transformation from X to Y* is a linear function $T: X' \to Y'$, where X' and Y' are linear subspaces of X and Y respectively. A transformation from X to X is called an *operator in X*.

We shall identify a transformation from X to Y with its graph, which is a linear subspace of the direct sum Hilbert space $X \oplus Y$. A linear subspace S of $X \oplus Y$ is thus a transformation if and only if $\eta \in Y$, $\langle 0, \eta \rangle \in S \Rightarrow \eta = 0$.

B3.   For any subset S of $X \oplus Y$, we define:

(i)   domain$(S) = \{\xi \in X : \langle \xi, \eta \rangle \in S$ for some $\eta\}$,

(ii)   range$(S) = \{\eta \in Y : \langle \xi, \eta \rangle \in S$ for some $\xi\}$,

(iii) $S^{\perp}$ = the orthogonal complement of S in $X \oplus Y$,

(iv) $S^{\sim} = \{\langle \eta, \xi \rangle : \langle -\xi, \eta \rangle \in S\}$,

(v) $S^* = (S^{\sim})^{\perp} = (S^{\perp})^{\sim}$ (the *adjoint of* S),

(vi) $S^{-1} = \{\langle \eta, \xi \rangle : \langle \xi, \eta \rangle \in S\}$ (the *inverse of* S).

Notice that $S^{\sim}$, S^*, and S^{-1} are subsets of $Y \oplus X$. We have $\langle \eta', \xi' \rangle \in S^*$ if and only if

$$(\xi, \xi') = (\eta, \eta') \qquad \text{for all } \langle \xi, \eta \rangle \text{ in } S.$$

If S is linear, then

$$(S^{-1})^* = (S^*)^{-1},$$

$$S^{**} = \text{the closure of } S \text{ in } X \oplus Y.$$

If S is linear, $S^{\sim}$, S^*, and S^{-1} are always linear subspaces of $Y \oplus X$, but in general they will not be transformations.

B4.   A transformation T from X to Y is said to be *closed* if it is closed as a subset of $X \oplus Y$. It is *densely defined* if domain(T) is dense in X. If the closure of T in $X \oplus Y$ is a transformation, we say that T *admits a closed extension*. T is *bounded* if it is in $\mathcal{O}(X, Y)$, that is, if it is continuous (i.e., $\{\|T\xi\| : \xi \in \text{do-}$ main$(T), \|\xi\| = 1\}$ is bounded) and domain$(T) = X$.

Definition.   We write $\mathcal{U}(X, Y)$ for the family of all closed densely defined transformations from X to Y, abbreviating $\mathcal{U}(X, X)$ to $\mathcal{U}(X)$ if $Y = X$.

Evidently $\mathcal{O}(X, Y) \subset \mathcal{U}(X, Y)$, $\mathcal{O}(X) \subset \mathcal{U}(X)$. It is easy to check that a continuous element of $\mathcal{U}(X, Y)$ is automatically in $\mathcal{O}(X, Y)$.

B5.   Proposition.   *If $T \in \mathcal{U}(X, Y)$ and* domain$(T) = X$, *then* $T \in \mathcal{O}(X, Y)$.
This is just the Closed Graph Theorem.

B6.   Proposition.   *Let T be a transformation from X to Y. Then: (A) T^* is a transformation (from Y to X) if and only if T is densely defined. (B)* domain(T^*) *is dense in Y if and only if T has a closed extension.*

Proof.

(A) $\xi \perp$ domain$(T) \Leftrightarrow \langle \xi, 0 \rangle \perp \langle \xi', \eta' \rangle$ for all $\langle \xi', \eta' \rangle$ in $T \Leftrightarrow \langle \xi, 0 \rangle \perp T \Leftrightarrow \langle 0, \xi \rangle \in T^*$. Thus domain$(T)$ is dense if and only if $\langle 0, \xi \rangle \in T^* \Rightarrow \xi = 0$.

(B) $\eta \perp \text{domain}(T^*) \Leftrightarrow \langle \eta, 0 \rangle \perp T^* \Leftrightarrow \langle \eta, 0 \rangle \in T^{\sim \perp \perp} = (T^\sim)^- = \bar{T}^\sim \Leftrightarrow$
  $\langle 0, \eta \rangle \in \bar{T}$. Thus $\text{domain}(T^*)$ is dense if and only if $\langle 0, \eta \rangle \in \bar{T} \Rightarrow \eta = 0$,
  that is, $\bar{T}$ is a transformation. ∎

B7. Corollary. *If* $T \in \mathcal{U}(X, Y)$, *then* $T^* \in \mathcal{U}(Y, X)$, *and* $T^{**} = T$.
  As this corollary suggests, the most important (unbounded) transforma-
tions are those which belong to $\mathcal{U}(X, Y)$.

Addition and Composition

B8. Definitions. Let S and T be transformations from X to Y, and λ a
complex number. The transformations λS and $S + T$ from X to Y are defined
in the natural way:

$$\text{domain}(\lambda S) = \text{domain}(S),$$

$$(\lambda S)(\xi) = \lambda S(\xi) \qquad\qquad (\xi \in \text{domain}(S)),$$

$$\text{domain}(S + T) = \text{domain}(S) \cap \text{domain}(T),$$

$$(S + T)(\xi) = S\xi + T\xi \qquad (\xi \in \text{domain}(S + T)).$$

Let Z be another Hilbert space, and V a transformation from Y to Z. The
composition $V \circ S$ is defined as usual:

$$\text{domain}(V \circ S) = \{\xi \in X : \xi \in \text{domain}(S), S\xi \in \text{domain}(V)\},$$

$$(V \circ S)(\xi) = V(S\xi) \qquad\qquad (\xi \in \text{domain}(V \circ S)).$$

Thus $V \circ S$ is a transformation from X to Z. Often the composition symbol $\circ$
will be omitted.

B9. Proposition. *Let* T, T_1, T_2, T_3 *be transformations and* λ, μ *complex
numbers.* (*The domains and ranges of these transformations lie in whatever
Hilbert spaces are appropriate to each relationship below.*) *Then:*

(i) $T_1 + T_2 = T_2 + T_1$;

(ii) $T_1 + (T_2 + T_3) = (T_1 + T_2) + T_3$;

(iii) $T_1(T_2 T_3) = (T_1 T_2)T_3$;

(iv) $(T_1 + T_2)T_3 = T_1 T_3 + T_2 T_3$;

(v) $T_1(T_2 + T_3) \supset T_1 T_2 + T_1 T_3$ (*equality holding if* $\text{domain}(T_1)$ *is all of
its Hilbert space*);

(vi) *if* T_1^{-1} *and* T_2^{-1} *are transformations, so is* $(T_1 T_2)^{-1}$, *and* $(T_1 T_2)^{-1} = T_2^{-1} T_1^{-1}$;

(vii) *if* T_1^* *and* T_2^* *are transformations, then* $(T_1 + T_2)^* \supset T_1^* + T_2^*$
(*equality holding if either* T_1 *or* T_2 *is bounded*);

(viii) $(T_1 T_2)^* \supset T_2^* T_1^*$ *provided* T_1^* *and* T_2^* *are transformations (equality holding if* T_1 *is bounded);*

(ix) $T_1 \subset T_2 \Rightarrow T_2^* \subset T_1^*$;

(x) $\lambda(T_1 + T_2) = \lambda T_1 + \lambda T_2$;

(xi) $(\lambda + \mu)T = \lambda T + \mu T$;

(xii) $(\lambda T)^* = \bar{\lambda} T^*$ *if* T^* *is a transformation and* $\lambda \neq 0$.

The proof of these statements is left to the reader.

Note: A statement such as $T_1 \subset T_2$ means of course that the transformation T_2 is an extension of T_1.

B10.   Proposition. *Let* T *be* *in* $\mathcal{U}(X, Y)$; *and* *put* $N_T = \{\xi \in \mathrm{domain}(T): T\xi = 0\}$, *and* $R_T = $ *the closure of range* (T). *Then (see* B7)

$$X = N_T \oplus R_{T^*}, \tag{1}$$

$$Y = N_{T^*} \oplus R_T. \tag{2}$$

Proof. If $\xi \in N_T$ and $\eta \in \mathrm{domain}(T^*)$, then $0 = (T\xi, \eta) = (\xi, T^*\eta)$. So $N_T \perp R_{T^*}$. On the other hand, let $\xi \perp R_{T^*}$; then $\eta \in \mathrm{domain}(T^*) \Rightarrow (T^*\eta, \xi) = 0 = (\eta, 0)$. So $\xi \in \mathrm{domain}(T^{**})$ and $T^{**}\xi = 0$. Since $T^{**} = T$ (by B7), this says that $\xi \in N_T$. Thus $(R_{T^*})^\perp \subset N_T$, giving (1). Relation (2) follows similarly. ∎

B11.   The following fundamental result shows how an (unbounded) transformation in $\mathcal{U}(X, Y)$ is determined by two bounded transformations. We denote the identity operator on X by I.

Theorem. *Let* T *be in* $\mathcal{U}(X, Y)$; *and put* $B = (I + T^*T)^{-1}$, $C = TB$. *Then* $B \in \mathcal{O}(X)$ *and* $C \in \mathcal{O}(X, Y)$; B *is a positive operator; and* $\|B\| \leq 1$, $\|C\| \leq 1$. *Furthermore, the restriction* T' *of* T *to* $\mathrm{domain}(T^*T)$ *is dense in* T *(i.e.,* T *is the closure of* T' *in* $X \oplus Y$*).*

Proof. Since $X \oplus Y = T \oplus T^\perp = T \oplus (T^*)^\sim$, to every ξ in X there is a unique ζ' in $\mathrm{domain}(T)$ and a unique η' in $\mathrm{domain}(T^*)$ such that

$$\langle \xi, 0 \rangle = \langle \zeta', T\zeta' \rangle + \langle T^*\eta', -\eta' \rangle, \tag{3}$$

that is,

$$\left. \begin{array}{l} \xi = \zeta' + T^*\eta', \\ T\zeta' = \eta', \end{array} \right\} \tag{4}$$

whence

$$\xi = (I + T^*T)\xi'. \tag{5}$$

Now define

$$B\xi = \xi', \; C\xi = \eta'.$$

Thus B and C are transformations with domain X (to X and Y respectively). From (4) and (5) we get

$$C = TB, \tag{6}$$

$$\text{range}(I + T^*T) = X. \tag{7}$$

Now the uniqueness of ξ' and η' gives

$$I + T^*T \text{ is one-to-one.} \tag{8}$$

Further, by (5)

$$(I + T^*T)B = I. \tag{9}$$

By (7) and (8), $(I + T^*T)^{-1}$ is an operator with domain X, and by (9)

$$B = (I + T^*T)^{-1} \tag{10}$$

Now the two terms on the right side of (3) are orthogonal in $X \oplus Y$; so

$$\|\xi\|^2 = \|\xi'\|^2 + 2\|\eta'\|^2 + \|T^*T\xi'\|^2,$$

whence in particular

$$\|\xi'\| \le \|\xi\|, \; \|\eta'\| \le 2^{-1/2}\|\xi\|.$$

From this we obtain

$$\|B\| \le 1, \; \|C\| \le 2^{-1/2}. \tag{11}$$

In particular B and C are bounded.

We shall now show that B is positive. For any ξ in X, we have by (9) and B9(iv)

$$\xi = B\xi + T^*TB\xi, \tag{12}$$

so that

$$(B\xi, \xi) = (B\xi, B\xi) + (B\xi, T^*TB\xi)$$

$$= (B\xi, B\xi) + (TB\xi, TB\xi) \ge 0. \tag{13}$$

Thus B is positive.

For future use we note that

$$(B\xi, \xi) = 0 \Rightarrow \xi = 0. \tag{14}$$

Indeed: If $(B\xi, \xi) = 0$, then $B\xi = 0$ by (13), so $\xi = 0$ by (12).

Finally, we must show that T' is dense in T. Since T is closed, it is enough to take a vector ξ in domain(T) such that $\langle \xi, T\xi \rangle \perp T'$, and show that $\xi = 0$. But by hypothesis, for all $\xi' \in$ domain(T^*T)

$$0 = (\xi, \xi') + (T\xi, T\xi') = (\xi, \xi') + (\xi, T^*T\xi')$$

$$= (\xi, (I + T^*T)\xi').$$

By (7) this implies $\xi = 0$, completing the proof. ∎

Commutation of Operators

B12.   Definition.   Let T be an operator in X and B an element of $\mathcal{O}(X)$. We say that B *commutes with* T (in symbols B comm T) if $BT \subset TB$.

If T also is bounded, then of course $BT \subset TB$ implies $BT = TB$.

We do *not* define commutation of two arbitrary (unbounded) operators.

B13.   Proposition.   *Let $B, B_1, B_2 \in \mathcal{O}(X)$ and let T, T_1, T_2 be operators in X. Then*:

(i) *If B comm T_1 and B comm T_2, then B comm($T_1 + T_2$) and B comm($T_1 T_2$).*

(ii) *If B_1 comm T and B_2 comm T, then $(B_1 + B_2)$ comm T and $(B_1 B_2)$ comm T.*

The proof follows from B9.

B14.   Proposition.   *Let $B \in \mathcal{O}(X)$, and let T be a one-to-one operator in X. Then*:

$$B \text{ comm } T \Rightarrow B \text{ comm } T^{-1}.$$

Proof.   If $\xi \in$ domain(T^{-1}), then $B\xi = BT(T^{-1}\xi) = TBT^{-1}\xi$ (since $BT \subset TB$); so $B\xi \in$ range(T) = domain(T^{-1}), and $T^{-1}B\xi = BT^{-1}\xi$. Thus $BT^{-1} \subset T^{-1}B$. ∎

B15.   Proposition.   *Let $B \in \mathcal{O}(X)$, and let T be an operator in X such that T^* is an operator. Then*

$$B \text{ comm } T \Rightarrow B^* \text{ comm } T^*.$$

Proof.   If $\xi \in \text{domain}(T^*)$, then for all η in domain(T):

$$(T\eta, B^*\xi) = (BT\eta, \xi) = (TB\eta, \xi)$$

$$= (B\eta, T^*\xi) = (\eta, B^*T^*\xi);$$

so $B^*\xi \in \text{domain}(T^*)$ and $T^*B^*\xi = B^*T^*\xi$. Therefore $B^*T^* \subset T^*B^*$.   ∎

Direct Sums of Transformations

B16.   For each i in an index set I, let X_i and Y_i be Hilbert spaces and T_i a transformation from X_i to Y_i. Form the direct sum Hilbert spaces $X = \sum_{i \in I}^{\oplus} X_i$, $Y = \sum_{i \in I}^{\oplus} Y_i$.

Definition.   By the *direct sum transformation* $\sum_{i \in I}^{\oplus} T_i$ we mean the transformation T from X to Y given by: (I) domain(T) is the set of those ξ in X such that (a) $\xi_i \in \text{domain}(T_i)$ for each i, and (b) $\sum_{i \in I} \|T_i\xi_i\|^2 < \infty$; (II) if $\xi \in \text{domain}(T)$, then $(T\xi)_i = T_i\xi_i$ $(i \in I)$.

If I is finite, say $\{1, \dots, n\}$, we may write $T_1 \oplus \cdots \oplus T_n$ instead of $\sum_{i=1}^{n \oplus} T_i$.

B17.   Proposition.   *In B16, if all the T_i are closed, so is T. If all the T_i are densely defined, so is T. Thus, if $T_i \in \mathcal{U}(X_i, Y_i)$ for each i, then $T \in \mathcal{U}(X, Y)$.*

Proof.   Assume that the T_i are all closed. Let $\{\xi^\alpha\}$ be a net of elements of domain(T) such that $\xi^\alpha \to \xi$ in X and $T\xi^\alpha \to \eta$ in Y. Thus $\xi_i^\alpha \to \xi_i$ in X_i and $T_i\xi_i^\alpha \to \eta_i$ in Y_i for each i; and so, since T_i is closed, $\xi_i \in \text{domain}(T_i)$ and $T_i\xi_i = \eta_i$ for each i. Since $\eta \in Y$, $\infty > \sum_i \|\eta_i\|^2 = \sum_i \|T_i\xi_i\|^2$, whence $\xi \in \text{domain}(T)$ and $T\xi = \eta$. So T is closed.

The second and third statements of the Proposition are now evident.   ∎

B18.   Proposition.   *Suppose in B16 that T_i^* is a transformation for each i. Then T^* is a transformation; and in fact $T^* = \sum_{i \in I}^{\oplus} T_i^*$.*

Proof.   Let $T' = \sum_{i \in I}^{\oplus} T_i^*$. If $\eta \in \text{domain}(T')$, we have $\eta_i \in \text{domain}(T_i^*)$ for each i; so, for each $\xi \in \text{domain}(T)$, $(T\xi, \eta) = \sum_i(T_i\xi_i, \eta_i) = \sum_i(\xi_i, T_i^*\eta_i) = (\xi, T'\eta)$. So $T' \subset T^*$.

Conversely, let $\langle \eta, \zeta \rangle \in T^*$; that is, $(T\xi, \eta) = (\xi, \zeta)$ for all ξ in domain(T). In particular, for each i in I this implies that $(T_i\xi_i, \eta_i) = (\xi_i, \zeta_i)$ for all ξ_i in domain(T_i), whence $\eta_i \in \text{domain}(T_i^*)$, $T_i^*\eta_i = \zeta_i$. But then $\infty > \sum_i\|\zeta_i\|^2 = \sum_i\|T_i^*\eta_i\|^2$, so that $\eta \in \text{domain}(T')$, $\zeta = T'\eta$. We have shown that $T^* \subset T'$. Combined with the preceding paragraph this completes the proof.   ∎

B19. Suppose that in B16 $X_i = Y_i$ for each i, so that $X = Y$. Let P_i be the projection of X onto X_i. Then it is clear that

$$P_j \operatorname{comm} \sum_{i \in I}^{\oplus} T_i \qquad \text{for each } j \text{ in } I. \tag{15}$$

Definition. If S is an operator in a Hilbert space Z, a closed subspace W of Z is said to *reduce* S if $S = S_1 \oplus S_2$, where S_1 and S_2 are operators in W and $W^{\perp}$ respectively.

It is easy to verify that W reduces S if and only if Q comm S, where Q is the projection of Z onto W.

Thus, in the context of (15), each X_i reduces $\sum_i^{\oplus} T_i$.

Metrical Equivalence and Normality

B20. Let X, Y, Z be any fixed Hilbert spaces.

Definition. Let T_1 and T_2 be transformations from X to Y and from X to Z respectively. Then T_1 and T_2 are *metrically equivalent* if $\operatorname{domain}(T_1) = \operatorname{domain}(T_2)$ and $\|T_1 \xi\| = \|T_2 \xi\|$ for all ξ in $\operatorname{domain}(T_1)$.

Suppose that T_1 and T_2 are metrically equivalent. If T_1 has a closed extension $\bar{T}_1$, then clearly T_2 has a closed extension $\bar{T}_2$, and $\bar{T}_1$ and $\bar{T}_2$ are metrically equivalent. In particular, T_2 is closed if T_1 is.

B21. Proposition. *If $T_1 \in \mathcal{U}(X, Y)$, $T_2 \in \mathcal{U}(X, Z)$, and $T_1^* T_1 = T_2^* T_2$, then T_1 and T_2 are metrically equivalent.*

Proof. If $\xi \in \operatorname{domain}(T_1^* T_1)$ ($= \operatorname{domain}(T_2^* T_2)$), then

$$\|T_1 \xi\|^2 = (\xi, T_1^* T_1 \xi) = (\xi, T_2^* T_2 \xi) = \|T_2 \xi\|^2. \tag{16}$$

Let $T_i' = T_i | \operatorname{domain}(T_i^* T_i)$. By hypothesis T_1' and T_2' have the same domain, and so by (16) they are metrically equivalent. By B11 T_1 and T_2 are the closures of T_1' and T_2' respectively. Thus it follows from the last paragraph of B20 that T_1 and T_2 are metrically equivalent. ∎

B22. *Definition.* An operator T in X is *normal* if $T \in \mathcal{U}(X)$ and $T^* T = T T^*$.

This certainly coincides with the usual definition of normality if T is bounded.

If T is normal then so is T^*.

B23. Proposition. *An operator T in X is normal if and only if T is densely defined and T and T^* are metrically equivalent.*

Proof.   Assume that T is normal. Then T and T^* are both in $\mathcal{U}(X)$ by B7; and $T^*T = TT^* = (T^*)^*T^*$. So T and T^* are metrically equivalent by B21.

Conversely, suppose that T is densely defined, and that T and T^* are metrically equivalent. Since T^* is closed, T is also closed (see B20). Thus $T \in \mathcal{U}(X)$. From the polarization identity applied to $\|T\xi\|^2 = \|T^*\xi\|^2$, we get

$$(T\xi, T\eta) = (T^*\xi, T^*\eta) \qquad \text{for all } \xi, \eta \text{ in domain}(T).$$

If $\eta \in \text{domain}(TT^*)$, this gives

$$(T\xi, T\eta) = (\xi, TT^*\eta) \qquad \text{for all } \xi \text{ in domain}(T),$$

showing that $T\eta \in \text{domain}(T^*)$ and $T^*T\eta = TT^*\eta$. Thus $TT^* \subset T^*T$. Hence by symmetry $T^*T \subset TT^*$. So $TT^* = T^*T$, and T is normal.   ∎

B24.   Corollary.   *If S and T are normal operators in X satisfying $S \subset T$, then $S = T$.*

Proof.   Since $S \subset T$, we have $T^* \subset S^*$. So, by B23, $\text{domain}(S) \subset \text{domain}(T) = \text{domain}(T^*) \subset \text{domain}(S^*) = \text{domain}(S)$. Hence $\text{domain}(S) = \text{domain}(T)$, and $S = T$.   ∎

B25.   *Definition.*   An operator T in X is *self-adjoint* (or *Hermitian*) if $T^* = T$.

Since $T^* = T \Rightarrow T \in \mathcal{U}(X)$ by B6, a self-adjoint operator is normal.

B26.   Proposition.   *If T is a self-adjoint [normal] operator in X, then so is $T + \lambda I$ (I being the identity operator on X) for all real [complex] numbers λ.*

Proof.   The first statement holds by B9(vii).

Assume T normal. Since $(T + \lambda I)^* = T^* + \bar{\lambda} I$ by B9(vii), we have by B23 for $\xi \in \text{domain}(T)$

$$\|(T + \lambda I)^*\xi\|^2 = \|T^*\xi\|^2 + |\lambda|^2\|\xi\|^2$$
$$+ \bar{\lambda}(\xi, T^*\xi) + \lambda(T^*\xi, \xi)$$
$$= \|T\xi\|^2 + |\lambda|^2\|\xi\|^2 + \bar{\lambda}(T\xi, \xi) + \lambda(\xi, T\xi)$$
$$= \|(T + \lambda I)\xi\|^2.$$

Now apply B23 to conclude that $T + \lambda I$ is normal.   ∎

B27.   If T is self-adjoint [normal], then clearly λT is self-adjoint [normal] for any non-zero real [complex] number λ.

B28. If T is a self-adjoint [normal] operator in X, and W is a closed linear subspace of X which reduces T, then evidently $T \,|\, W$ is self-adjoint [normal] in W. As a converse to this, we have:

Proposition. *For each i in an index set I, let T_i be a normal operator in a Hilbert space X_i. Then $T = \sum_{i \in I}^{\oplus} T_i$ is normal in the direct sum Hilbert space $X = \sum_{i \in I}^{\oplus} X_i$. In fact T is the only normal operator whose restriction to X_i is T_i for each i in I. If each T_i is self-adjoint, T is self-adjoint.*

Proof. By B23 T_i and T_i^* are metrically equivalent for each i. By B18 and the definition of $\sum_i^{\oplus} T_i$ (B16), this implies that T and T^* are metrically equivalent. So by B23 T is normal.

Let T' be another normal operator in X coinciding with T_i in X_i for each i. Since T' is closed, it is clear that $T \subset T'$. Therefore $T' = T$ by B24.

Finally, if each T_i is self-adjoint, T is self-adjoint by B18. ∎

B29. Proposition. *If T is a one-to-one self-adjoint operator in X, then T^{-1} is self-adjoint, and* range (T) *is dense in X.*

Proof. By B3 $(T^{-1})^* = (T^*)^{-1} = T^{-1}$; so T^{-1} is self-adjoint, and range (T) $= \operatorname{domain}(T^{-1})$ is dense. ∎

B30. Proposition. *For any T in $\mathscr{U}(X, Y)$, T^*T is self-adjoint.*

Proof. By B11, $B = (I + T^*T)^{-1}$ is self-adjoint; so $B^{-1} = I + T^*T$ is self-adjoint by B29. Thus T^*T is self-adjoint by B26. ∎

B31. Proposition. *Suppose T is a normal operator in X, and put $B = (I + T^*T)^{-1}$, $C = TB$ as in B11. Then B comm T, B comm C, C comm T, and C is normal.*

Proof. (I) First we show that B comm T.

By B9(v) $T(I + T^*T) \supset T + TT^*T$. But both sides of this have the same domain, namely $\operatorname{domain}(TT^*T)$; so

$$T(I + T^*T) = T + TT^*T. \tag{18}$$

Further, by B11,

$$(I + T^*T)B = I, \quad B(I + T^*T) \subset I. \tag{19}$$

From (18), (19), and B9(iii), (iv),

$$BT = BT(I + T^*T)B = B(T + TT^*T)B$$

$$= B(I + TT^*)TB$$

$$= B(I + T^*T)TB \qquad \text{(by normality of } T\text{)}$$

$$\subset TB.$$

So B comm T.

(II) We have

$$BC = BTB \subset TBB \qquad \qquad \text{(by (I))}$$

$$= CB.$$

Likewise

$$CT = TBT \subset TTB = TC.$$

Therefore B comm C and C comm T.

(III) Next we show that $C^* = T^*B$.

For all ξ in X, $B\xi \in \text{domain}(T)$ (by (19)) $= \text{domain}(T^*)$ (by B23). So, for all ξ in X and η in $\text{domain}(T)$,

$$(\eta, C^*\xi) = (C\eta, \xi) = (TB\eta, \xi)$$

$$= (BT\eta, \xi) \qquad \qquad \text{(by (I))}$$

$$= (T\eta, B\xi) \qquad \qquad \text{(since } B \text{ is self-adjoint)}$$

$$= (\eta, T^*B\xi).$$

Since the η run over a dense set, this gives $C^*\xi = T^*B\xi$.

(IV) For any ξ in X,

$$\|C\xi\|^2 = (TB\xi, TB\xi)$$

$$= (B\xi, T^*TB\xi) \qquad \qquad \text{(see (19))}$$

$$= (B\xi, TT^*B\xi) \qquad \qquad \text{(since } T \text{ is normal)}$$

$$= (T^*B\xi, T^*B\xi) \qquad \qquad \text{(see (III))}$$

$$= \|C^*\xi\|^2 \qquad \qquad \text{(by (III))}.$$

So by B23 C is normal. The proof is complete. ∎

The Operator of Differentiation

B32. The most useful and important examples of unbounded operators are
the operators of differentiation. As an illustration of the preceding theory we
give a brief discussion of the simplest such operator—the operation of
differentiation of a function of one variable on a bounded interval.

As Hilbert space we take $X = \mathscr{L}_2(\lambda)$, where λ is Lebesgue measure on the
closed interval $[0, 1]$. Let $\mathscr{F}$ be the subspace of $\mathscr{C}([0, 1])$ consisting of all
those f whose first derivatives are continuous on $[0, 1]$; and let $E: \mathscr{F} \to X$ be
the operator given by

$$(Ef)(t) = i\frac{d}{dt}f(t) \qquad\qquad (f \in \mathscr{F}; 0 \le t \le 1).$$

Evidently E is densely defined in X. It is not bounded, since $\{\|Ef\|_X : f \in \mathscr{F},$
$\|f\|_X = 1\}$ is clearly unbounded. E is not closed as it stands; but it has a
closed extension D, namely the operator $i(d/dt)$ applied to those absolutely
continuous complex functions on $[0, 1]$ whose derivatives lie in X. To avoid
presupposing the theory of absolutely continuous functions, let us define D to
be the set of all pairs $\langle f, g \rangle$, where $g \in X$, $f \in \mathscr{C}([0, 1])$, and

$$f(x) - f(0) = -i\int_0^x g(t)dt \qquad\qquad (20)$$

for all x in $[0, 1]$. (Notice that a function g in X is summable over any
subinterval of $[0, 1]$, and that $\int_0^x g(t)dt$ is continuous in x.) One verifies that
the g of (20) is uniquely determined by f. Therefore D is an operator in X
which clearly extends E. In fact, we claim that D is just the closure of E (in
$X \oplus X$). We shall leave the straightforward proof of this fact as an exercise to
the reader. In particular D is a closed operator, and hence belongs to $\mathscr{U}(X)$.

It would seem that D is the natural differentiation operator to consider.
However, we shall show that D as it stands fails to be normal, though suitable
restrictions of D are actually self-adjoint.

Let us determine D^*. Since $D = \bar{E}$, we have $D^* = E^*$; and therefore D^*
consists of those pairs $\langle g, h \rangle$ in $X \times X$ such that

$$i\int_0^1 f'(t)\overline{g(t)}dt = \int_0^1 f(t)\overline{h(t)}dt \qquad \text{for all } f \text{ in } \mathscr{F}. \qquad (21)$$

Suppose $\langle g, h \rangle \in D^*$. On putting

$$q(x) = \int_0^x h(t)dt$$

(so that $q \in \mathscr{C}([0, 1])$), and integrating the right side of (21) by parts, (21) becomes

$$i \int_0^1 f'(t)\overline{g(t)}dt = f(1)\overline{q(1)} - \int_0^1 f'(t)\overline{q(t)}dt$$

(since $q(0) = 0$), or

$$\int_0^1 f'(t)\overline{(q(t) - ig(t))}dt = f(1)\overline{q(1)} \tag{22}$$

for all f in $\mathscr{F}$. Taking $f \equiv 1$ in (22), we get

$$q(1) = 0, \tag{23}$$

which combines with (22) to give

$$\int_0^1 f'(t)\overline{(q(t) - ig(t))}dt = 0 \qquad \text{for all } f \text{ in } \mathscr{F}. \tag{24}$$

Since the f' in (24) runs over the dense subset $\mathscr{C}([0, 1])$ of X, (24) implies that

$$q = ig. \tag{25}$$

Now (23) and (25) imply that

$$D^* \subset C, \tag{26}$$

where $C = \{\langle g, h\rangle \in D : g(0) = g(1) = 0\}$.

Conversely, if $\langle g, h\rangle \in C$, a reversal of the above argument shows that (21) holds. Thus $C \subset D^*$, implying by (26) that

$$D^* = C. \tag{27}$$

Thus the adjoint of D is properly contained in D. In particular, D fails to be self-adjoint, or even normal (see B23).

There exist self-adjoint operators lying between D and D^*. For example, let us define D_0 as the restriction of D to $\{f \in \text{domain}(D): f(1) = f(0)\}$. Thus D_0 is an operator satisfying $D \supset D_0 \supset D^*$. In the same way that we showed that D is the closure of E, we can show that D_0 is the closure of the restriction E_0 of E to $\mathscr{F}_0 = \{f \in \mathscr{F}: f(1) = f(0)\}$. Thus $D_0^* = E_0^*$; and D_0^* consists of all those pairs $\langle g, h\rangle$ in $X \times X$ which satisfy (21) for all f in $\mathscr{F}_0$. Putting $q(x) = \int_0^x h(t)dt$ and arguing as before, we conclude that $\langle g, h\rangle \in D_0^*$ if and only if $q(1) = 0$ and (24) holds for all f in $\mathscr{F}_0$. But the latter is equivalent to the assertion that $q - ig$ is a constant function. It follows that $D_0^* = D_0$; and D_0 is a self-adjoint operator lying between D and D^*.

D_0 is the operator $i(d/dt)$ applied to those absolutely continuous functions g whose derivatives lie in X and which satisfy $g(0) = g(1)$.

Here is another description of D_0. For $n \in \mathbb{Z}$ let $e_n(t) = e^{2\pi int}$ ($t \in [0, 1]$). We have seen in A11 that the $\{e_n\}$ ($n \in \mathbb{Z}$) form an orthonormal basis of X. Each e_n belongs to $\mathscr{F}_0$ and so to domain(D_0), and

$$D_0 e_n = -2\pi n e_n.$$

From this and B28 it follows that D_0 is the unique self-adjoint operator in X whose domain contains all the e_n and which coincides on each e_n with the operator $i(d/dt)$.

Remark. D_0 is not the only self-adjoint operator lying between D and D^*. Indeed, let α be any real number, and let U be the unitary operator on X which multiplies each function in X by $t \mapsto e^{2\pi i \alpha t}$. By B9, the operator

$$D_\alpha = U D_0 U^{-1} - 2\pi\alpha I$$

(where I is the identity operator on X) is self-adjoint. Its domain is $U(\text{domain}(D_0))$, that is, the space of all those absolutely continuous functions g on $[0, 1]$ whose derivatives lie in X and which satisfy $g(1) = e^{2\pi i\alpha}g(0)$; and it coincides on its domain with D. Clearly $D_\alpha = D_\beta$ if and only if $\alpha \equiv \beta$ mod 1. Thus the D_α (α real) form a set of infinitely many self-adjoint operators lying between D and D^*.

If we fix α and set $e_n^\alpha(t) = e^{2\pi i(n+\alpha)t}$ ($0 \le t \le 1$), then $\{e_n^\alpha : n \in \mathbb{Z}\}$ (being the image of $\{e_n : n \in \mathbb{Z}\}$ under U) is an orthonormal basis of X. Since $e_n^\alpha \in \text{domain}(D_\alpha)$, we conclude from B28 that D_α is the unique self-adjoint operator in X whose domain contains e_n^α for all n in $\mathbb{Z}$, and which coincides on each e_n^α with $i(d/dt)$.

Appendix C

The Existence of Continuous Cross-Sections of Banach Bundles

C1. In this appendix we shall prove the following remarkable result due to A. Douady and L. dal Soglio-Herault [1]: A Banach bundle over a base space X has enough continuous cross-sections whenever X is either paracompact or locally compact. The proof given here is their proof.

C2. Let $\mathscr{B} = \langle B, \pi \rangle$ be a fixed Banach bundle over the Hausdorff space X. In addition to keeping the notation of II.13.4, we will write D for $\{\langle b, c \rangle \in B \times B : \pi(b) = \pi(c)\}$. If $U \subset X$ and $\varepsilon > 0$, we define $B(U, \varepsilon) = \{b \in B : \pi(b) \in U, \|b\| < \varepsilon\}$. Postulate (iv) of II.13.4 says that, if ε runs over all positive numbers and U runs over a basis of neighborhoods of a point x of X, then the $B(U, \varepsilon)$ run over a basis of neighborhoods of 0_x in B.

C3. As a first step toward constructing continuous cross-sections of $\mathscr{B}$, we shall construct cross-sections whose discontinuities are "small". To see what this means, we make a definition.

Definition. Let ε be a positive number. A subset U of B is ε-*thin* if $\|b - b'\| < \varepsilon$ whenever $b, b' \in U$ and $\pi(b) = \pi(b')$.

C4.   Proposition.   *If $\varepsilon > 0$ and $b \in B$, then b has an ε-thin neighborhood.*

Proof.   Since $\sigma : \langle b', b'' \rangle \mapsto b' - b''$ is continuous on D to B, there is a D-neighborhood W of the pair $\langle b, b \rangle$ such that

$$\sigma(W) \subset B(X, \varepsilon). \tag{1}$$

Since D carries the relativized topology of $B \times B$, we may as well assume that

$$W = (U \times U) \cap D, \tag{2}$$

where U is a neighborhood of b in B. The combination of (1) and (2) shows that U is ε-thin.   ∎

C5.   Proposition.   *Let b be an element of B; and let $\{U_i\}$ be a decreasing net of neighborhoods of b with the following two properties:*

(i)   *$\{\pi(U_i)\}$ shrinks to $\pi(b)$ (that is, $\{\pi(U_i)\}$ is a basis of neighborhoods of $\pi(b)$);*

(ii) *U_i is ε_i-thin, where $\{\varepsilon_i\}$ is a net of positive numbers such that $\lim_i \varepsilon_i = 0$.*

Then $\{U_i\}$ is a basis of neighborhoods of b.

Proof.   Put $x = \pi(b)$, $V_i = \pi(U_i)$. Thus the $B(V_i, \varepsilon_i)$ form a basis of neighborhoods of O_x.

Now let W be any neighborhood of b. Since $\rho : \langle b', b'' \rangle \mapsto b' + b''$ is continuous on D, in particular at $\langle b, O_x \rangle$, we can find a neighborhood W' of b and an index i such that

$$\rho[(W' \times B(V_i, \varepsilon_i)) \cap D] \subset W. \tag{3}$$

Noting that $\pi(W' \cap U_i)$ is a neighborhood of x, we next choose an index $j \succ i$ such that

$$V_j \subset \pi(W' \cap U_i). \tag{4}$$

We now claim that

$$U_j \subset W. \tag{5}$$

Indeed: Since $j \succ i$ and the $\{U_k\}$ are decreasing, (4) gives

$$U_j \subset U_i \cap \pi^{-1}\pi(W' \cap U_i). \tag{6}$$

Let $c \in U_j$. By (6) $c \in U_i$ and there exists an element c' of $W' \cap U_i$ such that $\pi(c') = \pi(c)$. Since U_i is ε_i-thin, this says that $\|c - c'\| < \varepsilon_i$. So $c - c' \in B(V_i, \varepsilon_i)$, and $c = c' + (c - c') \in \rho[(W' \times B(V_i, \varepsilon_i)) \cap D] \subset W$ by (3).

Since c was any element of U_j, we have proved (5). Thus, by the arbitrariness of W, $\{U_i\}$ is a basis of neighborhoods of b.   ■

Remark.   The above proposition fails if we omit the hypothesis that the $\{U_i\}$ are decreasing. Consider for example the simple Banach bundle $\langle \mathbb{C} \times \mathbb{R}, \pi \rangle$ over $\mathbb{R}$, where $\pi(z, t) = t$. One can easily construct a sequence $\{U_n\}$ of neighborhoods of $\langle 0, 0 \rangle$ such that U_n is n^{-1}-thin and $\{\pi(U_n)\}$ shrinks down to 0, while the image of U_n under the projection $\langle z, t \rangle \mapsto z$ contains $\{z : |z| \leq 1\}$ for all n. Of course the $\{U_n\}$ will not be decreasing in n.

C6.   We are now ready to define what it means for a cross-section to have only "small" discontinuities.

Definition.   Let $\varepsilon > 0$. A cross-section f of $\mathscr{B}$ is called *ε-continuous* at a point x of X if there is a neighborhood V of x and an ε-thin neighborhood U of $f(x)$ such that $f(V) \subset U$. If f is ε-continuous at all points of X, it is called simply *ε-continuous*.

Taking U and V to be open and replacing U by $U \cap \pi^{-1}(V)$, we can rephrase the above definition as follows: f is ε-continuous at x if and only if there is an open ε-thin neighborhood U of $f(x)$ such that $f(\pi(U)) \subset U$.

C7.   Proposition.   *If $\varepsilon > 0$ and f is a cross-section of $\mathscr{B}$ which is ε-continuous at a point x of X, then $y \mapsto \|f(y)\|$ is bounded on some neighborhood of x.*

Proof.   Choose U and V as in the definition of ε-continuity; and set

$$U' = \{b \in U : \|b\| < \|f(x)\| + 1\},$$

$$V' = \pi(U') \cap V.$$

Since U' is an open neighborhood of $f(x)$, V' is an open neighborhood of x. Let $y \in V'$. Then $f(y) \in U$ (since $V' \subset V$); and there is an element b of U' such that $\pi(b) = y$. Since $U' \subset U$ and U is ε-thin, this implies that $\|f(y) - b\| < \varepsilon$. Also $\|b\| < \|f(x)\| + 1$ (since $b \in U'$). Therefore $\|f(y)\| \leq \|b\| + \|f(y) - b\| < \|f(x)\| + 1 + \varepsilon$; and this is true for all y in the neighborhood V' of x.   ■

C8.   Proposition.   *Let f be a cross-section of $\mathscr{B}$ which is ε-continuous for all $\varepsilon > 0$. Then f is continuous.*

Proof.   Fix a point x of X. For each positive integer n let U_n be an n^{-1}-thin open neighborhood of $f(x)$ such that

$$f(\pi(U_n)) \subset U_n \tag{7}$$

(see C6). We notice that $U'_n = U_n \cap U_{n-1}$ is also n^{-1}-thin and satisfies (7). (Indeed, to see that U'_n satisfies (7), observe that $y \in \pi(U'_n) \Rightarrow y \in \pi(U_n)$ and $y \in \pi(U_{n-1}) \Rightarrow f(y) \in U_n \cap U_{n-1} = U'_n$.)   Thus,   replacing U_n by $U_1 \cap \cdots \cap U_n$, we may as well assume from the beginning that the $\{U_n\}$ are decreasing. From this and Proposition C5 it follows that the $U_n \cap \pi^{-1}(V)$, where $n = 1, 2, \ldots$ and V is an X-neighborhood of x, form a basis of neighborhoods of $f(x)$.

Let W be any neighborhood of $f(x)$. By the preceding paragraph,

$$U_n \cap \pi^{-1}(V) \subset W \tag{8}$$

for some $n = 1, 2, \ldots$ and some open neighborhood V of x. Thus, if $y \in V \cap \pi(U_n)$, it follows from (7) and (8) that $f(y) \in W$. Since $V \cap \pi(U_n)$ is a neighborhood of x, this proves the proposition.   ∎

C9.   Lemma.   *Let V be an open subset of X; and let $U_1, \ldots, U_n$ be open subsets of B such that $\pi(U_i) = V$ for all $i = 1, \ldots, n$. Furthermore, let $\phi_1, \ldots, \phi_n$ be continuous complex functions on X; and assume that $\phi_1(x) \neq 0$ for all x in V. Define W to be the set of all points of B of the form*

$$\sum_{i=1}^{n} \phi_i(x) b_i,$$

where $x \in V$ and $b_i \in U_i \cap B_x$ for each $i = 1, 2, \ldots, n$. Then W is open in B.

Proof.   Set $E = \{\langle b_1, \ldots, b_n \rangle \in B^n : \pi(b_1) = \pi(b_2) = \cdots = \pi(b_n) \in V\}$; and define $\rho : E \to E$ as follows:

$$\rho(b_1, \ldots, b_n) = \left\langle \sum_{i=1}^{n} \phi_i(x) b_i, b_2, b_3, \ldots, b_n \right\rangle$$

(where $x = \pi(b_1) = \cdots = \pi(b_n)$). Evidently ρ is continuous on E and has a continuous inverse ρ^{-1}:

$$\rho^{-1}(b_1, \ldots, b_n) = \left\langle \phi_1(x)^{-1} \left[b_1 - \sum_{i=2}^{n} \phi_i(x) b_i \right], b_2, \ldots, b_n \right\rangle$$

(where $x = \pi(b_1) = \cdots = \pi(b_n)$). So ρ is a homeomorphism of E onto itself.

Furthermore, let us define $\beta : E \to \pi^{-1}(V)$ and $\gamma : E \to \pi^{-1}(V)$:

$$\beta(b_1, \ldots, b_n) = \sum_{i=1}^{n} \phi_i(x) b_i \qquad (x = \pi(b_1) = \cdots = \pi(b_n)),$$

$$\gamma(b_1, \ldots, b_n) = b_1.$$

Notice that $\gamma: E \to \pi^{-1}(V)$ is an open surjection. Also $\gamma \circ \rho = \beta$. Since ρ is a homeomorphism, it follows that β is also open on E. Now $W = \beta((U_1 \times \cdots \times U_n) \cap E)$. Therefore W is open. ∎

C10. Proposition. *Let $\varepsilon > 0$ and $x \in X$. Let $\phi_1, \ldots, \phi_n$ be continuous non-negative real functions on X such that $\sum_{i=1}^n \phi_i(y) \equiv 1$. Let $f_1, \ldots, f_n$ be cross-sections of $\mathcal{B}$ which are ε-continuous at x. Then $f = \sum_{i=1}^n \phi_i f_i$ is ε-continuous at x.*

Proof. Since $\phi_i(x) \neq 0$ for some i, we may as well assume that $\phi_1(x) \neq 0$, and then restrict attention to an open neighborhood of x on which ϕ_1 is never 0.

For each i we take an open ε-thin neighborhood U_i of $f_i(x)$ such that

$$f_i(\pi(U_i)) \subset U_i. \tag{9}$$

Cutting down the U_i, we may assume that the $\pi(U_i)$ are all the same open neighborhood V of x. Using these U_i, we define W as in Lemma C9. By C9 W is an open neighborhood of $f(x)$. Further, if $b, b' \in W$ and $\pi(b) = \pi(b') = y$, we can write

$$b = \sum_{i=1}^n \phi_i(y) b_i,$$

$$b' = \sum_{i=1}^n \phi_i(y) b_i',$$

where $b_i, b_i' \in U_i \cap B_y$; and hence $\|b - b'\| \leq \sum_{i=1}^n \phi_i(y)\|b_i - b_i'\| < \sum_{i=1}^n \phi_i(y)\varepsilon = \varepsilon$ (since $\sum_i \phi_i \equiv 1$ and U_i is ε-thin). Therefore W is ε-thin. From (9) it follows that $f(V) \subset W$. Consequently f is ε-continuous at x. ∎

C11. Recall that X is *completely regular* if, given a neighborhood U of a point x of X, there is a continuous real function ϕ on X such that $\phi(x) = 1$ and $\phi \equiv 0$ outside U.

Proposition. *Assume that X is completely regular. Given $\varepsilon > 0$ and $b \in B$, there exists an ε-continuous cross-section f of $\mathcal{B}$ such that $f(\pi(b)) = b$.*

Proof. By C4 b has an open ε-thin neighborhood U. Put $x = \pi(b)$, $V = \pi(U)$. Now there obviously exists a cross-section g of the reduced bundle $\mathcal{B}_V$ such that $g(y) \in U$ for all y in V and $g(x) = b$. By the complete regularity of X

there is an open neighborhood W of x such that $\bar{W} \subset V$, and a continuous function $\phi: X \to [0, 1]$ such that $\phi(x) = 1$ and $\phi \equiv 0$ outside $\bar{W}$. Now define

$$f(y) = \begin{cases} \phi(y)g(y) & \text{for } y \in V, \\ 0_y & \text{for } y \notin V. \end{cases}$$

Since g is ε-continuous on V, it follows from C10 that $f\ (= \phi g + (1 - \phi)0)$ is ε-continuous on V. Being identically 0 on $X \setminus \bar{W}$, f is trivially ε-continuous on $X \setminus \bar{W}$. So f is ε-continuous on X. Evidently $f(x) = b$.   ∎

C12.   We recall that the topological space X is *paracompact* if it is Hausdorff and every open covering of X has a locally finite open refinement (see Bourbaki, General Topology [13], Chap. I, §9, no. 10). Every paracompact space is normal (see Bourbaki, General Topology [13], Chap. IX, §4, no. 4), hence completely regular.

C13.   We come now to the crucial step in the development of this appendix.

Proposition.   *Assume that X is paracompact. Suppose that $\varepsilon > 0$ and $x_0 \in X$; and let f be an ε-continuous cross-section of $\mathscr{B}$. Then there exists an $\varepsilon/2$-continuous cross-section f' of $\mathscr{B}$ satisfying:*

$$\|f'(x) - f(x)\| \leq \tfrac{3}{2}\varepsilon \qquad \text{for all } x \text{ in } X, \tag{10}$$

$$f'(x_0) = f(x_0). \tag{11}$$

Proof.   For the moment we fix any element x of X. By C11 there is a $\varepsilon/2$-continuous cross-section $\tilde{f}$ of $\mathscr{B}$ such that $\tilde{f}(x) = f(x)$. We claim that there exists a neighborhood U of x such that

$$\|\tilde{f}(y) - f(y)\| < \tfrac{3}{2}\varepsilon \qquad \text{for all } y \text{ in } U. \tag{12}$$

Indeed: Choose an open ε-thin neighborhood V of $f(x)$ such that $f(\pi(V)) \subset V$, and an open $\varepsilon/2$-thin neighborhood W of $\tilde{f}(x) = f(x)$ such that $\tilde{f}(\pi(W)) \subset W$. Let $U = \pi(V \cap W)$. If $y \in U$, there exists an element b of $V \cap W$ such that $\pi(b) = y$; and thus $\|f(y) - b\| < \varepsilon$ and $\|\tilde{f}(y) - b\| < \tfrac{1}{2}\varepsilon$, giving (12). This proves the claim.

By the above claim we can find an open covering $\{U_i\}$ $(i \in I)$ of X and for each i an $\varepsilon/2$-continuous cross-section g_i of $\mathscr{B}$ satisfying

$$\|g_i(y) - f(y)\| < \tfrac{3}{2}\varepsilon \qquad \text{for all } y \text{ in } U_i. \tag{13}$$

Suppose that the point x_0 belongs to some fixed U_k. From the preceding construction we can suppose that

$$g_k(x_0) = f(x_0). \tag{14}$$

Choose an open neighborhood Z of x_0 with $\bar{Z} \subset U_k$; and replace each U_i with $i \neq k$ by $U_i \setminus \bar{Z}$. The result is an open covering with the same properties as before, and with the additional property that

$$x_0 \notin \bar{U}_i \qquad \text{for } i \neq k. \tag{15}$$

Now by Bourbaki, General Topology [13], Chap. IX, §4, no. 4, Corollary 1, there is a continuous partition of unity $\{\phi_i\}$ $(i \in I)$, subordinate to $\{U_i\}$. We can therefore form the cross-section

$$f' = \sum_{i \in I} \phi_i g_i.$$

Proposition C10, applied to a small neighborhood of each point on which only finitely many of the ϕ_i do not vanish, shows that f' is $\varepsilon/2$-continuous.

Given x in X, we have

$$\|f'(x) - f(x)\| = \left\| \sum_{i \in I} \phi_i(x)(g_i(x) - f(x)) \right\|$$

$$\leq \sum_{i \in I} \phi_i(x)\|g_i(x) - f(x)\|. \tag{16}$$

Now for each index i, either $x \notin U_i$, in which case $\phi_i(x) = 0$, or $x \in U_i$, in which case $\|g_i(x) - f(x)\| < \frac{3}{2}\varepsilon$ by (13). So for all i, $\phi_i(x)\|g_i(x) - f(x)\| \leq \frac{3}{2}\varepsilon\phi_i(x)$; and (16) gives $\|f'(x) - f(x)\| \leq \sum_i \frac{3}{2}\varepsilon\phi_i(x) = \frac{3}{2}\varepsilon$, which is (10).

Finally, in view of (14) and (15) we have $\phi_k(x_0) = 1$ and $f'(x_0) = g_k(x_0) = f(x_0)$; and this is (11).   ∎

C14.   Two more simple propositions complete the preparation for the main theorem.

Proposition.   *Let $\varepsilon > 0$; and let $\{f_n\}$ be a sequence of ε-continuous cross-sections of $\mathcal{B}$ converging uniformly on X to a cross-section f. Then f is ε'-continuous for every $\varepsilon' > \varepsilon$.*

Proof.   Let $\varepsilon' > \varepsilon$. By hypothesis there is a positive integer n such that

$$\|f_n(x) - f(x)\| < \tfrac{1}{2}(\varepsilon' - \varepsilon) \qquad \text{for all } x. \tag{17}$$

Fix $x_0 \in X$; and choose an open ε-thin neighborhood U of $f_n(x_0)$ such that

$$f_n(V) \subset U, \text{ where } V = \pi(U). \tag{18}$$

Now   set $W = U + B(V, \frac{1}{2}(\varepsilon' - \varepsilon)) = \{b + c : \langle b, c \rangle \in D, \quad b \in U, \quad c \in B(V, \frac{1}{2}(\varepsilon' - \varepsilon))\}$. By C9 W is an open neighborhood of $f_n(x_0)$. Also, W is

ε'-thin; for if b, $c \in U$ and u, $v \in B(V, \frac{1}{2}(\varepsilon' - \varepsilon))$, where b, c, u, v are all in B_x $(x \in V)$, then

$$\|(b + u) - (c + v)\| \le \|b - c\| + \|u - v\|$$
$$< \varepsilon + 2 \cdot \tfrac{1}{2}(\varepsilon' - \varepsilon)$$
$$= \varepsilon'.$$

Further, if $x \in V$ we have

$$f(x) = f_n(x) + (f(x) - f_n(x)) \in U + B(V, \tfrac{1}{2}(\varepsilon' - \varepsilon)) \qquad \text{(by (17), (18))}$$
$$= W.$$

Thus $f(V) \subset W$; and f is ε'-continuous at x_0 for all x_0 in X. ∎

C15.   Proposition.   *Let $\{f_n\}$ be a sequence of cross-sections of $\mathscr{B}$ such that (i) $\{f_n\}$ converges uniformly on X to a cross-section f of $\mathscr{B}$, and (ii) f_n is ε_n-continuous, where $\lim_{n \to \infty} \varepsilon_n = 0$. Then the cross-section f is continuous.*

Proof.   By C14 f is ε-continuous for every $\varepsilon > 0$. So by C8 f is continuous. ∎

C16.   We are now ready for the main results.

Theorem.   *Assume that X is paracompact. Then $\mathscr{B}$ has enough continuous cross-sections.*

Proof.   Given b_0 in B, we must find a continuous cross-section f of $\mathscr{B}$ such that $f(x_0) = b_0$, where $x_0 = \pi(b_0)$.

Let f_0 be the cross-section of $\mathscr{B}$ defined by: $f_0(x_0) = b_0$, $f_0(x) = O_x$ for $x \ne x_0$. Thus f_0 is ε_0-continuous, where $\varepsilon_0 = 2\|b_0\| + 1$. Now set $\varepsilon_n = 2^{-n}\varepsilon_0$. Applying C13 repeatedly, we obtain a sequence $\{f_n\}$ of cross-sections of $\mathscr{B}$ with the following three properties: (i) f_n is ε_n-continuous for all n; (ii) $\|f_{n+1}(x) - f_n(x)\| \le \tfrac{3}{2}\varepsilon_n$ for all x and all n; (iii) $f_n(x_0) = b_0$ for all n. It follows from (ii) that $\{f_n\}$ converges uniformly on X to some cross-section f of $\mathscr{B}$. By (i) and C15 f is continuous. By (iii) $f(x_0) = b_0$. Thus the required continuous cross-section has been constructed. ∎

C17.   Theorem.   *Assume that X is locally compact. Then $\mathscr{B}$ has enough continuous cross-sections.*

Proof.   If X is compact it is paracompact, and so Theorem C16 is applicable. Assume then that X is not compact. Let $b_0 \in B$; and choose open neighborhoods U, V of $x_0 = \pi(b_0)$ such that $\bar{U} \subset V$ and $\bar{V}$ is compact. By C16 applied to the compact base space $\bar{V}$ there is a continuous cross-section g of the reduction of $\mathscr{B}$ to $\bar{V}$ satisfying $g(x_0) = b_0$. Now let ϕ be a real-valued continuous function on X with $\phi(x_0) = 1$ and vanishing outside of U. Then the cross-section f of $\mathscr{B}$ coinciding with ϕg on V and vanishing outside V is continuous on X and satisfies $f(x_0) = b_0$. ∎

C18. Remark. It has been observed by K. H. Hofmann ([4], Prop. 3.4) that the argument of this appendix, and hence also the truth of C16 and C17, remains valid if the norm-function in $\mathscr{B}$ is assumed to be merely upper semi-continuous rather than continuous. More precisely, let us define a *loose Banach bundle* over a topological space X to be a bundle over X satisfying the same axioms as in II.13.4 except that (i) is replaced by: (i') $s \mapsto \|s\|$ is upper semi-continuous on B (that is to say, $\{s \in B: \|s\| < r\}$ is open in B for every positive number r). Then every loose Banach bundle over a paracompact or locally compact space has enough continuous cross-sections.

Bibliography

AARNES, J. F.

[1] On the continuity of automorphic representations of groups, *Commun. Math. Phys.* **7**(1968), 332–336. MR 36 #6542.

[2] Physical states on C^*-algebras, *Acta Math.* **122**(1969), 161–172. MR 40 #747.

[3] Full sets of states on a C^*-algebra, *Math. Scand.* **26**(1970), 141–148. MR 41 #5978.

[4] Quasi-states on C^*-algebras, *Trans. Amer. Math. Soc.* **149**(1970), 601–625. MR 43 #8311.

[5] Continuity of group representations with applications to C^*-algebras, *J. Functional Anal.* **5**(1970), 14–36. MR 41 #393.

[6] Differentiable representations. I. Induced representations and Frobenius reciprocity, *Trans. Amer. Math. Soc.* **220**(1976), 1–35. MR 54 #5392.

[7] Distributions of positive type and representations of Lie groups, *Math. Ann.* **240**(1979), 141–156. MR 80e: 22017.

AARNES, J. F., EFFROS, E. G., and NIELSEN, O. A.

[1] Locally compact spaces and two classes of C^*-algebras, *Pacific J. Math.* **34**(1970), 1–16. MR 42 #6626.

AARNES, J. F., and KADISON, R. V.

[1] Pure states and approximate identities, *Proc. Amer. Math. Soc* **21**(1969), 749–752. MR 39 #1980.

ABELLANAS, L.

[1] Slices and induced representations, *Bull. Acad. Polon. Sci. Sér. Sci. Math. Astronom. Phys.* **19**(1971), 287–290. MR 44 #4142.

ALI, S. T.

[1] Commutative systems of covariance and a generalization of Mackey's imprimitivity theorem, *Canad. Math. Bull.* **27**(1984), 390–397. MR 86i: 22007.

AKEMANN, C. A.

[1] The general Stone-Weierstrass problem, *J. Functional Anal.* **4**(1969), 277–294. MR 40 #4772.

[2] Subalgebras of C^*-algebras and von Neumann algebras, *Glasgow Math. J.* **25**(1984), 19–25. MR 85c: 46059.

AKEMANN, C. A., and ANDERSON, J.

[1] The Stone-Weierstrass problem for C^*-algebras, *Invariant subspaces and other topics*, pp. 15–32, *Operator Theory: Adv. Appl.*, **6**, Birkhäuser, Basel-Boston, Mass., 1982. MR 85a: 46035.

AKEMANN, C. A., and WALTER, M. E.

[1] Nonabelian Pontryagin duality, *Duke Math. J.* **39**(1972), 451–463. MR 48 #3595.

ALBERT, A. A.

[1] Structure of algebras, *Amer. Math. Soc. Colloq. Publ.*, **24**, Providence, R. I. 1939. MR 1, 99.

ALESINA, A.

[1] Equivalence of norms for coefficients of unitary group representations, *Proc. Amer. Math. Soc.* **74**(1979), 343–349. MR 80e: 22008.

ALI, S. T.

[1] Commutative systems of covariance and a generalization of Mackey's imprimitivity theorem, *Canad. Math. Bull.* **27**(1984), 390–397. MR 86i: 22007.

AMBROSE, W.

[1] Spectral resolution of groups of unitary operators, *Duke Math. J.* **11**(1944), 589–595. MR 6, 131.

[2] Structure theorems for a special class of Banach algebras, *Trans. Amer. Math. Soc.* **57**(1945), 364–386. MR 7, 126.

ANDERSON, J., and BUNCE, J.

[1] Stone-Weierstrass theorems for separable C^*-algebras, *J. Operator Theory* **6**(1981), 363–374. MR 83b: 46077.

ANDLER, M.

[1] Plancherel pour les groupes algébriques complexes unimodulaires, *Acta Math.* **154**(1985), 1–104.

ANGELOPOULOS, E.

[1] On unitary irreducible representations of semidirect products with nilpotent normal subgroup, *Ann. Inst. H. Poincaré Sect. A(N.S.)* **18**(1973), 39–55. MR 49 #462.

ANKER, J. P.
[1] Applications de la p-induction en analyse harmonique, *Comment. Math. Helv.* **58**(1983), 622–645. MR 85c: 22006.

ANUSIAK, Z.
[1] On generalized Beurling's theorem and symmetry of L_1-group algebras, *Colloq. Math.* **23**(1971), 287–297. MR 49 #11147.

APOSTOL, C., FIALKOW, L., HERRERO, D., and VOICULESCU, D.
[1] *Approximation of Hilbert space operators. Volume II, Research Notes in Mathematics*, **102**, Pitman Books, London–New York, 1984. MR 85m: 47002.

APOSTOL, C., FOIAS, C., and VOICULESCU, D.
[1] Some results on non-quasitriangular operators. II, *Rev. Roumaine Math. Pures Appl.* **18**(1973); III, ibidem 309; IV, ibidem 487; VI, ibidem 1473. MR 48 #12109.
[2] On the norm-closure of nilpotents. II, *Rev. Roumaine Math. Pures Appl.* **19**(1974), 549–557. MR 54 #5876.

APOSTOL, C., HERRERO, D. A., and VOICULESCU, D.
[1] The closure of the similarity orbit of a Hilbert space operator, *Bull. New Series Amer. Math. Soc.* **6**(1982), 421–426. MR 83c: 47028.

APOSTOL, C., and VOICULESCU, D.
[1] On a problem of Halmos, *Rev. Roumaine Math. Pures Appl.* **19**(1974), 283–284. MR 49 #3574.

ARAKI, H., and ELLIOTT, G. A.
[1] On the definition of C^*-algebras, *Publ. Res. Inst. Math. Sci. Kyoto Univ.* **9**(1973), 93–112. MR 50 #8085.

ARENS, R.
[1] On a theorem of Gelfand and Naimark, *Proc. Nat. Acad. Sci. U.S.A.* **32**(1946), 237–239. MR 8, 279.
[2] Representation of *-algebras, *Duke Math. J.* **14**(1947), 269–283. MR 9, 44.
[3] A generalization of normed rings, *Pacific J. Math.* **2**(1952), 455–471. MR 14, 482.

ARMACOST, W. L.
[1] The Frobenius reciprocity theorem and essentially bounded induced representations, *Pacific J. Math.* **36**(1971), 31–42. MR 44 #6902.

ARVESON, W. B.
[1] A theorem on the action of abelian unitary groups, *Pacific J. Math.* **16**(1966), 205–212. MR 32 #6241.
[2] Operator algebras and measure-preserving automorphisms, *Acta Math.* **118**(1967), 95–109. MR 35 #1751.
[3] Subalgebras of C^*-algebras, *Acta Math.* **123**(1969), 141–224. MR 40 #6274.
[4] Subalgebras of C^*-algebras, II, *Acta Math.* **128**(1972), 271–308. MR 52 #15035.
[5] On groups of automorphisms of operator algebras, *J. Functional Anal.* **15**(1974), 217–243. MR 50 #1016.

[6] *An invitation to C*-algebras*, Graduate texts in math., **39**, Springer-Verlag, Berlin-Heidelberg-New York, 1976. MR 58 #23621.

ATIYAH, M. F.

[1] *K-Theory*, Benjamin, New York-Amsterdam, 1967. MR 36 #7130.

AUPETIT, B.

[1] *Propriétés spectrales des algèbres de Banach*, Lecture Notes in Math., **735**, Springer-Verlag, Berlin-Heidelberg-New York,1979. MR 81i: 46055.

AUSLANDER, L.

[1] *Unitary representations of locally compact groups—The elementary and type I theory*, Lecture Notes, Yale University, New Haven, Conn., 1962.

AUSLANDER, L., and KOSTANT, B.

[1] Polarization and unitary representations of solvable Lie groups, *Invent. Math.* **14**(1971), 255-354. MR 45 #2092.

AUSLANDER, L., and MOORE, C. C.

[1] *Unitary representations of solvable Lie groups*, Memoirs Amer. Math. Soc., **62**, Amer. Math. Soc., Providence, R. I., 1966. MR 34 #7723.

BACKHOUSE, N. B.

[1] Projective representations of space groups, III. Symmorphic space groups, *Quart. J. Math. Oxford Ser.* (2) **22**(1971), 277-290. MR 45 #2028.

[2] On the form of infinite dimensional projective representations of an infinite abelian group, *Proc. Amer. Math. Soc.* **41**(1973), 294-298. MR 47 #6940.

[3] Tensor operators and twisted group algebras, *J. Math. Phys.* **16**(1975), 443-447. MR 52 #546.

BACKHOUSE, N. B., and BRADLEY, C. J.

[1] Projective representations of space groups, I. Translation groups, *Quart. J. Math. Oxford Ser.* (2) **21**(1970), 203-222. MR 41 #5510.

[2] Projective representations of space groups, II. Factor systems, *Quart. J. Math. Oxford. Ser.* (2) **21**(1970), 277-295. MR 43 #7517.

[3] Projective representations of abelian groups, *Proc. Amer. Math. Soc.* **36**(1972), 260-266. MR 46 #7443.

[4] Projective representations of space groups, IV. Asymmorphic space groups, *Quart. J. Math. Oxford Ser.* (2) **23**(1972), 225-238. MR 51 #5734.

BACKHOUSE, N. B., and GARD, P.

[1] On induced representations for finite groups, *J. Math. Phys.* **17**(1976), 1780-1784. MR 54 #2775.

[2] On the tensor representation for compact groups, *J. Math. Phys.* **17**(1976), 2098-2100. MR 54 #4374.

BAGCHI, S. C., MATHEW, J., and NADKARNI, M. G.

[1] On systems of imprimitivity on locally compact abelian groups with dense actions, *Acta Math.* **133**(1974), 287-304. MR 54 #7690.

BAGGETT, L.
[1] Hilbert-Schmidt representations of groups. *Proc. Amer. Math. Soc.* **21**(1969), 502–506. MR 38 #5991.

[2] A note on groups with finite dual spaces, *Pacific J. Math.* **31**(1969), 569–572. MR 41 #3658.

[3] A weak containment theorem for groups with a quotient R-group, *Trans. Amer. Math. Soc.* **128**(1967), 277–290. MR 36 #3921.

[4] A description of the topology on the dual spaces of certain locally compact groups, *Trans. Amer. Math. Soc.* **132**(1968), 175–215. MR 53 #13472.

[5] A separable group having a discrete dual is compact, *J. Functional Anal.* **10**(1972), 131–148. MR 49 #10816.

[6] Multiplier extensions other than the Mackey extension, *Proc. Amer. Math. Soc.* **56**(1976), 351–356. MR 53 #13468.

[7] Operators arising from representations of nilpotent Lie groups, *J. Functional Anal.* **24**(1977), 379–396. MR 56 #536.

[8] Representations of the Mautner group, I, *Pacific J. Math.* **77**(1978), 7–22. MR 80e: 22014.

[9] A characterization of "Heisenberg groups"; When is a particle free? *Rocky Mountain J. Math.* **8**(1978), 561–582. MR 80a: 22014.

[10] On the continuity of Mackey's extension process, *J. Functional Anal.* **56**(1984), 233–250. MR 85e: 22011.

[11] Unimodularity and atomic Plancherel measure, *Math. Ann.* **266**(1984), 513–518. MR 86a: 22004.

BAGGETT, L., and KLEPPNER, A.
[1] Multiplier representations of Abelian groups, *J. Functional Anal.* **14**(1973), 299–324. MR 51 #791.

BAGGETT, L., and MERRILL, K.
[1] Representations of the Mautner group and cocycles of an irrational rotation, *Michigan Math. J.* **33**(1986), 221–229. MR 87h: 22011.

BAGGETT, L., MITCHELL, W., and RAMSAY, A,
[1] Representations of the discrete Heisenberg group and cocycles of irrational rotations, *Michigan Math. J.* **31**(1984), 263–273. MR 86k: 22017.

BAGGETT, L., and RAMSAY, A.
[1] Some pathologies in the Mackey analysis for a certain nonseparable group, *J. Functional Anal.* **39**(1980), 375–380. MR 83b: 22007.

[2] A functional analytic proof of a selection lemma, *Canad. J. Math.* **32**(1980), 441–448. MR 83j: 54016.

BAGGETT, L., and SUND, T.
[1] The Hausdorff dual problem for connected groups, *J. Functional Anal.* **43**(1981), 60–68. MR 83a: 22006.

BAGGETT, L., and TAYLOR, K.

[1] Groups with completely reducible representation, *Proc. Amer. Math. Soc.* **72**(1978), 593–600. MR 80b: 22009.

[2] A sufficient condition for the complete reducibility of the regular representation, *J. Functional Anal.* **34**(1979), 250–265. MR 81f: 22005.

[3] On asymptotic behavior of induced representations, *Canad. J. Math.* **34**(1982), 220–232. MR 84j: 22017.

BAKER, C. W.

[1] A closed graph theorem for Banach bundles, *Rocky Mountain J. Math.* **12**(1982), 537–543. MR 84h: 46010.

BALDONI-SILVA, M. W., and KNAPP, A. W.

[1] Unitary representations induced from maximal parabolic subgroups, *J. Functional Anal.* **69**(1986), 21–120.

BARGMANN, V.

[1] Irreducible unitary representations of the Lorentz group. *Ann. Math.* (2) **48**(1947), 568–640. MR 9, 133.

[2] On unitary ray representations of continuous groups, *Ann. Math.* (2) **59**(1954), 1–46. MR 15, 397.

BARGMANN, V., and WIGNER, E. P.

[1] Group theoretical discussion of relativistic wave equations, *Proc. Nat. Acad. Sci. U.S.A.* **34**(1948), 211–223. MR 9, 553.

BARIS, K.

[1] On induced representations of *p*-adic reductive groups, *Karadeniz Univ. Math. J.* **5**(1982), 168–177. MR 85f: 22016.

BARNES, B. A.

[1] When is a *-representation of a Banach *-algebra Naimark-related to a *-representation? *Pacific J. Math.* **72**(1977), 5–25. MR 56 #16385.

[2] Representations Naimark-related to a *-representation; a correction: "When is a *-representation of a Banach *-algebra Naimark-related to a *-representation?" [*Pacific J. Math.* **72**(1977), 5–25], *Pacific J. Math.* **86**(1980), 397–402. MR 82a: 46060.

[3] The role of minimal idempotents in the representation theory of locally compact groups, *Proc. Edinburgh Math. Soc.* (2) **23**(1980), 229–238. MR 82i: 22007.

[4] A note on separating families of representations, *Proc. Amer. Math. Soc.* **87**(1983), 95–98. MR 84k: 22006.

BARUT, A. O., and RACZKA, R.

[1] *Theory of group representations and applications*, Polish Scientific Publishers, Warsaw, 1977. MR 58 #14480.

BASS, H.

[1] *Algebraic K-theory*, Benjamin, New York, 1968. MR 40 #2736.

BECKER, T.

[1] A few remarks on the Dauns-Hofmann theorems for C^*-algebras, *Arch. Math. (Basel)* **43**(1984), 265–269.

BEHNCKE, H.

[1] Automorphisms of crossed products, *Tôhoku Math. J.* **21**(1969), 580–600. MR 42 #5056.

[2] C^*-algebras with a countable dual, *Operator algebras and applications*, Part 2, Kingston, Ont., 1980, pp. 593–595, Proc. Sympos. Pure. Math., **38**, Amer. Math. Soc., Providence, R. I., 1982. MR 83j: 46004b.

BEHNCKE, H., and LEPTIN, H.

[1] Classification of C^*-algebras with a finite dual, *J. Functional Anal.* **16**(1974), 241–257. MR 49 #9638.

BEKES, ROBERT A.

[1] Algebraically irreducible representations of $L_1(G)$, *Pacific J. Math.* **60**(1975), 11–25. MR 53 #10978.

BELFI, V. A., and DORAN, R. S.

[1] Norm and spectral characterizations in Banach algebras, *L'Enseignement Math.* **26**(1980), 103–130. MR 81j: 46071.

BENNETT, J. G.

[1] Induced representations and positive linear maps of C^*-algebras, Thesis, Washington University, 1976.

[2] Induced representations of C^*-algebras and complete positivity. *Trans. Amer. Math. Soc.* **243**(1978), 1–36. MR 81h: 46068.

BERBERIAN, S. K.

[1] *Baer *-rings*, Springer-Verlag, New York-Berlin, 1972. MR 55 #2983.

[2] *Lectures on functional analysis and operator theory*, Graduate Texts in Math. **15**, Springer-Verlag, Berlin-Heidelberg-New York, 1974. MR 54 #5775.

BEREZIN, F. A.

[1] Laplace operators on semi-simple Lie groups, *Trudy Moskov. Math. Obšč.* **6**(1957), 371–463 (Russian). MR 19, 867.

BEREZIN, F. A., and GELFAND, I. M.

[1] Some remarks on the theory of spherical functions on symmetric Riemannian manifolds, *Trudy Moskov. Mat. Obšč.* **5**(1956), 311–351 (Russian). MR 19, 152.

BERKSON, E.

[1] Some characterizations of C^*-algebras, *Illinois J. Math.* **10**(1966), 1–8. MR 32 #2922.

BERNAT, P., and DIXMIER, J.

[1] Sur le dual d'un groupe de Lie, *C. R. Acad. Sci. Paris* **250**(1960), 1778–1779. MR 27 #1536.

BERNSTEIN, I. N., and ZELEVINSKY, A. V.

[1] Induced representations of reductive p-adic groups. I, *Ann. Sci. École Norm. Sup.* (4) **10**(1977), 441–472. MR 58 #28310.

BERTRAND, J., and RIDEAU, G.

[1] Non-unitary representations and Fourier transform on the Poincaré group, *Rep. Math. Phys.* **4**(1973), 47–63. MR 53 #710.

BEURLING, A.

[1] *Sur les intégrales de Fourier absolument convergentes, et leur application à une transformation fonctionnelle*, Congrès des Math. Scand., Helsingfors (1938).

BICHTELER, K.

[1] On the existence of noncontinuous representations of a locally compact group, *Invent. Math.* **6**(1968), 159–162. MR 38 #4610.

[2] A generalization to the non-separable case of Takesaki's duality theorem for C^*-algebras, *Invent. Math.* **9**(1969), 89–98. MR 40 #6275.

[3] Locally compact topologies on a group and its corresponding continuous irreducible representations, *Pacific J. Math.* **31**(1969), 583–593. MR 41 #394.

BLACKADAR, B. E.

[1] Infinite tensor products of C^*-algebras, *Pacific J. Math.* **72**(1977), 313–334. MR 58 #23622.

[2] A simple unital projectionless C^*-algebra, *J. Operator Theory* **5**(1981), 63–71. MR 82h: 46076.

[3] K-*Theory for operator algebras*, Math. Sci. Research Institute Publications, Springer-Verlag, New York-Berlin-Heidelberg, 1986.

BLATTNER, R. J.

[1] On induced representations, *Amer. J. Math.* **83**(1961), 79–98. MR 23 #A2757.

[2] On induced representations. II. Infinitesimal induction, *Amer. J. Math.* **83**(1961), 499–512. MR 26 #2885.

[3] On a theorem of G. W. Mackey, *Bull. Amer. Math. Soc.* **68**(1962), 585–587. MR 25 #5135.

[4] A theorem on induced representations, *Proc. Amer. Math. Soc.* **13**(1962), 881–884. MR 29 #3894.

[5] Positive definite measures, *Proc. Amer. Math. Soc.* **14**(1963), 423–428. MR 26 #5095.

[6] Group extension representations and the structure space, *Pacific J. Math.* **15**(1965), 1101–1113. MR 32 #5785.

BLATTNER, R. J., COHEN, M., and MONTGOMERY, S.

[1] Crossed products and inner actions of Hopf algebras, *Trans. Amer. Math. Soc.* **298**(1986), 671–711.

BOCHNER, S.

[1] *Vorlesungen über Fouriersche Integrale*, Akad. Verlag, Leipzig, 1932.

[2] Integration von Funktionen, deren Werte die Elemente eines Vectorraumes sind, *Fund. Math.* **20**(1933), 262–276.

BOE, B. D.

[1] Determination of the intertwining operators for holomorphically induced representations of SU(p,q). *Math. Ann.* **275**(1986), 401–404.

BOE, B. D., and COLLINGWOOD, D. H.

[1] A comparison theory for the structure of induced representations, *J. Algebra* **94**(1985), 511–545. MR 87b: 22026a.

[2] Intertwining operators between holomorphically induced modules, *Pacific J. Math.* **124**(1986), 73–84.

BOERNER, H.

[1] *Representations of groups*, North-Holland Publishing Co., Amsterdam London, 1970. MR 42 #7792.

BOHNENBLUST, H. F., and KARLIN, S.

[1] Geometrical properties of the unit sphere of Banach algebras, *Ann. Math.* **62**(1955), 217–229. MR 17, 177.

BOIDOL, J.

[1] Connected groups with polynomially induced dual, *J. Reine Angew. Math.* **331**(1982), 32–46. MR 83i: 22010.

[2] *-regularity of some classes of solvable groups, *Math. Ann.* **261**(1982), 477–481. MR 84g: 22016.

[3] Group algebras with unique C*-norm, *J. Functional Anal.* **56**(1984), 220–232. MR 86c: 22006.

[4] A Galois-correspondence for general locally compact groups, *Pacific J. Math.* **120**(1985), 289–293. MR 87c: 22010.

[5] Duality between closed normal subgroups of a locally compact group G and hk-closed subduals of Ĝ (preprint).

BOIDOL, J., LEPTIN, H., SCHÜRMAN, J., and VAHLE, D.

[1] Räume primitiver Ideale von Gruppenalgebren, *Math. Ann.* **236**(1978), 1–13. MR 58 #16959.

BONIC, R. A.

[1] Symmetry in group algebras of discrete groups, *Pacific J. Math.* **11**(1961), 73–94. MR 22 #11281.

BONSALL, F. F., and DUNCAN, J.

[1] *Complete Normed Algebras*, Ergebnisse der Mathematik und ihrer Grenzgebiete. **80**. Springer-Verlag, Berlin-Heidelberg-New York, 1973. MR 54 #11013.

BORCHERS, H. J.

[1] On the implementability of automorphism groups, *Comm. Math. Phys.* **14**(1969), 305–314. MR 41 #4267.

BOREL, A.

[1] *Représentations de groupes localement compacts*, Lecture Notes in Math. **276**, Springer-Verlag, Berlin-Heidelberg-New York, 1972. MR 54 #2871.

[2] On the development of Lie group theory, *Niew Arch. Wiskunde* **27**(1979), 13–25. MR 81g: 01013.

BOREL, A., and WALLACH, N. R.

[1] *Continuous cohomology, discrete subgroups, and representations of reductive groups,* Annals of Mathematics Studies, **94**, Princeton University Press, Princeton, N. J.; University of Tokyo Press, Tokyo, 1980. MR 83c: 22018.

BOURBAKI, N.

The first twelve references are in the series *Éléments de Mathématique, Actualités Sci. et. Ind.,* Hermann et cie, Paris; each title is identified by a serial number.

[1] *Théorie des Ensembles,* Chapitres I, II, No. 1212(1966). MR 34 #7356.

[2] *Théorie des Ensembles,* Chapitre III, No. 1243(1963). MR 27 #4758.

[3] *Topologie générale,* Chapitres I, II, No. 1142(1965). MR 39 #6237.

[4] *Topologie générale,* Chapitres III, IV, No. 1143(1960). MR 25 #4021.

[5] *Algèbre,* Chapitre II, No. 1032(1967). MR 9, 406.

[6] *Algèbre,* Chapitres IV, V, No. 1102(1967). MR 30 #4576.

[7] *Algèbre,* Chapitre VIII, No. 1261(1958). MR 20, #4576.

[8] *Espaces Vectoriels Topologiques,* Chapitres I, II, No. 1189(1966). MR 34 #3277.

[9] *Espaces Vectoriels Topologiques,* Chapitres III, IV, V, No. 1229(1964). MR 17 1062.

[10] *Intégration,* Chapitres I–IV, V, VI, Nos. 1175(1965), 1244(1967), 1281(1959). MR 39 #6237.

[11] *Intégration,* Chapitres VII–VIII, No. 1306(1963). MR 31 #3539.

[12] *Théories Spectrales,* Chapitres I, II, No. 1332(1967). MR 35 #4725.

[13] *General topology,* Parts I, II, Hermann, Paris, Addison-Wesley Pub. Co., Reading, Mass., 1966. MR 34 #5044.

[14] *Theory of sets,* Hermann, Paris, Addison-Wesley Pub. Co., Reading, Mass., 1968.

[15] *Algebra,* Part I, Hermann, Paris, Addison-Wesley Pub. Co., Reading, Mass., 1974. MR 50 #6689.

[16] *Lie groups and Lie algebras,* Part I, Hermann, Paris, Addison-Wesley Pub. Co., Reading, Mass., 1975.

BOYER, R., and MARTIN, R.

[1] The regular group C^*-algebra for real-rank one groups, *Proc. Amer. Math. Soc.* **59**(1976), 371–376. MR 57 #16464.

[2] The group C^*-algebra of the de Sitter group, *Proc. Amer. Math. Soc.* **65**(1977), 177–184. MR 57 #13381.

BRATTELI, O.

[1] Inductive limits of finite dimensional C^*-algebras, *Trans. Amer. Math. Soc.* **171**(1972), 195–234. MR 47 #844.

[2] Structure spaces of approximately finite-dimensional C^*-algebras, *J. Functional Anal.* **16**(1974), 192–204. MR 50 #1005.

[3] Crossed products of UHF algebras by product type actions, *Duke Math. J.* **46**(1979), 1–23. MR 82a: 46063.

[4] *Derivations, dissipations and group actions on C*-algebras*, Lecture Notes in Math. 1229, Springer-Verlag, Berlin-New York, 1986.

BRATTELI, O., and ELLIOTT, G. A.

[1] Structure spaces of approximately finite-dimensional *C*-algebras, II., J. Functional Anal.* **30**(1978), 74–82. MR 80d: 46111.

BRATTELI, O., and ROBINSON, D. W.

[1] *Operator algebras and quantum statistical mechanics, I*, Texts and Monographs in Physics, Springer-Verlag, New York-Heidelberg, 1979. MR 81a: 46070.

[2] *Operator algebras and quantum statistical mechanics, II*, Texts and Monographs in Physics, Springer-Verlag, New York-Heidelberg, 1981. MR 82k: 82013.

BREDON, G. E.

[1] *Introduction to compact transformation groups*, Academic Press, New York-London, 1972. MR 54 #1265.

BREZIN, J., and MOORE, C. C.

[1] Flows on homogeneous spaces: a new look, *Amer. J. Math.* **103**(1981), 571–613. MR 83e: 22009.

BRITTON, O. L.

[1] Primitive ideals of twisted group algebras, *Trans. Amer. Math. Soc.* **202**(1975), 221–241. MR 51 #11011.

BRÖCKER, T., and TOM DIECK, T.

[1] *Representations of compact Lie groups*, Graduate Texts in Math. **98**, Springer-Verlag, New York-Berlin-Heidelberg-Tokyo, 1985. MR 86i: 22023.

BROWN, I. D.

[1] Representations of finitely generated nilpotent groups, *Pacific J. Math.* **45**(1973), 13–26. MR 50 #4811.

[2] Dual topology of a nilpotent Lie group, *Ann. Sci. École Norm. Sup.* (4) **6**(1973), 407–411. MR 50 #4813.

BROWN, I. D., and GUIVARC'H, Y.

[1] Espaces de Poisson des groupes de Lie, *Ann. Sci. École Norm. Sup.* (4) **7**(1974), 175–179(1975). MR 55 #570a.

BROWN, L. G.

[1] Extensions of topological groups, *Pacific J. Math.* **39**(1971), 71–78. MR 46 #6384.

[2] Locally compact abelian groups with trivial multiplier group (preprint 1968).

BROWN, L. G., GREEN, P., and RIEFFEL, M. A.

[1] Stable isomorphism and strong Morita equivalence of *C*-algebras, Pacific J. Math.* **71**(1977), 349–363. MR 57 #3866.

BRUHAT, F.

[1] Sur les représentations induites des groupes de Lie, *Bull. Soc. Math. France* **84**(1956), 97–205. MR 18, 907.

[2] Distributions sur un groupe localement compact et applications à l'étude des représentations des groupes p-adiques, *Bull. Soc. Math. France* **89**(1961), 43–75. MR 25 #4354.

[3] Sur les représentations de groupes classiques p-adiques. I, *Amer. J. Math.* **83**(1961), 321–338. MR 23 #A3184.

[4] Sur les représentations de groupes classiques p-adiques. II, *Amer. J. Math.* **83**(1961), 343–368. MR 23 #A3184.

[5] *Lectures on Lie groups and representations of locally compact groups.* Notes by S. Ramanan, Tata Institute of Fundamental Research Lectures on Math., **14**, Bombay, 1968. MR 45 #2072.

BUNCE, J. W.

[1] Representations of strongly amenable C^*-algebras, *Proc. Amer. Math. Soc.* **32**(1972), 241–246. MR 45 #4159.

[2] Characterizations of amenable and strongly amenable C^*-algebras, *Pacific J. Math.* **43**(1972), 563–572. MR 47 #9298.

[3] Approximating maps and a Stone-Weierstrass theorem for C^*-algebras, *Proc. Amer. Math. Soc.* **79**(1980), 559–563. MR 81h: 46082.

[4] The general Stone-Weierstrass problem and extremal completely positive maps, *Manuscripta Math.* **56**(1986), 343–351.

BUNCE, J. W., and DEDDENS, J. A.

[1] C^*-algebras with Hausdorff spectrum, *Trans. Amer. Math. Soc.* **212**(1975), 199–217. MR 53 #8911.

BURCKEL, R. B.

[1] *Weakly almost periodic functions on semigroups*, Gordon and Breach, New York, 1970. MR 41 #8562.

[2] *Characterizations of $C(X)$ among its subalgebras*, Marcel-Dekker, New York, 1972. MR 56 #1068.

[3] Averaging a representation over a subgroup, *Proc. Amer. Math. Soc.* **78**(1980), 399–402. MR 80m: 22005.

BURNSIDE, W.

[1] On the condition of reducibility for any group of linear substitutions, *Proc. London Math. Soc.* **3**(1905), 430–434.

BURROW, M.

[1] *Representation theory of finite groups*, Academic Press, New York-London, 1965. MR 38 #250.

BUSBY, R. C.

[1] On structure spaces and extensions of C^*-algebras, *J. Functional Anal.* **1**(1967), 370–377. MR 37 #771.

[2] Double centralizers and extensions of C^*-algebras, *Trans. Amer. Math. Soc.* **132**(1968), 79–99. MR 37 #770.

[3] On a theorem of Fell, *Proc. Amer. Math. Soc.* **30**(1971), 133–140. MR 44 #814.

[4] Extensions in certain topological algebraic categories, *Trans. Amer. Math. Soc.* **159**(1971), 41–56. MR 43 #7937.

[5] On the equivalence of twisted group algebras and Banach *-algebraic bundles, *Proc. Amer. Math. Soc.* **37**(1973), 142–148. MR 47 #4018.

[6] Centralizers of twisted group algebras, *Pacific J. Math.* **47**(1973), 357–392. MR 48 #11920.

BUSBY, R. C., and SCHOCHETMAN, I.

[1] Compact induced representations, *Canad. J. Math.* **24**(1972), 5–16. MR 45 #2495.

BUSBY, R. C., SCHOCHETMAN, I., and SMITH, H. A.

[1] Integral operators and the compactness of induced representations, *Trans. Amer. Math. Soc.* **164**(1972), 461–477. MR 45 #4167.

BUSBY, R. C., and SMITH, H. A.

[1] Representations of twisted group algebras, *Trans. Amer. Math. Soc.* **149**(1970), 503–537. MR 41 #9013.

CALKIN, J. W.

[1] Two-sided ideals and congruences in the ring of bounded operators in Hilbert space, *Ann. Math.* **42**(1941), 839–873. MR 3, 208.

CAREY, A. L.

[1] Square integrable representations of non-unimodular groups, *Bull. Austral. Math. Soc.* **15**(1976), 1–12. MR 55 #3153.

[2] Induced representations, reproducing kernels and the conformal group, *Commun. Math. Phys.* **52**(1977), 77–101. MR 57 #16465.

[3] Some infinite-dimensional groups and bundles, *Publ. Res. Inst. Math. Sci.* **20**(1984), 1103–1117.

CAREY, A. L., and MORAN, W.

[1] Some groups with T_1 primitive ideal spaces, *J. Austral. Math. Soc. (Ser. A)* **38**(1985), 55–64. MR 86c: 22007.

[2] Cocycles and representations of groups of CAR type, *J. Austral. Math. Soc. (Ser. A)* **40**(1986), 20–33. MR 87d: 22010.

CARLEMAN, T.

[1] Zur Theorie der linearen Integralgleichungen, *Math. Zeit.* **9**(1921), 196–217.

CARTAN, E.

[1] Les groupes bilinéaires et les systèmes de nombres complexes, *Ann. Fac. Sc. Toulouse*, 1898, *Oeuvres complètes, pt. II*, t. 1, pp. 7–105, Gauthier-Villars, Paris, 1952.

CARTAN, H.

[1] Sur la mesure de Haar, *C. R. Acad. Sci. Paris* **211**(1940), 759–762. MR 3, 199.

[2] Sur les fondements de la théorie du potentiel, *Bull. Soc. Math. France* **69**(1941), 71–96. MR 7, 447.

CARTAN, H., and GODEMENT, R.

[1] Théorie de la dualité et analyse harmonique dans les groupes abéliens localement compacts, *Ann. Sci. École. Norm. Sup.* (3) **64**(1947), 79–99. MR 9, 326.

CASTRIGIANO, D. P. L., and HENRICHS, R. W.

[1] Systems of covariance and subrepresentations of induced representations, *Lett. Math. Phys.* **4**(1980), 169–175. MR 81j: 22010.

CATTANEO, U.

[1] Continuous unitary projective representations of Polish groups: The BMS-group, Group theoretical methods in physics, Lecture Notes in Physics, Vol. 50, pp. 450–460, Springer, Berlin, 1976. MR 58 #6051.

[2] On unitary/antiunitary projective representations of groups. *Rep. Mathematical Physics* **9**(1976), 31–53. MR 54 #10478.

[3] On locally continuous cocycles, *Rep. Mathematical Physics* **12**(1977), 125–132. MR 57 #523.

[4] Splitting and representation groups for Polish groups, *J. Mathematical Physics* **19**(1978), 452–460. MR 57 #16460.

[5] On Mackey's imprimitivity theorem, *Comment Math. Helv.* **54**(1979), 629–641. MR 81b: 22009.

CAYLEY, A.

[1] On the theory of groups, as depending on the symbolic equation $\theta^n = 1$, *Phil. Mag.* **7**(1854), 40–47. Also, Vol. II, Collected Mathematical Papers, Cambridge, 1889, pp. 123–130.

CHOJNACKI, W.

[1] Cocycles and almost periodicity, *J. London Math. Soc.* (2) **35**(1987), 98–108.

CHRISTENSEN, E.

[1] On nonselfadjoint representations of C^*-algebras, *Amer. J. Math.* **103**(1981). 817–833. MR 82k: 46085.

CIVIN, P. and YOOD, B.

[1] Involutions on Banach algebras, *Pacific J. Math.* **9**(1959), 415–436. MR 21 #4365.

CLIFFORD, A. H.

[1] Representations induced in an invariant subgroup, *Ann. of Math.* **38**(1937), 533–550.

COHEN, P. J.

[1] Factorization in group algebras, *Duke Math. J.* **26**(1959), 199–205. MR 21 #3729.

COIFMAN, R. R. and WEISS, G.

[1] Representations of compact groups and spherical harmonics, *L'Enseignement Math.* **14**(1969), 121–173. MR 41 #537.

COLEMAN, A. J.

[1] *Induced and subinduced representations, group theory and its applications*, E. M. Loebl, Ed., 57–118, Academic Press, New York, 1968.

[2] Induced representations with applications to S_n and GL(n), Queens Papers in Pure and Applied Mathematics 4 (Queens Univ., Kingston), 91 pp. 1966. MR 34 #2718.

COLOJOARĂ, I.
[1] Locally convex bundles. (Romanian). *An. Univ. Craiova Mat. Fiz.-Chim.* No. 4(1976), 11–21. MR 58 #23648.

COMBES, F.
[1] Crossed products and Morita equivalence, *Proc. London Math. Soc.* (3), **49**(1984), 289–306. MR 86c: 46081.

COMBES, F. and ZETTL, H.
[1] Order structures, traces and weights on Morita equivalent C^*-algebras, *Math. Ann.*, **265**(1983), 67–81. MR 85f: 46106.

CONNES, A.
[1] Classification of injective factors. Cases II_1, II_∞, III_λ, $\lambda \neq 1$, *Ann. Math.* (2)**104**(1976), 73–115. MR 56 #12908.
[2] On the cohomology of operator algebras, *J. Functional Anal.* **28**(1978), 248–253. MR 58 #12407.
[3] On the spatial theory of von Neumann algebras, *J. Functional Anal.* **35**(1980), 153–164. MR 81g: 46083.

CONWAY, J. B.
[1] *A course in functional analysis*, Graduate Texts in Math. no. 96, Springer-Verlag, Berlin-Heidelberg-New York, 1985. MR 86h: 46001.

CORWIN, L.
[1] Induced representations of discrete groups, *Proc. Amer. Math. Soc.* **47**(1975), 279–287. MR 52 #8329.
[2] Decomposition of representations induced from uniform subgroups and the "Mackey machine," *J. Functional Anal.* **22**(1976), 39–57. MR 54 #468.

CORWIN, L., and GREENLEAF, F. P.
[1] Intertwining operators for representations induced from uniform subgroups. *Acta Math.* **136**(1976), 275–301. MR 54 #12967.

COURANT, R. and HILBERT, D.
[1] *Methods of mathematical physics*. Vol. I, Interscience Publishers, Inc., New York, 1953. MR 16, 426.

CUNTZ, J.
[1] K-theoretic amenability for discrete groups, *J. Reine Angew. Math.* **344**(1983), 180–195. MR 86e: 46064.

CURTIS, C. W., and REINER, I.
[1] *Representation theory of finite groups and associative algebras*, Interscience Publishers, 1962. MR 26 #2519.
[2] *Methods of representation theory (with applications to finite groups and orders)*. Vol. I, Wiley-Interscience (pure & applied Math.) 1981. MR 82i: 20001.

CURTO, R., MUHLY, P. S., and WILLIAMS, D. P.
[1] Crossed products of strongly Morita equivalent C^*-algebras, *Proc. Amer. Math. Soc.* **90**(1984), 528–530. MR 85i: 46083.

DADE, E. C.

[1] Compounding Clifford's Theory, *Ann. of Math.* **91**(1970), 236–290. MR 41 #6992.

DAGUE, P.

[1] Détermination de la topologie de Fell sur le dual du groupe de Poincaré, *C.R. Acad. Sci., Paris* **283**(1976), 293–296. MR 54 #12981.

DANG-NGOC, N.

[1] Produits croisés restreints et extensions de groupes, Mai 1975 (unpublished preprints).

DANIELL, P. J.

[1] A general form of integral, *Ann. of Math.* (2) **19**(1917–1918), 279–294.

VAN DANTZIG, D.

[1] Zur topologischen Algebra, II, Abstrakte b_v-adische Ringe, *Composit. Math.* **2**(1935), 201–223.

DAUNS, J.

[1] The primitive ideal space of a C^*-algebra, *Canad. J. Math.* **26**(1974), 42–49. MR 49 #1131.

DAUNS, J., and HOFMANN, K. H.

[1] *Representations of rings by sections*, Memoirs Amer. Math. Soc. **83**, Amer. Math. Soc., Providence, R. I., 1968. MR 40 #752.

[2] Spectral theory of algebras and adjunction of identity, *Math. Ann.* **179**(1969), 175–202. MR 40 #734.

DAY, M. M.

[1] *Normed linear spaces*, Ergebnisse der Math. **21**, Springer-Verlag, Berlin, Göttingen, Heidelberg, 1958. MR 20 #1187.

DEALBA, L. M. and PETERS, J.

[1] Classification of semicrossed products of finite-dimensional C^*-algebras, *Proc. Amer. Math. Soc.* **95**(1985), 557–564. MR 87e: 46088.

DELIYANNIS, P. C.

[1] Holomorphic imprimitivities, *Proc. Amer. Math. Soc.* **16**(1965), 228–233. MR 31 #625.

DERIGHETTI, A.

[1] Sur certaines propriétés des représentations unitaires des groupes localement compacts, *Comment. Math. Helv.* **48**(1973), 328–339. MR 48 #8686.

[2] Sulla nozione di contenimento debole e la proprietà di Reiter, *Rend. Sem. Mat. Fis. Milano* **44**(1974), 47–54. MR 54 #10476.

[3] Some remarks on $L^1(G)$, *Math Z.* **164**(1978), 189–194. MR 80f: 43009.

DIEUDONNÉ, J.

[1] *Treatise on Analysis. Vol. VI, Harmonic analysis* **10**, Pure and Applied Mathematics. Academic Press, New York, 1978. MR 58 #29825b.

[2] *Special functions and linear representations of Lie groups*. CBMS Regional Conf. Ser. Math. **42**, Amer. Math. Soc., Providence, R. I., 1980. MR 81b: 22002.

[3] *History of functional analysis*, North Holland Mathematics Studies, **49**, North-Holland Publishing Co., Amsterdam, 1981. MR 83d: 46001.

DIESTEL, J., and UHL, JR, J. J.

[1] *Vector measures, Math. Survey*, **15**, Amer. Math. Soc. Providence, R. I., 1977. MR 56 #12216.

DINCULEANU, N.

[1] *Vector measures*, Pergamon Press, Oxford, London, 1967. MR 34 #6011b.

[2] *Integration on locally compact spaces*, Noordhoff International Publishing Co., Leyden, 1974. MR 50 #13428.

DISNEY, S., and RAEBURN, I.

[1] Homogeneous C^*-algebras whose spectra are tori, *J. Austral. Math. Soc. Ser. A* **38**(1985), 9–39. MR 86i: 46057.

DIXMIER, J.

[1] Sur la réduction des anneaux d'opérateurs, *Ann. Sci. École Norm. Sup.* (3) **68**(1951), 185–202. MR 13, 471.

[2] Algèbres quasi-unitaires, *Comment. Math. Helv.* **26**(1952), 275–322. MR 14, 660.

[3] On unitary representations of nilpotent Lie groups, *Proc. Nat. Acad. Sci. U.S.A.* **43**(1957), 958–986. MR 20 #1927.

[4] Sur les représentations unitaires des groupes de Lie algébriques, *Ann. Inst. Fourier (Grenoble)* **7**(1957), 315–328. MR 20 #5820.

[5] Sur les représentations unitaires des groupes de Lie nilpotents. I, *Amer. J. Math.* **81**(1959), 160–170. MR 21 #2705.

[6] Sur les représentations unitaires des groupes de Lie nilpotents. II, *Bull. Soc. Math. France* **85**(1957), 325–388. MR 20 #1928.

[7] Sur les représentations unitaires des groupes de Lie nilpotents. III, *Canad. J. Math.* **10**(1958), 321–348. MR 20 #1929.

[8] Sur les représentations unitaires des groupes de Lie nilpotents. IV, *Canad. J. Math.* **11**(1959), 321–344. MR 21 #5693.

[9] Sur les représentations unitaires des groupes de Lie nilpotents. V, *Bull. Soc. Math. France* **87**(1959), 65–79. MR 22 #5900a.

[10] Sur les C^*-algèbres, *Bull. Soc. Math. France* **88**(1960), 95–112. MR 22 #12408.

[11] Sur les représentations unitaires des groupes de Lie nilpotents. VI, *Canad. J. Math.* **12**(1960), 324–352. MR 22 #5900b.

[12] Sur les structures boréliennes du spectre d'une C^*-algèbre, *Inst. Hautes Études Sci. Publ. Math.* **6**(1960), 297–303. MR 23 #A2065.

[13] Points isolés dans le dual d'un groupe localment compact, *Bull. Soc. Math. France* **85**(1961), 91–96. MR 24 #A3237.

[14] Points séparés dans le spectre d'une C^*-algèbre, *Acta Sci. Math. Szeged* **22**(1961), 115–128. MR 23 #A4030.

[15] Sur le revêtement universel d'un groupe de Lie de type I, *C. R. Acad. Sci. Paris* **252**(1961), 2805–2806. MR 24 #A3241.

[16] Représentations intégrables du groupe de DeSitter, *Bull. Soc. Math. France* **89**(1961), 9–41. MR 25 #4031.

[17] Dual et quasi dual d'une algèbre de Banach involutive, *Trans. Amer. Math. Soc.* **104**(1962), 278–283. MR 25 #3384.

[18] Traces sur les C^*-algèbres, *Ann. Inst. Fourier (Grenoble)* **13**(1963), 219–262. MR 26 #6807.

[19] Représentations induites holomorphes des groups resolubles algébriques, *Bull. Soc. Math. France* **94**(1966), 181–206. MR 34 #7724.

[20] Champs continus d'espaces hilbertiens et de C^*-algèbres. II, *J. Math. Pures Appl.* **42**(1963), 1–20. MR 27 #603.

[21] *Les C^*-algèbres et leurs représentations*, Gauthier-Villars, Paris, 1964. MR 30 #1404.

[22] *Les algèbres d'opérateurs dans l'espace hilbertien (algèbres de von Neumann)*, 2nd ed., Gauthier-Villars, Paris, 1969. *MR* 20 #1234.

[23] *Bicontinuité dans la méthode du petit groupe de Mackey*, Bull. Sci. Math. (2) **97**(1973), 233–240. MR 53 #3187.

[24] *C^*-algebras*, 15, North-Holland Publishing Co., Amsterdam, 1977. MR 56 #16388.

[25] *Von Neumann algebras*, 27, North-Holland Publishing Co., Amsterdam, 1981. MR 50 #5482.

DIXMIER, J., and DOUADY, A.

[1] Champs continus d'espaces hilbertiens et de C^*-algèbres, *Bull. Soc. Math. France* **91**(1963), 227–283. MR 29 #485.

DÔ, NGOK Z'EP.

[1] The structure of the group C^*-algebra of the group of affine transformations of the line (Russian), *Funkcional Anal. Priložen.* **9**(1974), 63–64. MR 51 #793.

[2] Quelques aspects topologiques en analyse harmonique, *Acta Math. Vietnam.* **8**(1983), 35–131 (1984). MR 86j: 22005.

DOPLICHER, S., KASTLER, D., and ROBINSON, D.

[1] Covariance algebras in field theory and statistical mechanics, *Commun. Math. Phys.* **3**(1966), 1–28. MR 34 #4930.

DORAN, R. S.

[1] Construction of uniform CT-bundles, *Notices Amer. Math. Soc.* **15**(1968), 551.

[2] *Representations of C^*-algebras by uniform CT-bundles*, Ph.D. Thesis, Univ. of Washington, Seattle, 1968.

[3] A generalization of a theorem of Civin and Yood on Banach *-algebras, *Bull. London Math. Soc.* **4**(1972), 25–26. MR 46 #2442.

DORAN, R. S., and BELFI, V. A.

[1] *Characterizations of C^*-algebras: the Gelfand-Naimark Theorems*, Pure and Applied Mathematics, 101, Marcel-Dekker Pub. Co., New York, 1986.

DORAN, R. S., and TILLER, W.

[1] Extensions of pure positive functionals on Banach *-algebras, *Proc. Amer. Math. Soc.* **82**(1981), 583–586. MR 82f: 46062.

[2] Continuity of the involution in a Banach *-algebra, *Tamkang J. Math.* **13**(1982), 87–90. MR 84d: 46086.

DORAN, R. S., and WICHMANN, J.

[1] The Gelfand-Naimark theorems for C^*-algebras, *Enseignement Math.* (2) **23**(1977), 153–180. MR 58 #12395.

[2] *Approximate identities and factorization in Banach modules.* Lecture Notes in Mathematics, **768**. Springer-Verlag, Berlin-Heidelberg-New York, 1979. MR 81e: 46044.

DORNHOFF, L.

[1] *Group representation theory, Part A, Ordinary theory*, Marcel-Dekker, New York, 1971. MR 50 #458a.

[2] *Group representation theory, Part B, Modular representation theory*, Marcel-Dekker, New York, 1972. MR 50 #458b.

DOUADY, A., and DAL SOGLIO-HÉRAULT, L.

[1] Existence de sections pour un fibré de Banach au sens de Fell, unpublished manuscript (see J. M. G. Fell [15], Appendix).

DOUGLAS, R. G.

[1] *Banach algebra techniques in operator theory*, Academic Press, New York and London, 1972. MR 50 #14335.

DUFLO, M.

[1] *Harmonic analysis and group representations*, Cortona, 1980, 129–221, Liguori, Naples 1982.

[2] Théorie de Mackey pour les groupes de Lie algébriques, *Acta Math.* **149**(1982), 153–213. MR 85h: 22022.

DUFLO, M., and MOORE, C. C.

[1] On the regular representation of a nonunimodular locally compact group, *J. Functional Anal.* **21**(1976), 209–243. MR 52 #14145.

DUNFORD, N.

[1] Resolution of the identity for commutative B^*-algebras of operators, *Acta Sci. Math. Szeged Pars B* **12**(1950), 51–56. MR 11, 600.

DUNFORD, N., and PETTIS, B. J.

[1] Linear operations on summable functions, *Trans. Amer. Math. Soc.* **47**(1940), 323–392. MR 1, 338.

DUNFORD, N., and SCHWARTZ, J. T.

[1] *Linear operators, Part I: General theory*, Interscience Publishers, New York and London, 1958. MR 22 #8302.

[2] *Linear operators, Part II: General theory*, Interscience Publishers, New York and London, 1963. MR 32 #6181.

DUNKL, C. F., and RAMIREZ, D. E.

[1] *Topics in harmonic analysis*, Appleton-Century Crofts, New York, 1971.

DUPONCHEEL, L.

[1] How to use induced representations. *Proceedings of the Conference on p-adic analysis*, pp. 72–77, Report, 7806, Math. Inst., Katolieke Univ., Niumegen, 1978. MR 80c: 22010.

[2] Non-archimedean induced representations of compact zerodimensional groups, *Composito Math.* **57**(1986), 3–13.

DUPRÉ, M. J.

[1] Classifying Hilbert bundles, *J. Functional Anal.* **15**(1974), 244–278. MR 49 #11266.

[2] Classifying Hilbert bundles. II, *J. Functional Anal.* **22**(1976), 295–322. MR 54 #3435.

[3] Duality for C*-algebras, Proc. Conf., Loyola Univ., New Orleans, La., pp. 329–338, Academic Press, New York, 1978. MR 80a: 46034.

[4] *The classification and structure of C*-algebra bundles*, Memoirs Amer. Math. Soc. **21**, (222), 1–77, Amer. Math. Soc. Providence, R. I., 1979. MR 83c: 46069.

DUPRÉ, M. J., and GILLETTE, R. M.

[1] *Banach bundles, Banach modules and automorphisms of C*-algebras*, **92**, Pitman's Research Notes in Mathematics Series, Pitman Pub. Co., New York, 1983. MR 85j: 46127.

DURBIN, J. R.

[1] On locally compact wreath products. *Pacific J. Math.* **57**(1975), 99–107. MR 51 #13125.

DYE, H.

[1] On groups of measure preserving transformations I, *Amer. J. Math.* **81**(1959), 119–159. MR 24 #A1366.

[2] On groups of measure preserving transformations II, *Amer. J. Math.* **85**(1963), 551–576. MR 28 #1275.

EDWARDS, C. M.

[1] C*-algebras of central group extensions. I, *Ann. Inst. H. Poincaré (A)* **10**(1969), 229–246. MR 40 #1536.

[2] The operational approach to algebraic quantum theory. I, *Commun. Math. Phys.* **16**(1970), 207–230. MR 42 #8819.

[3] The measure algebra of a central group extension, *Quart. J. Math. Oxford Ser.* (2) **22**(1971), 197–220. MR 46 #609.

EDWARDS, C. M., and LEWIS, J. T.

[1] Twisted group algebras I, II, *Commun. Math. Phys.* **13**(1969), 119–141. MR 40 #6279.

EDWARDS, C. M., and STACEY, P. J.

[1] On group algebras of central group extensions, *Pacific J. Math.* **56**(1975), 59–75. MR 54 #10480.

EDWARDS, R. E.

[1] *Functional analysis: theory and applications*, Holt, Rinehart, and Winston, New York-Toronto-London, 1965. MR 36 #4308.

[2] *Integration and harmonic analysis on compact groups*, London Mathematical Society Lecture Note Series. **8**, Cambridge University Press, London, 1972. MR 57 #17116.

EFFROS, E.

[1] A decomposition theory for representations of C^*-algebras, *Trans. Amer. Math. Soc.* **107**(1963), 83–106. MR 26 #4202.

[2] Transformation groups and C^*-algebras, *Ann. Math.* **81**(1965), 38–55. MR 30 #5175.

[3] *Dimensions and C^*-algebras*, CBMS Regional Conference Series in Mathematics, **46**, Conference Board of the Mathematical Sciences, Washington, D. C., 1981. MR 84k: 46042.

EFFROS, E. G., and HAHN, F.

[1] *Locally compact transformation groups and C^*-algebras*, Memoirs Amer. Math. Soc. **75**, Amer. Math. Soc., Providence, R. I., 1967. MR 37 #2895.

EFFROS, E. G., HANDELMAN, D. E., and SHEN, C.-L.

[1] Dimension groups and their affine representations, *Amer. J. Math.* **102**(1980), 385–407. MR 83g: 46061.

EHRENPREIS, L., and MAUTNER, F. I.

[1] Some properties of the Fourier transform on semi-simple Lie groups. I, *Ann. Math.* (2) **61**(1955), 406–439. MR 16, 1017.

[2] Some properties of the Fourier transform on semi-simple Lie groups. II, *Trans. Amer. Math. Soc.* **84**(1957), 1–55. MR 18, 745.

[3] Some properties of the Fourier transform on semi-simple Lie groups. III, *Trans. Amer. Math. Soc.* **90**(1959), 431–484. MR 21 #1541.

ELLIOTT, G. A.

[1] A characterization of compact groups, *Proc. Amer. Math. Soc.* **29**(1971), 621. MR 43 #2155.

[2] Another weak Stone-Weierstrass theorem for C^*-algebras, *Canad. Math. Bull.* **15**(1972), 355–358. MR 47 #4011.

[3] An abstract Dauns-Hofmann-Kaplansky multiplier theorem, *Canad. J. Math.* **27**(1975), 827–836. MR 53 #6334.

[4] On the classification of inductive limits of sequences of semi-simple finite dimensional algebras, *J. Algebra* **38**(1976), 29–44. MR 53 #1279.

[5] On the K-theory of the C^*-algebra generated by a projective representation of a torsion-free discrete abelian group, *Operator algebras and group representations*, 1, pp. 157–184, Pitman, London, 1984.

ELLIOTT, G. A., and OLESEN, D.

[1] A simple proof of the Dauns-Hofmann theorem, *Math. Scand.* **34**(1974), 231–234. MR 50 #8091.

EMCH, G. G.

[1] *Algebraic methods in statistical mechanics and quantum field theory*, Wiley-Interscience, John Wiley & Sons, New York, 1972.

ENFLO, P.

[1] Uniform structures and square roots in topological groups. I., *Israel J. Math.* **8**(1970), 230–252. MR 41 #8568.

[2] Uniform structures and square roots in topological groups. II., *Israel J. Math.* **8**(1970), 253–272. MR 41 #8568.

[3] A counterexample to the approximation problem in Banach spaces, *Acta Math.* **130**(1973), 309–317. MR 53 #6288.

ENOCK, M., and SCHWARTZ, J. M.

[1] Produit croisé d'une algèbre de von Neumann par une algèbre de Kac, *J. Functional Anal.* **26**(1977), 16–47. MR 57 #13513.

[2] Kac algebras and crossed products, *Algèbres d'opérateurs et leurs applications en physique mathematique*, Proc. Colloq., Marseille, 1977, pp. 157–166, Colloques Internat. CNRS, **274**, CNRS, Paris, 1979. MR 81e: 46051.

ENOCK, M., and SCHWARTZ, J.-M.

[1] Une dualité dans les algèbres de von Neumann, *Bull. Soc. Math. France Suppl. Mem.* **44**(1975), 1–144. MR 56 #1091.

[2] Produit croisé d'une algèbre de von Neumann par une algèbre de Kac. II, *Publ. Res. Inst. Math. Sci.* **16**(1980), 189–232. MR 81m: 46084.

[3] Algèbres de Kac moyennables, *Pacific J. Math.* **125**(1986), 363–379.

ERNEST, J.

[1] Central intertwining numbers for representations of finite groups, *Trans. Amer. Math. Soc.* **99**(1961), 499–508. MR 23 #A2467.

[2] A decomposition theory for unitary representations of locally compact groups, *Bull. Amer. Math. Soc.* **67**(1961), 385–388. MR 24 #A784.

[3] A decomposition theory for unitary representations of locally compact groups, *Trans. Amer. Math. Soc.* **104**(1962), 252–277. MR 25 #3383.

[4] A new group algebra for locally compact groups. I, *Amer. J. Math.* **86**(1964), 467–492. MR 29 #4838.

[5] Notes on the duality theorem of non-commutative, non-compact topological groups, *Tôhoku Math. J.* **16**(1964), 291–296. MR 30 #192.

[6] A new group algebra for locally compact groups. II, *Canad. J. Math.* **17**(1965), 604–615. MR 32 #159.

[7] The representation lattice of a locally compact group, *Illinois J. Math.* **10**(1966), 127–135. MR 32 #1288.

[8] Hopf-Von Neumann algebras, *Functional Analysis*, pp. 195–215, Academic Press, New York, 1967. MR 36 #6956.

[9] The enveloping algebra of a covariant system, *Commun. Math. Phys.* **17**(1970), 61–74. MR 43 #1553.

[10] A duality theorem for the automorphism group of a covariant system, *Commun. Math. Phys.* **17**(1970), 75–90. MR 42 #8298.

[11] A strong duality theorem for separable locally compact groups, *Trans. Amer. Math. Soc.* **156**(1971), 287–307. MR 43 #7555.

[12] On the topology of the spectrum of a C^*-algebra, *Math. Ann.* **216**(1975), 149–153. MR 53 #8913.

EVANS, B. D.

[1] C^*-*bundles and compact transformation groups*, Memoirs Amer. Math. Soc. **269**, Amer. Math. Soc., Providence, R. I., 1982. MR 84a: 46148.

EYMARD, P.

[1] L'algèbre de Fourier d'un groupe localement compact, *Bull. Soc. Math. France.* **92**(1964), 181–236. MR 37 #4208.

[2] *Moyennes invariantes et représentations unitaires*, Lecture Notes in Math., **300**, Springer-Verlag, Berlin-New York, 1972. MR 56 #6279.

FABEC, R. C.

[1] A theorem on extending representations, *Proc. Amer. Math. Soc.* **75**(1979), 157–162. MR 80f: 22002.

[2] Cocycles, extensions of group actions and bundle representations, *J. Functional Anal.* **56**(1984), 79–98. MR 85k: 22017.

[3] Induced group actions, representations and fibered skew product extensions, *Trans. Amer. Soc.* **301**(1987), 489–513.

FACK, T., and SKANDALIS, G.

[1] Structure des idéaux de la C^*-algèbre associée à un feuilletage, *C. R. Acad. Sci. Paris Sér. A-B* **290**(1980), A1057–A1059. MR 81h: 46088.

[2] Sur les représentations et idéaux de la C^*-algèbre d'un feuilletage, *J. Operator Theory* **8**(1982), 95–129. MR 84d: 46101.

FAKLER, R. A.

[1] On Mackey's tensor product theorem, *Duke Math. J.* **40**(1973), 689–694. MR 47 #8764.

[2] Erratum to: "On Mackey's tensor product theorem," *Duke Math. J.* **41**(1974), 691. MR 49 #5224.

[3] Representations induced from conjugate subgroups, *Indiana J. Math.* **19**(1977), 167–171. MR 82e: 22012.

[4] An intertwining number theorem for induced representations, *Nanta Math.* **11**(1978), 164–173. MR 80h: 22006.

FARMER, K. B.

[1] A survey of projective representation theory of finite groups, *Nieuw Arch. Wiskunde* **26**(1978), 292–308. MR 58 #16860.

FELDMAN, J.

[1] Borel sets of states and of representations, *Michigan Math. J.* **12**(1965), 363–366. MR 32 #375.

FELDMAN, J., and FELL, J. M. G.

[1] Separable representations of rings of operators, *Ann. Math.* **65**(1957), 241–249. MR 18, 915.

FELDMAN, J., HAHN, P., and MOORE, C. C.

[1] Orbit structure and countable sections for actions of continuous groups, *Adv. Math.* **28**(1978), 186–230. MR 58 #11217.

FELIX, R.

[1] Über Integralzerlegungen von Darstellungen nilpotenter Lie-gruppen (English summary), *Manuscripta Math.* **27**(1979), 279–290. MR 81d: 22007.

[2] When is a Kirillov orbit a linear variety? *Proc. Amer. Math. Soc.* **86**(1982), 151–152. MR 83h: 22017.

FELIX, R., HENRICHS, R. W., and SKUDLAREK, H.

[1] Topological Frobenius reciprocity for projective limits of Lie groups, *Math. Z.* **165**(1979), 19–28. MR 80e: 22010.

FELL, J. M. G.

[1] C^*-algebras with smooth dual, *Illinois J. Math.* **4**(1960), 221–230. MR 23 #A2064.

[2] The dual spaces of C^*-algebras, *Trans. Amer. Math. Soc.* **94**(1960), 365–403. MR 26 #4201.

[3] The structure of algebras of operator fields, *Acta. Math.* **106**(1961), 233–280. MR 29 #1547.

[4] A Hausdorff topology on the closed subsets of a locally compact non-Hausdorff space, *Proc. Amer. Math. Soc.* **13**(1962), 472–476. MR 25 #2573.

[5] Weak containment and induced representations of groups, *Canad. J. Math.* **14**(1962), 237–268. MR 27 #242.

[6] A new proof that nilpotent groups are CCR, *Proc. Amer. Math. Soc.* **13**(1962), 93–99. MR 24 #A3238.

[7] Weak containment and Kronecker products of group representations, *Pacific J. Math.* **13**(1963), 503–510. MR 27 #5865.

[8] Weak containment and induced representations of groups II, *Trans. Amer. Math. Soc.* **110**(1964), 424–447. MR 28 #3114.

[9] The dual spaces of Banach algebras, *Trans. Amer. Math. Soc.* **114**(1965), 227–250. MR 30 #2357.

[10] Non-unitary dual spaces of groups, *Acta. Math.* **114**(1965), 267–310. MR 32 #4210.

[11] Algebras and fiber bundles, *Pacific J. Math.* **16**(1966), 497–503. MR 33 #2674.

[12] Conjugating representations and related results on semisimple Lie groups, *Trans. Amer. Math. Soc.* **127**(1967), 405–426. MR 35 #299.

[13] An extension of Mackey's method to algebraic bundles over finite groups, *Amer. J. Math.* **91**(1969), 203–238. MR 40 #735.

[14] *An extension of Mackey's method to Banach *-algebraic bundles*, Memoirs Amer. Math. Soc. **90**, Providence, R. I., 1969. MR 41 #4255.

[15] *Banach *-algebraic bundles and induced representations*, Actes du Congrès International des Mathématiciens, Nice, 1970, Tome 2, pp. 383–388, Gauthier-Villars, Paris 1971. MR 54 #8315.

[16] A new look at Mackey's imprimitivity theorem. *Conference on Harmonic Analysis*, Univ. Maryland, College Park, Md., 1971, pp. 43-58. Lecture Notes in Math., **266**, Springer, Berlin, 1972. MR 53 #13471.

[17] *Induced representations and Banach *-algebraic bundles*, Lecture Notes in Math., **582**, Springer-Verlag, Berlin-Heidelberg-New York, 1977. MR 56 #15825.

FELL, J. M. G., and KELLEY, J. L.

[1] An algebra of unbounded operators, *Proc. Nat. Acad. Sci. U.S.A.* **38**(1952), 592-598. MR 14, 480.

FELL, J. M. G., and THOMA, E.

[1] Einige Bemerkungen über vollsymmetrische Banachsche Algebren, *Arch. Math.* **12**(1961), 69-70. MR 23 #A2067.

FIGÀ-TALAMANCA, A., and PICARDELLO, M.

[1] *Harmonic analysis on free groups*, 87, Marcel Dekker, New York, 1983. MR 85j: 43001.

FONTENOT, R. A., and SCHOCHETMAN, I.

[1] Induced representations of groups on Banach spaces, *Rocky Mountain J. Math.* **7**(1977), 53-82.. MR 56 #15824.

FORD, J. W. M.

[1] A square root lemma for Banach *-algebras, *J. London Math. Soc.* **42**(1967), 521-522. MR 35 #5950.

FOURIER, J. B. J.

[1] La théorie analytique de la chaleur, 1822, trans. A. Freeman, Cambridge, 1878.

FOURMAN, M. P., MULVEY, C. J., and SCOTT, D. S. (eds.)

[1] *Applications of sheaves*, Lecture Notes in Math., **753**, Springer-Verlag, Berlin-Heidelberg-New York, 1979. MR 80j: 18001.

FOX, J.

[1] Frobenius reciprocity and extensions of nilpotent Lie groups, *Trans. Amer. Math. Soc.* **298**(1986), 123-144.

FRENCH, W. P., LUUKKAINEN, J., and PRICE, J. F.

[1] The Tannaka-Krein duality principle, *Adv. Math.* **43**(1982), 230-249. MR 84f: 22011.

FREUDENTHAL, H., and DE VRIES, H.

[1] *Linear Lie groups*, Academic Press, New York-London, 1969. MR 41 #5546.

FROBENIUS, G.

[1] Über lineare Substitutionen und bilineare Formen, *J. Creele*, **84**(1878), 1-63.

[2] Über Gruppencharaktere, *Berl. Sitz.* (1896), 985-1021.

[3] Über Primfaktoren der Gruppendeterminant, *Berl. Sitz.* (1896), 1343-1382.

[4] Darstellung der Gruppen durch lineare Substitutionen. *Berl. Sitz.* (1897), 994-1015.

[5] Über Relationen zwischen den Charakteren einer Gruppe und denen ihrer Untergruppen, *Sitz. Preuss. Akad. Wiss.* (1898), 501-515.

FROBENIUS, G., and SCHUR, I.

[1] Über die Äquivalenz der Gruppen linearer Substitutionen, *Sitz. Preuss. Akad. Wiss.* (1906), 209–217.

FUJIMOTO, I., and TAKAHASI, S.

[1] Equivalent conditions for the general Stone-Weierstrass problem, *Manuscripta Math.* **53**(1985), 217–224.

FUJIWARA, H.

[1] On holomorphically induced representations of exponential groups, *Proc. Japan Acad.* **52**(1976), 420–423. MR 57 #12778.

[2] On holomorphically induced representations of exponential groups, *Japan J. Math.* (n.s.) **4**(1978), 109–170. MR 80g: 22008.

FUKAMIYA, M.

[1] On B*-algebras, *Proc. Japan Acad.* **27**(1951), 321–327. MR 13, 756.

[2] On a theorem of Gelfand and Naimark and the B*-algebra, *Kumamoto J. Sci. Ser. A*-1, 1 (1952), 17–22. MR 14, 884; MR 15, 1139.

FUNAKOSI, S.

[1] Induced bornological representations of bornological algebras, *Portugal. Math.* **35**(1976), 97–109. MR 56 #12924.

[2] On representations of non-type I groups, *Tôhoku Math. J.* (2) **31**(1979), 139–150. MR 81b: 22006.

GAAL, S. A.

[1] *Linear analysis and representation theory*, Grundlehren der Math. Wiss, Band 198, Springer-Verlag, New York-Heidelberg, 1973. MR 56 #5777.

GAMELIN, T. W.

[1] *Uniform algebras*, Prentice Hall, Englewood Cliffs, N. J., 1969. MR 53 #14137.

[2] *Uniform algebras and Jensen measures*, London Mathematical Society Lecture Note Series, **32**, Cambridge University Press, Cambridge-New York, 1978. MR 81a: 46058.

GANGOLLI, R.

[1] On the symmetry of L_1-algebras of locally compact motion groups and the Wiener Tauberian theorem, *J. Functional Anal.* **25**(1977), 244–252. MR 57 #6284.

GARDNER, L. T.

[1] On the "third definition" of the topology of the spectrum of a C^*-algebra, *Canad. J. Math.* **23**(1971), 445–450. MR 43 #6730.

[2] On the Mackey Borel Structure, *Canad. J. Math.* **23**(1971), 674–678. MR 44 #4532.

[3] An elementary proof of the Russo-Dye theorem, *Proc. Amer. Math. Soc.* **90**(1984), 171. MR 85f: 46017.

GELBAUM, B. R.

[1] Banach algebra bundles, *Pacific J. Math.* **28**(1969), 337–349. MR 39 #6077.

[2] *Banach algebra bundles. II, Troisième Colloque sur l'Analyse Fonctionnelle*, Liège, 1970, pp. 7–12, Vander, Louvain, 1971. MR 53 #14132.

[3] *Group algebra bundles, Problems in analysis*, Papers dedicated to Salomon Bochner, 1969, pp. 229–237. Princeton Univ. Press, Princeton, N. J., 1970. MR 50 #997.

GELBAUM, B. R., and KYRIAZIS, A.

[1] Fibre tensor product bundles, *Proc. Amer. Math. Soc.* **93**(1985), 675–680. MR 86g: 46107.

GELFAND, I. M.

[1] Sur un lemme de la théorie des espaces linéaires, *Comm. Inst. Sci. Math. Kharkoff* **13**(1936), 35–40.

[2] On normed rings, *Dokl. Akad. Nauk. SSSR* **23**(1939), 430–432.

[3] Normierte Ringe, *Mat. Sb.* **9**(1941), 3–24. MR 3, 51.

[4] Über absolut konvergente trigonometrische Reihen und Integrale, *Mat. Sb. N.S.* (51) **9**(1941), 51–66. MR 3, 51.

[5] The center of an infinitesimal group ring, *Mat. Sb. N.S.* (68) **26**(1950), 103–112(Russian). MR 11, 498.

[6] Unitary representations of the real unimodular group (principal nondegenerate series), *Izv. Akad. Nauk. SSSR* **17**(1953), 189–248 (Russian).

[7] Spherical functions in symmetric Riemann spaces, *Dokl. Akad. Nauk. SSSR* **70**(1956), 5–8 (Russian).

[8] The structure of a ring of rapidly decreasing functions on a Lie group, *Dokl. Akad. Nauk. SSSR* **124**(1959), 19–21 (Russian). MR 22 #3987.

[9] Integral geometry and its relation to the theory of representations, *Uspehi Mat. Nauk* **15**(1960), 155–164 (Russian). MR 26 #1903. [Translated in *Russian Math. Surveys* **15**(1960), 143–151.]

GELFAND, I. M., and GRAEV, M. I.

[1] Analogue of the Plancherel formula for the classical groups, *Trudy Moskov. Mat. Obšč.* **4**(1955), 375–404 (Russian). MR 17, 173. [*Amer. Math. Soc. Trans.* **9**(1958), 123–154. MR 19, 1181.]

[2] Expansion of Lorenz group representations into irreducible representations on spaces of functions defined on symmetrical spaces, *Dokl. Akad. Nauk. SSSR* **127**(1959), 250–253 (Russian). MR 23 #A1238.

[3] Geometry of homogeneous spaces, representations of groups in homogeneous spaces and related questions of integral geometry, *Trudy Moskov. Mat. Obšč.* **8**(1959), 321–390 (Russian). MR 23 #A4013.

GELFAND, I. M., GRAEV, M. I., and VERSIK, A. M.

[1] Representations of the group of smooth mappings of a manifold X into a compact Lie group, *Compositio Math.* **35**(1977), 299–334.

GELFAND, I. M., and NAIMARK, M. A.

[1] On the embedding of normed rings into the ring of operators in Hilbert space, *Mat. Sb.* **12**(1943), 197–213. MR 5, 147.

[2] Unitary representations of the group of linear transformations of the straight line, *Dokl. Akad. Nauk. SSSR* **55**(1947), 567–570 (Russian). MR 8, 563.

[3] Unitary representations of the Lorenz group, *Izv. Akad. Nauk. SSSR* **11**(1947), 411–504. (Russian). MR 9, 495.

[4] Normed rings with involution and their representations, *Izv. Akad. Nauk SSSR, Ser-math.* **12**(1948), 445–480. MR 10, 199.

[5] Unitary representations of the classical groups, *Trudy Mat. Inst. Steklov* **36**(1950), 1–288. (Russian). MR 13, 722.

GELFAND, I. M., and PYATETZKI-SHAPIRO, I.

[1] Theory of representations and theory of automorphic functions, *Uspehi Mat. Nauk* **14**(1959), 171–194 (Russian).

[2] Unitary representation in homogeneous spaces with discrete stationary subgroups, *Dokl. Akad. Nauk SSSR* **147**(1962), 17–20 (Russian).

GELFAND, I. M., and RAIKOV, D. A.

[1] Irreducible unitary representations of arbitrary locally bicompact groups, *Mat. Sb. N.S.* **13**(55) (1943), 301–316 (Russian). MR 6, 147.

GELFAND, I. M., RAIKOV, D. A., and ŠILOV, G. E.

[1] Commutative normed rings, *Uspehi Mat. Nauk* **1**: 2(12) (1946), 48–146. MR 10, 258.

[2] *Commutative normed rings* (Russian), Sovremennye Problemy Matematiki Gosudarstv. Izdat. Fiz.-Mat. Lit. Moscow, 1960. MR 23 #A1242. [Translated from the Russian, Chelsea Publishing Co., New York, 1964. MR 34 #4940.]

GELFAND, I. M., and ŠILOV, G. E.

[1] Über verschiedene Methoden der Einführung der Topologie in die Menge der maximalen Ideale eines normierten Rings, *Mat. Sb.* **51**(1941), 25–39. MR 3, 52.

GELFAND, I. M., and VILENKIN, N. JA.

[1] *Generalized functions, Vol. IV. Some applications of harmonic analysis: Rigged Hilbert spaces*, Gos. Izd., Moscow, 1961. MR 26 #4173; MR 35 #7123.

[2] *Integral geometry and connections with questions in the theory of representations*, Fitmatgiz, Moscow, 1962 (Russian).

GELFAND, I. M., MINLOS, R. A., and SHAPIRO, Z. YA.

[1] *Representations of the rotation and Lorentz groups and their applications*, Macmillan, New York, 1963. MR 22 #5694.

GHEZ, P., LIMA, R., and ROBERTS, J. E.

[1] W*-categories, *Pacific J. Math.* **120**(1985), 79–109. MR 87g: 46091.

GIERZ, G.

[1] *Bundles of topological vector spaces and their duality* (with an appendix by the author and K. Keimel). Lecture notes in Mathematics, **955**, Springer-Verlag, Berlin-New York, 1982. MR 84c: 46076.

GIL DE LAMADRID, J.

[1] Extending positive definite linear forms, *Proc. Amer. Math. Soc.* **91**(1984), 593–594. MR 85g: 46068.

GINDIKIN, S. G., and KARPELEVIC, F. I.

[1] Plancherel measure of Riemannian symmetric spaces of non-positive curvature, *Dokl. Akad. Nauk* **145**(1962), 252–255 (Russian). MR 27 #240.

GIOVANNINI, M.

[1] Induction from a normal nilpotent subgroup, *Ann. Inst. H. Poincaré Sect. A N.S.* **26**(1977), 181–192. MR 57 #535.

[2] *Induction from a normal nilpotent subgroup*, pp. 471–480. Lecture Notes in Physics, **50**, Springer, Berlin, 1976. MR 57 #12784.

GLASER, W.

[1] Symmetrie von verallgemeinerten L^1-Algebren, *Arch. Math. (Basel)* **20**(1969), 656–660. MR 41 #7448.

GLEASON, A. M.

[1] Measures on the closed subspaces of a Hilbert space, *J. Math. Mech.* **6**(1957), 885–894. MR 20 #2609.

GLICKFELD, B. W.

[1] A metric characterization of $C(X)$ and its generalization to C^*-algebras, *Illinois J. Math.* **10**(1966), 547–556. MR 34 #1865.

GLIMM, J. G.

[1] On a certain class of operator algebras, *Trans. Amer. Math. Soc.* **95**(1960), 318–340. MR 22 #2915.

[2] A Stone-Weierstress theorem for C^*-algebras, *Ann. Math.* **72**(1960), 216–244. MR 22 #7005.

[3] Type I C^*-algebras, *Ann. Math.* **73**(1961), 572–612. MR 23 #A2066.

[4] Locally compact transformation groups, *Trans. Amer. Math. Soc.* **101**(1961), 124–138. MR 25 #146.

[5] Families of induced representations, *Pacific J. Math.* **12**(1962), 855–911. MR 26 #3819.

[6] *Lectures on Harmonic analysis (non-abelian)*, New York Univ. Courant Institute of Math. Sciences, New York, 1965.

GLIMM, J. C., and KADISON, R. V.

[1] Unitary operators in C^*-algebras, *Pacific J. Math.* **10**(1960), 547–556. MR 22 #5906.

GODEMENT, R.

[1] Sur une généralization d'un théorème de Stone, *C. R. Acad. Sci Paris* **218**(1944), 901–903. MR 7, 307.

[2] Sur les relations d'orthogonalité de V. Bargmann. I. Résultats préliminaires, *C. R. Acad. Sci Paris* **225**(1947), 521–523. MR 9, 134.

[3] Sur les relations d'orthogonalité de V. Bargmann. II. Démonstration générale, *C. R. Acad. Sci. Paris* **225**(1947), 657–659. MR 9, 134.

[4] Les fonctions de type positif et la théorie des groupes, *Trans. Amer. Math. Soc.* **63**(1948), 1–84. MR 9, 327.

[5] Sur la transformation de Fourier dans les groupes discrets, *C. R. Acad. Sci. Paris* **228**(1949), 627–628. MR 10, 429.

[6] Théorie générale des sommes continues d'espaces de Banach, *C. R. Acad. Sci Paris* **228**(1949), 1321–1323. MR 10, 584.

[7] Sur la théorie des représentations unitaires, *Ann. Math.* (2) **53**(1951), 68–124. MR 12, 421.

[8] Mémoire sur la théorie des caractères dans les groupes localement compacts unimodulaires, *J. Math. Pures Appl.* **30**(1951), 1–110. MR 13, 12.

[9] A theory of spherical functions, I. *Trans. Amer. Math. Soc.* **73**(1952), 496–556. MR 14, 620.

[10] Théorie des caractères. I. Algèbres unitaires, *Ann. Math.* (2) **59**(1954), 47–62. MR 15, 441.

[11] Théorie des caractères. II. Définition et propriétés générales des caractères, *Ann. Math.* (2) **59**(1954), 63–85. MR 15, 441.

GOLDIN, G. A., and SHARP, D. H.

[1] Lie algebras of local currents and their representations, *Group Representations in Mathematics and Physics*, Battelle, Seattle, 1969, Rencontres, Lecture notes in Physics, **6**, Springer, Berlin, 1970, pp. 300–311.

GOLODETS, V. YA.

[1] Classification of representations of the anticommutation relations, *Russian Math. Surveys* **24**(1969), 1–63.

GOODEARL, K. R.

[1] *Notes on real and complex C*-algebras*, Birkhäuser, Boston, 1982. MR 85d: 46079.

GOODMAN, R.

[1] Positive-definite distributions and intertwining operators, *Pacific J. Math.* **48**(1973), 83–91. MR 48 #6319.

[2] *Nilpotent Lie groups: structure and applications to analysis*, Lecture Notes in Mathematics, **562**, Springer-Verlag, Berlin-New York 1976. MR 56 #537.

GOOTMAN, E. C.

[1] Primitive ideals of C*-algebras associated with transformation groups, *Trans. Amer. Math. Soc.* **170**(1972), 97–108. MR 46 #1961.

[2] The type of some C*- and W*-algebras associated with transformation groups, *Pacific J. Math.* **48**(1973), 93–106. MR 49 #461.

[3] Local eigenvectors for group representations, *Studia Math.* **53**(1975), 135–138. MR 52 #8327.

[4] Weak containment and weak Frobenius reciprocity, *Proc. Amer. Math. Soc.* **54**(1976), 417–422. MR 55 #8246.

[5] Induced representations and finite volume homogeneous spaces, *J. Functional Anal.* **24**(1977), 223–240. MR 56 #532.

[6] Subrepresentations of direct integrals and finite volume homogeneous spaces, *Proc. Amer. Math. Soc.* **88**(1983), 565–568. MR 84m: 22009.

[7] On certain properties of crossed products, *Proc. Symp. Pure Math.* **38**(1982), Part 1, 311–321.

[8] Abelian group actions on type I C*-algebras, *Operator algebras and their connections with topology and ergodic theory*, Busteni, 1983, pp. 152–169, Lecture Notes in Mathematics **1132**, Springer, Berlin-New York 1985.

GOOTMAN, E. C., and ROSENBERG, J.
  [1] The structure of crossed product C^*-algebras,: a proof of the generalized Effros-Hahn conjecture, *Invent. Math.* **52**(1979), 283–298. MR 80h: 46091.

GRAEV, M. I.
  [1] Unitary representations of real simple Lie groups, *Trudy Moskov. Mat. Obšč.* **7**(1958), 335–389 (Russian). MR 21 #3510.

GRANIRER, E. E.
  [1] On group representations whose C^*-algebra is an ideal in its von Neumann algebra, *Ann. Inst. Fourier (Grenoble)* **29**(1979), 37–52. MR 81b: 22007.
  [2] A strong containment property for discrete amenable groups of automorphisms on W^*-algebras, *Trans. Amer. Math. Soc.* **297**(1986), 753–761.

GRASSMANN, H.
  [1] Sur les différents genres de multiplication, *J. Crelle* **49**(1855), 199–217; Leipzig, Teubner, 1904.

GREEN, P.
  [1] C^*-algebras of transformation groups with smooth orbit space, *Pacific J. Math.* **72**(1977), 71–97. MR 56 #12170.
  [2] The local structure of twisted covariance algebras, *Acta Math.* **140**(1978), 191–250. MR 58 #12376.
  [3] Square-integrable representations and the dual topology *J. Functional Anal.* **35**(1980), 279–294. MR 82g: 22005.
  [4] The structure of imprimitivity algebras, *J. Functional Anal.* **36**(1980), 88–104. MR 83d: 46080.
  [5] Twisted crossed products, the "Mackey machine," and the Effros-Hahn conjecture, *Operator algebras and applications*, Part 1, Kingston, Ont., 1980, pp. 327–336, Proc. Sympos. Pure Math., **38**, Amer. Math. Soc., Providence, R. I., 1982. MR 85a: 46038.

GREENLEAF, F. P.
  [1] Norm decreasing homomorphisms of group algebras, *Pacific J. Math.* **15**(1965), 1187–1219. MR 29 #2664.
  [2] Amenable actions of locally compact groups, *J. Functional Anal.* **4**(1969), 295–315. MR 40 #268.
  [3] *Invariant means on topological groups and their applications*, Van Nostrand-Reinhold Co., New York, 1969. MR 40 #4776.

GREENLEAF, F. P., and MOSKOWITZ, M.
  [1] Cyclic vectors for representations of locally compact groups, *Math. Ann.* **190**(1971), 265–288. MR 45 #6978.
  [2] Cyclic vectors for representations associated with positive definite measures: Nonseparable groups, *Pacific J. Math.* **45**(1973), 165–186. MR 50 #2389.

GREENLEAF, F. P., MOSKOWITZ, M., and ROTHSCHILD, L. P.
  [1] Central idempotent measures on connected locally compact groups, *J. Functional Anal.* **15**(1974), 22–32. MR 54 #5741.

GROSS, K. I.

[1] On the evolution of noncommutative harmonic analysis, *Amer. Math. Monthly* **85**(1978), 525–548. MR 80b: 01016.

GROSSER, S., MOSAK, R., and MOSKOWITZ, M.

[1] Duality and harmonic analysis on central topological groups. I., *Indag. Math.* **35**(1973), 65–77. MR 49 #5225a.

[2] Duality and harmonic analysis on central topological groups. II. *Indag. Math.* **35**(1973), 78–91. MR 49 #5225b.

[3] Correction to "Duality and harmonic analysis on central topological groups. I." *Indag. Math.* **35**(1973), 375. MR 49 #5225c.

GROSSER, S., and MOSKOWITZ, M.

[1] Representation theory of central topological groups, *Trans. Amer. Math. Soc.* **129**(1967), 361–390. MR 37 #5327.

[2] Harmonic analysis on central topological groups, *Trans. Amer. Math. Soc.* **156**(1971), 419–454. MR 43 #2165.

[3] Compactness conditions in topological groups, *J. Reine Angew. Math.* **246**(1971), 1–40. MR 44 #1766.

GROTHENDIECK, A.

[1] Un résultat sur le dual d'une *C**-algèbre, *J. Math. Pures Appl.* **36**(1957), 97–108. MR 19, 665.

GUICHARDET, A.

[1] Sur un problème posé par G. W. Mackey, *C. R. Acad. Sci. Paris* **250**(1960), 962–963. MR 22 #910.

[2] Sur les caractères des algèbres de Banach à involution, *C. R. Acad. Sci. Paris* **252**(1961), 2800–2862..

[3] Caractères des algèbres de Banach involutives, *Ann. Inst. Fourier (Grenoble)* **13**(1963), 1–81. MR 26 #5437 MR 30, 1203.

[4] Caractères et représentations des produits tensoriels de *C**-algèbres, *Ann. Sci. École Norm. Sup.* (3) **81**(1964), 189–206. MR 30 #5176.

[5] Utilisation des sous-groupes distingués ouverts dans l'etude des représentations unitaires des groupes localement compacts, *Compositio Math.* **17**(1965), 1–35. MR 32 #5787.

[6] *Théorie générale des représentations unitaires.* Summer school on representation of Lie groups, Namur, 1969, pp. 1–59, Math. Dept., Univ. of Brussels, Brussels, 1969. MR 58 #11214.

[7] *Représentations de G^x selon Gelfand et Delorme*, Seminaire Bourbaki 1975/76, no. 486, pp. 238–255. Lecture Notes in Math. **567**, Springer, Berlin, 1977. MR 57 #9910.

[8] Extensions des représentations induites des produits semi-directs. *J. Reine Angew. Math.* **310**(1979), 7–32. MR 80i: 22017.

GURARIE, D.

[1] Representations of compact groups on Banach algebras, *Trans. Amer. Math. Soc.* **285**(1984), 1–55. MR 86h: 22007.

HAAG, R., and KASTLER, D.
[1] An algebraic approach to quantum field theory, *J. Math. Phys.* **5**(1964), 848–861. MR 29 #3144.

HAAGERUP, U.
[1] The standard form of von Neumann algebras, *Math. Scand.* **37**(1975), 271–283. MR 53 #11387.
[2] Solution of the similarity problem for cyclic representations of C*-algebras, *Ann. Math.* (2) **118**(1983), 215-240. MR 85d: 46080.
[3] All nuclear C*-algebras are amenable, *Invent. Math.* **74**(1983), 305–319. MR 85g: 46074.

HAAR, A.
[1] Der Messbegriff in der Theorie der kontinuerlichen Gruppen, *Ann. Math.* (2) **34**(1933), 147–169.

HADWIN, D. W.
[1] Nonseparable approximate equivalence, *Trans. Amer. Math. Soc.* **266**(1981), 203–231. MR 82e: 46078.
[2] Completely positive maps and approximate equivalence, *Indiana Univ. Math. J.*, **36**(1987), 211–228.

HAHN, P.
[1] Haar measure for measure groupoids, *Trans. Amer. Math. Soc.* **242**(1978), 1–33. MR 82a: 28012.
[2] The regular representation of measure groupoids, *Trans. Amer. Math. Soc.* **242**(1978), 34–72. MR 81f: 46075.

HAHN, P., FELDMAN, J., and MOORE, C. C.
[1] Orbit structure and countable sections for actions of continuous groups, *Adv. Math.* **28**(1978), 186–230. MR 58 #11217.

HALMOS, P. R.
[1] *Measure theory*, D. Van Nostrand Co., New York, 1950. MR 11, 504.
[2] *Introduction to Hilbert space and the theory of spectral multiplicity*, Chelsea Publ. Co., New York, 1951. MR 13, 563.
[3] *A Hilbert space problem book*, Graduate Texts in Mathematics, **19**, Springer-Verlag, Berlin-Heidelberg-New York, 1974. MR 34 #8178.

HALPERN, H.
[1] A generalized dual for a C*-algebra, *Trans. Amer. Math. Soc.* **153**(1971), 139–156. MR 42 #5058.
[2] Integral decompositions of functionals on C*-algebras, *Trans. Amer. Math. Soc.* **168**(1972), 371–385. MR 45 #5769.
[3] Mackey Borel structure for the quasi-dual of a separable C*-algebra, *Canad. J. Math.* **26**(1974), 621–628. MR 52 #3973.

HALPERN, H., KAFTAL, V., and WEISS, G.
[1] The relative Dixmier property in discrete crossed products, *J. Functional Anal.* **69**(1986), 121–140.

HAMILTON, W. R.

[1] *Lectures on quaternions*, Dublin, 1853.

HANNABUSS, K.

[1] Representations of nilpotent locally compact groups, *J. Functional Anal.* **34**(1979), 146–165. MR 81c: 22016.

HARISH-CHANDRA

[1] On some applications of the universal enveloping algebra of a semi-simple Lie algebra, *Trans. Amer. Math. Soc.* **70**(1951), 28–96. MR 13, 428.

[2] Representations of semi-simple Lie groups. II, *Proc. Nat. Acad. Sci. U.S.A.* **37**(1951), 362–365. MR 13, 107.

[3] Plancherel formula for complex semi-simple Lie groups, *Proc. Nat. Acad. Sci. U.S.A.* **37**(1951), 813–818. MR 13, 533.

[4] The Plancherel formula for complex semi-simple Lie groups, *Trans. Amer. Math. Soc.* **76**(1954), 485–528. MR 16, 111.

[5] Representations of a semi-simple Lie group on a Banach space, I, *Trans. Amer. Math. Soc.* **75**(1953), 185–243. MR 15, 100.

[6] Representations of semi-simple Lie groups, II, *Trans. Amer. Math. Soc.* **76**(1954), 26–65. MR 15, 398.

[7] Representations of semi-simple Lie groups. III, *Trans. Amer. Math. Soc.* **76**(1954), 234–253. MR 16, 11.

[8] Representations of semi-simple Lie groups. IV, *Amer. J. Math.* **77**(1955), 743–777. MR 17, 282.

[9] Representations of semi-simple Lie groups. V, *Amer. J. Math.* **78**(1956), 1–41. MR 18, 490.

[10] Representations of semi-simple Lie groups. VI, *Amer. J. Math.* **78**(1956), 564–628. MR 18, 490.

[11] On the characters of a semi-simple Lie group, *Bull. Amer. Math. Soc.* **61**(1955), 389–396. MR 17, 173.

[12] The characters of semi-simple Lie groups, *Trans. Amer. Math. Soc.* **83**(1956), 98–163. MR 18, 318.

[13] On a lemma of F. Bruhat, *J. Math. Pures. Appl.* (9) **35**(1956), 203–210. MR 18, 137.

[14] A formula for semi-simple Lie groups, *Amer. J. Math.* **79**(1957), 733–760. MR 20, #2633.

[15] Differential operators on a semi-simple Lie algebra, *Amer. J. Math.* **79**(1957), 87–120. MR 18, 809.

[16] Fourier transforms on a semi-simple Lie algebra. I, *Amer. J. Math.* **79**(1957), 193–257. MR 19, 293.

[17] Fourier transforms on a semi-simple Lie algebra. II, *Amer. J. Math.* **79**(1957), 653–686. MR 20, #2396.

[18] Spherical functions on a semi-simple Lie group. I, *Amer. J. Math.* **80**(1958), 241–310. MR 20, #925.

[19] Spherical functions on a semi-simple Lie group. II, *Amer. J. Math.* **80**(1958), 553–613. MR 21, #92.

[20] Automorphic forms on a semi-simple Lie group, *Proc. Nat. Acad. Sci. U.S.A.* **45**(1959), 570–573. MR 21, #4202.

[21] Invariant eigendistributions on semi-simple Lie groups, *Bull. Amer. Math. Soc.* **69**(1963), 117–123. MR 26, #2545.

HARTMAN, N. N., HENRICHS, R. W., and LASSER, R.

[1] Duals of orbit spaces in groups with relatively compact inner automorphism groups are hypergroups, *Monatsh. Math.* **88**(1979), 229–238. MR 81b: 43006.

HAUENSCHILD, W.

[1] Der Raum Prim$C^*(G)$ für eine abzählbare, lokal-endliche Gruppe G, *Arch. Math.* (Basel) **46**(1986), 114–117.

HAUENSCHILD, W., KANIUTH, E., and KUMAR, A.

[1] Harmonic analysis on central hypergroups and induced representations, *Pacific J. Math.* **110**(1984), 83–112. MR 85e: 43015.

HAWKINS, T.

[1] *Lebesgue's theory of integration: Its origins and development*, University of Wisconsin Press, 1970.

[2] The origins of the theory of group characters, *Arch. History Exact Sci.* **7**(1970-71), 142–170.

[3] Hypercomplex numbers, Lie groups and the creation of group representation theory, *Arch. History Exact Sci.* **8**(1971-72), 243–287.

[4] New light on Frobenius' creation of the theory of group characters, *Arch. History Exact Sci.* **12**(1974), 217–243.

VAN HEESWIJCK, L.

[1] Duality in the theory of crossed products, *Math. Scand.* **44**(1979), 313–329. MR 83d: 46082.

HELGASON, S.

[1] *Differential geometry and symmetric spaces*, Pure and Applied Mathematics, **12**, Academic Press, New York, 1962. MR 26, #2986.

[2] *Differential geometry, Lie groups, and symmetric spaces*, Pure and Applied Mathematics, **80**, Academic Press, New York-London, 1978. MR 80k: 53081.

[3] *Topics in harmonic analysis on homogeneous spaces*, Progress in Mathematics, **13**, Birkhäuser, Boston, Mass., 1981. MR 83g: 43009.

[4] *Groups and geometric analysis*, Pure and Applied Mathematics, **113**, Academic Press, New York-London, 1984. MR 86c: 22017.

HELSON, H.

[1] Analyticity on compact abelian groups, *Algebras in Analysis*, Birmingham, 1973, pp.1–62, Academic Press, London, 1975. MR 55 #989.

[2] *Harmonic analysis*, Addison Wesley Publishing Co., Reading, Mass., 1983. MR 85e: 43001.

HENNINGS, M. A.

[1] Fronsdal *-quantization and Fell inducing, *Math. Proc. Cambridge Philos. Soc.* **99**(1986), 179–188. MR 87f: 46097.

HENRARD, G.

[1] A fixed point theorem for C^*-crossed products with an abelian group, *Math. Scand.* **54**(1984), 27–39. MR 86b: 46111.

HENRICHS, R. W.

[1] Die Frobeniuseigenschaft FP für diskrete Gruppen, *Math. Z.* **147**(1976), 191–199. MR 53 #8324.

[2] On decomposition theory for unitary representations of locally compact groups, *J. Functional Anal.* **31**(1979), 101–114. MR 80g: 22002.

[3] Weak Frobenius reciprocity and compactness conditions in topological groups, *Pacific J. Math.* **82**(1979), 387–406. MR 81e: 22003.

HENSEL, K.

[1] *Zahlentheorie*, Berlin, 1913.

HERB, R.

[1] Characters of induced representations and weighted orbital integrals, *Pacific J. Math.* **114**(1984), 367–375. MR 86k: 22030.

HERB, R., and WOLF, J. A.

[1] The Plancherel theorem for general semisimple groups, *Compositio Math.* **57**(1986), 271–355. MR 87h: 22020

HERSTEIN, I. N.

[1] Group rings as *-algebras, *Publ. Math. Debrecen* **1**(1950), 201–204. MR 21, 475.

[2] *Noncommutative rings*, Carus Monographs **15**, Mathematical Association of America, Menascha, Wis., 1968. MR 37 #2790.

HEWITT, E.

[1] The ranges of certain convolution operators, *Math. Scand.* **15**(1964), 147–155. MR 32 #4471.

HEWITT, E., and KAKUTANI, S.

[1] A class of multiplicative linear functionals on the measure algebra of a locally compact abelian group, *Illinois J. Math.* **4**(1960), 553–574. MR 23 #A527.

HEWITT, E., and ROSS, K. A.

[1] *Abstract harmonic analysis, I. Structure of topological groups, integration theory, group representations*, Grundlehren der Math. Wiss. **115**, Springer-Verlag, Berlin-Göttingen-Heidelberg, 1963. MR 28 #158.

[2] *Abstract harmonic analysis, II. Structure and analysis for compact groups, analysis on locally compact abelian groups*, Grundlehren der Math. Wiss. **152**, Springer-Verlag, New York-Berlin, 1970. MR 41 #7378.

HIGGINS, P. J.

[1] *An introduction to topological groups*, Cambridge University Press, Cambridge, 1974. MR 50 #13355.

HIGMAN, D. G.

[1] Induced and produced modules, *Canad. J. Math.* **7**(1955), 490–508. MR 19, 390.

HILBERT, D.

[1] *Grundzüge einer allgemeinen Theorie der linearen Integralgleichungen*, Leipzig and Berlin, 1912.

HILDEBRANDT, T. H.

[1] Integration in abstract spaces, *Bull. Amer. Math. Soc.* **59**(1953), 111–139. MR 14, 735.

HILL, V. E.

[1] *Groups, representations, and characters*, Hafner Press, New York, 1975. MR 54 #7596.

HILLE, E.

[1] *Functional analysis and semigroups*, Amer. Math. Soc. Coll. Publ. 31, New York, 1948. MR 9, 594.

HILLE, E., and PHILLIPS, R. S.

[1] *Functional analysis and semigroups*, Amer. Math. Soc. Coll. Publ. 31, rev. ed., Amer. Math. Soc., Providence, R. I., 1957. MR 19, 664.

HILLE, E., and TAMARKIN, J. D.

[1] On the characteristic values of linear integral equations, *Acta Math.* **57**(1931), 1–76.

HIRSCHFELD, R.

[1] Représentations induites dans les espaces quasi-hilbertiens parfaits, *C. R. Acad. Sci. Paris Ser. A.-B* **272**(1971), A104–A106. MR 42 #7829.

[2] Projective limits of unitary representations, *Math. Ann.* **194**(1971), 180–196. MR 45 #6979.

[3] Duality of groups: a narrative review, *Nieuw Arch. Wisk.* (3) **20**(1972), 231–241. MR 49 #5226.

HOCHSCHILD, G.

[1] Cohomology and representations of associative algebras, *Duke. Math. J.* **14**(1947), 921–948. MR 9, 267.

HOFFMAN, K.

[1] *Banach spaces of analytic functions*, Prentice-Hall, Englewood Cliffs, N.J. 1962. MR 24 #A2844.

HOFMANN, K. H.

[1] *The duality of compact semigroups and C*-bigebras*, Lecture Notes in Mathematics, **129**, Springer-Verlag, Berlin-Heidelberg-New York, 1970. MR 55 #5786.

[2] Representations of algebras by continuous sections, *Bull. Amer. Math. Soc.* **78**(1972), 291–373. MR 50 #415.

[3] Banach bundles, Darmstadt Notes, 1974.

[4] Bundles and sheaves are equivalent in the category of Banach spaces, pp. 53–69, Lecture Notes in Mathematics, **575**, Springer-Verlag, Berlin-Heidelberg-New York, 1977.

HOFMANN, K. H., and LIUKKONEN, J. R., EDS.

[1] *Recent advances in the representation theory of rings and C*-algebras by continuous sections*, Memoirs Amer. Math. Soc. **148**, Amer. Math. Soc. Providence, R. I., 1974. MR 49 #7063.

HOLZHERR, A. K.

[1] Discrete groups whose multiplier representations are type I, *J. Austral. Math. Soc. Ser. A* **31**(1981), 486–495. MR 83g: 22003.

[2] Groups with finite dimensional irreducible multiplier representations, *Canad. J. Math.* **37**(1985), 635–643. MR 87a: 22010.

HOWE, R.

[1] On Frobenius reciprocity for unipotent algebraic groups over Q, *Amer. J. Math.* **93**(1971), 163–172. MR 43 #7556.

[2] Representation theory for division algebras over local fields (tamely ramified case), *Bull. Amer. Math. Soc.* **77**(1971), 1063–1066. MR 44 #4146.

[3] The Brauer group of a compact Hausdorff space and n-homogeneous C*-algebras, *Proc. Amer. Math. Soc.* **34**(1972), 209–214. MR 46 #4218.

[4] On representations of discrete, finitely generated, torsion-free, nilpotent groups, *Pacific J. Math.* **73**(1977), 281–305. MR 58 #16984.

[5] The Fourier transform for nilpotent locally compact groups, I. *Pacific J. Math.* **73**(1977), 307–327. MR 58 #11215.

[6] On the role of the Heisenberg group in harmonic analysis, *Bull. Amer. Math. Soc. N.S.* (2) **3**(1980), 821–843. MR 81h: 22010.

HOWE, R., and MOORE, C. C.

[1] Asymptotic properties of unitary representations, *J. Functional Anal.* **32**(1979), 72–96. MR 80g: 22017.

HULANICKI, A.

[1] Groups whose regular representation weakly contains all unitary representations, *Studia Math.* **24**(1964), 37–59. MR 33 #225.

HULANICKI, A., and PYTLIK, T.

[1] On cyclic vectors of induced representations, *Proc. Amer. Math. Soc.* **31**(1972), 633–634. MR 44 #6905.

[2] Corrigendum to: "On cyclic vectors of induced representations," *Proc. Amer. Math. Soc.* **38**(1973), 220. MR 48 #6318.

[3] On commutative approximate identities and cyclic vectors of induced representations, *Studia Math.* **48**(1973), 189–199. MR 49 #3024.

HURWITZ, A.

[1] Über die Erzeugung der Invarianten durch Integration, *Nachr. Ges. Gott. Math.-Phys.* K1. 1897, 71–90 (also Math. Werke. Vol. II, pp. 546–564. Birkhäuser, Basel, 1933).

HUSEMOLLER, D.

[1] *Fiber bundles*, Mcgraw-Hill, New York, 1966. MR 37 #4821.

HUSAIN, T.

[1] *Introduction to topological groups*, W. B. Saunders, Philadelphia-London, 1966. MR 34 #278.

IMAI, S., and TAKAI, H.

[1] On a duality for C^*-crossed products by a locally compact group, *J. Math. Soc. Japan* **30**(1978), 495–504. MR 81h: 46090.

IORIO, V. M.

[1] Hopf-C^*-algebras and locally compact groups, *Pacific J. Math.* **87**(1980), 75–96. MR 82b: 22007.

ISMAGILOV, R. C.

[1] On the unitary representations of the group of diffeomorphisms of the space R^n, $n > 2$, *J. Functional Anal.* **9**(1975), 71–72. MR 51 #14143.

JACOBSON, N.

[1] Structure theory of simple rings without finiteness assumptions, *Trans. Amer. Math. Soc.* **57**(1945), 228–245. MR 6, 200.

[2] The radical and semi-simplicity for arbitrary rings, *Amer. J. Math.* **67**(1945), 300–320. MR 7, 2.

[3] A topology for the set of primitive ideals in an arbitrary ring, *Proc. Nat. Acad. Sci. U.S.A.* **31**(1945), 333–338. MR 7, 110.

[4] *Structure of rings*, Amer. Math. Soc. Colloq. Publ. **37**, Amer. Math. Soc. Providence, R. I. 1956. MR 18, 373.

JACQUET, H., and SHALIKA, J.

[1] The Whittaker models of induced representations, *Pacific J. Math.* **109**(1983), 107–120. MR 85h: 22023.

JENKINS, J. W.

[1] An amenable group with a nonsymmetric group algebra, *Bull. Amer. Math. Soc.* **75**(1969), 359–360. MR 38 #6366.

[2] Free semigroups and unitary group representations, *Studia Math.* **43**(1972), 27–39. MR 47 #395.

JOHNSON, B. E.

[1] An introduction to the theory of centralizers. *Proc. Lond. Math. Soc.* **14**(1964), 229–320. MR 28 #2450.

[2] *Cohomology in Banach algebras*, Memoirs Amer. Math. Soc. **127**, Amer. Math. Soc. Providence, R. I. 1972. MR 51 #11130.

JOHNSON, G. P.

[1] Spaces of functions with values in a Banach algebra, *Trans. Amer. Math. Soc.* **92**(1959), 411–429. MR 21 #5910.

JORGENSEN, P. E. T.

[1] Perturbation and analytic continuation of group representations, *Bull. Amer. Math. Soc.* **82**(1976), 921–924. MR 58 #1026.

[2] Analytic continuation of local representations of Lie groups, *Pacific J. Math.* **125**(1986), 397–408.

JOY, K. I.

[1] A description of the topology on the dual space of a nilpotent Lie group, *Pacific J. Math.* **112**(1984), 135–139. MR 85e: 22013.

KAC, G. I.

[1] Generalized functions on a locally compact group and decompositions of unitary representations, *Trudy Moskov. Mat. Obšč.* **10**(1961), 3–40 (Russian). MR 27 #5863.

[2] Ring groups and the principle of duality, I. *Trudy Moskov. Mat. Obšč.* **12**(1963), 259–301 (Russian). MR 28 #164.

[3] Ring groups and the principle of duality, II. *Trudy Moskov. Mat. Obšč.* **13**(1965), 84–113 (Russian). MR 33 #226.

KADISON, R. V.

[1] *A representation theory for commutative topological algebra*, Memoirs Amer. Math. Soc., 7, Amer. Math. Soc., Providence, R. I., 1951. MR 13, 360.

[2] Isometries of operator algebras, *Ann. Math.* **54**(1951), 325–338. MR 13, 256.

[3] Order properties of bounded self-adjoint operators, *Proc. Amer. Math. Soc.* **2**(1951), 505–510. MR 13, 47.

[4] Infinite unitary groups, *Trans. Amer. Math. Soc.* **72**(1952), 386–399. MR 14, 16.

[5] Infinite general linear groups, *Trans. Amer. Math. Soc.* **76**(1954), 66–91. MR 15, 721.

[6] On the orthogonalization of operator representations, *Amer. J. Math.* **77**(1955), 600–621. MR 17, 285.

[7] Operator algebras with a faithful weakly closed representation, *Ann. Math.* **64**(1956), 175–181. MR 18, 54.

[8] Unitary invariants for representations of operator algebras, *Ann. Math.* **66**(1957), 304–379. MR 19, 665.

[9] Irreducible operator algebras, *Proc. Nat. Acad. Sci. U.S.A.* **43**(1957), 273–276. MR 19, 47.

[10] States and representations, *Trans. Amer. Math. Soc.* **103**(1962), 304–319. MR 25 #2459.

[11] Normalcy in operator algebras, *Duke. Math. J.* **29**(1962), 459–464. MR 26 #6814.

[12] Transformations of states in operator theory and dynamics, *Topology* **3**(1965), 177–198. MR 29 #6328.

[13] Strong continuity of operator functions, *Pacific J. Math.* **26**(1968), 121–129. MR 37 #6766.

[14] Theory of operators, Part II: operator algebras, *Bull. Amer. Math. Soc.* **64**(1958), 61–85.

[15] Operator algebras—the first forty years, in *Operator algebras and applications*, pp. 1–18, Proc. Symp. Pure Math., Amer. Math. Soc. Providence, R. I. 1982.

KADISON, R. V., and PEDERSEN, G. K.
  [1] Means and convex combinations of unitary operators, *Math. Scand.* **57**(1985), 249–266. MR 87g: 47078.
KADISON, R. V., and RINGROSE, J. R.
  [1] *Fundamentals of the theory of operator algebras*, vol. 1, *Elementary theory*, Pure and Applied Mathematics, **100**, Academic Press, New York-London, 1983. MR 85j: 46099.
  [2] *Fundamentals of the theory of operator algebras*, vol. 2, *Advanced theory*, Pure and Applied Mathematics, **100**, Academic Press, New York-London, 1986.
KAJIWARA, T.
  [1] Group extension and Plancherel formulas, *J. Math. Soc. Japan* **35**(1983), 93–115. MR 84d: 22011.
KAKUTANI, S., and KODAIRA, K.
  [1] Über das Haarsche Mass in der lokal bikompakten Gruppe, *Proc. Imp. Acad. Tokyo* **20**(1944), 444–450. MR 7, 279.
KALLMAN, R. R.
  [1] A characterization of uniformly continuous unitary representations of connected locally compact groups, *Michigan Math. J.* **16**(1969), 257–263. MR 40 #5787.
  [2] Unitary groups and automorphisms of operator algebras, *Amer. J. Math.* **91**(1969), 785–806. MR 40 #7825.
  [3] A generalization of a theorem of Berger and Coburn, *J. Math. Mech.* **19**(1969/70), 1005–1010. MR 41 #5982.
  [4] Certain topological groups are type I., *Bull. Amer. Math. Soc.* **76**(1970), 404–406. MR 41 #385.
  [5] Certain topological groups are type I. II., *Adv. Math.* **10**(1973), 221–255.
  [6] A theorem on the restriction of type I representations of a group to certain of its subgroups, *Proc. Amer. Math. Soc.* **40**(1973), 291–296. MR 47 #5177.
  [7] The existence of invariant measures on certain quotient spaces, *Adv. Math.* **11**(1973), 387–391. MR 48 #8682.
  [8] Certain quotient spaces are countably separated, I., *Illinois J. Math.* **19**(1975), 378–388. MR 52 #650.
  [9] Certain quotient spaces are countably separated, II, *J. Functional Anal.* **21**(1976), 52–62. MR 54 #5384.
  [10] Certain quotient spaces are countably separated, III. *J. Functional Anal.* **22**(1976), 225–241. MR 54 #5385.
KAMPEN, E. R. VAN.
  [1] Locally bicompact Abelian groups and their character groups, *Ann. Math.* (2) **36**(1935), 448–463.
KANIUTH, E.
  [1] Topology in duals of SIN-groups, *Math. Z.* **134**(1973), 67–80. MR 48 #4197.
  [2] On the maximal ideals in group algebras of SIN-groups, *Math. Ann.* **214**(1975), 167–175. MR 52 #6325.

[3] A note on reduced duals of certain locally compact groups, *Math. Z.* **150**(1976), 189–194. MR 54 #5386.

[4] On separation in reduced duals of groups with compact invariant neighborhoods of the identity, *Math. Ann.* **232**(1978), 177–182. MR 58 #6053.

[5] Primitive ideal spaces of groups with relatively compact conjugacy classes, *Arch. Math. (Basel)* **32**(1979), 16–24. MR 81f: 22009.

[6] Ideals in group algebras of finitely generated FC-nilpotent discrete groups, *Math. Ann.* **248**(1980), 97–108. MR 81i: 43005.

[7] Weak containment and tensor products of group representations, *Math. Z.* **180**(1982), 107–117. MR 83h: 22013.

[8] On primary ideals in group algebras, *Monatsh. Math.* **93**(1982), 293–302. MR 84i: 43003.

[9] Weak containment and tensor products of group representations, II, *Math. Ann.* **270**(1985), 1–15. MR 86j: 22004.

[10] On topological Frobenius reciprocity for locally compact groups, *Arch. Math.* **48**(1987), 286–297.

KAPLANSKY, I.

[1] Topological rings, *Amer., J. Math.* **69**(1947), 153–183. MR 8, 434.

[2] Locally compact rings. I, *Amer. J. Math.* **70**(1948), 447–459. MR 9, 562.

[3] Dual rings, *Ann. of Math.* **49**(1948), 689–701. MR 10, 7.

[4] Normed algebras, *Duke Math. J.* **16**(1949), 399–418. MR 11, 115.

[5] Locally compact rings. II, *Amer. J. Math.* **73**(1951), 20–24. MR 12, 584.

[6] The structure of certain operator algebras, *Trans. Amer. Math. Soc.* **70**(1951), 219–255. MR 13, 48.

[7] Group algebras in the large, *Tohoku Math. J.* **3**(1951), 249–256. MR 14, 58.

[8] Projections in Banach algebras, *Ann. Math.* (2) **53**(1951), 235–249. MR 13, 48.

[9] A theorem on rings of operators, *Pacific J. Math.* **1**(1951), 227–232. MR 14, 291.

[10] Modules over operator algebras, *Amer. J. Math.* **75**(1953), 839–858. MR 15, 327.

KARPILOVSKY, G.

[1] *Projective representations of finite groups*, Pure and Applied Mathematics, 84, Marcel-Dekker, New York, 1985.

KASPAROV, G. G.

[1] Hilbert C^*-modules: theorems of Stinespring and Voiculescu, *J. Operator Theory*, **4**(1980), 133–150. MR 82b: 46074.

KATAYAMA, Y.

[1] Takesaki's duality for a non-degenerate co-action, *Math. Scand.* **55**(1984), 141–151. MR 86b: 46112.

KATZNELSON, Y., and WEISS, B.

[1] The construction of quasi-invariant measures, *Israel J. Math.* **12**(1972), 1–4. MR 47 #5226.

KAWAKAMI, S.

[1] Irreducible representations of nonregular semi-direct product groups, *Math. Japon.* **26**(1981), 667–693. MR 84a: 22015.

[2] Representations of the discrete Heisenberg group, *Math. Japon.* **27**(1982), 551–564. MR 84d: 22015.

KAWAKAMI, S., and KAJIWARA, T.

[1] Representations of certain non-type I *C**-crossed products, *Math. Japon.* **27**(1982), 675–699. MR 84d: 22013.

KEHLET, E. T.

[1] A proof of the Mackey-Blattner-Nielson theorem, *Math. Scand.* **43**(1978), 329–335. MR 80i: 22019.

[2] A non-separable measurable choice principle related to induced representations, *Math. Scand.* **42**(1978), 119–134. MR 58 #16954.

[3] On extensions of locally compact groups and unitary groups, *Math. Scand.* **45**(1979), 35–49. MR 81d: 22009.

[4] Cross sections for quotient maps of locally compact groups, *Math. Scand.* **55**(1984), 152–160. MR 86k: 22011.

KELLEY, J. L.

[1] *General topology*, D. Van Nostrand Co., Toronto-New York-London, 1955. MR 16, 1136.

KELLEY, J. L., NAMIOKA, I., ET AL.

[1] *Linear topological spaces*, D. Van Nostrand Co., Princeton, 1963. MR 29 #3851.

KELLEY, J. L., and SRINAVASAN, T. P.

[1] *Measures and integrals*, Springer Graduate Series, Springer, New York, in Press.

KELLEY, J. L., and VAUGHT, R. L.

[1] The positive cone in Banach algebras, *Trans. Amer. Math. Soc.* **74**(1953), 44–55. MR 14, 883.

KEOWN, R.

[1] *An introduction to group representation theory*, Academic Press, New York, 1975. MR 52 #8230.

KHOLEVO, A. S.

[1] Generalized imprimitivity systems for abelian groups. *Izv. Vyssh. Uchebn. Zaved. Mat.* (1983), 49–71 (Russian). MR 84m: 22008.

KIRCHBERG, E.

[1] Representations of coinvolutive Hopf-*W**-algebras and non-abelian duality, *Bull. Acad. Polon. Sci. Sér. Sci. Math. Astronom. Phys.* **25**(1977), 117–122. MR 56 #6415.

KIRILLOV, A. A.

[1] Induced representations of nilpotent Lie groups, *Dokl. Akad. Nauk SSSR* **128**(1959), 886–889 (Russian). MR 22 #740.

[2] On unitary representations of nilpotent Lie groups, *Dokl Akad. Nauk. SSSR* **130**(1960), 966–968 (Russian). MR 24 #A3240.

[3] Unitary representations of nilpotent Lie groups, Dokl. Akad. Nauk. SSSR **138**(1961), 283–284 (Russian). MR 23 #A3205.

[4] Unitary representations of nilpotent Lie groups, *Uspehi Mat. Nauk* **106**(1962), 57–110 (Russian). MR 25 #5396.

[5] Representations of certain infinite-dimensional Lie groups, *Vestnik Moskov . Univ. Ser. I. Mat. Meh.* **29**(1974), 75–83 (Russian). MR 52 #667.

[6] *Elements of the theory of representations*, 2nd Ed. Nauka. Moscow, 1978 (Russian). MR 80b: 22001. [English translation of first ed.: Grundlehren der math. Wiss. 220, Springer-Verlag, Berlin-Heildelberg-New York, 1976].

KISHIMOTO, A.

[1] Simple crossed products of C^*-algebras by locally compact abelian groups (preprint), University of New South Wales, Kensington, 1979.

KITCHEN, J. W., and ROBBINS, D. A.

[1] Tensor products of Banach bundles, *Pacific J. Math.* **94**(1981), 151–169. MR 83f: 46078.

[2] Sectional representations of Banach modules, *Pacific J. Math.* **109**(1983), 135–156. MR 85a: 46026.

[3] Internal functionals and bundle duals, *Int. J. Math. Sci.* **7**(1984), 689–695. MR 86i: 46077.

KLEPPNER, A.

[1] The structure of some induced representations, *Duke Math. J.* **29**(1962), 555–572. MR 25 #5132.

[2] Intertwining forms for summable induced representations, *Trans. Amer. Math. Soc.* **112**(1964), 164–183. MR 33 #4186.

[3] Multipliers on abelian groups, *Math. Ann.* **158**(1965), 11–34. MR 30 #4856.

[4] Representations induced from compact subgroups, *Amer. J. Math.* **88**(1966), 544–552. MR 36 #1577.

[5] Continuity and measurability of multiplier and projective representations, *J. Functional Anal.* **17**(1974), 214–226. MR 51 #790.

[6] Multiplier representations of discrete groups, *Proc. Amer. Math. Soc.* **88**(1983), 371–375. MR 84k: 22007.

KLEPPNER, A., and LIPSMAN, R. L.

[1] The Plancherel formula for group extensions. I. II., *Ann. Sci. École. Norm. Sup.* (4) **5**(1972), 459–516; ibid. (4) **6**(1973), 103–132. MR 49 #7387.

KNAPP, A. W.

[1] *Representation theory of semisimple groups, an overview based on examples*, Princeton Mathematical Series, **36**, Princeton University Press, Princeton, N. J., 1986.

KOBAYASHI, S.

[1] On automorphism groups of homogeneous complex manifolds, *Proc. Amer. Math. Soc.* **12**(1961), 359–361. MR 24 #A3664.

KOOSIS, P.

[1] An irreducible unitary representation of a compact group is finite dimensional, *Proc. Amer. Math. Soc.* **8**(1957), 712–715. MR 19, 430.

KOPPINEN, M., and NEUVONEN, T.

[1] An imprimitivity theorem for Hopf algebras, *Math. Scand.* **41**(1977), 193–198. MR 58 #5758.

KOORNWINDER, T. H., and VAN DER MEER, H. A.

[1] Induced representations of locally compact groups, *Representations of locally compact groups with applications*, Colloq. Math. Centre, Amsterdam, 1977/78, pp. 329–376, MC syllabus, **38**, Math. Centrum, Amsterdam, 1979. MR 81k: 22001.

KOTZMANN, E., LOSERT, V., and RINDLER, H.

[1] Dense ideals of group algebras, *Math. Ann.* **246**(1979/80), 1–14. MR 80m: 22007.

KRALJEVIĆ, H.

[1] Induced representations of locally compact groups on Banach spaces, *Glasnik Math.* **4**(1969), 183–196. MR 41 #1928.

KRALJEVIĆ, H., and MILIČIĆ, D.

[1] The C*-algebra of the universal covering group of *SL*(2, *R*), *Glasnik Mat. Ser.* III (27) **7**(1972), 35–48. MR 49 #5222.

KRIEGER, W.

[1] On quasi-invariant measures in uniquely ergodic systems, *Invent. Math.* **14**(1971), 184–196. MR 45 #2139.

KUGLER, W.

[1] Über die Einfachheit gewisser verallgemeinerter L^1-Algebren, *Arch. Math. (Basel)* **26**(1975), 82–88. MR 52 #6326.

[2] On the symmetry of generalized L^1-algebras, *Math. Z.* **168**(1979), 241–262. MR 82f: 46074.

KUMJIAN, A.

[1] On C*-diagonals, *Can. J. Math.* **38**(1986), 969–1008.

KUNZE, R. A.

[1] L_p Fourier transforms on locally compact unimodular groups, *Trans. Amer. Math. Soc.* **89**(1958), 519–540. MR 20 #6668.

[2] On the irreducibility of certain multiplier representations, *Bull. Amer. Math. Soc.* **68**(1962), 93–94. MR 24 #A3239.

[3] *Seminar: Induced representations without quasi-invariant measures.* Summer School on Representations of Lie Groups, Namur, 1969, pp. 283–289. Math. Dept., Univ. of Brussels, Brussels, 1969. MR 57 #9904.

[4] On the Frobenius reciprocity theorem for square-integrable representations, *Pacific J. Math.* **53**(1974), 465–471. MR 51 #10530.

[5] Quotient representations, *Topics in modern harmonic analysis*, Vol. I, II, Turin/Milan, 1982, pp. 57–80, Ist. Naz. Alta Mat. Francesco Severi, Rome, 1983. MR 85j: 22009.

KUNZE, R. A., and STEIN, E. M.

[1] Uniformly bounded representations and harmonic analysis on the $n \times n$ real unimodular group, *Amer. J. Math.* **82**(1960), 1–62. MR 29 #1287.

[2] Uniformly bounded representations. II. Analytic continuation of the principal series of representations of the $n \times n$ complex unimodular group, *Amer. J. Math.* **83**(1961), 723–786. MR 29 #1288.

KURATOWSKI, K.

[1] *Introduction to set theory and topology*, Pergamon Press, Oxford-Warsaw, 1972. MR 49 #11449.

KUSUDA, M.

[1] Crossed products of C^*-dynamical systems with ground states, *Proc. Amer. Math. Soc.* **89**(1983), 273–278.

KUTZKO, P. C.

[1] Mackey's theorem for nonunitary representations, *Proc. Amer. Math. Soc.* **64**(1977), 173–175. MR 56 #533.

LANCE, E. C.

[1] Automorphisms of certain operator algebras, *Amer. J. Math.* **91**(1969), 160–174. MR 39 #3324.

[2] On nuclear C^*-algebras, *J. Functional Anal.* **12**(1973), 157–176. MR 49 #9640.

[3] Refinement of direct integral decompositions, *Bull. London Math. Soc.* **8**(1976), 49–56. MR 54 #8307.

[4] Tensor products of non-unital C^*-algebras, *J. London Math. Soc.* **12**(1976), 160–168. MR 55 #11059.

LANDSTAD, M.

[1] Duality theory for covariant systems, Thesis, University of Pennsylvania, 1974.

[2] Duality for dual covariant algebras, *Comm. Math. Physics* **52**(1977), 191–202. MR 56 #8750.

[3] Duality theory for covariant systems, *Trans. Amer. Math. Soc.* **248**(1979), 223–269. MR 80j: 46107.

[4] Duality for dual C^*-covariance algebras over compact groups, University of Trondheim (preprint), 1978.

LANDSTAD, M. B., PHILLIPS, J., RAEBURN, I., and SUTHERLAND, C. E.

[1] Representations of crossed products by coactions and principal bundles, *Trans. Amer. Math. Soc.* **299**(1987), 747–784.

LANG, S.

[1] *Algebra*, Addison-Wesley Publishing Co., Reading, Mass., 1965. MR 33 #5416.

[2] $SL_2(R)$, Addison-Wesley Publishing Co., Reading, Mass., 1975. MR 55 #3170.

LANGE, K.

[1] A reciprocity theorem for ergodic actions, *Trans. Amer. Math. Soc.* **167**(1972), 59–78. MR 45 #2085.

LANGE, K., RAMSAY, A., and ROTA, G. C.

[1] Frobenius reciprocity in ergodic theory, *Bull. Amer. Math. Soc.* **77**(1971), 713–718. MR 44 #1769.

LANGLANDS, R.
[1] The Dirac monopole and induced representations, *Pacific J. Math.* **126**(1987), 145–151.

LANGWORTHY, H. F.
[1] Imprimitivity in Lie groups, Ph.D. thesis, University of Minnesota, 1970.

LARSEN, R.
[1] *An introduction to the theory of multipliers*, Springer-Verlag New York-Heidelberg, 1971. MR 55 #8695.

LEBESGUE, H.
[1] Sur les séries trigonométriques, *Ann. École. Norm. Sup.* (3) **20**(1903), 453–485.

LEE, R. Y.
[1] On the C^*-algebras of operator fields, *Indiana Univ. Math. J.* **25**(1976), 303–314. MR 53 #14150.
[2] Full algebras of operator fields trivial except at one point, *Indiana Univ. Math. J.* **26**(1977), 351–372. MR 55 #3812.
[3] On C^*-algebras which are approachable by finite-dimensional C^*-algebras, *Bull. Inst. Math. Acad. Sinica* **5**(1977), 265–283. MR 58 #17864.

LEE, T. Y.
[1] Embedding theorems in group C^*-algebras, *Canad. Math. Bull.* **26**(1983), 157–166. MR 85a: 22012.

DE LEEUW, K., and GLICKSBERG, I.
[1] The decomposition of certain group representations, *J. d'Analyse Math.* **15**(1965), 135–192. MR 32 #4211.

LEINERT, M.
[1] Fell-Bündel und verallgemeinerte L^1-Algebren, *J. Functional Anal.* **22**(1976), 323–345. MR 54 #7694.

LEJA, F.
[1] Sur la notion du groupe abstract topologique, *Fund. Math.* **9**(1927), 37–44.

LEPTIN, H.
[1] Verallgemeinerte L^1-Algebren, *Math. Ann.* **159**(1965), 51–76. MR 34 #6545.
[2] Verallgemeinerte L^1-Algebren und projektive Darstellungen lokal kompakter Gruppen. I, *Invent. Math.* **3**(1967), 257–281. MR 37 #5328.
[3] Verallgemeinerte L^1-Algebren, und projektive Darstellungen lokal kompakter Gruppen. II, *Invent. Math.* **4**(1967), 68–86. MR 37 #5328.
[4] Darstellungen verallgemeinerter L^1-Algebren, *Invent. Math.* **5**(1968), 192–215. MR 38 #5022.
[5] *Darstellungen verallgemeinerter L^1-Algebren.* II, pp. 251–307, Lecture Notes in Mathematics **247**, Springer-Verlag, Berlin and New York, 1972. MR 51 #6433.
[6] The structure of $L^1(G)$ for locally compact groups, *Operator Algebras and Group Representations*, pp. 48–61, Pitman Publishing Co., Boston-London, 1984. MR 85d: 22015.

LIPSMAN, R. L.

[1] *Group representations*, Lecture Notes in Mathematics **388**, Springer-Verlag, Berlin and New York, 1974. MR 51 #8333.

[2] Non-Abelian Fourier analysis, *Bull. Sci. Math.* (2) **98**(1974), 209–233. MR 54 #13467.

[3] Harmonic induction on Lie groups, *J. Reine Angew. Math.* **344**(1983), 120–148. MR 85h: 22021.

[4] Generic representations are induced from square-integrable representations, *Trans. Amer. Math. Soc.* **285**(1984), 845–854. MR 86c: 22022.

[5] An orbital perspective on square-integrable representations, *Indiana Univ. Math.* **34**(1985), 393–403.

LITTLEWOOD, D. E.

[1] *Theory of group characters and matrix representations of groups*, Oxford University Press, New York, 1940. MR 2, 3.

LITVINOV, G. L.

[1] Group representations in locally convex spaces, and topological group algebras, *Trudy Sem. Vektor. Tenzor. Anal.* **16**(1972), 267–349 (Russian). MR 54 #10479.

LIUKKONEN, J. R.

[1] Dual spaces of groups with precompact conjugacy classes, *Trans. Amer. Math. Soc.* **180**(1973), 85–108. MR 47 #6937.

LIUKKONEN, J. R., and MOSAK, R.

[1] The primitive ideal space of [FC]⁻ groups, *J. Functional Anal.* **15**(1974), 279–296. MR 49 #10814.

[2] Harmonic analysis and centers of group algebras, *Trans. Amer. Math. Soc.* **195**(1974), 147–163. MR 50 #2815.

LONGO, R.

[1] Solution of the factorial Stone-Weierstrass conjecture. An application of the theory of standard split W^*-inclusions, *Invent. Math.* **76**(1984), 145–155. MR 85m: 46057a.

LONGO, R. and PELIGRAD, C.

[1] Noncommutative topological dynamics and compact actions on C^*-algebras, *J. Functional Anal.* **58**(1984), 157–174. MR 86b: 46114.

LOOMIS, L. H.

[1] Haar measure in uniform structures, *Duke Math. J.* **16**(1949), 193–208. MR 10, 600.

[2] *An introduction to abstract harmonic analysis*, D. Van Nostrand Co., Toronto-New York-London, 1953. MR 14, 883.

[3] Positive definite functions and induced representations, *Duke Math. J.* **27**(1960), 569–579. MR 26 #6303.

LOSERT, V., and RINDLER, H.

[1] Cyclic vectors for $L^p(G)$, *Pacific J. Math.* **89**(1980), 143–145. MR 82b: 43006.

LUDWIG, J.
  [1] Good ideals in the group algebra of a nilpotent Lie group, *Math. Z.* **161**(1978), 195–210. MR 58 #16958.
  [2] Prime ideals in the *C**-algebra of a nilpotent group, *Mh. Math.* **101**(1986), 159–165.

MACKEY, G. W.
  [1] On a theorem of Stone and von Neumann, *Duke Math. J.* **16**(1949), 313–326. MR 11, 10.
  [2] Imprimitivity for representations of locally compact groups, I, *Proc. Nat. Acad. Sci. U.S.A.* **35**(1949), 537–545. MR 11, 158.
  [3] Functions on locally compact groups, *Bull. Amer. Math. Soc.* **56**(1950), 385–412 (survey article). MR 12, 588.
  [4] On induced representations of groups, *Amer. J. Math.* **73**(1951), 576–592. MR 13, 106.
  [5] Induced representations of locally compact groups. I, *Ann. Math.* **55**(1952), 101–139. MR 13, 434.
  [6] Induced representations of locally compact groups. II, The Frobenius theorem, *Ann. Math.* **58**(1953), 193–220. MR 15, 101.
  [7] Borel structure in groups and their duals, *Trans. Amer. Math. Soc.* **85**(1957), 134–165. MR 19, 752.
  [8] Unitary representations of group extensions. I, *Acta Math.* **99**(1958), 265–311. MR 20 #4789.
  [9] *Induced representations and normal subgroups*, Proc. Int. Symp. Linear Spaces, Jerusalem, pp. 319–326, Pergamon Press, Oxford, 1960. MR 25 #3118.
  [10] Infinite dimensional group representations, *Bull. Amer. Math. Soc.* **69**(1963), 628–686. MR 27 #3745.
  [11] Ergodic theory, group theory, and differential geometry, *Proc. Nat. Acad. Sci. U.S.A.* **50**(1963), 1184–1191. MR 29 #2325.
  [12] *Group representations and non-commutative harmonic analysis with applications to analysis, number theory, and physics*, Mimeographed notes, University of California, Berkeley, 1965.
  [13] Ergodic theory and virtual groups, *Math. Ann.* **166**(1966), 187–207. MR 34 #1444.
  [14] *Induced representations of groups and quantum mechanics*, W. A. Benjamin, New York-Amsterdam, 1968. MR 58 #22373.
  [15] *The theory of unitary group representations*. Based on notes by J. M. G. Fell and D. B. Lowdenslager of lectures given at University of Chicago, 1955, University of Chicago Press, Chicago, 1976. MR 53 #686.
  [16] Products of subgroups and projective multipliers, *Colloq. Math.* **5**(1970), 401–413. MR 50 #13370.
  [17] *Ergodicity in the theory of group representations*, Actes du Congrès International des Mathématiciens, Nice, 1970, Tome 2, pp. 401–405, Gauthier-Villars, Paris, 1971. MR 55 #8247.

[18] *Induced representations of locally compact groups and applications*: *Functional analysis and related fields*, Proc. Conf. M. Stone, University of Chicago, 1968, pp. 132–166, Springer, New York, 1970. MR 54 #12968.

[19] *On the structure of the set of conjugate classes in certain locally compact groups*, Symp. Math. **16**, Conv. Gruppi Topol. Gruppi Lie, INDAM, Rome, 1974, pp. 433–467, Academic Press, London, 1975. MR 53 #8323.

[20] Ergodic theory and its significance for statistical mechanics and probability theory, *Adv. Math.* **12**(1974), 178–268. MR 49 #10857.

[21] *Unitary group representations in physics, probability, and number theory*, Benjamin/Cummings, Reading, Mass., 1978. MR 80i: 22001.

[22] *Origins and early history of the theory of unitary group representations*, pp. 5–19, London Math. Soc., Lecture Note Series, **34**, Cambridge University Press, Cambridge, 1979. MR 81j: 22001.

[23] Harmonic analysis as the exploitation of symmetry—a historical survey, *Bull. Amer. Math. Soc.* **3**(1980), 543–698. MR 81d: 01019.

MAGYAR, Z., and SEBESTYÉN, Z.

[1] On the definition of *C**-algebras, II, *Canad. J. Math.* **37**(1985), 664–681. MR 87b: 46061.

MARÉCHAL, O.

[1] Champs mesurables d'espaces hilbertiens, *Bull. Sci. Math.* **93**(1969), 113–143. MR 41 #5948.

MARTIN, R. P.

[1] Tensor products for *SL*(2, *k*) *Trans. Amer. Math. Soc.* **239**(1978), 197–211. MR 80i: 22033.

MASCHKE, H.

[1] Beweis des Satzes, dass diejenigen endlichen linearen Substitutionsgruppen, in welchen einige durchgehends verschwindende Koeffizienten auftreten, intransitiv sind, *Math. Ann.* **52**(1899), 363–368.

MATHEW, J., and NADKARNI, M. G.

[1] On systems of imprimitivity on locally compact abelian groups with dense actions, *Ann. Inst. Fourier (Grenoble)* **28**(1978), 1–23. MR 81j: 22002.

MAUCERI, G.

[1] Square integrable representations and the Fourier algebra of a unimodular group, *Pacific J. Math.* **73**(1977), 143–154. MR 58 #6054.

MAUCERI, G., and RICARDELLO, M. A.

[1] Noncompact unimodular groups with purely atomic Plancherel measures, *Proc. Amer. Math. Soc.* **78**(1980), 77–84. MR 81h: 22005.

MAURIN, K.

[1] Distributionen auf Yamabe-Gruppen, Harmonische Analyse auf einer Abelschen l.k. Gruppe, *Bull. Acad. Polon. Sci. Sér. Sci. Math. Astronom. Phys.* **9**(1961), 845–850. MR 24 #A2839.

MAURIN, K., and MAURIN, L.

[1] Duality, imprimitivity, reciprocity, *Ann. Polon. Math.* **29**(1975), 309–322. MR 52 #8330.

MAUTNER, F. I.

[1] Unitary representations of locally compact groups. I, *Ann. Math.* **51**(1950), 1–25. MR 11, 324.

[2] Unitary representations of locally compact groups. II, *Ann. Math.* **52**(1950), 528–556. MR 12, 157.

[3] The structure of the regular representation of certain discrete groups, *Duke Math. J.* **17**(1950), 437–441. MR 12, 588.

[4] The regular representation of a restricted direct product of finite groups, *Trans. Amer. Math. Soc.* **70**(1951), 531–548. MR 13, 11.

[5] Fourier analysis and symmetric spaces, *Proc. Nat. Acad. Sci. U.S.A.* **37**(1951), 529–533. MR 13, 434.

[6] A generalization of the Frobenius reciprocity theorem, *Proc. Nat. Acad. Sci. U.S.A.* **37**(1951), 431–435. MR 13, 205.

[7] Spherical functions over p-adic fields, *Amer. J. Math.* **80**(1958), 441–457. MR 20, #82.

MAYER, M. E.

[1] *Differentiable cross sections in Banach *-algebraic bundles*, pp. 369–387, Cargése Lectures, 1969, D. Kastler, ed., Gordon & Breach, New York, 1970. MR 58 #31188.

[2] Automorphism groups of C^*-algebras, Fell bundles, W^*-bigebras, and the description of internal symmetries in algebraic quantum theory, *Acta Phys. Austr. Suppl.* **8**(1971), 177–226, Springer-Verlag, 1971. MR 48 #1596.

MAYER-LINDENBERG, F.

[1] Invariante Spuren auf Gruppenalgebren, *J. Reine. Angew. Math.* **310**(1979), 204–213. MR 81b: 22008.

[2] Zur Dualitätstheorie symmetrischer Paare, *J. Reine Angew. Math.* **321**(1981), 36–52. MR 83a: 22009.

MAZUR, S.

[1] Sur les anneaux linéaires, *C. R. Acad. Sci. Paris* **207**(1938), 1025–1027.

MICHAEL, E. A.

[1] *Locally multiplicatively convex topological algebras*, Memoirs Amer. Math. Soc. **11**, Amer. Math. Soc., Providence, R. I., 1952. MR 14, 482.

MILIČIĆ, D.

[1] Topological representation of the group C^*-algebra of $SL(2, R)$, *Glasnik Mat. Ser. III (26)* **6**(1971), 231–246. MR 46 #7909.

MINGO, J. A., and PHILLIPS, W. J.

[1] Equivalent triviality theorems for Hilbert C^*-modules. *Proc. Amer. Math. Soc.* **91**(1984), 225–230. MR 85f: 46111.

MITCHELL, W. E.

[1] The σ-regular representation of $Z \times Z$, Michigan Math. J. **31**(1984), 259–262. MR 86h: 22008.

MONTGOMERY, D., and ZIPPIN, L.

[1] Topological transformation groups, Interscience, New York, 1955. MR 17, 383.

MOORE, C. C.

[1] On the Frobenius reciprocity theorem for locally compact groups, Pacific J. Math. **12**(1962), 359–365. MR 25 #5134.

[2] Groups with finite-dimensional irreducible representations, Trans. Amer. Math. Soc. **166**(1972), 401–410. MR 46 #1960.

[3] Group extensions and cohomology for locally compact groups. III, Trans. Amer. Math. Soc. **221**(1976), 1–33. MR 54 #2867.

[4] Group extensions and cohomology for locally compact groups. IV, Trans. Amer. Math. Soc. **221**(1976), 34–58. MR 54 #2868.

[5] Square integrable primary representations, Pacific J. Math. **70**(1977), 413–427. MR 58 #22381.

MOORE, C. C., and REPKA, J.

[1] A reciprocity theorem for tensor products of group representations, Proc. Amer. Math. Soc. **64**(1977), 361–364. MR 56 #8749.

MOORE, C. C., and ROSENBERG, J.

[1] Comments on a paper of I. D. Brown and Y. Guivarc'h, Ann. Sci. École Norm. Sup. (4) **8**(1975), 379–381. MR 55 #570b.

[2] Groups with T_1 primitive ideal spaces, J. Functional Anal. **22**(1976), 204–224. MR 54 #7693.

MOORE, C. C., and ZIMMER, R. J.

[1] Groups admitting ergodic actions with generalized discrete spectrum, Invent. Math. **51**(1979), 171–188. MR 80m: 22008.

MOORE, R. T.

[1] Measurable, continuous and smooth vectors for semigroups and group representations, Memoirs Amer. Math. Soc. **78**, Amer. Math. Soc. Providence, R. I., 1968. MR 37 #4669.

MORITA, K.

[1] Duality for modules and its applications to the theory of rings with minimum condition, Tokyo Kyoiku Daigaku Sec. A **6**(1958), 83–142. MR 20 #3183.

MORRIS, S. A.

[1] Pontryagin duality and the structure of locally compact abelian groups, Cambridge University Press, Cambridge, 1977. MR 56 #529.

[2] Duality and structure of locally compact abelian groups ... for the layman, Math. Chronicle **8**(1979), 39–56. MR 81a: 22003.

MOSAK, R. D.

[1] A note on Banach representations of Moore groups, Resultate Math. **5**(1982), 177–183. MR 85d: 22012.

Moscovici, H.
[1] Generalized induced representations, *Rev. Roumaine Math. Pures. Appl.* **14**(1969), 1539–1551. MR 41 #3671.
[2] Topological Frobenius properties for nilpotent Lie groups, *Rev. Roumaine Math. Pures Appl.* **19**(1974), 421–425. MR 50 #540.

Moscovici, H., and Verona, A.
[1] Harmonically induced representations of nilpotent Lie groups, *Invent. Math.* **48**(1978), 61–73. MR 80a: 22011.
[2] Holomorphically induced representations of solvable Lie groups, *Bull. Sci. Math.* (2) **102**(1978), 273–286. MR 80i: 22025.

Moskalenko, Z. I.
[1] On the question of the locally compact topologization of a group, *Ukrain. Mat. Ž.* **30**(1978), 257–260, 285 (Russian). MR 58 #11208.

Mostert, P.
[1] Sections in principal fibre spaces, *Duke Math. J.* **23**(1956), 57–71. MR 57, 71.

Mueller-Roemer, P. R.
[1] A note on Mackey's imprimitivity theorem, *Bull. Amer. Math. Soc.* **77**(1971), 1089–1090. MR 44 #6907.
[2] Kontrahierende Erweiterungen und Kontrahierbare Gruppen, *J. Reine Angew. Math.* **283/284**(1976), 238–264. MR 53 #10977.

Muhly, P. S., and Williams, D. P.
[1] Transformation group C*-algebras with continuous trace. II, *J. Operator Theory*, **11**(1984), 109–124. MR 85k: 46074.

Muhly, P. S., Renault, J. N., and Williams, D. P.
[1] Equivalence and isomorphism for groupoid C*-algebras, *J. Operator Theory* **17**(1987), 3–22.

Murnaghan, F.
[1] *The theory of group representations*, Dover, New York, 1963 (reprint).

Murray, F. J., and von Neumann, J.
[1] On rings of operators I, *Ann. Math.* **37**(1936), 116–229.
[2] On rings of operators II, *Trans. Amer. Math. Soc.* **41**(1937), 208–248.
[3] On rings of operators IV, *Ann. Math.* **44**(1943), 716–808. MR 5, 101.

Nachbin, L.
[1] *The Haar integral*, D. Van Nostrand, New York, 1965. MR 31 #271.

Nagumo, M.
[1] Einige analytische Untersuchungen in linearen metrischen Ringen, *Jap. J. Math.* **13**(1936), 61–80.

Naimark, M. A.
[1] Positive definite operator functions on a commutative group, *Izv. Akad. Nauk SSSR* **7**(1943), 237–244 (Russian). MR 5, 272.
[2] Rings with involution, *Uspehi Mat. Nauk* **3**(1948), 52–145 (Russian). MR 10, 308.

[3] On a problem of the theory of rings with involution, *Uspehi Mat. Nauk* **6**(1951), 160–164 (Russian). MR 13, 755.

[4] On irreducible linear representations of the proper Lorentz group, *Dokl. Akad. Nauk SSSR* **97**(1954), 969–972 (Russian). MR 16, 218.

[5] *Normed rings*, Gosudarstv. Izdat. Tehn.-Teor. Lit., Moscow, 1956 (Russian). MR 19, 870. MR 22 #1824.

[6] *Linear representations of the Lorentz group*, Gosudarstv. Izdat. Fiz.-Mat. Lit., Moscow, 1958 (Russian). MR 21 #4995.

[7] Decomposition into factor representations of unitary representations of locally compact groups, *Sibirsk. Mat. Ž.* **2**(1961), 89–99 (Russian). MR 24 #A187.

[8] *Normed algebras*, Wolters-Noordhoff Publishing Co., Groningen, 1972. MR 55 #11042.

NAIMARK, M. A., and STERN, A. I.

[1] *Theory of group representations*, Springer-Verlag, New York-Heidelberg-Berlin, 1982. MR 58 #28245. MR 86k: 22001.

NAKAGAMI, Y., and TAKESAKI, M.

[1] *Duality for crossed products of von Neumann algebras*, Lecture Notes in Mathematics **731**, Springer-Verlag, Berlin-Heidelberg-New York, 1979. MR 81e: 46053.

NAKAYAMA, T.

[1] Some studies on regular representations, induced representations and modular representations, *Ann. Math.* **39**(1938), 361–369.

VON NEUMANN, J.

[1] Allgemeine Eigenwerttheorie Hermitescher Funktionaloperatoren, *Math. Ann.* **102**(1929), 49–131.

[2] Zur Algebra der Funktionaloperatoren und Theorie der normalen Operatoren, *Math. Ann.* **102**(1929), 370–427.

[3] Die Eindeutigkeit der Schrödingerschen Operatoren, *Math. Ann.* **104**(1931), 570–578.

[4] Über adjungierte Functionaloperatoren, *Ann. Math.* **33**(1932), 294–310.

[5] Zum Haarschen Mass in topologischen Gruppen, *Compositio Math.* **1**(1934), 106–114.

[6] The uniqueness of Haar's measure, *Mat. Sb. N.S.* **1**(43) (1936), 721–734.

[7] On rings of operators. III, *Ann. Math.* **41**(1940), 94–161. MR 1, 146.

[8] On rings of operators, Reduction theory, *Ann. Math.* **50**(1949), 401–485. MR 10, 548.

[9] Über einen Satz von Herrn M. H. Stone, *Ann. Math.* (2) **33**(1932), 567–573.

NIELSEN, O. A.

[1] The failure of the topological Frobenius property for nilpotent Lie groups, *Math. Scand.* **45**(1979), 305–310. MR 82a: 22009.

[2] The Mackey-Blattner theorem and Takesaki's generalized commutation relations for locally compact groups, *Duke Math. J.* **40**(1973), 105–114.

[3] *Direct integral theory*, Marcel-Dekker, New York, 1980. MR 82e: 46081.

[4] The failure of the topological Frobenius property for nilpotent Lie groups, II., *Math. Ann.* **256**(1981), 561–568. MR 83a: 22011.

OKAMOTO, K.

[1] On induced representations, *Osaka J. Math.* **4**(1967), 85–94. MR 37 #1519.

OLESEN, D.

[1] A classification of ideals in crossed products, *Math. Scand.* **45**(1979), 157–167. MR 81h: 46083.

[2] A note on free action and duality in C^*-algebra theory (preprint), 1979.

OLESEN, D., LANDSTAD, M. B., and PEDERSEN, G. K.

[1] Towards a Galois theory for crossed products of C^*-algebras, *Math. Scand.* **43**(1978), 311–321. MR 81i: 46074.

OLESEN, D., and PEDERSEN, G. K.

[1] Applications of the Connes spectrum to C^*-dynamical systems, *J. Functional Anal.* **30**(1978), 179–197. MR 81i: 46076a.

[2] Applications of the Connes spectrum to C^*-dynamical systems, II, *J. Functional Anal.* **36**(1980), 18–32. MR 81i: 46076b.

[3] On a certain C^*-crossed product inside a W^*-crossed product, *Proc. Amer. Math. Soc.* **79**(1980), 587–590. MR 81h: 46075.

[4] Partially inner C^*-dynamical systems, *J. Functional Anal.* **66**(1986), 262–281.

OLSEN, C. L., and PEDERSEN, G. K.

[1] Convex combinations of unitary operators in von Neumann algebras, *J. Functional Anal.* **66**(1986), 365–380.

O'RAIFEARTAIGH, L.

[1] Mass differences and Lie algebras of finite order, *Phys. Rev. Lett.* **14**(1965), 575–577. MR 31 #1047.

ØRSTED, B.

[1] Induced representations and a new proof of the imprimitivity theorem, *J. Functional Anal.* **31**(1979), 355–359. MR 80d: 22007.

PACKER, J. A.

[1] K-theoretic invariants for C^*-algebras associated to transformations and induced flows, *J. Functional Anal.* **67**(1986), 25–59.

PALMER, T. W.

[1] Characterizations of C^*-algebras, *Bull. Amer. Math. Soc.* **74**(1968), 538–540. MR 36 #5709.

[2] Characterizations of C^*-algebras. II, *Trans. Amer. Math. Soc.* **148**(1970), 577–588. MR 41 #7447.

[3] Classes of nonabelian, noncompact, locally compact groups, *Rocky Mountain Math. J.* **8**(1978), 683–741. MR 81j: 22003.

PANOV, A. N.

[1] The structure of the group C^*-algebra of a Euclidean motion group. *Vestnik Moskov. Univ. Ser. I Mat. Meh.* (1978), 46–49 (Russian). MR 80a: 22008.

PARRY, W., and SCHMIDT, K.

[1] A note on cocycles of unitary representations, *Proc. Amer. Math. Soc.* **55**(1976), 185–190. MR 52 #14146.

PARTHASARATHY, K. R.

[1] *Probability measures on metric spaces*, Academic Press, New York-London, 1967. MR 37 #2271.

[2] *Multipliers on locally compact groups*, Lecture Notes in Mathematics, **93**, Springer-Verlag, Berlin-New York, 1969. MR 40 #264.

PASCHKE, W. L.

[1] Inner product modules over B^*-algebras, *Trans. Amer. Math. Soc.* **182**(1973), 443–468. MR 50 #8087.

[2] The double B-dual of an inner product module over a C^*-algebra B, *Canad. J. Math.* **26**(1974), 1272–1280. MR 57 #10433.

[3] Inner product modules arising from compact automorphism groups of von Neumann algebras, *Trans. Amer. Math. Soc.* **224**(1976), 87–102. MR 54 #8308.

PASCHKE, W. L., and SALINAS, N.

[1] C^*-algebras associated with free products of groups, *Pacific J. Math.* **82**(1979), 211–221. MR 82c: 22010.

PATERSON, A. L. T.

[1] Weak containment and Clifford semigroups, *Proc. Roy. Soc. Edinburgh. Sect. A* **81**(1978), 23–30. MR 81c: 43001.

PAULSEN, V. I.

[1] *Completely bounded maps and dilations*, Pitman Research Notes in Math., Vol. 146, John Wiley & Sons, New York, 1986.

PEDERSEN, G. K.

[1] Measure theory for C^*-algebras. I, *Math. Scand.* **19**(1966), 131–145. MR 35 #3453.

[2] Measure theory for C^*-algebras. II, *Math. Scand.* **22**(1968), 63–74. MR 39 #7444.

[3] Measure theory for C^*-algebras. III, *Math. Scand.* **25**(1969), 71–93. MR 41 #4263.

[4] Measure theory for C^*-algebras. IV, *Math. Scand.* **25**(1969), 121–127. MR 41 #4263.

[5] *C^*-algebras and their automorphism groups*, Academic Press, New York, 1979. MR 81e, 46037.

PEDERSEN, N.

[1] Duality for induced representations and induced weights, Copenhagen (preprint), 1978.

PENNEY, R. C.

[1] Abstract Plancherel theorems and a Frobenius reciprocity theorem, *J. Functional Anal.* **18**(1975), 177–190. MR 56 #3191.

[2] Rational subspaces of induced representations and nilmanifolds, *Trans. Amer. Math. Soc.* **246**(1978), 439–450. MR 80b: 22010.

[3] Harmonically induced representations on nilpotent Lie groups and auto-morphic forms on manifolds, *Trans. Amer. Math. Soc.* **260**(1980), 123–145. MR 81h: 22008.

[4] Lie cohomology of representations of nilpotent Lie groups and holomorphi-cally induced representations, *Trans. Amer. Math. Soc.* **261**(1980), 33–51. MR 81i: 22005.

[5] Holomorphically induced representations of exponential Lie groups, *J. Functional Anal.* **64**(1985), 1–18. MR 87b: 22018.

PERLIS, S.

[1] A characterization of the radical of an algebra, *Bull. Amer. Math. Soc.* **48**(1942), 128–132. MR 3, 264.

PETERS, F., and WEYL, H.

[1] Die Vollständigkeit der primitiven Darstellungen einer geschlossenen kontin-uerlichen Gruppe, *Math. Ann.* **97**(1927), 737–755.

PETERS, J.

[1] Groups with completely regular primitive dual space, *J. Functional Anal.* **20**(1975), 136–148. MR 52 #652.

[2] On traceable factor representations of crossed products, *J. Functional Anal.* **43**(1981), 78–96. MR 83b: 46086.

[3] Semi-crossed products of *C**-algebras, *J. Functional Anal.* **59**(1984), 498–534. MR 86e: 46063.

[4] On inductive limits of matrix algebras of holomorphic functions, *Trans. Amer. Math. Soc.* **299**(1987), 303–318.

PETERS, J., and SUND, T.

[1] Automorphisms of locally compact groups, *Pacific J. Math.* **76**(1978), 143–156. MR 58 #28263.

PETTIS, B. J.

[1] On integration in vector spaces, *Trans. Amer. Math. Soc.* **44**(1938), 277–304.

PHILLIPS, J.

[1] A note on square-integrable representations, *J. Functional Anal.* **20**(1975), 83–92. MR 52 #15026.

[2] Automorphisms of *C**-algebra bundles, *J. Functional Anal.* **51**(1983), 259–267. MR 84i: 46067.

PHILLIPS, J., and RAEBURN, I.

[1] Crossed products by locally unitary automorphism groups and principal bundles, *J. Operator Theory* **11**(1984), 215–241. MR 86m: 46058.

PIARD, A.

[1] Unitary representations of semi-direct product groups with infinite dimen-sional abelian normal subgroup, *Rep. Math. Phys.* **11**(1977), 259–278. MR 57 #12772.

PICARDELLO, M. A.

[1] Locally compact unimodular groups with atomic duals, *Rend. Sem. Mat. Fis. Milano* **48**(1978), 197–216 (1980). MR 82e: 22015.

PIER, J. P.

[1] *Amenable locally compact groups*, Pure and Applied Mathematics, John Wiley and Sons, New York, 1984. MR 86a: 43001.

PIMSNER, M. and VOICULESCU, D.

[1] K-groups of reduced crossed products by free groups, *J. Operator Theory* **8**(1982), 131-156. MR 84d: 46092.

PLANCHEREL, M.

[1] Contribution à l'étude de la représentation d'une fonction arbitraire par les intégrales définies, *Rend. Pal.* **30**(1910), p. 289.

POGUNTKE, D.

[1] Epimorphisms of compact groups are onto, *Proc. Amer. Math. Soc.* **26**(1970), 503-504. MR 41 #8577.

[2] Einige Eigenschaften des Darstellungsringes kompakter Gruppen, *Math. Z.* **130**(1973), 107-117. MR 49 #3020.

[3] Decomposition of tensor products of irreducible unitary representations, *Proc. Amer. Math. Soc.* **52**(1975), 427-432. MR 52 #5862.

[4] Der Raum der primitiven Ideale von endlichen Erweiterungen lokalkompakter Gruppen, *Arch. Math. (Basel)* **28**(1977), 133-138. MR 55 #12862.

[5] Nilpotente Liesche Gruppen haben symmetrische Gruppenalgebren, *Math. Ann.* **227**(1977), 51-59. MR 56 #6283.

[6] Symmetry and nonsymmetry for a class of exponential Lie groups, *J. Reine Angew. Math.* **315**(1980), 127-138. MR 82b: 43013.

[7] Einfache Moduln über gewissen Banachschen Algebren: ein Imprimitivitäts-satz, *Math. Ann.* **259**(1982), 245-258. MR 84i: 22008.

PONTRYAGIN, L. S.

[1] *Der allgemeine Dualitätssatz für abgeschlossene Mengen.* Verhandlungen des Internat. Math.-Kongr., Zürich, 1932. Zürich u. Leipzig: Orell Füssli, Bd. II, pp. 195-197.

[2] Sur les groupes topologiques compacts et cinquième problème de D. Hilbert, *C. R. Acad. Sci. Paris* **198**(1934), 238-240.

[3] Sur les groupes abéliens continus, *C. R. Acad. Sci. Paris* **198**(1934), 328-330.

[4] The theory of topological commutative groups, *Ann. Math.* (2) **35**(1934), 361-388.

[5] *Topological groups*, Princeton University Press, Princeton, N. J., 1939. MR 1, 44, MR 19, 867, MR 34 #1439.

[6] *Topological groups*, 2nd ed., Gordon and Breach, Science Publishers, New York, 1966.

POPA, S.

[1] Semiregular maximal abelian *-subalgebras and the solution to the factor Stone-Weierstrass problem, *Invent. Math.* **76**(1984), 157-161. MR 85m: 46057b.

POVZNER, A.
[1] Über positive Funktionen auf einer Abelschen Gruppe, *Dokl. Akad. Nauk SSSR, N.S.* **28**(1940), 294–295.

POWERS, R. T.
[1] Simplicity of the C^*-algebra associated with the free group on two generators, *Duke Math. J.* **42**(1975), 151–156. MR 51 #10534.

POZZI, G. A.
[1] Continuous unitary representations of locally compact groups. Application to Quantum Dynamics. Part I. Decomposition Theory. *Suppl. Nuovo Cimento* **4**(1966), 37–171. MR 36 #6406.

PRICE, J. F.
[1] On positive definite functions over a locally compact group, *Canad. J. Math.* **22**(1970), 892–896. MR 41 #8593.

[2] *Lie groups and compact groups*, London Mathematical Society Lecture Note Series, **25**, Cambridge University Press, Cambridge-New York-Melbourne, 1977. MR 56 #8743.

PROSSER, R. T.
[1] *On the ideal structure of operator algebras*, Memoirs Amer. Math. Soc. **45**, Amer. Math. Soc. Providence, R. I., 1963. MR 27 #1846.

PTÁK, V.
[1] Banach algebras with involution, *Manuscripta Math.* **6**(1972), 245–290. MR 45 #5764.

PUKANSZKY, L.
[1] *Leçons sur les représentations des groupes*, Monographies de la Société Mathématique de France, Dunod, Paris, 1967. MR 36 #311.

[2] Unitary representations of solvable Lie groups, *Ann. Sci. École Norm. Sup.* (4) **4**(1971), 457–608. MR 55 #12866.

[3] The primitive ideal space of solvable Lie groups, *Invent. Math.* **22**(1973), 75–118. MR 48 #11403.

[4] Characters of connected Lie groups, *Acta Math.* **133**(1974), 81–137. MR 53 #13480.

PYTLIK, T.
[1] L^1-harmonic analysis on semi-direct products of Abelian groups, *Monatsh. Math.* **93**(1982), 309–328. MR 83m: 43006.

PYTLIK, T., and SZWARC, R.
[1] An analytic family of uniformly bounded representations of free groups, *Acta Math.* **157**(1986), 287–309.

QUIGG, J. C.
[1] On the irreducibility of an induced representation, *Pacific J. Math.* **93**(1981), 163–179. MR 84j: 22007.

[2] On the irreducibility of an induced representation II, *Proc. Amer. Math. Soc.* **86**(1982), 345–348. MR 84m: 22010.

[3] Approximately periodic functionals on C^*-algebras and von Neumann algebras, *Canad. J. Math.* **37**(1985), 769–784. MR 86k: 46088.

[4] Duality for reduced twisted crossed products of C^*-algebras, *Indiana Univ. Math. J.* **35**(1986), 549–572.

RAEBURN, I.

[1] On the Picard group of a continuous trace C^*-algebra, *Trans. Amer. Math. Soc.* **263**(1981), 183–205. MR 82b: 46090.

[2] On group C^*-algebras of bounded representation dimension, *Trans. Amer. Math. Soc.* **272**(1982), 629–644. MR 83k: 22018.

RAEBURN, I., and TAYLOR, J. L.

[1] Continuous trace C^*-algebras with given Dixmier-Douady class, *J. Austral. Math. Soc. Ser. A* **38**(1985), 394–407. MR 86g: 46086.

RAEBURN, I., and WILLIAMS, D. P.

[1] Pull-backs of C^*-algebras and crossed products by certain diagonal actions, *Trans. Amer. Math. Soc.* **287**(1985), 755–777. MR 86m: 46054.

RAIKOV, D. A.

[1] Positive definite functions on commutative groups with an invariant measure, *Dokl. Akad. Nauk SSSR, N.S.* **28**(1940), 296–300.

[2] The theory of normed rings with involution, *Dokl. Akad. Nauk SSSR* **54**(1946), 387–390. MR 8, 469.

RAMSAY, A.

[1] Virtual groups and group actions, *Adv. Math.* **6**(1971), 253–322. MR 43 #7590.

[2] Boolean duals of virtual groups, *J. Functional Anal.* **15**(1974), 56–101. MR 51 #10586.

[3] Nontransitive quasi-orbits in Mackey's analysis of group extensions, *Acta Math.* **137**(1976), 17–48. MR 57 #524.

[4] Subobjects of virtual groups, *Pacific J. Math.* **87**(1980), 389–454. MR 83b: 22009.

[5] Topologies on measured groupoids, *J. Functional Anal.* **47**(1982), 314–343. MR 83k: 22014.

READ, C. J.

[1] A solution to the invariant subspace problem, *Bull. London Math. Soc.* **16**(1984), 337–401. MR 86f: 47005.

RENAULT, J.

[1] *A groupoid approach to C^*-algebras*, Lecture Notes in Mathematics, **793**, Springer-Verlag, Berlin-New York, 1980. MR 82h: 46075.

REPKA, J.

[1] A Stone-Weierstrass theorem for group representations, *Int. J. Math. Sci.* **1**(1978), 235–244. MR 58 #6055.

RICKART, C. E.

[1] Banach algebras with an adjoint operation, *Ann. Math.* **47**(1946), 528–550. MR 8, 159.

[2] *General theory of Banach algebras*, D. Van Nostrand Co., Princeton, 1960. MR 22 #5903.

RIEDEL, N.
[1] Topological direct integrals of left Hilbert algebras. I., *J. Operator Theory* **5**(1981), 29–45. MR 83e: 46049.
[2] Topological direct integrals of left Hilbert algebras. II., *J. Operator Theory* **5**(1981), 213–229. MR 83e: 46050.

RIEFFEL, M. A.
[1] On extensions of locally compact groups, *Amer. J. Math.* **88**(1966), 871–880. MR 34 #2771.
[2] Induced Banach representations of Banach algebras and locally compact groups, *J. Functional Anal.* **1**(1967), 443–491. MR 36 #6544.
[3] Unitary representations induced from compact subgroups, *Studia Math.* **42**(1972), 145–175. MR 47 #398.
[4] On the uniqueness of the Heisenberg commutation relations, *Duke Math. J.* **39**(1972), 745–752. MR 54 #466.
[5] Induced representations of C^*-algebras, *Adv. Math.* **13**(1974), 176–257. MR 50 #5489.
[6] Morita equivalence for C^*-algebras and W^*-algebras, *J. Pure Appl. Algebra* **5**(1974), 51–96. MR 51 #3912.
[7] Induced representations of rings, *Canad. J. Math.* **27**(1975), 261–270. MR 55 #3004.
[8] Strong Morita equivalence of certain transformation group C^*-algebras, *Math. Ann.* **222**(1976), 7–22. MR 54 #7695.
[9] *Unitary representations of group extensions; an algebraic approach to the theory of Mackey and Blattner.* Studies in analysis, pp. 43–82. Advances in Mathematics, Suppl. Studies, **4**, Academic Press, New York 1979. MR 81h: 22004.
[10] Actions of finite groups on C^*-algebras, *Math. Scand.* **47**(1980), 157–176. MR 83c: 46062.
[11] C^*-algebras associated with irrational rotations, *Pacific J. Math.* **93**(1981), 415–429. MR 83b: 46087.
[12] *Morita equivalence for operator algebras*, Proc. Sympos. Pure Math., pp. 285–298, **38**, Part 1, Amer. Math. Soc., Providence, R. I. 1982.
[13] *Applications of strong Morita equivalence to transformation group C^*-algebras*, *Proc. Sympos. Pure Math.*, pp. 299–310, 38, Part 1, *Amer. Math. Soc., Prov., R.I.* 1982.
[14] K-theory of crossed products of C^*-algebras by discrete groups, Group actions on rings, Proc. AMS-IMS-SIAM Summer Res. Conf., 1984, *Contemp. Math.* **43**(1985), 227–243.

RIESZ, F.
[1] Sur les opérations fonctionnelles linéaires, *C. R. Acad. Sci. Paris* **149**(1909), 974–977. (Also in Gesammelte Arbeiten, vol. I., pp. 400–402, Budapest, 1960).
[2] *Les systèmes d'equations linéaires à une infinité d'inconnus*, Paris, 1913.

[3] Über lineare Funktionalgleichungen, *Acta Math.* **41**(1918), 71–98.

[4] Sur la formule d'inversion de Fourier, *Acta Sci. Math. Szeged* **3**(1927), 235–241.

RIESZ, F., and SZ-NAGY, B.

[1] *Leçons d'analyse fonctionelle*, Académie des Sciences de Hongrie, Budapest, 1952. MR 14, 286.

[2] *Functional analysis*, Frederick Ungar Publishing Co., New York, 1955. MR 17, 175.

RIGELHOF, R.

[1] Induced representations of locally compact groups, *Acta Math.* **125**(1970), 155–187. MR 43 #7550.

RINGROSE, J. R.

[1] On subalgebras of a C^*-algebra, *Pacific J. Math.* **15**(1965), 1377–1382. MR 32 #4561.

ROBERT, A.

[1] Exemples de groupes de Fell (English summary), *C. R. Acad. Sci. Paris Ser. A-B* (8) **287**(1978), A603–A606. MR 83k: 22015.

[2] *Introduction to the representation theory of compact and locally compact groups*, London Mathematical Society Lecture Note Series, **80**, Cambridge University Press, 1983. MR 84h: 22012.

ROBINSON, G. DE B.

[1] *Representation theory of the symmetric group*, University of Toronto Press, Toronto, 1961. MR 23 #A3182.

ROELCKE, W., and DIEROLF, S.

[1] *Uniform structures on topological groups and their quotients*, McGraw-Hill, New York, 1981. MR 83a: 82005.

ROSENBERG, A.

[1] The number of irreducible representations of simple rings with no minimal ideals, *Amer. J. Math.* **75**(1953), 523–530. MR 15, 236.

ROSENBERG, J.

[1] The C^*-algebras of some real and p-adic solvable groups, *Pacific J. Math.* **65**(1976), 175–192. MR 56 #5779.

[2] Amenability of crossed products of C^*-algebras, *Comm. Math. Phys.* **57**(1977), 187–191. MR 57 #7190.

[3] A quick proof of Harish-Chandra's Plancherel theorem for spherical functions on a semisimple Lie group, *Proc. Amer. Math. Soc.* **63**(1977), 143–149. MR 58 #22391.

[4] Frobenius reciprocity for square-integrable factor representations, *Illinois J. Math.* **21**(1977), 818–825. MR 57 #12771.

[5] Square-integrable factor representations of locally compact groups *Trans. Amer. Math. Soc.* **237**(1978), 1–33. MR 58 #6056.

[6] Appendix to: "Crossed products of UHF algebras by product type actions" [*Duke Math. J.* **46**(1979), 1–23] by O. Bratteli, *Duke Math. J.* **46**(1979), 25–26. MR 82a: 46064.

ROSENBERG, J., and VERGNE, M.

[1] Hamonically induced representations of solvable Lie groups, *J. Functional Anal.* **62**(1985), 8–37. MR 87a: 22017.

ROUSSEAU, R.

[1] The left Hilbert algebra associated to a semi-direct product, *Math. Proc. Cambridge Phil. Soc.* **82**(1977), 411–418. MR 56 #16386.

[2] An alternative definition for the covariance algebra of an extended covariant system, *Bull. Soc. Math. Belg.* **30**(1978), 45–59. MR 81i: 46081.

[3] Un système d'induction pour des groupes topologiques, *Bull. Soc. Math. Belg.***31**(1979), 191–195. MR 81m: 22008.

[4] The covariance algebra of an extended covariant system, *Math. Proc. Cambridge Phil. Soc.* **85**(1979), 271–280. MR 80e: 46040.

[5] The covariance algebra of an extended covariant system. II, *Simon Stevin* **53**(1979), 281–295. MR 82g: 46105.

[6] Le commutant d'une algèbre de covariance, *Rev. Roum. Math. Pures Appl.* **25**(1980), 445–471. MR 81m: 46095.

[7] A general induction process, *Quart. J. Math. Oxford Ser.* (2) **32**(1981), 453–466. MR 83a: 22005.

[8] Tensor products and the induction process, *Arch. Math.* (*Basel*) **36**(1981), 541–545. MR 83h: 22014.

[9] Crossed products of von Neumann algebras and semidirect products of groups, *Nederl. Akad. Wetensch. Indag. Math.* **43**(1981), 105–116. MR 82i: 46103.

[10] Quasi-invariance and induced representations, *Quart. J. Math. Oxford* (2) **34**(1983), 491–505. MR 86i: 22008.

[11] A general induction process. II., *Rev. Roumaine Math. Pures Appl.* **30**(1985), 147–153. MR 86i: 22009.

RÜHL, W.

[1] *The Lorentz group and harmonic analysis*, W. A. Benjamin, New York, 1970. MR 43 #425.

RUDIN, W.

[1] *Fourier analysis on groups*, J. Wiley, New York, 1962. MR 27 #2808.

[2] *Functional analysis*, Mcgraw Hill, New York, 1973. MR 51 #1315.

RUSSO, R., and DYE, H. A.

[1] A note on unitary operators in C^*-algebras, *Duke Math. J.* **33**(1966), 413–416. MR 33 #1750.

SAGLE, A. A., and WALDE, R. E.

[1] *Introduction to Lie groups and Lie algebras*, Academic Press, New York-London, 1973. MR 50 #13374.

SAKAI, S.

[1] C^*-algebras and W^*-algebras, Ergebnisse der Math. **60**, Springer-Verlag, Berlin, Heidelberg, New York 1971. MR 56 #1082.

SALLY, P. J.

[1] *Harmonic analysis on locally compact groups*, Lecture Notes, University of Maryland, 1976.

[2] Harmonic analysis and group representations, *Studies in Harmonic analysis*, **13**, pp. 224-256, MAA Studies in Mathematics, Mathematical Association of America, Washington D. C., 1976.

SANKARAN, S.

[1] Representations of semi-direct products of groups, *Compositio Math.* **22**(1970), 215-225. MR 42 #722.

[2] Imprimitivity and duality theorems, *Boll. Un. Mat. Ital.* (4) **7**(1973), 241-259. MR 47 #5176.

SAUVAGEOT., J. L.

[1] Idéaux primitifs de certains produits croisés, *Math. Ann.* **231**(1977/78), 61-76. MR 80d: 46112.

[2] Idéaux primitifs induits dans les produits croisés, *J. Functional Anal.* **32**(1979), 381-392. MR 81a: 46080.

SCHAAF, M.

[1] *The reduction of the product of two irreducible unitary representations of the proper orthochronous quantum-mechanical Poincaré group*, Lecture Notes in Physics, **5**, Springer-Verlag, Berlin-New York, 1970. MR 57 #16471.

SCHAEFFER, H. H.

[1] *Topological vector spaces*, Macmillan, New York, 1966. MR 33 #1689.

SCHEMPP, W.

[1] *Harmonic analysis on the Heisenberg nilpotent Lie group, with applications to signal theory*, Pitman Research Notes in Mathematics, 147, John Wiley & Sons, New York, 1986.

SCHLICHTING, G.

[1] Groups with representations of bounded degree, Lecture Notes in Mathematics **706**, pp. 344-348. Springer, Berlin, 1979. MR 80m: 22006.

SCHMIDT, E.

[1] Entwicklung willkürlicher Funktionen nach Systemen vorgeschriebener, *Math. Ann.* **63**(1907), 433-476.

SCHMIDT, K.

[1] *Cocycles on ergodic transformation groups*, Macmillan Lectures in Mathematics 1, Macmillan Co. of India, Delhi 1977. MR 58 #28262.

SCHOCHETMAN, I. E.

[1] Topology and the duals of certain locally compact groups, *Trans. Amer. Math. Soc.* **150**(1970), 477-489. MR 42 #422.

[2] Dimensionality and the duals of certain locally compact groups, *Proc. Amer. Math. Soc.* **26**(1970), 514-520. MR 42 #418.

[3] Nets of subgroups and amenability, *Proc. Amer. Math. Soc.* **29**(1971), 397-403. MR 43 #7551.

[4] Kernels of representations and group extensions, *Proc. Amer. Math. Soc.* **36**(1972), 564–570. MR 47 #8761.

[5] Compact and Hilbert-Schmidt induced representations, *Duke Math. J.* **41**(1974), 89–102. MR 48 #11392.

[6] Induced representations of groups on Banach spaces, *Rocky Mountain J. Math.* **7**(1977), 53–102. MR 56 #15824.

[7] *Integral operators in the theory of induced Banach representations*, Memoirs Amer. Math. Soc. **207**, Amer. Math. Soc. Providence, R. I., 1978. MR 58 #6060.

[8] The dual topology of certain group extensions, *Adv. Math.* **35**(1980), 113–128. MR 81e: 22005.

[9] Integral operators in the theory of induced Banach representations. II. The bundle approach, *Int. J. Math. Sci.* **4**(1981), 625–640. MR 84c: 22009.

[10] Generalized group algebras and their bundles, *Int. J. Math. Sci.* **5**(1982), 209–256. MR 84f: 46094.

SCHREIER, O.

[1] Abstrakte kontinuerliche Gruppen, *Abh. Math. Sem. Univ. Hamburg* **4**(1926), 15–32.

SCHUR, I.

[1] Über die Darstellung der endlichen Gruppen durch gebrochene lineare Substitutionen, *J. Crelle* **127**(1904), 20–50.

[2] Neue Begründung der Theorie der Gruppencharaktere, *Sitz. preuss. Akad. Wiss.* (1905), 406–432.

[3] Neue Anwendungen der Integralrechnung auf Probleme der Invariantentheorie. I, II, III, *Sitz. preuss. Akad. Wiss. phys.-math. Kl.* (1924), 189–208, 297–321, 346–355.

SCHWARTZ, J.-M.

[1] Sur la structure des algèbres de Kac, I, *J. Functional Anal.* **34**(1979), 370–406. MR 83a: 46072a.

[2] Sur la structure des algèbres de Kac, II, *Proc. London Math. Soc.* **41**(1980), 465–480. MR 83a: 46072b.

SCUTARU, H.

[1] Coherent states and induced representations, *Lett. Math. Phys.* **2**(1977/78), 101–107. MR 58 #22406.

SEBESTYÉN, Z.

[1] Every C^*-seminorm is automatically submultiplicative, *Period. Math. Hungar.* **10**(1979), 1–8. MR 80c: 46065.

[2] On representability of linear functionals on *-algebras, *Period. Math. Hungar.* **15**(1984), 233–239. MR 86a: 46068.

SEDA, A. K.

[1] A continuity property of Haar systems of measures, *Ann. Soc. Sci. Bruxelles* **89**(1975), 429–433. MR 53 #6555.

[2] Haar measures for groupoids, *Proc. Roy. Irish. Sect.* **A76**(1976), 25–36. MR 55 #629.

[3] Quelques résultats dans la catégorie des groupoids d'opérateurs, *C. R. Acad. Sci. Paris* **288**(1979), 21–24. MR 80b: 22005.

[4] Banach bundles and a theorem of J. M. G. Fell, *Proc. Amer. Math. Soc.* **83**(1981), 812–816. MR 84d: 22006.

[5] Banach bundles of continous functions and an integral representation theorem, *Trans. Amer. Math. Soc.* **270**(1982), 327–332. MR 83f: 28009.

[6] Sur les espaces de fonctions et les espaces de sections, *C. R. Acad. Sci. Paris Sér. I Math.* **297**(1983), 41–44. MR 85a: 46041.

SEGAL, I. E.

[1] The group ring of a locally compact group. I, *Proc. Nat. Acad. Sci. U.S.A.* **27**(1941), 348–352. MR 3, 36.

[2] Representation of certain commutative Banach algebras, *Bull. Amer. Math. Soc.* Abst. 130, **52**(1946), 421.

[3] The group algebra of a locally compact group, *Trans. Amer. Math. Soc.* **61**(1947), 69–105. MR 8, 438.

[4] Irreducible representations of operator algebras, *Bull. Amer. Math. Soc.* **53**(1947), 73–88. MR 8, 520.

[5] Two-sided ideals in operator algebras, *Ann. Math.* **50**(1949), 856–865. MR 11, 187.

[6] The two-sided regular representation of a unimodular locally compact group, *Ann. Math.* (2) **51**(1950), 293–298. MR 12, 157.

[7] An extension of Plancherel's formula to separable unimodular groups, *Ann. Math.* (2) **52**(1950), 272–292. MR 12, 157.

[8] *Decompositions of operator algebras*, I *and* II, Memoirs Amer. Math. Soc. **9**, Amer. Math. Soc. Providence, R. I., 1951. MR 13, 472.

[9] A non-commutative extension of abstract integration, *Ann. Math.* (2) **57**(1953), 401–457. MR 14, 991. MR 15, 204.

[10] *Caractérisation mathématique des observables en théorie quantique des champs et ses conséquences pour la structure des particules libres*, Report of Lille conference on quantum fields, pp. 57–163, C.N.R.S., Paris, 1959.

[11] An extension of a theorem of L. O'Raifeartaigh, *J. Functional Anal.* **1**(1967), 1–21. MR 37 #6079.

SELBERG, A.

[1] Harmonic analysis and discontinuous groups in weakly symmetric Riemannian spaces with applications to Dirichlet series, *J. Indian Math. Soc.* **20**(1956), 47–87. MR 19, 531.

SEN, R. N.

[1] Bundle representations and their applications, pp. 151–160, Lecture Notes in Mathematics, **676**, Springer-Verlag, 1978. MR 80e: 22003.

[2] Theory of symmetry in the quantum mechanics of infinite systems. I. The state space and the group action, *Phys. A* **94**(1978), 39–54. MR 80h: 81031a.

[3] Theory of symmetry in the quantum mechanics of infinite systems. II. Isotropic representations of the Galilei group and its central extensions, *Phys. A* **94**(1978), 55–70. MR 80h: 81031b.

SERIES, C.
[1] Ergodic actions on product groups, *Pacific J. Math.* **70**(1977), 519–547. MR 58 #6062.
[2] An application of groupoid cohomology, *Pacific J. Math.* **92**(1981), 415–432. MR 84f: 22014.

SERRE, J. P.
[1] *Linear representations of finite groups*, Graduate Texts in Mathematics, **42**, Springer-Verlag, Berlin-Heidelberg-New York, 1977. MR 80f: 20001.

SHERMAN, S.
[1] Order in operator algebras, *Amer. J. Math.* **73**(1951), 227–232. MR 13, 47.

SHIN'YA, H.
[1] *Spherical functions and spherical matrix functions on locally compact groups*, Lectures in Mathematics, Department of Mathematics, Kyoto University, **7**, Kinokuniya Book Store Co., Tokyo, 1974. MR 50 #10149.
[2] Irreducible Banach representations of locally compact groups of a certain type, *J. Math. Kyoto Univ.* **20**(1980), 197–212. MR 82c: 22009.
[3] On a Frobenius reciprocity theorem for locally compact groups, *J. Math. Kyoto Univ.* **24**(1984), 539–555. MR 86a: 22008.
[4] On a Frobenius reciprocity theorem for locally compact groups. II, *J. Math. Kyoto Univ.* **25**(1985), 523–547. MR 87b: 22006.

SHUCKER, D. S.
[1] Square integrable representations of unimodular groups, *Proc. Amer. Math. Soc.* **89**(1983), 169–172. MR 85d: 22013.

SHULTZ, F. W.
[1] Pure states as a dual object for C^*-algebras, *Comm. Math. Phys.* **82**(1981/82), 497–509. MR 83b: 46080.

SINCLAIR, A. M.
[1] *Automatic continuity of linear operators*, Lecture Notes Series **21**, London Math. Soc., London, 1977. MR 58 #7011.

SINGER, I. M.
[1] *Report on group representations*, National Academy of Sciences – National Research Council, Publ. **387** (Arden House), Washington, D.C. (1955), 11–26.

SKUDLAREK, H. L.
[1] On a two-sided version of Reiter's condition and weak containment, *Arch. Math. (Basel)* **31**(1978/79), 605–610. MR 80h: 43010.

SMITH, H. A.
[1] Commutative twisted group algebras, *Trans. Amer. Math. Soc.* **197**(1974), 315–326. MR 51 #792.
[2] Characteristic principal bundles, *Trans. Amer. Math. Soc.* **211**(1975), 365–375. MR 51 #13128.

[3] Central twisted group algebras, *Trans. Amer. Math. Soc.* **238**(1978), 309–320. MR 58 #7093.

SMITH, M.

[1] Regular representations of discrete groups, *J. Functional Anal.* **11**(1972), 401–406. MR 49 #9641.

SMITHIES, F.

[1] The Fredholm theory of integral equations, *Duke Math. J.* **8**(1941), 107–130. MR 3, 47.

SOLEL, B.

[1] Nonselfadjoint crossed products: invariant subspaces, cocycles and subalgebras, *Indiana Univ. Math. J.* **34**(1985), 277–298. MR 87h: 46140.

SPEISER, A.

[1] *Die Theorie der Gruppen von endlicher Ordnung*, Grundlehren der Mathematischen Wiss., **5**, Berlin, 1937.

STEEN, L. A.

[1] Highlights in the history of spectral theory, *Amer. Math. Monthly* **80**(1973), 359–381. MR 47 #5643.

STEIN, E. M.

[1] Analytic continuation of group representations, *Adv. Math.* **4**(1970), 172–207. MR 41 #8584.

ŠTERN, A. I.

[1] The connection between the topologies of a locally bicompact group and its dual space, *Funkcional. Anal. Priložen.* **5**(1971), 56–63 (Russian). MR 45 #3639.

[2] Dual objects of compact and discrete groups, *Vestnik Moskov. Univ. Ser. I Mat. Meh.* **2**(1977), 9–11 (Russian). MR 56 #8748.

STEWART, J.

[1] Positive definite functions and generalizations, an historical survey, *Rocky Mountain J. Math.* **6**(1976), 409–434. MR 55 #3679.

STINESPRING, W. F.

[1] Positive functions on C^*-algebras, *Proc. Amer. Math. Soc.* **6**(1955), 211–216. MR 16, 1033.

[2] A semi-simple matrix group is of type I, *Proc. Amer. Math. Soc.* **9**(1958), 965–967. MR 21 #3509.

STONE, M. H.

[1] Linear transformations in Hilbert space. III. Operational methods and group theory, *Proc. Nat. Acad. Sci. U.S.A.* **16**(1930), 172–175.

[2] *Linear transformations in Hilbert space and their applications to analysis*, Amer. Math. Soc. Coll. Publ. **15**, New York, 1932.

[3] Notes on integration. I, II, III, *Proc. Nat. Acad. Sci. U.S.A.* **34**(1948), 336–342, 447–455, 483–490. MR 10, 24. MR 10, 107. MR 10, 239.

[4] Applications of the theory of Boolean rings to general topology, *Trans. Amer. Math. Soc.* **41**(1937), 375–481.

[5] The generalized Weierstrass aproximation theorem, *Math. Magazine* **21**(1948), 167–184, 237–254. MR 10, 255.

[6] On one-parameter unitary groups in Hilbert space, *Ann. Math.* (2) **33**(1932), 643–648.

[7] A general theory of spectra, I, *Proc. Nat. Acad. Sci. U.S.A.*, **26**(1940), 280–283.

STØRMER, E.

[1] Large groups of automorphisms of C^*-algebras, *Commun. Math. Phys.* **5**(1967), 1–22. MR 37 #2012.

STOUT, E. L.

[1] *The theory of uniform algebras*, Bogden & Quigley, New York, 1971. MR 54 #11066.

STRASBURGER, A.

[1] Inducing spherical representations of semi-simple Lie groups, *Diss. Math. (Rozpr. Mat.)* **122**(1975), 52 pp. MR 53 #8332.

STRATILA, S.

[1] *Modular theory in operator algebras*, Abacus Press, Tunbridge Wells, 1981. MR 85g: 46072.

STRATILA, S., and ZSIDÓ, L.

[1] *Lectures on von Neumann algebras*, Abacus Press, Tunbridge Wells, 1979. MR 81j: 46089.

SUGIURA, M.

[1] *Unitary representations and harmonic analysis. An introduction*, Kodansha, Tokyo; Halstead Press, John Wiley & Sons, New York-London-Sydney, 1975. MR 58 #16977.

SUND, T.

[1] Square-integrable representations and the Mackey theory, *Trans. Amer. Math. Soc.* **194**(1974), 131–139. MR 49 #9115.

[2] Duality theory for groups with precompact conjugacy classes. I, *Trans. Amer. Math. Soc.* **211**(1975), 185–202. MR 53 #8332.

[3] Isolated points in duals of certain locally compact groups, *Math. Ann.* **224**(1976), 33–39. MR 54 #10477.

[4] A note on integrable representations, *Proc. Amer. Math. Soc.* **59**(1976), 358–360. MR 55 #3154.

[5] Duality theory for groups with precompact conjugacy classes. II, *Trans. Amer. Math. Soc.* **224**(1976), 313–321. MR 55 #12863.

[6] Multiplier representations of exponential Lie groups, *Math. Ann.* **232**(1978), 287–290. MR 57 #16467.

[7] Multiplier representations of nilpotent Lie groups (preprint) 1976.

SUTHERLAND, C.

[1] Cohomology and extensions of von Neumann algebras. I, *Publ. Res. Inst. Math. Sci.* **16**(1980), 105–133. MR 81k: 46067.

[2] Cohomology and extensions of von Neumann algebras, II, *Publ. Res. Math. Sci.* **16**(1980), 135–174. MR 81k: 46067.

[3] Induced representations for measured groupoids (preprint), 1982.

SZMIDT, J.

[1] On the Frobenius reciprocity theorem for representations induced in Hilbert module tensor products, *Bull. Acad. Polon. Sci. Sér. Sci. Math. Astronom. Phys.* **21**(1973), 35–39. MR 47 #6941.

[2] The duality theorems. Cyclic representations. Langlands conjectures, *Diss. Math. (Rozpr. Mat.)* **168**(1980), 47 pp. MR 81j: 22012.

SZ.-NAGY, B.

[1] *Spektraldarstellung linearer Transformationen des Hilbertschen Raumes,* Ergebnisse Math. V. **5**, Berlin (1942). MR 8 #276.

TAKAHASHI, A.

[1] Hilbert modules and their representation, *Rev. Colombiana Mat.* **13**(1979), 1–38. MR 81k: 46056a.

[2] A duality between Hilbert modules and fields of Hilbert spaces, *Rev. Colombiana Mat.* **13**(1979), 93–120. MR 81k: 46056b.

TAKAI, H.

[1] The quasi-orbit space of continuous C^*-dynamical systems, *Trans. Amer. Math. Soc.* **216**(1976), 105–113. MR 52 #6444.

TAKENOUCHI, O.

[1] Sur une classe de fonctions continues de type positif sur un groupe localement compact, *Math. J. Okayama Univ.*, **4**(1955), 153–173. MR 16, 997

[2] Families of unitary operators defined on groups, *Math. J. Okayama Univ.*, **6**(1957), 171–179. MR 19, 430.

[3] Sur la facteur représentation d'un group de Lie résoluble de type (E), *Math. J. Okayama Univ.* **7**(1957), 151–161. MR 20 #3933.

TAKESAKI, M.

[1] Covariant representations of C^*-algebras and their locally compact automorphism groups, *Acta Math.* **119**(1967), 273–303. MR 37 #774.

[2] A duality in the representation of C^*-algebras, *Ann. Math.* (2) **85**(1967), 370–382. MR 35 #755.

[3] A characterization of group algebras as a converse of Tannaka-Stinespring-Tatsuuma duality theorem, *Amer. J. Math.* **91**(1969), 529–564. MR 39 #5752.

[4] A liminal crossed product of a uniformly hyperfinite C^*-algebra by a compact Abelian group, *J. Functional Anal.* **7**(1971), 140–146. MR 43 #941.

[5] *Duality and von Neumann algebras,* Lecture Notes in Mathematics, **247** Springer, New York, 1972, 665–786. MR 53 #704.

[6] Duality for crossed products and the structure of von Neumann algebras of type III, *Acta Math.* **131**(1973), 249–308. MR 55 #11068.

[7] *Theory of operator algebras.* I, Springer-Verlag, New York-Heidelberg-Berlin, 1979. MR 81e: 46038.

TAKESAKI, M., and TATSUUMA, N.

[1] Duality and subgroups, *Ann. Math.* **93**(1971), 344–364. MR 43 #7557.

[2] Duality and subgroups. II. *J. Functional Anal.* **11**(1972), 184-190. MR 52 #5865.

TAMAGAWA, T.

[1] On Selberg's trace formula, *J. Fac. Sci. Univ. Tokyo Sect.* (1) **8**(1960), 363-386. MR 23 #A958.

TATSUUMA, N.

[1] A duality theorem for the real unimodular group of second order, *J. Math. Japan* **17**(1965), 313-332. MR 32 #1290.

[2] A duality theorem for locally compact groups, *J. Math. Kyoto Univ.* **6**(1967), 187-293. MR 36 #313.

[3] Plancherel formula for non-unimodular locally compact groups, *J. Math. Kyoto Univ.* **12**(1972), 179-261. MR 45 #8777.

[4] Duality for normal subgroups, *Algèbres d'opérateurs et leurs applications en physique mathématique*, Proc. Colloq., Marseille, 1977, pp. 373-386, Colloques Int. CNRS, 274, CNRS, Paris, 1979. MR 82b: 22009.

TAYLOR, D. C.

[1] Interpolation in algebras of operator fields, *J. Functional Anal.* **10**(1972), 159-190. MR 51 #13700.

[2] A general Hoffman-Wermer theorem for algebras of operator fields, *Proc. Amer. Math. Soc.* **52**(1975), 212-216. MR 52 #6455.

TAYLOR, K. F.

[1] The type structure of the regular representation of a locally compact group, *Math. Ann.* **222**(1976), 211-224. MR 54 #12965.

[2] Group representations which vanish at infinity, *Math. Ann.* **251**(1980), 185-190. MR 82a: 22006.

TAYLOR, M. E.

[1] *Noncommutative harmonic analysis*, Mathematical Surveys and Monographs **22**, Amer. Math. Soc., Providence, R. I., 1986.

TERRAS, A.

[1] *Harmonic analysis on symmetric spaces and applications*, Springer-Verlag, New York-Berlin, 1985. MR 87f: 22010.

THIELEKER, E.

[1] On the irreducibility of nonunitary induced representations of certain semidirect products, *Trans. Amer. Math. Soc.* **164**(1972), 353-369. MR 45 #2097.

THOMA, E.

[1] Über unitäre Darstellungen abzählbarer, diskreter Gruppen, *Math. Ann.* **153**(1964), 111-138. MR 28 #3332.

[2] Eine Charakterisierung diskreter Gruppen vom Typ I, *Invent. Math.* **6**(1968), 190-196. MR 40 #1540.

TITCHMARSH, E. C.

[1] *The theory of functions*, 2nd ed, Oxford University Press, Cambridge, 1939.

[2] *Introduction to the theory of Fourier integrals*, Oxford University Press, Cambridge, 1948.

TOMIYAMA, J.

[1] Topological representation of C^*-algebras, *Tôhoku Math. J.* **14**(1962), 187-204. MR 26 #619.

[2] *Invitation to C^*-algebras and topological dynamics*, World Scientific Advanced Series in Dynamical Systems, Vol. 3, World Scientific, Singapore, 1987.

TOMIYAMA, J., and TAKESAKI, M.

[1] Applications of fibre bundles to a certain class of C^*-algebras, *Tôhoku Math. J.* **13**(1961), 498-523. MR 25 #2465.

TURUMARU, T.

[1] Crossed product of operator algebra, *Tôhoku Math. J.* (2) **10**(1958), 355-365. MR 21 #1550.

VALLIN, J. M.

[1] C^*-algèbres de Hopf et C^*-algèbres de Kac, *Proc. London Math. Soc.* (3) **50**(1985), 131-174. MR 86f: 46072.

VAINERMAN, L. I.

[1] A characterization of objects that are dual to locally compact groups, *Funkcional. Anal. Priložen* **8**(1974), 75-76 (Russian) MR 49 #463.

VAINERMAN, L. I., and KAC, G. I.

[1] Non-unimodular ring groups and Hopf-von Neumann algebras, *Mat. Sb.* **94**(1974), 194-225. MR 50 #536.

VAN DAELE, A.

[1] *Continuous crossed products and type III von Neumann algebras*, Lecture Notes London Math. Soc. **31**, Cambridge University Press (1978). MR 80b: 46074.

VAN DIJK, G.

[1] On symmetry of group algebras of motion groups, *Math. Ann.* **179**(1969), 219-226. MR 40 #1782.

VARADARAJAN, V. S.

[1] *Geometry of quantum theory.* I, Van Nostrand, Princeton, N. J., 1968. MR 57 #11399.

[2] *Geometry of quantum theory.* II, Van Nostrand, Princeton, N. J., 1970. MR 57 #11400.

[3] *Harmonic analysis on real reductive groups*, Lecture Notes In Mathematics, **576**, Springer-Verlag, Berlin-New York, 1977. MR 57 #12789.

VARELA., J.

[1] Duality of C^*-algebras, *Memoirs Amer. Math. Soc.* **148**(1974), 97-108. MR 50 #5490.

[2] Sectional representation of Banach modules, *Math. Z.* **139**(1974), 55-61. MR 50 #5473.

[3] Existence of uniform bundles, *Rev. Colombiana Mat.* **18**(1984), 1-8. MR 86i: 46076.

VAROPOULOS, N. T.
  [1] Sur les formes positives d'une algèbre de Banach, C. R. *Acad. Sci. Paris* **258**(1964), 2465–2467. MR 33 #3121.

VASIL'EV, N. B.
  [1] C^*-algebras with finite-dimensional irreducible representations, *Uspehi Mat. Nauk* **21**(1966), 135–154 [English translation: *Russian Math. Surveys* **21**(1966), 137–155]. MR 34 #1871.

VESTERSTRØM, J., and WILS, W.
  [1] Direct integrals of Hilbert spaces. II., *Math. Scand.* **26**(1970), 89–102. MR 41 #9011.

VOGAN, D. A., JR.
  [1] *Representations of real reductive Lie groups*, Progress in Mathematics, **15**, Birkhauser, Boston, Mass., 1981. MR 83c: 22022.
  [2] Unitarizability of certain series representations, *Ann. Math.* (2) **120**(1984), 141–187. MR 86h: 22028.
  [3] The unitary dual of GL(n) over an Archimedean field, *Invent. Math.* **83**(1986), 449–505. MR 87i: 22042.

VOICULESCU, D.
  [1] Norm-limits of algebraic operators, *Rev. Roumaine Math. Pure Appl.* **19**(1974), 371–378. MR 49 #7826.
  [2] Remarks on the singular extension in the C^*-algebra of the Heisenberg groups, *J. Operator Theory* **5**(1981), 147–170. MR 82m: 46075.
  [3] Asymptotically commuting finite rank unitary operators without commuting approximants, *Acta Sci. Math.* (*Szeged*), **45**(1983), 429–431. MR 85d: 47035.
  [4] Dual algebraic structures on operator algebras related to free products, *J. Operator Theory* **17**(1987), 85–98.

VOWDEN, B. J.
  [1] On the Gelfand-Naimark theorem, *J. London Math. Soc.* **42**(1967), 725–731. MR 36 #702.

WAELBROECK, L.
  [1] Le calcule symbolique dans les algèbres commutatives, *J. Math. Pures Appl.* **33**(1954), 147–186. MR 17, 513.
  [2] Les algèbres à inverse continu, *C. R. Acad. Sci. Paris* **238**(1954), 640–641. MR 17, 513.
  [3] Structure des algèbres à inverse continu, *C. R. Acad. Sci. Paris* **238**(1954), 762–764. MR 17, 513.

VAN DER WAERDEN, B. L.
  [1] *Die gruppentheoretische Methode in der Quantenmechanik*, Grundlagen der Math. Wiss., **36** Springer, Berlin, 1932.
  [2] *Moderne Algebra*, Leipzig, 1937.

WALLACH, N. R.

[1] *Harmonic analysis on homogeneous spaces*, Pure and Applied Mathematics, **19**, Marcel-Dekker, New York, 1973. MR 58 #16978.

WALTER, M. E.

[1] Group duality and isomorphisms of Fourier and Fourier-Stieltjes algebras from the *W**-algebra point of view, *Bull. Amer. Math. Soc.* **76**(1970), 1321–1325. MR 44 #2047.

[2] The dual group of the Fourier-Stieltjes algebra, *Bull. Amer. Math. Soc.* **78**(1972), 824–827. MR 46 #611.

[3] *W**-algebras and nonabelian harmonic analysis, *J. Functional Anal.* **11**(1972), 17–38. MR 50 #5365.

[4] A duality between locally compact groups and certain Banach algebras, *J. Functional Anal.* **17**(1974), 131–160. MR 50 #14067.

[5] On the structure of the Fourier-Stieltjes algebra, *Pacific J. Math.* **58**(1975), 267–281. MR 54 #12966.

WANG, S. P.

[1] On isolated points in the dual spaces of locally compact groups, *Math. Ann.* **218**(1975), 19–34. MR 52 #5863.

[2] On integrable representations, *Math. Z.* **147**(1976), 201–203. MR 53 #10975.

WARD, H. N.

[1] The analysis of representations induced from a normal subgroup, *Michigan Math. J.* **15**(1968), 417–428. MR 40 #4384.

WARNER, G.

[1] *Harmonic analysis on semi-simple Lie groups*. I, Die Grundlehren der mathematischen Wissenschaften, **188**, Springer-Verlag, New York-Heidelberg, 1972. MR 58 #16979.

[2] *Harmonic analysis on semi-simple Lie groups*. II, Die Grundlehren der mathematischen Wissenschaften, **189**, Springer-Verlag, New York-Heidelberg, 1972. MR 58 #16980.

WARNER, S.

[1] Inductive limits of normed algebras, *Trans. Amer. Math. Soc.* **82**(1956), 190–216. MR 18, 52.

WAWRZYŃCZYK, A.

[1] On the Frobenius-Mautner reciprocity theorem, *Bull. Acad. Polon. Sci. Ser. Sci. Math. Astronom. Phys.* **20**(1972), 555–559.

[2] Reciprocity theorems in the theory of representations of groups and algebras, *Diss. Math. (Rozp. Mat.)* **126**(1975), 1–60. MR 56 #534.

[3] *Group representations and special functions*, C. Reidel Publishing Co., Dordrecht-Boston, Mass., PWN-Polish Scientific Publishers, Warsaw, 1984.

WEDDERBURN, J. M.

[1] On hypercomplex numbers, *Proc. London Math. Soc.* (2) **6**(1908), 77–118.

WEIERSTRASS, K.

  [1] Über die analytische Darstellbarkeit sogenannter willkürlicher Funktionen reeller Argumente, *Sitz. Preuss. Akad. Wiss.* (1885), 633–640, 789–906.

WEIL, A.

  [1] *L'intégration dans les groupes topologiques et ses applications*, Actualités Sci. et Ind. **869**, Hermann et Cie., Paris 1940. MR 3, 198.

  [2] Sur certains groupes d'opérateurs unitaires, *Acta Math.* **111**(1964), 143–211. MR 29 #2324.

WEISS, G.

  [1] Harmonic analysis on compact groups, *Studies in Harmonic Analysis*, **13**, pp. 198–223, MAA Studies in Mathematics, Mathematical Association of America, Washington, D. C., 1976. MR 57 #13383.

WENDEL, J. G.

  [1] Left centralizers and isomorphisms of group algebras, *Pacific J. Math.* **2**(1952), 251–261. MR 14, 246.

WERMER, J.

  [1] Banach algebras and analytic functions, *Adv. Math.* **1**(1961), 51–102. MR 26 #629.

WESTMAN, J. J.

  [1] Virtual group homomorphisms with dense range, *Illinois J. Math.* **20**(1976), 41–47. MR 52 #14235.

WEYL, H.

  [1] Theorie der Darstellung kontinuerlicher halb-einfacher Gruppen durch lineare Transformationen. I, II, III, *Nachtrag. Math. Z.* **23**(1925), 271–309; **24**(1926), 328–376, 377–395, 789–791.

  [2] *The classical groups, their invariants and representations*, Princeton University Press, Princeton, N. J., 1939. MR 1, 42.

  [3] *Theory of groups and quantum mechanics*, Dover, New York, 1964 (translated from German, 2nd ed., published 1931).

WEYL, H., and PETER, F.

  [1] Die Vollständigkeit der primitiven Darstellungen einer geschlossenen kontinuerlichen Gruppe, *Math. Ann.* **97**(1927), 737–755.

WHITNEY, H.

  [1] On ideals of differentiable functions, *Amer. J. Math.* **70**(1948), 635–658. MR 10, 126.

WHITTAKER, E. T., and WATSON, G. N.

  [1] *Modern analysis*, Cambridge Press, London, 1927.

WICHMANN, J.

  [1] Hermitian *-algebras which are not symmetric, *J. London Math. Soc.* (2) **8**(1974), 109–112. MR 50 #8088.

  [2] On the symmetry of matrix algebras, *Proc. Amer. Math. Soc.* **54**(1976), 237–240. MR 52 #8947.

[3] The symmetric radical of an algebra with involution, *Arch. Math.* (*Basel*), **30**(1978), 83–88. MR 58 #2313.

WIENER, N.

[1] Tauberian theorems, *Ann. Math.* **33**(1932), 1–100, 787. (Also Selected Papers of N. Wiener, M.I.T. Press, 1964, pp. 261–360).

[2] *The Fourier integral and certain of its applications*, Cambridge University Press, Cambridge, 1933.

WIGHTMAN, A. S.

[1] *Quelques problèmes mathématiques de la théorie quantique relativiste*, Report of Lille conference on quantum fields, C.N.R.S., Paris, 1959, pp. 1–35.

WIGNER, E. P.

[1] On unitary representations of the inhomogeneous Lorentz group, *Ann. Math.* **40**(1939), 149–204.

[2] Unitary representations of the inhomogeneous Lorentz group including reflections, *Group theoretical concepts and methods in elementary particle physics* (Lectures Istanbul Summer School Theoretical Physics, 1962), pp. 37–80, Gordon and Breach, New York, 1964. MR 30 #1210.

WILCOX, T. W.

[1] A note on groups with relatively compact conjugacy classes, *Proc. Amer. Math. Soc.* **42**(1974), 326–329. MR 48 #8685.

WILLIAMS, D. P.

[1] The topology on the primitive ideal space of transformation group C^*-algebras and CCR transformation group C^*-algebras, *Trans. Amer. Math. Soc.* **226**(1981), 335–359. MR 82h: 46081.

[2] Transformation group C^*-algebras with continuous trace, *J. Functional Anal.* **41**(1981), 40–76. MR 83c: 46066.

[3] Transformation group C^*-algebras with Hausdorff spectrum, *Illinois J. Math.* **26**(1982), 317–321. MR 83g: 22004.

WILLIAMS, F. L.

[1] *Tensor products of principal series representations*, Lecture Notes in Mathematics **358**, Springer-Verlag, Berlin-New York, 1973. MR 50 #2396.

WILLIAMSON, J. H.

[1] A theorem on algebras of measures on topological groups, *Proc. Edinburgh Math. Soc.* **11**(1958/59), 195–206. MR 22 #2851.

[2] *Lebesgue integration*, Holt, Rinehart & Winston, New York, 1962. MR 28 #3135.

WILS, W.

[1] Direct integrals of Hilbert spaces. I., *Math. Scand.* **26**(1970), 73–88. MR 41 #9010.

WORONOWICZ, S.

[1] On a theorem of Mackey, Stone and von Neumann, *Studia Math.* **24**(1964-65), 101–105. MR 28 #4815, MR 30 #1205.

Wulfsohn, A.

[1] The reduced dual of a direct product of groups, *Proc. Cambridge Phil. Soc.* **62**(1966), 5-6. MR 32 #5791.

[2] The primitive spectrum of a tensor product of C^*-algebras, *Proc. Amer. Math. Soc.* **19**(1968), 1094-1096. MR 37 #6771.

[3] A compactification due to Fell, *Canad. Math. Bull.* **15**(1972), 145-146. MR 48 #3004.

Yamagami, S.

[1] The type of the regular representation of certain transitive groupoids, *J. Operator Theory* **14**(1985), 249-261.

Yang, C. T.

[1] Hilbert's fifth problem and related problems on transformation groups. *Proc. Symp. Pure Math.*, 28, Part I, pp. 142-146, Amer. Math. Soc., Providence, R.I., 1976. MR 54 #13948.

Yosida, K.

[1] On the group embedded in the metrical complete ring, *Japan. J. Math.* **13**(1936), 7-26.

[2] *Functional analysis*, Grundlehren der Math. Wiss., **123**, Springer-Verlag, Berlin, 1965. MR 31 #5054.

Zaitsev, A. A.

[1] Holomorphically induced representations of Lie groups with an abelian normal subgroup. *Trudy Moskov. Mat. Obšč.* **40**(1979), 47-82 (Russian). MR 80m: 22025.

[2] Equivalence of holomorphically induced representations of Lie groups with abelian normal subgroups. *Mat. Sb. N.S.* (160) **118**(1982), 173-183, 287 (Russian). MR 84j: 22019.

Żelazko, W.

[1] *Banach algebras*, Elsevier, Amsterdam, 1973. MR 56 #6389.

Zelevinsky, A. V.

[1] Induced representations of reductive p-adic groups II. On irreducible representations of GL(n), *Ann. Sci. École. Norm. Sup.* (4) **13**(1980), 165-210. MR 83g: 22012.

Zeller-Meier, G.

[1] Produits croisés d'une C^*-algèbre par un groupe d'automorphismes, *C. R. Acad. Sci. Paris*, **263**(1966), A20-23. MR 33 #7877.

[2] Produits croisés d'une C^*-algèbre par un groupe d'automorphismes, *J. Math. Pures Appl.* (9) **47**(1968), 101-239. MR 39 #3329.

Želobenko, D. P.

[1] A description of a certain class of Lorentz group representations, *Dokl. Akad. Nauk SSSR* **125**(1958), 586-589 (Russian). MR 21 #2920.

[2] Linear representations of the Lorentz group, *Dokl. Akad. Nauk SSSR.* **126**(1959), 935-938. MR 22 #906.

[3] *Compact Lie groups and their representations*, Translations of Mathematical Monographs, **40**, Amer. Math. Soc. Providence, R. I., 1973. MR 57 #12776b.

ZETTL, H.

[1] Ideals in Hilbert modules and invariants under strong Morita equivalence of C^*-algebras, *Arch. Math.* **39**(1982), 69–77. MR 84i: 46060.

ZIMMER, R. J.

[1] Orbit spaces of unitary representations, ergodic theory, and simple Lie groups, *Ann. Math.* (2) **106**(1977), 573–588. MR 57 #6286.

[2] Induced and amenable ergodic actions of Lie groups, *Ann. Sci. École Norm. Sup.* (4) **11**(1978), 407–428. MR 81b: 22013.

[3] Amenable actions and dense subgroups of Lie groups, *J. Functional Anal.* **72**(1987), 58–64.

ZYGMUND, A.

[1] *Trigonometric series*, I, II, Cambridge University Press, New York, 1959. MR 21 #6498.

Name Index

Subject Index

Index of Notation

Bundles, cross-sections, and fibers

Direct sums

Equivalences and orderings

Positive functionals and representation theory

Special spaces

Tensor products

Topology

PURE AND APPLIED MATHEMATICS

* Presently out of print

Printed and bound by CPI Group (UK) Ltd, Croydon, CR0 4YY

13/10/2024

01773514-0004